RADIO COMMUNICATION

Radio Communication

J. H. Reyner

B.SC., A.C.G.I., D.I.C., C.ENG., F.I.E.E., LIFE MEMBER I.E.E.E.

and

P. J. Reyner

M.A. (CANTAB.), C.ENG., M.I.E.E.

Third Edition

SI Units

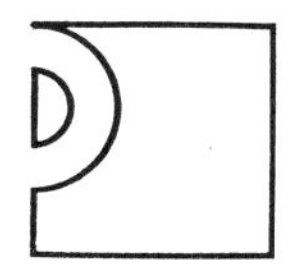

Pitman Publishing

Third edition 1972

SIR ISAAC PITMAN AND SONS LTD.
Pitman House, Parker Street, Kingsway, London, WC2B 5PB
P.O. Box 46038, Portal Street, Nairobi, Kenya

SIR ISAAC PITMAN (AUST.) PTY. LTD.
Pitman House, 158 Bouverie Street, Carlton, Victoria 3053, Australia

PITMAN PUBLISHING COMPANY S.A. LTD.
P.O. Box 11231, Johannesburg, South Africa

PITMAN PUBLISHING CORPORATION
6 East 43rd Street, New York, N.Y. 10017, U.S.A.

SIR ISAAC PITMAN (CANADA) LTD.
495 Wellington Street West, Toronto, 135, Canada

THE COPP CLARK PUBLISHING COMPANY
517 Wellington Street West, Toronto, 135, Canada

Cased edition: ISBN 0 273 36164 3
Paperback edition: ISBN 0 273 36165 1

Text set in 10/11 pt. Monotype Modern printed by letterpress, and bound in Great Britain at the Pitman Press, Bath
G2—(T.5426/1352:74

Preface to Third Edition

THE RESPONSE to the earlier editions has encouraged us to enlarge the scope by the inclusion of a treatment of logic circuitry and similar developments of interest to the practising electronics engineer.

The basic coverage remains that required for the City and Guilds Telecommunications Technicians' Course, with a slight change of emphasis. Despite the almost universal use of solid-state techniques in many fields, thermionic devices still play an important role in certain applications, and will continue to do so for many years. The treatment of valves in Part I has therefore been maintained, though certain obsolete matter has been omitted and the space devoted to an extended discussion of semiconductor usage, including field-effect transistors.

The chapters in Part II, dealing with practical applications, treat valve and transistor circuitry separately, as before, but the discussions of transistor usage have been appropriately extended, while Chapter 16 has been enlarged to cover colour television.

Two new chapters have been added. The first of these deals with solid-state switching devices, including logic circuitry and the use of integrated-circuit techniques, while the second discusses briefly some of the specialized developments in space communications.

We have retained the wide selection of examples from examination papers which we believe to be of value. A study of recent papers shows that they do not, in the main, include any material not already covered in previous years. The fundamental basis of the art is in fact stable; hence although some 1970 examples are included, extensive updating appears neither necessary nor desirable.

Finally, Appendix 2 has been modified to cover the recently agreed SI units, which have been used throughout the book in place of the former MKS units, while a new Appendix 4 has been added summarizing the principal techniques of transistor construction.

LITTLE MISSENDEN
March, 1972

J. H. R.
P. J. R.

Contents

Part II. Practical Applications

Plates

Part I

Basic Principles

I

Fundamentals

1.1. Basic Concepts

Telecommunications engineering is concerned with the transmission of information by utilizing the effects produced when an electric current is varied. The variations may be transmitted by means of wires, as is done with telegraph and telephone systems, or by means of electromagnetic waves as used in radio communication. It is this second category with which this book is primarily concerned, though some discussion of the simpler aspects of telegraph and telephone techniques will necessarily be required.

The fundamental requirement for a proper study of the subject is a clear understanding of the nature of electricity, which is no longer regarded as a mysterious invisible fluid pumped through wires by a battery, but rather as the movement or displacement of fundamental particles called *electrons* in the atoms of the materials themselves. Hence the study and utilization of electrical phenomena constitute the twin sciences known as electronics and electronic engineering.

Atomic Structure

All matter is believed to be composed of atoms which are like miniature solar systems containing a central positively-charged nucleus surrounded by negatively-charged electrons revolving in definite orbits.

These atoms do not normally exist alone but associate themselves with other atoms, either of their own kind or of some other element, to form *molecules* in which a number of atoms group together to provide an equilibrium between the various internal forces, under conditions of minimum energy.

In gases these molecules remain as independent entities moving at random through the space in which they are confined. If two molecules happen to collide they bounce apart like elastic balls. In solid materials, however, the molecules have much less freedom, and generally associate themselves with other molecules to form symmetrical assemblies of either crystalline or chain-like formation.

It should be realized that all matter is very largely open space. The diameter of an atomic nucleus is of the order of 10^{-11} mm, while the electrons, which are of similar size, revolve in orbits of the order of 10^{-7} mm diameter, i.e. some 10,000 times the size of the nucleus. It is the powerful internal forces resulting from the relatively compact grouping of the molecules in a solid which cause it to appear impervious and reflect light waves so that it looks solid.

Liquids occupy a position between solids and gases. The internal forces are weaker so that the molecules have less restraint and the substance will behave in some ways like a solid and in others more like a gas.

Free Space

Since the structure of the atom involves the interplay of internal forces, there must be a medium through which they operate. It is clearly non-physical since the forces are found to exist in a complete vacuum, and 19th-century physics postulated the existence of what was called the *ether* to fulfil this function. This has been replaced today by the concept of *free space*, which is said to possess certain specific electrical properties, as is discussed later.

One can then regard the sub-atomic particles as local distortions or condensations of free space, which create a sphere of influence in their immediate environment. They are said to carry an *electric charge*, which may be of two kinds. A particle carrying a negative charge is called an *electron*, while a positively charged particle is called a *proton**. The nature of the force is such that similarly-charged particles will repel one another, while dissimilar charges will be mutually attracted.

The nucleus of an atom thus contains an appropriate number of protons surrounded (normally) by an equal number of electrons rotating in orbits at distances and speeds which produce a centrifugal force sufficient to counteract the electrical attraction between the two opposite charges, and hence maintain a state of equilibrium. (The nucleus may also contain some additional particles called *neutrons*, which carry no charge but merely add to the mass, as discussed in Appendix 1.)

* The electron is assigned a negative charge because in the classic early experiments with minute oil-drops carrying an accumulation of electrons it was found that the force exhibited was in opposition to that of gravity.

Electric Charge

Basically, it is disturbances of this state of equilibrium which give rise to the whole range of electrical (and magnetic) phenomena. The electrons in the outermost orbits of many atoms can be detached by the application of a suitable force and thereby become free. One of the simplest examples of this is the behaviour of certain insulating materials under the influence of friction. If a glass rod is rubbed smartly with a silken cloth some of the electrons in the material are transferred to the cloth leaving the rod with a deficiency.

The rod is then said to be positively charged and it will actually exercise appreciable mechanical forces. If, for example, it is brought near some small pieces of tissue paper it will attract them like a magnet. The word electricity, in fact, is derived from the Greek word *ηλεκτρον* (electron), meaning amber, because it had been observed that amber had this property of becoming "electrified."

The mechanism of this attraction is itself based on disturbance of the equilibrium. The atoms on the surface of the rod, being deficient in electrons, exhibit an invisible force seeking to attract electrons from any neighbouring material to restore the equilibrium. This force acts on the atoms in the paper attracting electrons to the surface and if the pieces of paper are light enough the mechanical attraction between these electrons and the positively-charged (deficient) atoms on the surface of the rod is sufficient to lift the paper to the rod.

Electromotive Force

This displacement of electrons constituting an electric charge can be produced by other means than surface friction. In fact, except for certain specialized types of equipment, frictional methods are rarely used.

There are, in fact, several ways of producing a force which will cause an electron displacement. One is the familiar electric battery in which chemical interaction between two different materials produces an *electromotive force*, or *e.m.f.* Alternatively an e.m.f. may be produced by mechanical means, as in a dynamo, which converts mechanical forces into electrical energy. Then, finally, there are arrangements which will convert heat or light into electrical energy.

The mechanism of these various processes is discussed later. All that need be noted at the moment is that, by suitable techniques, it is possible to produce an e.m.f. which will cause the displacement of electrons which constitutes an electric charge or current.

Electric Current; Conductors and Insulators

The movement of electrons produced by an e.m.f. is called an electric current. In the effects so far described the movement has only been momentary because the freedom of the electrons is limited. Materials of this type are known as *insulators* or *dielectrics*. There are, however, materials, mostly metallic in nature, in which the internal forces are appreciably less strong so that electrons are continually escaping from their parent atom and drift at random. They are soon captured by another atom which is deficient of an electron but there is always a quantity of stray electrons in transit.

If an e.m.f. can be introduced into the system, however, these random electrons will all be subjected to a force tending to move them in the same direction, producing a continuous flow of electrons through the material. In this case the current is sustained, and materials of this type are called *conductors*.

Hence it will be seen that there are two possible kinds of current:

1. Conduction current which is a continuous movement of electrons through the material.

2. Displacement current which is a momentary movement of electrons within the material. This might appear to be of limited importance but, as will be seen later, if the direction of the e.m.f. is continually reversed the electrons can be caused to surge to and fro in the material so that a sustained oscillating current is obtained.

In either case the intensity or strength of the current depends on the rate of flow of electrons; the more electrons passing a given point in a given time, the greater the current; and since an accumulation of electrons at rest constitutes an electric charge we can define the intensity of an electric current, whether momentary or sustained, as the rate of change of charge.

Gases normally behave as insulators but can under certain conditions behave as conductors. Liquids are usually (but not necessarily) insulators except for aqueous solutions (salts dissolved in water) which conduct in a special way, to be discussed later.

There are also certain materials known as *semiconductors* which occupy an intermediate position, behaving partly like insulators and partly like conductors. An understanding of their behaviour requires a more detailed discussion of the mechanism of conduction than is necessary at this juncture. It will suffice to note that developments in the use of these semiconductors have produced an entirely new range of techniques which are discussed in detail in Chapter 6.

Units

We may conveniently conclude this section with a brief reference to the units adopted for the measurement of the various quantities so far discussed. It might appear that the unit of charge could be the charge on a single electron but apart from the fact that this is far too small for practical purposes it is not commensurate with the units normally adopted. All engineering quantities can be expressed in terms of the three fundamental measurements of length, mass, and time and for many years the centimetre, the gramme, and the second were used as primary units (the CGS system). From these a series of derived units can be obtained but unfortunately there were two schools of thought, one based on electrostatic phenomena and the other on electromagnetic effects. In 1950 it was agreed to bring into use a common system of units known as the MKS system. This was based on the metre, kilogramme, and second as the fundamental units, which did not, in itself, make a significant change; but by assigning certain specific values to the characteristics of free space, as explained in Appendix 2, a simplified and uniform system was developed.

This is, in the main, still in use but its scope has been widened to include other than electrical quantities to form what is known as the International System of Units, and these SI units have been adopted in the present text.

In the International System the fundamental electrical quantity is not the charge but the current, which can be more conveniently measured in terms of the physical forces developed. The unit of current is the *ampere*, which is defined as the value of a constant current which, when maintained in two straight, parallel conductors of infinite length and negligible circular cross-section, separated by a distance of one metre in vacuo, would produce between the conductors a force of 2×10^{-7} newton per metre length.

The unit of charge is the *coulomb* which is the quantity of electricity which results when a current of one ampere flows for one second. This is actually $6{\cdot}24 \times 10^{18}$ times the charge on an electron, but this figure is not of importance in engineering, though it is of significance in electron physics.

The unit of e.m.f. is the *volt* which is derived from the relationship between current and resistance as explained in the next section.

1.2. Ohmic Conduction; Resistance

We have seen that a movement of electrons constitutes an electric current, but this can take many forms. The simplest form is that which occurs in those materials in which there is a proportion of relatively free electrons which can be caused to move through the material by the application of a suitable e.m.f. Such materials are known as conductors and are, in the main, metallic in nature. The atomic structure of metals (as explained in Appendix 1) is such that the outermost orbit only contains one or two electrons which in consequence are relatively loosely held. From time to time an electron may leave its parent atom, but in the absence of any directing force any such free electrons will circulate at random through the material.

In the presence of an e.m.f., however, this movement will be co-ordinated into a steady drift towards any point or points of positive potential, which may be the result of a battery or other source external to the material, or may be the result of potentials actually produced within the material itself.

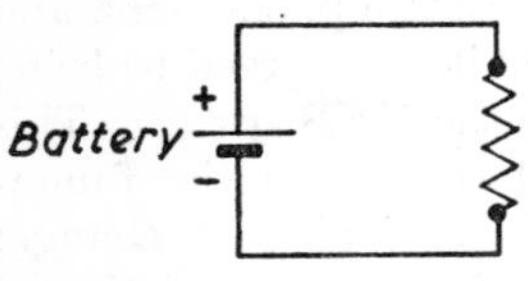

Fig. 1.1. Simple Electric Circuit

Consider the effect of a battery connected to a length of wire, as in Fig. 1.1. The electrons in the wire next to the positive terminal will be drawn out of the wire, leaving behind them a deficiency which will be filled by some of the random free electrons and so in turn along the wire, the deficiency at the negative end being made up by electrons supplied (by chemical action) by the battery itself. There will thus be a continuous flow of current which will persist as long as the battery maintains its e.m.f.

(Note that this current is an electron flow towards the point of positive potential. The accepted convention of current as flowing from positive to negative originated before the mechanism of current flow was understood.)

Resistance

The magnitude of the current depends on the material of the conductor. If an atom loses one of its outer electrons it becomes positively charged or *ionized*, and any free electrons are soon recaptured by an ionized atom. The pressure of an e.m.f., however, increases the velocity (and hence the energy) of the free electrons, some of which may thereby overcome the attraction of the ionized atoms and continue on their journey. The number of electrons which successfully evade recapture (and hence the current) is determined by

(*a*) The applied e.m.f., and

(*b*) The strength of the internal forces.

The second factor is clearly dependent on the internal structure of the material. It is, in fact, a definite and measurable physical property of the material, which is called the *resistivity*, and a particular length of conductor or circuit is said to possess a certain *resistance*.

The actual resistance of a piece of material depends upon its length, for obviously the greater the distance which the electrons have to travel the more the chance of their recapture and hence the less the current produced by a given e.m.f. It is also inversely proportional to the cross-sectional area since the larger the area the greater the number of free electrons available. This is expressed mathematically by saying that $R = \rho l/A$ where l and A are the length and cross-sectional area and ρ is the resistivity, which is the resistance of a cube of the material of unit dimensions. Typical values of ρ are given in Table 4.1, p. 153.

The reciprocal of resistivity is called the *conductivity*, for clearly the less the resistance the more easily will the material conduct.

The resistivity is dependent upon the temperature. The atoms of the material are in a state of vibration which increases with temperature so that the chances of recapture are increased. Hence the resistivity normally increases slightly as the temperature is raised, though there are some materials, notably carbon, in which the resistivity decreases with temperature.* Conversely, if the temperature is reduced, the resistance normally decreases and in fact with certain metals a point is reached, near to absolute zero (−273°C) at which the resistance almost disappears and a current, once started, will continue *after the removal of the e.m.f.* for several days. This is known as *superconductivity*, and is utilized in certain special applications.

Ohm's Law

In normal circumstances the current, as we have seen, is directly proportional to the e.m.f. and inversely proportional to the resistance. This relationship is known as Ohm's Law, which states that $I = kE/R$, where I is the current, E is the e.m.f., R is the resistance, and k is a constant. In practice the units of e.m.f. and resistance are so chosen as to make $k = 1$, so that $I = E/R$.

The unit of resistance is the *ohm*, named after the scientist who first formulated the relationship. It is a practical unit, being that resistance in which a steady current of one ampere generates heat energy at the rate of one joule per second (1 watt). (See page 12.)

* Negative temperature coefficients are frequently found in materials of the semiconductor class, wherein the mechanism of conduction is somewhat different, as explained in Chapter 6; and although carbon is not normally regarded as a semiconductor, its atomic structure is of the same form.

The unit of e.m.f., the *volt*, is thus the e.m.f. which will produce a current of one ampere in a resistance of one ohm.

The metals used for conductors all obey this strict proportionality and are therefore said to be ohmic conductors. There are certain materials and alloys which do not behave in this way and these are known as non-linear or non-ohmic conductors.

Kirchhoff's Laws; Concept of Back E.M.F.

In the circuit of Fig. 1.1, $I = E/R$. This may be rewritten in the form $E - IR = 0$, indicating that the applied e.m.f. is exactly offset by the voltage drop produced by the passage of the current I through the resistance R. The relationship, however, may be expressed slightly differently by suggesting that the passage of the current through the resistance has developed a *back e.m.f.*, $-IR$, so that the *total e.m.f. in the circuit is zero.*

The advantage of expressing the relationship in this way is that, as will be seen shortly, there are other forms of circuit element in which the passage of current develops a voltage drop, particularly when the current is varying; moreover there are certain forms of "active" circuit element which contain or generate e.m.f.s of their own. If any voltage drop is regarded as producing a back e.m.f., we have a consistent form of relationship which applies to any form of circuit.

The relationship is known as Kirchhoff's Second Law, which states that the *total e.m.f. in a circuit at any instant is zero.*

Kirchhoff's First Law relates to the current distribution in a circuit. Circuits are rarely as simple as that of Fig. 1.1. There is often more than one path for the current, as for example in Fig. 1.3. In such cases Kirchhoff's First Law states that at any point in a circuit *the sum of the currents flowing at any instant is zero.* Thus at the point A in Fig. 1.3 there will be a current I flowing towards A and two currents I_1 and I_2 flowing away from it. Kirchhoff's First Law says that

$$I - I_1 - I_2 = 0$$

so that $I = I_1 + I_2$.

In this simple example this may appear self-evident but with more complex circuits the answer may not be so readily apparent. It should be noted that in circuits where the current is varying it is the instantaneous values of current or e.m.f. which must be used.

Series and Parallel Connection

Practical circuits nearly always contain more than one element, and it is necessary to know how the effects of the various parts may be added together. Fig. 1.2 is a circuit containing two resistances connected end to end, i.e. *in series*. Since there is only one path for the current, it must be the same throughout the circuit. If this current is I, the back e.m.f. on the resistances will be $-IR_1$ and $-IR_2$ respectively so that $E - IR_1 - IR_2 = 0$. Hence $E = I(R_1 + R_2)$, which means that R_1 and R_2 in series are equivalent to a single resistance of value equal to the sum of R_1 and R_2. Moreover this same reasoning applies to any number of resistances.

In Fig. 1.2 the resistances have been shown as separate units. It is often required in practice to provide units having a given value of resistance. Such an arrangement is called a *resistor* and the construction of typical forms of resistor is discussed in Section 4.1.

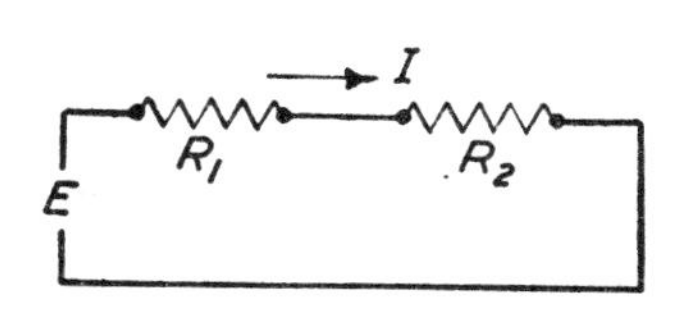

FIG. 1.2. RESISTANCES IN SERIES

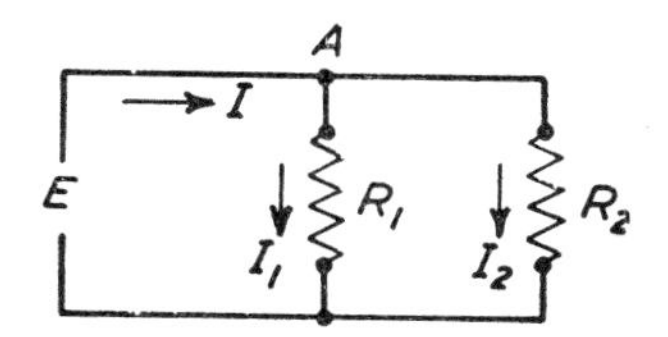

FIG. 1.3. RESISTANCES IN PARALLEL

In any circuit, however, the connecting wires possess some resistance which will normally have to be taken into account. In Fig. 1.2, for example, R_1 might be a resistor while R_2 might represent the resistance of the wiring, assumed to be concentrated for convenience.

The e.m.f. will, in fact, be expended in overcoming the total circuit resistance and will be divided across the circuit in proportion to the resistance of the various portions. This is sometimes expressed by referring to the voltage drop across the circuit. Thus if $R_1 = 9$ ohms and $R_2 = 1$ ohm then an e.m.f. of 10 volts will produce a current of 1 ampere and there will be a voltage drop of 9 volts across R_1 and 1 volt across R_2.

Fig. 1.3 shows two resistors *in parallel*. Here it is the e.m.f. across the circuit which is the same, so that $I_1 = E/R_1$ and $I_2 = E/R_2$. But we have seen that $I = I_1 + I_2$, so that if R is the effective resistance of the two resistors R_1 and R_2 together, $I = E/R = E/R_1 + E/R_2$. Hence $1/R = 1/R_1 + 1/R_2$.

This again may be applied to any number of resistors. The reciprocal of the resistance, $1/R$, is called the *conductance*; hence for any number of resistors in parallel the total conductance is equal to the sum of the conductances of the individual parts.

With only two resistors it is sometimes convenient to rewrite the above expression as follows:

$$\frac{1}{R} = \frac{1}{R_1} + \frac{1}{R_2}$$

$$= \frac{(R_2 + R_1)}{R_1 R_2}$$

Hence $$R = \frac{R_1 R_2}{(R_1 + R_2)}$$

Heating Effect of a Current; Energy and Power

We have seen that the electron drift which constitutes an electric current is not unrestricted. Collisions occur between the electrons and the atoms and these absorb energy. In simple terms, a collision between an electron and an atom displaces one or more of the constituent electrons from their normal orbits. In due course the atom reverts to its former condition and the surplus energy is released in the form of electromagnetic vibrations, usually as heat, though it may take other forms such as light or X-rays.

The energy developed by these collisions is proportional to the quantity of electrons and their velocity, which is determined by the e.m.f. Hence the energy $W \propto EQ$, and the unit of energy, which is called the *joule*, is so chosen as to make this an exact relationship, so that

$$\text{Energy (joules)} = \text{E.M.F. (volts)} \times \text{Quantity (coulombs)}$$

The effect of this energy depends upon the rate at which it is expended. It is more exhausting to run up-hill than to amble up at a leisurely pace. The rate of doing work (expending energy) is called the *power*, and the unit of power, which is called the *watt*, is the expenditure of one joule in one second*.

Hence $$\text{Power} = \text{E.M.F.} \times \text{Quantity per second}$$
$$= \text{E.M.F.} \times \text{Current}$$

In symbols P (watts) $= E$ (volts) $\times\ I$ (amps). But by Ohm's Law $I = E/R$, so that we can write $P = EI = E^2/R = I^2R$, any of which forms may be used according to convenience.

These expressions only apply when the current is steady. If the current is varying they are still fundamentally correct, but have to

* Mechanical power was formerly expressed in different units, but the watt is now used for all power. See Appendix 2.

be modified to express the average power, as explained in Section 1.5 (page 48).

Concept of Load

The resistance of a practical circuit is not simply that of the conductors. The object is to pass the current through some device from which useful work can be obtained. This useful part of the circuit is called the *load* and since energy is expended therein it can be and usually is regarded as a resistance, called the *load resistance*.

Any practical circuit thus consists of two parts, the conductor resistance r and the load resistance R. A current I flowing in the circuit will develop a power I^2R in the load and a (smaller) power I^2r which is dissipated (mainly as heat) in the conductors. This latter power is not useful and is called the *conductor loss*.

The form of the load varies. It may be a device deliberately designed to produce heat as in an electric fire or furnace, or light, as in an electric lamp. It may be a device designed to convert electrical energy into mechanical energy, such as an electric motor, while yet again it may be an arrangement to convert one form of electrical energy into another, as in a radio set.

1.3. Electrostatics; Capacitance

It has been shown that the disturbance of equilibrium which constitutes an electric charge results in a force. The charged body is said to produce an *electric field* which will influence any other charge in the vicinity. Such a field is usually represented diagrammatically as a number of *lines of force*.

Consider a uniform field such as is shown in Fig. 1.4. If a charged body (or simply an electric charge) is brought into this field there will be a force acting upon it depending on the strength of the charge and the *intensity* of the field. The force will, in fact, be given by the expression $F = Eq$ where E is the intensity of the field and q is the charge. The intensity E is sometimes called the *field strength*.

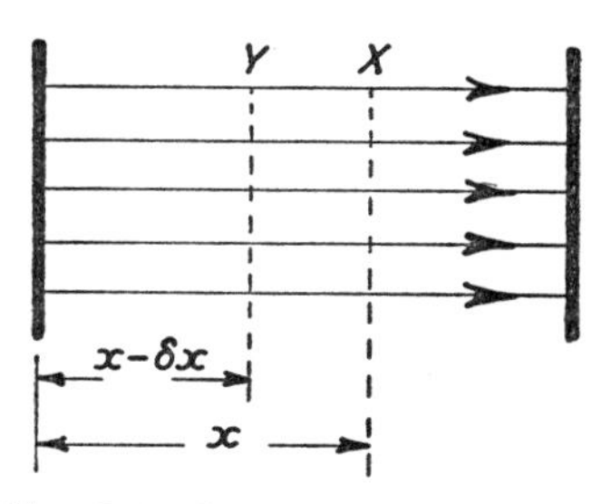

FIG. 1.4. LINES OF FORCE REPRESENTING AN ELECTRIC FIELD BETWEEN TWO PLATES

Potential

To move the charge against the field, work will have to be done in overcoming this force. In fact, in order to bring the charge into the field at all some work will have had to be done; this is called the *potential* at the particular point.

If a weight is lifted against gravity, work has to be done and the weight acquires some potential energy which is given back as kinetic energy if the weight is allowed to fall to the ground. Similarly the work done in bringing a unit electric charge from infinity to any point in an electric field is called the electric potential at that point and if we consider two points X and Y in Fig. 1.4 there will be a *potential difference* or p.d. between them.

If the distances of the points X and Y from the source of the field are x and $x - \delta x$ respectively and the potentials are V and $V + \delta V$, then by definition the rise in potential δV will be equal to the work done in moving a unit charge over the distance δx against the field which will be $-E\delta x$. Hence $E = -\delta V/\delta x$ which as $\delta x \rightarrow 0$ becomes $-dV/dx$, the rate of change of potential or *potential gradient*. Thus the intensity of the field at any point is equal to the potential gradient and this applies whether the field is uniform or varying.

Potential gradients vary widely in practice from a few microvolts/metre in an electric wave to many thousands of volts/metre in the insulation of high-voltage transformers.

Flux Density

We can regard a charged body as being surrounded by an *electrostatic flux*. In the case of an isolated charge this flux radiates uniformly in all directions as in Fig. 1.5, while the flux between two bodies (such as the pair of parallel plates in Fig. 1.4) is confined mainly to the space between them.

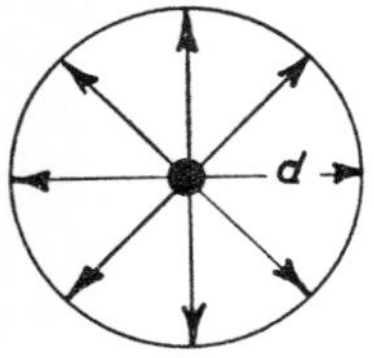

FIG. 1.5. THE ELECTRIC FIELD FROM A CHARGE RADIATES UNIFORMLY IN ALL DIRECTIONS

Now the influence of this electric field is determined not by the amount of the flux but by its relative concentration, i.e. the *flux density* (sometimes called the displacement), which is the flux per unit area. The flux is regarded as being numerically equal to the charge, so that the flux density $D = Q/A$ where A is the area over which the flux is distributed.

Permittivity

What is the relationship between this flux density and the intensity? The answer depends on what is called the *permittivity* of the medium. The work done in moving a charge against the field is not the same for different materials, some of which permit the flux to be established more readily. Mica, for example, is five times more permissive than air so that only one-fifth of the energy is required to move a charge against the field. Hence for a given flux density the intensity is dependent on the permittivity so that if we call the permittivity ε, $E = D/\varepsilon$ and $D = E\varepsilon$.

Since the intensity E is equal to the voltage gradient this means that for a given voltage the flux, and hence the charge, are proportional to the permittivity.

Inverse Square Law

Let us consider how the intensity varies with distance. Consider an isolated charge. The total flux will be distributed uniformly through the surrounding space and at a distance, d, the effective area over which this flux is distributed will be the surface of a sphere of radius d, which is $4\pi d^2$. Hence the flux density will be $Q/4\pi d^2$ and the intensity or field strength $Q/4\pi\varepsilon d^2$. This may be written $E = E_0/d^2$, where $E_0 = Q/4\pi\varepsilon$, the field strength at unit distance.

We can now consider the force developed between two charges separated by a distance d. If the charges are Q_1 and Q_2 the intensity of the field produced by Q_1 will be $E = D/\varepsilon = Q_1/4\pi d^2\varepsilon$. The force produced by this field on the charge Q_2 is $EQ_2 = Q_1Q_2/4\pi\varepsilon d^2$. Hence the force is dependent on the product of the two charges and is inversely proportional to the square of the distance between them. This is known as the *inverse square law*.

Capacitance

From what has been said it is clear that there is a connection between electric charge and potential. In fact, any variation in the amount of charge is accompanied by a change of potential. This relationship is called the *capacitance* of the system, which is the charge (in coulombs) necessary to raise the potential by one volt. Theoretically any body possesses a certain capacitance but in practice we are more concerned with the capacitance between two bodies, which, by definition, is the charge which must be acquired by one body to achieve a potential difference of one volt relative to the other.

In some circumstances, this capacitance is a hindrance which has to be circumvented while in other applications it is useful, for which purpose special arrangements are made to obtain the maximum capacitance in the space available. Such a device is called a *capacitor* or in older parlance, a *condenser*, and in its simplest form consists of two parallel plates as shown in Fig. 1.6. If a source of e.m.f., such as a battery, is connected to the plates, there will be a momentary flow of electrons resulting in an accumulation at the (negative) plate and a deficiency at the other, so that the arrangement becomes electrically charged and there will be an electric field between the plates. If, as in a practical case, there is some insulation or *dielectric* between the plates the electric field will cause a corresponding displacement of the electrons in the dielectric, which will become similarly charged.

This condition will persist even if the battery is removed, and the capacitor will remain "charged" until some conducting path is

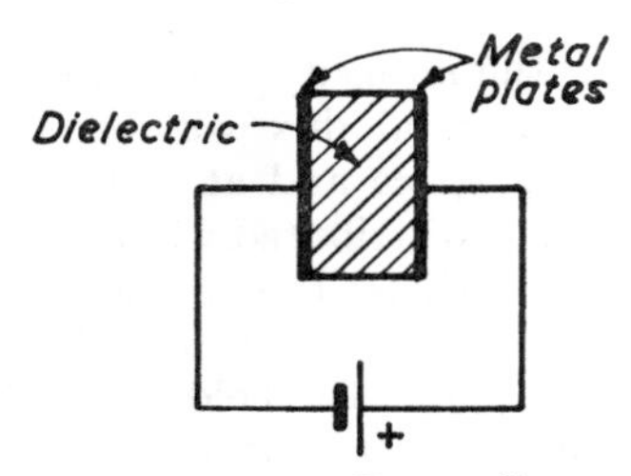

FIG. 1.6. PARALLEL PLATE CAPACITOR

established between the plates which will allow the accumulated electrons to return to their normal condition. In practice the capacitor would be discharged in time by leakage currents which would flow through the dielectric and even the surrounding air (particularly if the air were moist) since no material is a perfect insulator.

It should be noted, however, that the action of a capacitor does not require the presence of a dielectric. Provided the plates become charged, by the application of a suitable e.m.f., an electric field is set up between them. The presence of a dielectric merely modifies the amount of charge produced. Conversely if we have two plates as in Fig. 1.6 and we introduce between them a thin sheet of metal, the electrons in this sheet will be displaced by the influence of the electric field, producing charges of $+Q$ and $-Q$ on the opposite faces, where $Q = Da$, D being the flux density of the field and a the area of the (inserted) plate.

Capacitance Relationships

The capacitance of a parallel-plate capacitor is readily calculable. Since by definition the capacitance is the charge acquired for a p.d. of one volt we can write $C = Q/V$ where C is the capacitance, Q is the charge, and V the applied (or acquired) p.d.

Now assuming a uniform electric field between the plates the flux density D will be Q/A where A is the area of the plates. So we can write

$$Q = AD = AE\varepsilon$$

Again assuming a uniform field, the p.d. across the plates is d times the voltage gradient, where d is the distance between them, so that $V = Ed$.

Hence $$C = Q/V = AE\varepsilon/Ed = A\varepsilon/d$$

The undetermined factor here is the permittivity ε, which is itself made up of two factors. As already stated, the capacitor of Fig. 1.6 does not require any material (even air) between the plates. The presence of any material between the plates will simply modify the amount of charge to an extent depending upon the relative permittivity of the material.

Hence $\varepsilon = \varepsilon_0\varepsilon_r$ where ε_0 is the *permittivity of free space*, sometimes called the *electric space constant*, and ε_r is the relative permittivity of the dielectric used. In SI units the value of ε_0 is

$$1/(36\pi \times 10^9) = 8{\cdot}854 \times 10^{-12}$$

so we can write the expression for the capacitance between two parallel plates in the form $C = (8{\cdot}854\, A\varepsilon_r/d) \times 10^{-12}$ farads. In practice this unit is too large and capacitances are specified in *microfarads* for which the symbol μF is used, while for many purposes even this is too large and the capacitance is expressed in *picofarads* (pF), sometimes called *micromicrofarads* ($\mu\mu F$)

$$1 \text{ microfarad } (\mu\text{F}) = 10^{-6} \text{ farad}$$

$$1 \text{ picofarad (pF)} = 10^{-12} \text{ farad} = 10^{-6} \text{ microfarad}$$

Practical Forms of Capacitor

For most requirements the capacitance of a single pair of plates is insufficient. For example, two plates 10 cm $\times$ 10 cm would have an area of 10^{-2} m^2. If they were 1 mm apart in air the capacitance would be

$$C = \frac{8{\cdot}85 \times 10^{-2}}{10^{-3}} \,.\, 10^{-12} = 88{\cdot}5 \text{ pF only}$$

(In fact it would be slightly less because there is some dispersion of the electric flux at the edges of the plates so that the flux density is not quite uniform.)

It is customary, therefore, to use assemblies of plates connected together alternately. Such an arrangement is equivalent to $(n - 1)$ capacitors in parallel, where n is the number of plates.

For many purposes variable capacitors are required. These may be constructed by assemblies of plates so arranged that either the area or the separation can be varied. The construction of practical forms of capacitor, both fixed and variable, is discussed in Section 4.2.

Relative Permittivity

The value of the relative permittivity ε_r is determined experimentally and is known for a wide range of materials. It is found that the capacitance with air dielectric is not appreciably different from the capacitance in a vacuum so that ε_r for air is taken as unity. The relative permittivity of some other materials in common use is given in Chapter 4.

The relative permittivity was formerly known by various other names, notably *specific inductive capacity* or *dielectric constant.*

Growth of Charge

We have seen that if a capacitor is connected across a source of e.m.f. there will be a displacement of the electrons in the dielectric and the capacitor will become charged. The amount of the charge will adjust itself so that the back e.m.f. Q/C is equal and opposite to the applied e.m.f. V.

This, however, is not an instantaneous process. Time is always required for any circuit to readjust itself to changed conditions. Let us consider what happens more closely. When the e.m.f. is first applied the capacitor is not charged and therefore offers no back e.m.f. A current will therefore flow, limited only by the resistance of the connecting leads (or any other resistance in the circuit). This current, however, will produce a charge which will generate a back e.m.f. q/C. This will begin to limit the current, which thus becomes less and less as the charge builds up, until the condition of equilibrium is reached such that $Q = CV$.

At any instant we can say, by Kirchhoff's Law, that

$$E - iR - q/C = 0$$

where E is the applied e.m.f.,

i and q are the instantaneous current and charge, and

R and C are the resistance and capacitance.

But current = rate of change of charge. Hence $i = dq/dt$ so that we can write

$$E - Rdq/dt - q/C = 0$$

This is a differential equation of which the solution is

$$q = CE(1 - e^{-t/CR})$$

where e is the exponential* constant (2·7183) and t is the time elapsing.

Let us see what this means in physical terms. At the instant of connecting the e.m.f. to the capacitor $t = 0$, so that the term in the brackets becomes $(1 - e^0) = (1 - 1) = 0$. Hence $q = 0$. As the time increases the term $e^{-t/CR}$, which is $1/e^{t/CR}$, becomes smaller and smaller until finally it becomes negligible and $q =$ simply CE. The capacitor is then fully charged.

The voltage across the capacitor obeys a similar law since $v = q/C$. Hence

$$v = E(1 - e^{-t/CR})$$

Time Constant

How long does this operation take? The answer depends on the value of the product CR. When $t = CR$, $t/CR = 1$, and $q = CE(1 - 1/e) = CE\,(1 - 1/2{\cdot}7183) = 0{\cdot}6321\,CE$. In other words the capacitor has acquired 63·21 per cent of its full charge. When $t = 2CR$ the charge is 86·5 per cent of the final value and if we plot

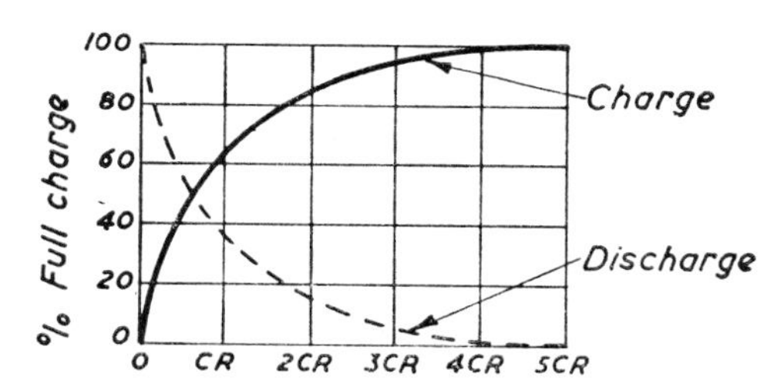

FIG. 1.7. EXPONENTIAL GROWTH AND DECAY

charge against time we obtain the curve shown in Fig. 1.7. This is known as an *exponential* curve and this type of variation is frequently encountered in electrical engineering.

* If a quantity a is to be multiplied by itself the process is expressed in the form a^2, and the expression a^x means a multiplied by itself x times. The term x is called the *exponent*.

Now if we differentiate a^x the result is xa^{x-1}. This may be more or less than a^x, but there is one particular value of a for which the derivative xa^{x-1} is equal to a^x. In other words its rate of change is exactly equal to its value at any instant. There are many physical processes which behave in just this way so that mathematical calculations frequently involve this fundamental value which is called the *exponential constant*. It is not an exact number, being actually 2·7183. . . ., and is written as e. Sometimes (but not in this book) ε is used.

The product CR is called the *time constant* of the circuit, which as we have just seen is the time required for the capacitor to acquire 63·21 per cent of its full charge. If the time constant is small the capacitor will charge up virtually immediately, but with a larger value appreciable time is involved. For example with $C = 1\ \mu F$ and $R = 1\ M\Omega$, making $CR = 1$, the capacitor would take nearly 5 seconds to acquire 99 per cent full charge.

Similar conditions obviously obtain when a capacitor is discharged. The charge does not immediately disappear but decays exponentially in inverse manner to the charging curve of Fig. 1.7, losing 63 per cent of its charge in time $t = CR$. In this case $v = E\ e^{-t/CR}$.

Charging Current

The current which flows during the charging (or discharging) process is determined by the rate of change of charge, $= dq/dt$. If we differentiate the expression for q just deduced we find that

$$i = (E/R)e^{-t/CR}$$

In physical terms again, this means that initially (when $t = 0$), $i = E/R$, and then falls off exponentially as shown by the dotted curve in Fig. 1.7.

Energy Stored in a Capacitor

Since work is done in raising the potential of a capacitor it follows that, as long as the capacitor remains charged, energy is stored in it. But as we have just seen, the charge in a capacitor due to the application of an e.m.f. is not immediately established but requires a certain (small) time during which the p.d. across the capacitor gradually builds up to the full value. The energy input into the capacitor may therefore be taken as the total charge Q multiplied by the average p.d. during the period, $= \frac{1}{2}V$, where V is the final voltage. Hence the energy stored is $\frac{1}{2}QV = \frac{1}{2}CV^2$ since $Q = CV$.

A more rigorous derivation of this expression is as follows. The work δw done in bringing an elementary charge δq up to a potential v is $v\delta q$. But $v = q/C$ so that $\delta w = q\delta q/C$. Hence if Q is the final charge on the capacitor the total work done is

$$W = \frac{1}{C}\int_0^Q q\,dq = \frac{1}{2C}\cdot Q^2 = \tfrac{1}{2}CV^2$$

Dielectric Loss

The distortion of the atomic structure in a dielectric which constitutes an electric charge is not a perfectly elastic process. Hence the work done in raising the potential of a capacitor is not completely given out again when the capacitor is discharged. The difference is called the *dielectric loss*, and when a capacitor is being used under conditions where the charge is continually varying (i.e. the capacitor is passing a current) the effects of this dielectric loss may be appreciable (See Chapter 4, p. 167).

Capacitances in Series and Parallel

Occasions often arise where two or more capacitors are connected together. There are two possible arrangements, one where the capacitors are *in series* as shown in Fig. 1.8 (*a*) and the other where

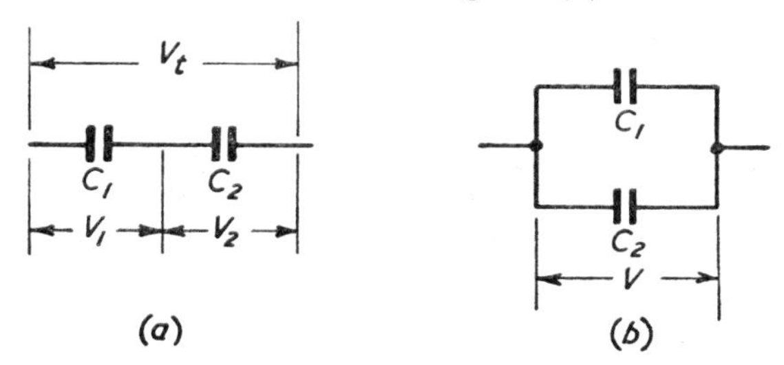

FIG. 1.8. CAPACITORS IN SERIES AND PARALLEL

they are all connected *in parallel* across the same p.d. as in Fig. 1.8 (*b*), while combinations of both are possible.

For the series case the quantity of electrons displaced by the application of a p.d. across the system (i.e. the charge Q) will clearly be the same whether we consider the whole system or the individual capacitors, but the p.d. will be divided across the capacitors, in inverse proportion to their capacitance. Hence, since $C = QV$, we have

$$V_t = V_1 + V_2$$

i.e. $$Q/C_t = Q/C_1 + Q/C_2$$

so that $$1/C_t = 1/C_1 + 1/C_2$$

This reasoning may be applied to any number of capacitors in series so that the reciprocal of the total capacitance is the sum of the reciprocals of the individual capacitances. It will be noted that the series capacitance is always less than either or any of the individual capacitances.

The above expression may be rearranged in the form

$$C_t = C_1C_2/(C_1 + C_2)$$

which is sometimes more convenient.

In the parallel case the p.d. across the capacitors is the same but the total charge is the sum of the charges on the individual capacitors. Hence

$$Q_t = Q_1 + Q_2$$

i.e.

$$C_t V = C_1 V + C_2 V$$

so that

$$C_t = C_1 + C_2$$

Again this can be applied to any number of capacitors, the total capacitance being the sum of the individual capacitances.

Non-uniform Dielectrics

The capacitance of a system having a mixed dielectric (e.g. ebonite and air) may be obtained by considering the system as made up of two capacitors in series, the distances between the plates being the thickness of the respective dielectrics.

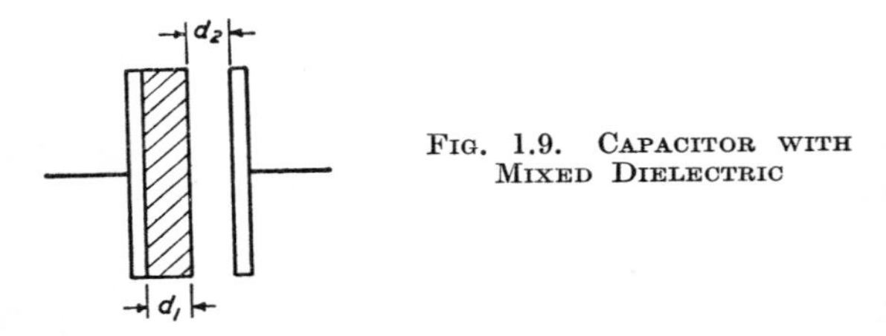

FIG. 1.9. CAPACITOR WITH MIXED DIELECTRIC

Fig. 1.9 illustrates a mixed dielectric. For the solid portion we have:

$$C_1 = A\varepsilon_r\varepsilon_0/d_1$$

where A = the area of the plate,
ε_r = the relative permittivity of the material,
d_1 = thickness of the material, and
ε_0 = the electric space constant.

For the air-spaced portion

$$C_2 = A\varepsilon_0/d_2$$

whence

$$C = A\varepsilon_0/[(d_1/\varepsilon_r) + d_2]$$

Special Configurations

It is sometimes necessary to allow for naturally existing capacitances. We may therefore conclude this section with a discussion of some of the more common forms.

TWO PARALLEL WIRES

Fig. 1.10 represents two long parallel wires of radius r separated by a distance d, large compared with r. Under these conditions the charges on the wires may be assumed to be uniformly distributed. Let the total charge per unit length be Q. At a distance x this will produce a flux operating through the surface of a cylinder of unit length and radius x, i.e. an area of $2\pi x$. Hence the flux density D will be $Q/2\pi x$ and the intensity of the field will be $Q/2\pi x\varepsilon$. Similarly the other wire carrying a charge of $-Q$ will produce an intensity $-Q/2\pi\varepsilon(x - d)$, (since we must measure our distance from the same point), $= Q/2\pi\varepsilon(d - x)$.

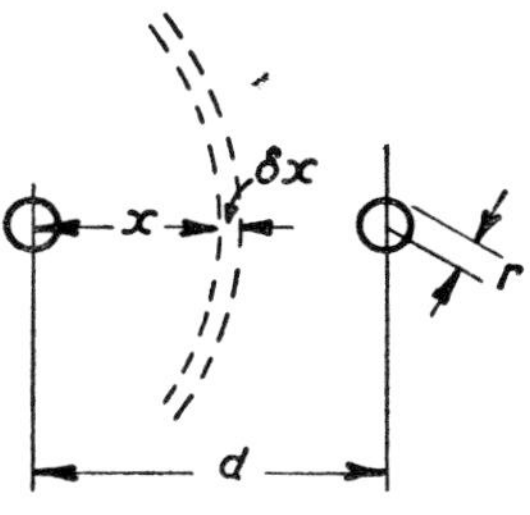

FIG. 1.10. TWO PARALLEL WIRES

The total intensity at x is the sum of these two and this is the potential gradient at the point. Hence

$$E = \frac{Q}{2\pi\varepsilon}\left(\frac{1}{x} + \frac{1}{d - x}\right) = \frac{dv}{dx}$$

This potential gradient is not constant so that the voltage between the two wires must be obtained by integrating dv/dx over the range $x = r$ to $x = d - r$, i.e.,

$$V = \frac{Q}{2\pi\varepsilon}\int_r^{d-r}\left(\frac{1}{x} + \frac{1}{d - x}\right)dx = \frac{Q}{\pi\varepsilon}\log_e\frac{d - r}{r}$$

and since r is small compared with d, this simplifies to

$$V = (Q/\pi\varepsilon)\log_e d/r$$

The capacitance $C = Q/V$, whence $C = \pi\varepsilon/\log_e(d/r)$. For air spacing $\varepsilon = \varepsilon_0$, which is $1/36\pi \times 10^9$, and substituting this in the expression for C we have

$$C = \frac{1}{36\log_e d/r}\cdot 10^{-3}\ \mu\text{F per metre}$$

The potential gradient at a distance x is, as stated above,

$$\frac{Q}{2\pi\varepsilon}\left(\frac{1}{x} + \frac{1}{d - x}\right)$$

This is a maximum when $x = r$, for which the expression becomes

$$\frac{Q}{2\pi\varepsilon}\left(\frac{1}{r} + \frac{1}{d - r}\right) = \frac{Q}{2\pi\varepsilon r}\ \text{very nearly,}$$

since $1/(d - r)$ is small compared with $1/r$. But from the expression for V above,

$$\frac{Q}{2\pi\varepsilon} = \frac{V}{2 \log_e (d - r)/r}$$

Hence the maximum potential gradient (which occurs at the surface of the wires) is

$$\frac{V}{2r \log_e (d - r)/r} \text{ volts/metre}$$

CONCENTRIC CYLINDERS

Fig. 1.11 represents two coaxial cylinders, such as would be found in a coaxial transmission line, of radii r_1 and r_2 respectively. By similar reasoning to the above it can be shown that at a radius x the intensity is

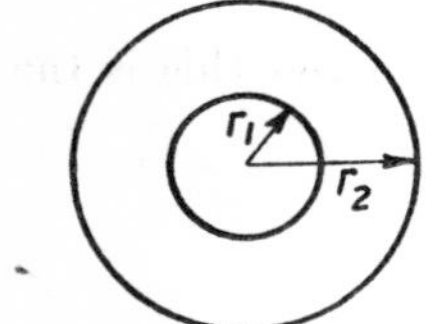

FIG. 1.11. TWO CONCENTRIC CYLINDERS

$$E = Q/2\pi\varepsilon x = dv/dx$$

whence
$$V = \frac{Q}{2\pi\varepsilon}\int_{r_1}^{r_2}\frac{1}{x}\,dx$$

$$= \frac{Q}{2\pi\varepsilon}\log_e (r_2/r_1) = Q/C$$

Concentric conductors may well have a solid dielectric; hence writing $\varepsilon = \varepsilon_0\varepsilon_r$ and substituting $1/36\pi \times 10^9$ for ε_0, the expression becomes

$$C = \frac{\varepsilon_r}{18 \log_e r_2/r_1} \cdot 10^{-3}\ \mu\text{F per metre}$$

The potential gradient at a radius x is, as stated above, $Q/2\pi\varepsilon x$. Substituting for Q this becomes

$$\frac{1}{x} \cdot \frac{V}{\log_e r_2/r_1}$$

This is a maximum when $x = r_1$, where the potential gradient is

$$\frac{V/r_1}{\log_e r_2/r_1} \text{ volts per metre}$$

1.4. Electromagnetism; Inductance

We have seen that the disturbance of equilibrium in a dielectric which results in an electric charge gives rise to an electric field, and that energy is stored in the system while this condition persists. If the electrons are free to move, as they are in a conductor, one may expect to find some similar manifestation and, in fact, electrons in motion do produce a force similar in some respects to that which exists between electric charges.

Two wires carrying an electric current exercise a definite force on each other which is evidence of the existence of some effect due to the current. This is known as a *magnetic field.* It does not radiate from the wires like an electric field but surrounds the wire as a sort of sheath of gradually decreasing intensity as the distance from the wire increases.

If we consider a cross-section of this sheath in a plane normal to

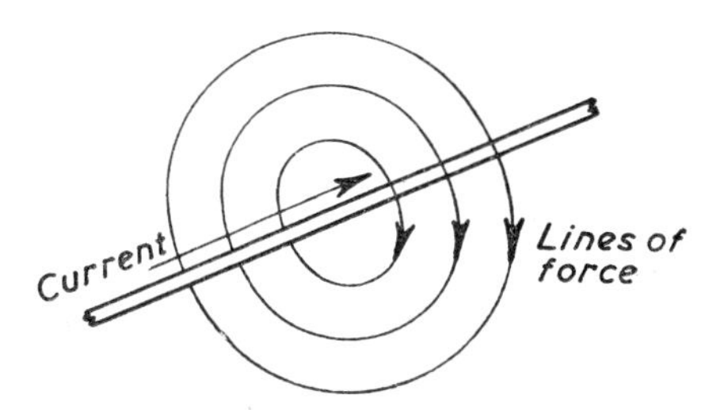

Fig. 1.12. Magnetic Field around a Wire

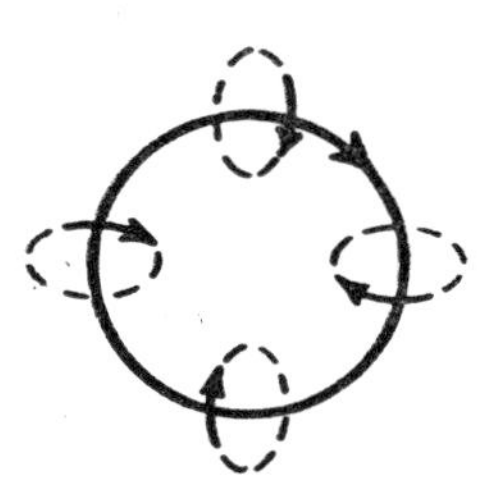

Fig. 1.13. Magnetic Field Produced by a Single Turn of Wire

the wire we can represent the field by a series of lines of force, as shown in Fig. 1.12. It will be seen that they form a series of concentric circles, the direction of the field being at right angles to the direction of the current. Using the conventional direction of current flow as being from positive to negative (which we have seen is actually the opposite direction to the electron flow) the direction of the field is clockwise when the current flow is away from one.

In practice we are usually more concerned with a complete circuit rather than an isolated portion and the simplest form of circuit would be a single turn of wire, as shown in Fig. 1.13. At various points round this turn we can draw the lines of force and it will be seen that they combine to give a force acting along the axis of the turn. It will also be seen that the clockwise rule still applies, the current in any segment of the turn producing a field at right-angles, as in Fig. 1.12.

If the wire takes the form of a number of turns in series the magnetic fields of the individual turns reinforce one another so that the total field is increased. Such an arrangement, called an *inductance coil* or *inductor*, is shown in Fig. 1.14.

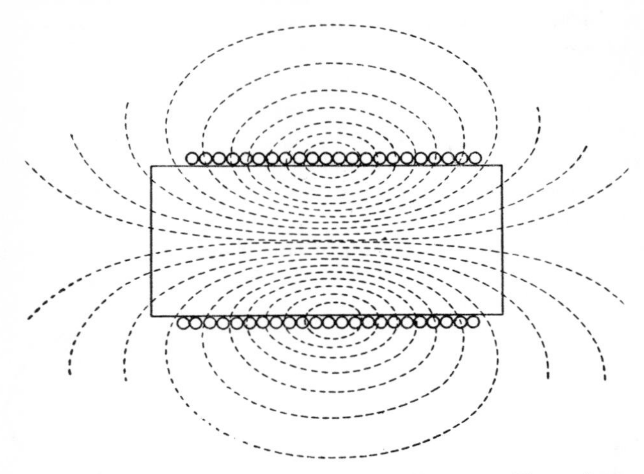

FIG. 1.14. ILLUSTRATING DISTRIBUTION OF MAGNETIC FIELD AROUND AN INDUCTOR

The Magnetic Circuit; M.M.F.

There is an important difference between electric and magnetic fields, namely that the magnetic field is always a complete loop, as is clear from the various diagrams. It forms, in fact, a magnetic circuit similar to a conducting circuit, and in many respects behaves similarly.

The lines of magnetic field are said to constitute a *magnetic flux* which is produced by a *magnetomotive force* (m.m.f.) analogous to the e.m.f. which produces current in an electrical circuit. There is, in fact, a relationship similar to Ohm's Law in a conducting circuit, namely

$$\Phi = M/S$$

where Φ = the flux,

M = the m.m.f., and

S = a quantity analogous to resistance, called the *reluctance*, which will be defined shortly.

Now, the m.m.f. is clearly proportional to the current in the coil (I) and also to the number of turns (N). It is, indeed, defined as the product of these two, so that $M = IN$.

Field Strength; Magnetic Path

For many purposes, however, we are more concerned with the *m.m.f. gradient* or *field strength*, for which the symbol H is used. Strictly speaking the m.m.f. gradient at any point is the change of m.m.f. over a small length of path at that point, i.e. $H = dM/dx$, but if the m.m.f. gradient is assumed to be uniform we can say simply that $H = M/l = IN/l$ where l is the total length of the magnetic path.

It is not always easy to decide what the length of the magnetic path is. Sometimes, as will be seen shortly, the magnetic field can be concentrated within certain limits (as in the toroid of Fig. 1.16) so that the path length can be defined. Otherwise it is necessary to state clearly what path is being considered. For example, the field strength at a distance r from the single wire of Fig. 1.12 can be calculated without difficulty, for at a radius r the magnetic path is clearly the distance round the circle of this radius $= 2\pi r$. The m.m.f. is simply I, since the wire is part of a single turn so that $N = 1$. Hence $H_r = I/2\pi r$.

Permeability

There is, however, another factor involved in determining the total magnetic flux produced, namely the *relative permeability* of the material. It is found, for example, that if a coil such as that shown in Fig. 1.14 is provided with an iron core the flux produced by a given current is increased many hundreds of times.

In any material the electrons revolving in their orbits produce heir own magnetic fields but these have a purely random orientation. In certain materials, an external m.m.f. can produce an alignment of some of these orbits, thus increasing the effective flux.

Materials which have a greater permeability than air are called *paramagnetic* though this term is usually applied only to materials in which the increase is relatively small. Materials which exhibit considerably increased relative permeability are called *ferromagnetic* (though as explained in Chapter 4 some do not actually include iron in their composition). There are also some materials which are called *diamagnetic*, having less permeability than air but the difference is only slight. The substance which exhibits the effect most strongly is bismuth which has a relative permeability of 0·9998.

Flux Density; Units

A given m.m.f., therefore, operating in a material of given permeability will produce a certain total flux. The concentration of this flux will depend on the area over which it is distributed, so that the *flux density* $B = \Phi/A$. The flux density, as in the electrostatic case, is proportional to the field strength and is, in fact, defined by the relationship $B = \mu H$, where μ is the permeability.

The flux density is measured in terms of the number of lines of force per unit area. These lines of force are, of course, an imaginary concept but a greater field strength will obviously produce a greater concentration of lines of force. In the old CGS system the unit chosen was 1 line/cm^2, which was called the *gauss*. The SI unit of flux is the *weber* (Wb) which is 10^8 CGS lines of force, and the unit of flux density is the *tesla* (T), which is 1 weber/metre^2.

The permeability, μ, is itself the product of the *permeability of free space*, or the *magnetic space constant*, μ_0, and the relative permeability μ_r, which for air is 1. The value of μ_0 in SI units is $4\pi/10^7$ so that

$$B = \mu H = 4\pi\mu_r H \,.\, 10^{-7} = 1{\cdot}257\ \mu_r(IN/l) \,.\, 10^{-6} \text{ teslas}$$

Reluctance

We are now able to define the *reluctance* of a magnetic circuit, to which reference was made earlier. It was stated that the flux = m.m.f./reluctance i.e. $\Phi = M/S$, so that $S = M/\Phi$.

But $M = Hl$ and $\Phi = HA\mu$ so that $S = l/A\mu$. Hence the reluctance (equivalent to resistance in a conducting circuit) is directly proportional to the length of the magnetic path and inversely proportional to the area and the permeability.

The reciprocal of the reluctance ($A\mu/l$) is called the *permeance*, analogous to the conductance in an electric circuit.

To evaluate the reluctance in any particular instance it is necessary to know the length and area of the magnetic circuit, and as has been stated this is not always easy to define. This may, however, be ignored for the present. It is discussed further in Chapter 4.

Induced E.M.F.

So far it has been assumed that the current in the wire or coil, and hence the magnetic field, is of constant strength. This is an equilibrium condition, the only back e.m.f. in the circuit being that due to the resistance. If the current changes, however, this equilibrium is disturbed and forces are brought into play which resist any change. The force takes the form of an additional e.m.f. which in magnitude and direction is such as to try to preserve the original conditions. This is an important principle, known as Lenz's Law, and the e.m.f. is called an *induced e.m.f.*

Faraday, in his classic research into electromagnetism, discovered that the induced e.m.f. was proportional to the rate at which the flux was varied, i.e. $e \propto d\Phi/dt$ and this is a fundamental relationship which applies whether the flux is varied mechanically as in a dynamo, or electrically as in an oscillating circuit.

In a coil such as that of Fig. 1.14 the e.m.f. will clearly depend on the influence of the flux changes on the turns of the coil itself, and hence it is called the *e.m.f. of self-induction* which is dependent upon the *flux linkage*, i.e. the manner in which the flux interacts with the turns of the coil. If all the flux linked with every turn the linkage would be simply $N\Phi$, where N is the number of turns, but it is clear that this is not necessarily the case. In Fig. 1.14, for example, it is only the turns in the middle portion of the winding which are linked with the total flux, so that in a practical case the linkage is $kN\Phi$, where k is a constant which depends on the configuration of the circuit. It is always less than unity, though with suitable arrangements it may be made very close to unity.

Inductance

The linkage $kN\Phi$ may be written $(kN\Phi/I)I$ and hence the induced e.m.f., which is the rate of change of linkage, may be written as $e = (kN\Phi/I) \times$ rate of change of current. The expression $kN\Phi/I$, which is the flux linkage per unit current, is called the *inductance* of the coil. The unit of inductance is the *henry*, which is defined as the inductance in which an e.m.f. of one volt is induced when the current changes at a rate of one ampere per second.

We have already seen that $\Phi = \mu ANI/l$, so that the inductance

$$L = kN\Phi/I = k\mu AN^2/l$$

where N = number of turns,
A = area of flux path,
l = length of flux path,
μ = permeability, and
k = linkage factor.

The e.m.f. of self-induction is the rate of change of linkage. Hence

$$e = kNd\Phi/dt = kN(\mu AN/l)\,di/dt = L\,di/dt$$

where i is the instantaneous current. (Note that when dealing with varying quantities it is customary to use small letters to represent the instantaneous values whereas capital letters are used for steady or average values.)

Since $\mu = \mu_r\mu_0$ we can substitute $4\pi/10^7$ for μ_0, so that if l and A are in metres and square metres,

$$L = 1{\cdot}257\, k\mu_r AN^2/l \times 10^{-6} \text{ henrys}$$

Cylindrical Coils

A coil of the form of Fig. 1.14 is called a *solenoid* and it will be seen that its inductance is considerably influenced by its shape. With a coil of short axial length the linkage factor k may approach unity, but as the length increases k becomes progressively less. The variation of this factor with relative dimensions is shown in Fig 1.15.

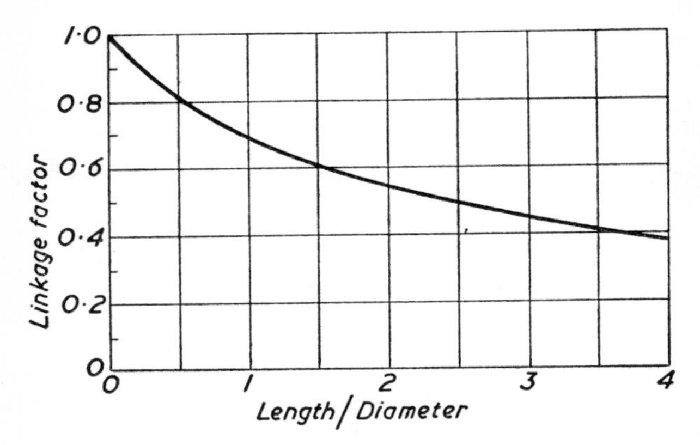

FIG. 1.15. VALUE OF LINKAGE FACTOR FOR SINGLE-LAYER COILS

If the coil is wound with more than one layer so that it has appreciable radial thickness the flux linkage is again reduced, and a further correction factor has to be introduced dependent on the ratio of radial thickness to mean diameter. Curves or abacs for calculating inductance, which allow for these various factors, are readily available today.

For a quick approximation the following empirical formula may be used

$$L = 0{\cdot}2\,\frac{n^2 D^2}{3{\cdot}5D + 8l} \times \frac{D - 2{\cdot}25d}{D} \text{ microhenrys}$$

where D = outside diameter of coil,
l = axial length of winding, and
d = radial thickness of winding.

The dimensions are in inches and the formula is valid within a few per cent within the limits $l/D = 2$ and $d/D = 0{\cdot}3$.

Rectangular Coils

Inductances are sometimes wound on rectangular formers. In such cases the inductance may be calculated to a sufficient accuracy for practical purposes by regarding the coil as equivalent to a cylindrical coil having a diameter equal to the mean of the inscribed and escribed circles. Thus with a former having sides a and b (a being the larger) the inscribed circle diameter is b, while the escribed circle diameter is $\sqrt{(a^2 + b^2)}$. The equivalent circular diameter is then $\frac{1}{2}[\sqrt{(a^2 + b^2)} + b]$.

Self-Capacitance

In practice, no coil is a pure inductance. Apart from the fact that the winding necessarily possesses resistance, there is also some capacitance between the various turns, and between the terminal connections with the material of the coil former as dielectric. This is known as the distributed or *self-capacitance.*

It is not readily calculable, and is normally quite small (a few pF only), but is obviously considerably larger with a multi-layer coil. Its influence is only of importance at frequencies where its reactance (see page 44) becomes comparable with that of the inductance.

It forms part of the unavoidable stray capacitances which are present in any practical circuit, as discussed later.

Toroidal Coils

We have seen that the magnetic flux has to form a complete circuit. If the turns of a coil could be arranged to enclose the flux throughout, the linkage would be virtually perfect. Such a coil is called a *toroid* (Fig. 1.16) and has the merit that the flux is almost

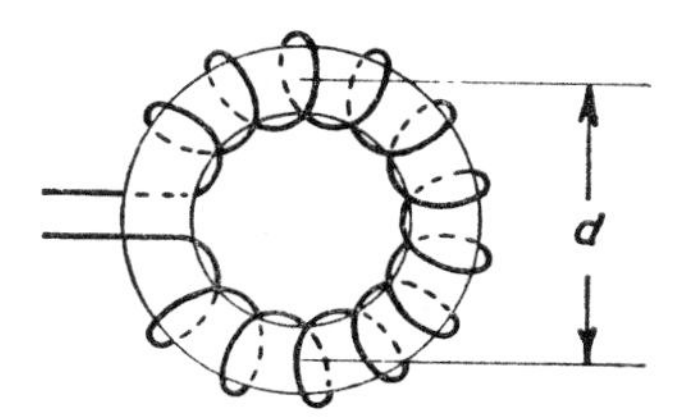

FIG. 1.16. TOROIDAL COIL

entirely confined within the winding, so that there is a minimum of "stray" field, which is often important, while the length of the magnetic path is definite, being πd, where d is the mean diameter of the ring.

The inductance of a toroid is thus simply $\mu AN^2/\pi d$, which, since $\mu = \mu_0\mu_r = 4\pi\mu_r/10^7$, reduces to $0{\cdot}4\ \mu_r AN^2/d$ microhenrys (where d is in metres and A in square metres). The ring may be of insulating material or of some magnetic material, according to requirements, but toroids are difficult to wind and are only used for special purposes.

The alternative method of confining the flux is to provide a path for the flux made of iron or other paramagnetic material; this is an important part of practical communication engineering and is discussed more fully in Chapter 4.

Mutual Inductance

If the flux produced by a coil is allowed to link with the turns of another coil as in Fig. 1.17 any variation in the flux will induce an e.m.f. in the second coil as well as the first. This is called an e.m.f. of *mutual induction* and there is said to be *mutual inductance* between

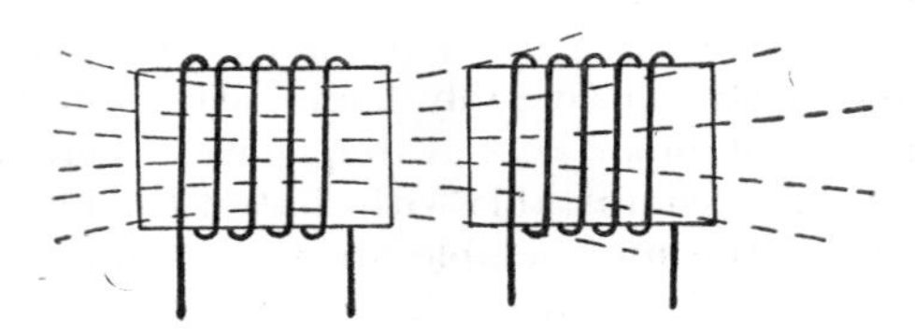

FIG. 1.17. ILLUSTRATING MUTUAL INDUCTANCE

the coils. The unit of mutual inductance is also the henry, which is the mutual inductance for which an e.m.f. of one volt is induced in the *secondary* if the current in the *primary* changes at a rate of one ampere per second.

In symbols $$e_2 = M\, di_1/dt$$

This possibility of inducing e.m.f. in an entirely separate circuit is of great value in practical engineering. Two coils so arranged that the flux from one links with the other are called a *transformer*, which is discussed more fully in Chapters 2 and 4.

Inductances in Series and Parallel

If two inductances L_1 and L_2 are connected together, their combined inductance will be the total linkage produced by unit current flowing through the combination.

If the coils are connected in series, *each* coil will have associated with it a flux due to its self-inductance, and also a flux due to the mutual inductance M from the other coil. The total linkage is therefore

$$L_{\text{series}} = (L_1 + M) + (L_2 + M) = L_1 + L_2 + 2M$$

If the coils are connected in parallel, each coil will only carry a portion of the total current. The appropriate proportion can be calculated and hence the combined linkage can be deduced as

$$L_{\text{parallel}} = \frac{L_1 L_2 - M^2}{L_1 + L_2 - 2M}$$

The above expressions assume that the coils are wound in the same direction so that M is positive. If the coils are in opposition $-M$ must be substituted for M.

Variable Inductors

The inductance of a coil (or circuit) may be varied by varying the number of turns or the permeability. Both methods are used, while a third method is to use two coils in series or parallel and to vary the mutual inductance between them. Practical forms of inductor, both fixed and variable, are discussed in Section 4.3.

Current in an Inductive Circuit

If an e.m.f. is applied across an inductor the magnetic field is not established immediately. As in the case of a capacitor, time is required for the circuit to adjust itself to the new conditions. When the current begins to flow a back e.m.f. will be produced proportional to the rate of change of current. This will limit the current which will thus build up to its final steady value in an exponential fashion similar to the growth of charge in a capacitor.

At any instant the current is given by the relationship

$$E - iR - L\, di/dt = 0$$

This is a differential equation of which the solution is

$$i = (E/R)(1 - e^{-Rt/L})$$

which is an exponential relationship similar to that of Fig. 1.7. At the instant of switching on, the rate of change of current is very large and the applied e.m.f. is almost entirely occupied in overcoming the back e.m.f. of self-induction. The term $e^{-Rt/L}$ is, in fact, unity and $i = 0$.

As the current begins to flow the rate of change decreases, the term $e^{-Rt/L}$ becomes smaller, until when the current finally reaches its steady value there is no inductive back e.m.f. and the current is simply limited by the resistance.

Similarly if the e.m.f. is removed the current does not immediately cease but decays exponentially. It may be noted here that, if the circuit is broken so that R becomes infinite, the attempt to maintain the current flow during this decay period will result in a very high momentary induced e.m.f. which may reach a dangerous value, causing breakdown of the insulation.

The term L/R is called the *time constant* and is the time taken for the current to reach 63 per cent of its final value since, when $t = L/R$, the current $i = (E/R)(1 - 1/e) = 0{\cdot}6321\ E/R$.

Energy Stored in an Inductor

From what has just been said it will be clear that work has to be done in establishing a magnetic field, and hence, in an inductor carrying current, energy will be stored in the magnetic field. During the process of establishing the field the current is varying exponentially and the total energy will thus be the sum of the instantaneous energies during the period.

Now over a small element of time δt the energy is $eq = ei\delta t$ and $e = L\,di/dt$. Hence if the final steady current is I, the total energy involved in establishing the field (and so stored in the inductor while the current continues to flow) is

$$W = \int_0^I Li\frac{di}{dt}\cdot dt = \tfrac{1}{2}LI^2$$

Permanent Magnets; Polarity

Reference has already been made to the increased permeability of ferromagnetic materials due to the internal realignment of the molecules under the influence of an m.m.f. With certain materials, such as steel, this realignment persists after the m.m.f. has been removed and the material is then said to be permanently magnetized.

Such permanent magnets are used to a considerable extent in electrical engineering practice and small magnetized rods are frequently employed to demonstrate some of the effects of magnetic fields. For example, if one magnet is brought near to another magnet it will be found that one end will be attracted and the other repelled. This is because of the interaction of the lines of force. If these are in the same direction they will combine, while if they are in opposite directions they will repel, as shown in Fig. 1.18. If the magnets are not directly in line, they will try to locate themselves so that each lies along the direction of the magnetic field at that point.

The two ends of a magnet, in fact, exhibit different polarities, usually referred to as N and S respectively, as explained in the next section; and, as with electric charges, like poles repel and unlike poles attract.

The poles, of course, are not isolated like electric charges though sometimes the arbitrary concept of a unit magnetic pole is employed. From such a concept it can be shown, by reasoning similar to that used in the electrostatic case, that the force exerted by a magnet falls off inversely as the square of the distance away, which is true

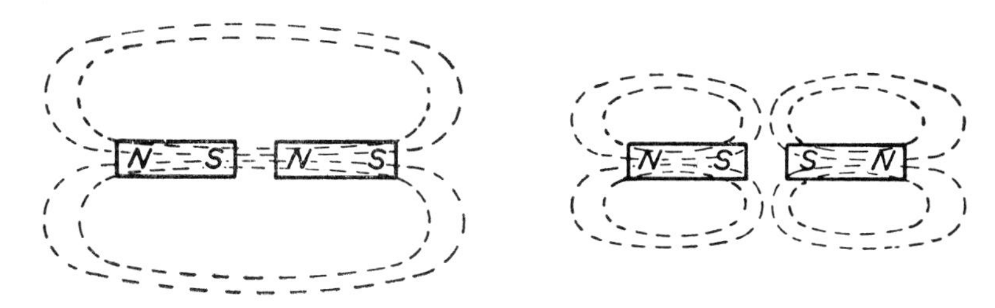

FIG. 1.18. BAR MAGNETS

so long as the field is not distorted. It is also possible to introduce the idea of magnetic potential as the work done in bringing a unit magnetic pole from infinity to the particular point in the field (analogous to the definition of electric potential). These concepts, however, are mainly of academic interest and are not employed in practical engineering.

Earth's Magnetic Field

It has long been known that the earth itself is a huge permanent magnet. Many hundreds of years ago mariners were assisted in their navigation by a property possessed by pieces of naturally magnetized iron ore known as lodestones which, if freely suspended, would take up a position with one end pointing North. Hence the poles of a magnet are termed N and S respectively.

Actually the earth's field is relatively small ($\simeq$ 14 A/m at European latitudes). It behaves as if produced by a magnet located approximately on the earth's axis. The actual axis is displaced some 12° from the spin axis but varies slightly from year to year. The effect is thought to be due to circulating currents in the liquid core of the earth but there are many subsidiary influences which also contribute.

The field is not parallel to the surface of the earth except near the equator. As the latitude increases the field becomes inclined at an increasingly steep angle, which in European latitudes is some 67°. Compass needles have therefore to be provided with a small weight on one pole to counteract the *magnetic dip*, as it is called.

Force on a Conductor

An important aspect of the magnetic field is the force exerted on a conductor. Consider a wire carrying a current situated in a uniform magnetic field as shown in Fig. 1.19. This is a perspective

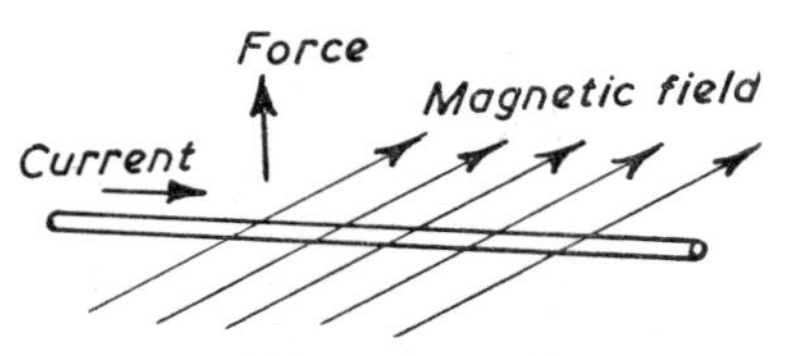

FIG. 1.19. FORCE ON A CONDUCTOR

view representing a horizontal magnetic field with a wire, also horizontal but at right angles to the flux, carrying a current I. The wire will experience a mechanical force tending to move it out of the field, at right angles to the directions of the field *and* the current. With the field and current in the directions shown the force will be vertically upwards.

The direction of motion can be memorized by using the left-hand rule. If the thumb, first and second fingers of the left hand are

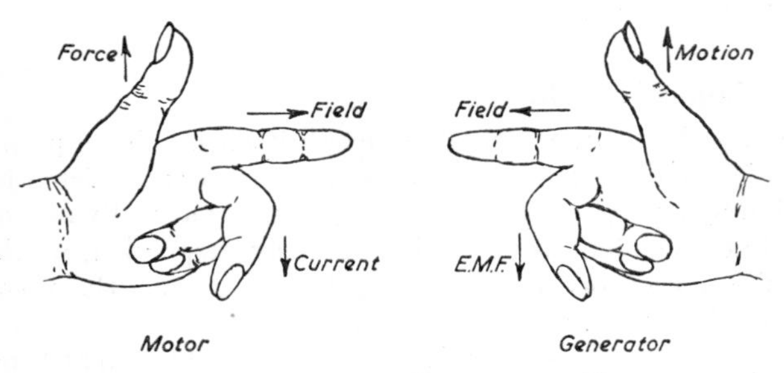

FIG. 1.20. LEFT- AND RIGHT-HAND RULES

arranged at right angles, then if the first and second fingers represent the direction of the field and the current respectively the thumb will indicate the direction of motion, as shown in Fig. 1.20.

The force produced is given by $F = BIl$, where B is the flux density, I is the current and l is the effective length of conductor within, and at right angles to, the field. (If the conductor is not at right angles, its effective length becomes $l \sin \theta$, where θ is the angle between the conductor and the magnetic field.)

Motors and Generators

The existence of this force is the basis of all electric motors and indicating meters, practical forms of which are discussed later. It is usual to construct the moving element in the form of a coil, in which case the force produced is increased N times, where N is the number of turns in the coil. From this the torque and power can be determined as explained in Chapter 4.5.

Conversely, if the wire is moved mechanically across (i.e. at right angles to) the field, an e.m.f. will be induced tending to oppose the movement. This e.m.f. will therefore be in the opposite direction to the current which would produce the same motion in a motor. In this case, therefore, the right-hand rule should be used, with the thumb and first finger representing the direction of motion and field, when the direction of the induced e.m.f. is indicated by the second finger (Fig. 1.20).

This is the basic principle of the electric generator, the induced e.m.f. being $e = Blv$ volts where v is the velocity of motion in metres/sec and B and l are the flux density and length of the conductor as before.

This expression may be deduced as follows. In one second the conductor will move v metres across the field; hence the area swept out will be lv and the flux enclosed during the period will be Blv. This is the change of flux in one second, i.e. the rate of change of flux which, as we have seen, is the e.m.f. generated.

Circuit Inductance

Although inductance is usually deliberately contrived in a concentrated form by winding turns of wire into a coil, it must be remembered that any closed circuit possesses inductance. Where the current is varying very rapidly the effect of the inductance of the circuit wiring itself may be appreciable.

Consider the field existing at a distance x from the centre of a wire carrying a current I (Fig. 1.12). This field will surround the wire in a concentric sheath of elemental thickness dx, the cross-sectional area of which, for unit length, will be dx. The field strength $B = \mu I/2\pi x$ per unit length so the flux $= \mu I dx/2\pi x$. The total flux will be the sum of all these elemental fluxes over the appropriate range of values for x.

The lower limit is clearly the radius of the wire itself, r, while the upper limit must be the distance d from the nearest return part of the circuit. (The inductance of an isolated wire, in fact, has no meaning. The wire only possesses inductance as part of a circuit.)

With this understanding we can evaluate the total flux, and hence the inductance which, by definition, is the flux per unit current, so that

$$L = \int_r^d (\mu/2\pi) \frac{1}{x} \, . \, dx = (\mu/2\pi) \log_e d/r$$

In air $\mu = \mu_0 = 4\pi/10^7$, so that

$$L = 2 \log_e d/r \, . \, 10^{-7} \text{ henrys/metre}$$

$$= 0{\cdot}2 \log_e d/r \text{ microhenrys/metre}$$

This is an expression which is easily memorized. There is, in fact, a slight correcting term which takes account of the flux inside the wire itself but this can usually be neglected.

Note that the larger the radius of the wire the less the inductance. Hence when a lead is required to be of low inductance it is best made of large diameter (or by using several wires symmetrically disposed round the circumference of a circle).

Two Parallel Wires

For transmission lines consisting of two parallel wires the above expression gives the inductance *per conductor*. The inductance of the pair is therefore twice as great, so that

$$L = 0{\cdot}4 \log_e d/r \text{ microhenrys/metre}$$

Coaxial Cable

For a coaxial cable, as in Fig. 1.11, the same expression applies. In this case d is the *radius* of the outer conductor, so that

$$L = 0{\cdot}2 \log_e R/r \text{ microhenrys/metre}$$

1.5. Alternating Currents

As has been stated, radio communication is largely concerned with currents which are not steady but flow first in one direction and then in the other. Such "alternating" currents, as they are called, can be produced in various ways, one of which is to rotate a coil in a magnetic field.

Fig. 1.21 illustrates such an arrangement. CC is the cross-section of a single-turn coil which is rotated at a steady speed in a uniform magnetic field. As the coil is rotated, the flux linked with the coil changes. Due to this changing linkage, e.m.f.s are set up according to the laws of magnetic induction, proportional to the rate of change of the linkage.

Consider the e.m.f. induced in the coil when it is uniformly rotated. The rate of change of linkage depends on the rate at which the coil moves across the field, i.e. the rate at which it *cuts* the magnetic lines of force. At the points DD the coil is not moving across the field at all, and the e.m.f. induced will be zero. As the coil is rotated, however, it begins to cut more and more lines of force until at the points AA it is moving across the field at the maximum rate and a maximum e.m.f. is induced. As the coil continues to rotate, the number of lines cut in a given time grows less and less until it reaches the position DD again and the e.m.f. is zero. It then proceeds to move across the field in the opposite direction, and at the point AA a maximum e.m.f. is again induced in the coil, this time, however, in the opposite direction, after which it once more dies away to zero.

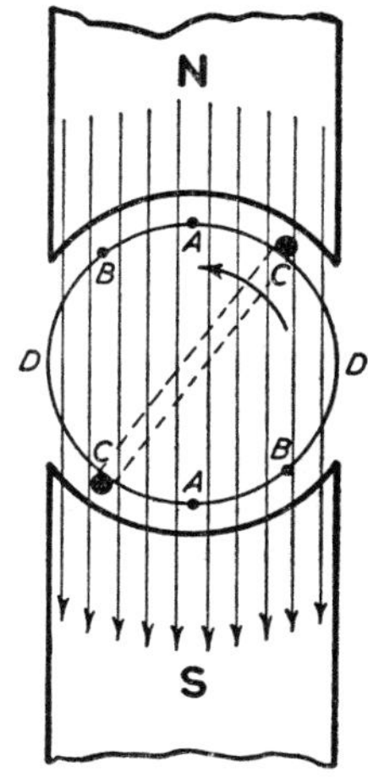

FIG. 1.21. CROSS-SECTION OF SIMPLE ALTERNATOR

Sine Waves; Vectors

The manner in which the e.m.f. will vary is illustrated in Fig. 1.22 where the line OA represents the position of the conductor in the

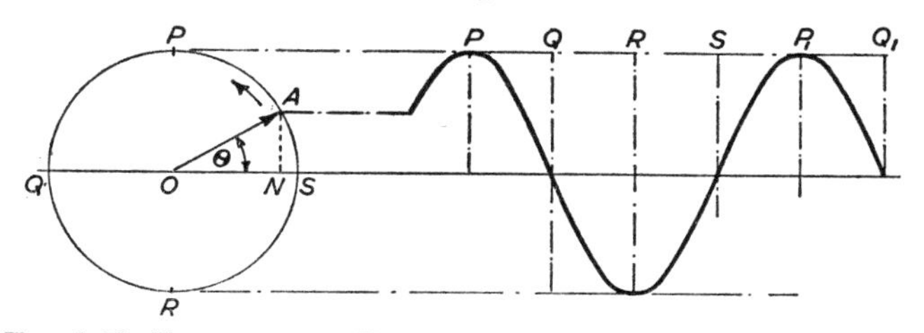

FIG. 1.22. ILLUSTRATING PRODUCTION OF ALTERNATING CURRENT

field. If OA is rotating anticlockwise with a uniform velocity the e.m.f. induced is proportional to the rate at which the inductor moves across the flux, i.e. the component of the velocity at right angles to the field. From simple geometry this can be shown to be proportional to the ordinate AN.

When $\theta = 90°$, $AN = OA$, so that if we say that OA represents the maximum e.m.f. E, the instantaneous e.m.f. $e = E \sin \theta$.

Hence the variation of e.m.f. with time will be a rhythmic change from positive to negative which is called a *sine wave*. The line OA is called the *vector* of the wave and, as will be seen, it is often convenient to consider a wave from the point of view of the vector producing it rather than in its actual sine wave form. For convenience and uniformity vectors are usually assumed to rotate in an anti-clockwise direction.

Phase Difference

Consider the wave produced by a second vector OB, at an angle of 90° with OA, as shown in Fig. 1.23 by the dotted line. This wave will pass through its maximum before the original wave, and for this reason is said to lead by 90° on the wave OA. Conversely the wave OA is said to lag 90° behind OB. A wave may lead or lag by less or more than 90°, however, and in fact any intermediate position

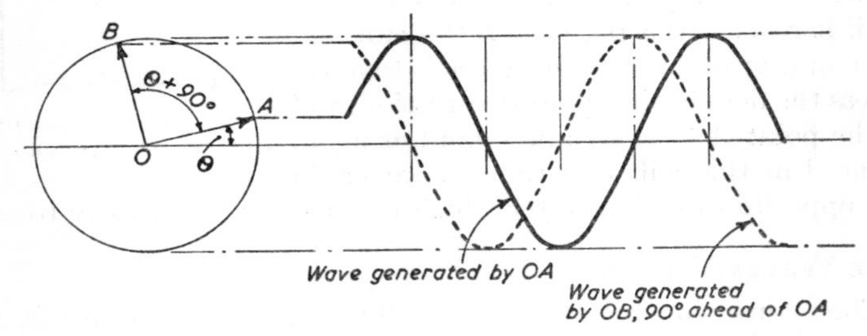

FIG. 1.23. ILLUSTRATING PHASE DIFFERENCE

between being *in phase* and 180° *out of phase* (i.e. exact opposition) may obtain, depending on the conditions.

The following points should be noted with respect to phase differences. If two waves are in phase they rise and fall exactly together. A *leading* wave reaches its maximum, in a given direction, *before* the second wave, whereas a *lagging* wave reaches its maximum *after* the other wave. If a wave leads or lags *by 90°* on a given wave it passes through its maximum when the original wave is passing through zero, and vice versa.

Fig. 1.24 illustrates this point; it also serves to indicate how two waves of different strengths may be represented, the length of the vector being made proportional to the actual strength of the wave in question. The wave marked voltage (full line) is due to the vector OE. The wave marked "current leading" is due to OI_2, which is 90° ahead of OE. The wave marked "current lagging," however, is due to the vector OI_1, which is only lagging 45° behind

OE. Here it will be observed it reaches its maxima *after* the full-line wave but *before* it has reached zero.

The question of phase difference between sine waves is very important, and the student will do well to study the subject until

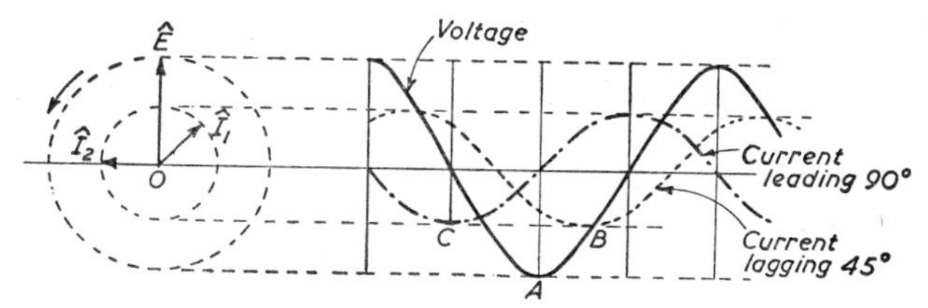

FIG. 1.24. VOLTAGE AND CURRENT MAY BE REPRESENTED ON THE SAME DIAGRAM

it is thoroughly understood. If any difficulty is experienced he should draw sine waves for himself. This may be done by drawing a circle and marking off points on the circumference every 30°. At each position the projection AN (Fig. 1.22) is measured. A horizontal base is then taken and 12 points $\frac{1}{4}$ in. apart are marked off. These points represent, to a time basis, the successive points on the circle spaced 30° apart, at each of which the appropriate value of AN is plotted, as in Fig. 1.22.

Frequency

Referring again to Fig. 1.22, it is obvious that each complete revolution of the vector OA will produce one complete wave. The number of such waves occurring in one second is termed the *frequency*. For many years the abbreviation c/s was used to denote the number of cycles per second, but this has now been replaced by the *hertz* (Hz) as the unit of frequency (= 1 c/s).

Since the range of frequencies encountered in practice is very wide, suitable prefixes are used to denote appropriate multiples,

$$\text{kHz (kilohertz)} = 10^3\,\text{Hz}$$
$$\text{MHz (megahertz)} = 10^6\,\text{Hz}$$
$$\text{GHz (gigahertz)} = 10^9\,\text{Hz}$$

The frequency of the normal alternating current (a.c.) supply is 50 Hz, though 25 Hz is occasionally found. In the Americas, and on H.M. Ships, 60 Hz is used. Generators for aircraft or radar often use higher frequencies (e.g. 400 Hz) as explained in Chapter 4.

Speech and music involve frequencies ranging broadly from 15 Hz to 20 kHz, while radio frequencies extend well into the gigahertz region, as detailed in Section 3.1 (page 134).

Equation to a Sine Wave

We have seen that the e.m.f. of a sine wave is $e = E \sin \theta$, where E is the maximum value, and θ is the angle through which the rotating vector OA in Fig. 1.22 has moved. Since OA is rotating uniformly, θ depends on the time which has elapsed since the beginning of the cycle. If n is the number of revolutions made in a given time t, then $\theta = n \times 360° = 2\pi n$ radians.* (Note that n need not be a whole number.)

If the frequency of the wave is f, OA will revolve f times per second, so that in time t the number of revolutions is $n = ft$.

Hence $$\theta = 2\pi ft \text{ radians}$$

The term $2\pi f$ is called the *angular velocity* of the vector, and is usually represented by ω. Hence

$$\theta = \omega t \quad \text{and} \quad e = E \sin \omega t$$

This is, therefore, a means of expressing the actual value of the e.m.f. at any particular instant, in terms of the maximum value of the wave, the frequency of the wave, and the time.

The time required for a complete cycle is called the *period* of the wave; for the wave just considered, the period is clearly $1/f$, so that the frequency f is sometimes called the *periodicity*.

A lagging or leading wave may be expressed in the same form. In Fig. 1.23 if θ is the angle through which OA has revolved, OB will have revolved through an angle of $\theta + 90°$. Hence the equation to the wave generated by OB is

$$e = E \sin (\theta + 90°)$$

Since $360° = 2\pi$ radians, $90° = \pi/2$ radians; $\theta = \omega t$, so that

$$e = E \sin (\omega t + \pi/2)$$

In general if a wave is leading or lagging by an angle ϕ, the wave may be expressed in the form $e = E \sin (\omega t \pm \phi)$, the $+$ sign being used when the wave is leading, and the $-$ sign when it is lagging.

Rate of Change of a Sine Wave

A sine wave is possessed of a very valuable property relating to the manner in which its value changes from instant to instant. Referring once more to Fig. 1.22 and considering successive values

* A radian is the angle through which the vector must move for the arc traced out to be equal to the radius of the vector. In one complete revolution the arc traced out is the complete circumference of a circle, which is 2π times the radius. Hence $360° = 2\pi$ radians.

of AN very close to each other, it will be seen that at the points P, R, P_1 the curve is horizontal; i.e. at that particular instant it is not changing at all and the rate of change is zero. At the points Q, S, Q_1, however, the value is changing rapidly and the rate of change is a maximum.

It is found that if the rate at which AN increases or decreases is plotted against the time, a second sine wave is obtained leading by 90° on the original wave. A little thought will show that the actual numerical value of this wave must be proportional to the frequency of the original wave, for the higher the frequency, the more rapid must be the actual rate of change, quite apart from any other considerations. Actually if $e = E \sin \omega t$ the rate of change of e.m.f. is

$$\frac{de}{dt} = E\omega \cos \omega t = E\omega \sin\left(\omega t + \frac{\pi}{2}\right)$$

This is a highly important property of sine waves which may be stated as follows:

The rate of change of a sine wave is a second sine wave leading by 90°, whose maximum value is ω times as great as that of the original wave.

Current in a Circuit

We have discussed the simple laws relating current and voltage in a circuit carrying direct current. We will now extend this investigation and consider the current in a circuit having an alternating e.m.f. applied across it.

Resistive Circuit

Consider first a circuit containing resistance only. At each instant, by Kirchhoff's Law, $v - iR = 0$, i.e. $i = v/R$. Hence if

$$v = V \sin \omega t, \quad i = \frac{V}{R} \sin \omega t$$

This is a second sine wave *in phase* with the first, so that in a circuit containing resistance only, the voltage and current rise and fall together.

Inductive Circuit

Consider the current in an inductance, neglecting the resistance for the time being. Due to the varying current in the coil, an e.m.f. of self-induction will be set up, proportional to the rate of change of the current.

Now, in Section 1.4 it was shown that the e.m.f. of self-induction is $L\,di/dt$. Hence $v - L\,di/dt = 0$. If

$$i = I \sin \omega t, \quad di/dt = I\omega \cos \omega t$$

Hence $\qquad v - IL\omega \cos \omega t = 0 \quad \text{and} \quad v = IL\omega \cos \omega t$

The term $IL\omega$ is clearly the maximum value of the voltage, V, so that we can write

$$v = V \cos \omega t = V \sin (\omega t + \pi/2)$$

Both the current and voltage, therefore, are sine waves, but are 90° out of phase, *the current lagging* 90° *behind the voltage.*

CAPACITIVE CIRCUIT

Consider, thirdly, the current in a capacitor when an alternating voltage is applied across the plates. Here, at any instant, if q = charge, C = capacitance, and v = voltage, then

$$q = Cv$$

by definition (Section 1.3).

The current in a capacitor is the rate at which the charge is varying (from the definition of current); i.e.

$$\text{Current} = \text{Rate of change of charge} = dq/dt$$

But $q = Cv$ and, since C is constant and only v is varying, we may write $i = C\,dv/dt$

Let $\qquad v = V \sin \omega t$

Then $\qquad dv/dt = V\omega \cos \omega t$

Therefore $\qquad i = VC\omega \cos \omega t = VC\omega \sin (\omega t + \pi/2)$

Hence the current in a capacitor having a sine wave voltage applied across it is also a sine wave *leading by* 90°.

As far as phase is concerned, therefore, the results may be summarized as follows:

(*a*) Resistance: Current and voltage in phase.
(*b*) Inductance: Current lagging 90° behind the voltage.
(*c*) Capacitance: Current leading 90° ahead of the voltage.

Reactance

In the inductive case we wrote V for $IL\omega$ so that $I = V/L\omega$. This is analogous to the resistive case where $I = V/R$, so that an inductance behaves (numerically) like a resistance of value $L\omega$.

This is called the *inductive reactance* and is usually given the symbol X.

Similarly in the capacitive case where $i = VC\omega \cos \omega t$, the term $VC\omega$ is the maximum value of the current $= I$. Hence

$$I = VC\omega = V/(1/C\omega)$$

The capacitance is thus equivalent numerically to a resistance of value $1/C\omega$, which is called the *capacitive reactance*.

It will be noted that the reactance is dependent upon frequency. Inductive reactance, X_L, increases as the frequency is increased while capacitive reactance, X_C, decreases.

The reciprocal of the reactance is called the *susceptance*.

Impedance

Consider a coil having resistance as well as inductance. Let the current be $i = I \sin \omega t$; then the voltage in the circuit must be equal to the sum of the back e.m.f.s due to both resistance and inductance. Therefore

$$\begin{aligned} v &= iR + L\, di/dt \\ &= IR \sin \omega t + IL\omega \sin(\omega t + \pi/2) \\ &= I\{R \sin \omega t + L\omega \sin(\omega t + \pi/2)\} \end{aligned}$$

This expression may be simplified by trigonometry to the following:

$$v = I\sqrt{[R^2 + (L\omega)^2]} \, . \sin(\omega t + \phi)$$

where ϕ is the phase difference between the current and voltage. It is less than 90°, being such that $\tan \phi = L\omega/R$.

The quantity $\sqrt{[R^2 + (L\omega)^2]}$ is called the *impedance* and is usually given the symbol Z.

It will be noted that the impedance $Z = \sqrt{(R^2 + X^2)}$, where X is the reactance. This is a fundamental relationship applying to any form of circuit. Thus the impedance of a capacitance in series with a resistance is

$$Z = \sqrt{[R^2 + (1/C\omega)^2]}$$

The reciprocal of the impedance is called the *admittance*, and it is sometimes more convenient to use admittance than impedance, as explained on page 86.

Phase Angle

As just said, the phase angle is determined by the ratio of reactance to resistance, so that in general terms $\tan \phi = X/R$. However, for many purposes it is convenient to express it, in inverted form, in terms of the ratio of the resistance to the total impedance, as explained on page 48. The phase angle is then given by the relationship $\cos \phi = R/Z$ as will be seen from the vector diagram in Fig. 2.2).

Circuit Containing Resistance, Inductance, and Capacitance

Consider the simple case of a coil in series with a capacitor. If we pass an alternating current through the circuit, the voltages across each will be out of phase with the current.

The current in an inductance lags 90° behind the voltage, so that, conversely, we may say that the voltage leads by 90° on the current. The capacitance, on the other hand, behaves in an exactly opposite manner, and the voltage lags 90° behind the current.

The voltages are thus in opposition, reaching their maxima at the same instant, but in opposite directions, and they will go through every intermediate point together, but in opposition. They will thus tend to cancel each other out, and in fact can be replaced by one single wave, whose value is the numerical difference of the two component waves.

The reactance of the circuit is thus $L\omega \sim 1/C\omega$, the sign $\sim$ signifying the numerical difference between the two quantities. If the former term is greater than the latter the inductive effect preponderates, and the whole circuit behaves like an inductance (of smaller value than that of the coil itself). If the second term is the larger, the reverse is the case, and the arrangement behaves as if it were a capacitance.

Where a circuit contains resistance, the impedance is obtained in an exactly similar manner to that for the simpler circuits already considered. For the reactance X, however, we must insert the expression deduced above, so that the whole becomes

$$Z = \sqrt{[R^2 + (L\omega \sim 1/C\omega)^2]}$$

It will be clear that by a suitable choice of values the inductive and capacitive reactances may be made to cancel. This happens when $L\omega = 1/C\omega$, a condition known as *resonance*.

R.M.S. Values

As already mentioned, it is not always necessary or convenient to use the full expression for a sine wave. We can, of course, simply use the maximum value and this is sometimes done, but it is clearly more useful to adopt a value which would produce the same results as a steady or direct current (usually abbreviated to d.c., while alternating current is abbreviated to a.c.).

This can be done in terms of energy, which is not dependent on the direction of the current. As was shown in Section 1.2, the energy is proportional to the square of the current or e.m.f., and hence if the current is varying, the energy developed during any cycle will be proportional to the mean value of the *square* of the current at successive instants, and the equivalent steady current would be one having a value equal to the square-root of this "mean square." This equivalent value is thus called the *root-mean-square* or r.m.s. value of the current (or e.m.f.) and it may be noted that it applies to any form of fluctuating waveform, and not merely to sine waves.

The mean-squared value of a sine wave $\hat{E} \sin \theta$ is

$$\frac{1}{2\pi}\int_0^{2\pi} \hat{E}^2 \sin^2 \theta \, d\theta$$

$$= \frac{\hat{E}^2}{4\pi}\int_0^{2\pi} (1 - \cos 2\theta)\, d\theta$$

$$= \frac{\hat{E}^2}{4\pi}\left[\theta - \frac{1}{2}\sin\theta\right]_0^{2\pi} = \frac{\hat{E}^2}{2}$$

Hence the r.m.s. value

$$E = \hat{E}/\sqrt{2} = 0{\cdot}707\hat{E}$$

conversely, the peak value

$$\hat{E} = 1{\cdot}414E$$

Form Factor

Alternatively, the value of an alternating waveform may be expressed in terms of its mean *rectified* value. The average value of a sine wave (or any symmetrical waveform) is clearly zero since

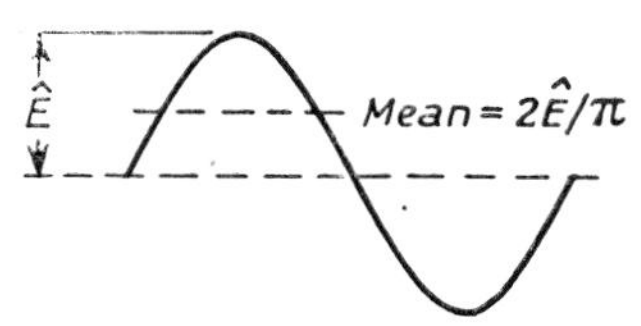

FIG. 1.25. ILLUSTRATING MEAN VALUE OF A SINE WAVE

each half cycle is followed by a corresponding half cycle in the reverse direction. Each half cycle, however, will have a mean value as shown in Fig. 1.25.

If $e = \hat{E} \sin \theta$ the mean value over a half cycle will be

$$\frac{\hat{E}}{\pi}\int_0^{\pi} \sin\theta \, d\theta = 2\hat{E}/\pi$$

The r.m.s. value, as just stated, is $\hat{E}/\sqrt{2}$. The ratio between the r.m.s. and mean values is called the *form factor* and is equal to

$$\pi/2\sqrt{2} = 1{\cdot}11$$

It must be remembered that these simple expressions only apply to pure sine waves. Many of the waveforms encountered in practice are complex, in which case the r.m.s. and mean values, and the form factor, are quite different as will be seen shortly.

Power

As explained in Section 1.2, the power in a circuit is obtained by multiplying the voltage by the current. As long as the current and voltage are in phase it is sufficient to multiply the r.m.s. values together, and this gives the power in watts. Thus a circuit carrying a current of 10 amps and having a voltage across it of 100 volts, will absorb a power of 1,000 watts or 1 kilowatt (kW).

If the current and voltage, however, are *not* in phase the power is not as great. Indeed, a pure inductance or capacitance (i.e. having no resistance) cannot absorb any power at all. In such cases the voltage and current are 90° out of phase. Thus at the instant when the voltage is a maximum, the current is zero and, therefore, the power is nothing. Similarly, when the current is a maximum the voltage is zero, so that once again the power is zero.

In the intervening periods energy is alternately absorbed from and returned to the circuit, but the average power consumed remains zero. Current which is 90° out of phase is often referred to as "wattless" current, for this reason.

Power Factor

In a practical circuit, therefore, power will only be developed in the resistance, and to calculate this one must multiply the current by that component of the voltage which is expended in overcoming the back e.m.f. across the resistance. Hence the apparent power EI must be multiplied by a *power factor*, which is the ratio of the resistance to the total impedance of the circuit. Hence

$$\text{Power factor} = R/\sqrt{(R^2 + X^2)}$$

It is shown in the next chapter (page 74) that this ratio is $\cos\phi$, where ϕ is the phase angle between the voltage and current, and the power factor is sometimes expressed in this form.

In circuits concerned with the generation or utilization of energy one endeavours to maintain the power factor large, so that the actual power is as nearly as possible equal to the apparent power. Clearly, the power factor can never exceed unity, which is the value obtained when there is no inductance or capacitance in the circuit, or when the effect of one has been neutralized by the other.

There are many circuits, however, particularly those using capacitors, in which it is desirable to keep the effective resistance as low as possible, which requires the power factor to be small. In such circumstances R is small compared with $L\omega$ or $1/C\omega$ and we can write the power factor as R/X, that is to say, the ratio of the resistance to the reactance in the circuit.

Complex Waves; Harmonics

Many of the waveforms encountered in practice are far from being pure sine waves. Fig. 1.26 shows a wave made up of two sine waves, one having only 10 per cent of the amplitude but twice the

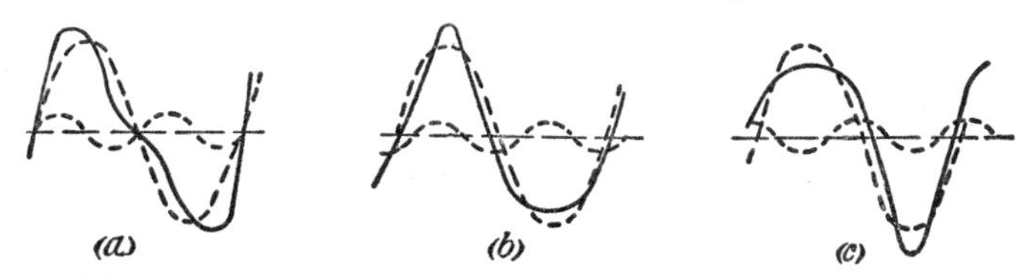

FIG. 1.26. WAVEFORMS CONTAINING SECOND HARMONIC

frequency of the other. The two waves are known as the *fundamental* and the *second harmonic* respectively. Harmonics may be present having any multiple of the fundamental frequency and of variable amplitude. Few practical waveforms, in fact, are pure.

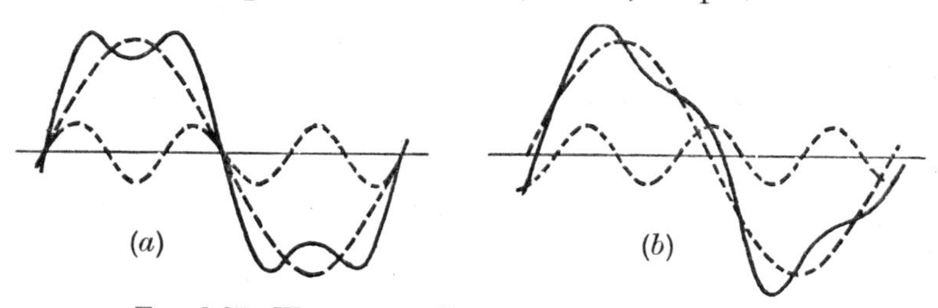

FIG. 1.27. WAVEFORMS CONTAINING THIRD HARMONIC

The harmonics are not necessarily in phase with the fundamental, and the actual form of the wave will in fact depend upon the relative phase of the harmonics. This is shown in Fig. 1.26, where three different waveforms are illustrated, each containing the same amount of second harmonic, but of different phase.

Fig. 1.27 illustrates the effect of a third harmonic. The form of the wave again depends on the phase of the harmonic, but in this case it will be seen that the waveform is symmetrical, the negative half wave being an inverted copy of the positive half wave. This is characteristic of waves containing odd harmonics, whereas even harmonics produce asymmetry, as shown in Fig. 1.26.

If the percentage of harmonics is small, the r.m.s. value of the voltage or current is only slightly affected (although the form factor may be appreciably different). Suppose we have a voltage having an r.m.s. value of 10 volts, and which we know to contain 5 per cent of second harmonic, to be applied across a circuit containing 200 ohms resistance and an inductance of 1 henry. Let the fundamental frequency be 50 Hz.

The reactance of the circuit at the fundamental frequency will be $\sqrt{(200^2 + 314^2)} = 372{\cdot}3$ ohms and the current will thus be 26·9 mA.

The reactance of the circuit to the second harmonic will be $\sqrt{(200^2 + 628^2)} = 659{\cdot}1$ ohms. The voltage is 0·5 volt and the

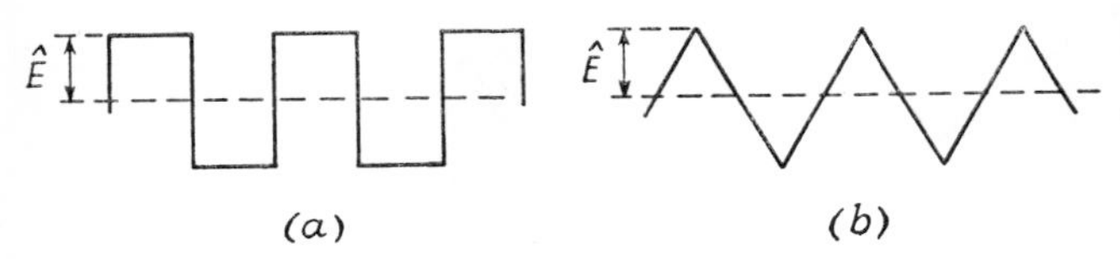

FIG. 1.28. TWO TYPICAL NON-SINUSOIDAL WAVEFORMS

current is thus 0·76 mA, which is only 2·8 per cent of the fundamental.

Hence an inductive circuit tends to suppress harmonics to an extent which increases as the "order" of the harmonic increases, while a capacitive circuit will tend to increase the percentage of harmonic, as the reader can easily verify for himself.

Where a larger harmonic content is present the approximate treatment above is not accurate enough and it is necessary to analyse the voltage wave to determine the effective fundamental and harmonic voltages.

Few of the waveforms encountered in communications work are pure sine waves. The majority are complex in form and hence contain a high proportion of harmonics. They may be of a regularly recurring nature, as in the vowel sounds of speech illustrated in Fig. 1.37 or they may be single irregular pulses, known as transients. In addition, many circuits used for control or test purposes may be of a specifically non-sinusoidal form, such as are illustrated in Figs. 1.28 to 1.30.

Equipment to handle such waveforms must clearly be able to respond adequately not only to the fundamental but also to the highest significant harmonic frequency. It can be shown that any waveform, however complex, can be analysed into a fundamental component and a series of harmonics. For example, the square wave of Fig. 1.28 (*a*) is not greatly dissimilar to that of Fig. 1.27 (*a*) containing a fundamental and an in-phase third harmonic. The addition of suitably proportioned fifth and seventh harmonics will produce an even closer approximation as will be seen shortly.

Fourier Series

The exact analysis of waveforms was developed by Fourier, who showed that any periodic function may be written in the form

$$f(x) = a_0 + a_1 \cos x + a_2 \cos 2x + \ldots$$
$$+ b_1 \sin x + b_2 \sin 2x + \ldots$$

The values of the constants a_0, $a_1 \ldots$, $b_1 \ldots$ are obtained by noting the actual value of the function at suitable points in the cycle.

Specifically, if the period of $f(x)$ is 2π,

$$a_0 = \frac{1}{2\pi}\int_0^{2\pi} f(x)\,dx$$

$$a_n = \frac{1}{\pi}\int_0^{2\pi} f(x)\cos nx\,dx$$

$$b_n = \frac{1}{\pi}\int_0^{2\pi} f(x)\sin nx\,dx$$

The Fourier analysis for the square wave of Fig. 1.28 (*a*) gives a fundamental and a series of odd harmonics, of amplitude inversely proportional to the order of the harmonic, i.e. $33\frac{1}{3}$ per cent third harmonic, 20 per cent fifth harmonic, and so on. The equation is

$$e = \frac{4\hat{E}}{\pi}(\sin \omega t + \tfrac{1}{3}\sin 3\omega t + \tfrac{1}{5}\sin 5\omega t + \ldots)$$

The triangular wave of Fig. 1.28 (*b*) may similarly be built up from a fundamental and odd harmonics (since it is symmetrical), the expression being

$$e = \frac{8\hat{E}}{\pi}\left(\sin \omega t - \frac{1}{3^2}\sin 3\omega t + \frac{1}{5^2}\sin 5\omega t - \ldots\right)$$

Fig. 1.29 illustrates a saw-tooth wave, frequently used in oscillograph work. This is an asymmetrical wave and will thus contain *even* harmonics. It is, in fact, similar to an inverted wave of the form of Fig. 1.26 (*a*), but a certain proportion of odd harmonics will be required in order to produce the sharp transitions from maximum to minimum. Also since the e.m.f. does not reverse, being always above the zero line, it is necessary to add a constant d.c. component (equal to half the peak amplitude), the complete expression being

$$e = \hat{E}\left[\frac{1}{2} - \frac{1}{\pi}\left(\sin \omega t + \tfrac{1}{2} \sin 2\omega t + \tfrac{1}{3} \sin 3\omega t + \ldots\right)\right]$$

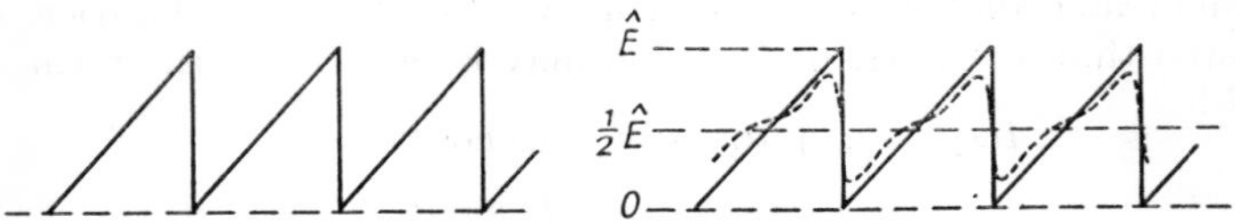

FIG. 1.29. SAW-TOOTH WAVE

The dotted waveform is Fig. 1.26 (*a*) inverted

A further interesting case is the rectified sine wave of Fig. 1.30. Fig. 1.30 (*a*) shows a sine wave in which the negative half-cycles are suppressed. The equation to this is

$$e = \hat{E}\left[\frac{1}{\pi} + \tfrac{1}{2} \sin \omega t - \frac{2}{\pi}\left(\frac{1}{1.3} \cos 2\omega t + \frac{1}{3.5} \cos 4\omega t + \ldots + \frac{\cos 2n\omega t}{(2n-1)(2n+1)} + \ldots\right)\right]$$

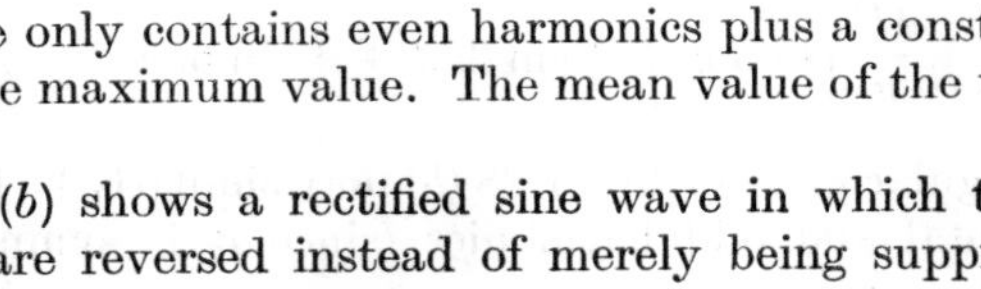

(*a*) (*b*)

FIG. 1.30. RECTIFIED SINE WAVES

This wave only contains even harmonics plus a constant term of $1/\pi$ times the maximum value. The mean value of the wave is thus $\hat{E}/\pi$.

Fig. 1.30 (*b*) shows a rectified sine wave in which the negative half-cycles are reversed instead of merely being suppressed. The equation to this is similar, being

$$e = \hat{E}\left[\frac{2}{\pi} - \frac{4}{\pi}\left(\frac{1}{1.3} \cos 2\omega t + \frac{1}{3.5} \cos 4\omega t + \ldots\right)\right]$$

Here, as one would expect, the mean value is $2\hat{E}/\pi$ (as was stated earlier on page 47).

Average Values of Complex Waves

The r.m.s. and mean values of complex waves are evaluated by adding the appropriate values for each of the components. Thus for a wave

$$e = E_1 \sin \omega t + E_2 \sin 2\omega t + E_3 \sin 3\omega t + \ldots$$

the r.m.s. value is $\sqrt{[\frac{1}{2}(E_1^2 + E_2^2 + E_3^2 + \ldots)]}$

If the wave contains a zero-frequency (d.c.) component, the mean and r.m.s. values of this component are the same as its actual value, i.e. E_0. Hence in this case the r.m.s. value is

$$\sqrt{[E_0^2 + \frac{1}{2}(E_1^2 + E_2^2 + \ldots)]}$$

Similarly the mean (rectified) value is

$$E_0 + \frac{2}{\pi}(E_1 + E_2 + \ldots)$$

These expressions apply to any form of complex wave, provided that its Fourier components are known. In the case of square waves or pulses, however, the Fourier series is apt to be lengthy and it is easier to base the calculations on simple proportion as explained in Chapter 17 (page 715).

1.6. Conduction in Gases and Liquids

As stated in Section 1.1, gases are normally insulators. The molecules are of simple structure and do not readily part with an electron (though a notable exception is water vapour, in which the electrons have some freedom, so that water vapour is slightly conducting). Normally, therefore, the only current which will flow in a gas is a displacement current caused by a varying or alternating e.m.f.

Ionization

It is possible, however, for occasional molecules of gas to lose an electron, in which case the gas contains a number of free electrons and a corresponding number of *ions*, i.e. molecules each deficient in an electron. The gas is then said to be *ionized*.

Each free electron drifts through the gas until it encounters an ionized molecule when the two will recombine, and at normal pressure the rate of recombination is high, so that equilibrium is quickly restored. As the pressure is reduced, however, the distance between the molecules increases so that a free electron can travel relatively long distances before being captured by an ionized molecule. The average distance which an electron can travel is called the *mean free path* and, as the pressure is reduced, the mean free path increases so that the gas remains ionized for a longer period.

If we have two electrodes in a glass bulb or other container with an e.m.f. applied across them, then, if the gas is ionized, a conduction current will flow. The electrons will move towards the positive electrode, which is called the *anode* (meaning literally "way to"), while the positively-charged ions drift towards the negative electrode, called the *cathode* ("way from").

Glow Discharge

At any instant the current will be equal to the e.m.f. divided by the resistance; in a gas, however, the resistance is not constant but varies with the e.m.f. Consider the effect of an e.m.f. applied across two electrodes in a glass bulb filled with gas. The electric field produced by the e.m.f. will tend to displace some of the electrons in the gas molecules, but as explained in Appendix 1 any actual displacement can only occur in jumps from one orbit to another.

This can happen in two ways:

(*a*) An electron may jump from one orbit to another (of higher energy level). Generally it then returns almost immediately to its original state, giving up its excess energy in the form of light waves. There are, however, certain orbits from which the electron cannot revert to normal. This is known as a *metastable* state and may last for an appreciable fraction of a second until a collision occurs with another atom.

(*b*) An electron may jump out of orbit altogether, leaving the atom ionized. The e.m.f. required to produce this result is called the *ionization potential*, and varies with different gases, typical values being shown in Table 1.1.

TABLE 1.1 IONIZATION POTENTIALS

Gas	*Ionization Potential*
Hydrogen	13·5/15·9*
Helium	24·5
Nitrogen	14·5/16·7*
Oxygen	13·6/15·8*
Neon	21·5
Argon	15·7
Mercury vapour	10·4/19*

* The two values are for the gas in atomic or molecular state respectively.

Ionization by Collision

As the e.m.f. is still further increased the velocity of the electrons becomes so high that any encounter with a gas molecule displaces an electron, producing still further ionization. This is known as ionization by collision and is clearly cumulative. The current increases very rapidly by what is called *avalanche effect* up to a limiting value where the gas is completely ionized. This limiting value depends on the number of gas molecules available, and hence increases with the pressure.

The potential at which this effect occurs is called the *breakdown potential*, and depends on the gas pressure, the shape of the electrodes and the distance between them.

Any detailed discussion of gaseous conduction is beyond the scope of the present work. Practical usage falls into two distinct classes. In the first the gas is at low pressure in a suitable container, usually of glass, the current being fairly small. Such devices, usually known as glow tubes or gas-filled valves, are discussed further in Chapter 5.

The second usage is concerned with gases at atmospheric pressure, in which case the current flows in the form of an arc or spark.

Conduction in Liquids

The behaviour of liquids under the influence of an e.m.f. falls into two distinct classes depending upon whether the substance is a pure liquid or a solution. Certain substances, such as oil, exist naturally in liquid form. They are dielectrics and behave in the same manner as solid dielectrics, being capable of carrying a displacement current but not passing any appreciable conduction current. A notable exception, of course, is mercury which is a liquid metal and behaves as such, while any molten metal still behaves as a metal even though liquid in form.

Electrolysis

The conditions are entirely different, however, with solutions, particularly aqueous solutions, i.e. solutions consisting of chemical salts dissolved in water. As explained in Appendix 1, chemical combination between two elements is due to what may be called economy of possibilities. Thus common salt, which is sodium chloride, is a combination of sodium (which has only one electron in its outer shell as against a possible maximum of 8), with chlorine which has 7 electrons in its outer shell. If the two elements join forces the chlorine atom can absorb the isolated sodium electron, so that between them they fill up the outer shell completely.

If the salt is dissolved in water, however, the sodium and chlorine atoms separate but the chlorine still retains the electron which it borrowed from the sodium. It will therefore have an excess of negative charge, and is called a chlorine *ion* ($Cl-$). Similarly the sodium, now deficient in an electron, becomes a positively charged sodium ion ($Na+$).*

Consider now a solution of salt in water with two electrodes connected to a source of e.m.f. as shown in Fig. 1.31. The electric

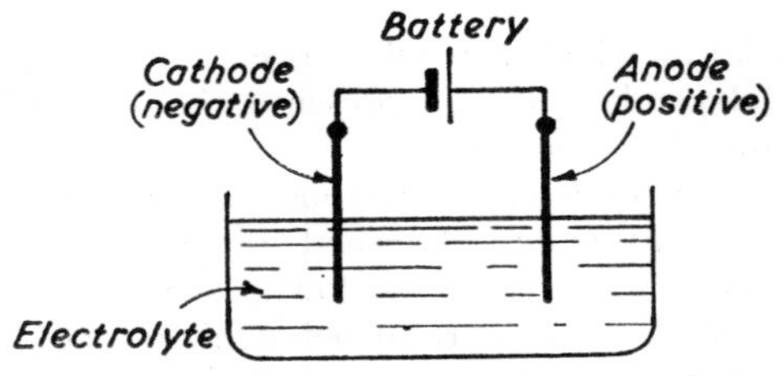

FIG. 1.31. ILLUSTRATING ELECTROLYSIS

field existing between the electrodes will cause the negative ions to move towards the positive electrode (the anode) and the positive ions to move towards the cathode, so that a current will flow. When a $Cl-$ ion arrives at the anode it will give up its surplus electron and become a normal chlorine atom again, while similarly an $Na+$ ion arriving at the cathode will collect an electron and become neutral, and there will thus be a flow of electrons through the connecting wires (including the source of e.m.f.).

The neutral atoms released at the electrodes are in a very active condition, called a *nascent* state, which they cannot maintain. The chlorine atom will combine with another to form a chlorine molecule and bubbles of chlorine gas will appear at the surface of the anode (some of which will dissolve in the water). The nascent sodium will combine with the water to form sodium hydroxide, releasing hydrogen, which will appear as bubbles at the cathode.

This process is known as *electrolysis* and the liquid is called the *electrolyte*. It will be seen that the passage of current through an aqueous solution produces a breakdown of the dissolved chemical into its constituent parts. This has many practical implications, some useful, some troublesome. For example, if the chemical is a solution of a suitable metal salt (such as nickel sulphate) the nascent metal forms a metallic coating on the cathode, this being the basis of electroplating.

* It should be noted that in aqueous solutions both positive and negative ions exist, whereas an ionized gas molecule is always positive.

Faraday's Laws

The first quantitative study of electrolytic action was made in 1830 by Faraday, who formulated two basic laws which are known by his name. These are:

1. The amount of electrolytic action which takes place is independent of the size of the electrodes but is proportional to the quantity of electricity which passes through the solution (i.e. the product of the current and the time).
2. The amount of substance liberated at the electrodes is proportional to the equivalent (chemical) weight of the substance.

The equivalent weight of an element is the relative weight of the element which would combine with approximately 1 gramme of hydrogen (actually 1·008 g as explained in Appendix 1), and hence in order to determine the quantity of material liberated by electrolytic action it is simply necessary to find (experimentally) the appropriate factor which will express the result in terms of the quantity of electricity involved.

Now, it is found that if a current of one ampere is passed through a solution of silver nitrate for one second the mass of silver deposited at the cathode is always 1·118 mg. This relationship, in fact, is so precise that it has been used as the practical standard of current. The equivalent weight of silver is 107·88 so that the factor we are seeking is $1{\cdot}036 \times 10^{-8}$ in SI units or $1{\cdot}036 \times 10^{-2}$ if the result is expressed in mg/coulomb. The mass of material liberated by one coulomb is called the *electrochemical equivalent* and can be deduced for any element by multiplying the equivalent weight by the above factor. Some typical values are shown in the table.

Note that some elements, such as copper, can combine in more than one way. They are said to have different valencies. Thus copper

TABLE 1.2. ELECTROCHEMICAL EQUIVALENTS

Element	*Equivalent Weight*	*Electrochemical Equivalent (mg/coulomb)*
Hydrogen	1·008	0·01045
Oxygen	8·0	0·0829
Sodium	23·0	0·2383
Nickel	29·35	0·3041
Chlorine	35·46	0·3674
Copper (monovalent)	63·57	0·6588
Copper (divalent)	31·79	0·3294
Silver	107·88	1·118

can have two equivalent weights, one twice the other, depending upon the particular chemical combination used.

1.7. Vibrations and Waves

The transmission of intelligence from one point to another is accomplished by means of waves produced by suitable forms of vibration. There are actually two types of wave, longitudinal and transverse.

Longitudinal waves are of a material nature and operate by the movement of particles along the direction of travel of the disturbance, the most common example of this being sound waves.

In a transverse wave the movement takes place at right angles to the direction of travel. Such waves can be material, e.g. ripples on a pond, but are mainly electromagnetic in character, such as light or radio waves.

The two types of wave, though similar in some respects, behave essentially differently in others and are best considered separately.

Sound Waves

Consider the sound produced by a tuning fork as in Fig. 1.32. If this is set in vibration, the prongs will move to and fro some hundreds of times a second, and this movement will displace the surrounding

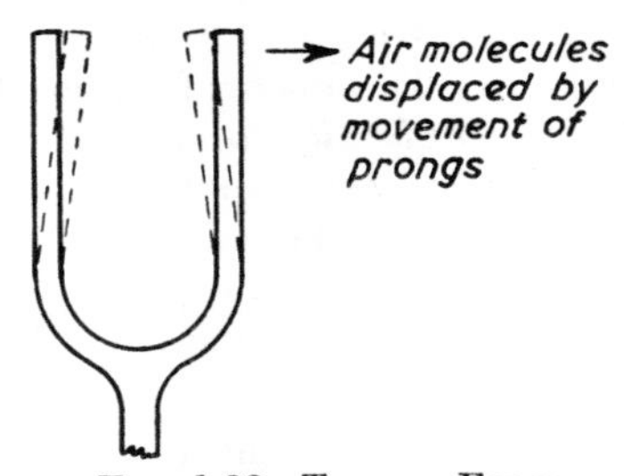

FIG. 1.32. TUNING FORK

air. When the prong moves outwards the molecules of air will be pushed away: when it moves inwards it will create a partial vacuum which will be filled by molecules of air rushing back again.

But the molecules of air thus displaced will in their turn displace other molecules farther away so that the disturbance is transmitted by this jostling of adjacent molecules for an appreciable distance. The farther one goes from the source the greater the volume of air available so that the jostling becomes less pronounced. The actual displacement, in fact, decreases as the square of the distance, as one would expect.

This effect is called a *sound wave*, and it is because of this that sound can be heard at a distance. The human ear is designed to be very sensitive to slight differences of air pressure so that the tiny displacement of the air molecules may be detected at a considerable distance from the source and we say that we "hear" the sound.

Velocity of Propagation; Wavelength

The transmission of the sound obviously takes a finite time. Consider a section of the air (or other medium) a short distance away from the source of the sound as in Fig. 1.33 (*a*). The white circles represent molecules moving from left to right, having been displaced by the sound source, while the black circles are molecules as yet unmoved. These in turn, will be displaced to the right but with a slight time lag because of their inertia, and they will then

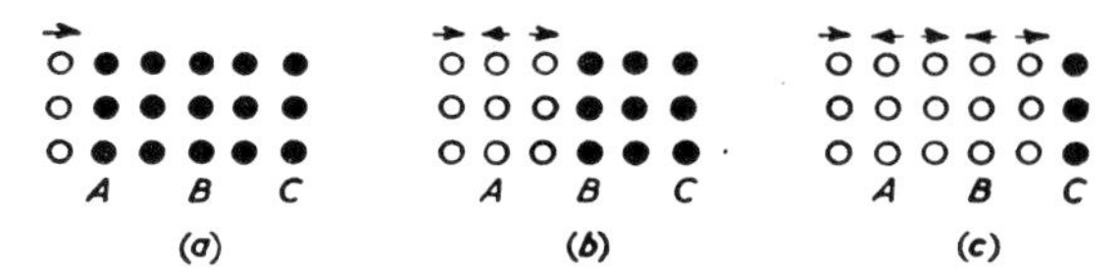

FIG. 1.33. ILLUSTRATING TRAVEL OF SOUND WAVE

displace further molecules so that the wavefront will gradually travel to the right.

Meanwhile the "white" molecules will move back in the reverse direction and then forward again reaching the original position one cycle later, during which time the wavefront will have travelled a certain distance depending on the velocity with which disturbances can travel in the particular medium—called the *velocity of propagation*. The situation will then be as in Fig. 1.33 (*b*), and at B there is clearly a repetition of the conditions at A.

During the next cycle the wave front will travel a further similar distance as in Fig. 1.33 (*c*) and so the disturbance travels outwards, the space between the source and the wavefront being composed of a series of compressions and rarefactions an equal distance apart, moving outwards with a constant velocity.

It is important to note that the air *as a whole* does not move; it is only the individual molecules which are momentarily displaced and so communicate the intelligence to neighbouring molecules. Certain physical disturbances may produce actual movement of large masses of air (e.g. the blast from an explosion) and as a result a transient sound may be produced, but this is not a true sound wave.

Wavelength

We have assumed that the molecules in Fig. 1.33 were at the condition of maximum displacement, corresponding to the maximum movement of the prong of the tuning fork, but it will be clear that a similar process takes place over the whole cycle from maximum in one direction to maximum in the opposite direction and back again. Hence at any point in space the air pressure will gradually vary between these two maxima in opposite directions but the

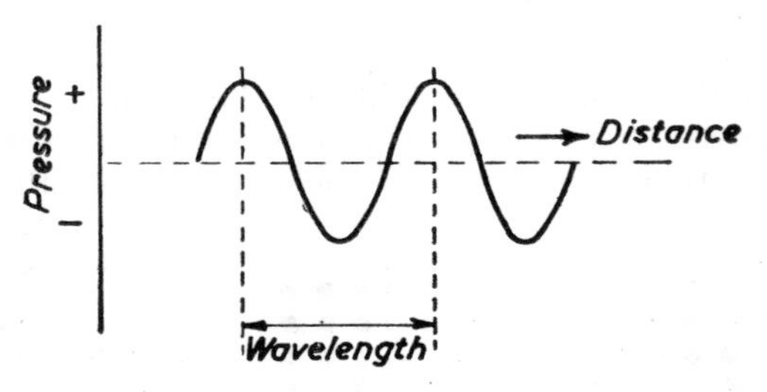

FIG. 1.34. ILLUSTRATING WAVELENGTH

points of maximum (or any intermediate) pressure will travel outwards as illustrated in Fig. 1.34.

The arrangement constitutes what is called a *travelling wave* and the distance between any two similar points (e.g. two points of maximum pressure in the same direction) is called the *wavelength*.

Now it will be clear that this distance will depend upon

(*a*) The velocity of travel of the wave.

(*b*) The frequency of the vibrations, for if a complete cycle of vibration is completed in half the time the wavefront will only have travelled half the distance.

In fact we can say that

$$\text{Frequency} \times \text{Wavelength} = \text{Velocity of propagation}$$

and this is a fundamental property of any wave motion.

The velocity of propagation depends on the medium in which the wave is transmitted for clearly the transmission, depending as it does upon physical displacement of adjoining molecules, can take place in any material; and the greater the density (i.e. the closer the molecules) the more rapidly will any disturbance be transmitted. Sound waves can thus be transmitted through liquid or solid materials just as well as—in fact better than—in air.

Approximate values of the velocity of sound are shown below. As just explained the exact value depends on density and temperature, but the figures indicate the relative values.

TABLE 1.3. VELOCITY OF SOUND AT NORMAL TEMPERATURES

Air . . .	330 metres/sec
Water . . .	1,500 metres/sec
Steel . . .	5,000 metres/sec

Frequency of Sound Waves

Longitudinal vibrations of this type can occur at frequencies from a few hertz up to tens of thousands of hertz. The range of normal sound to which the human ear is sensitive is from about 15 to

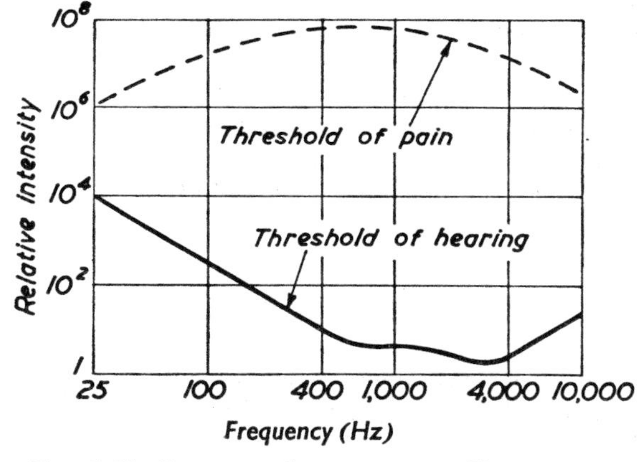

FIG. 1.35. RELATIVE SENSITIVITY OF HUMAN EAR

15,000 Hz, though some people, particularly in youth, can hear frequencies of 20,000 Hz or more. The ear is not uniformly sensitive. Fig. 1.35 illustrates the minimum audible sound level at different frequencies, from which it will be seen that the ear is some 5,000 times more sensitive around 2,000 Hz than at 25 Hz.

The top curve shows the threshold of pain. Air pressures in excess of this will cause distress.

There are vibrations beyond the limit of human hearing. The bat, for example, "sees" by emitting vibrations at a frequency of some 50,000 Hz and detects the echoes from any obstacles by the reflections produced. (Actually it emits these waves in a series of short bursts about 20,000 times a second and it is this pulse frequency which can be heard as the bat's "squeak.")

In the industrial field supersonic vibrations in the range 20,000 to 50,000 Hz are used for a variety of purposes such as drilling, cleaning, echo-sounding, etc.

Nature of Sound

The various sounds with which we are familiar, including speech and music, are the result of air vibrations of this type, but these vibrations are of widely differing character. There are two essential qualities in any sound, namely, the *pitch* and tone or *timbre.* The pitch is determined by the fundamental frequency. A small number of vibrations per second gives a low or deep note, while as the frequency is increased the note becomes higher in pitch. It is found that two vibrations, one of which is exactly twice the frequency of the other, produce notes which are similar in quality but of different pitch. The higher-frequency vibration is said to be the *octave* of the other.

The term "octave" is derived from musical parlance for the major scale in music consists of eight notes related by small (but uneven) ratios, the eighth note being of exactly twice the frequency. The actual intervals (with the names assigned to the different notes) are:

Note	Do	Re	Mi	Fa	Sol	La	Si	Do
Relative frequency . .	1	9/8	5/4	4/3	3/2	5/3	15/8	2

Musical notes cover a range of some 8 or 9 octaves while speech covers about half this range, the actual frequencies being shown in Fig. 1.36.

The quality or timbre of any note, however, depends on the nature of the vibration. Very few sounds consist of a pure sine wave. There is nearly always an admixture of *harmonics* or *overtones.* These are other vibrations, smaller in amplitude than the fundamental, but having a frequency which is an integral multiple thereof. The same note played on a flute, a violin and a piano, for example, will have quite a different quality though the pitch will be the same. The flute note is nearly a pure sine wave, but the violin and piano notes will contain many (different) harmonics which will give them entirely different and distinctive qualities.

Speech is similarly composed of vibrations which are rarely pure but which contain very many harmonics. Fig. 1.37 shows two simple vowel sounds, but even these sounds uttered by two different people will not be identical. The quality of voice, in fact, depends on the nature of the harmonics produced.

More extensive examples of sound waves, showing their relative harmonic content, are given in Chapter 15 of *Telecommunication Principles* by R. N. Renton (Pitman).

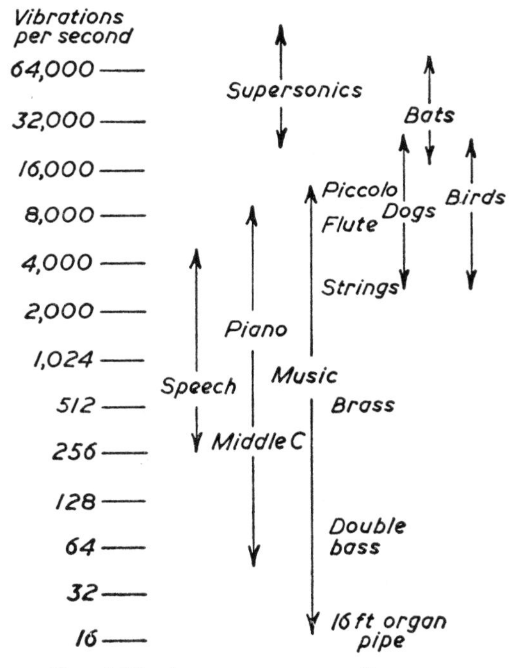

FIG. 1.36. AUDIO-FREQUENCY SPECTRUM

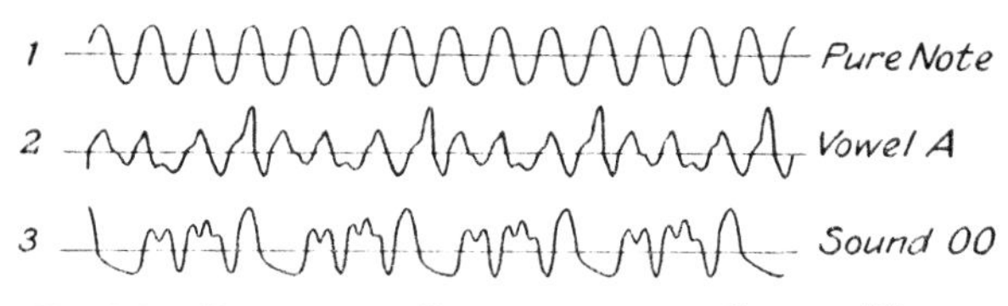

FIG 1.37. DIAGRAM OF REPRESENTATIVE SPEECH WAVES

Transients

Not all sounds are repetitive. An object falling to the ground, for example, will produce a sudden sharp displacement of the air which we hear as a thud or bang. The beat of a drum or clash of cymbals in music produces a similar single pulse, while many of the sounds in speech, such as the labials p and b, or dentals t and d are short explosive pulses.

These single pulses are called *transients*. They can be analysed into a fundamental and a series of harmonics, the harmonics in this case being large and of a high order (i.e. many times the fundamental frequency). The more important aspect is the wavefront, i.e., the relative speed with which the air pressure builds up, and when it is required to reproduce such sounds electrically the system has to be able to handle the extremely rapid changes required.

Telephony

It is, indeed, the translation of sound waves into electrical currents and the subsequent reconversion into sound with which communication engineering is largely concerned. The first process, i.e. the translation into electric currents, is accomplished by a

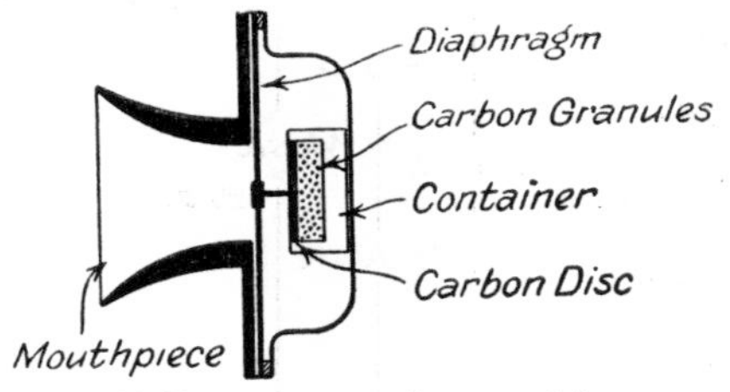

FIG. 1.38. DIAGRAM OF CARBON MICROPHONE

device called a *microphone* which consists essentially of an arrangement in which variation of air pressure causes a change in electrical characteristics. One of the simplest (and earliest) of such devices is the carbon microphone illustrated in Fig. 1.38.

Carbon, though not a metal, is a conductor of electricity; but carbon exists in many forms, from lampblack to diamonds, and the electrical performance varies very widely according to the form. In particular it is found that if a number of pieces of carbon are in contact the electrical resistance depends on the pressure between them. A carbon microphone, therefore, consists of a container filled with small granules of a suitable grade of carbon, one side of the container being flexible and hence able to yield to the varying air pressure of the sound wave. This in turn varies the resistance of the carbon granules so that if an e.m.f. is applied across the microphone the current flowing will vary in accordance with the sound impinging on the diaphragm, thus providing an electric current of the same form as the sound wave.

For many purposes (e.g. the normal house telephone) this simple arrangement is quite adequate but it only responds satisfactorily over a limited range of frequencies. For special requirements different forms of microphone are used which utilize electromagnetic or piezo-electric effects. These are discussed in more detail in Section 4.7.

The electric current into which the sound waves have been translated can now be transmitted by wire or radio to a distant

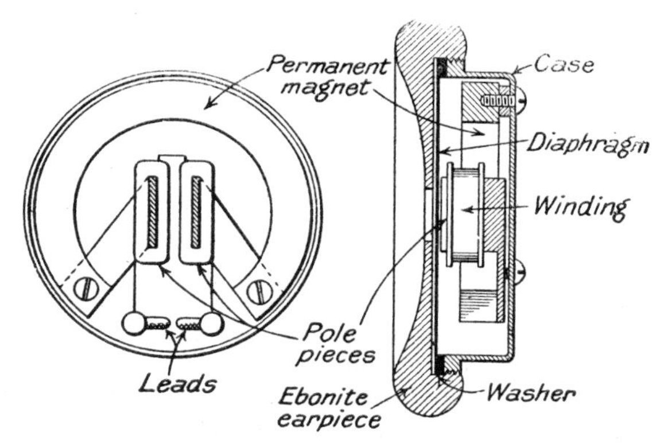

FIG. 1.39. TELEPHONE EARPIECE

point where it can be re-converted into sound. This is done by causing the current to move a diaphragm the vibrations of which set up air waves which will reproduce the original sound. One of the simplest of such arrangements is the telephone receiver illustrated in Fig. 1.39. Two coils of wire are mounted on iron pole pieces which are located very close to a flexible iron diaphragm. The varying currents through the coils produce varying attraction of the diaphragm which causes it to vibrate and reproduce the sound waves.

Again, this arrangement is only suitable over a limited frequency range; moreover the air vibrations produced are relatively small. By making the diaphragm larger, and providing for greater actual movement, sound waves of considerable intensity can be produced. Such a device is called a *loudspeaker*, and while basically the same in principle it incorporates a number of modifications in detail, as discussed in Section 4.8.

Reflection; Standing Waves

If a sound wave travelling through the air meets a solid obstacle there will be a sharp change in the conditions. The air molecules will be unable to continue to move smoothly forward and will have to transmit their energy to other molecules around, and mainly backwards. Some small part of the energy will be absorbed and transmitted through the solid material but most of it will be *reflected*.

This will produce a series of sound waves travelling back towards the source where they may arrive some seconds later to produce an echo. On the return journey, however, they will meet the forward-going waves so that the air pressure at any point will be the difference between the two, and there will be points at which the two cancel each other out leaving only a small residual due to the fact that the reflected wave, having travelled farther, will be slightly less strong than the forward wave.

These points of cancellation will, in fact, be separated by exactly half a wavelength, while in between there will be places where the waves reinforce each other to produce a maximum. In between the cancellation points, or *nodes*, there will, in fact, be a gradual increase in the sound up to a maximum followed by a fall to zero again.

This is called a *standing wave* because, although the forward and reflected waves are travelling through the air, the interference between them always produces the same intensity at any point in the path between the source and the reflector.

Transverse Vibrations

We now come to the second type of vibration, the transverse vibration in which the displacement of the medium is at right angles to the direction of the wave. A simple demonstration of this type of wave may be obtained by taking a length of about ten feet of heavy cord or thin rope and fixing the far end to a suitable anchorage. The near-end is held in the hand with the rope slack.

If the hand is moved slowly from side to side no unusual effect occurs but if it is jerked sharply sideways a ripple will travel along the rope. This arises because the inertia of the rope does not permit it to change its position suddenly so that the disturbance is communicated from one piece of rope to the next with a small time lag.

Two things will be noted from this experiment.

1. If the hand is only moved slowly there is no ripple and the far end does not move appreciably, but (if the rope is slack) a rapid movement will produce a ripple which reaches the far end with considerable amplitude.

2. The ripple which travels *along* the rope is a displacement of the appropriate portion *at right angles* to the direction of travel.

The ripple is, in fact, a form of transverse vibration and if the hand is moved rapidly from side to side a definite wave motion will be noted in the rope. It will also be noticed that the more rapid the movement the closer together the waves become. This is similar to the longitudinal vibrations, and in fact the same rule applies, viz: frequency × wavelength = velocity. As before, the wavelength is the distance between successive points of similar displacement—e.g. the distance between the peaks in Fig. 1.34.

The velocity in the crude experiment just described is not very definite. It will depend on the tension in the rope and will probably not even be uniform along the length of the rope, but this is unimportant since the experiment is merely to demonstrate the essential difference between transverse and longitudinal vibrations.

In fact there is a very wide range of transverse vibrations in nature which make use of electrical "ropes." It has already been pointed out that an electron has associated with it an electric field which can be represented as a series of lines of force radiating from it. These lines of force possess inertia just like a rope so that if the electron is caused to change its position rapidly, ripples will be produced in the lines of force associated with it. These will travel through free space in what are called *electromagnetic waves*. Radio waves are of this type, as also is the vast range of natural phenomena from heat and light to X-rays and cosmic rays.

Since the "tension" in the lines of force is constant it is found that all these waves travel with a constant velocity $c = 3 \times 10^8$ metres/second and hence the wavelength and frequency of electromagnetic waves are connected by the relationship

$$\text{Frequency (Hz)} \times \text{Wavelength (metres)} = c = 3 \times 10^8\,\text{m/s}$$

The mechanism by which electromagnetic waves are produced is discussed more fully in Chapter 3, where it is explained that the waves are produced by sudden changes in the position of electrons. These changes may be relatively slow, produced by causing electric currents to travel up and down elevated structures called aerials, in which case we have the range of radio waves of gradually decreasing wavelength as the frequency is increased. Ultimately a region is reached where the necessary movement of the electrons becomes comparable with molecular dimensions, and the waves which are produced in this region are of a physical character which we know as heat and infra-red waves. With still shorter wavelengths comparable with atomic dimensions we come into the region of visible light.

As the wavelength becomes shorter we pass through the region of ultra-violet light, i.e. waves which are similar to those of light but beyond the limit of visibility of the human eye, while at still shorter wavelengths we find X-rays, gamma rays and cosmic rays, all of which are of the same basic character but are produced by very special types of physical disturbance. The complete range of vibrations known to science is shown in Fig. 1.40.

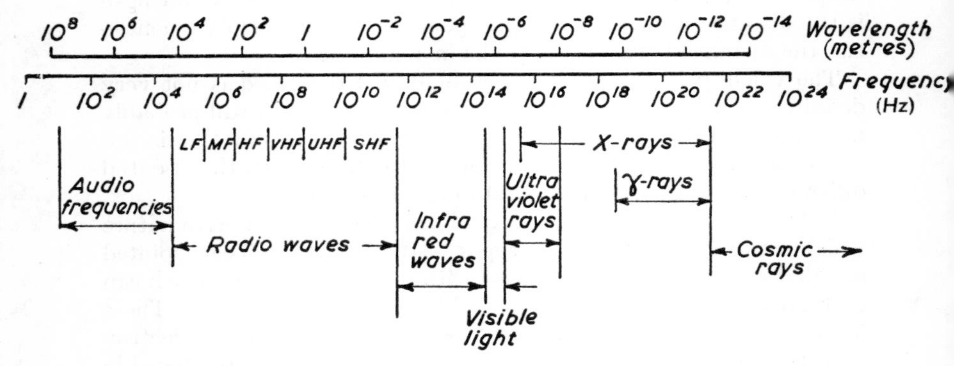

Fig. 1.40. ELECTROMAGNETIC SPECTRUM

PHYSICAL TRANSVERSE VIBRATIONS

Transverse vibrations are also found in physical structures, but because some part of the structure has to be fixed the vibrations only appear as standing waves. A violin or piano string, for example, may be caused to vibrate in a direction at right angles to its length. The vibrations will travel along the wire to one end, where they will be reflected back to the other end and again reflected and so on. This will happen very quickly and the energy will be dissipated almost immediately except at the particular frequency for which the length of the string is equal to an exact wavelength or suitable fraction thereof. This condition is known as (mechanical) *resonance*, and vibrations at this natural frequency can be produced relatively easily and may persist for an appreciable time.

1.8. Sources of E.M.F.

We have seen the effects of e.m.f. in various forms of circuit, but little has been said of the ways in which such e.m.f.s can be produced in practice. There are six main methods, which will be discussed in more detail at the appropriate time in the later chapters, but a brief summary will be helpful here. They are as follows:

1. *Friction.* We have seen that a dielectric or insulated body may be charged by the physical removal of surface electrons by friction. This process is not used to any extent in practice, mainly

because the quantity of electricity so extracted is very small, but it is employed for certain special purposes, notably the production of very high potentials for particular requirements. There are also occasions where the production of static electricity by friction constitutes a hazard. For example, in paper mills the rapidly moving paper acquires a charge which can build up to a dangerous value and protective devices have to be installed to remove this charge. The air friction on the skin of an aircraft will also cause it to become charged and tyres of conducting rubber are used to allow this charge to leak away when the aircraft lands.

2. *Electromagnetic induction.* By far the most common method of producing e.m.f. is electromagnetic induction, in which the linkage between a coil and a magnetic field is varied. This produces an e.m.f. of self or mutual induction which by suitable design can be of almost any desired character, steady or varying, low or high voltage, etc.

A machine in which the change of linkage is produced by mechanical rotation is called a *dynamo*, or if the e.m.f. generated is alternating, an *alternator*. The principal forms of generator are discussed in Section 4.5.

3. *Electrochemical e.m.f.* All substances possess a certain potential so that if two elements are brought into contact a potential difference exists between them. With certain elements in suitable chemical solutions this p.d. can be made available for external use. Such arrangements are called batteries and are discussed in Section 4.6.

In this case the electrical energy results from a chemical change in the structure of the materials used. There are, however, occasions where this electrochemical action is a hazard and precautions have to be taken to avoid or minimize its effect.

4. *Thermo-electric e.m.f.* The potential of any element depends on temperature. Hence if two metals are brought into contact and the temperature of the junction is increased the potential difference between the two metals will be changed. By choosing suitable materials this thermal e.m.f. can be sufficient to be of practical use in which case heat energy is transformed into electrical energy.

Typical values of e.m.f. are shown in Table 1.4.

Since the e.m.f.s are potential differences they must be relative to some datum, for which platinum has been taken, but this does not mean that platinum must be one of the elements used. A common combination is copper-constantan which gives a p.d. of 41·8 μV per °C.

Such arrangements are called *thermo-couples* and may be used singly or as an assembly of a number of elements connected in series

TABLE 1.4 THERMO-ELECTRIC E.M.F.

Metal	P.D. against Platinum per °C (microvolts)
Aluminium . . .	3·8
Antimony . . .	47·0
Bismuth . . .	−65·0
Copper	7·4
Iron	16·0
Constantan . . .	−34·4

to form a *thermopile.* As will be seen, however, the e.m.f. generated is small, so that thermo-couples are used mainly for measurement or control purposes since they provide a convenient means of converting temperatures into electrical terms.

5. *Photo-electric e.m.f.* Certain elements and compounds are found to undergo a change of structure when exposed to light or other electromagnetic waves in this region. This may take the form of a change of electrical resistance, as in the case of selenium, or an actual emission of electrons from the material, which thus behaves as a source of e.m.f. With proper precautions and within suitable limits the e.m.f. may be made proportional to the incident light. Further details are given in Chapter 5.

6. *Piezo-electric effect.* If a crystal of quartz is subjected to mechanical pressure or twisting it is found that a small e.m.f. is developed in a plane at right angles to the strain. This is known as piezo-electric action and is utilized for translating mechanical vibrations into electrical voltages.

The e.m.f., though small, is directly proportional to the mechanical stress. Appreciably greater response can be obtained from Rochelle salt, but this is more fragile, while a variety of other substances have now been found which exhibit the effect.

The action is reversible, an e.m.f. across the crystal causing a mechanical deformation. The effect is normally very small, but if the frequency of the e.m.f. coincides with the mechanical resonant frequency of the crystal the crystal can be maintained in a state of vibration of extremely constant frequency, and this technique is used in the construction of precision oscillators.

2

Circuit Theory

2.1. Vectors

WE have discussed the basic forms of e.m.f. and the various ways in which an applied e.m.f is opposed by the back e.m.f.s developed. Practical radio engineering is concerned with the correct choice of circuitry to obtain the required currents and voltages at the requisite points, and the present chapter deals with the methods principally used in such calculations.

In general, practical circuits are never simple arrangements of a single impedance operated on by a single e.m.f. There is always a combination of impedances in series and/or parallel and it is necessary to be able to calculate the currents and voltages in the various portions and to assess what the combined effect will be. Moreover, since communication of any sort necessarily involves varying quantities, the currents and voltages will be alternating and therefore the calculation of the combined effect is not a matter of simple addition or subtraction but will need to take account of the phase.

This aspect has already been discussed briefly in Section 1.5, where it was shown, for example, that the voltage across an inductance and resistance in series is not simply the sum of the voltage across each but is $\sqrt{(V_L^2 + V_R^2)}$. This is often called the *vector* sum, because it is the sum of two vectors V_L and V_R at right angles.

It is, indeed, frequently helpful to consider a circuit in terms of the current and voltage vectors. The method by which a sine wave may be deduced from a uniformly rotating rod or vector was indicated in Chapter 1, where it was also explained how sine waves out of phase with each other may be represented by two vectors at a suitable angle.

For example, consider the simple cases of pure resistance, inductance, or capacitance. In the first case, if OV is the vector of the voltage, then OI will be the vector of the current, the length of OI being proportional to V/R as shown in Fig. 2.1 (*a*).

Similarly for an inductance, $OI = V/L\omega$ and will be lagging 90° behind the voltage, while the current in a capacitance will be $OI = VC\omega$ leading by 90° (Figs. 2.1 (*b*) and (*c*) respectively).*

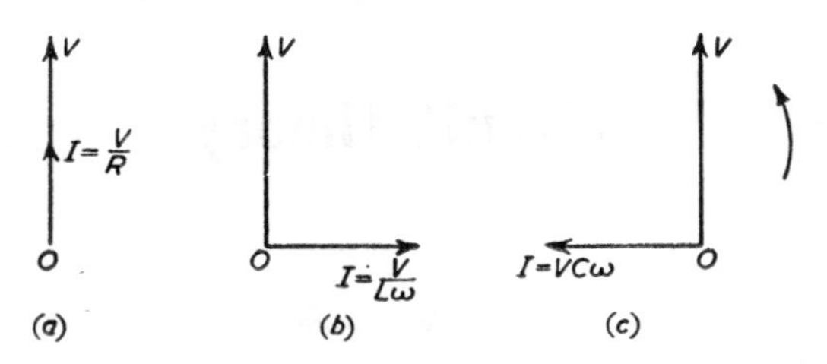

FIG. 2.1. VOLTAGE AND CURRENT VECTORS

Addition of Vectors

Consider now a circuit as in Fig. 2.2 containing inductance and resistance. The current flowing through the two portions is the same; the voltage on the resistance, however, is in phase with the current, while that on the inductance is leading by 90°. Hence, if OI is the current vector, OA will represent the voltage across the resistance, which is equal to IR and is in phase with OI. The voltage on the inductance will be equal to $IL\omega$ and will be represented

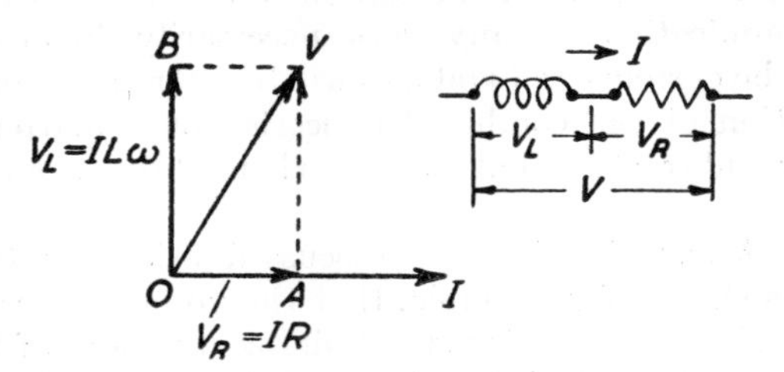

FIG. 2.2. INDUCTIVE CIRCUIT

by OB which is leading by 90° on OI (which is the same as saying that the current is lagging 90° behind the voltage). The total voltage across the whole circuit will obviously be the sum of the voltages across the two separate portions. These two voltages, however, are out of phase and hence must be added "vectorially."

* It should be noted that, whereas in Figs. 1.22 to 1.24 the vectors were made to represent the maximum values of the sine waves, they may equally well represent the r.m.s. values.

By simple geometry this vector sum is OV. To the distance OA representing V_R we have to add a distance at right angles equal to OB, representing V_L, which we can do by drawing a line from A at right angles, of length $AV = OB$. This will terminate at V which could be reached directly by the line OV; hence OV represents the voltage across R and L in series.

This construction can be used to add any two vectors, whatever the angle between them. Thus in Fig. 2.3, if OA and OB represent

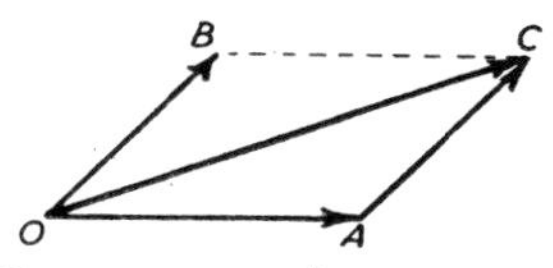

FIG. 2.3. ILLUSTRATING ADDITION OF VECTORS

two sine waves, then the sum of the two will be a third sine wave whose vector is OC. The vector OC is obtained by drawing from A a line AC parallel and equal to OB and joining OC.

This simple method enables any two vectors to be added together provided they are of the same frequency, and also, of course, provided that they can legitimately be added together. For example a current cannot be added to a voltage, but only to another current.

Use of Scale

It should be observed that if the current and voltage vectors have been drawn to scale, OV in Fig. 2.2 will actually represent the resultant voltage across the circuit in magnitude as well as phase.

According to the well-known laws of right-angled triangles

$$OA^2 + AV^2 = OV^2$$

In other words,

$$OV^2 = (IR)^2 + (IL\omega)^2 = I^2(R^2 + (L\omega)^2)$$

Now, if Z is the impedance of the circuit, the voltage

$$OV = IZ$$

so that $$OV^2 = I^2Z^2$$

Hence, $R^2 + (L\omega)^2 = Z^2$ and $Z = \sqrt{[R^2 + (L\omega)^2]}$

which was the result obtained mathematically in Section 1.5 (page 45).

Moreover it will be seen that the angle ϕ between the voltage vector OV and the current vector OI is given by

$$\tan \phi = AV/OA = IL\omega/IR = L\omega/R = X/R$$

while
$$\cos \phi = OA/OV = IR/IZ = R/Z.$$

Resonance

Consider next the circuit shown in Fig. 2.4, which contains capacitance as well as inductance and resistance. Here if I is the current, V_L is the voltage across the inductance, and V_C is the voltage across the capacitance. The resultant of these two depends on whether V_L or V_C is the greater. In this case V_C is assumed greater than V_L and the resultant is V_X, as shown. In addition to this there is the voltage on the resistance V_R, the voltage across the whole circuit being V, the resultant of V_R and V_X. The net effect therefore in the circuit considered is that the current leads on the voltage by a small angle, the effect of the capacitance overwhelming that of the inductance.

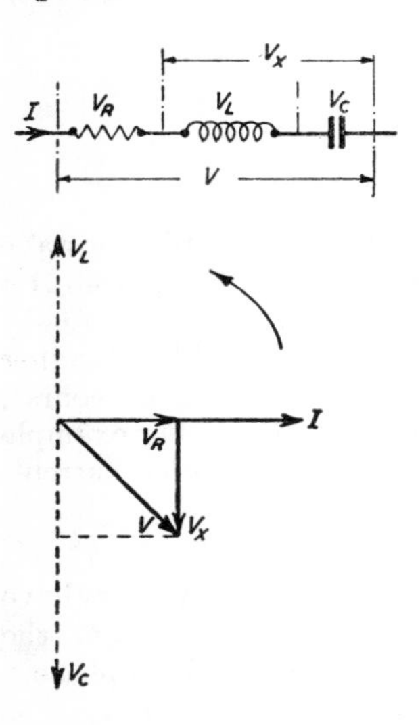

Fig. 2.4. Circuit Containing Resistance, Inductance and Capacitance

It will be seen that the voltage on the capacitance is directly opposite to that on the inductance, and hence the resultant will be the *numerical* difference between them. The combined reactance is therefore

$$X = (L\omega - 1/C\omega)$$

and the impedance

$$Z = \sqrt{[R^2 + (L\omega - 1/C\omega)^2]}$$

It will be observed that if $L\omega$ is made exactly equal to $1/C\omega$ the effects of the inductance and capacitance will cancel out and the circuit will behave as if it was a pure resistance. This effect, which only holds for the particular frequency which makes $L\omega = 1/C\omega$, is called *resonance*, and is considered further later.

The vector diagram of a resonant circuit is given in Fig. 2.5. Here V_L is equal and opposite to V_C, so that the only voltage left in the circuit is V_R, the voltage on the resistance. Actually, since the resistance is part of the inductance, the voltage on the inductance will be V_{LR} as shown, but the resultant of V_{LR} and V_C is V_R, which is in phase with I.

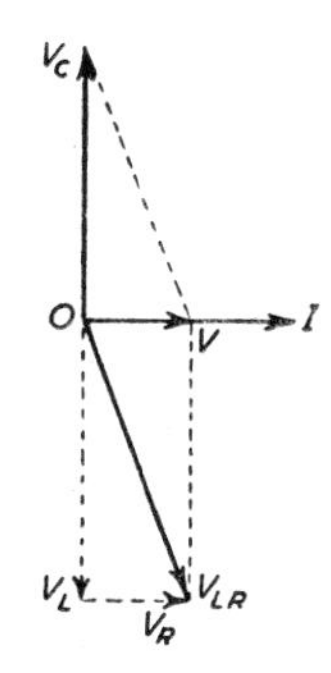

FIG. 2.5. VECTOR DIAGRAM OF RESONANT CIRCUIT

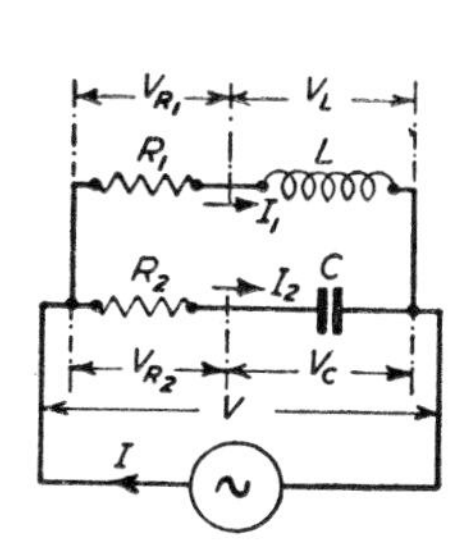

FIG. 2.6. PARALLEL CIRCUITS

Parallel Circuits

A different type of circuit is shown in Fig. 2.6, which shows an inductance and capacitance in parallel. The primary fact here is that the voltages across the two portions are the same, but the currents will be different and, of course, out of phase. The total current will be the (vector) sum of the currents in the two paths.

Now assume that the voltage across the whole is V, and let the currents in the top and bottom parts of the circuit be I_1 and I_2 respectively. Consider the top path first. The voltage across the resistance R_1 will be in phase with I_1, while the voltage across L will be 90° ahead. The total voltage V will therefore be the resultant of these two.

In the bottom path the voltage is the resultant of the voltage on the resistive portion R_2, which is in phase with I_2, and that on the capacitor C, which is 90° behind. (The current in a capacitance leads on the voltage, and hence the voltage will lag behind the current.)

These two resultant voltages, as has been explained, must be the same. Fig. 2.7 (*a*) shows the voltage V and the currents I_1 and I_2 lagging and leading respectively. The component voltages across the resistances and the inductance and capacitance are also indicated.

Fig. 2.7 (*b*) shows the two currents and the resultant current I with their relation to the voltage V. It will be seen that I leads on V by a small angle. Obviously a resonance condition could be obtained here, the values of L and C being adjusted so that I is in phase with V.

If the total resistance $R_1 + R_2$ is small compared with the reactances the condition for resonance is the same as for the series case, viz. $L\omega = 1/C\omega$, but the subject is one of particular importance in radio engineering and is therefore discussed in detail on page 90.

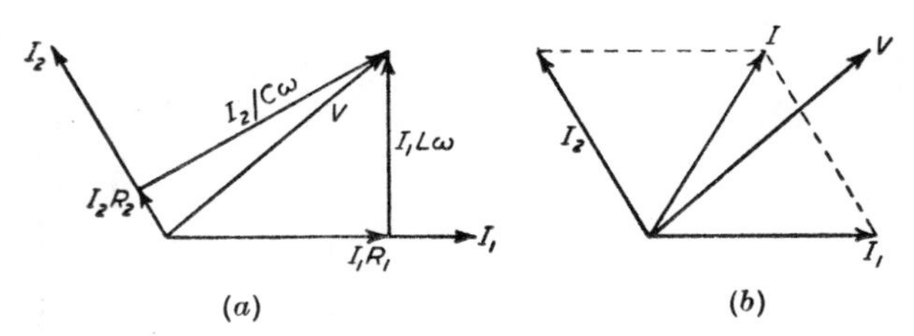

FIG. 2.7. VECTOR DIAGRAMS OF FIG. 2.6 CIRCUIT

Vector Notation

We have seen that the numerical or "scalar" value of any impedance can be evaluated by adding together the squares of the resistance and reactance, and taking the square root. This does not completely define the impedance, however, for it does not specify the phase angle between the resistive and reactive components.

Impedances are therefore sometimes specified in terms of scalar magnitude *and* phase angle, a special sign being used to indicate the quadrant in which the *voltage vector* lies, the current vector being assumed horizontal and to the right of the origin. For example, consider a resistance of 56·6 ohms in series with a capacitor of reactance also 56·6 ohms. The scalar impedance of the two in series would be $56{\cdot}6\sqrt{2} = 80$ ohms and since $R = X$ the phase angle would be 45°. The voltage would lag behind the current so that the voltage vector will lie in the fourth quadrant, as shown in Fig. 2.8. The impedance would then be written $80/\overline{45^\circ}$. Two such impedances may be added graphically by drawing lines of length proportional to the scalar value at the correct angle and constructing the usual parallelogram. Thus, in Fig. 2.8, if we wish to add the vectors $80/\overline{45^\circ}$ and $40\underline{/30^\circ}$ we draw a line equal and parallel to the $40\underline{/30^\circ}$ vector from the end of the first vector. The resultant is then the line joining the origin to the extremity, which may be measured for length and angle as before, e.g. in the case shown the resultant is $98{\cdot}5/\overline{22^\circ}$.

If one of the two quantities has to be subtracted from the other, then one of the vectors must be reversed as shown dotted in Fig. 2.8 and the resultant again worked out.

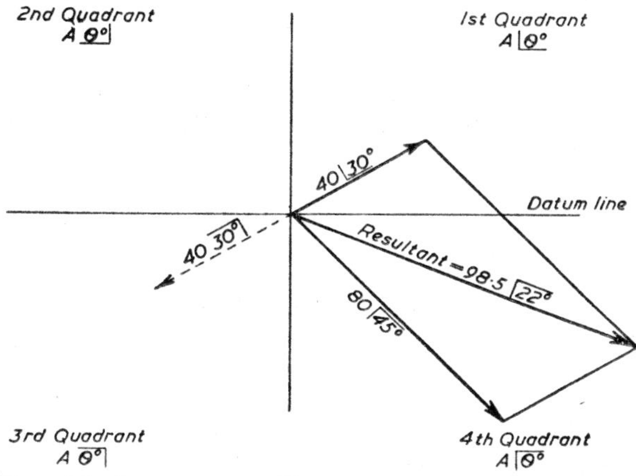

FIG. 2.8. ILLUSTRATING NOMENCLATURE BY QUADRANTS

The "j" Operator

An alternative method, which is in many ways more convenient, involves the division of the expression into "real" and "imaginary" parts. If we have a quantity a which can be represented by a vector

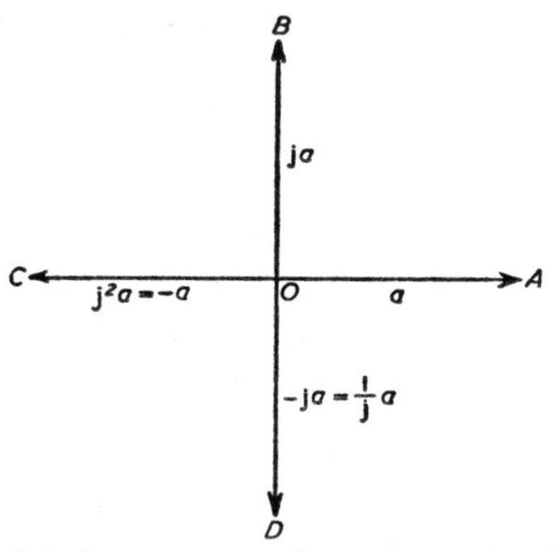

FIG. 2.9. ILLUSTRATING CONCEPT OF $j = \sqrt{(-1)}$

in a given direction such as OA in Fig. 2.9, then if we multiply this quantity by -1 we represent it graphically by a vector in the opposite direction,

$$OC = -a$$

Let us suppose that the vector has rotated from the position OA to OC through the position OB, where it is at right angles to the original position. How are we to represent the vector when it is in this position?

Let us represent this motion by j, so that the vector in the position OB is ja. Now suppose we rotate the vector a further 90° so that it comes into the position OC. Then its motion will be represented by $j \times j = j^2$, and $OC = j^2a$. But we have seen that $OC = -a$, so that $j^2 = -1$, and $j = \sqrt{(-1)}$.

From the usual mathematical standpoint $\sqrt{(-1)}$ has no meaning since the square of any quantity, positive or negative, is always positive. The square root of a negative quantity is therefore imaginary and is frequently spoken of as such. It will be clear, however, that it can have a meaning if we regard it as representing a quantity at right angles to the normal.

This is particularly convenient in electrical engineering calculations where the voltages and currents in reactive components are actually at right angles. Thus the vector $80\overline{/45°}$ in Fig. 2.8 would be written in the form $56{\cdot}6 - j\,56{\cdot}6$. The minus sign is used because the reactive part of the vector is below the datum line. If we are depicting a vector which is leading, so that the reactive part will be above the datum line, we use the positive sign.

Consider the case of a resistance in series with an inductance. The voltage on the resistance is IR. The voltage on the inductance is $IL\omega$, at right angles to the resistive component. Then the voltage developed across the whole circuit is

$$IR + IjL\omega = I(R + jL\omega)$$

The circuit shown in Fig. 2.4 can be treated in the same way. The voltage across the resistance is IR as before. That across the inductance is $jL\omega I$, while that across the capacitance is $-j\dfrac{I}{C\omega}$, the minus sign being used because the reactive part is below the datum line. Now, when using this form of notation it is possible to add together all the resistive components and all the reactive components, and therefore we can rewrite the expression

$$V = I[R + j(L\omega - 1/C\omega)]$$

It should be noted that this gives us exactly the same result as we obtained with the other method, namely that the total impedance is made up of resistance R and the reactance $(L\omega - 1/C\omega)$.

With this notation, therefore, we have a simple and convenient method of representing impedances both in magnitude and phase.

Moreover the combined effect of several impedances may readily be assessed by separating the resistive and reactive components as just mentioned. For example, the two impedances $80\overline{/45°}$ and $40\underline{/30°}$ in Fig. 2.8 would be written $56{\cdot}6 - j\,56{\cdot}6$ and $34{\cdot}6 + j\,20$. The resultant would then be $91{\cdot}2 - j\,36{\cdot}6$.

Modulus

The scalar value of a vector quantity can be evaluated by simple geometry. The expression $91{\cdot}2 - j\,36{\cdot}6$ represents two vectors at right angles and we can add them together by adding the squares and taking the square root. Thus, the scalar value of the vector, or the "modulus" as it is called, is

$$\sqrt{91{\cdot}2^2 + (-\,36{\cdot}6)^2} = 98{\cdot}5$$

which is the same result as before.

Vectors may be multiplied or divided by the use of this j notation equally well, applying the ordinary laws of algebra and remembering that $j = \sqrt{-1}$. For example, let us multiply $(a + jb)$ by $(c + jd)$:

$$(a + jb)(c + jd) = ac + jbc + jad + j^2bd$$

But $j^2 = -1$ so that the expression $= (ac - bd) + j\,(bc + ad)$.

It is important, however, to note that before converting a vector expression into scalar form *all* the reactive terms must be segregated from the resistive terms. Thus the quantity $(a + jb) + jc$ is *not* equal to $\sqrt{[(a^2 + b^2) + c^2]}$. It must be written $[a + j(b + c)] = \sqrt{[a^2 + (b + c)^2]} = \sqrt{[a^2 + b^2 + c^2 + 2bc]}$.

Rationalizing

Suppose we wish to add two vector quantities $a + jb$ and $\dfrac{1}{c + jd}$. Proceed as follows:

$$a + jb + \frac{1}{c + jd} = \frac{(a + jb)(c + jd) + 1}{c + jd}$$

$$= \frac{(ac - bd) + j(bc + ad) + 1}{c + jd}$$

We now resort to a mathematical trick and multiply both top and bottom by $c - jd$. This leaves the value of the expression unaltered, but the denominator is

$$(c + jd)(c - jd) = c^2 - j^2d^2 = c^2 + d^2$$

which is a real quantity. The numerator can thus be separated into resistive and reactive components and each divided by $c^2 + d^2$, leaving a simple vector in the form $A + jB$, which can be completely evaluated both as regards magnitude and phase angle.

It is convenient to remember, in working with these vector quantities that $-j = 1/j$. For example, $1/jC\omega = -j(1/C\omega)$.

2.2. Network Theory

Most practical circuits contain several impedances in various combinations. Such arrangements are called *networks* and may be quite elaborate.

The behaviour of such networks can be analysed by assigning values i_1, i_2, etc., to the currents in selected portions or *meshes* and then expressing the relationships between these currents by the application of Kirchhoff's laws. These laws were stated on page 10 in relation to simple ohmic circuits, but they apply universally to any type of circuit if the instantaneous values of current and e.m.f. are used.

The two laws are

1. The algebraic sum of the (instantaneous) currents meeting at any point of a network is zero. This is an obvious requirement in a closed circuit as otherwise there would be an accumulation of current at the point, which is impossible.

2. In any mesh or part of a network the total (instantaneous) e.m.f. is zero. This means that the total e.m.f. present is exactly offset by the (algebraic) sum of the back e.m.f.s developed by the product of the circuit impedances and the current flowing in them.

(It may be noted that the first law does not apply if the circuit is not continuous, as in the charging of a capacitor from a battery. In such a case the law has to be expressed in terms of the quantity of electricity, but the second law still applies.)

By the application of these laws the performance of any circuit may be analysed to determine the current (or voltage) or the impedance at any part of the network. With complex networks the calculations may sometimes be simplified by adopting certain transformations discussed shortly, but the basic principle is unaltered.

The procedure may be illustrated in simple terms by considering the circuit of Fig. 2.10. We can assign values i_a, i_b and i_c to the currents in R_1, R_2 and R_3 respectively, from which three equations could be written down and the values of the respective currents determined. The calculations may be simplified, however, by the use of *cyclic currents*. If we assume a current i_1 in the mesh R_1R_3

and i_2 in the mesh R_2R_3, as shown, we only need two equations. Thus we can write

$$e = i_1R_1 + (i_1 - i_2)R_3$$
$$0 = i_2R_2 + (i_2 - i_1)R_3$$

since i_1 and i_2 flow through R_3 in opposite directions.

This supplies two simultaneous equations which can, after simplification, be solved in the usual way, giving

$$i_1 = e(R_2 + R_3)/\Sigma$$
$$i_2 = eR_3/\Sigma$$

where $\Sigma = R_1R_2 + R_2R_3 + R_3R_1$.

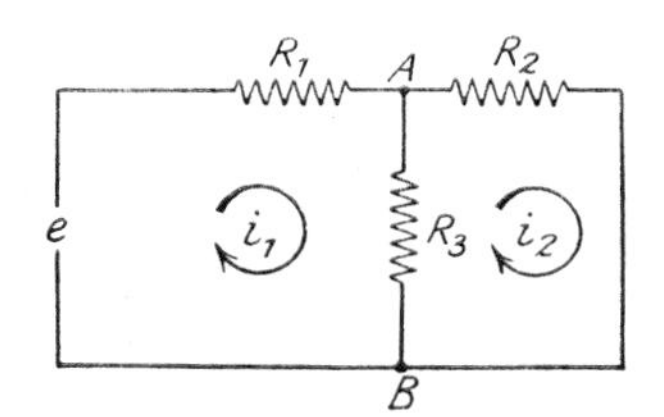

FIG. 2.10. SIMPLE T NETWORK

Principle of Superposition

As just said, the calculations on more complex networks may often be simplified by the use of certain ancillary principles. One of these is that the various sources of e.m.f. in a network act independently of each other and hence the action of each may be considered separately. This is known as the *principle of superposition*, the full statement being as follows.

The current at any point (or the voltage between any two points) in a linear network due to the simultaneous action of a number of e.m.f.s distributed throughout the network is the sum of the currents (or voltages) which would exist at these points if the e.m.f.s were acting separately.

The remaining sources are therefore considered as delivering no e.m.f. *but any internal impedance which they possess must still be considered as part of the circuit.*

Note that the theorem only applies to linear networks, i.e. networks in which the current in each element is proportional to the voltage across it. It does not apply to networks containing non-linear elements such as an iron-cored inductance or a rectifier or unilateral elements such as valves and transistors.

Reciprocity Theorem

Another network theorem, known as the *reciprocity theorem*, states that if any source of e.m.f. located at a given point produces a certain current at some other point, then the same source of e.m.f. acting at the second point will produce a similar current at the first point.

Thus for either position of e.m.f. and current their ratio, which is called the *transfer impedance*, is the same and the circuit need only be analysed in one direction.

As before, the theorem is limited to networks containing linear bilateral impedances, i.e. impedances which pass current equally in both directions in proportional fashion.

Compensation Theorem

When it is desired to assess the effect of a change in some part of a network the calculations may sometimes be simplified by use of the *compensation theorem*.

This states that, if the impedance Z of a branch of a network carrying a current I is changed by ΔZ, the effect on the currents in

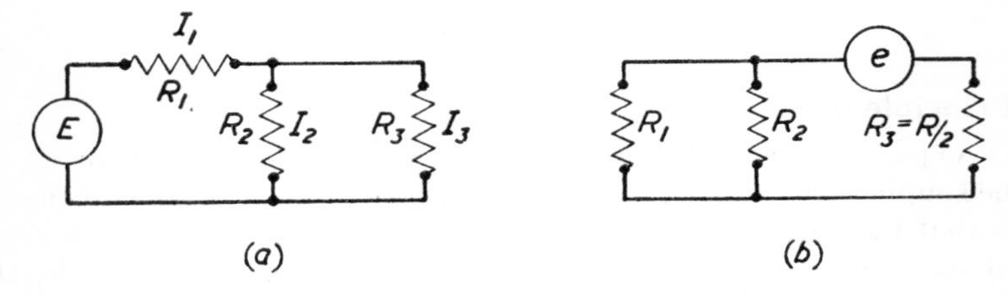

FIG. 2.11. ILLUSTRATING COMPENSATION THEOREM

all the other branches of the network is that which would be produced by an e.m.f. of $-I\Delta Z$ in series with the changed branch.

In other words, we can assume that the current in the particular branch which is being modified is not changed, but that to compensate for the altered impedance we introduce an e.m.f. equal to $-I\Delta Z$.

A simple example will serve to show how this theorem can be applied. In Fig. 2.11(a) we have an e.m.f. E in series with a network of three resistances R_1, R_2, and R_3. Let us assume these are all equal. Then clearly, by simple calculation,

$$I_1 = 2E/3R \quad \text{and} \quad I_2 = I_3 = E/3R$$

Suppose R_3 is reduced to $R/2$. Again by simple calculation we find that $I_1 = 3E/4R$, an increase of $E/12R$.

The compensation theorem states that we could re-write the circuit as in Fig. 2.11 (*b*), where $e = I_3Z = I_3R/2 = E/6$.

The additional current flowing in the network will be that due to the e.m.f. e acting on $R/2$ in series with R_1 and R_2 in parallel, which in this instance is $R/2$. Hence the additional current is $E/6R$, which will divide equally through R_1 and R_2. The additional current in R_1 is thus $E/12R$, which is the result already found.

In this instance, because the resistances were all made equal, the use of the compensation theorem has not provided any advantage, but in a practical and less straightforward network it is often of value.

Bridge Networks

A particular form of network often used is the "bridge" arrangement of Fig. 2.12. Here an e.m.f. is applied across two corners and current is taken from the other two. The feature of the network is

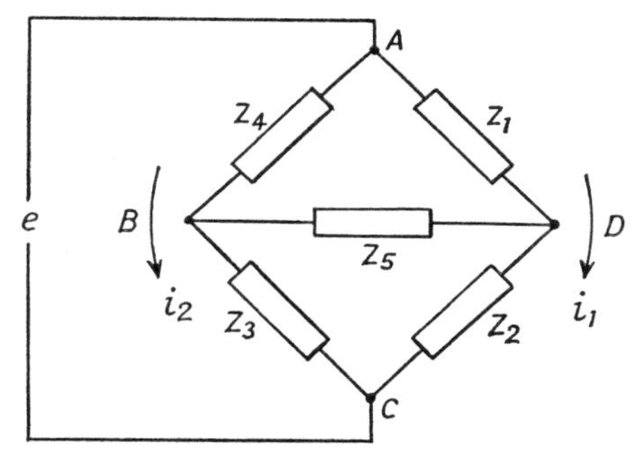

FIG. 2.12. BRIDGE NETWORK

that if certain conditions are complied with the output current is zero and the bridge is said to be balanced.

Clearly the current from the source of e.m.f. divides, part flowing through Z_1Z_2 and the remainder through Z_4Z_3. There will also be some current through Z_5 unless the bridge is balanced. This may flow via Z_1 or Z_4 depending on the conditions but we are mainly interested in the balance conditions. If the currents are i_1 and i_2 then $i_1(Z_1 + Z_2) = i_2(Z_4 + Z_3)$. Also if the bridge is balanced, $i_1Z_2 = i_2Z_3$, making the potentials at the points B and D equal. Hence $i_1/i_2 = Z_3/Z_2$. Substituting i_2Z_3/i_1 for Z_2 in the first equation we get $i_1Z_1 = i_2Z_4$ so that $i_1/i_2 = Z_4/Z_1$.

Hence $Z_3/Z_2 = Z_4/Z_1$ or $Z_1Z_3 = Z_2Z_4$. In other words, the condition for balance is that the products of the impedances of the opposite arms shall be equal.

If the arms are resistive we have the familiar Wheatstone Bridge in which Z_1 and Z_4 are fixed "ratio arms," Z_2 an unknown resistance and Z_3 is varied until the bridge is balanced, when Z_2 can be calculated from the expression just derived. Similar bridges using impedance arms are used to measure reactances or impedances, while special forms can be devised for particular requirements, as for instance a bridge which will only balance at one particular frequency.

Equivalent Networks

It is often required to replace one form of network with another of simpler or more convenient form. Such transformations can be made by arranging that the impedances of the two networks are the same under the following three conditions:

1. Looking from the input end with the output open-circuited.
2. Looking from the input end with the output short-circuited.
3. Looking from the output end with the input open-circuited.

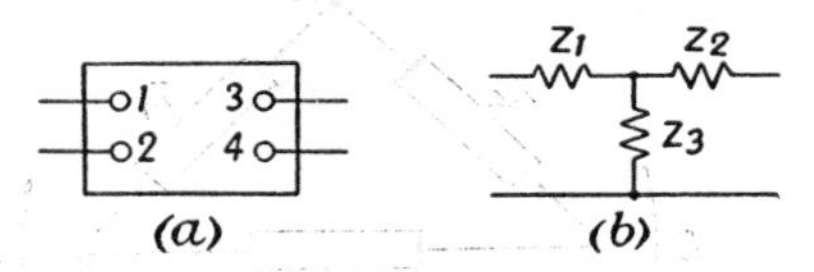

FIG. 2.13. FOUR-TERMINAL NETWORK AND EQUIVALENT T NETWORK

Let us apply this to the network of Fig. 2.13 (*a*) and derive the equivalent T network of Fig. 2.13 (*b*). The original network can be any four-terminal network, such as the bridge of Fig. 2.12, and the first step is to calculate the three impedances mentioned above, which we will call Z_{oc}, Z_{sc}, and Z'_{oc}.

The same three impedances for the equivalent T network may be written in terms of Z_1, Z_2, and Z_3 so that, equating the two sets of impedances, we have

$$Z_{oc} = Z_1 + Z_3$$
$$Z_{sc} = Z_1 + Z_2Z_3/(Z_2 + Z_3)$$
$$Z'_{oc} = Z_2 + Z_3$$

Rearranging these equations we obtain values of Z_1, Z_2, and Z_3 as under:

$$Z_3 = \sqrt{[(Z_{oc} - Z_{sc})Z'_{oc}]}$$
$$Z_1 = Z_{oc} - Z_3$$
$$Z_2 = Z'_{oc} - Z_3$$

Sometimes one requires an equivalent π network. This may be derived by applying the same basic rules, but the resulting expressions are more elaborate than for the equivalent T. It is often simpler to derive the equivalent T first and then deduce the equivalent π by use of the star-mesh transformation.

Star-Mesh Transformation

This is a transformation by which a network consisting of a number of impedances meeting at a point, as at Fig. 2.14 (*a*), can

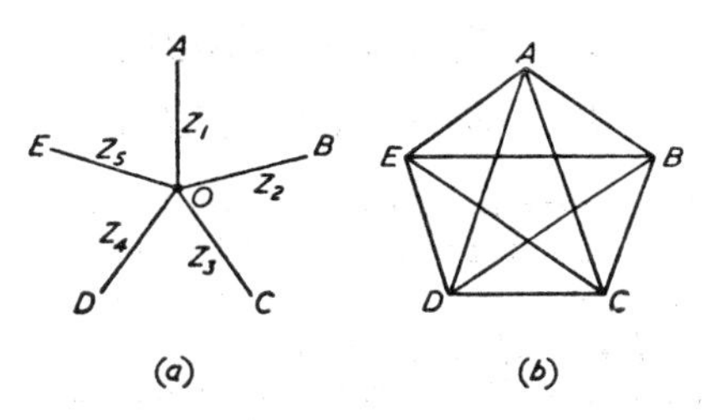

Fig. 2.14. Star Network with Equivalent Mesh

be replaced by equivalent impedances joining all the points, calculated as follows.

Let the original star impedances be Z_1, Z_2, Z_3, etc.

Then the mesh impedances between any two points is the product of the star impedances at those points multiplied by

$$[(1/Z_1) + (1/Z_2) + (1/Z_3) + \ldots] = \Sigma(1/Z)$$

For example, in Fig. 2.14 (*b*), the

$$\text{mesh impedance } AB = Z_1Z_2[\Sigma(1/Z)]$$

Similarly, $\quad$ impedance $AC = Z_1Z_3[\Sigma(1/Z)]$

and the general impedance $= Z_mZ_n[\Sigma(1/Z)]$

Let us apply this to transform the T network of Fig. 2.15 (*a*) into the equivalent π network of Fig. 2.15 (*b*).

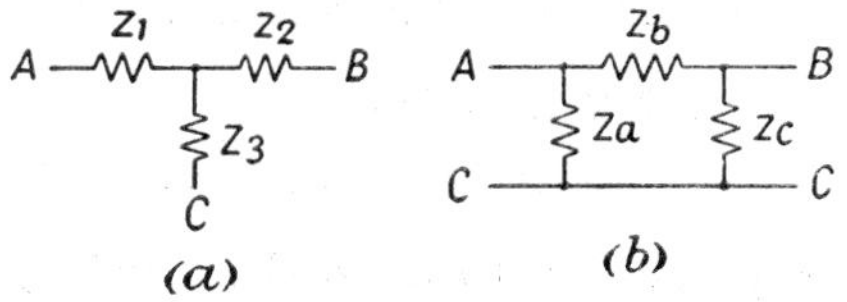

Fig. 2.15. T Network and Equivalent π Network

Z_a, which is the mesh impedance between points A and C, is given by

$$Z_a = Z_1Z_3[1/Z_1 + 1/Z_2 + 1/Z_3]$$
$$= Z_1Z_3\left[\frac{Z_1Z_2 + Z_2Z_3 + Z_3Z_1}{Z_1Z_2Z_3}\right]$$
$$= \frac{Z_1Z_2 + Z_2Z_3 + Z_3Z_1}{Z_2} = Z_1 + Z_3 + Z_3Z_1/Z_2$$

Similarly $$Z_b = Z_1 + Z_2 + Z_1Z_2/Z_3$$

and $$Z_c = Z_2 + Z_3 + Z_2Z_3/Z_1$$

It may be more convenient to use admittance (= 1/impedance) because with two conductors in parallel the admittances are simply added together whereas the impedance has to be evaluated by the usual expression $1/Z = 1/Z_1 + 1/Z_2$.

If the admittances of the various branches are Y_1, Y_2, Y_3, etc. ($= 1/Z_1, 1/Z_2, 1/Z_3$, etc.), the mesh admittance is simply $Y_mY_n/\Sigma(Y)$.

Thévenin's Theorem

There is one other transformation which is often of considerable use. This is Thévenin's Theorem, which states quite simply:

The current in any branch of a network is that which would result from the application of an e.m.f. E equal to the e.m.f. available across the specified points *when the branch is removed*, operating on the impedance of the particular branch plus the impedance of the remainder of the network between the points in question.

The procedure will be clear if we consider a simple example. Suppose we wish to know the current in R_3 of Fig. 2.10.

If R_3 is removed, the voltage across the points AB will be

$$E = eR_2/(R_1 + R_2)$$

The impedance of the network at these points, still with R_3 removed, will be that of R_1 and R_2 in parallel, i.e. $R_1R_2/(R_1 + R_2) = Z$.

Then by Thévenin's Theorem the current in R_3 will be

$$E/(R_3 + Z) = \frac{eR_2/(R_1 + R_2)}{R_3 + R_1R_2/(R_1 + R_2)} = \frac{eR_2}{R_1R_2 + R_2R_3 + R_3R_1}$$

This is a speedy solution to a problem which, though simple, would still require appreciably more calculation by ordinary methods, and in more complex networks the saving can be even more pronounced.

Matrices and Determinants

Expressions of the form $A_1B_2 - B_1A_2$ often occur in circuit calculations. Some simplification can result if this is written in the form $\begin{vmatrix} A_1B_1 \\ A_2B_2 \end{vmatrix}$, which can be expanded by writing down the diagonal from left to right and subtracting the diagonal from right to left.

This is a simple example of a technique known as *matrix algebra,* the specific expressions being called *determinants.* It is a form of mathematical shorthand which often enables the solution of equations to be written down on sight.

Suppose we have two equations

$$A_1x + B_1y + C_1 = 0$$

$$A_2x + B_2y + C_2 = 0$$

In matrix notation, this can be written

$$\frac{x}{\begin{vmatrix} B_1C_1 \\ B_2C_2 \end{vmatrix}} = \frac{y}{\begin{vmatrix} C_1A_1 \\ C_2A_2 \end{vmatrix}} = \frac{1}{\begin{vmatrix} A_1B_1 \\ A_2B_2 \end{vmatrix}}$$

Note the symmetry of the middle term, i.e. the sequence $ABCA$ is maintained. Otherwise the sign will be wrong as explained later.

The value of x (or y) can then be written down by treating items 1 and 3 (or 2 and 3) as algebraic fractions which can be rationalized by cross-multiplication. Thus

$$\frac{x}{\begin{vmatrix} B_1C_1 \\ B_2C_2 \end{vmatrix}} = \frac{1}{\begin{vmatrix} A_1B_1 \\ A_2B_2 \end{vmatrix}}$$

whence $x = (B_1C_2 - C_1B_2)/(A_1B_2 - B_1A_2)$.

The simplicity of this form will be apparent particularly if A, B, and C are not simple numbers but complex expressions.

As an example let us solve the expressions for the circuit of Fig. 2.10. The equations, suitably rearranged, are

$$i_1(R_1 + R_3) - i_2R_3 - e = 0$$

$$-i_1R_3 + i_2(R_2 + R_3) = 0$$

Then

$$\frac{i_1}{\begin{vmatrix} -R_3 & -e \\ (R_2+R_3) & 0 \end{vmatrix}} = \frac{i_2}{\begin{vmatrix} -e & (R_1+R_3) \\ 0 & -R_3 \end{vmatrix}}$$

$$= \frac{1}{\begin{vmatrix} (R_1+R_3) & -R_3 \\ -R_3 & (R_2+R_3) \end{vmatrix}}$$

Whence

$$i_1 = \frac{e(R_2+R_3)}{(R_1+R_3)(R_2+R_3)-R_3^2} = \frac{e(R_2+R_3)}{R_1R_2+R_2R_3+R_3R_1}$$

Similarly,

$$i_2 = \frac{eR_3}{R_1R_2+R_2R_3+R_3R_1}$$

which is the same as was obtained by Thévenin's Theorem.

Third-order Determinants

For a three-variable expression we obtain a third-order determinant of the form

$$\begin{vmatrix} A_1 & B_1 & C_1 \\ A_2 & B_2 & C_2 \\ A_3 & B_3 & C_3 \end{vmatrix}$$

This is handled by expanding it into three second-order determinants, namely

$$A_1\begin{vmatrix} B_2C_2 \\ B_3C_3 \end{vmatrix} + A_2\begin{vmatrix} B_3C_3 \\ B_1C_1 \end{vmatrix} + A_3\begin{vmatrix} B_1C_1 \\ B_2C_2 \end{vmatrix}$$

Note again the sequence of the second term to preserve the sign. As an example, let us take the equations

$$A_1x + B_1y + C_1z = 0$$
$$A_2x + B_2y + C_2z = 0$$
$$A_3x + B_3y + C_3z = e$$

We rewrite these, mentally, in a form involving a constant term D such that the equation $= 0$ in each case. In the first two equations $D = 0$ while in the third it is $-e$. Then

$$\frac{x}{\begin{vmatrix} B_1 & C_1 & 0 \\ B_2 & C_2 & 0 \\ B_3 & C_3 & -e \end{vmatrix}} = \frac{y}{\begin{vmatrix} C_1 & 0 & A_1 \\ C_2 & 0 & A_2 \\ C_3 & -e & A_3 \end{vmatrix}} = \frac{z}{\begin{vmatrix} 0 & A_1 & B_1 \\ 0 & A_2 & B_2 \\ -e & A_3 & B_3 \end{vmatrix}} = \frac{1}{\begin{vmatrix} A_1B_1C_1 \\ A_2B_2C_2 \\ A_3B_3C_3 \end{vmatrix}}$$

which determines x, y, and z. Usually the individual solutions will simplify either directly or on expansion. For example, the third determinant resolves itself into $-e\begin{vmatrix} A_1B_1 \\ A_2B_2 \end{vmatrix}$, the remaining terms going out.

Similarly the first determinant can be expanded into

$$B_1\begin{vmatrix} C_2 & 0 \\ C_3 & -e \end{vmatrix} + B_2\begin{vmatrix} C_3 & -e \\ C_1 & 0 \end{vmatrix} + B_3\begin{vmatrix} C_1 & 0 \\ C_2 & 0 \end{vmatrix} = -B_1C_2e + B_2C_1e$$

It is not practicable to discuss the subject further. There are, however, certain simple rules which may be stated. These are:

1. Interchanging two columns (or rows) of a determinant changes its sign.

Thus if in the determinant originally cited on page 87 we interchange the columns, we get

$$\begin{vmatrix} B_1A_1 \\ B_2A_2 \end{vmatrix} = B_1A_2 - A_1B_2 = -\begin{vmatrix} A_1B_1 \\ A_2B_2 \end{vmatrix}$$

2. Multiplying any column (or row) by a constant multiplies the whole determinant by that factor.

Thus $$\begin{vmatrix} mA_1B_1 \\ mA_2B_2 \end{vmatrix} = m\begin{vmatrix} A_1B_1 \\ A_2B_2 \end{vmatrix}$$

as the reader may easily verify.

3. If each constituent of any column or row comprises two terms the determinant may be expressed as the sum of two simple determinants.

Thus $$\begin{vmatrix} (A_1 + a_1)B_1 \\ (A_2 + a_2)B_2 \end{vmatrix} = \begin{vmatrix} A_1B_1 \\ A_2B_2 \end{vmatrix} + \begin{vmatrix} a_1B_1 \\ a_2B_2 \end{vmatrix}$$

which again may be simply verified.

4. If the constituents of any column or row are increased or diminished by equal multiples of another column or row the result is unaltered.

Thus $$\begin{vmatrix} (A_1 + mB_1)B_1 \\ (A_2 + mB_2)B_2 \end{vmatrix} = \begin{vmatrix} A_1B_1 \\ A_2B_2 \end{vmatrix}$$

which is often useful in effecting simplifications.

2.3. Tuned Circuits

We saw in the first section of this chapter that inductive and capacitive reactances have opposite effects, and it was mentioned that this gives rise to a type of circuit having very wide applications.

Consider the circuit shown in Fig. 2.16. This circuit shows a resistance, an inductance, and a capacitance, all in series with a generator producing an alternating e.m.f. The resistance represents the total resistance in the circuit, made up of the resistance of the wire in the inductance coil, the resistance of the connecting leads,

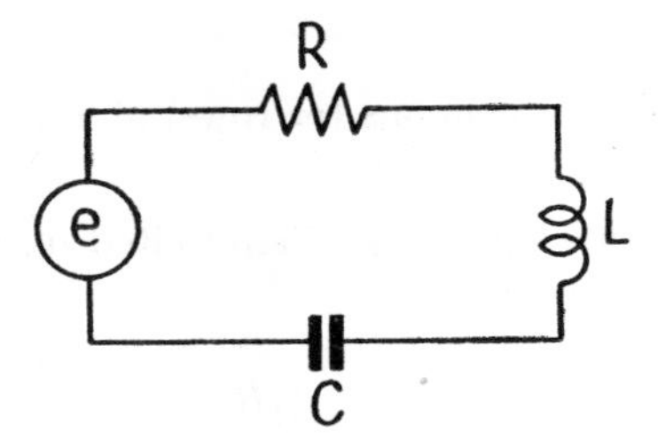

FIG. 2.16. SERIES RESONANT CIRCUIT

the losses in the capacitor (to which we shall refer later) and the internal resistance of the generator. Lumping all the resistances together in this manner is a convenient practice widely adopted in circuit design, since it enables us to consider the generator, the inductance, and the capacitance as being loss-free.

Let us assume that there is an alternating current of value I flowing through the circuit. The voltage developed across the inductance will be $j\omega LI$, 90° ahead of the current. Similarly, the voltage across the capacitance is $I/j\omega C$ or $-jI/\omega C$, the minus sign indicating that this voltage is in the opposite direction to that across the inductance.

Since the voltage across the inductance is exactly 90° ahead of the current and that across the capacitance is exactly 90° behind the current, the two voltages are in direct opposition to one another. There is obviously some condition at which the voltages are not only in opposition but are also exactly equal, with the result that the total voltage across the inductance and capacitance together is zero. Under these conditions the only back e.m.f. in the circuit is that developed across the resistance and the generator would only have to overcome this resistive voltage drop. Expressed in another way, this means that the current in the circuit, for a given generator e.m.f., will be limited only by the resistance in the circuit, being in fact

$$I = \frac{E}{R + j(\omega L - 1/\omega C)} = \frac{E}{R}$$

since $(\omega L - 1/\omega C) = 0$.

This condition is known as *resonance* and it clearly arises when we make the reactance of the inductance equal to the reactance of the capacitance, i.e. we make $\omega L = 1/\omega C$, which may be rewritten in the form $\omega^2 LC = 1$.

Hence for resonance $\omega = 1/\sqrt{(LC)}$ and the resonant frequency is $f = 1/2\pi\sqrt{(LC)}$ hertz, where L is in henrys and C is in farads.

There are two ways of obtaining resonance in practice. One is to vary the value of the inductance or the capacitance, or both, and the other is to vary the frequency, and both methods are used. In the tuning circuit of a radio receiver the capacitance is usually varied until the condition of resonance is obtained (as indicated by the signal strength becoming a maximum), whereas in a superheterodyne receiver the inductances and capacitances in the i.f. amplifier are fixed and the frequency of the input signal is varied until the maximum response is obtained.

Resonance Curves

The behaviour of the circuit at different frequencies may be plotted to obtain what is known as a *resonance curve*. Let us suppose that we have a constant voltage but that the frequency is varied continuously over a region around the actual resonant point. At resonance, as we have seen, the current which will flow is limited only by the resistance in the circuit since the inductive and capacitive reactances cancel out. As we reduce the frequency, the inductive reactance decreases but the capacitive reactance increases and the two do not cancel. The impedance of the circuit is, therefore, the net reactance in quadrature with the resistance, and as this is necessarily greater than the resistance alone the current will fall.

The response falls off rapidly, and at a frequency only slightly removed from resonance the impedance is almost completely reactive. As we get still further away from the resonant point the effect of the inductance becomes negligible and the circuit behaves almost in the same manner as a capacitance, so that the current merely falls off steadily with decreasing frequency.

A similar state of affairs occurs if we increase the frequency beyond the resonant point. Here it is the inductive reactance which becomes larger and the capacitive reactance which decreases.

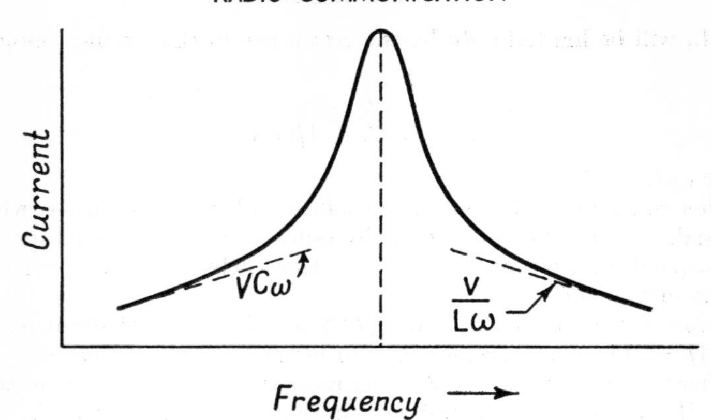

FIG. 2.17. RESONANCE CURVE

The variation of response with frequency is, in fact, approximately symmetrical in form, as shown in Fig. 2.17.

The sharpness or steepness of this curve is a property of considerable interest to the communication engineer. The skirts of the curve are determined by the type of circuit, but the peak is controlled by the amount of resistance present. As the resistance is reduced, the maximum current at the resonant point becomes larger, and in fact if there were no resistance in the circuit the current would be infinite. The effect of reduced resistance, therefore, is to increase the height of the peak, causing the curve to grow, as it were, as in Fig. 2.18. In many cases it is more convenient to show the curves having the same maximum height, in which case the curves

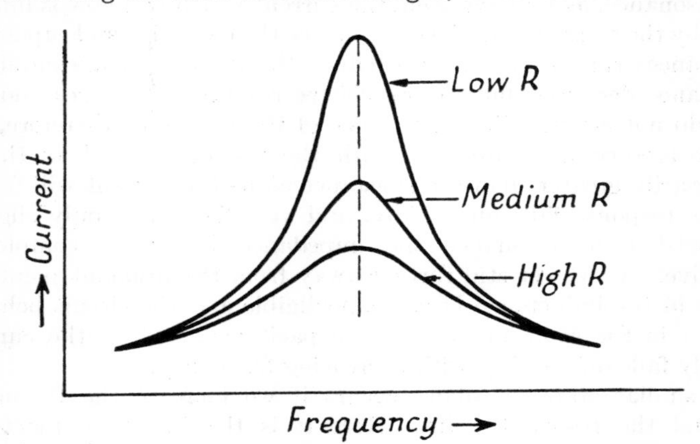

FIG. 2.18. EFFECT OF RESISTANCE ON RESONANCE CURVE

could be redrawn as shown in Fig. 2.19, but it should be remembered that this apparent narrowing of the skirts of the curve only arises from the change of scale and that the real effect of reducing the resistance is to increase the peak value of the current.

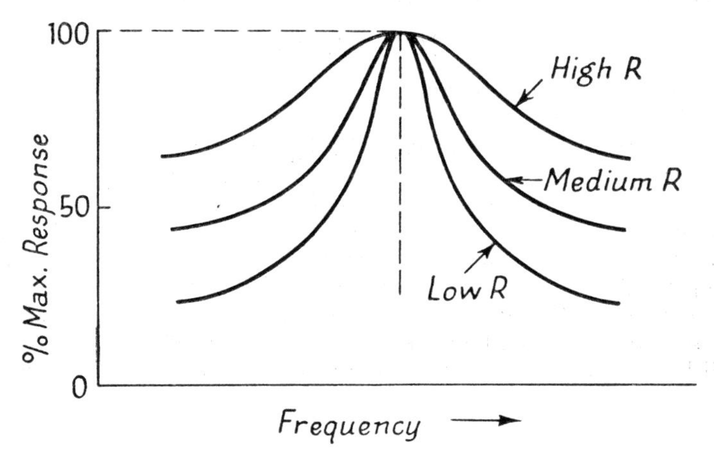

FIG. 2.19. CURVES OF FIG. 2.18 REDRAWN ON A PERCENTAGE BASIS

Magnification Factor

The property of a resonant circuit whereby the current circulating round the circuit is a maximum at a given frequency has obvious applications to the tuning required in communication equipment. Sometimes, however, one is more concerned with the voltage in the circuit, and it will be clear that voltage developed either across the coil or the capacitor is many times greater than voltage across the generator. At resonance we have seen that the reactances cancel out and the circuit adjusts itself so that:

$$I = E/[R + j(\omega L - 1/\omega C)] = E/R$$

The voltage across the inductance, however, is ωLI. The generator voltage $E = IR$, and hence the ratio of the voltage across the coil to that supplied by the generator is $\omega L/R$.

This ratio is known as the *Q factor*. In normal radio circuits its value is of the order of 100 to 300. In audio-frequency circuits the Q factor may be only 10 to 15. Circuits can be devised for special applications, having Q factors of 1,000 or more, but the usual order of Q is as stated above.

Since the voltage across the capacitor is equal and opposite to the voltage across the inductor at resonance it is clear that we should obtain the same answer if we calculated the Q factor in terms of the capacitance.

The voltage across the capacitor $= I/C\omega$, so that

$$Q = (I/C\omega)/IR = 1/RC\omega = \omega L/R$$

since $1/C\omega = \omega L$ at resonance.

It is possible to draw a universal resonance curve which applies to any circuit, by making use of this Q factor. Such a curve is shown in Fig. 2.20. The vertical scale represents the ratio of the current at resonance to that at some frequency off resonance, while the horizontal scale represents the amount of mistuning (i.e. the departure from resonance) in terms of a factor which involves the percentage of mistuning and the Q of the circuit. Obviously the greater the Q, the greater will be the ratio of the current at resonance to the current in a specified condition of mistune, and it will be seen that this result is obtained with the curve of Fig. 2.20.

Phase Angle

It will be clear that as the circuit departs from its resonant condition both the amplitude *and* the phase of the current will change. At resonance the current is in phase with the e.m.f. Below resonance the circuit is capacitive while above resonance it is inductive. The phase angle in fact is arc tan $(\omega L - 1/\omega C)/R$, so that $\phi \to -\pi/2$ when $1/\omega C \gg \omega L$, and $\to +\pi/2$ when $\omega L \gg 1/\omega C$.

The following rules may be quoted, as they are useful in dealing with resonant circuits.

1. If the circuit is mistuned by an amount $1/2Q$ of the resonant frequency the current is reduced to 0·71 times the resonant value and the phase angle is 45°.

2. If the mistuning is $1/Q$ times the resonant frequency the current is reduced to 0·45 times the resonant value, and the phase angle is 63·5°.

These two rules are, of course, merely particular cases of the general calculations regarding resonant circuits. For the sake of completeness, the mathematical expressions are given here.

$$\frac{\text{Actual current}}{\text{Current at resonance}} = \frac{R}{R + j(\omega L - 1/\omega C)}$$

$$= \frac{R}{R + (1/j\omega C)(1 - \omega^2 LC)} \quad . \quad . \quad . \quad (1)$$

Now let the mistuning be represented by $\delta = (\omega - \omega_0)/\omega_0$, where ω_0 is the resonant value. Then $\omega = \omega_0(1 + \delta)$.

If we now put $-j/\omega C$ for $1/j\omega C$ and divide top and bottom by R, the expression becomes

$$I/I_0 = \frac{1}{1 - \dfrac{j}{RC\omega_0(1+\delta)}[1 - \omega_0^2 LC(1+\delta)^2]} \quad . \quad (2)$$

But $1/RC\omega_0 = Q$ and $\omega_0^2 LC = 1$, so that the expression simplifies to

$$\frac{I}{I_0} = \frac{1}{1 - \dfrac{jQ}{1+\delta}[-2\delta - \delta^2]} = \frac{1}{1 + jQ\delta\left[\dfrac{2+\delta}{1+\delta}\right]} \quad . \quad (3)$$

This expression neglects any variation in the resistance of the circuit with frequency which will be small as long as δ is small. The curve of Fig. 2.20 is plotted in terms of a factor $\alpha = Q\delta$.

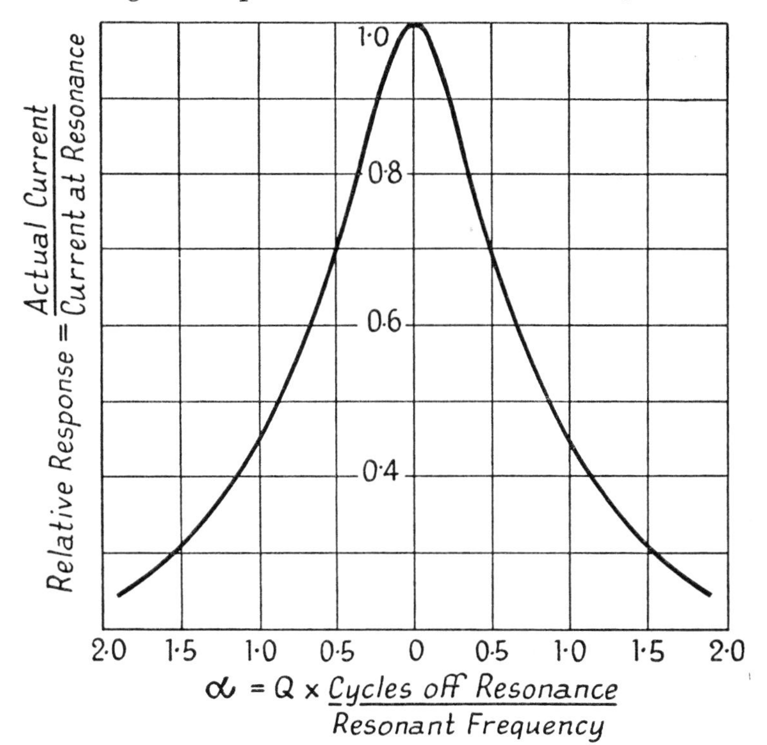

FIG. 2.20. UNIVERSAL RESONANCE CURVE

If δ is small, the expression simplifies to

$$I/I_0 = 1/(1 + 2jQ\delta) = 1/\sqrt{(1 + 4Q^2\delta^2)} \quad . \quad . \quad (4)$$

It will be noted that if $\delta = 1/2Q$, $I/I_0 = 1/\sqrt{2}$, while if $\delta = 1/Q$, $I/I_0 = 1/\sqrt{5} = 0{\cdot}45$, as quoted earlier.

If δ is appreciable, $Q\delta$ becomes $\gg 1$. The expression (3) then simplifies to

$$I/I_0 = (1 + \delta)/Q\delta(2 + \delta) \text{ (numerically)}$$

This may be written in terms of the actual frequency ratio $\omega/\omega_0 = \gamma$. Then $\gamma = 1 + \delta$ and $\delta = \gamma - 1$. Making these substitutions we get

$$\frac{I}{I_0} = \frac{\gamma}{Q(\gamma - 1)(\gamma + 1)} = \frac{\gamma}{Q(\gamma^2 - 1)} \quad . \quad . \quad (5)$$

This gives a negative answer when $\gamma < 1$, indicating that the current changes sign when going through resonance.

Parallel Resonance

The circuit of Fig. 2.16 is called a *series resonant circuit* because the three elements are all in series with the source of e.m.f. A circuit which is often encountered, however, is the *parallel resonant circuit* in Fig. 2.21. Here the coil and the capacitor are in parallel

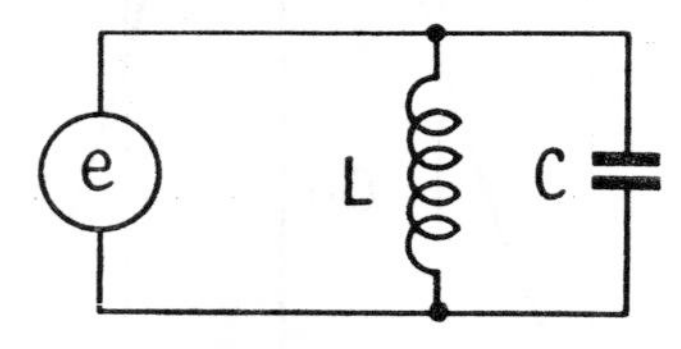

FIG. 2.21. PARALLEL RESONANT CIRCUIT

and the e.m.f. is applied across the two. The current supplied by the generator therefore passes partly through the coil and partly through the capacitor. This circuit is not markedly different from the other, the essential difference being that the roles of the current and voltage are interchanged.

The current through the coil will lag 90° behind the voltage, while the current through the capacitor will lead 90° ahead of the voltage. Thus, as before, we have two currents, one lagging and the other leading, so that they are in opposition and the net current which has to be supplied by the e.m.f. will be the difference between the two. Clearly, under the particular conditions which make these two currents equal, the net current is zero, and the circuit has an infinite impedance.

This, of course, is an ideal case where there is no resistance. In practice, there is some resistance in the coil and some loss in the capacitor, which we represent by small resistances in series with each arm. For a strictly accurate treatment it is necessary to show resistances in each arm, but in the majority of cases the resistance in the inductive arm is many times greater than the effective resistance due to losses in the capacitive arm, and therefore we can consider the circuits as of the form in Fig. 2.22 (*b*), without being seriously in error.

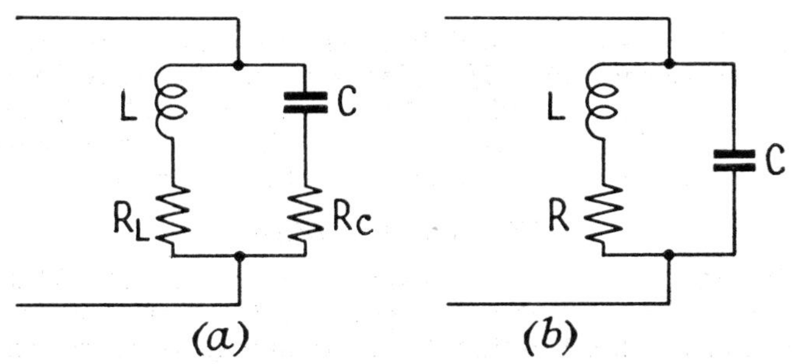

Fig. 2.22. Parallel Resonant Circuits having Resistance

Here the presence of the resistance in series with the inductance causes the current in the coil to lag by an angle which is slightly less than 90°. Consequently, the inductive current does not entirely cancel out the capacitive current even at resonance, and there is a small resistive component which has to be supplied by the generator. As shown on page 98, the network of Fig. 2.22 (*b*) behaves, at resonance, as if it were a resistance of value L/CR. Clearly the smaller the resistance, the larger does this expression become, so that if R is made zero the expression becomes infinite, as we have just seen.

The calculations on a parallel circuit such as this are more conveniently done in terms of the admittances, which are the reciprocals of the impedances. The admittance of the capacitance is $j\omega C$, while the admittance of the inductive arm is $1/(R + j\omega L)$. Rationalizing by multiplying top and bottom by $R - j\omega L$, this becomes

$$(R - j\omega L)/(R^2 + \omega^2 L^2)$$

The total admittance is the sum of the admittances of the two arms

$$Y = j\omega C + (R - j\omega L)/(R^2 + \omega^2 L^2)$$

Separating resistive and reactive terms

$$Y = \frac{R}{R^2 + \omega^2 L^2} + j\left[\omega C - \frac{\omega L}{R^2 + \omega^2 L^2}\right]$$

and the impedance $= 1/Y$.

Resonance occurs when the reactive term is zero, i.e. when

$$\omega C = \omega L/(R^2 + \omega^2 L^2)$$

which can be written

$$\omega^2 = \frac{1}{LC} - \frac{R^2}{L^2}$$

This will be seen to be slightly different from the series case, for which, at resonance, $\omega^2 = 1/LC$, but in practical circuits the second term R^2/L^2 is usually negligible by comparison with $1/LC$, in which case the expression becomes simply $\omega^2 = 1/LC$, as for the series case.

At resonance the effective impedance of the circuit is

$$1/Y = (R^2 + \omega^2 L^2)/R$$

Substituting $\omega^2 = 1/LC$ this reduces to $R + L/CR$. With the values normally encountered in practice, L/CR is very large (of the order of 10^5 ohms) so that the first term may be omitted and the effective impedance at resonance (often called the *dynamic resistance*) is simply L/CR.

It should be noted, however, that the term R^2/L^2 in the expression for ω is not necessarily negligible. If the Q of the circuit is below about 10 this additional term may be appreciable, in which case there will be two possible "resonant" conditions. The first is the condition for which the "feed" current supplied from the generator is in phase with the e.m.f. This condition is satisfied when $\omega^2 = 1/LC - R^2/L^2$. The second condition is that for minimum feed current (maximum impedance), which will occur at a slightly higher frequency given by $\omega^2 = 1/LC$ but the current will not be quite in phase.

Normally these two conditions are indistinguishable, but the difference should be remembered. Parallel resonant circuits are often employed in circumstances where the effective circuit resistance is intentionally made large, in which case the difference is appreciable. Fig. 2.23 (*a*) shows the vector diagram for an arbitrary example in which V, $L\omega$ and $1/C\omega$ are taken as 1 and $R = \frac{1}{2}$. The full lines represent the resonant condition for which $\omega = \sqrt{(1/LC - R^2/L^2)} = 0{\cdot}865$. The resultant current is in phase with the voltage, and has a value of $0{\cdot}5$ $(= L/CR)$. The dotted lines show the conditions when $\omega = 1/\sqrt{(LC)} = 1$. Here the resultant current is slightly less (0·445) and leads by 30°.

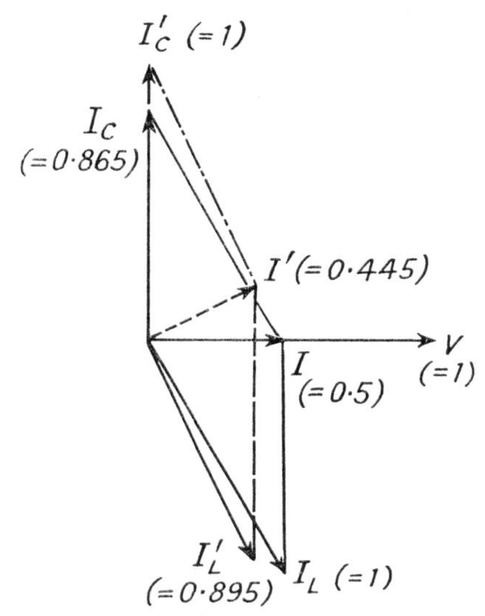

FIG. 2.23 (*a*). VECTOR DIAGRAMS FOR A PARALLEL CIRCUIT WITH APPRECIABLE RESISTANCE ($R = \frac{1}{2}X_L$)
Full-line vectors—feed current in phase with voltage; $\omega = 0{\cdot}865\omega_0$
Dotted vectors—$X_L = X_C$; $\omega = \omega_0 = 1/\sqrt{(LC)}$

Dynamic Resistance

As already mentioned the resonant impedance L/CR is often called the dynamic resistance of the circuit. This only applies at resonance, and as the frequency is varied the impedance falls off rapidly in an exactly similar manner to the variation of current in a series resonant circuit. Indeed, we can use the universal resonance curve of Fig. 2.20 to give the ratio of the actual circuit impedance to that at resonance.

This, as we shall see, is very useful in amplifier design, where the stage gain is dependent upon the impedance of the load. If this load is a parallel-resonant circuit of the type just discussed it is clear that the impedance and hence the gain of the circuit would be a maximum at the resonant point and will fall off steeply on either side.

Most of the r.f. and i.f. circuits in communication equipment utilize resonant circuits of the type just described. In the audio-frequency portions the circuits are not usually tuned though some resonant action is often present.

Circuit Losses

As explained in Chapter 1 it is not only the resistance of the conductor which absorbs energy in a circuit. There are various other causes of loss of energy, but for most purposes it is sufficient to assume that the effect of all the losses in the circuit is equivalent to a series resistance, so that the total resistance in the circuit is made up of the ohmic resistance of the wire plus a small loss resistance due to the various causes already specified.

Sometimes the circuit is of such a nature that the loss can most easily be envisaged in the form of a resistance across the circuit. A valve connected across a tuned circuit has this sort of effect. It is, however, a simple matter by applying the laws of a.c. theory to convert parallel resistance into an equivalent series resistance.

Consider the circuit of Fig. 2.23 (*b*). The current I supplied by the generator will flow through L and R, and then through C and P in

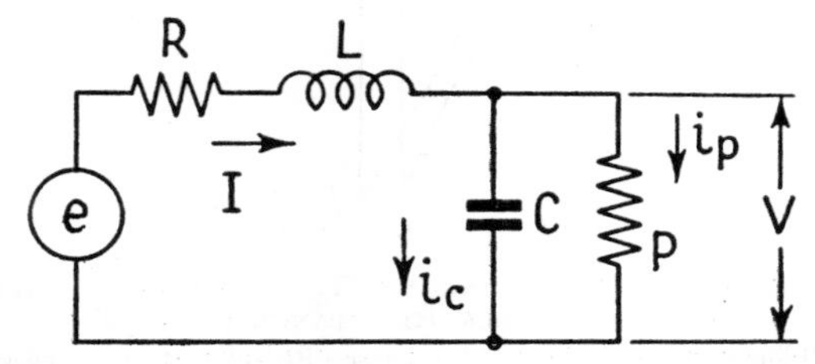

FIG. 2.23 (*b*). CURRENTS AND VOLTAGES IN A LOSSY RESONANT CIRCUIT

parallel. The voltage across C and P is the same so that, if i_c and i_p are the respective currents, we can write

$$V = i_c/j\omega C = i_p P$$

Therefore
$$i_c = i_p j\omega CP$$

The current I is the sum of these two currents, so that

$$I = i_c + i_p = i_p(1 + j\omega CP)$$

The impedance of C and P in parallel is V/I and since $V = i_p P$, we have

$$Z_{CP} = P/(1 + j\omega CP) = P(1 - j\omega CP)/(1 + \omega^2 C^2 P^2)$$

Considering now the whole circuit, and writing G for $1 + \omega^2 C^2 P^2$, we have

$$Z = R + j\omega L + Z_{CP} = R + j\omega L + P(1 - j\omega CP)/G$$
$$= [R + P/G] + j\omega[L - CP^2/G]$$

We see, therefore, that the parallel resistance P is equivalent to an increased series resistance, while it also modifies the reactance of the circuit.

If the circuit is tuned the expression can be simplified. In the first place, for any radio-frequency circuit ωCP is usually much greater than unity, so that G becomes simply $\omega^2C^2P^2$.

If also $\omega^2LC = 1$ we can write

$$\begin{aligned} Z &= [R + P/\omega^2C^2P^2] + j\omega[L - CP^2/\omega^2C^2P^2] \\ &= [R + L/CP] + j\omega[L - L] \end{aligned}$$

In other words the reactance vanishes—the normal tuning effect—and the effective resistance is increased by L/CP—a simple result, useful to remember.

Natural Oscillations

So far we have only considered resonance as a special case of ordinary a.c. theory, and in modern transmission and reception the majority of problems will fall within this category. It is necessary, however, to be familiar with the behaviour of resonant circuits in the condition of free oscillation.

Suppose we have a circuit comprising resistance, inductance, and capacitance in series, and in some way we charge the capacitor and then let it discharge through the circuit. The rush of current through the coil will build up a magnetic field. This field will not collapse immediately but will prolong the current in accordance with Lenz's Law. Current will continue to flow after the discharge is completed and the capacitor will charge up in the opposite direction.

When this process has been completed another discharge will occur in the reverse direction from the first one and the whole cycle of events will be repeated. The process will continue indefinitely, although the current at each successive oscillation will obviously grow weaker due to the dissipation of energy in the resistance of the circuit, just as a pendulum will ultimately come to rest due to air friction.

The behaviour of the circuit can be deduced from first principles. At any instant we know, by Kirchhoff's second law, that

$$L\,di/dt + Ri + q/C = 0$$

Rewriting this in terms of q gives

$$L\,d^2q/dt^2 + R\,dq/dt + q/C = 0$$

This is a differential equation of which the solution is

$$i = (E/\beta L)e^{-\alpha t} \sinh \beta t$$

where $\alpha = R/2L$, $\beta = \sqrt{(\alpha^2 - 1/LC)}$ and $\sinh \beta t = \frac{1}{2}(e^{\beta t} - e^{-\beta t})$.

If $\alpha > 1/LC$, β is real and the circuit is non-oscillatory, the current simply rising to a maximum and then falling to zero.

On the other hand if $\alpha^2 < 1/LC$, β becomes imaginary and equal

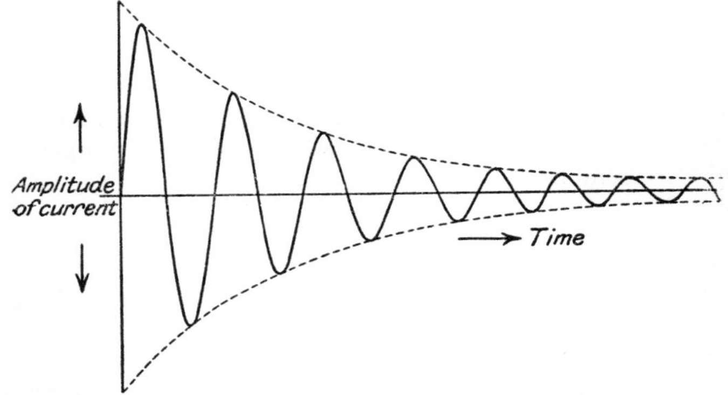

FIG. 2.24. ILLUSTRATING DECAY OF CURRENT IN OSCILLATORY CIRCUIT

to $j\sqrt{[(1/LC - \alpha^2)]}$; but $\sinh jx = j \sin x$, so that the expression for the current becomes

$$i = \frac{E}{\omega L} e^{-(R/2L)t} \sin \omega t$$

where E is the voltage to which the capacitor is charged,

L and R are the inductance and resistance in the circuit,

t is the time since the discharge started, and

$$\omega = 2\pi f = \sqrt{\left[\frac{1}{LC} - \frac{R^2}{4L^2}\right]}.$$

This current is of the form of an oscillation of gradually decreasing amplitude as shown in Fig. 2.24. The decay of the current is controlled by the *damping factor* $R/2L$, which clearly becomes larger as the resistance in the circuit increases. In fact, if R is such that $R^2/4L^2 = 1/LC$, the circuit ceases to oscillate as already stated.

Decrement

The frequency of the oscillation is known as the *natural frequency* of the circuit. It must not be confused with the expression for parallel resonance. As before, however, when the resistance is low, the second term becomes negligible, and the expression reduces to $f = 1/2\pi\sqrt{LC}$, which is the *resonant frequency*, and is the same as that obtained for maximum current in a series circuit with steady applied e.m.f.

The peak value of the current (i.e. the first maximum) is obtained by putting $t = \pi/2\omega$. Then

$$i_{max_1} = \frac{E}{\omega L}\, e^{-\pi R/4L\omega} \left(\sin \frac{\pi}{2}\right)$$

Sin $\pi/2 = 1$, so that the term in the brackets may be omitted.

The next peak is obtained by putting $t = 5\pi/2\omega$, giving

$$i_{max_2} = \frac{E}{\omega L}\, e^{-5\pi R/4L\omega}$$

The ratio of i_{max_2} to i_{max_1} (or strictly the ratio of any peak to the preceding peak in the same direction) is called the *decrement* of the circuit, and is constant if the resistance is constant.

It obviously equals $e^{-\pi R/L\omega}$, and it is customary to quote the logarithm (to base e) of this quantity which is

$$-\frac{\pi R}{L\omega} = -\pi R\sqrt{\frac{C}{L}}, \quad \text{assuming} \quad \omega = \frac{1}{\sqrt{LC}}$$

Hence to obtain low decrement, so that the oscillation shall die away slowly, R must be small and the ratio C/L should be small also.

Damping

The earliest forms of radio communication were by means of *damped* waves of this sort, but such methods are now obsolete. The term damping, however, is still used to indicate the effect of resistance. We say, for example, that certain forms of connection may introduce damping into the circuit, meaning that the effective resistance has been increased so that the circuit is not so sharply tuned as it was.

As we have seen, the transition from non-oscillatory to oscillatory condition occurs when $R^2/4L^2 = 1/LC$. This is called *critical damping* and the discharge is said to be *dead beat*, which means that the current rises to a maximum and then falls rapidly to zero without actually reversing. The maximum current is $E\sqrt{(C/L)}$, which occurs when $t = 2L/R = \sqrt{(LC)}$.

Energy in a Resonant Circuit

The energy in a resonant circuit is partly electromagnetic and partly electrostatic, being, at any instant, $\frac{1}{2}Li^2 + \frac{1}{2}Cv^2$. At the moment when the capacitor is completely discharged the energy is stored entirely in the inductor, while a quarter of a cycle later, when the current is zero, the energy is stored entirely in the capacitor, and continues to change from one form to the other during the continuance of the oscillation.

2.4. Transformers; Coupled Circuits

In many applications in communication engineering it is desired to transfer energy from one circuit to another, the two circuits often being quite separate from each other. The most common method of achieving this transfer is by use of a transformer. The basic action of a transformer was discussed in Section 1.4, where it was shown that, if two inductances are placed in such a position that

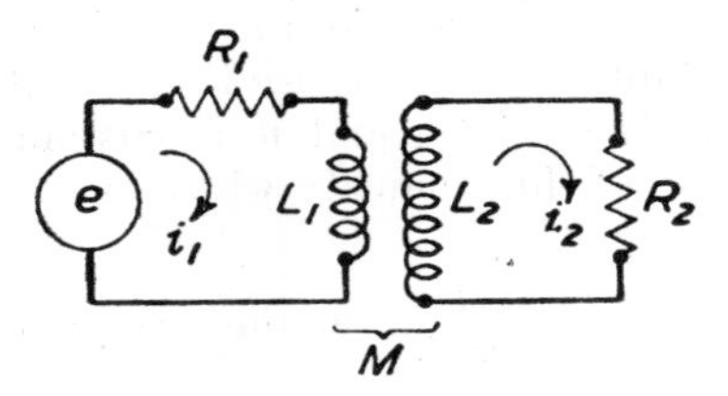

FIG. 2.25. SIMPLE TRANSFORMER

the magnetic field from one links with the other, there is said to be mutual inductance between them and any variation of the current in one coil will induce an e.m.f. in the other.

Fig. 2.25 shows a simple transformer. There is a source of e.m.f. connected across the inductance L_1 which is "coupled" to a second inductance L_2. The resistances R_1 and R_2 represent the circuit resistances of the primary and secondary respectively; for the moment it is assumed that there is no capacitance present. We require to know the currents and voltages in the two circuits.

The technique of the treatment is interesting because the two circuits interact on one another. The presence of the secondary circuit affects what happens in the primary and vice versa. Therefore, we assume arbitrary currents in the two circuits and write down the appropriate equations which we can then solve by reference to known conditions. Let us assume that the current in the primary circuit is i_1, and the current in the secondary circuit is i_2.

From ordinary a.c. theory we can write

$$i_1(R_1 + jX_1) + jM\omega i_2 = e$$

$$i_2 Z_2 + jM\omega i_1 = 0$$

where $Z_2 = R_2 + jX_2$.

Thus $$i_2 = -jM\omega i_1/Z_2$$

Substituting this in the first expression we have

$$i_1(R_1 + jX_1) - j^2(M^2\omega^2/Z_2)i_1 = e$$

Therefore the effective primary impedance

$$Z_1' = e/i_1 = R_1 + jX_1 + M^2\omega^2/Z_2$$

since $j^2 = -1$.

But $$\frac{M^2\omega^2}{Z_2} = \frac{M^2\omega^2}{R_2 + jX_2}$$

Rationalizing this we have

$$\frac{M^2\omega^2}{R_2 + jX_2} = \frac{M^2\omega^2(R_2 - jX_2)}{R_2^2 + X_2^2} = \frac{M^2\omega^2}{Z_2^2}(R_2 - jX_2)$$

Hence $$Z_1' = R_1 + jX_1 + (M^2\omega^2/Z_2^2)(R_2 - jX_2)$$
$$= \left[R_1 + \frac{\omega^2 M^2}{Z_2^2} . R_2\right] + j\left[X_1 - \frac{\omega^2 M^2}{Z_2^2} . X_2\right]$$

This is a most interesting result which shows that the presence of the secondary circuit has increased the effective resistance of the primary by an amount $\frac{\omega^2 M^2}{Z_2^2} . R_2$, while the reactance of the primary circuit has been *decreased* by an amount $\frac{\omega^2 M^2}{Z_2^2} . X_2$. These additional amounts of resistance and reactance are said to be "reflected" from the secondary into the primary.

In an exactly similar manner we can derive expressions showing the effect of the primary on the secondary and this gives us for the equivalent secondary impedance

$$Z_2' = \left[R_2 + \frac{\omega^2 M^2}{Z_1^2} \cdot R_1\right] + j\left[X_2 - \frac{\omega^2 M^2}{Z_1^2} \cdot X_1\right]$$

Transformation Ratio

The secondary e.m.f. $e_2 = M\omega i_1$ which may be more or less than the primary e.m.f. e_1 depending on the circuit conditions. The ratio e_2/e_1 is called the *transformation ratio* and, other things being equal, is equal to the ratio of secondary to primary turns. But in radio-frequency transformers this simple relationship may not apply and there are, in fact, two classes of transformers, namely

(*a*) *Loose-coupled transformers.* These are usually (but not necessarily) air cored, in which case only a part of the primary flux links with the secondary, and the coupling factor k is small.

(*b*) *Close-coupled transformers.* Here the windings are housed on a ferromagnetic core so that virtually all the primary flux links with the secondary and k approaches unity.

It should be noted that radio-frequency transformers are often wound on closed (ferrite) cores but the primary turns only form a small part of the primary circuit. This reduces the effective value of k, so that they behave like a loose-coupled system.

Loose-coupled Transformers

In a radio-frequency transformer the requirement is usually to obtain maximum secondary current. It is customary therefore to tune one or both of the windings to the frequency of the signal.

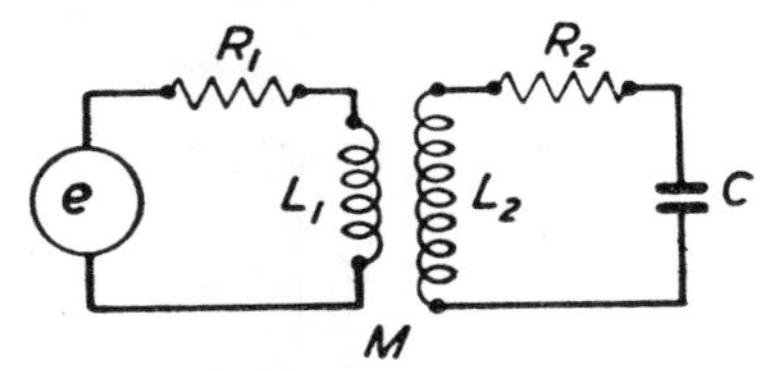

FIG. 2.26. TUNED TRANSFORMER

When both windings are tuned the arrangement becomes what is called a coupled-circuit network, which is discussed later.

A simpler circuit is that having only the secondary tuned as in Fig. 2.26. Such transformers are used in aerial circuits and inter-stage coupling networks. The same basic formulae apply, but since the secondary is tuned, $X_2 = 0$ and $Z_2 = R_2$.

Hence the effective primary impedance becomes

$$Z_1' = R_1 + \omega^2 M^2/R_2 + jX_1 = \omega^2 M^2/R_2 \text{ approximately}$$

since both R_1 and X_1 are small by comparison.

Let us now write $M = k\sqrt{(L_1 L_2)}$, where k is the *coupling factor*; $\omega^2 = 1/L_2 C$, so that

$$Z_1' = k^2 L_1/CR_2$$

This is very similar to the dynamic resistance of the secondary, which is L_2/CR_2, and indicates that the apparent impedance of the primary is some fraction of the effective impedance of the whole (tuned) secondary. The effect is, in fact, as if the primary were tapped across part of the circuit, and this is sometimes done instead of using a separate primary winding.

It is sometimes convenient to express the various formulae in terms of the transformation ratio. To a first approximation the inductances of the two windings on the transformer will be proportional to the square of the number of turns. Hence we can write $L_2/L_1 = n^2$, where n is the ratio of the number of turns on the secondary compared with the primary.

Hence for $M^2\omega^2$ we can write $k^2L_1{}^2\omega^2n^2 = k^2L_2{}^2\omega^2/n^2$. In particular, the equivalent primary impedance just deduced reduces to

$$Z_1' = k^2L_1/CR_2 = (k^2/n^2)(L_2/CR_2)$$

The apparent input impedance is thus k^2/n^2 of that across the secondary. As a corollary, any impedance connected across the input will be equivalent to an impedance n^2/k^2 times as great across the secondary.

Coupled Circuits

If both the primary and secondary of a transformer are tuned we find that the interaction of the two circuits is very marked, and

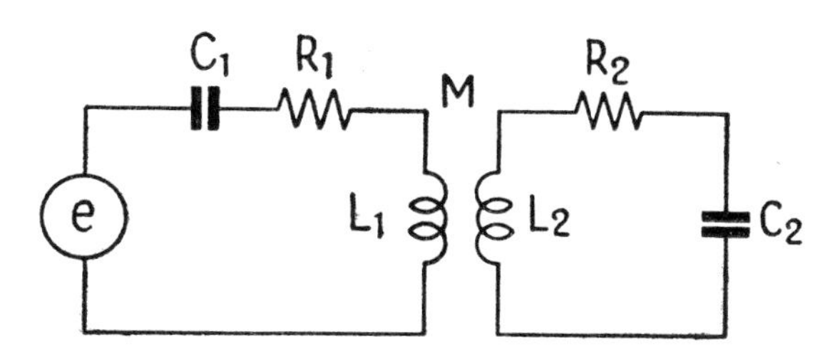

Fig. 2.27. Coupled Circuit

unless the coupling between the circuits is very weak the resonance curve is severely distorted.

Let us refer to the circuit of Fig. 2.27 and let us assume that both primary and secondary circuits are tuned so that $X_1 = X_2 = 0$. X_1' is also zero. The primary current, however, will not be a maximum because of the additional resistance which has been reflected into the circuit from the secondary. The actual additional resistance will be $(M^2\omega^2/Z_2^2)R_2$.

If we vary the frequency, however, so that the circuits are no longer tuned, the impedance of the secondary, Z_2, rises rapidly, with the result that the equivalent resistance reflected into the

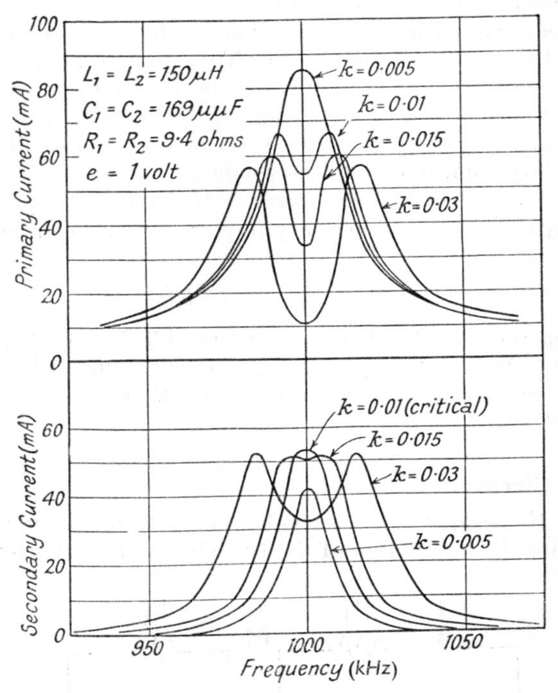

FIG. 2.28. PRIMARY AND SECONDARY RESONANCE CURVES FOR VARYING DEGREES OF COUPLING

primary is *less* than before. This allows the primary current to increase, and at frequencies close to the resonant point this effect will more than offset the increased impedance of the primary itself. The result therefore is that the primary current shows two maxima, one on each side of the true resonant point, as shown in Fig. 2.28.

If the coupling is increased the same effect begins to occur in the secondary current so that we have double humps both in the primary and the secondary resonance curves, and with quite a small coupling the peaks may be pronounced and widely separated.

Secondary Current

From Fig. 2.27 we can write

$$e = Z_1 i_1 - jM\omega i_2$$

and
$$0 = jM\omega i_1 + Z_2 i_2$$

where i_1 and i_2 are the primary and secondary currents. Solving these two equations gives

$$i_2 = -jM\omega e/(Z_1 Z_2 + M^2\omega^2) \quad . \quad . \quad . \quad (6)$$

At resonance $X_1 = X_2 = 0$ and the expression simplifies to

$$i_2 = -jM\omega e/(R_1 R_2 + M^2\omega^2) \quad . \quad . \quad . \quad (7)$$

This is a maximum when the two terms in the denominator are equal so that $M^2\omega^2 = R_1 R_2$, a condition which is termed *critical coupling*. The coupling factor

$$k = M/\sqrt{(L_1 L_2)} = M\omega/\sqrt{(L_1\omega \,.\, L_2\omega)}$$

If $M\omega = \sqrt{(R_1 R_2)}$, $k_c = 1/\sqrt{(Q_1 Q_2)} = 1/Q$ if the circuits are identical.

It can be shown that up to this critical coupling the secondary resonance curve only shows one tuning point. Beyond this value double humps appear in the secondary current. Moreover, the maximum value of the current, even at these peaks, can never exceed the value with critical coupling, which is the maximum possible.

The maximum value of the secondary current is clearly

$$-jM\omega e/2R_1 R_2 \text{ (making } R_1 R_2 = M^2\omega^2)$$
$$= -je/2\sqrt{(R_1 R_2)} \text{ (writing } \sqrt{(R_1 R_2)} \text{ for } M\omega)$$

If $R_1 = R_2 = R$, this becomes, numerically, simply $e/2R$, i.e. one-half the current which would flow if the e.m.f. were applied to either circuit alone.

Selectivity

At frequencies away from resonance the reactances are not zero. It is convenient to express them in terms of the factor $\alpha = Q\delta$, where $Q = L\omega/R = 1/\omega CR$, and δ is $(\omega - \omega_0)/\omega_0$. Then as was shown on page 95, $X = jQR\delta(2 + \delta)/(1 + \delta) \simeq j2QR\delta$ if $\delta \ll 1$. Hence

$$Z_1 = R_1(1 + j2Q_1\delta) \qquad \text{and} \qquad Z_2 = R_2(1 + j2Q_2\delta)$$

Substituting these values in the expression for secondary current, and assuming that the circuits are identical so that $R_1 = R_2 = R$, and $X_1 = X_2 = X$, we have

$$i_2 = -jM\omega e/[M^2\omega^2 + R^2(1 + j2Q\delta)^2]$$

At resonance $\delta = 0$, so that the ratio of the current off tune to that at resonance becomes

$$i_2/i_0 = [M^2\omega^2 + R^2(1 + j2Q\delta)^2]/(M^2\omega^2 + R^2) \quad . \quad (8)$$

If we now apply critical coupling so that $M^2\omega^2 = R^2$, the expression simplifies to

$$i_2/i_0 = 1/(1 + 2jQ\delta - 2Q^2\delta^2)$$
$$= 1/\sqrt{(1 + 4Q^4\delta^4)} \quad . \quad . \quad . \quad . \quad (9)$$

It will be noted that this involves the fourth power of α ($= Q\delta$), as against the square with a single circuit, as is to be expected since there are now two circuits in cascade.

Bandwidth

In practice one is concerned with the ability of a circuit to respond to a band of frequencies around the resonant point in order to accommodate the sidebands which convey the intelligence in a modulated transmission, as explained in Chapter 3. This is assessed in terms of the *bandwidth*, which is the total frequency spread between the points on the sides of the resonance curve at which the response falls by a specified amount. The usual criterion is a fall of 3 dB, which corresponds to a reduction of $1/\sqrt{2}$, though the bandwidth may be specified in terms of a smaller loss (e.g. 1 dB). Now, $\delta = (\omega - \omega_0)/\omega_0 = (f - f_0)/f_0$, on either side of the resonant point. Hence the bandwidth is twice $(f - f_0)$, so that $B = 2\delta f_0$.

Eqn. (9) shows that for a double-tuned circuit the response falls to $1/\sqrt{2}$ of the maximum when $4Q^4\delta^4 = 1$, which requires $Q\delta = 1/\sqrt{2}$. Thus $\delta = 1/Q\sqrt{2}$, and the bandwidth $= 2\delta f_0 = \sqrt{2}f_0/Q$. It was shown on page 96 that with a single circuit the response is 3 dB down when $\delta = 1/2Q$, providing a bandwidth $B = f_0/Q$. Hence the use of a double-tuned circuit increases the bandwidth for a given value of Q by $\sqrt{2}$ times. This is the "flat-top" effect mentioned earlier.

Cascading

Beyond this point the response falls off more steeply than for a single circuit. At twice the 3 dB bandwidth i/i_0 is 0·24 as against 0·45 for a single circuit, while at three times the 3 dB width i/i_0 is 0·11, which is very nearly the square of the corresponding value (0·316) for a single circuit.

In practice r.f. amplifiers use a series of tuned stages and provided that there is no interaction (cf. page 413) the overall response is the product of those of the individual stages. The denominator of eqns. (5) and (9) is the attenuation of the circuit for a given mistune, and if this is expressed in decibels the overall attenuation is the sum of the individual attenuations. Thus a single circuit at three times the 3 dB width would have an attenuation of 3·16, or 10 dB. Two such circuits in cascade would have an attenuation of 20 dB. (With a coupled circuit there is still a slight interaction so that the actual attenuation is 9, or 19 dB).

Because of this it may be necessary to calculate the appropriate Q value to provide less than 3 dB attenuation at the stipulated bandwidth. Thus with three stages in cascade an overall attenuation of 3 dB would require each stage to have an attenuation of only 1 dB. This is discussed more fully in Chapter 9, but Table 2.1 shows the values of Q for attenuations of 1 and 1·5 dB.

TABLE 2.1. Q VALUES FOR SPECIFIED ATTENUATIONS

Attenuation (dB)	*Q value*	
	Single circuit	*Double circuit*
3	f_0/B	$1{\cdot}41f_0/B$
1·5	$0{\cdot}65f_0/B$	$1{\cdot}15f_0/B$
1	$0{\cdot}50f_0/B$	f_0/B

Other Forms of Coupling

The expressions so far developed apply to any form of circuit provided that the term $M\omega$ is replaced by an equivalent reactance. Fig. 2.29 shows four types of circuit in common use.

Circuit (*b*) is a common-capacitance coupled arrangement in which $M\omega$ is replaced by $1/C\omega$. An alternative form is shown in circuit (*c*). The coupling reactance here is not so easy to specify, but the circuit can be converted to the form of circuit (*b*) by using the transformations quoted in Section 2.2. If this is done we arrive at the following:

$$C_1 = C'/C_4$$
$$C_2 = C'/C_3$$
$$C = C'/C_o$$

where $C' = C_3C_4 + C_3C_o + C_oC_4$.

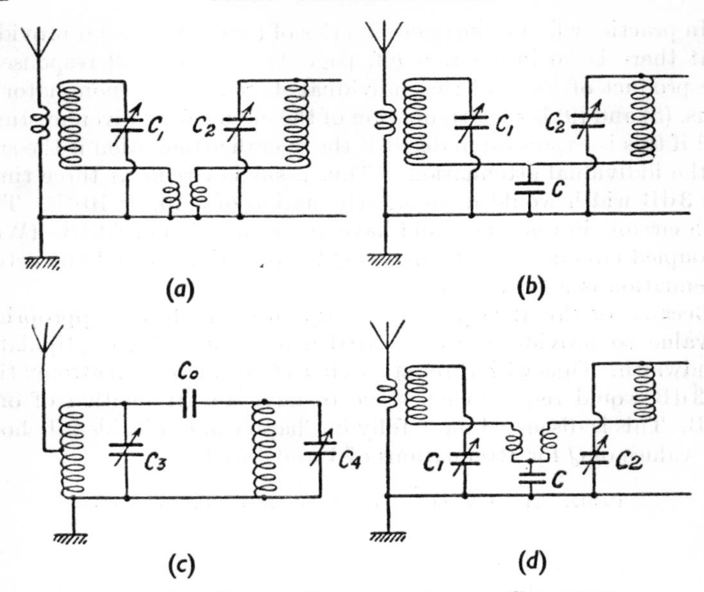

FIG. 2.29. TYPICAL BAND-PASS AERIAL COUPLINGS
(*a*) Mutual coupling; (*b*) Common-capacitance coupling;
(*c*) Top-capacitance coupling; (*d*) Mixed coupling

Free Oscillation

Where the coupled circuits are not provided with a source of e.m.f. but are allowed to oscillate naturally, e.g. in a valve oscillator, similar effects occur with certain important reservations. If the coupling is weak, the energy in the primary is partially transferred to the secondary, the remainder being dissipated in primary losses.

As the coupling is increased the energy transfer increases to an optimum after which an unstable condition occurs in which the energy surges backwards and forwards from one circuit to the other producing a complex double-frequency oscillation. The critical conditions to avoid this may best be arrived at by considering the frequencies at which the circuit will oscillate.

The theory is simplified if we assume that both primary and secondary circuits are tuned to the same angular frequency ω. Let us then assume that the two frequencies of the system are $\omega + \delta\omega$ and $\omega - \delta\omega$.

At the upper frequency

$$X_1 = L_1(\omega + \delta\omega) - 1/C_1(\omega + \delta\omega) = \frac{L_1C_1(\omega + \delta\omega)^2 - 1}{C_1(\omega + \delta\omega)}$$

Now $L_1C_1\omega^2 = 1$, and since $\delta\omega$ is small, $\omega + \delta\omega = \omega$ nearly. The expression then simplifies to

$$X_1 = 2L_1\delta\omega$$

Similarly, $$X_2 = 2L_2\delta\omega$$

The primary reactance $X_1' = X_1 - (M^2\omega^2/Z_2^2)X_2$, and if the circuit is self-oscillating it will choose a frequency such that $X_1' = 0$.

Therefore $$2L_1\delta\omega - \frac{M^2\omega^2 2L_2\delta\omega}{R_2^2 + 4L_2^2\delta\omega^2} = 0$$

i.e. $$8L_1L_2^2\delta\omega^3 + (2L_1R_2^2 - 2L_2M^2\omega^2)\delta\omega = 0$$

This has three roots, namely

$$\delta\omega = 0 \quad \text{and} \quad \delta\omega = \pm \tfrac{1}{2}\sqrt{\left[\frac{M^2\omega^2}{L_1L_2} - \frac{R_2^2}{L_2^2}\right]}$$

If the circuit is to be mono-oscillatory the second two roots must be imaginary, which means that the expression under the root sign must be negative. Hence

$$M^2\omega^2 < R_2^2L_1/L_2 \quad \text{or} \quad M < (R_2/\omega L_2)\sqrt{(L_1L_2)}$$

This requirement is equivalent to saying that the secondary volt-amperes ($i_2^2L_2\omega$) must be less than the primary volt-amperes, which is sometimes a more convenient way of stating the conditions.

The existence of three possible frequencies may come as a surprise to the reader who is accustomed to consider a coupled circuit as having only two modes of oscillation. It should be noted, however, that if M is greater than the critical value the frequency given by $\delta\omega = 0$, $(f = f_o)$, is unstable.

Suppose f increases slightly. Then X_2 is no longer zero and X_1' is reduced. The resonant frequency rises accordingly and the current rapidly slides into the upper stable frequency corresponding to the positive value of $\delta\omega$. Similarly if the frequency falls the circuit immediately takes up the lower value of oscillation, and stable conditions only obtain with one or other of these two modes.

Close-coupled Transformers

We now come to the consideration of the second class of transformer, the close-coupled type. Here the arrangements are such that practically all the flux produced by the primary links with the secondary so that the coupling is virtually perfect. This is only possible when the coils are wound together on a core of magnetic material as discussed in Chapter 4 and even then some small leakage is inevitable though this can often be ignored.

In such circumstances it is possible (and more convenient) to rewrite the various expressions in terms of the turns ratio or transformation ratio, in which case the mutual inductance, which may not be known, does not appear in the formulae.

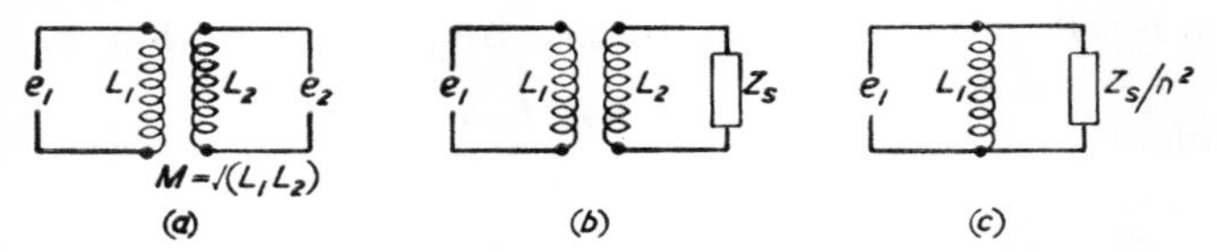

FIG. 2.30. CLOSE-COUPLED TRANSFORMER WITH RESISTANCELESS WINDINGS

Fig. 2.30 (*a*) represents a simple close-coupled transformer with no load across the secondary and in which the windings are assumed to have no resistance.

We can write

$$i_1 = e_1/L_1\omega \quad \text{and} \quad e_2 = M\omega i_1$$

Now $M = k\sqrt{(L_1L_2)}$ and k is assumed to be unity. Hence

$$e_2/e_1 = \omega\sqrt{(L_1L_2)}/\omega L_1 = \sqrt{(L_2/L_1)}$$

But the inductance of the windings is proportional to the square of the number of turns so that $\sqrt{(L_2/L_1)} = T_2/T_1 = n$. Hence

$$e_2/e_1 = n$$

This is a fundamental relationship in a closed-core transformer, namely that the primary and secondary e.m.f.s are in direct proportion to the number of turns on the respective windings.

Let us now consider the effective impedance of the transformer with a load connected across the secondary as in Fig. 2.30 (*b*). From basic theory

$$Z_1' = j\omega L_1 + M^2\omega^2/Z_2$$

But since k is unity, $M^2 = L_1L_2$ while $Z_2 = R_2 + j(\omega L_2 + X_2)$, so that

$$Z_1' = j\omega L_1 + \frac{\omega^2 L_1 L_2}{R_2 + j(\omega L_2 + X_2)} = \frac{j\omega L_1 R_2 - \omega L_1 X_2}{R_2 + j(\omega L_2 + X_2)}$$

which by a slight rearrangement

$$= \frac{j\omega L_1(R_2 + jX_2)}{j\omega L_2 + (R_2 + jX_2)}$$

Now $L_2 = n^2L_1$ and $(R_2 + jX_2) = Z_s$, the secondary load. The expression thus becomes

$$Z_1' = \frac{j\omega L_1 Z_s}{n^2 j\omega L_1 + Z_s} = \frac{j\omega L_1 \,.\, Z_s/n^2}{j\omega L_1 + Z_s/n^2}$$

This is the impedance of a load Z_s/n^2 in parallel with the primary inductance L_1 so that the equivalent circuit is as shown in Fig. 2.30 (*c*), consisting simply of the primary inductance shunted by the reflected secondary load Z_s/n^2.

Leakage Inductance

In a practical transformer the coupling is not perfect, so that k is not unity though it may be nearly so (0·98 or even higher). This means that some part of the flux produced by the primary does not link with the secondary, which has the same effect as a small

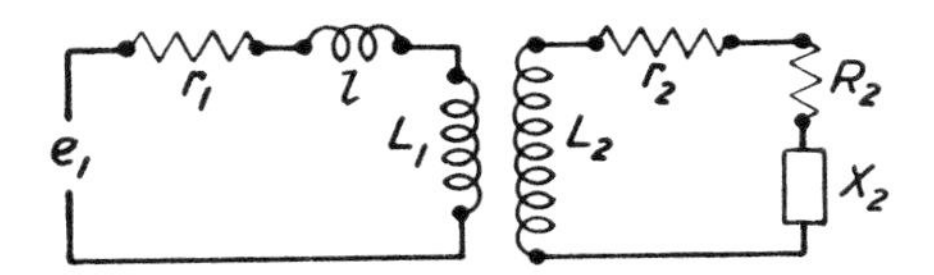

FIG. 2.31. TRANSFORMER WITH WINDING RESISTANCE AND LEAKAGE

separate inductance in series with a perfectly-coupled transformer as shown in Fig. 2.31, which also includes the resistance r_1 to represent the resistance of the primary winding.

The small series inductance is called the *leakage inductance* and can be shown to be equal to $(1 - k^2)L_1$. Strictly speaking the presence of leakage inductance affects the transformation ratio but this may be ignored here.

The simple circuit of Fig. 2.31 in fact does not take into account all the factors, but as these differ in importance according to the circumstances in which the transformer is used, more detailed discussion will be given at the appropriate place, notably in Chapters 4 and 10.

Insertion Gain

If an e.m.f. E having a source resistance R_S feeds a load R_L, the maximum power in the load is obtained when $R_L = R_S$. In general, this will not be the case, but the condition may be satisfied (or approached) by introducing a suitable impedance-changing network between the source and the load. The ratio of the power developed in the load with this network inserted to that without it is called the *insertion gain.*

The most usual form of impedance-changing network is a transformer. If this has a ratio n, the equivalent primary impedance R_L' (neglecting losses) is R_L/n^2, which must equal R_S for correct matching. Then the load power without the transformer

$$= I^2R_L = E^2R_L/(R_S + R_L)^2$$

With the transformer in circuit the power is

$$E^2R_L/n^2(R_S + R_L/n^2)^2$$

Hence the insertion gain $A_i = (R_S + R_L)^2/n^2(R_S + R_L/n^2)^2$

$$= (R_S + R_L)^2/4n^2R_S^2 \text{ if } R_L/n^2 = R_S$$

This is true for any value of n. Thus if $R_S = 200$ and $R_L = 2{,}000$, $n^2 = 10$ and $A_i = 2{,}200^2/(40 \times 200^2) = 3{\cdot}025$. If $R_L = 20$, $n^2 = 0{\cdot}1$ and $A_i = 220^2/(0{\cdot}4 \times 200^2) = 3{\cdot}025$ as before.

If the impedances are reactive, the current will be E/Z, but the condition for maximum power in the load is still that $R_L' = R_S$, so that the above calculations remain valid.

If some other form of network is introduced, such as an attenuator or filter, A_i may become fractional, indicating an *insertion loss.* This is discussed further in Chapter 14 (page 625).

Transfer Function

The ratio of the values of any particular function at the output and input of a network is called the *transfer function.* If one is interested in the relative voltages the transfer function is simply V_{out}/V_{in}. This, in general, will not be a scalar quantity but a vector quantity of the form $|V_2/V_1|\angle\theta$. Another form of transfer function is the impedance transfer which is the ratio of the input voltage to the output current. The input voltage here is not the source e.m.f. but the voltage actually developed across the input to the network. In the example just given $V_{in} = E/2$ (if $R/n^2 = R_S$) while $I_{out} = I_{in}/n = E/2nR_S$. Hence the impedance transfer function is nR_S.

If the matching is not correct and/or the impedances are reactive the values of V_{in} and I_{out} must be calculated from the circuit constants. Use is also made in filter calculations of a *transfer coefficient* which involves the ratio of the input and output volt-amperes, as explained on page 627.

3

Principles of Radio Communication

RADIO communication involves three essential requirements. First it is necessary to produce electromagnetic waves which, as explained in Chapter 1, are a particular kind of disturbance set up when electrons are caused to change their position rapidly. The first part of this chapter therefore deals with the mechanism by which these waves are produced and the manner in which they are propagated through space.

The mere presence of these waves, however, does not communicate any intelligence (apart from the fact of their presence) so that it is necessary to alter the waves in some way, in a predetermined manner which will convey the required information. We can, for example, start and stop the waves in accordance with the dots and dashes of the Morse code or, alternatively, we can produce variations corresponding to the sound waves of speech or music. This is called *modulation* and the second section of this chapter deals with the principal methods by which modulation can be achieved.

Finally it is necessary at the distant point to provide equipment which will detect the presence of the waves and convert the minute energy in the waves into electric currents of practical magnitude. This receiving equipment, moreover, must not only detect the waves but must also convert the modulation into suitable variations of current so as to reproduce the original information in the required form. This process is known as *detection* or *demodulation*, and the third part of the chapter will deal with the basic principles of this process.

3.1. Production of Radio Waves

An electromagnetic wave may be considered as a disturbance which arises from a sudden change in the condition of an electrical system. Sometimes this system is the atomic structure of a material which is disturbed by an increase of temperature, as in the case of light waves. Radio waves arise from disturbances in electrical circuits which are of much greater physical dimensions than atomic structures. Hence the waves produced here are of much longer wavelength but the form of the disturbance is similar.

A simple exposition of wave motion was given in Section 1.7, but we shall now consider the process in greater detail, particularly

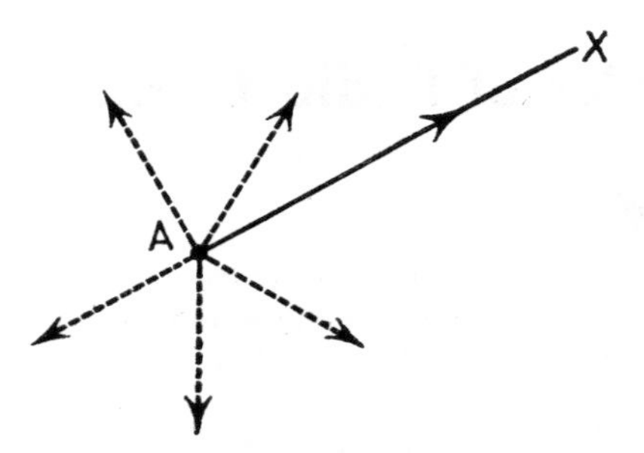

FIG. 3.1. LINES OF FORCE RADIATING FROM AN ELECTRON

in respect of the electromagnetic waves used in radio communication. Let us first consider a single electron as in Fig. 3.1. This electron has associated with it an electric field which surrounds it in all directions and which we usually represent as a series of lines of force radiating outwards in a straight line from the electron itself.

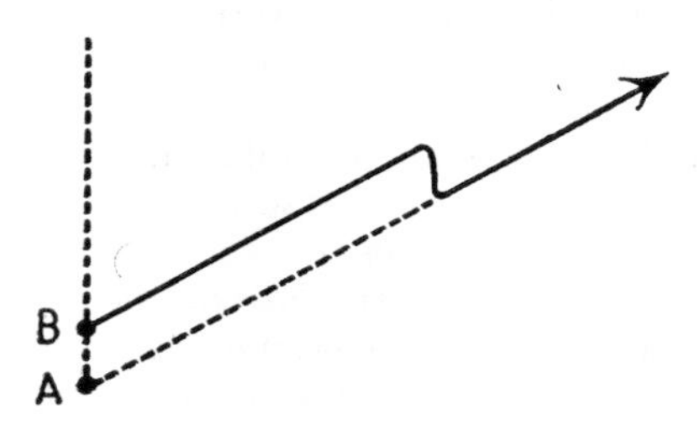

FIG. 3.2. KINK PRODUCED BY SUDDEN CHANGE OF POSITION OF THE ELECTRON

The intensity of the field is dependent on the relative concentration of these lines of force, which we saw in Section 1.3 to be inversely proportional to the square of the distance. Hence at a comparatively short distance the field strength becomes negligible.

However, consider the effect on one of these imaginary lines of force, such as AX in Fig. 3.1., if the position of the electron is suddenly changed, as indicated in Fig. 3.2, the electric field cannot suddenly change its position. It will, in fact, take time for the information that the electron has changed its position to be communicated along the line of force and there will thus be a kink in the line. Before this kink the line of force radiates from the electron in its new position B. Beyond it, the line of force is still in its former condition corresponding to the original position A of the electron and the kink itself represents the change from one position to the other due to the sudden movement of the electron.

This kink constitutes the electric wave. It travels outwards at a high, but not infinite velocity—actually a velocity which is found to be the same for all electromagnetic waves, namely 3×10^8 m/sec, and the interesting fact is that, although the static field strength very quickly becomes negligible as the distance increases, the strength of the wave only falls off inversely as the distance so that the effect of the disturbance can be noticed at much greater distances.

Strength of Field

Let us examine this process more precisely. Consider an electron moving upwards with a velocity v. Suppose that for a short time δt the electron is subjected to an acceleration α, so that its velocity

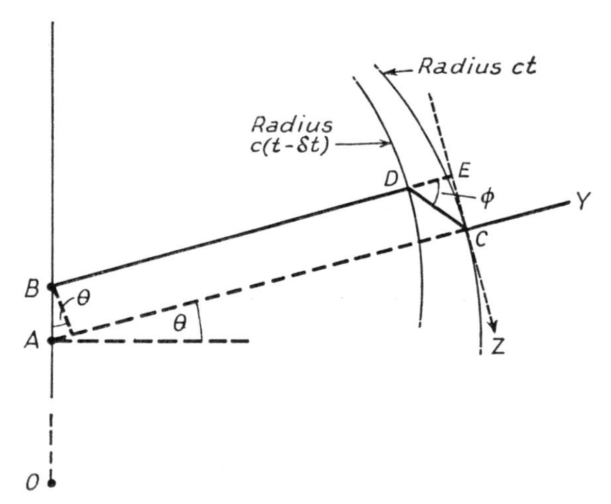

Fig. 3.3. Illustrating Formation of Radio Wave

becomes $v + \alpha\delta t$. Fig. 3.3 represents the situation at a time t after the change of conditions. If the velocity had remained unchanged, the electron would have arrived at A where $OA = vt$; but because of the momentary acceleration it actually arrives at B, where $OB = (v + \alpha\delta t)t$. Hence $AB = \alpha t\delta t$.

However, as just said, the news of this changed position cannot be transmitted instantaneously but travels outwards with a velocity c, the velocity of light. Hence at the time t the field from the electron will be undisturbed beyond a radius $r = ct$ and will continue to radiate from A. On the other hand, within a slightly smaller radius $c(t - \delta t)$ the field radiates from the new position B. In between the two there is a sudden transition from one condition to the other, and it is this which constitutes the radio wave.

Actually since the electric field is radiating in all directions the information regarding the change of state is contained within two rapidly expanding spheres of radii ct and $c(t - \delta t)$, and the diagram of Fig. 3.3 is a cross-section of these spheres. The portion we are interested in is the component DC, which can be resolved into two further components, one radial, represented by DE, and the other tangential represented by EC.

Now as was shown in Section 1.3 the field strength at a distance r from a charge q is E_0/r^2 where $E_0 = q/4\pi\varepsilon$. This is the field strength in a radial direction and hence the field strength along the line DC will be $E_0/r^2 \cos\phi$.

Resolving this radially and tangentially we have

$$\text{Radial component} = (E_0/r^2 \cos\phi)\cos\phi = E_0/r^2$$

Hence the radial component is the same as for the steady-state condition. There is now, however, a tangential component, given by

$$\text{Tangential component} = (E_0/r^2 \cos\phi)\sin\phi = (E_0/r^2)\tan\phi$$

Now $CE = AB\cos\theta = \alpha t \,.\, \delta t \,.\, \cos\theta$, where θ is the "angle of azimuth," i.e. the angle which AC makes with the horizontal, while if r is large, DE is very nearly equal to the difference in radius of the two spheres $= ct - c(t - \delta t) = c\delta t$. Hence

$$\tan\phi = \frac{CE}{DE} = \frac{\alpha t \,.\, \delta t \,.\, \cos\theta}{c \,.\, \delta t}$$

and since $r = ct$, so that $t = r/c$,

$$\tan\phi = \left(\frac{\alpha r}{c^2}\right)\cos\theta$$

so that the tangential component is

$$\frac{E_0}{r^2}\cdot\frac{\alpha r}{c^2}\cdot\cos\theta = \frac{E_0}{r}\cdot\frac{\alpha}{c^2}\cdot\cos\theta$$

It is this tangential component which is abnormal and which, in fact, constitutes the electric wave, and it should be noted that the field strength of the wave is:

(*a*) proportional to the acceleration α, so that the more rapidly the electrons are accelerated, or in other words the higher the frequency of the oscillation of the currents in the aerial, the greater is the field strength produced.

(*b*) inversely proportional to the distance from the origin (as opposed to the inverse square law of ordinary electrostatics), and

(*c*) proportional to $\cos \theta$; i.e. it is greatest in the horizontal direction (assuming that the electron movement is vertical) and is zero in the vertical direction.

Aerials

A practical form of radiator therefore must be such as to provide this rapid electron movement, which is accomplished by feeding into the system an alternating current of high frequency—usually many millions of hertz. Such a device is called an *aerial*, practical forms of which will be discussed later.

With such a system carrying an alternating current the movement of the electrons in one direction is immediately followed by a corresponding reverse movement, backwards and forwards. Consequently if we take the cross-section of the whole of the space surrounding the aerial, and add together all the components of the field strength, we shall obtain a series of bean-shaped loops of electric field travelling outwards and continually expanding in size with a velocity of 3×10^8 m/sec. This is represented in Fig. 3.4. The actual direction of the electric field at any point is at right angles to the line joining that point to the aerial.

Wavelength

The strength of the field falls off as the angle of azimuth increases, as we have just seen. For the majority of cases, only those waves which leave at a small angle to the horizontal are important. For small values of θ (up to 10° or so), $\cos \theta$ is practically unity. Hence near the ground level the field strength is uniform, so that the practical manifestation of a wireless wave is as shown in Fig. 3.5, which represents a succession of belts of electric field moving rapidly along the direction of travel. These belts are sinusoidal in form having a maximum intensity at one point. The strength then decreases to zero, reverses and reaches a maximum in the opposite direction and then reverses again so that we have alternate bands of positive and negative field. The distance between two points of maximum strength in the same direction is known as the *wavelength* of the wave.

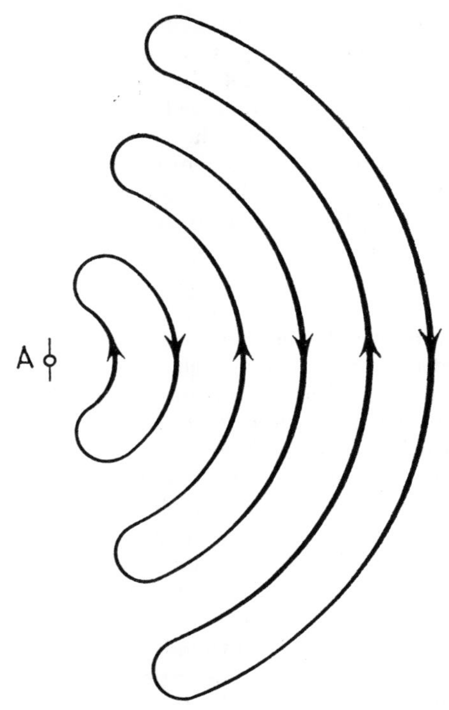

FIG. 3.4. FORM OF RADIATION FROM AN AERIAL

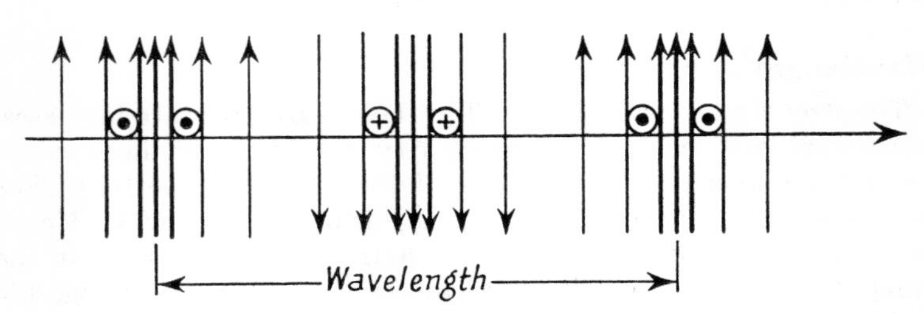

FIG. 3.5. FORM OF RADIATION NEAR GROUND LEVEL

Remembering that the disturbance travels outwards with a constant velocity it is clear that the more rapid the frequency of oscillation of the electrons generating the wave the closer will the disturbances follow one upon the other. Hence the wavelength will be shorter and, in fact, the wavelength and frequency are related by the simple expression

$$\lambda f = c$$

where λ is the wavelength in metres,
f is the frequency in hertz,
$c = 3 \times 10^8$ m/sec.

Magnetic Field

Now, the movement of the electric field through space generates a magnetic field, and it is sometimes convenient to regard the disturbance from a magnetic point of view. The magnetic field produced will be at right angles to the electric field and also at right angles to the direction of motion of the waves, so that it will be at right angles to the plane of the paper in Fig. 3.5, where it is represented by the circles with the crosses and the dots in the centre, the crosses representing a field going into the paper and the dots a field coming out of it.

It should particularly be noted, however, that this magnetic field is generated by the movement of the electric field and is, in fact, merely another way of regarding the same phenomenon. According to the convenience of the problem to be considered the wave may be regarded *either* as a moving belt of electric field *or* a moving belt of magnetic field, but not both. The condition is, indeed, quite different from that which exists in a circuit. In a simple resonant circuit the energy is first stored in the form of electrostatic energy in the capacitor, and one quarter cycle later is stored entirely in the form of magnetic energy in the coil. There is a phase lag of 90° between these two states and the energy is continually transformed from one form to the other. The conditions in an electromagnetic wave are essentially different from this, since the electric and magnetic fields are in phase.

Practical Aerials

Consider now the field strength produced by a practical aerial of length l carrying a current i.

The current $$i = q/t = \frac{q}{l} \cdot \frac{l}{t} = \frac{q}{l} v$$

where v is the velocity of the electrons.

Hence $v = (l/q)i$ and the acceleration $\alpha = dv/dt = (l/q)\, di/dt$.

Now, if $i = I \sin \omega t, \quad \alpha = (l/q)I\omega \cos \omega t$

Substituting this in the expression for field strength deduced earlier and writing $q/4\pi\varepsilon$ for E_0, we have

$$E_r = \frac{Il\omega}{4\pi\varepsilon rc^2} \cos \theta \cos \omega t$$

But we have seen that $\omega = 2\pi f$ and $f = c/\lambda$. Making this substitution and inserting the appropriate values for $c = 3 \times 10^8$ m/sec and $\varepsilon = 1/(36\pi \times 10^9)$ we have (in free space)

$$E_r = 188{\cdot}5 \frac{Il}{\lambda r} \cos \theta \cos \omega t \text{ volts/metre}$$

We have assumed that the current at any instant is the same throughout the whole length l of the aerial. In practice this is not true, and hence instead of the total length l we substitute a length h which is the length of a fictitious aerial in which the current is uniform throughout. This is known as the *equivalent height* of the aerial and is discussed in greater detail in Chapter 8 where various practical forms of aerial are considered.

Moreover, in a practical aerial for long and medium waves it is customary to earth the mid-point. When this is done, it behaves like a free aerial of twice the actual (effective) height, as explained on page 363; and since we are interested only in the radiation at ground level, $\theta = 0$, and $\cos \theta = 1$. The radiated field is then simply

$$E_r = 377\ Ih/\lambda r \text{ volts/metre}$$

Reception

To receive the signal at the distant point a receiving aerial is erected. This is made similar in form to that at the transmitter (though for long and medium waves it is not so large), and the electric field in its passage across the aerial induces an e.m.f. equal to the product of the field strength and the effective height of the aerial. Thus, if at a given point the field strength of the wave is 10 microvolts/metre and the effective height of the aerial is 3 metres, the voltage induced will be 30 microvolts. This will produce a current which is made a maximum by tuning the aerial (i.e. adjusting the circuit to resonate at the frequency in question) and the resulting signal is then amplified as required. The technique is discussed in detail in Section 3.3.

Propagation

These fundamental expressions refer to the generation of waves in free space. In practice, the transmitting aerial is located on the surface of the earth and because of the earth's curvature the waves will very soon leave the surface, as shown in Fig. 3.6, so that in the absence of other factors there would be no reception at a point B an appreciable distance away. The radiation from A is, of course, not confined to the direction at right angles to the transmitting aerial, so that there will be some radiation in the direction AB, but

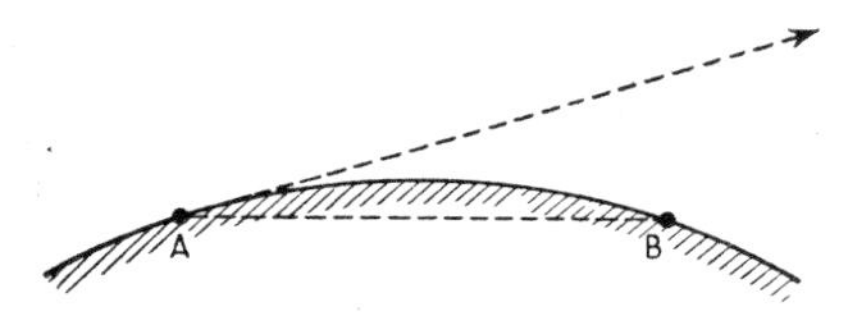

FIG. 3.6. EFFECT OF EARTH'S CURVATURE

this will have to pass through the earth and its energy will be rapidly absorbed.

Reception at considerable distances is, in fact, only possible because of reflection or refraction in the earth's atmosphere which causes the waves to return to the earth, as will be discussed shortly. There are, in fact, three main types of propagation, namely:

(*a*) Ground waves, which travel over the surface of the earth,
(*b*) Sky waves, which are reflected back to the earth by the upper atmosphere,
(*c*) Space waves, which travel in the space close to the earth for a limited distance.

Ground Waves

Consider first the ground wave. This would appear to be limited to the relatively short distance over which a line joining the transmitter and receiver does not pass through the earth to any serious extent. In practice the range is rather more than this because of the phenomenon known as *wave drag*.

The velocity with which an electromagnetic wave travels is proportional to $1/\sqrt{(\mu\varepsilon)}$ where μ is the permeability and ε is the permittivity $= \varepsilon_0\varepsilon_r$. In free space $\varepsilon = \varepsilon_0$ and the velocity is the velocity of light, c, but if ε_r is greater than unity, as it is with a solid or liquid dielectric, the velocity is reduced. That portion of the wave which travels in the surface of the earth thus moves slower than the main part of the wave which causes the "feet" of the waves to drag as shown in Fig. 3.7. The difference in velocity is only small but it is sufficient to permit the waves to creep round the earth for an appreciable distance.

Energy is, however, absorbed from the wave by the losses in the earth, so that the field strength at a distance is less than the value for free space, so that in fact

$$E_d = AE_0/d$$

where E_d = field strength of ground wave at a distance d from the transmitter,

E_0 = theoretical (free-space) field strength at unit distance from the transmitter.

The value of A depends on the nature of the terrain, and also on the frequency. At low frequencies (up to about 1 MHz) the earth is mainly resistive and the losses are relatively small. As one would expect the losses are less over sea than over land because of the better conductivity of sea water. At higher frequencies, the earth behaves more as an inefficient dielectric and the attenuation is appreciably greater.

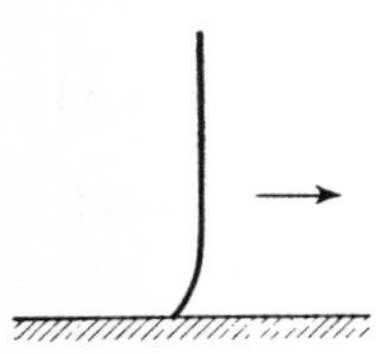

Fig. 3.7. Wave Drag

There is extensive literature on the subject which cannot be discussed further here. It may be noted, however, that at distances exceeding $0{\cdot}02\lambda^2\sigma$, where λ is the wavelength in use (in metres) and σ is the conductivity (in S/cm), the factor A becomes practically proportional to $1/d$, so that the received field strength is proportional to $1/d^2$ and not to $1/d$ as in the ideal case.

For normal land σ is of the order of 10^{-4} S/cm but may be as high as $4{\cdot}5 \times 10^{-2}$ for sea water.

Sky Waves

It will be clear that, because of the combined effect of the earth's curvature and losses, the range of the ground wave is limited. Transmission can be achieved over considerably greater distances, however, by means of the sky wave which is reflected from the upper atmosphere.

If the waves strike this reflecting layer at too steep an angle they are not reflected, but below a certain critical angle, which varies with conditions as will be seen later, the waves are returned to earth, from which they are again reflected and can, therefore, travel onwards by successive reflections from the electrified layer and the earth's surface. At each reflection some energy is lost but this is usually small and the signal strength is still approximately subject to the inverse distance law. It is, indeed, quite common with short waves to detect signals which have been completely round the earth, and under some conditions a curious effect known as *one-seventh-second echo* is obtained in which the normal signal is followed by another much fainter signal one-seventh of a second later. This is the same signal which has travelled onwards completely round the earth and again operated the receiver, taking one-seventh

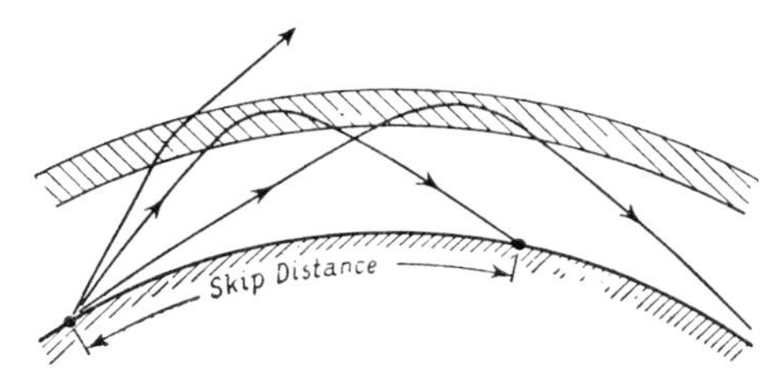

FIG. 3.8. ILLUSTRATING REFLECTION FROM IONOSPHERE

of a second in its transit and, of course, being much fainter due to the additional attenuation.

As we have seen, the reflection is obtained as long as the angle of incidence is not too steep. There is, however, a blank patch over which the ground wave has ceased to operate and the reflected wave has not yet arrived because the angle at which the waves would have to reach the reflecting layer in order to come back within this specified distance would be too steep. This blank period is known as the *skip distance.* It may vary from a few hundred to several thousand miles depending upon the wavelength being used and the conditions at the time. The effect is illustrated in Fig. 3.8.

The Ionosphere

The reflections from the upper atmosphere arise because of the existence of layers of ionized gas and the region has, therefore, become known as the *ionosphere.* Its existence was first predicted by Kennelly and Heaviside independently in 1902, the height of the suggested layer being of the order of 50 to 100 km above the earth's surface.

As the height above the earth increases, the pressure of the gases constituting the atmosphere decreases. For a height of about 8 km we have the ordinary atmosphere or *troposphere.* Here we can have varying degrees of temperature, pressure and humidity, with varying air currents and cloud formations. In general, the air pressure is fairly high, and except under thunder conditions the air is not electrified or ionized.

Above the troposphere there is a much more extensive region, known as the *stratosphere*. In this there are no air currents and the atmosphere is at a more or less constant temperature, but at a gradually decreasing pressure. The atmosphere, therefore, tends to separate itself into layers of constituent gases in order of density, which is the reason for the name stratosphere. Because of this the upper limits of this region contain light gases at a very low pressure and such gases are easily ionized.

Under the influence of the ultra-violet rays in the sunlight these gases in the upper regions are more or less permanently electrified, though the extent of the ionization depends upon the time of day.

Experiments to determine the exact height of the ionosphere were undertaken by Sir Edward Appleton and Dr. Barnett, in 1924,* as a result of which it was shown that there are two layers of ionization, one at a height of around 110 to 130 km above the earth's surface and the other at an average height of a little over 300 km. The lower level is known as the E layer and the higher region as the F layer. These layers are much more well defined in darkness than in daylight. The presence of the sun's rays causes considerable ionization to take place and the electrified belt is more extensive and less well defined during the daytime. Indeed during daylight there is not only an increase in the ionization but also a penetration downwards nearer to the earth, resulting in the formation of an almost separate layer known as the D layer which is little more than 90 km up. Moreover, during daylight the F layer becomes very much more extended and appears to divide itself into two more or less well defined regions, known as the F_1 and F_2 layers, at heights of approximately 250 and 400 km.

At night the disappearance of the sun's rays causes the gases to recombine into two more or less well defined layers, the E and F layers as already explained. Therefore, the reflection which is obtained is much more sharply defined at night than by day.

Mechanism of Reflection

Let us now see what happens if an electromagnetic wave passes into an electrified belt, such as the E or F layer. As already mentioned, the velocity of propagation of an electromagnetic wave is

* These experiments were conducted by radiating short pulses of waves skywards, and noting the time taken for them to be returned a short distance away by reflection from the ionosphere. It was from this technique that radar was evolved.

proportional to $1/\sqrt{(\mu\varepsilon)}$, where μ is the permeability and ε is the permittivity of the medium. Ordinarily both these factors are unity, but when we reach an electrified medium the permittivity decreases so that the wave will move with an increased velocity.

Now, the electric field in a wave is not confined to a small region but spreads over a considerable distance, particularly by the time it has reached a point some 100 km above the surface of the earth. Consequently the top of the wave reaches the electrified belt before the middle and bottom portions, and the top therefore begins to travel faster than the bottom so that a bending is produced as illustrated in Fig. 3.9, which causes the wave to be turned round

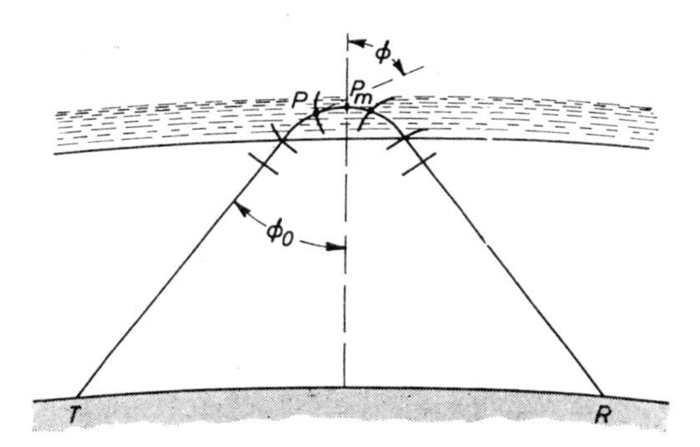

FIG. 3.9. ILLUSTRATING ACTION OF IONIZED LAYER

and directed back towards the earth again. The amount of bending produced depends upon the frequency of the wave, as also does the attenuation or loss of energy in the bending process.

The behaviour is calculable from ordinary optical laws, one of which, known as Snell's Law, states that the effective direction of travel at a point P is such that

$$n \sin \phi = \sin \phi_0$$

where n is the refractive index of the medium and ϕ and ϕ_0 are as shown in Fig. 3.9. Now, the refractive index, which is the square root of the relative permittivity, depends upon the degree of ionization, and can be expressed in the form

$$n = \sqrt{\varepsilon_r} = \sqrt{(1 + 81N^2/f^2)}$$

where f is the frequency of the wave and N is the number of free electrons per cubic centimetre in the medium.

The top of the wave path, P_m, sometimes called the point of reflection, occurs when $\phi = 90°$, so that $\sin \phi = 1$, and $n = \sin \phi_0$. Hence the smaller the angle of incidence, the smaller is the refractive index required if the wave is to be reflected. With vertical incidence ($\phi_0 = 0$), n must be zero, which occurs when $f_v = 9N$.

If the frequency exceeds this value, the refractive index becomes imaginary and no reflection results. The transition occurs at a frequency of the order of 2·6 MHz (corresponding to a wavelength of 115 metres) and is accompanied by considerable absorption of energy because the time of mean free travel of the electrons in the ionized layer corresponds roughly with the period of oscillation of the wave.

At angles of incidence less than 90°, $\sin \phi_0$ is less than unity. Hence the critical frequency f_c is greater than the value for vertical incidence, f_v, and is, in fact, given by

$$f_c = f_v/\cos \phi_0 = f_v \sec \phi_0$$

This is known as the *secant law*, from which for any given transmission (for which ϕ_0 can be estimated) it is possible to determine the *maximum usable frequency* (m.u.f.) with which satisfactory results can be expected. This depends, of course, on a knowledge of f_v, which varies with time of day and year, but has been established by prolonged observations under normal conditions, from which suitable frequencies and angles of path can be determined to meet specified requirements.

Since the ionosphere is by no means uniform, either in height or properties, allowance has to be made in practice for a variety of secondary effects. For example, a long-range transmission may span regions of transition from light to darkness involving quite different ionospheric conditions, and it is usually necessary to change the frequency according to the time of day. These and other special requirements are discussed more fully in Chapter 13.

Waves that are not sufficiently bent round to be returned to earth by the E layer will continue outwards into space until they encounter the F layer where they will receive further bending. The density of the electrons in the F layer is much less than in the E layer so that the additional bending produced is only slight. On the other hand, the extent of the F layer is rather greater than the E layer and the effect is that waves which would otherwise be lost have their reflection completed and are returned to the earth again.

Within the range of usable frequencies the attenuation of the wave is very small so that the field strength is substantially that given by the expression for free space, and is inversely proportional to the length of the path from the transmitter.

Fading

The lower surface of the ionosphere is not sharply defined and in consequence the plane of polarization of the wave may become twisted. The electromagnetic waves which have so far been considered are vertically polarized, which means that the electric fields all lie substantially in the vertical plane. This need not necessarily be the case, however, because the electric fields can be in any direction at right angles to the direction of the propagation of the wave. If the transmitting aerial is arranged vertically then the electric fields in the wave will all be in the vertical plane and the receiving aerial must also be erected vertically. A receiving aerial arranged horizontally at right angles to the direction of travel of the wave will not receive any signal because there is no component of the electric field along the length of the aerial; therefore, the electrons in the conductor cannot be set in motion.

If the plane of polarization of the electric wave is changed, however, by reflection at the upper atmosphere, then its effect on the receiving aerial will be reduced, and if the plane of polarization is twisted through 90° so that the wave arrives horizontally polarized the receiving aerial will not respond at all.

In practice some of the waves are twisted more than others and the conditions are continually changing, with the result that the wave in travelling down to earth again behaves as if its effective plane of polarization were gradually rotating. Consequently at one instant the receiving aerial will derive the full signal from the wave, while a short time later when the effective plane of polarization has rotated through 90° it will receive no signal at all. This gives rise to the effect known as *fading* whereby the strength of the received signal is not steady but continually varying. The fade may be slow or rapid, it may be complete or partial, and the effect will be familiar to every reader.

Special aerial arrangements can be used to minimize the effect of fading, as discussed in Chapter 8, while the receiving equipment is usually provided with automatic gain control to provide still further compensation.

Space Waves

The third type of propagation which applies principally to the very short wavelengths is that in which the received signal is derived partly by a direct transmission along the path between transmitter and receiver, and partly by a wave reflected *from the ground.* At low frequencies any waves travelling into the ground are absorbed but as the frequency increases an increasing proportion of the energy is reflected. The two paths are as shown in Fig. 3.10.

The range of such waves is still limited to an approximately optical path, being in general about twice the length of the direct line-of-sight path. The direct and reflected waves, however, will have slightly different path lengths and there will thus be a phase

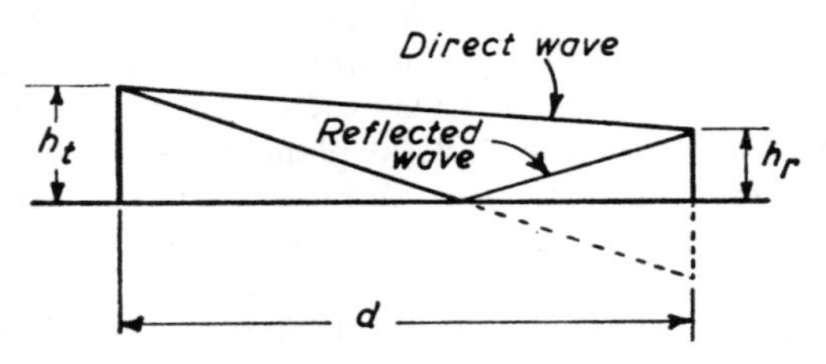

Fig. 3.10. Illustrating Space Waves

difference between them. From the geometry of the figure the difference in path length d' is $2h_th_r/d$. The phase difference is dependent on the ratio d'/λ, being in fact $4\pi h_th_r/\lambda d$.

The sum of the two waves is proportional to the sine of the phase angle, but since this is small $\sin\theta = \theta$ very nearly so that the received field strength is

$$\frac{E_0}{d}\cdot\frac{4\pi h_t h_r}{\lambda d} = \frac{kE_0}{\lambda d^2}$$

where E_0 is the field strength in free space at unit distance, and $k = 4\pi h_t h_r$. Hence the received field strength is proportional to $1/d^2$ and not to $1/d$.

Reception can be obtained over much greater distances than the optical path under certain atmospheric conditions. The relative permittivity of the atmosphere near the ground is slightly greater than unity because of the presence of gas molecules (mainly water vapour, which has a high permittivity). The gas concentration, however, decreases as the height increases, so that the permittivity decreases, producing a refraction of the rays similar to ionospheric reflection, which is the reason why the effective range is rather greater than the direct optical path.

Normally the permittivity decreases linearly with increasing height, but when air masses of differing temperature and moisture content overlie each other, conditions often arise where, for a while, the permittivity decreases at a greater rate, which produces increased refraction and consequent increased range. The waves can, in fact, follow the earth's curvature in a sort of duct, so long as the abnormal atmospheric conditions persist.

Classification of Radio Waves

The range and behaviour of a radio wave is thus considerably dependent on the wavelength, which must be chosen to suit the particular requirements.

The general classification and characteristics of radio waves in modern use are given in Table 3.1 overleaf.

3.2. Modulation

No information can be communicated except by a change of conditions. For example, if a continuous wave were radiated the only information which could be received at a given point would be that the wave was present. In order to transmit some intelligence it is necessary to change the state in some way or other. The simplest way of doing this is to stop and re-start the wave at appropriate intervals so that information can be transmitted by the use of some suitable code, but as we shall see, there are various other ways in which the conditions can be altered, such as, for example, altering the frequency.

This change in the state or conditions of the wave is termed *modulation* and a wave which has been modified in this manner is called a *modulated wave*.

Types of Modulation

The most obvious form of modulation is to start and stop the wave at appropriate intervals to produce the dots and dashes of the Morse code. Alternatively, the amplitude may be varied in such a manner as to conform with the relatively slow vibrations of speech or music to transmit telephonic information.

Both these types of modulation produce the effect by altering the general level or amplitude of the wave being transmitted: consequently this system is known as *amplitude modulation*. The wave of which the amplitude is varied is known as the carrier wave or often just the carrier, because its function is merely to carry from the transmitter to the receiver the changes of amplitude in which the real intelligence lies. When we have, at the receiver, made a note of the changes of amplitude, the carrier wave itself is no longer required and may be, and actually is, removed.

One disadvantage of amplitude modulation is that the system does not respond readily to sudden changes. Just as a mechanical

TABLE 3.1. CLASSIFICATION OF RADIO WAVES

Class	*Abbreviation*	*Frequency*	*Wavelength*	*Characteristics*	*Usage*
Very-low frequency	V.L.F.	10–30 kHz	3–10 km	Good and constant propagation at all times and seasons but requires large aerials and input powers.	Global communication
Low frequency	L.F.	30–300 kHz	10–1 km	Similar to V.L.F. but slightly more variable performance.	Broadcasting (L.F. and M.F.)
Medium frequency	M.F.	300–3000 kHz	1000–100 m	Attenuation low at night but high in day. Only require medium powers.	
High frequency	H.F.	3–30 MHz	100–10 m	Transmission depends on ionosphere and hence varies from very good to very poor with time and season.	Long-range communication
Very-high frequency	V.H.F.	30–300 MHz	10–1 m	Not reflected by ionosphere so that range is normally limited to direct (visual) path. Transmission within these limits good and constant. (V.H.F. to E.H.F.)	T.V. and radar
Ultra-high frequency	U.H.F.	300–3000 MHz	100–10 cm		Microwave radar and communication links
Super-high frequency	S.H.F.	3–30 GHz	10–1 cm		
Extra-high frequency	E.H.F.	30–300 GHz	10–1 mm		
Long infra-red		300–3000 GHz	1–0·1 mm	These waves merge into the region of natural oscillations and are at present largely experimental. (Long and short infra-red)	Space communication
Short infra-red		3–30 THz	0·1–0·01 mm		

There is also a "band" classification:

Band	L	S	C	X	J	K	Q
Frequency range (GHz)	1–2	2–4	4–8·2	8·2–12·4	12·4–18	18–26	26–40

flywheel possesses inertia which resists any sudden change in speed, so an oscillatory circuit will not change its frequency instantaneously. Some early forms of transmitter therefore were arranged to radiate continuously, but at slightly different frequencies in the key-up and key-down conditions, the receiver being tuned to discriminate between the "marking" and "spacing" waves.

This method of communicating intelligence is known as *frequency modulation*. At first sight it does not appear to be applicable to the transmission of speech or music but it will be clear that if the chang in frequency could be made proportional to the instantaneous amplitude of the speech we could modulate the frequency of the carrier wave instead of its amplitude.

This system, though more complex, has certain advantages, the most important being the discrimination which it provides against random amplitude modulation.

Atmospheric disturbances and even locally generated "noise" in the receiver itself result in a fluctuation of the strength of the signal. With frequency modulation, however, the required intelligence is all contained in the change in frequency of signal and it is, therefore, possible to pass the signals through a limiting stage immediately prior to the detector as a result of which the signal is of constant amplitude irrespective of any fluctuations which may have been introduced as a result of the interference. This results in a very marked improvement in the signal/noise ratio.

A somewhat similar form of modulation can be achieved by varying the phase of the carrier wave. This is merely a variant of frequency modulation since any change in frequency is necessarily accompanied by a change of phase, as is discussed on p. 144.

Finally there is *pulse modulation*. This arose as a corollary to the development of pulse technique in radar. For radar it is necessary to generate very short bursts of waves which start and stop extremely rapidly and are separated by relatively long intervals. These pulses of waves travel outwards from the aerial and are reflected from the target, and the direction from which the reception is obtained, and the time interval which elapses, provide the information as to the bearing and range of the target.

Such pulses are a special form of amplitude modulation. The requirements are that the amplitude shall build up very rapidly to its value, remaining constant for a short period and then falling to zero equally rapidly. The time-constants of the circuits necessarily prevent an instantaneous rise and fall, but the technique has now been so developed that pulses can be generated lasting for only a fraction of a microsecond.

The development of this technique has led to the introduction of various forms of pulse modulation for the transmission of speech. The instantaneous value of the speech amplitude is "sampled," and this information is then translated into a code involving a series of pulses separated by varying time intervals.

These various forms of modulation will now be considered in detail.

Amplitude Modulation

Fig. 3.11 represents an amplitude-modulated wave. It will be clear that there are two essential requirements.

(*a*) The modulation frequency must be only a small fraction of the carrier frequency.

(*b*) The changes in carrier amplitude must be faithful reproductions of the modulation frequency.

It is clear that the carrier wave can be modulated to varying depths. When the variation in amplitude, as a result of the modulation, is just sufficient to reduce the carrier wave to zero the modulation is said to be complete and in Fig. 3.11 we have three

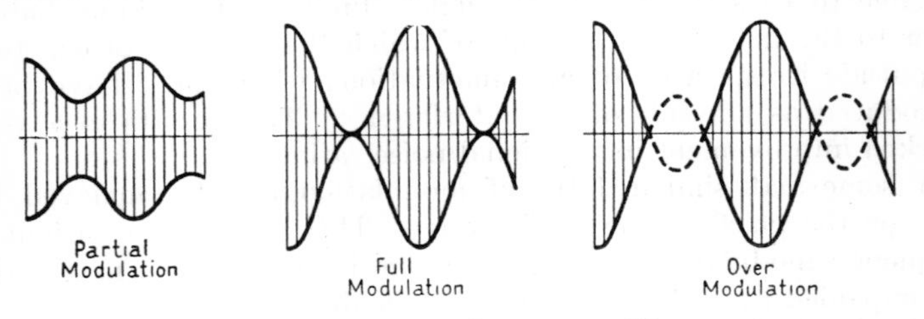

Fig. 3.11. Degrees of Amplitude Modulation

different depths of modulation. On the left is partial modulation where the amplitude of the carrier is altered but not completely reduced to zero. In the centre we have full modulation where the intensity is varied between zero and twice the normal, and on the right is a condition known as over-modulation where the maximum modulation is greater than the maximum amplitude of the carrier. Here the carrier wave is reduced to zero before the modulation wave has completed its cycle so that there is a blank portion during which no carrier is radiated, which is not a permissible condition because it does not comply with (*b*) above.

Indeed, it is found in practice that both the transmitter and receiver begin to introduce distortion of their own, due to non-linearity of the circuit parameters, when the depth of modulation exceeds about 80 per cent, so that it is usual to restrict the maximum modulation depth to between 80 per cent and 90 per cent, as a result of which the average modulation is only of the order of 30 per cent. With telegraph signals, of course, where we are concerned merely with starting and stopping the carrier, the modulation can be complete though even in this case it is often found convenient to leave the carrier oscillating faintly in the key-up position.

Sidebands: Bandwidth

The carrier wave can be represented by the expression $e = E \sin \omega t$, where E is the peak carrier amplitude and ω is 2π times the carrier frequency. When we modulate the wave the peak value of the carrier is subject to an additional variation at the frequency of the modulation. Hence the expression for an amplitude-modulated wave is

$$e = (E_0 + E_1 \sin pt) \sin \omega t$$

Here p is the angular modulation frequency ($= 2\pi$ times the actual modulation frequency) and the term E_1 represents the peak modulation. Clearly for full modulation E_1 must be equal to E_0. If E_1 is less than E_0 the modulation is partial while if it is greater than E_0 we have over-modulation, a condition which we have seen is to be avoided. The ratio of E_1/E_0 is known as the *depth of modulation* and is often represented by the symbol m.

Let us now expand this expression:

$$\begin{aligned} e &= (E_0 + E_1 \sin pt) \sin \omega t \\ &= E_0 \sin \omega t + E_1 \sin pt \sin \omega t \\ &= E_0 \sin \omega t + \tfrac{1}{2}E_1 \cos (\omega - p)t - \tfrac{1}{2}E_1 \cos (\omega + p)t \end{aligned}$$

It will be seen that the expression has been rewritten in the form of three separate sine waves. One is the original carrier wave itself. In addition we have two fresh oscillations, one at a frequency equal to the original carrier frequency minus the modulation frequency, and the other at the original carrier frequency plus the modulation frequency. These two frequencies are known as *side frequencies* and there is one such pair of side frequencies for every modulation frequency. Consequently when the modulation is varying in frequency over a wide range there is a large number of these side frequencies extending over a range on either side of the carrier to an extent equal to the maximum modulation frequency. These are known as *sidebands*, and it is the presence of these sidebands which makes it necessary to separate the frequencies of different transmitters by an appreciable amount so that there may be room in the frequency spectrum for the various communications existing simultaneously. It is, in fact, necessary to allocate various transmissions into channels, these channels being wide enough in frequency to accommodate the whole range of the sidebands required.

Consider, for example, a telephony transmission operating on a frequency of 1 MHz and modulated with speech or music up to a maximum frequency of 5 kHz. The sidebands of the transmission will then extend over a range of 5 kHz above or below the carrier frequency so that the whole transmission will have a *bandwidth* of 10 kHz.

Practical methods of producing the required modulation are discussed in Section 7.3.

Power in a Modulated Wave

The power in a modulated sine wave is proportional to the square of the amplitude. Consequently, referring to the expression deduced above it will be seen that the power in a modulated wave is proportional to $E_0^2 + (\frac{1}{2}E_1)^2 + (-\frac{1}{2}E_1)^2 = E_0^2 + \frac{1}{2}E_1^2$. Thus, with 100 per cent modulation ($E_1 = E_0$) the total power is increased by 50 per cent but only one-third of the power is in the sidebands, two-thirds being wasted in the carrier, and if the modulation is less than 100 per cent this discrepancy becomes even more noticeable. If we write $E_1 = mE_0$ then the expression for the power may be written

$$P \propto E_0^2(1 + m^2/2)$$

When $m = 1$ the expression inside the brackets becomes 1·5 as we have just seen. Suppose, however, that our modulation depth is 30 per cent so that $m = 0{\cdot}3$. The expression in the brackets then becomes 1·045 so that the power in the sidebands, which is the only part effective in communicating the intelligence, is only a little over 4 per cent of the total power.

Carrier Suppression

It is clearly advantageous to remove the carrier wave. This can be achieved by the use of certain types of balanced modulator which are discussed in Chapter 7 (see Fig. 7.17), with considerable saving of power. If the carrier is then re-introduced at the receiving end, and adjusted to an amplitude slightly greater than the peak sideband amplitude, the effect will (in theory) be the same as if the carrier had been transmitted in the first place.

Complete carrier suppression, however, is not practicable because the operation requires the re-introduced carrier to be of the same frequency *and phase* as that at the transmitter. This is impossible in practice, but the difficulty can be overcome by radiating a small proportion of the carrier only, and re-inforcing this at the receiving end, as will be discussed shortly.

Effect of Carrier Phase Shift

The reason for the phase requirement will be understood by reference to Fig. 3.12, which shows vector diagrams representing the carrier and the two sidebands. The carrier frequency can be

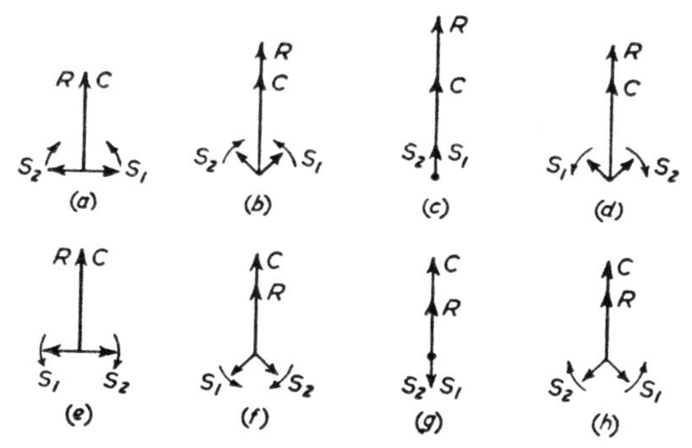

FIG. 3.12. VECTOR DIAGRAM OF MODULATED CARRIER WITH SIDEBANDS IN PHASE

represented by a vector C which will normally be considered as rotating and which will, therefore, generate the carrier wave in the ordinary manner. If this carrier is modulated we have to add two further vectors each having an amplitude relative to the carrier proportional to the depth of modulation. These vectors are represented in the diagram by S_1 and S_2. Now, since one side frequency is higher than the carrier and the other is lower, the vector S_1 will be rotating relative to the carrier vector in the same direction, while the vector S_2 will be rotating in the opposite direction. In effect, therefore, the whole diagram in each of the positions shown in Fig. 3.12 is rotating but we are only looking at it in an instantaneous stationary position.

Now, in Fig. 3.12 (*a*) the two side frequencies are in such a position that they are momentarily in opposition. Their effects thus cancel each other out and the amplitude of the carrier is not altered. The resultant R therefore is of the same length as the carrier C. In Fig. 3.12 (*b*) the condition of affairs is shown a fraction of a second later where the two side frequencies have each rotated 45°. They now produce a resultant which is added to the carrier so that the resultant R is greater than the carrier vector. In other words the amplitude of the carrier has increased. At (*c*) the two side frequencies have come into phase with the carrier and this is the condition of maximum amplitude. The succeeding figures show the progressive conditions as the amplitude of the carrier falls to a minimum and then begins to increase again.

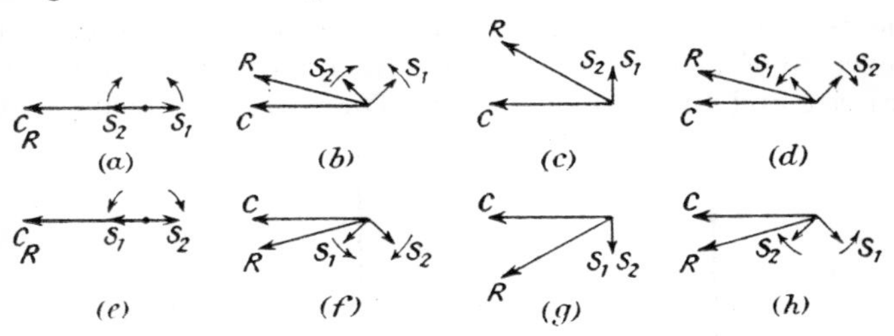

FIG. 3.13. VECTOR DIAGRAM OF MODULATED CARRIER WITH SIDEBANDS 90° OUT OF PHASE

Fig. 3.13 shows the same sequence, but in this case the carrier vector is rotated 90° relative to the modulation. In each of the figures the vector R represents the sum of the carrier C and the two side vectors S_1 and S_2, and it will be seen that in this case the effect of the side frequencies has very little influence on the amplitude of the resultant but merely causes it to swing in phase a little behind and a little ahead of the carrier. We have, in fact, produced a little phase modulation but a negligible amount of amplitude modulation.

Single-sideband Operation

This is why it is essential to preserve the phase of the carrier relative to the modulation, and because this is too difficult it is not usual in practice to suppress the carrier completely. It may be reduced in amplitude to some small fraction of its normal value and then restored to its full value at the receiving end, or the filtering arrangement may be such as to eliminate one of the sidebands only, leaving the carrier and the remaining sideband unaffected. This latter form of working is known as *single-sideband* working and results in some economy of power and also an appreciable economy of bandwidth since the total channel required is now only some 55 per cent of that required for a full double-sideband transmission. Single-sideband working also has the advantage that it can be received with a normal type of receiving circuit, though this is only true as long as the carrier is transmitted at its full amplitude.

If the carrier is suppressed in part then special arrangements must be made at the receiver in order to amplify the carrier frequency more than the side frequencies so that the normal modulation depth is restored. With superheterodyne receivers this is automatically achieved so that many transmissions today, including v.h.f. radio and t.v., use this *vestigial sideband* system.

The method of suppressing the carrier and sidebands is considered in more detail in Chapter 7.

Intermodulation

Amplitude modulation can also result owing to interference between two oscillations of slightly differing frequencies. This effect is sometimes used deliberately in the technique known as *heterodyning* which is discussed in Section 9.4, but it can also arise from interaction between a wanted signal and an undesired signal (usually of much smaller amplitude) operating on an adjacent frequency.

The combined signal may be written

$$\begin{aligned} e &= A \sin \omega t + B \sin (\omega + \omega_1)t \\ &= A \sin \omega t + B [\sin \omega t \cos \omega_1 t + \cos \omega t \sin \omega_1 t] \end{aligned}$$

Periodically $\cos \omega_1 t$ and $\sin \omega_1 t$ will pass through the values 1, 0, −1, 0, etc.

When $\sin \omega_1 t = 0$, $\cos \omega_1 t$ is ± 1. Then

$$e = A \sin \omega t \pm B \sin \omega t = (A \pm B) \sin \omega t$$

The amplitude of the signal will thus vary between $(A + B)$ and $(A - B)$, so that there is an amplitude modulation.

When $\cos \omega_1 t = 0$, $\sin \omega_1 t$ is ± 1. Then

$$e = A \sin \omega t \pm B \cos \omega t$$

Now, if B/A is small (and equal to p), then

$$e = A \sin (\omega t \pm p\pi/2) \text{ approximately}$$

indicating a small (periodic) change of phase.

In practice these variations of amplitude and phase are usually of minor importance because B is normally much less than A. In such circumstances the output from the subsequent detector stage is determined almost entirely by the modulation of the wanted signal, as is explained in Section 9.5.

Intermodulation should not be confused with *cross-modulation,* which is a different, and sometimes more serious, form of interference which arises if the amplifier and/or detector stages are not strictly linear, as discussed in Section 9.5.

Frequency Modulation

Let us now consider frequency modulation. Reference has already been made to the simplest form, as used for telegraph transmission, in which the marking and spacing waves are transmitted with the same amplitude but different frequencies. A more complicated arrangement is necessary to deal with speech and music because here the modulation intelligence to be transmitted is itself varying not only in amplitude but in frequency, i.e. both the strength and the pitch of the various tones are continually varying. With amplitude modulation this is quite a straightforward business for it is only necessary to vary the amplitude of the carrier wave in faithful reproduction of the variations in intensity and frequency of the modulation itself.

If we wish to use frequency modulation there are two factors which can be varied. One is the actual change of frequency from its mean value and the other is the rate at which the change is made. These two possibilities can be utilized to convey the required intelligence. The amount by which the frequency is changed is made proportional to the amplitude of the modulation so that, if we choose, say, a maximum deviation of 75 kHz from the mean value, the frequency of the carrier will be made to swing between limits plus or minus 75 kHz about mean value, for full modulation. For any intensity less than the maximum the deviation would be correspondingly reduced until with no modulation at all the carrier frequency would not be altered.

The carrier would then be caused to swing above and below the mean frequency to the required extent *at a rate* depending upon the modulation frequency. To convey a modulation of 100 Hz, for example, the carrier would be caused to vary between the prescribed limits 100 times a second.

The reception of a frequency-modulated wave requires a special form of receiver in which changes of frequency are converted into proportional changes of amplitude. The technique is discussed further in Sections 3.3 and 9.6.

Analysis of Frequency-modulated Wave

At first sight it might appear that the bandwidth of a frequency-modulated wave was simply the maximum deviation of the frequency on either side of the carrier, but this is not the case. Actually it contains a number of high-order side frequencies which occupy a band appreciably wider than the maximum deviation.

The equation to a frequency-modulated wave can be written in the form

$$e = E_0 \sin(\omega t + m \sin pt)$$

where $\omega = 2\pi \times$ carrier frequency

$p = 2\pi \times$ modulation frequency

$m =$ modulation index

$$= \frac{\text{deviation of carrier from mean value}}{\text{modulation frequency}}$$

$$= \Delta f / f_m$$

The *modulation index m* can obviously vary in a complex manner, since both its constituents depend on the modulation conditions. The deviation Δf is made proportional to the amplitude of the modulating signal and can thus vary from a (chosen) maximum down to a value approaching zero, while the modulation frequency can have any value within the required (a.f.) spectrum.

Thus with a small-amplitude signal, Δf may be less than f_m, so that m is fractional, while with full modulation and a low modulation frequency, m may be 20 or more. Hence it is not a convenient factor in design calculations, a more significant parameter being the frequency deviation, as defined later.

The equation can be expanded into the form

$$e = E[A_0 \sin \omega t + A_1\{\sin(\omega + p)t - \sin(\omega - p)t\} + A_2\{\sin(\omega + 2p)t + \sin(\omega - 2p)t\} + \ldots]$$

where A_0, A_1, A_2, etc., are a particular form of constant, depending on the value of m, known as Bessel functions. Thus the wave contains components of ω, $\omega \pm p$, $\omega \pm 2p$, $\omega \pm 3p$, . . . , and so occupies a considerably greater bandwidth than an amplitude-modulated wave. In practice it is not necessary to calculate the extent of these sidebands in detail because it can be shown that if side-frequency amplitudes less than 1 per cent of the carrier are neglected the total bandwidth may be taken as $2(\Delta f + f_m)$.

Phase Modulation

As has already been mentioned, a frequency-modulated wave is inherently phase-modulated and vice versa. We have seen that the equation for a frequency-modulated wave is

$$e = E_0 \sin(\omega t + m \sin pt) = E_0 \sin(\omega t + \theta)$$

which is the expression for a sine wave with a phase angle θ relative to $E_0 \sin \omega t$. In this case θ is not constant, but is itself a sine wave $m \sin pt$.

Clearly therefore we can use the same expression to denote either phase or frequency modulation, and as explained in Chapter 7 it is often more convenient to modulate the phase rather than the frequency. There is, however, an important difference between the two systems, namely that whereas with phase modulation the modulation index m is simply proportional to the amplitude of the modulating signal, with frequency modulation it is *also* inversely proportional to the modulation frequency.

Equations for Frequency and Phase Modulation

This distinction does not arise from any mysterious difference between the two systems, but simply because, *by definition*, the modulation index of a frequency-modulated wave is $\Delta f/f_m$. This will be understood if we write the equation for the wave in the form

$$e = E_0 \sin \theta_t$$

where θ_t is the total angular (phase) displacement at time t. The instantaneous angular frequency, ω_i, is then $d\theta_t/dt$.

Now, by definition, a frequency-modulated wave is one in which the instantaneous angular frequency

$$\omega_i = \omega_0 + \Delta\omega \cos pt$$

where $\Delta\omega = 2\pi\Delta f = kE_m$, where E_m is the maximum amplitude of the modulating signal, and $p = 2\pi f_m$, f_m being the modulation frequency.

$\Delta\omega$ is thus, by definition, independent of f_m. The second term is a cosine term because with $t = 0$ the rate of increase of θ_t is a maximum.

But $\omega_i = d\theta_t/dt$, so that $\theta_t = \int \omega_i \, dt$.

Hence $$e = E_0 \sin \theta_t = E_0 \sin\left(\omega_0 t + \frac{\Delta\omega}{p} \sin pt\right)$$

$$= E \sin(\omega_0 t + m_f \sin pt)$$

where $m_f = \Delta\omega/p = \Delta f/f_m$.

A phase-modulated wave, on the other hand, is by definition a wave in which it is the phase angle θ_t, and not the angular frequency, which is proportional to the modulation amplitude, so that

$$\theta_t = \omega_0 t + A \sin pt$$

where $A = k'E_m$, E_m being the maximum amplitude of the modulating signal, as before.

Hence if we write m_p for A, the equation to a phase-modulated wave becomes

$$e = E_0 \sin (\omega_0 t + m_p \sin pt)$$

This is of the same form as the expression for a frequency-modulated wave but the two will only be identical if $m_p = m_f = \Delta f/f_m$. Hence $\Delta f = m_p f_m = k'E_m f_m$.

Thus in a phase-modulated wave the frequency deviation is proportional both to the amplitude *and frequency* of the modulating signal.

It is sometimes stated that a frequency-modulated wave is the integral of a phase-modulated wave. This is an over-simplification which is not correct. What is meant is that, as stated above,

$$\theta_t = \int \omega_i \, dt$$

Frequency-Modulation Parameters

As said earlier, the deviation in a frequency-modulated transmission is proportional to the amplitude of the modulating signal, and for efficient operation this deviation should be several times the maximum modulation frequency. In practice, a specific maximum deviation is chosen (usually 75 kHz) to correspond with full modulation. This is called the "rated" deviation, and the ratio of this to the maximum modulation frequency is called the

$$\text{Deviation ratio} = \frac{\text{Rated deviation}}{\text{Rated maximum modulation frequency}}$$

This is a specific value of the modulation index, m, corresponding to the rated maximum values of Δf and f_m.

The actual deviation is then made proportional to the amplitude of the modulating signal, from which is derived the

$$\text{Modulation index} = \frac{\text{Actual deviation}}{\text{Actual modulation frequency}}$$

These definitions are as specified in BS. 204 (items 14029-14032), which also defines Δf as

$$\text{Frequency deviation} = \begin{cases} \text{Peak difference between instantaneous} \\ \text{frequency of modulated wave and} \\ \text{carrier frequency in a cycle of modu-} \\ \text{lation} \end{cases}$$

Thus in a system having a rated deviation of 75 kHz and a maximum modulation frequency of 15 kHz, the deviation ratio would be $75/15 = 5$.

If the maximum modulation amplitude is, say, 20 V (which would be arranged to produce a deviation of 75 kHz), then a signal of 12 V would produce a deviation of $(12/20) \times 75 = 45$ kHz. At the maximum modulation frequency (15 kHz) the modulation index will then be $45/15 = 3$.

Since the amplitude of a frequency-modulated wave is constant, the concept of modulation depth does not apply. Its equivalent would be the ratio of actual/rated deviation, but this has no practical significance.

Practical forms of frequency-modulated circuit are discussed in Chapter 7. They utilize special forms of oscillator in which the frequency (or phase) is made to depend on the amplitude of a control voltage, which is arranged to be proportional to the instantaneous value of the modulating signal. The frequency then swings on either side of the mean value at a rate determined by the modulation frequency, the extent of the excursion being determined by the peak value of the modulation amplitude.

As said earlier, the bandwidth is $2(\Delta f + f_m)$.

Pulse Modulation

Pulse modulation is a special form of amplitude modulation based on a sampling technique, of which the essential principle is illustrated in Fig. 3.14. The instantaneous value of the wave is sampled at regular intervals and this information is then used to modulate the carrier. It is found that if the number of samples per second is greater than twice the highest frequency contained in the signal wave, the original wave can be successfully reconstituted from the succession of pulses.

The sampled information may be utilized in several ways. One is to transmit a succession of pulses of constant width having an amplitude controlled by the appropriate sample, as in Fig. 3.14 (*b*). This is known as *pulse-amplitude* modulation.

A second method is to transmit the pulses at a constant (full-carrier) amplitude, but to vary the duration in accordance with the sample information. This may be achieved by a modified multivibrator circuit as described in Chapter 17. This is called *pulse-width* modulation, and is illustrated in Fig. 3.14 (*c*).

A modification of this method is to derive from the variable-width pulses a series of short pulses of constant amplitude and width whose

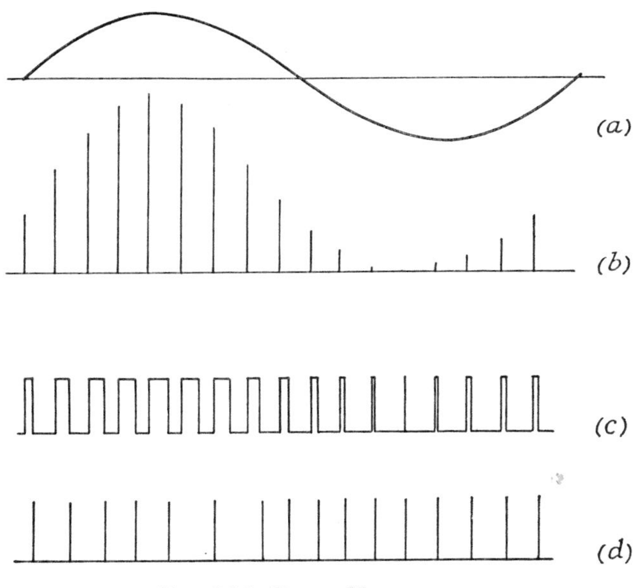

FIG. 3.14. PULSE MODULATION

(*a*) Original wave
(*b*) Pulse-amplitude modulation
(*c*) Pulse-width modulation
(*d*) Pulse-position modulation

position is controlled by the width of the original pulses (and hence by the initial sampling information). This is called *pulse-position* modulation, as shown in Fig. 3.14 (*d*). The derivation of this variable positioning is also discussed in Chapter 17.

Pulse-width and pulse-position modulation possess the same advantages as frequency modulation in signal/noise ratio. Provided that the received pulse is sufficiently large, variations in amplitude do not affect the performance to any serious extent. The steep wavefront of the pulses, however, necessitates a wide bandwidth (of several MHz) so that the system is mainly used at ultra-high frequencies.

Multiplexing

With pulse modulation there can be relatively large idle periods between the pulses, and hence, since the pulse-repetition rate is constant, it is possible to transmit a further series of pulses in these intervals. Thus with a pulse rate of 10 kHz the time interval is 100 μsec. If the basic pulse width is 1 μsec, which is varied by the modulation to 10 μsec, there is still 90 μsec available in which five or six additional channels could be accommodated with a reasonable margin of separation. It is only necessary to delay the second series by, say, 15 μsec, and so on, to obviate any risk of interference, and this process, which is known as *multiplexing*, is extensively used to increase the amount of information which can be transmitted on a given carrier.

Pulse-code Modulation

A development of pulse modulation is what is known as *pulse-code* modulation. Here the input is sampled as before but the relative amplitudes are then classified in relation to a series of standard amplitudes. Thus if eight standards are chosen from 0–100 per cent, each 14·3 per cent apart, a 30 per cent signal would be rated as 2 (28·6 per cent) and so on. This introduces a small error, which becomes less as the number of standard levels is increased. The process is called *quantization*.

Hence in the example chosen the information would be given by one of 8 possible amplitudes. This in itself is no advantage, but by expressing the 8 possible values in a binary code only 3 digits are required, as explained in Appendix 3, while a seven-digit code will handle 128 standard amplitudes.

This type of coded information is not only economical but is in a suitable form for the operation of computers.

It may be noted that pulse modulation techniques are not confined to the modulation of a radio-frequency carrier. They can be used as a convenient method of transmitting intelligence over a land line.

3.3. Reception

For the reception of radio waves it is first necessary to provide an aerial which consists essentially of a length of wire so arranged that the electric fields in the travelling electric wave induce e.m.f.s in the wire as they pass. These e.m.f.s are normally very small, being of the order of a few microvolts; moreover, apart from any special directional or other properties, such an aerial will respond equally well to signals from any source and it is therefore necessary to be able to select the wanted signal and reject any others.

This is accomplished by tuning the system to the particular frequency required by the use of suitable resonant circuits. Except in the simplest cases, a single circuit is insufficient and the signal is usually magnified by means of amplifiers which are themselves coupled by further tuned circuits so that a selective discrimination is obtained resulting in an e.m.f. of practicable magnitude (usually of the order of a volt) free of interfering signals.

This e.m.f., however, is at the frequency of the carrier wave and it is now necessary to extract from it the intelligence contained in the modulation. The most practical method of achieving this is to

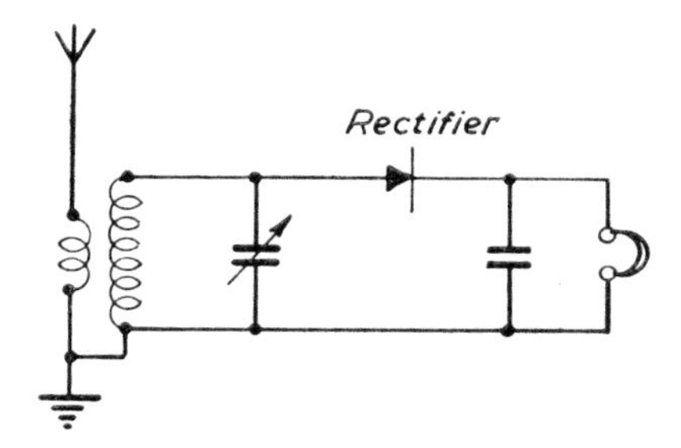

Fig. 3.15. Detector Circuit

rectify the signal which is done by passing the current through some form of non-linear element which conducts more readily in one direction than the other. Such a device is called a *detector* or *demodulator*.

Fig. 3.15 illustrates the essentials of a receiving circuit. The incoming signal is picked up on the aerial and the induced e.m.f. is transferred to a tuned secondary circuit which develops across the capacitor a substantially larger e.m.f. because of the resonant magnification. This signal, amplified if necessary in a tuned r.f. amplifier, is then applied to the detector circuit consisting of a rectifier in series with a telephone receiver.

Now the mean value of a modulated carrier wave, as will be seen from Fig. 3.11, is zero irrespective of the modulation, but if the signal is rectified so that the negative half-cycles are suppressed as shown in Fig. 3.16 it will be seen that the mean value is no longer zero but has a positive value, the amplitude of which varies in accordance with the modulation. It is, in fact, equivalent to a steady d.c. current plus an a.c. current corresponding exactly with the modulation. This is the basic process of detection for an amplitude-modulated wave. In practice the telephones would be replaced by a resistor across which the modulation would develop an audio-frequency voltage which would then be fed into an amplifying stage, while in all but the simplest type of receiver the r.f. signal would be suitably amplified before being applied to the detector.

With frequency modulation it is necessary first to convert the

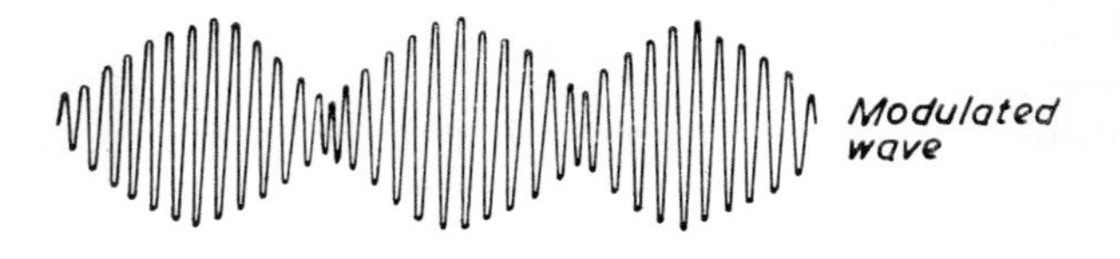

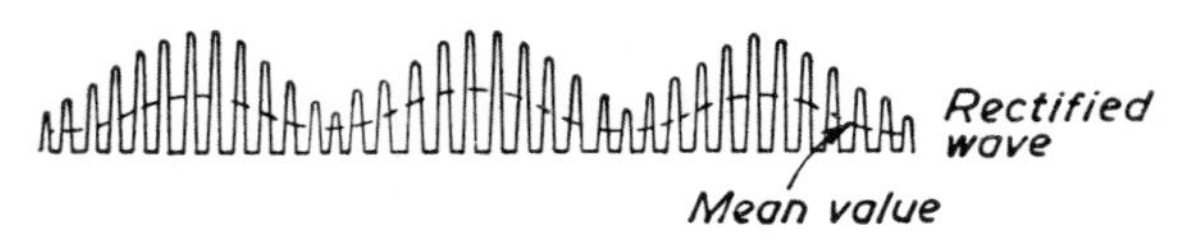

FIG. 3.16. PRINCIPLE OF DETECTION

variations in the frequency of the carrier into amplitude variations and then to follow this with a rectifier as just described. The device responsive to the variations in frequency is called a *discriminator* and may take various forms. One method would be to set the

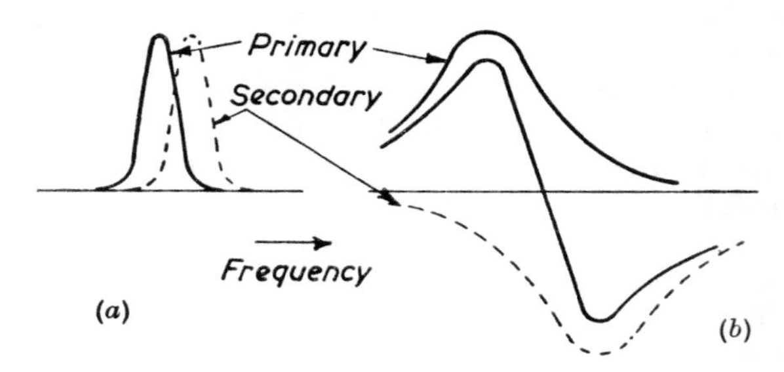

FIG. 3.17. ILLUSTRATING ACTION OF DISCRIMINATOR CIRCUIT

capacitor in Fig. 3.15 slightly off tune so that its operating point would be on the side of the resonance curve. Any variation in the frequency of the carrier would then produce a corresponding variation in the voltage across the capacitor.

Such an arrangement, however, is critical to operate and the more usual procedure is to use a loose-coupled transformer having both primary and secondary tuned. With such a circuit, using critical coupling, the maximum current in the primary does not occur at the same frequency as that in the secondary, as shown in Fig. 3.17 (*a*). If the detector is now made to respond to the difference between the primary and secondary e.m.f.s a curve of the form shown in Fig. 3.17 (*b*) is obtained. The frequency scale has been expanded for convenience, and it will be seen that around the mid-frequency point the output is proportional to the frequency, and so converts frequency variations into changes of amplitude.

Practical forms of detector circuit are discussed in Sections 9.5 and 9.6.

Bandwidth

It was shown in the previous section that the modulation of a carrier wave produces a range of side frequencies, and that the relevant information is contained entirely within these sidebands. Hence it is essential that the circuits involved, in both receiver *and* transmitter, shall respond adequately to a range of frequencies immediately adjacent to the carrier frequency, to an extent dependent upon the bandwidth of the particular transmission.

It is usually considered satisfactory if the *overall* response of the system is not more than 3 dB down (i.e. not less than $1/\sqrt{2}$ times the maximum) at the extremes of the band, though for high-fidelity reception a higher standard of performance may be desirable.

The requirements are discussed in detail in Part II (and particularly in Chapter 9), but it will be useful here to summarize the bandwidths involved in the different forms of transmission. These are briefly as follows:

Amplitude modulation. $B = 2f_m$, where f_m is the maximum modulation frequency. For audio reception f_m is of the order of 10 to 15 kHz, but for television it is several MHz, as discussed in Chapter 16.

Frequency modulation. $B = 2(\Delta f + f_m)$, where Δf is the rated deviation and f_m is the maximum modulation frequency.

Pulse modulation. Here the bandwidth is dependent on the system used, as discussed in Chapter 17, but is roughly given by

$$B = 0{\cdot}4/t \text{ where } t \text{ is the time of rise of the pulse.}$$

4

Practical Components

THE application of the basic principles discussed in the preceding chapters involves the use of components which exhibit the required properties to a known and controllable extent.

To treat this subject exhaustively is quite beyond the scope of the present work, but it is desirable to indicate the general methods employed. More detailed information will be found in *Telecommunication Principles* by R. N. Renton, and/or *Radio & Electronic Components* (*Vols.* 1–6) by G. W. A. Dummer, *et al.*

4.1. Resistors

In the conductors used to carry electric currents the inevitable resistance of the wires is a disadvantage and has to be kept to a minimum. On the other hand there are many occasions on which resistance can be utilized as, for example, in the various networks associated with amplifiers. In such circumstances it is desirable to use materials which have a large resistivity so that the required value of resistance can be produced in a reasonably small compass.

The resistance of any conductor is given by the expression

$$R = \rho l/A$$

where l is the length of conductor,

A is the cross-sectional area, and

ρ is the resistivity.

The *resistivity* or *specific resistance* is defined as the resistance of a cube of the material of unit dimensions. It is measured in ohm-metres and typical values are given in Table 4.1.

The reciprocal of the resistivity is known as the *conductivity*, and it will be seen that the metal which exhibits the highest conductivity is silver. This is too expensive for general use, though it is

TABLE 4.1. RESISTIVITIES AND TEMPERATURE COEFFICIENTS

Material	*Resistivity (at 20°C) in ohm-m* $\times 10^{-8}$	*Temperature Coefficient per °C* ($\times 10^{-4}$)
Silver	1·6	40
Copper	1·7	39
Aluminium	2·7	39
Iron (*according to grade*)	10–100	55–20
Transformer steel	55	—
Nickel-chromium (80/20)	103	1·0
Copper nickel (60/40)	48	negligible
Carbon	4,500–5,000	−0·2 to −0·5

employed occasionally. The material in more general use is copper which will be seen to be nearly as good, and electrical conductors are usually of copper in the form of wire or strip, covered where necessary with suitable insulation.

Aluminium is also a good conductor though it has only 63 per cent of the conductivity of copper, but because it is so much lighter (its specific gravity being only 31 per cent that of copper) an aluminium conductor will have only half the resistance of an equivalent weight of copper conductor. Such conductors therefore are used for aerial lines such as are employed in high-voltage transmission, usually in the form of a steel core for strength with aluminium conductors stranded round the core.

On the other hand, where a high resistance is required use is made of materials of high resistivity. For such purposes alloys of two or more metals are the most suitable. An alloy of 80 per cent nickel with 20 per cent chromium has a high resistivity and is used for the elements of electric fires and heaters. It is suitable for this application because it does not oxidize too easily at the dull red temperatures at which heater elements usually operate, but it has the disadvantage of a large temperature coefficient. An alloy of approximately 58 per cent copper and 42 per cent nickel has a somewhat lower resistivity but has the advantage of a very small temperature coefficient and is therefore suitable for the manufacture of resistors which are required to be constant in value under varying temperature conditions. Alloys such as this go by various trade names such as Eureka, Constantan, etc.

Temperature Coefficient

The resistance of a conductor, in general, depends upon temperature. With metallic conductors it is found that the resistance at any given temperature is proportional to the temperature above absolute zero. This is a theoretical zero, actually 273° below zero on the Celsius scale. This zero cannot actually be achieved in practice but many physical phenomena behave as if they originated at this point, though in practice as one approaches zero secondary effects are often introduced. As mentioned in Chapter 1, at temperatures approaching absolute zero certain materials begin to exhibit what is called superconductivity—i.e. they behave as if they had virtually no resistance at all.

It is more usual to express the resistance in terms of its value at convenient standard temperature, usually taken as 20°C. In this case the resistance can be expressed in the form

$$R_t = R_{20}(1 + \alpha t)$$

where R_t is the resistance at a temperature of t°C,

R_{20} is the resistance at 20°C, and

α is a constant for the particular material.

The term α is called the *temperature coefficient*, being in fact the relative change of resistance per degree and typical values for this temperature coefficient are also given in Table 4.1. It will be seen to vary with the material and it will also be noted that with the copper-nickel alloy the temperature coefficient is virtually zero, so that the resistance of a conductor of this material does not vary appreciably with temperature. It will also be noted that the temperature coefficient of a metal is positive so that the resistance increases as the temperature rises. With carbon, on the other hand, the temperature coefficient is negative so that the resistance falls with increasing temperature.

Carbon Resistors

For many purposes resistors are required having values of many thousands or even millions of ohms. To obtain such resistors with a wire conductor requires a very fine wire and a considerable length. Hence it is more convenient to utilize the much higher resistivity of carbon. Carbon exists in many forms from lampblack to diamonds, and it is possible by controlling the grade of carbon used, often admixed with suitable additives, to construct resistors in the form of short rods.

The current-carrying capacity of a resistor of this type is determined by its size. Energy will be absorbed by the resistor in the form of heat equal to I^2R, and this heat must be dissipated, usually by radiation from the surface of the material. The temperature of the resistor will, in fact, rise until an equilibrium condition is obtained where the energy loss is all satisfactorily radiated as heat. Practical resistors are therefore rated in terms of the wattage dissipation which they will handle, and they range from small units 8 mm long and 2·5 mm diameter which are rated at one-eighth of a watt up to units between 25 and 50 mm long and proportionately fatter which will dissipate 3 or 4 watts. The carbon rod itself is usually housed in a protective covering of moulded material, the actual

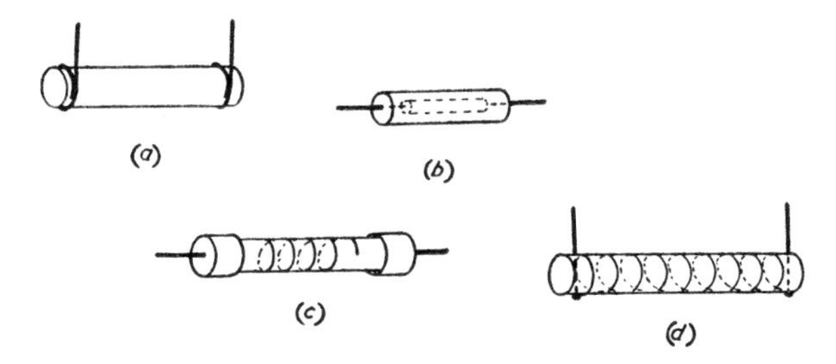

FIG. 4.1. TYPICAL RESISTOR CONSTRUCTIONS

(a) Carbon rod resistor. (b) Carbon rod in insulating sleeve.
(c) Thin film (high-stability) type. (d) Wire-wound resistor.

value of the resistance being indicated by small coloured rings in accordance with an agreed code.

Because of the substantial negative temperature coefficient of carbon these resistors must not be allowed to become overheated or the resistance decreases rapidly with consequent increase of current, so that the effect is cumulative. The resistance also is subject to variation after periods of use.

Thin-film Resistors

Greater stability can be obtained by the use of an alternative technique in which a thin film of carbon or metal is deposited on to a rod of insulating material. The material is then cut away, usually in the form of a spiral, so as to alter the effective length of the path until the required resistance value is obtained. Such resistors, known as high-stability resistors, maintain their resistance appreciably more constant.

Fig. 4.1. illustrates the various types of resistor construction.

Wire-wound Resistors

Where precision is required or large wattage dissipations are necessary, wire-wound resistors are used. The wire is of appropriately small diameter, often as little as $\frac{1}{1000}$th of an inch. The wire is wound as a single-layer solenoid on a suitable former and is then given a coat of protective lacquer. Where the resistor has to dissipate appreciable heat it is usual to coat it with a vitreous enamel and the whole resistor is then baked in a furnace. This holds the wire rigid and prevents appreciable movement due to the successive expansion and contraction due to the heating and cooling.

Resistors wound in this manner are, of course, inductive. For many purposes the inductance is so small that the inductive reactance is quite negligible by comparison with the resistance. However, if this is not the case some form of non-inductive winding has to be adopted. One of the methods used is what is known as the Ayrton–Perry winding which consists of two single-layer solenoids wound one on top of the other but in opposite directions as illustrated in Fig. 4.2.

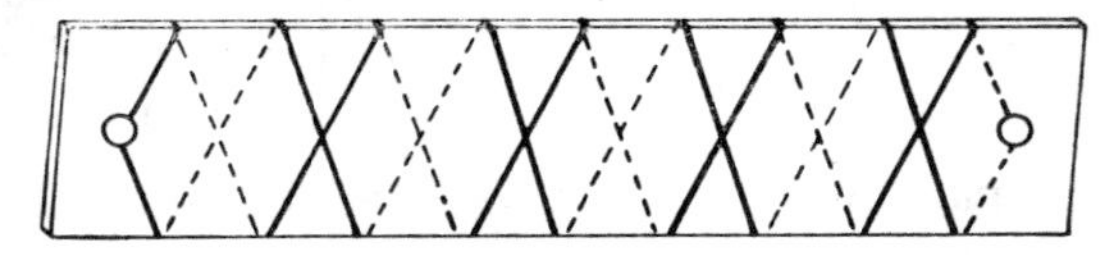

FIG. 4.2. AYRTON–PERRY NON-INDUCTIVE WINDING

Variable Resistors

For many purposes variable resistors are required. In these the resistance element is usually made up in a circular form with a light rotating "slider" to make contact with any desired point along the element. By connecting the element across a source of voltage and taking the output between the slider and one end, the device acts as a variable voltage divider, sometimes loosely called a potentiometer.

In a wire-wound potentiometer the element is wound on a thin flexible card of suitable insulating material which is then bent into a circle and mounted in a suitable housing. The slider rests on the edge of the winding from which the insulation is removed. This construction is only practicable for low resistance values. More usually the element is made of carbon composition.

The element may be made to provide a specific non-linear change of resistance with rotation of the slider (e.g. square-law, reciprocal law etc.). The grading is achieved in wire-wound devices either by tapering the width of the card or by changing the gauge of wire at suitable points (though this obviously does not provide a smoothly varying law). Carbon-composition elements may be graded by changing the dimensions and/or the composition.

Many circuits require the value of a resistance to be adjusted on test to give the required performance. For this purpose pre-set resistors are used, which are made up as small versions of the normal variable resistor in a form convenient for mounting directly in the circuit. Small wire-wound preset resistors are made with the element on a thin card with a sliding contact which is moved along the card until the required value is obtained.

More elaborate forms of variable resistor are made to provide a closer and more accurate control of the resistance, but these need not be discussed here.

Non-linear Resistors

For certain requirements it is convenient to use resistors having a non-linear current-voltage relationship. This is a characteristic property of semiconductor materials, of which the resistance depends on the current and also to a marked extent on the temperature. These effects may be utilized in two ways:

Varistors. Materials such a silicon carbide obey a law of the form $i = kv^x$, where x is between 3 and 5. If x is, say, 3·1 a 20 per cent increase in voltage will halve the effective resistance, and such a device connected across a circuit will provide a measure of surge protection.

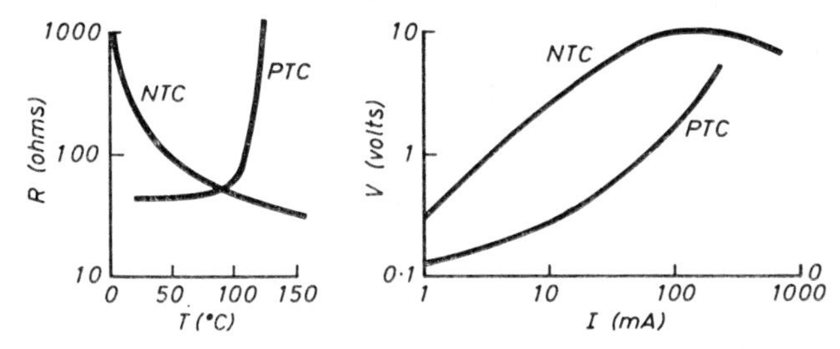

FIG. 4.3. TYPICAL THERMISTOR CHARACTERISTICS

Thermistors. These contain a small bead of semiconductor material so proportioned that the current flowing raises the temperature sufficiently to cause a significant change of resistance. If the material has a negative temperature coefficient (NTC) the resistance decreases with current, as in a varistor; if the coefficient is positive (PTC) the resistance increases with current. Fig. 4.3 illustrates typical characteristics.

Thermistors are used in a variety of ways to compensate for changes in circuit conditions. They are responsive to changes in the ambient temperature, but this is of secondary importance, though they are sometimes used as temperature-sensitive elements.

Barretters. An alternative form of non-linear resistor is the barretter, or ballast lamp, which contains a filament of iron wire enclosed in a glass bulb containing hydrogen. The operating temperature of the wire and the pressure of the gas are so arranged that over a limited range of voltage the current remains constant.

Skin Effect

A wire carrying an electric current produces a magnetic field. This field exists both inside as well as outside the wire and is, in fact, greatest at the centre of the wire. If the current is varying, an e.m.f. of self-induction is set up tending to oppose the flow of the current and this is again greatest at the centre of the wire. The effect, therefore, is that the impedance of the wire progressively increases towards the centre so that the current tends to flow in the outer skin of the wire.

In consequence of this the effective resistance of the conductor is increased. The increase is not important at low frequencies but it is quite considerable at radio frequencies. The ratio of the resistance at a frequency f to that at d.c. is given by the expression

$$R_f/R_0 = a \,.\, d\sqrt{(\mu f/\rho)} + b$$

where d = diameter of conductor,

μ = permeability of conductor,

ρ = resistivity of conductor,

and a and b are constants.

The effect is accentuated when the wire is wound into a coil as discussed in Section 4.3 because the magnetic field inside the coil itself also tends to drive the current to the outside edge of the winding. As will be seen from the expression above, the seriousness of the effect increases as the diameter of the wire increases. For many purposes it is found that even at radio frequencies the increase in resistance is not too serious if the wire used is of relatively fine gauge (of the order of 0·025 mm), but in transmitters where considerable radio-frequency currents have to be handled, the size of the conductor has to be much larger than this and special constructions are adopted to minimize the skin effect. One method is to use hollow tube for the conductor since if the current is flowing mainly in the outside of the conductor there is no point in filling the inside with expensive copper. The skin effect can also be minimized by using a rectangular strip for the conductor.

Still another method is to use stranded wire. If the conductor is made up of a number of fine-wire conductors in parallel, the wires being so arranged that each in turn comes to the surface of the conductor, then the increase in resistance due to skin effect can be very considerably minimized. Wires such as this are called *Litzendraht* (or litz) conductors and are considerably used where it is important to obtain the lowest possible effective resistance. The material is difficult to use because it is necessary to clean the ends of each individual strand and solder them all together at the beginning and end of the conductor: if any strand is not connected or is broken the eddy current loss induced in the dead strand may cause a considerable increase in the effective resistance.

Another device often used with high-frequency circuits is to silver-plate the conductors. Silver is a slightly better conductor than copper and therefore if the outer skin of the conductor, in which most of the current flows, is of silver, the effective resistance will be reduced.

Fuses

The radiation of energy from a body in the form of heat increases with the temperature. When a wire carries a current the energy used in overcoming the resistance, which is I^2Rt, is expended as heat and the temperature of the wire rises to a value such that this energy can be satisfactorily radiated. Hence the greater the current the higher the temperature until ultimately the conductor becomes white hot and melts.

A normal conductor is, of course, so proportioned that its temperature rise is not excessive, but this ability to produce melting is used to provide protection of circuits by what are known as *fuses*. These are short lengths of fine wire which are designed to reach a melting temperature at currents appreciably in excess of that for which the rest of the circuit has been designed. Then if the current becomes excessive due to a fault the fuse "blows" and breaks the circuit, thereby saving the rest of the circuit from damage.

The fusing current is roughly proportional to $d^{3/2}$, but the actual law is modified by the form in which the fuse is made up. A sudden surge of current may cause such a rapid rise of temperature that the fuse explodes, scattering molten metal around the vicinity. It is usual, therefore, to enclose the fuse in some form of protective housing. This may take the form of a removable unit of ceramic or other insulating material with contacts which plug into clips at each end. The fuse wire is then connected between these contacts, running through a hole or channel in the ceramic holder, so that a fresh length of wire can be inserted after the fault has been cleared.

Alternatively, the fuse wire can be housed inside a small tube of glass or ceramic material, making an expendable unit which is replaced by a new one when necessary. This form of "cartridge" fuse can be designed to have a small time lag in operation so that it will withstand momentary surges, of, say, five times normal current but will blow on a sustained current of two to three times normal.

4.2. Capacitors

The basic idea of capacitance was discussed in Chapter 1. In communication engineering frequent use is made of devices designed to produce a known value of capacitance in convenient form. Such a device is called a capacitor or, in older terms, a condenser.

The simplest form of capacitor is shown in Fig. 4.4 (*a*), which simply consists of a pair of parallel plates. The capacitance of such an arrangement is proportional to the area of the plates and

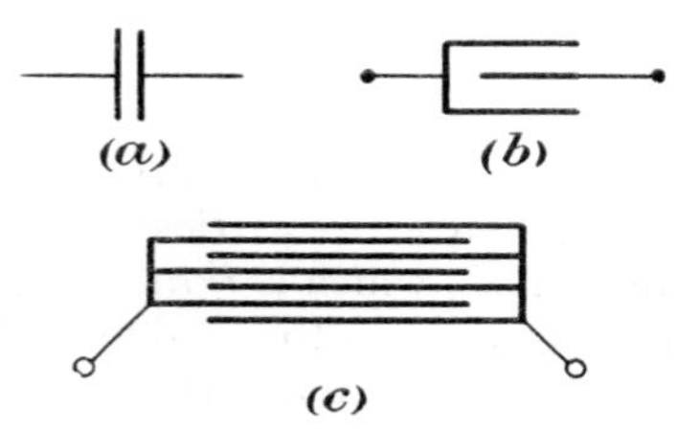

Fig. 4.4. Simple Fixed Capacitor Formations

inversely proportional to the distance between them. Hence the nearer the plates are together the larger the capacitance; obviously, there is a limit to the separation between the plates and also to the size, so that in order to increase the capacitance several pairs of plates are used.

Fig. 4.4 (*b*) shows the effect of adding another plate to the first pair. Here both sides of the middle plate are effective and double the capacitance is obtained. The process may be continued as in Fig. 4.4. (*c*), a series of plates being mounted up in a bank and alternate plates being connected together.

Practical Constructions

The construction of a capacitor depends on the work it is intended for. The capacitance may be fixed or may be made variable at will. In receiving equipment and similar low-voltage applications it is desirable to keep the physical size of the capacitor small. This may be done by using thin metal foil for the plates and separating them by a solid dielectric, preferably one having a high permittivity such as mica. Early small capacitors were made up in this way, and where the capacitor is required to carry appreciable current this construction is still used.

For most purposes, however, a cheaper and more stable capacitor can be produced by depositing a metal coating on the dielectric. This minimizes the small changes of capacitance which can arise from slight mechanical movements of the plates due to temperature or other causes. One such construction in common use, known as the *silver-mica* capacitor, uses silver deposited on mica, while there are various alternative forms in which a ceramic base is used. Such capacitors have a temperature coefficient depending only on the expansion of the base, the metal coating having sufficient elasticity to follow the changes in dimensions of the material on which it is deposited, and because of this intimate contact they are much less liable to erratic changes.

The main variation is thus that due to temperature. Silver-mica capacitors exhibit a very small positive coefficient, the capacitance increasing by about 0·0025 per cent per degree C. Calcium ceramic materials (e.g. porcelains) have a positive coefficient of the order of 0·01 per cent, while there are special ceramic materials which exhibit a marked negative coefficient (e.g. titanium dioxide, which has a coefficient of − 0·068 per cent).

Capacitors can be made up with a combination of two dielectrics so blended as to provide a zero temperature coefficient, while by deliberately using a suitable negative coefficient the variation in value of other parts of the circuit, such as the coils, which is almost invariably positive, may be offset over quite a wide range of temperature.

There are many variants of this basic technique, including monobloc units in which the element is encapsulated under pressure in an epoxy resin, which results in high stability with very small size to meet the requirements of modern miniaturized circuitry.

Foil Capacitors

For large capacitances a paper dielectric may be employed. The capacitor is made up from thin strips of metal foil and specially prepared thin paper. Alternate strips of foil and paper are rolled up together until the required capacitance is obtained. The voltage which the paper will withstand is limited and hence for high voltages two or more thicknesses of paper are used to each foil.

The completed unit is then vacuum-impregnated at a temperature sufficient to drive out any moisture and allow the impregnant to fill any voids, after which it is sealed in a suitable case, which may be partially or completely air-tight. Formerly wax was used as the impregnant, but this expands with heat and so tends to deplete the inner layers of the winding. Hence it is more usual to use petroleum jelly or a polyphenol resin.

The individual foils clearly constitute what is in effect a "pancake" coil and so possess inductance. To minimize this (and also the resistance) connections are made by inserting a number of foil strips at several points as the winding proceeds, so connecting the sections in parallel.

Non-inductive Capacitors

Where minimum inductance is required, the foils can be arranged to overlap on alternate sides, as shown in Fig. 4.5. When the winding is complete, connections can be taken to these projecting edges.

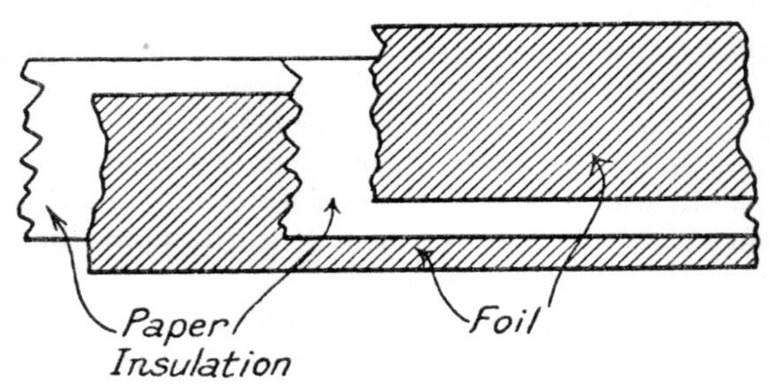

FIG. 4.5. ILLUSTRATING CONSTRUCTION OF NON-INDUCTIVE CAPACITOR

Even so, some residual inductance is present, of the order of 0·01 to 0·1 μH, depending on the construction.

The effect is only important at high frequencies, but even a small foil capacitor may resonate in the low megahertz region, which would render it useless as an r.f. bypass.

Large capacitors (up to about 8 μF) are normally housed in rectangular steel containers occupying about 40 cm^3 per microfarad. For smaller values the assembly is housed in a tube of metal or insulating board, with the connections brought out as wires at each end.

Paper-insulated capacitors do not have good long-term stability, though this is not usually of major importance. Considerable improvement can be obtained by using thin sheets of plastic material, such as polystyrene, for the dielectric.

In many instances the current which the capacitor has to handle is relatively small. This permits the use of metallized foil in the construction, the conducting surface being deposited direct onto

FIG. 4.6. TYPICAL FIXED CAPACITORS

4 μF paper dielectric; 50 μF electrolytic
5,000 pF silver mica; 1,000 pF silver ceramic
16 μF tubular electrolytic; 200 pF silver mica; 8 μF low-voltage electrolytic
0·01 μF disc-type silver ceramic; 0·25 μF polystyrene tubular; 0·1 μF paper tubular

FIG. 4.6 (*a*). A VARIABLE CAPACITOR

FIG. 4.7. TYPICAL RECEIVING COILS

Right: Single-layer solenoid.
Left: Multi-layer "honeycomb" coil with adjustable ferrite "slug" to provide a small variation of inductance.
Centre: Ferrite "pot" core with sectionalized former.

FIG. 4.8. TYPICAL CONSTRUCTIONS USING FERRITE MATERIAL

(*By courtesy of Mullard Ltd.*)

(*See page 202*)

an insulating foil of polyester or polycarbonate. This reduces the thickness, and hence the overall size. A typical polycarbonate 10 μF capacitor (63 V working) has a diameter of 16·5 mm and a length of 35 mm.

There is thus a wide range of foil capacitors from low-voltage units for transistor circuitry to high-voltage types for transmitting equipment. Further details may be found in *Modern Electronic Components* by G. W. A. Dummer (Pitman).

Electrolytic Capacitors

Paper capacitors, however, are inevitably bulky but it is possible, for certain applications, to obtain very high capacitance in small volume in what are called *electrolytic* capacitors. If a metal plate, usually of aluminium, is immersed in a suitable solution such as aluminium borate and a current is passed through the assembly with the aluminium plate positive, a very thin film of aluminium oxide forms on the surface of the plate. This film forms a dielectric layer between the aluminium plate and the electrolyte so that the arrangement exhibits the properties of a capacitor.

The oxide film is so thin (of the order of 10^{-5} cm) that the capacitance is far higher than could be obtained in the same space by ordinary methods. The maximum capacitance obtainable with a given anode surface area is inversely proportional to the voltage at which the capacitor is formed, since the film thickness is dependent on the forming voltage. On the other hand, if the voltage applied to the capacitor in use exceeds the forming voltage, or is in the reverse direction, partial breakdown of the dielectric film will occur. An electrolytic capacitor is thus limited in its application and it cannot be used on alternating currents. It can, however, be employed for smoothing out or by-passing alternating currents, provided there is a steady d.c. voltage across it, and this property is useful in many circumstances, as will be seen later.

Early forms of electrolytic capacitor were of the "wet" type in which an aluminium foil, corrugated to increase the surface area, was rolled up in cylindrical form and immersed in a metal can containing electrolyte. The electrolyte, however, is liable to leak, and ultimately dries out, particularly if the capacitor is allowed to get hot, but it has the advantage of being self-sealing.

It is more usual to use a "dry" construction. Here the capacitor is similar in form to a paper insulated capacitor, except that one foil is formed with the oxide film as just described. The formed foil is the anode and the plain foil is the contacting electrode to the electrolytic cathode, which consists of porous paper impregnated with electrolyte (usually a paste of glycol and ammonium tetraborate). The assembly of formed foil, impregnated separator and plain foil is then rolled up and housed in a suitable container.

A further increase in capacitance for a given size can be obtained by mildly etching the anode foil with acid before forming, so that an irregular surface of oxide is formed having a much greater area than that of the plain foil. Since the electrolyte is a paste it is able to make contact with the whole of the irregular surface area of the oxide film. An etched-foil capacitor can have a capacitance up to 8 times greater than a plain-foil capacitor of the same size, but the power factor is somewhat increased.

A "dry" electrolytic capacitor is not self-sealing and, therefore, must not be subjected to voltage in excess of the forming voltage. Nor should it be used under hot conditions for it is not truly dry and continued high temperature will cause rapid deterioration. Further disadvantages of electrolytic capacitors are the high power factor (about 10 times that of a paper capacitor), the high leakage current and the variation in capacitance with time and temperature. However, electrolytic capacitors have a great advantage in size for a given capacitance over all other forms of capacitor.

A greatly improved performance is obtainable by the use of tantalum, the oxide of which has a high permittivity ($\varepsilon_r = 27$). Such capacitors are not only smaller in size, but have a long storage life, improved stability and very low leakage current. They are more expensive than aluminium types and are therefore usually limited to low-voltage applications, but are extensively used in sophisticated transistor circuitry.

The conventional foil construction may be used, but "solid" types are made using a sintered pellet of tantalum powder. This is oxidized electrolytically, after which the electrolyte is removed, contact with the oxide being maintained by an admixture of manganese dioxide. A capacitor of this type will provide a capacitance of $5\,\mu$F, 20 V working, in a bead only 6·4 mm in diameter.

Some typical fixed capacitors are illustrated in Fig. 4.6.

Breakdown Voltage

Any dielectric will only withstand a certain potential gradient. If this is exceeded there is a momentary conduction current which in turn produces ionization resulting in a cumulative build-up of a current of such intensity that the dielectric may be punctured.

All solid dielectric capacitors, therefore, are rated to function at or below a specified working voltage. The permissible voltage depends on temperature, for the electric strength of an insulating material decreases rapidly as the temperature rises. A paper-insulated capacitor which can safely be used up to 750 V at 60°F would be limited to 600 V working at 75°F.

It must be remembered that the important factor is the peak value of the voltage. Capacitors, in general, will safely withstand momentary voltages higher than normal, though with electrolytic capacitors the margin is small (10 per cent or less). Capacitor ratings often include a "surge" voltage figure, and the circuit must be such that the peak value of voltage under any conditions does not exceed this figure.

Variable Capacitors

There are two methods of varying the capacitance of a capacitor. One is to vary the distance between the plates, and the other is to alter the effective area of the plates.

The second method is more common. A bank of plates is arranged as shown in Fig. 4.6 (*a*), the moving plates being capable of rotation into the spaces between the fixed plates. This gives a continuously variable capacitance, the maximum value depending on the number and distance apart of the plates.

The variation of capacitance with angular rotation depends on the shape of the plates. With plates of constant radius the capacitance variation is linear, but by suitable variation of the radius the capacitance change can be made to obey various other forms of law, such as:

(*a*) *Square Law.* This will give an approximately linear relation between the shaft rotation and the wavelength to which the circuit is tuned. As shown in Section 2.3, $\lambda = k\sqrt{(LC)}$, so that if the plates are so shaped as to make $C \propto \theta^2$, $\theta = k_1 \sqrt{(C)} = k_2\lambda$.

The relationship is only approximate since C must include the stray circuit capacitance which, in general, is not known beforehand.

(*b*) *S.L.F. Law.* By making $C \propto 1/\theta^2$ a linear relationship can be obtained between the frequency to which a circuit will tune and θ. This is known as a straight-line-frequency or S.L.F. Law.

(*c*) *Logarithmic Law.* $C \propto \log \theta$.

(*d*) *Constant* Δf *Law.* As explained in Chapter 9, superheterodyne receivers use a capacitor for the oscillator tuning which provides a constant difference in frequency from the input signal. This can be obtained by suitable shaping of the oscillator capacitor plates, though it is more usual to obtain the effect required in a slightly different manner.

The maximum capacitance depends on the size and number of plates and their relative spacing. Capacitors used in receiving equipment usually have a capacitance of the order of 200 to 500 pF.

Trimmers

For many purposes a capacitance is required which can be set to a given value (adjusted by actual trial in the circuit) and then locked. Such capacitors are known as trimmers or preset capacitors.

Various forms of trimmers exist, the principal ones being:

(*a*) Miniature versions of a normal variable capacitor, sometimes using a solid dielectric.

(*b*) Capacitors comprising a cylindrical inner element on a screw which can be moved inside an outer tube to any desired extent.

(*c*) Capacitors comprising two flat plates, usually separated by a thin mica sheet, with means for varying the distance between them by means of a screw.

The various types are illustrated in Fig. 4.9.

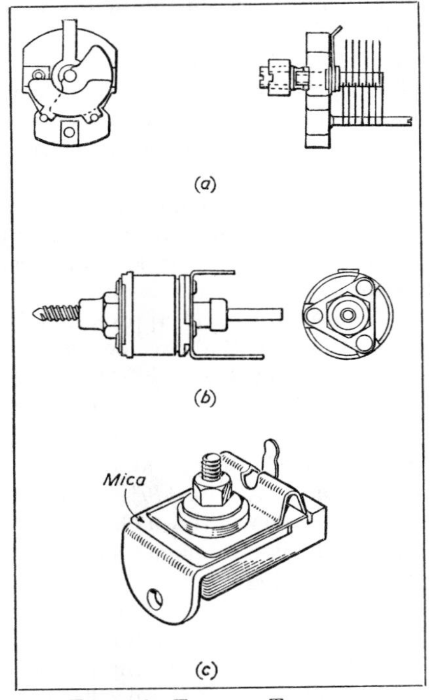

FIG. 4.9. TYPICAL TRIMMERS

Power Factor; Loss Angle

The ideal capacitor is simply a device for storing electrical energy and releasing it when required, and no energy should be absorbed during the process. With an air dielectric this condition is very nearly obtained, but with liquid or solid insulation there is a certain electrical friction occasioned by the movements of the electrons in the material, which results in a loss of energy. There is also some loss caused by the (ohmic) insulation resistance of the dielectric.

It is clear that we cannot support two sets of plates in mid-air without some solid dielectric at a suitable point, but by keeping this to a minimum the dielectric loss can be kept low.

The effect of the dielectric loss can be represented by assuming that a practical capacitor is composed of an ideal capacitor associated with a loss resistance which may be treated as being in series or in shunt, whichever is more convenient.

If we assume the resistance to be in series, then by the normal a.c. laws the current $I = V/(R + 1/j\omega C)$. This current will not lead by 90° but by some smaller angle given by $\phi = \text{arc tan } V_C/V_R = \text{arc tan } (1/RC\omega)$. Since in general the loss resistance is much smaller than the reactance, ϕ is nearly 90°, and it is more usual to specify the *loss angle* of the capacitor, which is the angle ψ by which the current fails to lead by 90°.

Since this angle is very small it may be expressed simply as $R/X(= RC\omega)$ radians (the radian being the measure of the ratio of an arc of a circle to its radius, as explained on page 42).

The power factor, which is the ratio of resistance to impedance, is in this case indistinguishable from R/X. Hence

$$\text{Power factor} = RC\omega$$

which is the same as the loss angle in radians.

Typical values of power factor for different dielectrics are given in Table 4.2 below. It will be seen that the power factor is in general very low, which is equivalent to saying that the Q of a capacitor (which is the ratio of reactance to resistance, $=1/RC\omega$) is very high. Nevertheless, loss does exist and has to be allowed for.

TABLE 4.2. POWER FACTOR OF TYPICAL DIELECTRICS
(average values)

Material	*Power Factor*	*Relative Permittivity*
Air	0	1
Mica	0·0004	5
Glass	0·009	5·5
Bakelite	0·05	6
Mycalex	0·002	8
Polystyrene	0·0002	2·5
Ceramics	0·002	6·5
Paper	0·005	2

If we represent the loss by a parallel resistance, the current is VY, where Y is the *admittance* of C and R_p in parallel,

$$= V(1/R_p + j\omega C) = V[(1 + j\omega CR_p)/R_p].$$

The loss angle ψ is then arc tan $I_R/I_C = 1/R_p\omega C$ radians. Hence in terms of parallel resistance,

$$\text{Power factor} = 1/R_p\omega C$$

By equating this to the expression for a series resistance we see that $R_p = 1/\omega^2C^2R_s$.

Variation of Loss with Frequency

Dielectrics, by their very nature, are not uniform in their behaviour, but those used as insulators in capacitors tend to exhibit a constant power factor.

Hence the equivalent loss resistance is inversely proportional to frequency.

The resistance of the plates and connections, of course, is virtually constant until skin effect begins to be appreciable, from which point $R \propto \sqrt{(f)}$.

There is a third source of loss known as *dielectric absorption*. The material appears to absorb a certain abnormal extra charge. If a capacitor is discharged and then set aside and again short-circuited a little later, a further discharge will occur showing that the charge was not all dispersed on the first occasion.

The energy loss from this cause is comparatively small, and is found to be inversely proportional to the frequency, i.e. it is greatest at low (power) frequencies, and grows less as the frequencies are raised.

Air is the most efficient dielectric in this respect. Good oil, free from moisture, is nearly as good. Compared with air, glass has an efficiency of about 60 per cent, and mica 40 to 90 per cent, depending on its quality. Synthetic materials, such as resin-impregnated paper, vary widely in performance, but in general introduce appreciable loss.

Factors Affecting Capacitor Performance

The effective series resistance of a capacitor is not only due to dielectric loss, but includes the actual resistance of the plates and leads. This is normally negligible but becomes of importance at high frequencies because of the increase due to skin effect. It may be noted also that the capacitor inevitably possesses some inductance, the effect again being negligible until frequencies in the short-wave region are attained.

The effect is to increase the apparent capacitance to a value

$$C' = C/(1 - \omega^2 LC)$$

The losses in an electrolytic capacitor are considerably higher than for other forms, and rise rapidly with increasing temperature. This rise with temperature is true of all dielectrics and becomes important at high frequencies when the increased losses cause a rise in temperature which in turn still further increases the losses. This necessitates a considerable reduction in the permissible working voltage so that the current through the capacitor may be kept within safe limits. A capacitor which has an electric strength sufficient to withstand 10 kV at low frequencies may not be able to handle more than a few hundred volts at 10 MHz.

Transmitting Capacitors

Capacitors for transmitters usually employ air dielectric, though liquid dielectrics such as paraffin are sometimes used. In any case the plate spacing has to be sufficient to prevent breakdown at the voltage in use.

As the distance between the plates is increased the breakdown voltage increases, but not in direct proportion; e.g. if the distance is doubled the breakdown voltage only increases about 1·75 times. Hence transmitting capacitors are often made up in several sections in series, each section having only a portion of the full voltage across it.

The potential gradient in which breakdown occurs is called the *electric strength*, formerly the dielectric strength. It is measured in volts per unit distance, but since, as has just been stated, the relationship is not linear and also the breakdown potential is dependent on the shape or disposition of the electrodes, it is important to specify the exact conditions as an indication of the order of potential gradient required to produce breakdown. Table 4.3 gives some representative figures.

TABLE 4.3. ELECTRIC STRENGTHS FOR PLATES 1 mm APART

Dielectric	*Electric Strength (volts/mm)*
Air	4,000
Oil	7,000
Glass	8,000
Bakelite	30,000
Mica (best ruby)	60,000

As voltages of from 10,000 to 50,000 volts are quite common in transmitting practice this question requires proper attention. One form of transmitting capacitor is made up by mounting a series of plates separated by air and supported at the corners on glass blocks.

Another method is to hang plates from rods of glass or other insulating material. The dielectric is thus principally air, the distance between the plates being made large enough to avoid any flashover. A more common method is to build up the plates in a rigid structure, and to use an oil or mica dielectric. This not only increases the capacitance but enables the plates to be assembled nearer together without any risk of the voltage sparking across. Such capacitors, however, are considerably more costly.

Dielectric Materials

The material for insulation, not only in capacitors but for general purposes, is a study in itself.

Paper products impregnated with suitable composition are widely used. Thin sheets of paper are either compressed together to form one thick slab, or are rolled up to form rods of circular or other section. The paper is impregnated with Bakelite varnish or other special insulating varnish, and in some cases linen is employed to give greater mechanical strength. This class of product goes by various trade names, such as Paxolin or Tufnol.

Mica is considerably used for small parts, but its brittleness seriously limits its application. There are, however, various mica products formed by bonding powdered mica with suitable materials and forming sheets or slabs. A particular example is a synthetic red-coloured material composed of rubber and mica, known by various trade names such as Keramot, Silvonite, etc. It is very good for medium radio frequencies, and combines low losses with high electric strength (55 kV per mm).

Hard-rubber compounds, such as ebonite, are not usually employed except at quite low frequencies, for the sulphur content of the material introduces considerable loss which in a transmitter may produce so much heat as to cause serious mechanical distortion.

A material which is much used, particularly for high-frequency work, is Mycalex, which is ground mica bonded with lead borate glass and moulded under pressure at about 675°C. It is a dull grey material which can be machined with special tools. Numerous other synthetic materials are continually being developed.

Porcelain is suitable for many applications. This is a ceramic material containing mainly china clay (kaolin), flint, and feldspar, which are finely ground and mixed, and then mixed with a small quantity of moisture to enable them to be moulded. They are then fired and, for some applications, glazed to provide a smooth surface. It is also possible to use a dry casting process in which the powder is formed into the required shape under pressure without firing.

Similar ceramic compounds have been highly developed, particularly for short-wave work. Magnesium silicate is also used instead of clay (which is an aluminium silicate), while steatite, which is a purified talc sometimes known as soapstone, is also utilized.

The ceramic insulators are undoubtedly the best from the point of view of dielectric properties, but because of their fragility they are only used in positions where the highest efficiency is essential.

4.3. Air-cored Inductors

Inductors are used in both transmitting and receiving equipment. In the latter case the chief consideration is the obtaining of the requisite inductance in a fairly compact form.

For low values of inductance the coils are usually wound in single layers upon cylindrical formers of bakelized paper or ceramic material. Such a coil is called a "solenoid." Another form of inductor sometimes employed is the spiral or pancake variety. Here the successive turns, instead of being side by side, are wound one on top of the other spiral fashion.

For higher values of inductance multi-layer coils are employed, i.e. when one layer is completed a further layer is wound on top. With this arrangement, however, there is appreciable stray capacitance between the turns, which may be minimized by using a "honeycomb" winding. Here the winding is arranged to travel from one side of the coil to the other every turn. If this is done, the successive layers of the coil cross each other at an angle (giving a criss-cross or honeycomb effect) and do not lie flat on top of one another.

It is possible to arrange that the successive turns of each layer are spaced by a small distance from each other, the result being that a large number of air spaces is obtained throughout the coil. This method, however, results in a rather bulky coil and it is more usual to use a wave-winding in which the traverse from side to side is so arranged that the wires lie close together without the large air spaces of a true honeycomb-weave. This gives a more compact coil having a performance not significantly worse. Wave-wound coils have the advantage of considerable rigidity, being quite self-supporting. They are often given a light dip in varnish and then baked at a low temperature.

Where the coil is not required to carry high-frequency currents a simple multi-layer winding may be used. This may be wound on a former having cheeks at each end to retain the wire, and there is usually an interleave of paper or other insulating material between layers in order to keep them apart. Otherwise, one turn several layers up may slip down and lie against a turn near the beginning of the coil, an arrangement which would not only create a high self-capacitance, but might result in a breakdown of the insulation.

An intermediate form of construction is to use a former having a number of slots. The winding is arranged to fill each slot in turn. Due to this sectionalizing, the self-capacitance is considerably minimized. Typical coil constructions are illustrated in Fig. 4.7.

All these constructions are subject to minor temperature effects due to changes in the physical dimensions, on which the inductance primarily depends. This applies both to air-cored coils and the ferrite assemblies described later, and inductors in general exhibit a small positive temperature coefficient.

Transmitting Inductors

For transmitting units the coils have to be wound with much thicker conductors to carry the heavy currents, and are usually wound with stranded wire, copper strip, or copper tube, to avoid losses due to eddy currents and skin effect.

The turns also are well spaced apart and are wound on a suitably-insulated former owing to the high voltages involved. Both cylindrical and spiral coils are employed, but multilayer coils are seldom used. The inductance in a high-power transmitting station has a voltage across it of the order of 50,000 volts, so that the need for careful insulation is at once apparent.

Magnetic Coupling

It has been seen that if one coil is placed near another which is carrying current, it is affected by the variations of the magnetic field of the latter. This condition of affairs is referred to as a "coupling" between the two coils. The extent of such coupling, i.e. the mutual inductance between the coils, depends on the proportion of the flux which links with the coil.

If the coil is placed at right angles to the first coil no flux will flow *through* the coil but only across it; in other words, there will be no linkage between the flux and the coil and no coupling will result (Fig. 4.10 (*a*)). As the secondary is turned on its axis it links with more and more of the flux until the maximum position is reached when the coils are in line (Fig. 4.10 (*b*)).

This coupling will not, of itself, produce any effect but if the magnetic field is varying, induced e.m.f. will appear in any coil which links with the field.

It will be observed that the magnetic field is not confined within the coil but spreads appreciably around the outside. Hence coupling

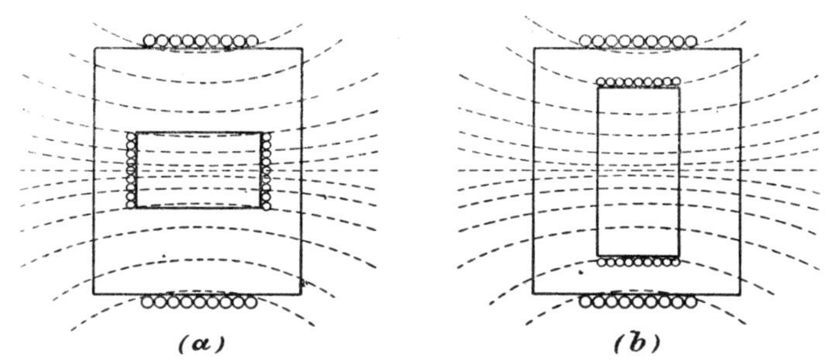

FIG. 4.10. ILLUSTRATING MAGNETIC COUPLING

exists between any two coils in proximity, and the mutual inductance may be varied by altering the distance between them and/or their relative orientation. This is utilized in various ways in practice, from the production of variable inductors as discussed later, to the transfer of energy from one circuit to another, as in radio-frequency transformers.

Coils with Closed Magnetic Circuits

The type of inductor so far considered has an "open" magnetic circuit. The magnetic flux generated within the winding has to complete its circuit through the surrounding medium and, while this is sometimes useful (as in an r.f. transformer), the presence of this "stray" field is often an inconvenience. Moreover the indeterminate length of the magnetic path makes inductance calculations difficult.

These disadvantages may be overcome by providing a closed magnetic circuit. This can be done in two ways, by using a toroidal winding, or by employing a closed core of magnetic material.

(*a*) TOROIDAL COILS

A toroidal winding was illustrated in Fig. 1.16. Here the coil is wound on a ring-shaped core and the magnetic field is very largely confined within the core so that not only is the stray field reduced but the length of the magnetic path is definite.

Special winding methods are necessary with a toroid because the spool holding the wire has to be small enough to pass through the centre aperture of the core. Consequently this type of coil is only used in special circumstances.

(*b*) Magnetic Cores

Inductors wound on closed magnetic cores are very frequently used in communication engineering. The design and construction however, involves appreciably different technique and they are therefore discussed at length in Section 4.4.

Eddy Currents

There is, however, another effect of great importance resulting from the presence of the magnetic field. Any metal object in the vicinity of the field will, if the field is varying, have an e.m.f. induced in it which will produce what are called *eddy currents*. The metal, in fact, behaves as if it were an assembly of a large number of short-circuited coils, in each of which currents will circulate depending on the resistivity of the material.

Now, it was shown in Section 2.4 that the presence of a secondary winding coupled to a coil affects the primary impedance, causing an increase in the effective resistance and a decrease in the inductance. Hence the presence of any metal near a coil carrying a varying current will absorb energy from the circuit. This is known as the *eddy-current loss* and can be represented by an additional resistance in series with the normal conductor resistance.

For this reason it is desirable to keep metal objects out of the magnetic field as far as possible. This is particularly important with transmitting inductors, for any metal in a powerful magnetic field may have induced in it eddy currents of sufficient intensity to make the metal red hot.

The induced e.m.f. is proportional to the magnetic field and the frequency and hence the eddy currents will be proportional to Bf/ρ where ρ is the resistivity of the material, and the energy loss will be proportional to the square of the current.

The effect thus becomes increasingly important as the frequency is increased. It is not large at low frequencies, though it does enter into the calculations on power and a.f. transformers, where it can be minimized by increasing the resistivity of any metal which has to be in the field, as explained in the next section.

At radio frequencies the eddy-current loss is appreciable with quite small fields and there is the further effect that the inductance of the coil is reduced by the eddy currents.

Shielding

There is, however, an important use to which such currents may be put. The eddy currents will themselves produce a magnetic field which we know, by Lenz's Law, will be in such a direction as to oppose the original field. If we enclose a coil completely in a metal box or can, the eddy currents induced in the can will produce a counter magnetic field. Moreover the strength of the eddy currents will automatically adjust itself so that this counter magnetic field virtually cancels the initial field at that point.

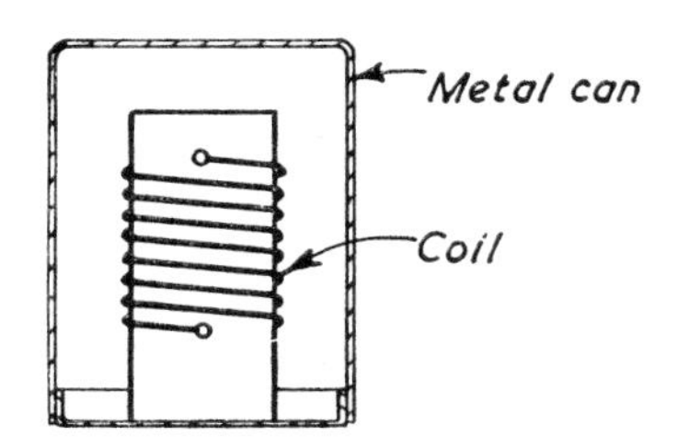

FIG. 4.11. SHIELDED COIL

Thus *outside the can* the effective magnetic field is reduced almost to zero so that any coupling between the original coil and any adjacent coils is largely eliminated. Inside the can, of course, the field is also reduced, but only by an amount equivalent to its value at the radius of the screening can. This is considerably less than the field inside the coil itself, so that the only effect is a slight reduction in the inductance and a slight increase in the effective resistance.

This process of shielding or screening is of great usefulness in the design of receiving equipment since it permits components to be placed appreciably closer together than would otherwise be possible. To be effective the shield must be such as to permit substantial eddy currents to exist; i.e. it must be of high-conductivity material, such as copper or aluminium, and it must have a thickness several times the skin-depth at the particular frequency in use (*see* Section 4.1.)

Variable Inductances

For many purposes it is desirable to be able to vary the inductance of a circuit. This is not so easily contrived as a variable capacitance, and is usually limited to relatively small variations.

The most obvious method is to alter the number of turns on the coil. This will normally only provide discrete changes and is little used except in transmitter inductors where the capacitance is fixed, so that tuning must be provided by varying the inductance (cf. Fig. 7.1). For such purposes the coil is constructed with bare wire or tube, with clip connections which can be adjusted as required.

With any form of tapped coil allowance must be made for possible "dead-end losses". The unused portion of the coil will constitute with its own self-capacitance a circuit which is closely coupled to

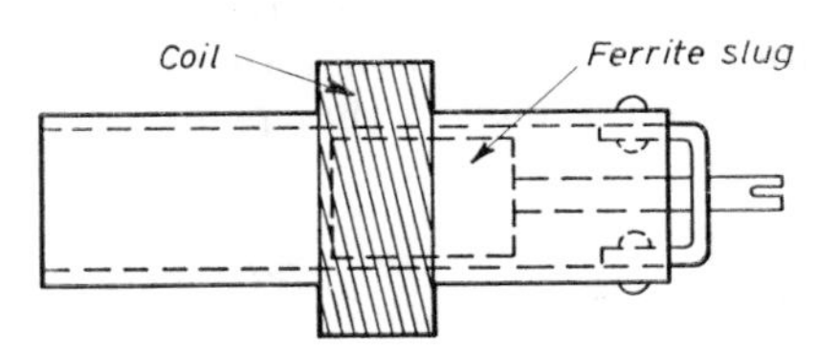

FIG. 4.12. PERMEABILITY-TUNED COIL

the active portion, and may produce undesirable effects. In some cases it is preferable to short-circuit the unused turns.

For receiving circuits it is useful to be able to vary the inductance within small limits to allow for stray circuit capacitances. This is usually achieved by altering the permeability of the magnetic circuit. With a single- or multi-layer solenoid this can be provided by fitting a small ferrite slug within the former, as shown in Fig. 4.12. A screwdriver adjustment allows the position of the slug to be adjusted and so provides the required (small) variation of inductance.

The variation obtained by this means is only a few per cent, which is sufficient for the purpose. Somewhat greater (and more precise) control of the inductance is obtainable with a completely closed ferrite core, as illustrated in Fig. 4.31, and this form of construction is frequently employed.

Certain special applications, usually associated with measuring equipment, require an inductance capable of being varied over an appreciable range. This can be arranged by providing two coils, so mounted that the coupling between them is variable. The inductance is then $L_1 + L_2 \pm 2M = L_1 + L_2 \pm 2k\sqrt{(L_1L_2)}$. Such a device is called a *variometer*, a simple form being illustrated in Fig. 4.13 (*a*). With this construction the coupling factor between stator and rotor is only about 0·5 so that the total variation obtainable is only about 3:1, but by using modified construction (e.g. spherical former) the coupling can be increased and a range of 10:1 or more can be obtained.

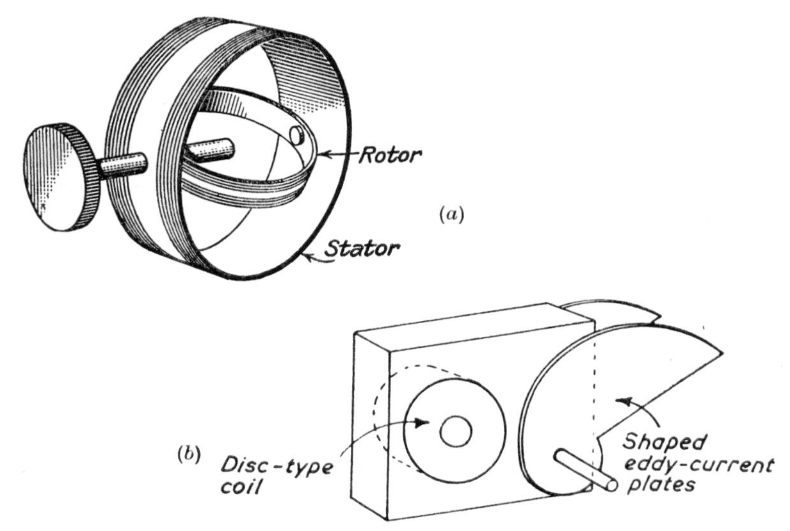

FIG. 4.13. TWO FORMS OF VARIOMETER

An alternative method is to utilize the change of inductance caused by eddy currents. This is known as spade tuning, suitably shaped metal plates being brought into proximity to the coil, which is usually of the disc type as shown in Fig. 4.13 (*b*). The range of variation with this type of variometer is only small, of the order of 10 to 20 per cent.

Losses in Coils

The main source of loss in an inductor is the ohmic resistance of the wire. This we have seen to be more than the d.c. resistance because of the skin effect mentioned in Section 4.1.

The efficiency of an inductor is rated in terms of the ratio of reactance to resistance which is known as the magnification factor $Q = L\omega/R$. Where the coil constitutes part of a tuned circuit or a filter network the object is to make this factor as high as practicable. Where the inductor is being used simply as an impedance (as in a choke) a high Q is relatively unimportant.

In addition to the purely ohmic losses there are dielectric losses occurring in the material of the former and the insulation on the wire itself, plus any eddy-current losses as just discussed.

We have seen that the eddy-current loss is proportional to the square of the frequency; the dielectric loss varies approximately as the cube of the frequency. The dielectric loss can be considered as due to a lossy capacitance connected across the coil and, as was shown in Section 4.2, the equivalent shunt resistance of a lossy capacitance $R_p = 1/\psi\omega C$ where ψ is the loss angle. A resistance R_p across a coil of inductance L is equivalent to a series resistance

$$R = \omega^2 L^2/(R_p + j\omega L)$$

$= \omega^2 L^2/R_p$ if R_p is large compared with $j\omega L$. Hence the equivalent loss resistance due to the dielectric is

$$\frac{\omega^2 L^2}{1/\psi\omega C} = \omega^3 L^2 C\psi$$

Hence for a given coil the Q will at first increase with frequency, though not directly, since the ohmic resistance will increase approximately as $\sqrt{f}$ due to skin effect. As the frequency is further increased dielectric losses begin to be appreciable and the Q begins to fall again.

If the coil has a core of iron or ferrous material additional losses will be caused by the presence of the iron in the magnetic field. These are discussed in the next section.

4.4. Iron-cored Inductors and Transformers

For many applications it is convenient to provide a substantially closed path for the magnetic flux in an inductor or transformer. This not only permits a better utilization of the material but also renders the arrangement more amenable to calculation in certain respects. The issue is complicated, however, by the fact that the permeability of ferrous material is not a constant factor, like the permittivity of a dielectric, but varies with the degree of magnetization.

The Magnetic Circuit

As explained in Section 1.4, the conditions existing in a magnetic circuit are in some respects analogous to the flow of current in a conducting circuit, the appropriate relationship being

$$\Phi = \frac{M}{S}$$

where Φ is the flux, M is the magnetomotive force (m.m.f.), S is the reluctance.

The reluctance, which is analogous to the resistance in a conducting circuit, is $S = l/A\mu$, where l and A are the length and cross-sectional area of the magnetic path, and μ is the permeability.

In an air-cored inductor, or indeed in any system having an

"open" magnetic circuit, it is difficult to define l and A, but where a substantially continuous magnetic path can be provided for the flux, much more precise calculations become possible.

These are the conditions which exist in iron-cored inductors and transformers. Fig. 4.14 illustrates a simple magnetic circuit and the meaning of l and A is clear.

The m.m.f., M, is, by definition, equal to IN, where I is the current in the coil and N is the number of turns. The magnetizing force, H, is the m.m.f. gradient, or m.m.f. per unit length. Hence the relevant fundamental expressions are:

$$\Phi = M/S$$
$$M = IN$$
$$H = IN/l$$

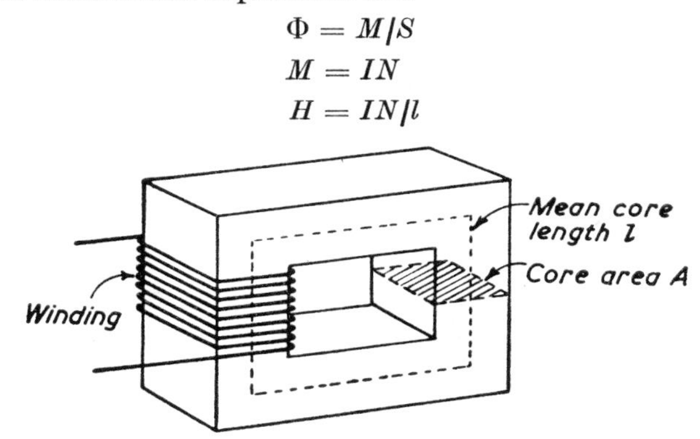

FIG. 4.14. IRON-CORED INDUCTOR

The flux $\Phi = HA\mu$, where μ is the permeability and equals $\mu_0\mu_r$. Substituting the value of $\mu_0 = 4\pi/10^7$ we have

$$\Phi = 1{\cdot}257\ HA\mu_r\ .\ 10^{-6} \text{ webers}$$
$$= 1{\cdot}257\ INA(\mu_r/l)\ .\ 10^{-6}$$

and
$$B = \Phi/A$$
$$= 1{\cdot}257\ (IN/L)\mu_r\ .\ 10^{-6} \text{ teslas}$$
$$= 1{\cdot}257\ H\mu_r\ .\ 10^{-6}$$

B-H Curve; Saturation

Now, as already mentioned, the permeability of iron and magnetic materials generally is not constant but varies with the value of B, which in turn depends on H. One thus needs to know the conditions under which the iron is working, and, to some extent, its past history.

Fig. 4.15 (a) shows the magnetization or B-H curve for soft iron, and also the variation of μ_r with H. It will be seen that the value of B rises rapidly at first and then much more slowly. This subsequent falling off is known as *saturation.*

Magnetism may be considered as the result of an orientation of the orbits of the electrons in the atoms each of which forms a small electromagnet. Normally the distribution is random and only a small magnetic effect results, known as residual magnetism or

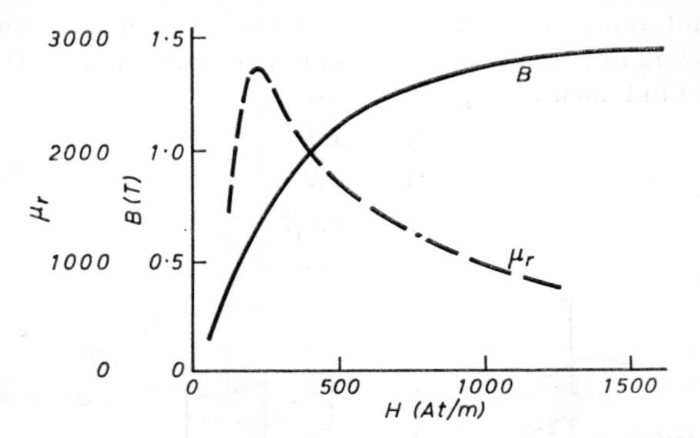

FIG. 4.15 (*a*). VARIATION OF B AND μ_r WITH H

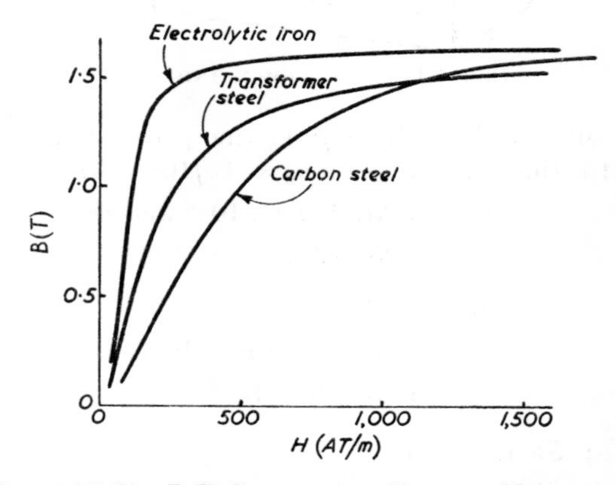

FIG. 4.15 (*b*). *B-H* CURVES FOR VARIOUS MATERIALS

remanence. The magnetizing force H causes the orbits to align themselves so that their magnetic effects all tend in the same direction. At first this action is rapid, but after the majority of the orbits have been correctly aligned, increase in H produces little effect and in the limit the only increase produced will be that due to the magnetizing force itself. The material would then be fully saturated. Such a condition is rarely attained in practice, but it is easy to reach a condition where doubling the value of H only produces a 5 per cent increase in B.

It will be seen that μ_r rises to a maximum and also then falls off rapidly. We shall see later that the effective value of μ_r is much below the static value of several thousand shown.

Fig. 4.15 (*b*) shows similar *B-H* curves for other materials.

Hysteresis

A further peculiarity is that, after iron has been magnetized, the removal of the magnetizing force does not result in complete disappearance of the magnetism. Even if the magnetizing force is only reduced slightly the magnetism will be greater than it was for the same value of H previously. If the magnetizing force is taken over a cycle from a given positive value to a similar negative value and back again, the value of B will follow a curious path as shown in Fig. 4.16. This is called *hysteresis* and plays an important part in the design of iron circuits.

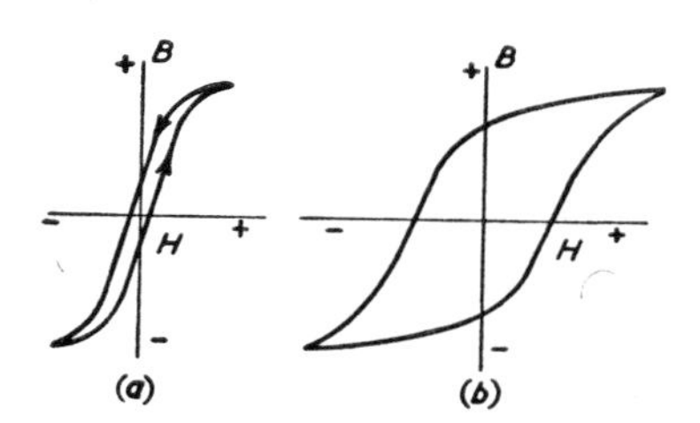

FIG. 4.16. HYSTERESIS CURVES

In an inductor carrying an alternating current the iron is taken through this cycle many times a second. It will be clear that the effect of hysteresis is that some of the energy originally applied to the iron is stored in the molecular structure and before we can magnetize the iron in the opposite direction we must dissipate this energy. There is thus a loss of energy every cycle, which is known as the hysteresis loss.

It is proportional to the maximum flux density B_p to which the iron is magnetized, rather more than in direct proportion but not as much as to the square, while it is also directly proportional to the number of cycles per second. The loss is, in fact, of the form $W = kfB_p^{1\cdot 6}$ where k is a constant depending on the material. It is found that the loss is reduced by incorporating from one to four per cent of silicon in the iron, and such material, known by various trade names such as Stalloy, Silcor, etc., is normally used for transformers and chokes.

It is put up in the form of thin sheets as explained later under the heading of *Iron Loss*.

Energy in a Magnetic Field

We saw in Chapter 1 that the work done in establishing a magnetic field is stored in the field and it was shown that the value of this stored energy in an inductance carrying a current I is $\frac{1}{2}LI^2$. We can rewrite this expression in terms which do not involve the inductance or the current, as such, and so obtain an expression in terms of the field strength. We know that $L = \mu N^2 A/l$, where A and l are the area and length of the magnetic circuit; also the magnetizing force $H = IN/l$. Hence

$$\tfrac{1}{2}LI^2 = \tfrac{1}{2}\,\mu N^2 A I^2/l = \tfrac{1}{2}\mu(N^2 I^2/l^2)Al = \tfrac{1}{2}\mu H^2.Al$$

But Al is the volume of the magnetic circuit, so we have simply

$$\text{Energy per unit volume} = \tfrac{1}{2}\mu H^2 = \tfrac{1}{2}B^2/\mu = \tfrac{1}{2}BH \text{ joules}$$

(In these expressions, of course, $\mu = \mu_0\mu_r$.)

Remanence; Permanent Magnets

It will be seen from Fig. 4.16 that if the magnetizing force, having been increased to a maximum, is now reduced to zero, the value of B, which follows the upper of the curves on the hysteresis loop when H is decreasing, will not fall to zero. This residual value of B is called the *remanence*. To reduce it to zero a negative value of H will have to be applied, and this is called the *coercive force*.

For normal core material both these quantities should be kept low because the area inside the hysteresis loop represents energy absorbed by the iron; but conversely, if one wants to utilize this property of remanence the material should have a large B-H loop. Carbon-steel is such a material, having a loop of the form shown in Fig. 4.16 (*b*). It requires a large coercive force, but after the removal of the magnetizing force it exhibits a large remanent magnetism. Such an arrangement is called a *permanent magnet*.

Various materials have been developed which have a very large hysteresis loop for use as permanent magnets. The effectiveness of a material for this purpose is not determined solely by its remanence, but by the energy stored, which is proportional to the area of the hysteresis loop. This is controlled not only by B_{rem} but also by the coercive force H_c, both of which should be large. The performance of a permanent magnet is thus dependent on that portion of the hysteresis loop lying between these two values, as shown in Fig. 4.17, which is known as the *demagnetization curve*. For any given value of B the energy, as just stated, is $\frac{1}{2}BH$ and it is customary to

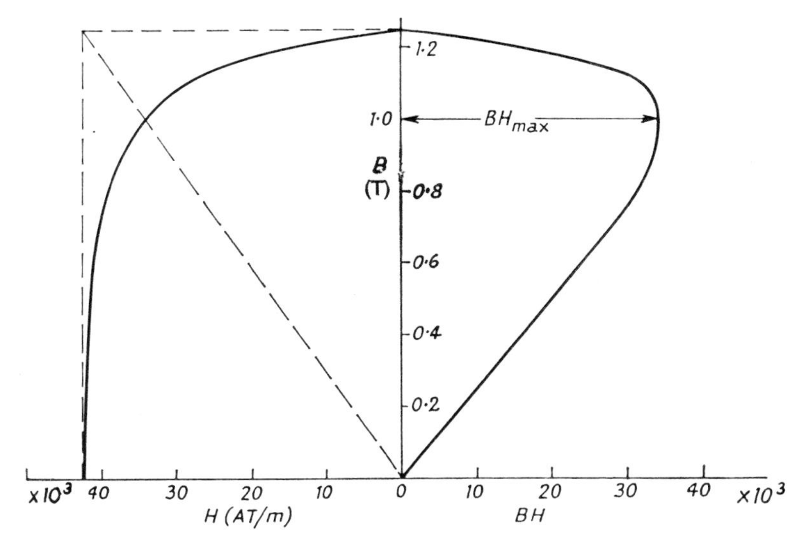

FIG. 4.17. DEMAGNETIZATION AND ENERGY PRODUCT CURVES FOR A PERMANENT MAGNET

rate the effectiveness of a magnetic material in terms of its *energy product BH*. This is shown on the right-hand side of Fig. 4.17 and will be seen to have a broad maximum at about 80% of B_{rem}. This value of BH_{max} is quoted as part of the specification of the material, some typical values being shown in Table 4.4. (The optimum value

TABLE 4.4. PERMANENT MAGNETIC MATERIALS

Material	*Coercive Force* At/m	*Remanence* T	*Energy Product* $(BH)_{max}$ joules
Tungsten steel (6% W, 0·7% C)	5,200	1·05	2,400
Cobalt steel (16% Co, 9% Cr, 1% C)	14,300	0·8	4,800
Alcomax (24% Co, 14% Ni, 8% Al, 3% Cu)	43,500	1·25	35,000

of B, for which BH is a maximum, is given approximately by the intersection between the demagnetization curve and a line from zero through $B_{rem}H_c$, as shown dotted in the figure.)

Design of Magnet Systems

Permanent magnets are mainly used to provide a magnetic field in various electronic devices. Thus in a moving-coil meter (see Fig. 4.43) the iron circuit consists of a permanent magnet supplying the field, a soft-iron core with associated pole-pieces, and an annular gap in which the coil moves. The design of such a system involves the determination of the size of magnet necessary to provide the required flux in the air gap.

This is assessed from the demagnetization curve. Suppose, for example, that it is required to produce a flux density of 0·8 T in a (total) air gap 2 mm long and 0·8 cm² cross-sectional area. The energy in the gap $= \frac{1}{2}B_gH_gA_gl_g$. This must equal the energy in the magnet $= \frac{1}{2}B_mH_mA_ml_m$. If $B_m = B_g$ and $A_m = A_g$, then

$$H_ml_m = H_gl_g = (B_g/\mu_0)l_g \quad (1)$$

From the curve of Fig. 4.17 it will be seen that to produce $B = 0{\cdot}8\,\mathrm{T}$, $H = 39{,}000$ At/m. Whence

$$l_m = \frac{0{\cdot}8 \times 2 \times 10^{-3} \times 10^7}{4\pi \times 39{,}000} = 3{\cdot}28\ \mathrm{cm}$$

Now, the magnet cross-section may not be the same as that of the air gap, in which case B_m will not equal B_g. In fact, $B_m = 0{\cdot}8$ is not the optimum for the material of Fig. 4.17, which occurs when $B_m = 1$. This higher value of B_m can be obtained by reducing A_m by 0·8, so that it becomes 0·64 cm². The length will then need to be increased, though not by 1/0·8 because of the more efficient condition. From the curve, if $B_m = 1{\cdot}0\,\mathrm{T}$, $H = 34{,}500$ At/m so that $H_ml_m = 34{,}500$ and

$$l_m = (0{\cdot}8 \times 2 \times 10^{-3} \times 10^7)/4\pi,$$
$$= 3{\cdot}7\ \mathrm{cm}$$

The volume of steel is $3{\cdot}7 \times 0{\cdot}64 = 2{\cdot}37$ cm³, as against $3{\cdot}28 \times 0{\cdot}8 = 2{\cdot}62$ cm³ for the original condition, so that there is a saving of approximately 9 per cent.

Actually some m.m.f. will be required to force the flux through the soft-iron pole pieces and the central cylinder. This can be allowed for by replacing l_g by $(l_g + l_i/\mu_r)$, where l_i is the total length of the soft-iron material (assumed to have the same cross section as the air gap) and μ_r is the relative permeability of the iron at the value of B_g (0·8 T). This correction is small. If $l_c = 50$ mm and $\mu_r = 1{,}000$, $l_i/\mu_r = 0{\cdot}05$, which is only 2·5 per cent of l_g in the example chosen. It is therefore customary to allow a small arbitrary increase of a few per cent in l_g to allow for the pole pieces.

Iron-cored Inductors

Let us now relate these data to the design of iron-cored inductors, often called chokes. For a given flux Φ the inductance, assuming no leakage, is $\Phi N/I$. We have seen that

$$\Phi = 1{\cdot}257\, INA\mu_r/l \,.\, 10^{-6}$$

so that

$$L = 1{\cdot}257 \,.\, N^2A\mu_r/l \,.\, 10^{-6} \text{ henrys}$$

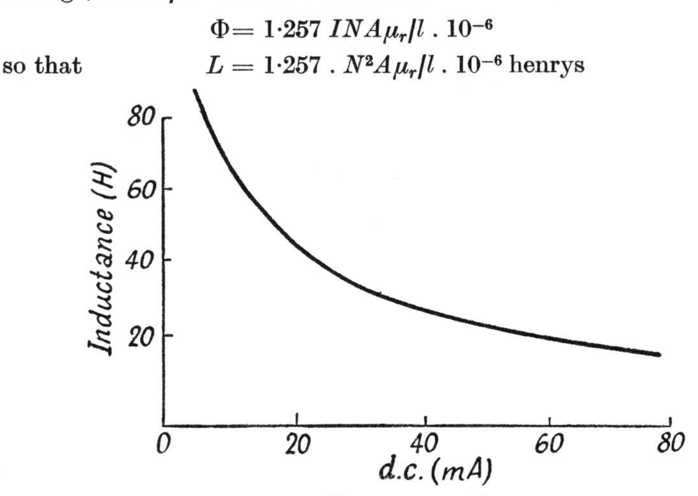

FIG. 4.18. ILLUSTRATING VARIATION OF INDUCTANCE WITH D.C. CURRENT

Thus the inductance is proportional to the square of the turns, the area of the core and inversely proportional to the length of the core. It is, however, directly dependent on μ_r, which we have seen to be a very variable quantity depending entirely on the degree of magnetization. Thus the inductance of a choke carrying an alternating current is varying from instant to instant and we can only assess some average value.

Moreover, a choke is often required to carry a steady d.c. (though this is not always the case). It thus has a steady magnetization with a ripple superposed. The permeability of the iron will vary with the amount of steady magnetization rather in the manner shown in Fig. 4.15 so that (except possibly for very low values of H) the inductance falls off steadily as the d.c. increases. This effect is known as *saturation* and is illustrated in Fig. 4.18. Sometimes this effect is of use (as in a "swinging" choke for use with a choke-input rectifier system) but as a rule it is preferable to avoid this dependence of inductance upon the current. This is the more so since the permeability is affected not only by the d.c. but also by the a.c. component, the inductance tending to increase as the ripple becomes larger.

Incremental Permeability

The effective permeability, in fact, is not the static value for the particular value of H. It is found that if an a.c. component of H is superposed on a steady value the iron goes through a small hysteresis loop as shown in Fig. 4.19, and the effective permeability

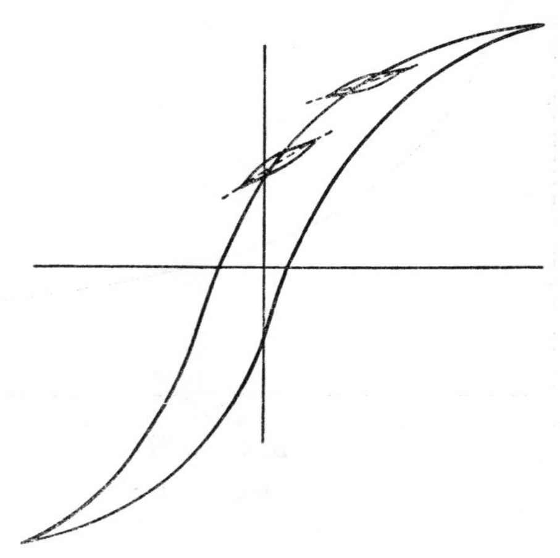

FIG. 4.19. ILLUSTRATING INCREMENTAL PERMEABILITY

of the iron is clearly dependent on the average slope of this loop (the dotted line in Fig. 4.19).

It will be seen that this *incremental permeability*, as it is called, depends upon both H_{dc} and H_{ac}. In general

(i) μ_i decreases as H_{dc} increases;

(ii) μ_i increases as H_{ac} increases.

The curves of Fig. 4.20 show values of μ_i for typical silicon-steel sheet. Similar figures are published by the makers for other materials including the special high-permeability alloys such as Radiometal and Mumetal.

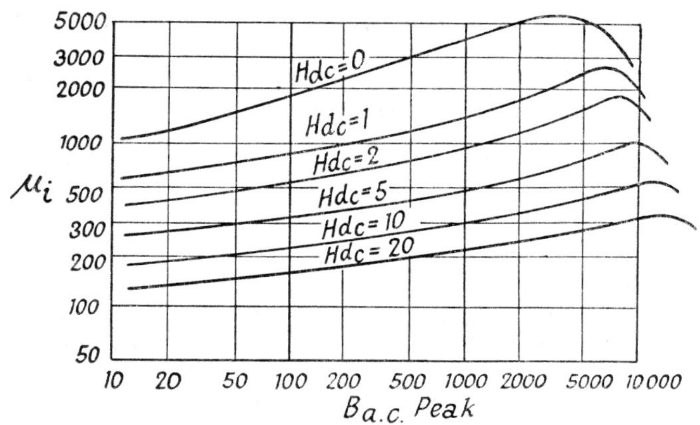

FIG. 4.20. TYPICAL VALUES OF INCREMENTAL PERMEABILITY

Use of Air Gap

The effects of varying permeability may be offset to a considerable extent by including in the iron circuit a small air gap. Even a small gap (of 0·25 to 1 mm in the average choke) has a reluctance many times the iron, so that the performance is determined mainly by the air gap, which is of constant permeability.

The flux is reduced because of the increased reluctance, but the permeability increases to an extent which may more than compensate for this, so that the inductance with an air gap is often greater, for a given d.c. through the winding, than with no gap.

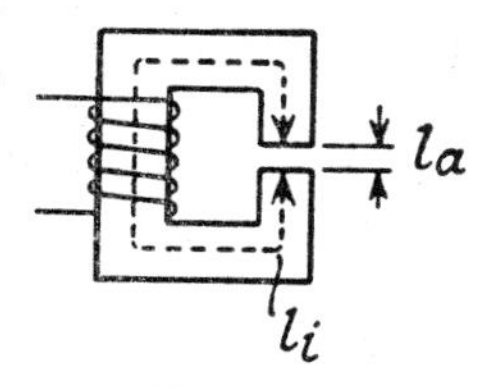

FIG. 4.21. GAPPED CORE

Fig. 4.21 shows a magnetic circuit having an iron path l_i and an air gap l_a. The reluctances of these paths are $l_i/A\mu_0\mu_i$ and $l_a/A\mu_0$ respectively. The m.m.f. $M = \Phi S$, so that

$$M_1 = \Phi l_i/A\mu_0\mu_i \quad \text{and} \quad M_2 = \Phi l_a/A\mu_0$$

Hence to obtain a given value of Φ the total ampere-turns required is $M_1 + M_2$.

The total m.m.f. $= M_1 + M_2$ may be written in the form

$$\frac{\phi}{A\mu_0\mu_i}(l_i + \mu_i l_a)$$

so that the effect of the air gap may be taken as increasing the length of the magnetic path by $\mu_i l_a$ (assuming A to be the same, which is usually valid).

It will be clear that if the number of turns and the current through the coil remain unchanged the flux will be reduced in the ratio $l_i/(l_i + \mu_i l_a)$. If we wish to maintain the flux at its original value the ampere-turns must be correspondingly increased. In the case of a choke we do not wish to do this—in fact, the object of introducing the gap is to reduce the flux—but with some air-gap circuits, such as the field of a loudspeaker "pot," we do wish to maintain the flux, and this method of calculation applies equally to either usage.

Optimum Air Gap

Now let the ratio $l_i/l_a = p$. Then the effective length of the magnetic circuit is $l_i(1 + \mu_i/p)$. We can thus write for the inductance of the choke

$$L = \frac{HAn\mu_i}{l_i(1 + \mu_i/p)} = \frac{1{\cdot}26An^2}{l_i} \cdot \frac{\mu_i p}{p + \mu_i}$$

The first term is constant, while the second depends on p. But as we reduce p (increasing air gap) μ_i increases and vice versa. In fact, over a wide range of practical values we are not greatly in error by assuming the product $\mu_i p$ to be constant.

On this assumption we may write $\mu_i = k/p$. The second term in the expression for inductance then becomes $1/(p + k/p)$. If we plot this expression in terms of p we find that it rises to a maximum and then begins to fall off again, and we find (either graphically or by calculus) that this maximum occurs when $p = \mu_i$.

This is another way of saying that the reluctances of the air and iron paths are equal, which the reader will recognize as a condition often found in electrical practice (e.g. maximum power when internal and external resistances are equal). The value of μ_i, of course, is the incremental permeability.

The best results therefore are obtained with a gap ratio approximately equal to μ_i. With a choke having $l_i = 20$ cm and $\mu_i = 400$ this would give a gap of

$$p = \mu_i = 400 = l_i/l_a = 20/l_a$$

whence $l_a = 0{\cdot}5$ mm. With such a gap the inductance, for a given H_{dc}, would not differ greatly from the value at the same H_{dc} without a gap. In most cases it will be found to exceed the ungapped figure and, of course, the inductance with the gap is much more constant despite variations both of the d.c. and the a.c. ripple current through the choke.

It is not practicable to express this improvement in simple terms because the various factors are so interdependent. The simple result just obtained for the optimum gap only arises from the assumption that $\mu_i p = k$, which is by no means rigidly true. The practical designer, however, is able to make his calculations with comparative ease by the use of tables or curves drawn up partly from more extended theory and partly from empirical data.

Pull of Electromagnet

A particular case of the gapped inductor is the relay or electromagnet. Here we are interested not in the inductance but in the mechanical force developed. This can be assessed as follows.

Consider two magnetic surfaces, of area A, separated by a distance x. If the flux density is B, the energy per unit volume is $\frac{1}{2}B^2/\mu_0$. If now the distance between the surfaces is increased by a small amount δx, the work done in separating them will be $F\delta x$. This must equal the increase in energy in the field. Hence

$$F\delta x = (\tfrac{1}{2}B^2/\mu_0) \times A\delta x$$

so that the magnetic pull will be

$$F = \tfrac{1}{2}B^2A/\mu_0 = \tfrac{1}{2}BHA \text{ newtons}$$

From the configuration of the magnet system and the relative lengths of the air and iron paths the flux density can be evaluated in terms of the m.m.f., and the magnetic pull is then immediately calculable.

Transformers

As explained in Section 2.4, a transformer consists essentially of an arrangement in which two coils are magnetically coupled to one another. If a varying current is passed through one coil, known as the primary, the changing flux linking with the second coil, called the secondary, induces a voltage therein.

In a perfect transformer all the magnetic flux produced by the primary should link with the secondary. This ideal state of affairs cannot be attained, but for power frequencies and audio frequencies it is possible to approach the ideal by arranging the coils around an iron core. Since the permeability of iron is many times greater than that of air the lines of magnetic force lie mainly within the iron, and if both coils are wound on the same core the greater part of the flux produced by the primary links with the secondary and vice versa.

Consider a simple transformer as in Fig. 4.22, consisting of a primary winding and a secondary winding both wound over the same iron core, with the primary connected to a source of alternating e.m.f. The alternating current flowing through the primary winding will produce an alternating magnetic flux. This flux linking with the turns of the secondary winding will induce e.m.f.s therein proportional to the rate of change of flux and the number of turns in the winding. If the flux is varying sinusoidally the secondary voltage will also be sinusoidal so that as long as we ensure that the

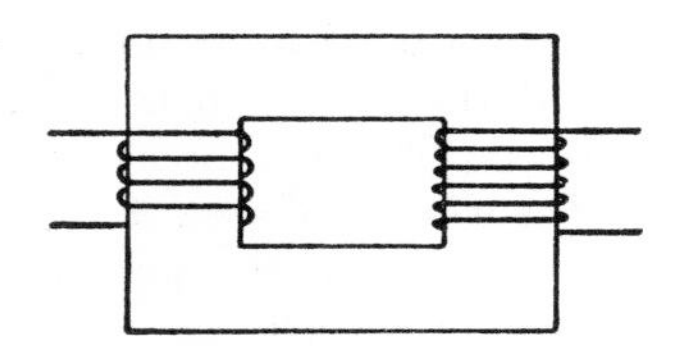

FIG. 4.22. DIAGRAM OF SIMPLE TRANSFORMER

flux wave is not distorted the transformer will deliver a sine wave output on its secondary.

It is convenient to work from this assumption of an alternating flux as a starting point. This flux will, of course, induce voltages in both primary and secondary windings. Therefore, in the first place if we are to maintain the flux we must apply across the primary a voltage equal and opposite to the back e.m.f. induced in the primary. As long as we continue to do this the flux wave will be maintained and this flux linking with the secondary will induce a corresponding voltage.

Let $\Phi = \Phi_p \sin \omega t$ be the flux, where Φ_p is the peak flux.

$$\text{Primary back e.m.f.} = e_1 = -N_1 d\Phi/dt = -N_1\Phi_p\omega \cos \omega t$$

$$\text{Secondary e.m.f.} = e_2 = -N_2\Phi_p\omega \cos \omega t$$

Thus the ratio of the e.m.f.s on the primary and secondary windings is equal to the turns ratio.

In practice we design the transformer so that e_1 is equal to the applied voltage v. Now e_1 depends on the turns, the frequency of supply $f(= \omega/2\pi)$ and the flux. If the first two factors are fixed the flux will adjust itself until $e_1 = v$ (numerically).

We know that $\Phi_p = B_pA$, where B_p is the maximum, or peak flux density; so we can write $e_1 = B_pAn_1\omega \cos \omega t$.

Hence the r.m.s. primary e.m.f. $E_1 = B_pAN_1\omega/\sqrt{2}$. Writing $\omega/\sqrt{2}$ as $2\pi f/\sqrt{2} = 4{\cdot}44f$, we have

$$\mathbf{E_1 = 4{\cdot}44B_pAn_1f}$$

This is the fundamental expression used in transformer design. We choose a suitable value of B_p, usually around 1 tesla, and then adjust the area of core A and the number of turns N_1 to make E_1 equal to the input voltage required (e.g. 240V). The secondary turns N_2 then $= E_2N_1/E_1$, subject to a correction for internal voltage drop as explained later.

Effect of Secondary Load

What happens if we connect a load across the secondary? We know, from Lenz's law, that the e.m.f.s of self and mutual induction are in such a direction as to oppose the change of flux which generates them. Hence if we connect a load across the secondary a current will flow $= E_2/Z_2$ which will be in such a direction as to *reduce* the flux in the core.

But once this happens, the primary e.m.f. E_1 is no longer equal to V and hence more current will flow in the primary until the flux is restored to its original value. The secondary ampere-turns producing the demagnetizing effect is I_2N_2, and this must be counteracted by an equal and opposite (additional) primary ampere-turns I_1N_1. Hence $I_1N_1 = - I_2N_2$, and the reflected load current

$$I_1 = - (N_2/N_1)I_2$$

Magnetizing Current

There must be some primary current, even when there is no load on the secondary, in order to maintain the flux. This initial current is termed the no-load or *magnetizing* current. It is nearly 90° out of phase with the primary voltage because the primary is almost wholly inductive. There is a small resistive component due to the winding resistance and the losses in the iron, as we shall see later, but in general this is small.

The total primary current is thus the vector sum of the magnetizing current and the reflected load current which is $-(N_2/N_1)I_2$ and is thus 180° out of phase with the secondary current. Fig. 4.23 shows the simplified vector diagrams neglecting losses, for resistive, inductive and capacitive secondary loads. The primary voltages and load currents are in opposition to the secondary quantities in each case.

Under no-load conditions there is no secondary current and the only primary current is the magnetizing current I_m. This produces a flux Φ in phase with I_m, and since this is common to both windings, e.m.f.s E_1 and E_2 are induced lagging 90° behind I_m. The primary voltage must be equal and opposite to E_1 as shown.

With a resistive load, I_2 is in phase with E_2, and I_1' is 180° out of phase with E_2 and hence is in phase with V. The total primary current I_1 is then the vector sum of I_m and I_1' as shown, lagging slightly behind V.

With an inductive load, I_1' lags behind V by the same amount as I_2 lags behind E_2, while I_1 lags slightly more because of I_m. With a capacitive load the reflected primary current leads on V, but at a slightly lesser angle than I_2 because I_m is lagging.

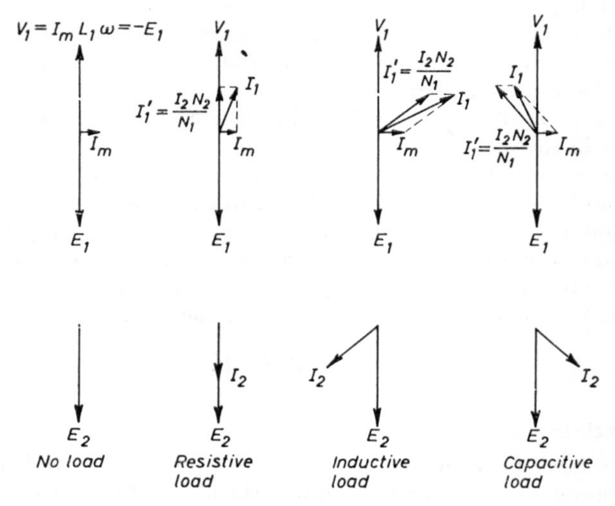

FIG. 4.23. TYPICAL TRANSFORMER VECTOR DIAGRAMS

Equivalent Circuit of Transformer

We see, therefore, that the connection of a load across the secondary of a transformer results in a secondary current which is reflected into the primary, the ratio of the currents being in the inverse ratio of the number of turns. This is only part of the story, however, because as soon as we begin to draw a current from the transformer the internal impedance of the device begins to take effect and we find that the output voltage is no longer in simple turns relation to the input voltage. Fig. 4.24 illustrates the equivalent circuit of a transformer. We commence with the voltage V_1, which is the input voltage. The primary winding, however, has a certain resistance so that the voltage actually applied across the transformer is $V_1 - I_1R_1$.

The secondary e.m.f. will be $E_2 = (V_1 - I_1R_1)N_2/N_1$ and the secondary output voltage will be less than this amount because of the voltage drop on the secondary resistance, so that

$$V_2 = E_2 - I_2R_2$$

It will be seen, therefore, that the turns ratio must be calculated on the internal e.m.f.s and is actually appreciably greater than V_2/V_1 to allow for the internal voltage drops.

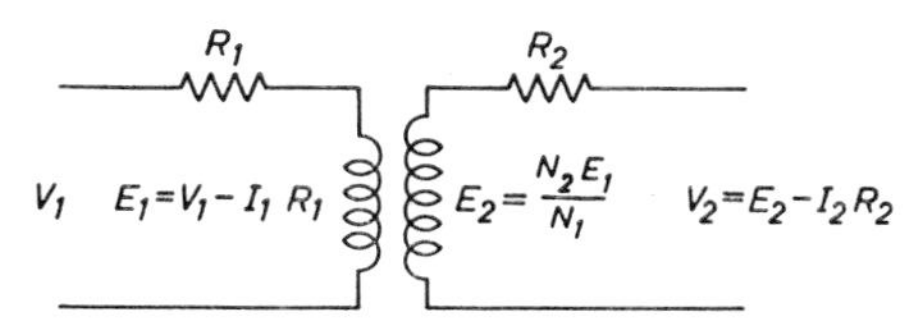

FIG. 4.24. EQUIVALENT CIRCUIT OF TRANSFORMER

Regulation

Because of these internal voltage drops, which are clearly dependent upon the current flowing, the voltage output from the secondary of the transformer is not constant but falls steadily as the load increases. The fall in voltage from no load to full load expressed as a fraction of the full-load voltage, is known as the *regulation* of the transformer. This figure varies with the size and quality of the transformer. A small radio set transformer may have a regulation of 15 to 20 per cent. As the size of transformer increases and the quality becomes better the regulation improves to something like 5 per cent, while with a large transformer the figure may be less than 1%.

It is obviously desirable to keep the regulation good because the resistance of the windings not only causes a voltage drop but actually occasions heat losses which are wasteful. If the transformer is a small one it may be that these losses can be tolerated, but as the transformer increases in size it is obviously necessary to restrict the losses as otherwise the heat generated becomes considerable.

Leakage Inductance

The resistance in the winding is not the only source of voltage drop. We have seen that the object of winding the coils on an iron core is to arrange that all the flux produced by the primary shall link with the secondary and vice versa. This object is not completely maintained and a certain amount of the flux on the primary does not link with the secondary, and vice versa.

This is known as the *leakage flux* and the effect will be to introduce additional e.m.f.s into the circuit. In the primary, for example, in addition to the main e.m.f. generated by the main flux cutting the turns there will be a subsidiary e.m.f. due to the leakage flux. The same applies to the secondary, and the simplest way of representing this effect is to introduce into the equivalent circuit a small inductance as shown in Fig. 4.25. Strictly speaking the leakage inductance should be introduced in series with both primary and secondary

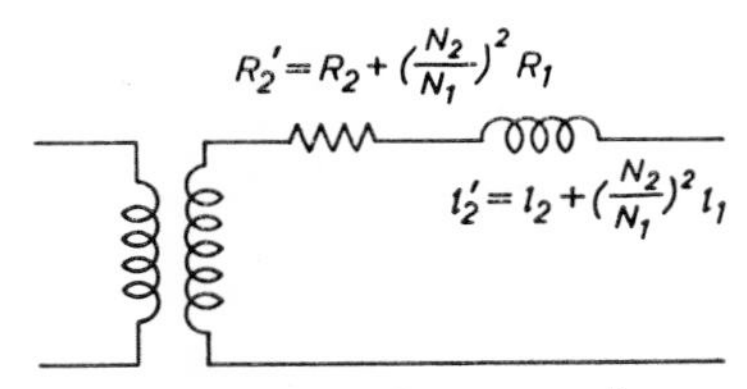

FIG. 4.25. EFFECT OF LEAKAGE INDUCTANCE

but it is usually adequate to represent the leakage flux by a leakage inductance in series with one winding only.

This has been done in Fig. 4.25 where both the leakage and the winding resistance have been represented as being in the secondary circuit. By the usual coupled-circuit laws

$$R_2' = R_2 + R_1 N_2^2/N_1^2$$

$$l_2' = l_2 + l_1 N_2^2/N_1^2$$

With this arrangement

$$V_1 = E_1$$

but

$$V_2 = E_2 - I_2(R_2' + j\omega l_2')^*$$

Transformer Construction

The amount of the leakage flux depends on the construction of the transformer. Fig. 4.26 (*a*) shows a type of transformer which is bad. Here the primary winding is placed on one limb of the core and the secondary winding is placed on the other. It is clear that there must be appreciable flux linking the primary which does not link the secondary.

We can improve matters by splitting the winding and putting

* This assumes a resistive load, which may not be valid. If I_2 makes an angle ϕ with V_2, the second term, representing the voltage drop, should be

$$I_2 R_2 \cos \phi + I l_2' \omega \sin \phi$$

Unless ϕ is large the difference is not serious.

half the primary on each limb of the core and winding half the secondary over each section of the primary. The sections may be in series or parallel as required. Thus, for a high-voltage transformer we might connect the primaries in parallel and the secondaries in series. Whichever connection is used we must ensure that the directions are correct so that the current at any instant is such as to produce a magnetic field which is in the same direction *around the core*.

This type of transformer is used for medium and high powers and is known as a *core-type* transformer. For small powers the arrangement of Figs. 4.26 (*c*) and (*d*) is more usual. Here a three-limb core

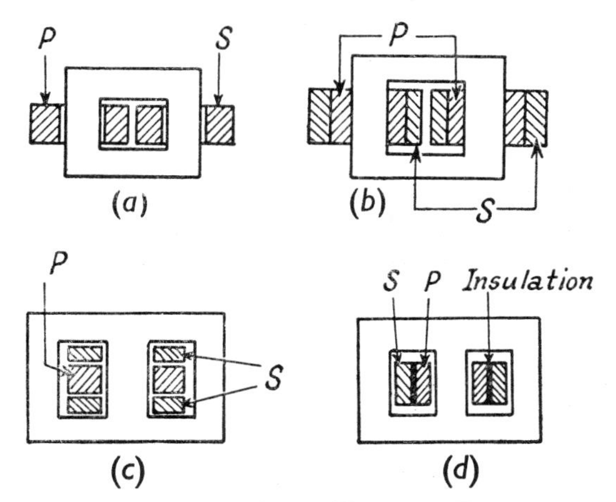

FIG. 4.26. CORE AND SHELL TYPES OF TRANSFORMER

is adopted, and since the flux has two alternative paths it is clear that the outer limbs, as well as the top and bottom limbs, need only be half the thickness of the centre limb.

This type of construction, known as the *shell type*, has the advantage that only one set of windings is needed, the secondary being wound over the primary as shown in Fig. 4.26 (*d*). Both these arrangements provide greatly reduced leakage inductance, but if still lower leakage is required still further precautions are required and it is necessary to interleave the windings in the manner shown in Fig. 4.26 (*c*). This interleaving or sandwiching process clearly limits the stray flux and thus reduces the leakage inductance.

Leakage inductance is difficult to calculate and in most design work practical experience is taken as a guide.

Transformer Losses

A practical transformer does not transfer all the primary energy to the secondary, some energy being lost in the process. The losses are of two main forms.

Copper Loss. The resistance of the windings causes heating, the loss due to this being $I_1^2R_1 + I_2^2R_2$.

Iron Loss. In addition, we have the loss in the iron core due to the varying magnetization. The effect of hysteresis was discussed on page 181, where it was shown that it produces a loss of the form $W_h = kfB_p^{1\cdot 6}$. There is a further loss due to the circulating eddy currents set up in the iron.

This eddy-current loss would be quite serious if a solid core were used. In a practical transformer, therefore, the core is built up of

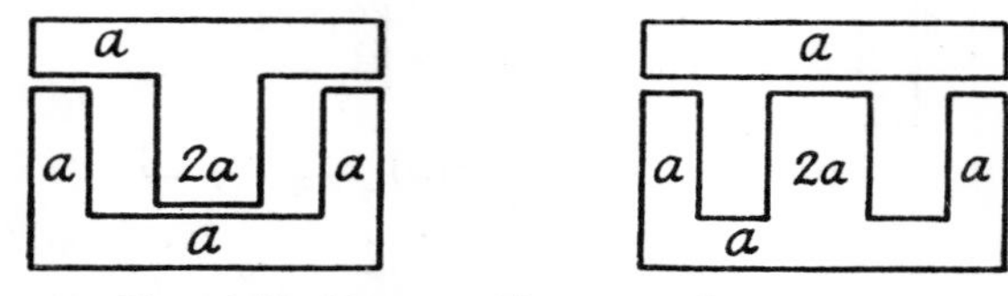

FIG. 4.27. TYPICAL FORMS OF LAMINATION

thin sheets or strips, insulated from each other by a thin facing of paper or paint on one side. For small transformers the *laminations* are stamped out in the form of T and U pieces or sometimes E and I pieces as shown in Fig. 4.27. They are assembled with the joints alternately at each end so that the effect of a substantially solid core is obtained, but built up of insulated thin laminae so that any eddy currents induced can only flow in restricted paths. In addition, the sheets are made of silicon steel, which not only reduces hysteresis but has a high electrical resistance so that the eddy currents are still further reduced.

Eddy-current loss is proportional to $B_p^2f^2$ and thus becomes increasingly important as the frequency is raised. It is also dependent on the thickness of the sheet. The usual thickness for 50 Hz is 0·014 in., though sometimes 0·020 in. is used. For higher frequencies or special circumstances requiring an unusually high value of B_p, the stampings may be 0·007 or even 0·005 in. thick.

The cost is greatly increased, for more stampings are required for a given area of core, while the material itself is more expensive and the stamping of the laminations is more costly owing to the higher precision required in the tools. Otherwise burrs are formed at the edges and these pierce the insulation when the laminations are stacked together, so defeating the object of the use of thin sheets. Even with normal sheets the presence of burrs is to be avoided.

The steel manufacturers market material under various trade names in which the hysteresis and eddy-current losses are of the same order at normal frequencies. The total iron loss for typical material is shown in Fig. 4.28.

There is also a small eddy-current loss due to circulating currents in the clamps used to hold the whole structure together but this is

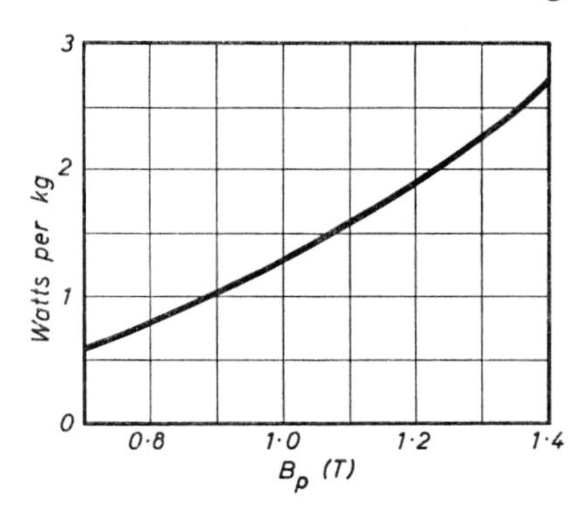

FIG. 4.28. TOTAL IRON LOSS FOR TYPICAL TRANSFORMER SHEET

usually negligible in small transformers. If there is reason to believe that it is of importance, insulated joints must be provided where the bolts come through the clamps and similar precautions taken wherever practicable to reduce the number of complete conducting paths which might permit eddy currents.

Analysis of Losses

The iron loss is present all the time, being a direct result of the magnetization of the iron which does not change with load. It may thus be measured by finding the power taken by the transformer on no load. (This is not VI_0, because I_0, the magnetizing current, is nearly 90° out of phase.) If it is desired to separate hysteresis and eddy-current losses this test must be repeated at two different frequencies. The measured loss is then $p = k_1 f + k_2 f^2$. If this is found for two values of f two equations are obtained which can be solved for k_1 and k_2. It is important that B_p shall be the same throughout so that the applied voltage must be proportional to frequency. Hence if we use $f_1 = 25$ and $f_2 = 50$, V_1 must be twice as great in the 50 Hz test.

The copper loss varies with I^2 and is taken at full load. One method is to short-circuit the secondary and to apply a reduced voltage to the primary of such a value that the secondary current is the required value. The power taken by the transformer in this condition is the copper loss because V_1 will be so small that B_p, and hence the iron loss, will be negligible.

These tests require a wattmeter, which may not be available. The losses, however, may be estimated with reasonable accuracy by calculation, basing the iron loss on data similar to that of Fig. 4.28 and calculating the copper loss from the resistance of the windings, either estimated or measured.

The *efficiency* of a transformer is the ratio of output watts/input watts. For small transformers it is around 80 to 90 per cent, rising to 95 per cent for medium transformers and 99 per cent or even higher for very large transformers. In many cases, in radio practice, the efficiency as such is not as important as ensuring that the losses shall not cause undue temperature rise. As a good practical rule it may be taken that a transformer can safely dissipate 0·05 watt per square cm of its surface. Thus a transformer having dimensions of $15 \times 12 \times 10$ cm would have a surface area of 900 cm^2, and thus could safely dissipate about 45 watts. Hence it could be used to handle a power of about 500 watts with an efficiency of 95 per cent.

The most efficient design is one in which the iron loss and copper loss are equal, but this may not be the most economical design. Iron is cheaper than copper and it may pay, say, to increase the area of the core 50 per cent, with corresponding increase in iron loss. This would permit the number of turns to be reduced with saving of copper. In such an instance the copper loss might be only half the iron loss.

Iron Distortion

The choice of the maximum flux density is not entirely determined by the losses, but is influenced by the permissible iron distortion. Reference to the magnetization curves of Fig. 4.16 shows that the relationship between B and H is not linear, though it is approximately so over the initial portion of the curves. We have assumed that the flux is sinusoidal but in fact a sinusoidal magnetizing current will produce a flat-topped flux wave, and if B_p is too large, giving serious distortion of the flux wave, the secondary e.m.f. is far from sinusoidal, as shown in Fig. 4.29.

It is customary to operate under conditions such that the maximum value of B is on the "knee" of the B–H curve. Too small a value of B_p does not utilize the iron to maximum advantage, requiring more turns on the winding. Too large a value of B_p introduces increasing distortion. The choice is thus a compromise, slightly higher values of B_p being permissible in a power transformer than in an audio-frequency transformer where negligible distortion is required.

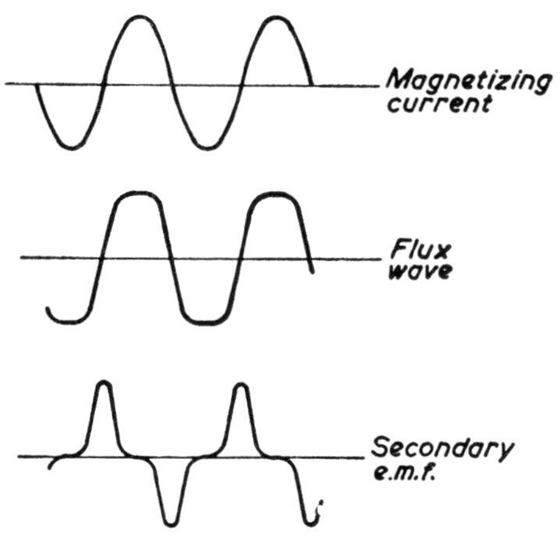

FIG. 4.29. ILLUSTRATING IRON DISTORTION

Effect of Frequency

The standard supply frequency in Britain is 50 Hz, while America has adopted 60 Hz. Power transformers are thus normally required to operate at frequencies of this order. From the fundamental transformer equation it will be seen that for a given maximum flux density in the iron the product ANf is constant for any particular value of voltage. Hence if the frequency is, say, 25 Hz the core area or the number of turns will need to be twice the value for 50 Hz (or both increased so that the product AN is doubled). Conversely, if the frequency is increased the volume of iron required is reduced. This is useful where weight is of importance, as in airborne equipment, which is usually operated at a frequency of 400 Hz, while frequencies as high as 4,000 Hz are sometimes adopted for special applications.

The iron loss, however, increases with frequency so that it is usually necessary to reduce the flux density somewhat. It is the eddy-current loss which preponderates, so that some improvement is obtained by reducing the thickness of the laminations; material

0·007 in. or 0·005 in. thick is frequently used in place of the 0·014 in. or 0·020 in. stampings customarily used at 50 Hz.

As already mentioned, the effect of leakage inductance increases directly as the frequency, but this is to some extent offset by the smaller physical size which of itself reduces the leakage inductance. It may, however, be necessary to sandwich the windings if good regulation is essential.

Audio-frequency Transformers

Transformers are used in audio-frequency amplifiers for inter-stage couplings and for matching the loudspeaker or other load to the output stage, as discussed in Chapter 10. Such transformers have to operate over a range of frequency which may run from 20 to

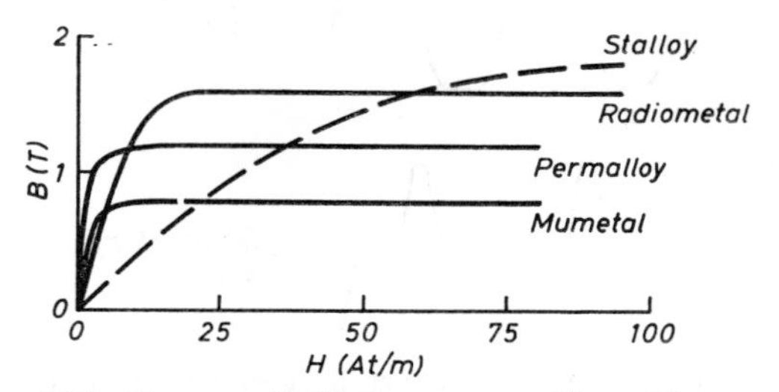

FIG. 4.30. TYPICAL *B–H* CURVES FOR CORE MATERIALS

20,000 Hz, though for many requirements a range of 50 to 10,000 Hz is sufficient.

The same basic considerations apply; the iron circuit is designed to handle the maximum voltage swing required at the lowest frequency, the value of B_p being chosen with regard to the permissible iron distortion.

The design of interstage transformers is affected by the fact that the primary will normally carry a steady direct current with the signal current superposed. This applies to both valve and transistor circuitry, as is discussed in later chapters, and its effect is to limit the available magnetization swing, the primary being designed to produce a value of B midway between zero and the knee of the B–H curve.

This necessitates a larger core size than would otherwise be required, and it is often convenient to remove the d.c. component by a parallel-feed arrangement, as in Fig. 10.9. (This limitation does not apply with a push-pull output stage, in which the currents in the two halves of the primary are in opposition.)

The physical size, however, can be considerably reduced by the use of special core materials. Certain nickel alloys exhibit permeabilities, at low values of H, 10 to 20 times greater than ordinary transformer steel. This permits the required performance to be achieved with a much smaller core size, so that although the material is more expensive its use is still economical. Moreover, the number of turns is reduced, which minimizes the leakage inductance.

Characteristics of some of these materials are shown in Fig. 4.30 and Table 4.5. Their performance is critically dependent on the heat treatment during manufacture and they should not be subjected to mechanical stress during assembly.

TABLE 4.5. HIGH-PERMEABILITY MATERIALS

Material	*Composition* (%)	*Permeability* (max)	*Coercive force* (At/m)	*Saturation induction* (T)
Stalloy	96 Fe:4 Si	7,000	50	1·95
Mumetal	76 Ni:17 Fe 5 Cu:2 Cr	100,000	2·4	0·8
Radiometal	50 Fe:45 Ni: 5 Cu	25,000	24	1·6
Permalloy A	78·5 Ni: 21·5 Fe	90,000	4	1·1
Perminvar	45 Ni:25 Co: 30 Fe	150,000	8	1·5

Radio-frequency Transformers

At frequencies above about 50 kHz the iron losses with conventional materials become prohibitive. Consequently for many years coils and transformers for radio frequencies were air cored, and had to be enclosed in screening cans to limit the radiation as explained on page 175.

Attempts were made to employ closed-core techniques by using iron-dust cores constructed of finely-divided iron particles bonded together under pressure with a suitable binding medium, but the permeability was low, of the order of 10–20 only.

Ferrites

A considerable improvement resulted from the development of materials known as *ferrites*. These are chemical compounds which exhibit appreciable permeability and at the same time have a high resistance so that the eddy-current losses are reduced.

In simple terms these ferrites are produced by replacing one of the iron atoms in ferric oxide, Fe_3O_4 (which is a magnetic iron ore, usually known as magnetite or lodestone) by an atom of a divalent metal such as copper, manganese, nickel or zinc, thus producing a compound of the form MFe_2O_4. It is found that such a material is still strongly magnetic but that its resistance is many thousands of times higher than pure iron. By sintering these materials at a high temperature (1,000–1,400°C) and allowing them to cool, they crystallize in a cubic system which is not only mechanically strong but magnetically homogeneous. With these materials, cores can be produced in a variety of shapes from simple rods to completely closed structures, and these cores are used for a variety of purposes either in place of conventional magnetic materials or in applications utilizing special unique properties.

Ferrite materials may be magnetically "soft" (i.e. having low remanence), or "hard" as in a permanent magnet. There are two basic types of magnetically-soft ferrites in common use today. These are the manganese-zinc ferrites and the nickel-zinc ferrites, and the properties of these two types depend upon the proportions of the principal oxides used in the composition and the manufacturing processes employed. The broad division between the two types is that the manganese-zinc ferrites have high permeability (600–1,000) and low losses below about 0·5 MHz, whereas the nickel-zinc ferrites have lower permeability but much higher resistivity and can be used up to 100 MHz.

Magnetically soft ferrites are most often found in the form of cylindrical "pot" cores for high-quality inductors, combining high Q-factor, high stability and the facility of simple inductance adjustment, but they are also used in large quantities for television line-scan transformers, picture-tube deflection yokes and rod aerials for domestic radio receivers. Fig. 4.8, opposite p. 163, illustrates some typical ferrite constructions. On the left are various forms of transformer core, while some ring and rod forms are shown on the right. In the centre is a selection of storage cores. Fig. 4.31 shows the construction of a typical r.f. transformer in more detail.

Storage Ferrites

Certain ferrites, usually magnesium-manganese ferrite and copper-manganese ferrite, may be made to have substantially rectangular hysteresis loops. When a magnetizing pulse $+H$, is applied to an inductor wound on such a core, the resulting flux is $+B$ and the core will remain in this state indefinitely. If a magnetizing pulse

$-H$ is now applied the flux will change to $-B$ and again remain in this condition indefinitely. Thus such a device has two stable states and can be used to store information in a binary code. In practice, the hysteresis loop is not truly rectangular but the ratio between the maximum flux B_{max} and the residual flux B_r is of the order of 0·95, so the ratio of flux change is adequate to give a clear

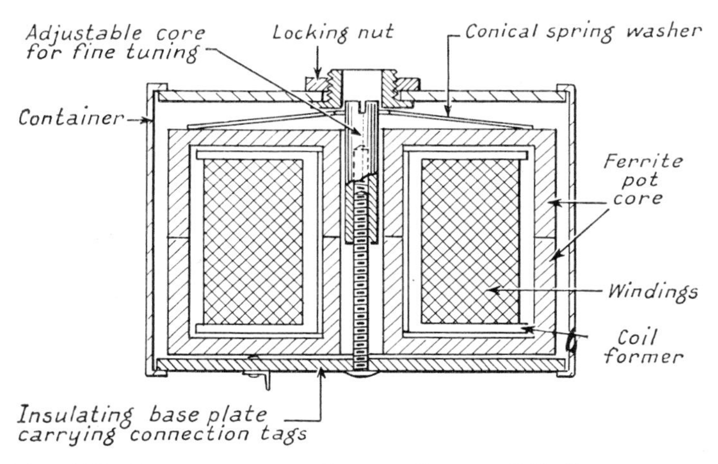

FIG. 4.31. CONSTRUCTION OF TYPICAL FERRITE R.F. TRANSFORMER
The connections are brought out to the tags in the base plate through slots in the sides of the ferrite core pieces (not shown).

indication of the state of the core, and small toroidal cores of these ferrites form the elements of many computing circuits.

Magnetically hard ferrites are also made usually consisting of barium ferrite $BaFe_{12}O_{19}$, which differs from the magnetically soft ferrites in having a hexagonal crystal structure. This ferrite has a much higher coercive force than metal magnets and can thus be used to make shorter magnets which have a greatly improved resistance to demagnetizing fields. Barium ferrite magnets are used in synchronous motors, as loudspeaker and headphone magnets and in many other applications where magnets are required to have high coercive force and high resistivity combined with small size and light weight.

Three-phase Transformers

Power transformers are often required to work on three-phase systems such as are described on page 209. Such transformers may be made up as three separate single-phase transformers connected in star or delta as required. Often, however, the three phases are accommodated on a special core having three limbs of equal cross-section.

Such a core is illustrated in Fig. 4.32. Let us assume that the red phase is at its maximum voltage. This will produce full flux in the middle limb and this flux will complete its circuit by passing half through each of the other two limbs. The back e.m.f. induced in

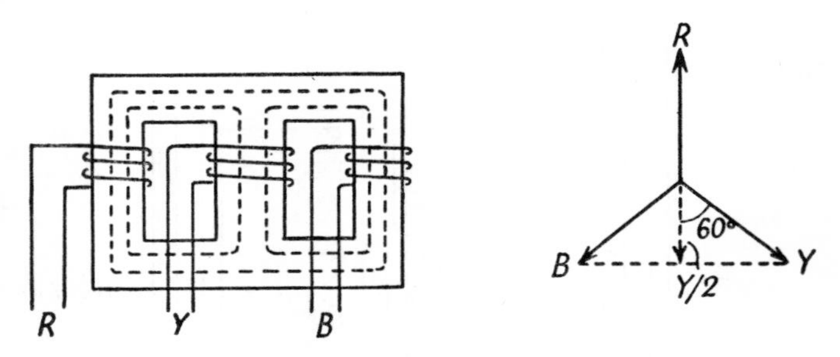

FIG. 4.32. THREE-PHASE CORE

these limbs, therefore, is half the value of the red phase voltage and in the opposite direction. It will be seen from the vector diagram, however, that the component of the yellow phase voltage in opposition to the red phase voltage is $Y/2$, and similarly for the blue phase, so that the flux in the second and third limbs is exactly what is required.

The area of each limb is therefore designed in accordance with the same principles as for a single-phase transformer and the windings are calculated accordingly.

4.5. Generators, Motors and Meters

By far the greatest source of power for communications equipment is the electricity supply mains. Here electricity is generated by dynamos or alternators and is then distributed throughout the country by transmission lines and ultimately brought into the user's premises in a convenient form. Alternatively, for mobile equipment, the generator may be located at the site or in the equipment itself, being driven by a suitable prime mover such as a petrol engine.

The simplest form of generator is a coil rotating in a magnetic field as shown in Fig. 4.33. The e.m.f. induced is

$$BlNv$$

where B = field strength,
l = length of the coil,
N = number of turns in the coil, and
v = velocity at which the conductor moves at right angles to the flux.

If the coil is rotating, $v = V \sin \omega t$, where ω is the angular velocity of the rotation. Hence an arrangement such as this will

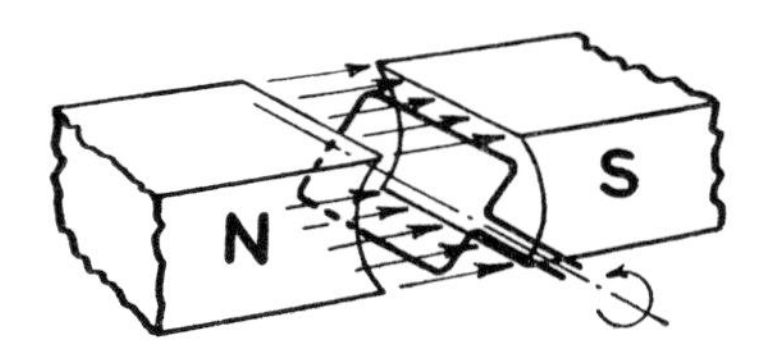

FIG. 4.33. DIAGRAMMATIC VIEW OF ALTERNATOR

produce a sinusoidal alternating e.m.f. proportional to the field strength, the length of the coil and the speed of rotation.

The speed is determined by the frequency required, and is therefore fixed, so that to increase the e.m.f. we must increase B or l or both. Increased field strength is obtained by filling the space between the pole-pieces with iron, leaving only a small clearance to permit the central *rotor* to revolve freely. The coil is then housed in a slot in this rotor, as shown in Fig. 4.34.

The simple arrangement of Fig. 4.33 does not make the best use of the available space. If a four-pole stator is used, as shown in Fig. 4.34, the rotor passes through a complete flux reversal in a quarter of a revolution only. It is then possible to include two coils, each occupying only half the rotor periphery. Both coils will have similar e.m.f.s induced in them and they may therefore be connected together, either in series or parallel as required. The frequency generated is then pn, where p is the number of pairs of poles and n is the rotor speed in r.p.s. Still better utilization of the rotor or *armature* is obtained by distributing the windings evenly round the the whole periphery in a series of coils housed in slots and suitably connected to form a continuous coil.

Small alternators are made up in this form. The poles on the stator may be permanent magnets but are more usually electromagnets, energized by *field windings* carrying d.c., which maintain the flux.

The connections to the rotor are brought out to *slip rings* against which brushes, usually of carbon, are held by springs. The d.c. for the field windings has to be supplied either by rectifying part of

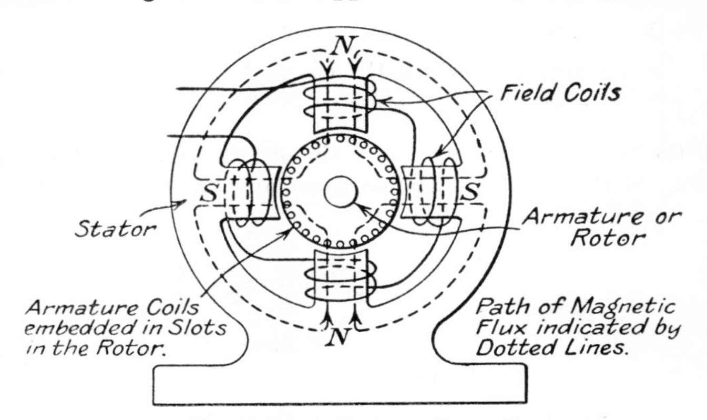

Fig. 4.34. Diagram of Salient-Pole Alternator

the a.c. output from the machine or from a d.c. generator mounted on the same shaft.

In larger machines it is more convenient to reverse the arrangement and use a fixed armature with a rotating field. This is done because the field current is only relatively small, and at a low voltage, so that the slip-rings and brush gear are simpler. The rotor then carries windings which are energized with d.c. in such a way as to provide alternate N and S poles at the periphery, and the rotation of this field induces e.m.f.s in the armature windings, which are housed in slots in the stator.

Regulation

When an alternator is supplying current the terminal voltage will be less than the generated e.m.f. because of the voltage drop on the internal impedance. As with a transformer, the ratio of the voltage drop at full load to the actual (full load) voltage is called the regulation and this may vary from 10 to 20 per cent for small machines down to a few per cent only with a large machine.

In some cases the field current is arranged to increase slightly as the main alternator load increases, in which case the generated e.m.f. will rise and substantially perfect regulation can be obtained.

Alternator Losses

The flux in an alternator is continually rotating and hence at any point the iron circuit will be subject to a varying flux. There will thus be iron losses set up as with a transformer, so the armature is built up of insulated stampings to reduce eddy-current losses, while the material itself is silicon steel to minimize hysteresis loss. The field system is similarly subjected to varying flux though only partially and is thus also laminated, though the thickness of the stampings can be somewhat greater.

Copper losses appear in the windings. The main loss is in the armature since this carries the full load current, but there is also some loss in the field coils which generates additional heat.

As with a transformer the optimum arrangement is usually that for which the iron and copper losses are equal.

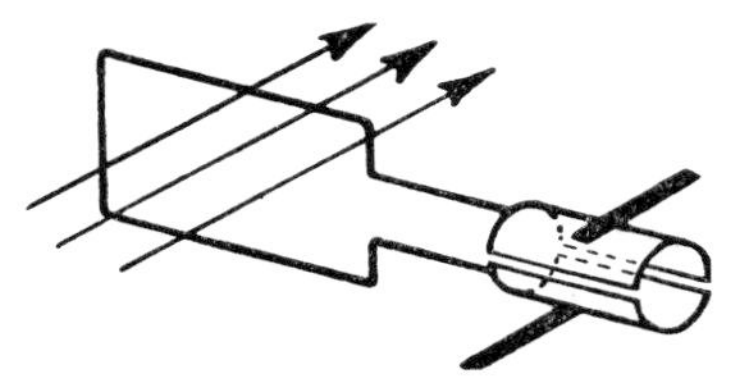

FIG. 4.35. SIMPLE COMMUTATOR

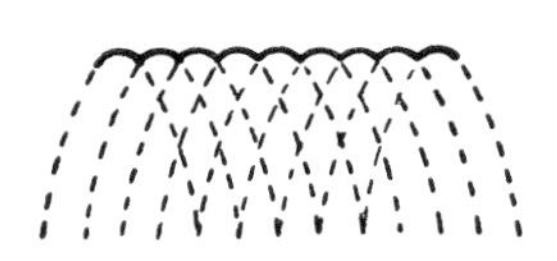

FIG. 4.36. OUTPUT WITH MULTI-SECTION COMMUTATOR

D.C. Generators

If the output is required to be unidirectional instead of alternating it is necessary to include some form of rotary switch. If the coil in a simple alternator is connected to a split slip-ring as shown in Fig. 4.35 the connections to the coil will be reversed every half cycle so that the output will be in the form of a series of half sine waves, all in the same direction. The split slip-ring is called a *commutator*.

A d.c. generator is arranged in this way, but to obtain a more regular output, a series of coils is arranged round the periphery of the rotor, each connected in turn to the output, which thus takes the form shown in Fig. 4.36. The commutator will not be a simple two-section affair but will consist of a series of insulated segments which may number several hundred in a large machine.

In practice the coils are not separate but are joined together to form a continuous winding. There are various ways of doing this which need not be discussed here, but we can say that the e.m.f. generated is $knz\Phi$, where n is the speed, z is the number of armature conductors, Φ is the flux, and k is a factor determined by the number of poles on the stator and the particular form of armature winding adopted.

The field coils in a d.c. generator can be energized by the machine itself. There are two possible arrangements. In the first illustrated in Fig. 4.37 (*a*), the field is connected across the output. When the machine supplies current its voltage will fall because of the internal voltage drop. This in turn will reduce the field current and the voltage variation will be accentuated.

The field coils, however, can be designed to be energized by the

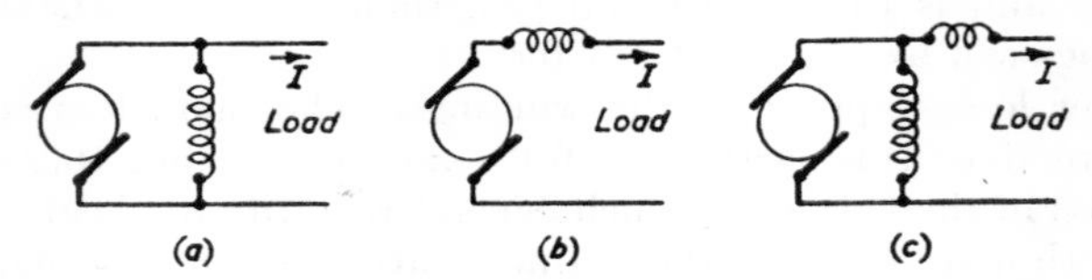

Fig. 4.37. Field Coil Connections for D.C. Generators

load current, being connected in series as in Fig. 4.37 (*b*), in which case the flux will increase as the load current increases, causing the voltage on load to rise. Fig. 4.37 (*c*) shows a compound winding in which there are two sets of field windings, one in shunt and the other in series. By suitably proportioning these two, the generator may be made to have any desired characteristic.

As with an alternator, the iron circuit of a d.c. generator is laminated to minimize losses due to the varying flux.

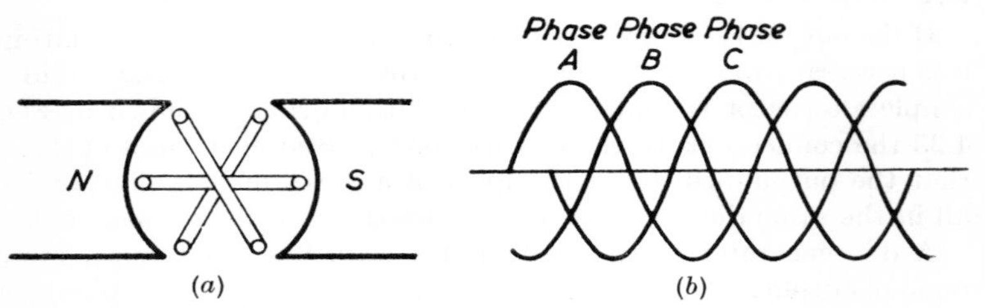

Fig. 4.38. Diagram of Simple Three-phase Alternator

Three-phase Generators

Fig. 4.38 (*a*) shows a modification of the simple alternator of Fig. 4.33 in which, instead of a single rotating coil, there are three separate coils symmetrically spaced. Each of these will have induced in it a sinusoidal e.m.f., but because of their angular displacement the three e.m.f.s will be spaced electrically by 120°, as shown in Fig. 4.38 (*b*).

Such an arrangement is called a three-phase alternator. Its practical form would, of course, be modified on the lines already discussed for single-phase alternators, but the provision of three separate phases in this manner has a number of advantages. It permits better utilization of the material not only in the generator itself but also in the transformers and cables of the distribution system, while it simplifies the construction of electric motors and other devices, as will be seen.

Modern electricity supply systems are all three-phase. The current is generated at a voltage of 33 or 66 kV, stepped-up in transformers to 132 or 275 kV and distributed on the overhead "grid" lines to suitable points where it is stepped down to a convenient voltage for industrial or domestic use. The high voltage is used for the transmission because the current is correspondingly reduced and hence the losses in the transmission line are minimized.

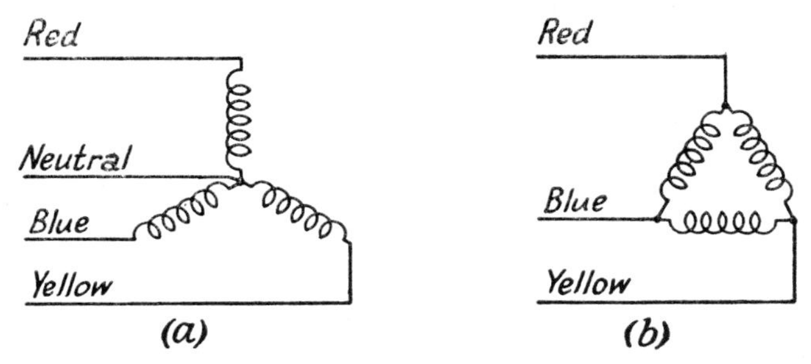

FIG. 4.39. STAR AND DELTA CONNECTIONS

Three-phase Relationships

The three phases, customarily referred to as the red, blue, and yellow phase respectively, may be connected in two ways, both of which are used according to convenience. In the *star connection* shown in Fig. 4.39 (*a*) one end of each phase is connected to a common *neutral* point, usually connected to earth. The voltage between the phases is then the vector *difference* between the individual phase voltages, which, as is shown in Fig. 4.40, is $\sqrt{3}$ times the phase voltage. To subtract OB from OA, add a vector $AB' = -OB$. This gives the resultant OB' at an angle of 30° to OA; whence $OB' = 2\ OA \cos 30° = (\sqrt{3})OA$. The line current, however, is clearly the same as the phase current.

The normal domestic supply is single phase and is taken from one of the phases and neutral, the voltage being transformed down to a value which is considered safe. The standard in Great Britain is 240 volts which means a voltage between phases of $240\sqrt{3} = 415$ volts. Domestic apparatus is thus designed to operate from 240 volts single phase, while industrial motors and similar devices run from a 415-volt three-phase supply.

In America and certain continental countries the domestic voltage is only 115, with a phase voltage of 200.

The alternative *delta connection* is illustrated in Fig. 4.39 (*b*). Here one end of each phase is joined to the beginning of the next, no part of the system being earthed. This connection is mainly used for motors and industrial devices. The line and phase voltages are the same, but the line current is the sum of the currents taken by two adjoining phases and is thus $\sqrt{3}$ times the phase current.

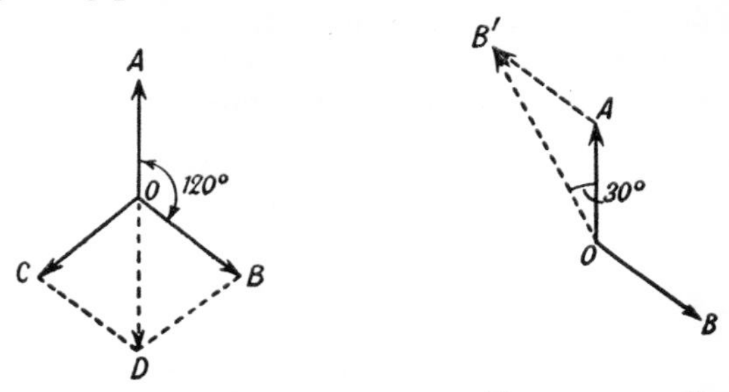

FIG. 4.40. VECTOR RELATIONSHIPS IN THREE-PHASE SYSTEM

Transformers may be designed for either form of connection and may have a delta-connected primary with star-connected secondary, or vice versa, if required.

When a three-phase system is supplying a normal three-phase load the sum of the voltages (or currents) in the three phases is zero. This will be clear from Fig. 4.38 and the vector diagram of Fig. 4.40. The sum of OB and OC is OD, which is equal and opposite to OA. This means that the neutral wire of a balanced star-connected system carries no current. In practice the need to supply single-phase systems for domestic use introduces a lack of balance. The local distribution systems are arranged so that, as far as possible, the loads on the three phases are equal so that the system remains balanced; but this cannot be performed exactly, so there is always some unbalance current in the neutral.*

Electric Motors

An electric motor is the converse of a generator, being used to translate electrical power into mechanical rotation. If we take a

* Note that the neutral wire at the actual point of supply has to carry the same current as the line wire. It is only when the three phases are combined at the local distribution point that the cancellation of currents in the neutral occurs.

simple d.c. generator and apply a d.c. voltage across the terminals a current will flow in the field coils which will establish a magnetic flux and the current which flows in the armature will produce a mechanical force equal to $BINl$, as explained in Section 1.4, which will cause the armature to rotate. As it does so the connections will be changed by the commutator, so that the maximum current is always flowing in that portion of the armature which is in the maximum field and continuous rotation will result.

As soon as this rotation has been set up the machine will behave like a generator, and a back e.m.f. will be induced in the armature proportional to the speed. Hence the armature current will be determined by the difference between the applied voltage and the back e.m.f., operating into the (effective) armature resistance. Since the back e.m.f. is proportional to the speed, the speed will automatically adjust itself such that the current flowing is sufficient to develop the mechanical power required.

Speed of D.C. Motors

Actually, when the motor is running, the difference between the applied voltage and the back e.m.f. is quite small. It was shown when discussing d.c. generators that $E = knz\Phi$ so that we can say that the speed $n = E/kz\Phi = V/kz\Phi$ very nearly, V being the applied voltage. Hence the speed is inversely proportional to the number of armature conductors and the flux.

It is therefore a simple matter to control the speed of a d.c. motor by varying the strength of the field, and this is one of its principal advantages. The field system of a d.c. motor may be connected across the input as with a shunt-field generator, in which case the speed for a given setting of field current is constant except for the small drop in speed necessary to allow the armature to draw sufficient current to supply the torque required.

Alternatively the field winding may be connected in series with the armature, having less turns but capable of carrying the full load current. With this arrangement the flux is proportional to the armature current and is therefore small when the motor is running light and increases as the load (and hence the current) increases. Hence the speed falls off as the load increases, but the torque developed is proportional to the square of the current since the flux has also increased, and this is useful in a motor which is required to develop a high torque at low speeds, such as a traction motor. A series motor, however, should never be run with no load because the field then is so weak that the speed may become dangerously high.

Motor Starters

When starting a d.c. motor there is initially no back e.m.f. and hence if the full voltage is applied the current will be limited only by the armature resistance, which is normally quite low. In all but the smallest motors therefore it is necessary to introduce a series resistance when first switching on. This is gradually cut out in steps until the motor is up to speed. Such a device is called a motor starter and usually takes the form of an arm which is moved slowly over a series of studs which connect gradually decreasing resistance in the circuit. At full travel the arm is held over by an electromagnet which releases it if the circuit is broken for any reason, so that the procedure has to be repeated each time the motor is started.

A.C. Motors

There is a wide variety of a.c. motors and reference can only be made to the more important. It is possible to operate a d.c. commutator motor on an a.c. supply, for although the current is alternating, there is at any instant a torque produced by the interaction of the armature current and the flux, and, since both of these change sign together when the current reverses, the resulting torque is always in the same direction and is the mean of the instantaneous torques throughout the cycle. This simple arrangement is not efficient and is usually only employed for small motors designed to operate on either d.c. or a.c. supply. For larger machines certain modifications are required which are beyond the present scope.

Induction Motors

The more usual form of a.c. motor is the induction motor which for industrial purposes is operated from a three-phase supply. Consider three field windings spaced equally round a stator and supplied with three-phase current. Because of the phase difference between the currents each of the field windings will receive maximum current in turn and the flux produced will thus move round the stator. The rotor is provided with a number of conductors which are all joined to a common end ring, forming a short-circuited secondary in which currents will be induced by the flux. The interaction between these currents and the flux will produce a torque causing the rotor to rotate, and since the flux itself is rotating it will drag the rotor round with it. This type of motor is often called a *squirrel-cage motor* because of the cage-like structure of the rotor conductors. The speed of the rotor is virtually that of the rotating flux though there must be a slight difference in order that the rotor conductors shall cut the flux. The greater this difference, the greater the induced current, so that the speed automatically adjusts itself to a value such that the current is sufficient to provide the torque required.

The speed is actually $60f/p$ r.p.m., where f is the frequency of the supply and p is the number of *pairs* of poles. Thus with a 2-pole motor and a 50 Hz supply the speed is 3,000 r.p.m. less the slight drop necessary to supply the torque. This small difference in speed is called the *slip*, and at full load is of the order of 5 per cent. As the load increases, however, the slip begins to increase disproportionately and the motor stops.

An induction motor is thus a substantially constant-speed machine with relatively small overload capacity (and a poor starting torque).

The necessary rotating flux can be produced by using two stator windings spaced 90° out of phase. This is the principle of the two-phase motor which is often used in control systems.

Single-phase Induction Motors

A single-phase supply will not produce a rotating field but it is possible to run an induction motor off a single-phase supply by two methods.

Split-phase Motor

If the stator is wound for a two-phase supply the current in the second phase may be fed through a capacitor which will produce a leading current and thus provide a phase difference. It is not

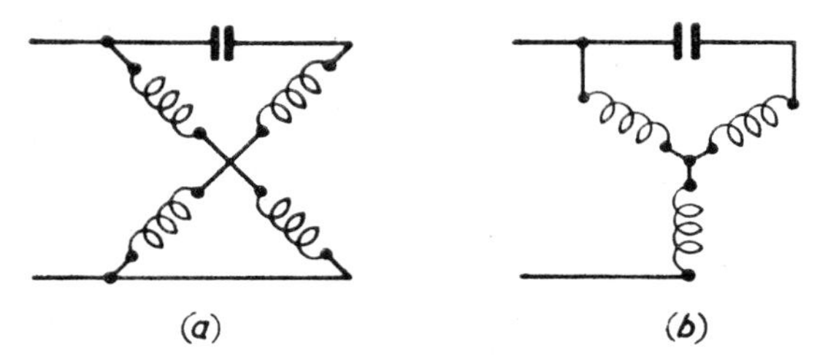

Fig. 4.41. Split-phase Field Connections

possible to obtain a full 90° difference, but between 60° and 75° is practicable and this is sufficient to produce a rotating field. The circuit is shown in Fig. 4.41 (*a*). A three-phase stator may also be connected for split-phase operation, as shown in Fig. 4.41 (*b*). Neither arrangement is as efficient as true 3-phase operation and split-phase working is normally only used in small machines.

SHADED-POLE MOTOR

An alternative arrangement is shown in Fig. 4.42. Here a portion of the stator pole pieces is enclosed by a short-circuited winding. Currents are induced in these coils, which produce a secondary flux which lags behind the main flux by 90°. By disposing these "shading" poles between the main poles the effect of a two-phase winding is obtained and a rotating field results. In practice the shading poles cannot be located the full 90° away from the main poles, but a sufficient displacement can be obtained for satisfactory operation. Again the system suffers in efficiency and is only used for small motors.

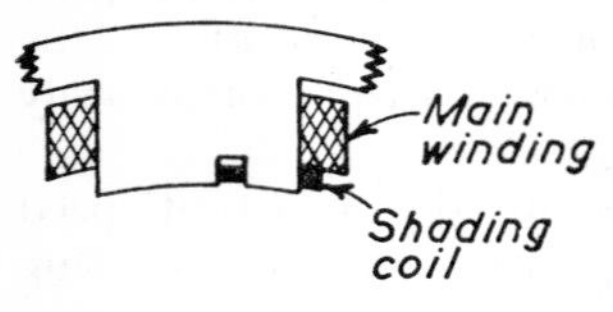

FIG. 4.42. SHADED POLE

Synchronous Motors

If the rotor of an induction motor is magnetized, the rotor will align itself with the rotating stator flux and will run at a truly constant speed determined only by the frequency of the current supplying the stator flux. This is sometimes useful, though the torque developed is less than with an induction motor.

The arrangement is not self-starting, it being necessary to run the rotor up to speed by some external means when it will pull into synchronism. However, if the rotor is provided with an additional squirrel cage or similar winding it will run up to speed as an induction motor and will then pull into synchronism provided the slip is small.

Another form of small synchronous machine is the hysteresis motor. This is provided with a solid steel rotor in which magnetism is induced by the stator flux. This causes a distortion of the flux path and the rotor takes up a position such as to reduce this distortion to a minimum, which occurs when the rotor revolves at synchronous speed. The stator may be two- or three-phase.

A type of synchronous motor used in small clock or timing mechanisms employs a stator provided with a series of poles and energized by a single-phase a.c. supply. The poles thus develop maximum flux every half cycle, and by arranging a similar (but not identical) series of poles on the rotor it will run at a speed which brings the poles into alignment at the appropriate instant. If the rotor is magnetized with alternate N and S poles the arrangement can be made of self-starting.

Stepping Motors

The toothed-rotor construction can be utilized to provide a device which will rotate in discrete steps on the application of suitable (square-topped) pulses to the stator. This is used for electrical repeating in control or telemetry systems. The stepping angle is usually $7\frac{1}{2}°$ or 15° with a maximum pulse rate of a few hundred per second. Typical torques range from 0·004 to 0·04 newton-metres according to size.

Meters

The mechanical forces produced by the interaction of magnetic or electric fields are utilized to provide indications of the strength of voltages and currents. Such instruments are of two main types, *moving-coil* and *moving-iron.*

Fig. 4.43 illustrates the essentials of a moving-coil meter. It

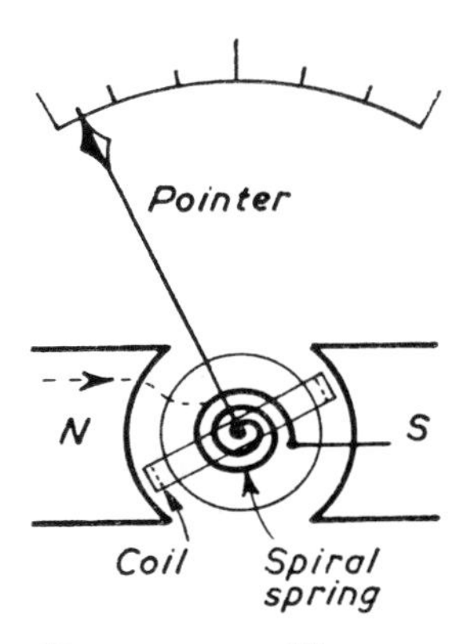

FIG. 4.43. ESSENTIALS OF MOVING-COIL METER

consists of a permanent magnet with a cylindrical air gap in which a coil is pivoted. If current is passed through the coil a torque is produced, as in an electric motor. The movement is restrained by a spiral spring which develops an increasing torque as the coil rotates. Hence the coil takes up a position proportional to the current through it, and if a light pointer is attached to the coil it will indicate the current against a suitable scale.

To obtain a linear relationship the magnetic flux must be constant and always at right angles to the coil. This is achieved by including a cylinder of soft iron inside the coil, as shown, leaving only a small annular gap in which the coil rotates.

The coil is of many turns of fine wire wound on a light former and held in jewelled pivots. Spiral hair springs serve to provide the restraining torque and also to lead the current into and out of the coil. The system is usually designed to provide a full-scale deflection with either 1 or 5 mA, though instruments are available requiring only 50 μA for f.s.d.

For measuring larger currents the meter is shunted with a resistor which by-passes the greater part of the current, while for voltage measurement a suitable series resistance is included such that the voltage to be measured passes the appropriate current through the meter.

Fig. 4.44 illustrates a moving-iron type of meter. In its simplest form this consists of a solenoid in which is pivoted a light soft-iron

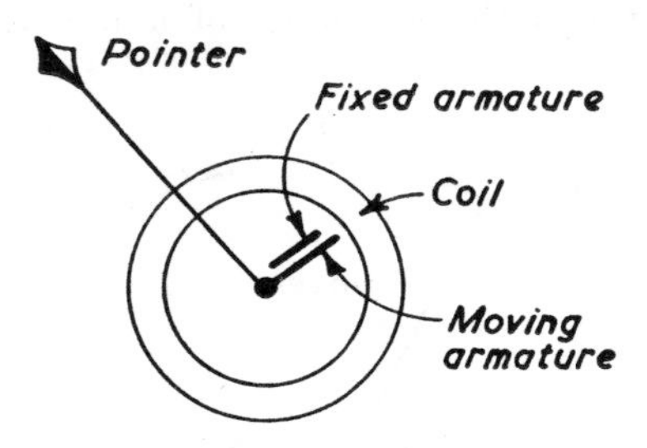

FIG. 4.44. ESSENTIALS OF REPULSION-TYPE MOVING-IRON METER

armature which will endeavour to align itself with the flux. The movement produced by such an arrangement, however, is not linear and improved performance, both in sensitivity and linearity, is obtained by introducing a second fixed armature. Since both are similarly magnetized by the current in the coil there is a repulsive force between them and this can be arranged to give a movement of the pointer nearly proportional to the current, except over the first 10° or so.

No permanent magnet is required, so that a moving-iron instrument is lighter and cheaper than the moving-coil type. On the other hand, the sensitivity is appreciably less, a current of 5 to 10 mA being required for full-scale deflection, and the power consumption is greater. This may be of importance, for it is an axiom in measurement that the measuring instrument must not appreciably influence the quantity being measured. This requirement is discussed more fully in Chapter 15.

Different current ranges are obtained in a moving-iron meter by suitable choice of the number of turns in the solenoid so that the use of a shunt is not necessary. For voltage measurements a series resistor is used as with the moving-coil type.

A.C. Meters

The moving-iron meter has the advantage that the deflection is independent of the direction of the current. Hence it will read a.c. as well as d.c. and with modern materials and constructions is equally accurate on both types of current up to frequencies of a few hundred hertz. Since the magnetic repulsion is proportional to the square of the current the reading obtained is a true r.m.s. reading and the same scale applies for both a.c. and d.c.

With a moving-coil meter the deflection is dependent upon the direction of the current and hence such a meter by itself will not give any indication on a.c. If the current is rectified, however, so that it consists of a series of half sine-waves all in the same direction, a moving-coil meter will read the mean value, which is $2/\pi = 0{\cdot}637$ times the peak. This will be 0·9 times the r.m.s. value, but this can be allowed for in the calibration, which can be made to indicate in terms of r.m.s. value.

Assemblies of small copper-oxide or other semiconductor rectifiers are made specifically for this purpose and are included inside the meter itself. Such instruments, known as *rectifier meters*, are in considerable use, but it should always be remembered that they do not read the true r.m.s. value and that the indication is only correct on a pure sine wave.

Thermal Meters

For certain applications use is sometimes made of the heating effect of a current. One method employs a linkage which converts the mechanical expansion of a fine wire due to heating, into a pointer movement. Such meters are sluggish and inaccurate. An alternative method is to use a thermo-couple which produces an e.m.f. proportional to the heat developed, and then apply this to a sensitive d.c. meter. This usage is discussed further in Chapter 15.

Electrostatic Meters

One other type of instrument which is sometimes used is the electrostatic meter. It consists of a light metal plate carried on a pivot which allows it to rotate between two fixed plates in a manner similar to a variable capacitor. If an e.m.f. is applied across the plates the electrostatic attraction will cause the moving plate to rotate and so provide an indication. The force is small so that such meters can only be used to measure voltages of 1,000 volts or more. The reading is proportional to the square of the voltage, so that the meter can be used on a.c. or d.c.

4.6. Batteries

Batteries are sources of electrochemical e.m.f. and are used where considerations of portability or convenience predominate. They are of two main types, known as *primary* and *secondary* cells respectively.

Primary Cells

A primary cell is one in which the e.m.f. is produced by the interaction of suitable chemicals which are themselves expended in the process. One of the most common types is the Leclanché cell in which two rods, of zinc and carbon respectively, are immersed in a solution of ammonium chloride (sal ammoniac). This arrangement exhibits a potential difference of approximately 1·4 volts between the carbon and the zinc (the carbon being positive) and, if the cell is connected to an external circuit, current will flow. An electrolytic action then takes place, the zinc reacting with the electrolyte to form zinc chloride and releasing ammonia at the negative (zinc) electrode and hydrogen at a positive (carbon) electrode. The ammonia dissolves in the water but the hydrogen is not soluble and forms a film round the carbon which rapidly prevents the passage of further current.

This action is known as *polarization* and to make effective use of the cell the hydrogen must be removed. This is done by surrounding the carbon rod with manganese dioxide. This gives up some of its oxygen to combine with the hydrogen, forming water, leaving a lower (and unstable) oxide of manganese which during the periods when the cell is not in use absorbs oxygen from the atmosphere and reverts to its normal state. The manganese dioxide *depolarizer* is usually mixed with graphite to improve the conductivity and is held round the carbon rod in a porous pot or bag. The effectiveness of the action depends on the amount of depolarizing agent available, but a typical "wet" cell will give currents of the order of one ampere. The zinc electrode is steadily consumed as long as current is passing and ultimately both the zinc and the electrolyte have to be replaced.

There are other forms of primary cell, using different chemicals and depolarizers but the principle remains the same and they need not be discussed here.

Dry Cells

To avoid the use of a liquid electrolyte, the Leclanché cell can be made up in a "dry" form. Here the carbon rod with its surrounding depolarizer is housed in a zinc can, the intervening space being filled with a jelly of sal ammoniac and glycerine, the top being sealed with pitch. Such cells can be made in small and compact form and can be assembled in series to provide any required e.m.f. from 1½ to several hundred volts.

The internal resistance of such a construction is necessarily higher than with the wet type, and becomes progressively higher in use so that the life of the dry battery is determined more by this rise in internal resistance than the dissolving of the zinc. There is also some chemical action taking place even when the battery is not in use so that it does not have an unlimited shelf life.

Other Forms of Primary Cell

There are various other types of cell of which only two are of importance in radio engineering. These are—

(*a*) *Weston standard cell.* This uses a cadmium cathode and a mercury anode in a solution of cadmium sulphate with mercurous sulphate depolarizer. It delivers a very constant e.m.f. of 1·0186 V provided that it is only required to supply negligible current. It is used as a standard of e.m.f. in potentiometer circuits (*See* Chapter 15).

A modified form of standard cell known as the *Clark cell* uses a zinc amalgam instead of cadmium for the cathode. It delivers 1·43 V.

(*b*) *Mallory cell.* This uses a zinc-mercury combination which can be made up as a dry cell having certain advantages, notably absence of deterioration when not in use. A zinc anode is employed with a self-depolarizing cathode of mercuric oxide mixed with graphite. The electrolyte is potassium hydroxide. The e.m.f. on load is between 1·3 and 1·35 V.

Fuel Cells

A current passed through an aqueous solution will cause the water to dissociate into its constituent elements by the electrolytic action described on page 55. Techniques have now been devised whereby this action can be reversed, so that hydrogen and oxygen can be induced to combine electrolytically to form water, and develop an e.m.f. in the process.

Such a device is called a fuel cell and is illustrated in simple form in Fig. 4.45. Two electrodes of porous nickel are immersed in a weak solution of potassium hydroxide. Oxygen supplied to electrode A combines with water to form hydroxyl ions (OH). At B these recombine with hydrogen to form water again. The process requires an interchange of electrons and hence only takes place when there is an external circuit, but if this is provided appreciable current can flow. A group of electrodes in a large cell can supply continuous currents of the order of tens of amperes. The e.m.f. is about 0·9 V.

The electrodes have to incorporate certain catalysts to enable the action to take place, but the principle is basically as described. The reaction produces an excess of water so that the strength of the solution has to be periodically increased.

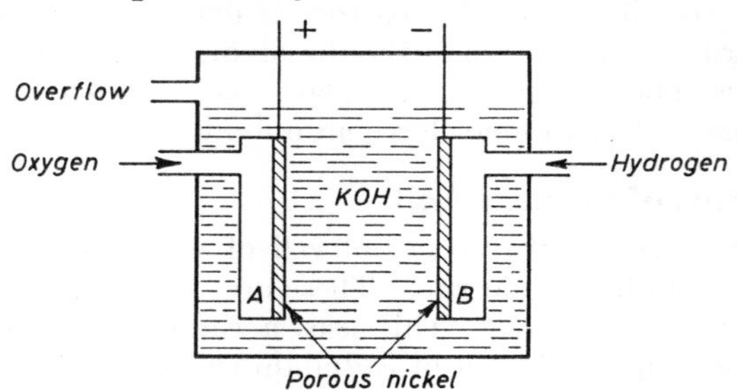

FIG. 4.45. SECTION THROUGH A FUEL CELL

Secondary Cells

There are certain chemical combinations in which the changes due to the passage of current are reversible, so that if current is passed in the opposite direction the original conditions can be restored. Such a device is called a *secondary cell*, the most common type being the lead-acid cell.

Here the plates are of lead and lead peroxide in a solution of dilute sulphuric acid. If current is drawn from the cell the positive plate gives up its oxygen, which combines with the hydrogen in the sulphuric acid to form water, while the sulphate ions combine with the lead to form lead sulphate.* If current is passed through the cell in the opposite direction, however, the process is reversed and the positive plate is reconverted to lead peroxide.

Thus it is possible to put back into the cell the energy which has been extracted from it, and the arrangement is therefore called an *accumulator*. The e.m.f. available is initially about 2·2 volts per cell. On discharge this drops rapidly to 2 volts and then falls slowly as the cell discharges to about 1·9 volts, at which point the discharge should be stopped and the cell recharged.

To recharge a cell it is necessary to apply an e.m.f. greater than that of the cell itself so that current is passed in the reverse direction. The e.m.f. of the cell itself increases as it becomes charged, becoming as high as 2·5 to 2·6 volts when fully charged, though this falls to 2·2 volts when the charging e.m.f. is removed.

Since the chemical action during discharge releases water it will

* Actually *both* plates become transformed into lead sulphate, but the process is complex and need not be discussed here.

be seen that the specific gravity of the acid decreases in the process. This can be used to determine the state of the battery, the specific gravity falling by about 10 per cent during discharge. The initial value depends on the type of cell, being of the order of 1·21 with a large stationary cell to 1·25 with a car battery.

A lead-acid cell must not be allowed to remain in a discharged condition, for the lead sulphate tends to change its character and will no longer recombine in the proper manner when recharging is attempted.

The plates of an accumulator are made in the form of a grid of lead alloy, the spaces of which are filled with a paste of lead oxide. The plates are then "formed" by passing current through the cell which reduces the paste on the negative plate to spongy lead and forms lead peroxide on the positive plate. It is also customary to include porous separators of insulating material between the plates to resist the mechanical deformation which might otherwise occur due to the force between the plates when passing a heavy current.

Secondary cells are rated in terms of their ampere-hour capacity. Thus a cell of 80 Ah capacity will supply 8 amps for 10 hours, or 4 amps for 20 hours, and *pro rata*, before requiring to be recharged. If the current is only taken intermittently the charge will last rather longer. The current permissible depends on the size and construction. An 80 Ah car battery will supply 20 to 30 amps for short periods while a large stationary battery may be required to supply several hundred amps.

Alkaline Cells

An alternative form of secondary cell uses plates of nickel hydroxide and iron in an electrolyte of dilute potassium hydroxide (caustic potash). In some cases cadmium is used instead of iron for the negative plate.

The e.m.f. with either type is approximately 1·2 volts and as with the lead cell the chemical changes due to the passage of current are reversible. The advantages of this type of cell are reduced weight and greater robustness, though the smaller e.m.f. necessitates a larger number of cells to obtain a given voltage.

A later type of alkaline cell uses electrodes of silver and zinc. In the charged condition the positive plate becomes oxidized, while, as the cell discharges, the oxygen is transferred to the zinc. The electrolyte is potassium hydroxide, which does not enter into the reaction. The e.m.f. on load is approximately 1·5 V.

Silver-zinc cells show considerable economy of weight. Lead-acid types develop approximately 20 ampere-hours per kg, whereas for silver-zinc cells the figure is about 80 Ah/kg.

Nickel-cadmium cells are now available in small sealed units, thus providing the convenience of a dry cell with the facility of rechargeability. The Mallory cell is also available in rechargeable form.

4.7. Microphones

A microphone is a device for turning sound energy into electrical energy, the ideal microphone being one which reproduces the waveform of the sound exactly. Microphones may be divided into two main groups: a *pressure* microphone is one in which the electrical response is caused by variations in pressure of the sound wave, whilst in a *velocity* microphone the electrical response corresponds to the particle velocity resulting from the propagation of the sound wave through an acoustic medium.

The Carbon Microphone

The simplest type of microphone depends for its operation on the fact that the resistance between two surfaces of carbon depends on the pressure between them. This effect can be utilized to create a

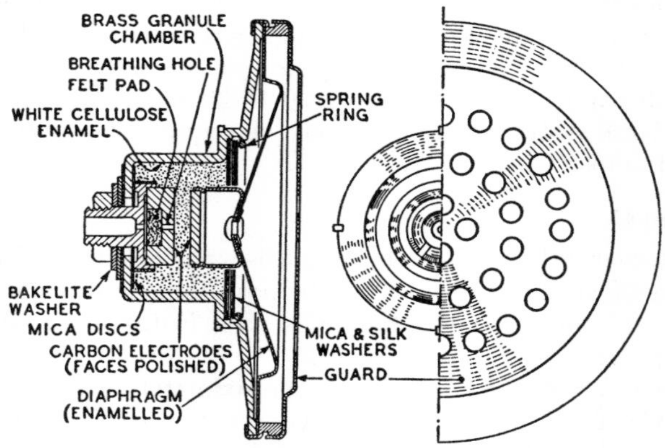

FIG. 4.46. DIAGRAM OF CARBON MICROPHONE
(*By courtesy of the Post Office*)

microphone by mounting two carbon blocks in a suitable housing and filling the intervening space with finely-divided carbon granules. If one of the blocks is connected to (or is itself in the form of) a diaphragm any variations in air pressure impinging on the diaphragm will cause a corresponding variation in the resistance. Hence if a current is passed through the microphone from a battery, the current will vary accordingly. This type of microphone is used in telephone handsets, a typical construction being illustrated in Fig. 4.46.

Such a microphone requires a source of low d.c. voltage to pass current through the system, and suffers considerable non-linear distortion for other than small displacements. The noise level is high and the frequency response is seriously peaked, but the sensitivity is good.

Where greater fidelity is required a more sophisticated system is required. One method is to use the piezo-electric effect described on page 70, whereby mechanical pressure can be converted into e.m.f. directly. Alternatively the air vibrations can be caused to produce variations in capacitance or inductance. Brief descriptions of these methods are given below. For more detailed information of the reader may refer to *Telecommunication Principles* by R. N. Renton (Pitman).

Crystal Microphones

A crystal microphone depends for its operation on the e.m.f. generated by a piezo-electric crystal when it is deformed. Rochelle salt is the most commonly used material owing to its high piezo-electric sensitivity.

The crystal microphone may be actuated directly or through a diaphragm. The directly actuated or sound-cell type has lower sensitivity but flatter frequency characteristics and is almost non-directional, whilst the diaphragm-operated type, though having higher sensitivity, has a less uniform response and is more directional at high frequency.

Crystal record-player pick-ups use a Rochelle salt crystal which is deformed by the motion of the needle in the record grooves (see page 510) and this produces a varying e.m.f.

Capacitor Microphones

A capacitor microphone operates by variations in capacitance between a thin stretched diaphragm and a rigid plate parallel to it. Such a system must be heavily damped to avoid serious resonance but can then be made to have a response which is nearly flat from 30 Hz to 10 kHz.

A potential of several hundred volts is applied across the plates through a high resistance, across which e.m.f. is developed by the variations in current due to the changing capacitance. The arrangement, however, is relatively insensitive, giving an output of a few millivolts only. Hence appreciable amplification is required, and because of the high impedance the first stage is made up as a pre-amplifier, mounted as close as possible to the microphone itself.

Moving-coil Microphones

The moving-coil microphone is, in essence, a small version of the moving-coil speaker described on page 226. The frequency response in the simplest construction is somewhat limited but is substantially flat over the operating band. However, by more elaborate construction the response can be extended to 60–10,000 Hz.

Pressure Ribbon Microphones

The pressure ribbon microphone consists of a light metallic ribbon suspended in a magnetic field, freely accessible to the atmosphere on one side and terminated in an acoustic resistance on the other. This resistance may take the form of a finite pipe damped with tufts of felt. Very high quality microphones may be made in this way and non-directional microphones with a flat response up to 15 kHz have been produced.

Velocity Ribbon Microphones

The free-ribbon microphone consists essentially of a loosely stretched ribbon suspended in the air gap between two pole pieces. In addition to supplying the magnetic flux, the pole pieces serve as a baffle for acoustically separating the two sides of the ribbon. Under the influence of a sound wave the ribbon is driven from its equilibrium position and the motion of the ribbon in the magnetic field induces a voltage between the two ends of the ribbon which is proportional to the particle velocity in the sound wave.

The resonance of a free-ribbon microphone is usually below the audible limit and the frequency response can be very flat over the range 30–15,000 Hz. The free-ribbon microphone is strongly bi-directional but reproduces transients faithfully and is accepted as the best type for high fidelity.

4.8 Loudspeakers

Loudspeakers may be divided into two principal groups. Horn loudspeakers are those in which the diaphragm is coupled to the air by means of a horn, whilst direct-radiation loudspeakers are those in which the cone or diaphragm is directly coupled to the air.

The earliest forms of loudspeaker were merely enlarged versions of the simple telephone earpiece described earlier (Fig. 1.39), fitted with a horn or trumpet to concentrate the sound. This simple arrangement is no longer used because of its limited range of frequency response, while the horn introduces undesirable resonances. Both these limitations, however, can be overcome with a properly designed horn and drive, as explained in the next paragraph.

One point may be noted here. The pole pieces in a telephone receiver are mounted on a permanent magnet so that there is a constant pull on the diaphragm, and the currents in the coils cause variations in this pull. This provides a marked increase in sensitivity because the pull on the diaphragm depends on the square of the magnetic flux density. Hence, if the permanent magnet contributes a flux density B and the currents in the coils cause this to vary by a small amount b, the total variation in the pull on the diaphragm is proportional to

$$(B + b)^2 - (B - b)^2 = 4Bb$$

Hence not only do we obtain greatly increased sensitivity (for B may be 10^6 times b) but the pull is now dependent on b and not b^2. If this were not so there would be a pull on the diaphragm every half cycle, producing a quite unacceptable frequency-doubling action.

Hence with all loudspeakers a permanent field is provided and to obtain the best sensitivity this is made as high as practicable.

Design of Horn

In order that the vibrations of the diaphragm may be converted into sound energy, it is necessary to set in motion a column of air. One way of doing this is to mount a horn over the orifice of the loudspeaker and this acts as an acoustic transformer, converting the small-amplitude high-pressure air pulsations at the diaphragm itself into low-pressure high-volume pulsations at the far end.

The design of the horn must be carried out carefully. The ordinary conical horn is not found satisfactory as phase differences are set up throughout its length which cause vibrations in certain parts of the column of air to interfere with those in other parts and restrict the effective radiation of sound. If this is to be avoided, it is necessary to design the horn to follow an exponential or logarithmic law, so that the area at any point equals $A_0 e^{bx}$, where A_0 is the initial area and x is the distance along the horn.

Such a horn will give practically uniform radiation down to a point where the horn suddenly ceases to load the diaphragm. In other words, the air currents fall out of phase very rapidly and a "cut-off" is obtained. This cut-off depends entirely upon the rate of expansion of the horn, and, in general, the larger and longer the horn, the lower the cut-off frequency.

Whereas a loudspeaker with an ordinary horn would give a very poor response below 200 Hz, it is possible with an exponential horn to obtain satisfactory radiation at frequencies as low as 20 Hz.

Moving-coil Speakers

Another way of achieving the desired results, namely the setting in motion of an appreciable column of air, is to increase the size of the diaphragm; this led to the development of cone speakers in which a simple reed mechanism similar to that of a telephone ear-piece is coupled to a cone of paper, buckram, or some similar material, having a diameter of 20 cm. or more.

However to be fully effective at low frequencies a considerable movement of the diaphragm is required. This led to the development of the moving-coil speaker, illustrated in Fig. 4.47. Here, in the

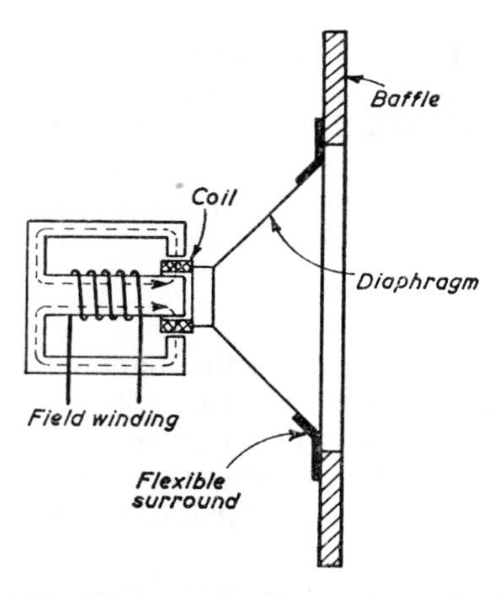

Fig. 4.47. Diagram of Moving-coil Speaker

centre of the diaphragm is mounted a small coil which is capable of lateral movement in and out of an air gap in a powerful magnetic field. This magnetic field may be produced by passing a current round the core of the field system as shown, or a somewhat similar arrangement may be constructed with a permanent magnet. The former is technically preferable owing to the fact that higher field strength can be obtained, but great progress has been made with permanent magnet systems, and most modern speakers use permanent-magnet fields.

It will be clear from the figure that the magnetic field flows across the gap and therefore cuts the turns of the coil at right angles. Any current flowing through the coil, therefore, will set up a force tending to move the coil out of the gap, the direction of motion being dependent upon the direction of the current. Hence, an alternating current will cause a continuous oscillation of the coil, and this will be transmitted to the diaphragm, setting up the desired radiation of sound waves.

Since the diaphragm is free to move and has very little restoring force, it is capable of moving a considerable distance. Now, it can be shown that, in order to radiate a uniform amount of sound energy, the motion of diaphragm should be inversely proportional to the frequency. Clearly, if we move a diaphragm a short distance a large number of times per second we produce the same agitation of the air as in moving the diaphragm a large distance a relatively small number of times per second. This is the principle on which the moving-coil speaker works, the vibration of the diaphragm at low frequencies being of considerable amplitude. Even so, however, it was found at first that the radiation of the low frequencies was disappointingly small, and it was not for some time that the real cause was discovered. This was found to be an interference between the air waves produced by the front and the back of the diaphragm.

Use of Baffle

At these very low frequencies, the wavelength of the sound is very much greater than the diameter of the diaphragm, and therefore air waves produced by the front and back of the diaphragm are sensibly in phase. If the diaphragm moves forward it pushes air away from it, but, at the same time, a partial vacuum is set up behind the diaphragm. All that happens, therefore, is that air rushes round the edge of the diaphragm from the front, where it has been displaced, to the back, where there is a partial vacuum.

The air vibrations are set up, but they are not radiated into the room. To obviate this difficulty, a baffle is used, consisting of a stout piece of wood, or other suitable material, arranged as a continuation of the diaphragm. This is illustrated in Fig. 4.47, and usually consists of a large square piece of wood with a hole in the centre through which the diaphragm operates. The sound waves coming off the front of the diaphragm are now unable to interfere with those coming off the back, and consequently the sound vibrations are radiated into the room.

In many cases the baffle is not a plain piece of wood but is made in the form of a box. This does not have quite the same effect, but does assist to a large extent in producing the proper radiation. The back of the box, however, must either be open or provided with large slots covered with light gauze so that air may have free passage in and out of the box. Otherwise, resonance will be set up at certain frequencies and the reproduction will sound unnaturally low-pitched.

Phase Difference

Difficulty still arises with the large diaphragm type of loudspeaker in the radiation of the upper frequencies. Let us suppose that we have a simple oscillating diaphragm AB, as shown in Fig. 4.48. Consider the radiation at a point X off the axis of the diaphragm. The radiation will reach this point from various points of the diaphragm, and if we consider the two extremes A and B, it will be seen that there is an appreciable difference in the length of the paths to the point X. If the diaphragm of the loudspeaker is comparable with the wavelength of the sound being reproduced, then it is clear that these two sound waves will arrive out of phase and will interfere with one another.

This condition of affairs will be obtained at about 1,500 to 2,000 Hz, and frequencies above this, therefore, are not radiated uniformly, but are strongly directional, being transmitted principally along the axis. The rest of the diaphragm is thus largely ineffective, and the result of these two factors is a cut-off in the upper frequencies, speech lacking crispness, and music being flat. To minimize this, the diaphragm is not usually a flat disc but is made conical in shape. This also tends to give the diaphragm greater rigidity, so that it is capable of moving as a whole and not flexing too much.

Even so, however, at the upper frequencies the diaphragm cannot

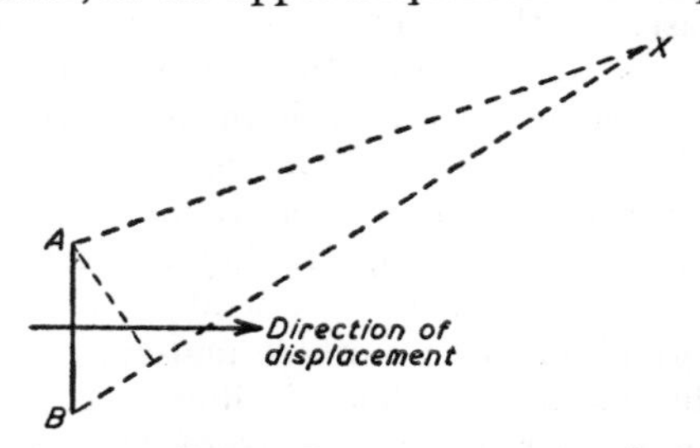

FIG. 4.48. ILLUSTRATING PHASE DIFFERENCE WITH LARGE DIAPHRAGM

vibrate as one unit, but breaks up into a number of subsidiary portions, each of which vibrates in its own way, and the design of the diaphragm is a matter of great difficulty. The modern tendency is to use diaphragms of comparatively soft material which are unable to break up into "paper" resonances at the upper frequencies. Such a system is not a good reproducer of the upper frequencies and must therefore be fed with an increased supply of energy in the upper registers in order to maintain a balance of tone. This arrangement, however, is preferable to a diaphragm which breaks up into resonances and accentuates certain upper frequencies unduly.

Tweeters

Where really high-fidelity reproduction is required two or more speakers are used. A conventional moving-coil unit is employed for the low and middle registers with small additional units, known as "tweeters", specifically designed to handle the upper registers. The art has become highly specialized because at the higher frequencies reflections from the walls of the room have a marked influence, and careful siting of the speakers is necessary.

Public Address Horn Speakers

Public address speakers utilize a moving-coil drive with a horn radiator as shown in Fig. 4.49. It will be seen that the moving-coil system is similar to that already shown, but instead of being

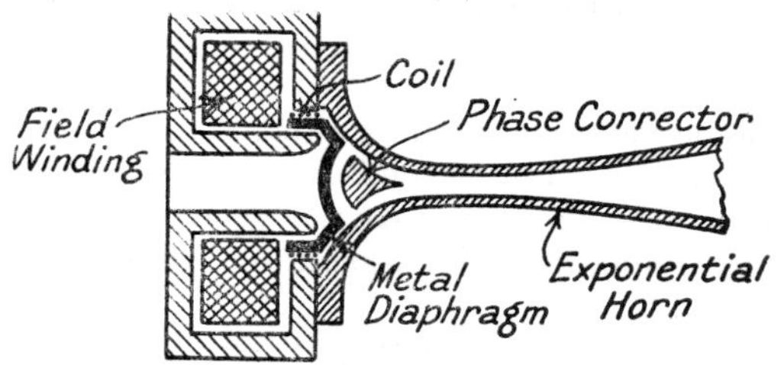

FIG. 4.49. MECHANISM OF COIL-DRIVEN HORN SPEAKER

attached to a large diaphragm a comparatiuely small diaphragm of about 5 cm in diameter is used, and an exponential horn is fitted over the diaphragm in order to transform the energy to low-pressure high-volume fluctuations as already explained.

An interesting point, however, is the phase corrector placed in front of the diaphragm, and shown shaded in the figure. This is a small conical-shaped obstruction, the function of which is to overcome the difficulty already mentioned of unequal radiation from the diaphragm. If this phase corrector were not there, the sound waves coming off the middle and edge of the diaphragm would be out of phase, and would therefore not be radiated. With the corrector in place, however, the radiations from the centre of the diaphragm are forced to take a longer route. If the distances from any part of the diaphragm to the throat of the horn are within a quarter wavelength at the (highest) frequency concerned, no serious cancellation occurs. This gives a considerable increase in response to the upper frequencies, and if the horn is correctly designed to obey an exponential law, these sound waves are transmitted down the horn by successive reflections without serious loss, and a good transmission is obtainable.

5

Thermionic Valves

THE successful development of the art of radiocommunication was entirely due to the evolution of the *thermionic valve*, and although this is being replaced to an increasing extent by solid-state devices there are many applications in which valve techniques continue to be used.

The thermionic valve, or vacuum tube, is essentially a device in which current is produced by the emission of electrons from a cathode under conditions of complete or partial vacuum, this current then being varied by voltages applied to suitable control electrodes.

5.1 Electron Emission

When a metal is heated the electrons in the atoms take up fresh orbits at greater distances, corresponding to a condition of greater energy. The process is neither uniform nor continuous, consisting of spasmodic jumps from one energy level to another. Some of the electrons fall back into more normal orbits emitting light or heat waves in the process, while others are so violently agitated that they leave the confines of the atom and even, if conditions permit, escape from the material altogether.

TABLE 5.1. TYPICAL EMISSION CONSTANTS

Material	*Work Function (electron-volts)**	*A*	*b*
Carbon	4·0	30	50,300
Tungsten	4·5	60	52,400
Thorium	3·4	60	38,900
Thoriated tungsten	2·6	3	30,500
Caesium	1·8	162	21,000
Strontium oxide	1·3	0·001	15,000 (approx.)
Barium oxide	1·0		

* An electron-volt (eV) is the energy required to raise the potential of an electron by one volt.

The energy which an electron has to acquire before it can escape varies with different materials. It is known as the *work function*, some typical values being given in Table 5.1.

The amount of the *electron emission*, as it is called, will obviously be largely determined by the work function of the material and also by the temperature, since the higher the temperature the greater the disturbance of the electron orbits. The relationship is customarily expressed in a form first postulated by S. Dushman which says that the emission in amps/cm² of emitting surface is

$$J_s = AT^2 \, \mathrm{e}^{-b/T}$$

where T is the absolute temperature (= °C + 273), and
A and b are constants for the particular material, b being proportional to the work function.

Values for A and b are given in Table 5.1.

Space Charge

The emitted electrons shoot out from the heated cathode and then fall back again. But they are repelled by further emitted electrons, which they in turn repel. The result is a cloud of electrons surrounding the cathode to form what is called a *space charge* and a condition of equilibrium is obtained. If the electrons, however, are subjected to a positive electric field by introducing another electrode connected to a source of (positive) e.m.f., electrons will be attracted to this anode which will destroy the equilibrium in the space charge so that further electrons will be emitted from the cathode. The net result will be that a current will flow across the space from cathode to anode and return to the cathode via the external circuit.*

Since the action only takes place if the anode is at a positive potential the device acts as a non-return valve and it was from this property that the generic name "valve" became used in British parlance. As we shall see, it is possible to introduce other electrodes into the valve structure, which exercise control functions of various kinds, and the simple diode is only used for certain limited applications. With the addition of these further electrodes the action becomes far more than that of a simple non-return valve, and the Americans have preferred to call such devices *vacuum tubes* or,

* The effective source of the electrons is thus not the cathode itself but some region within the space charge which is known as the *virtual cathode*. The distinction is only of importance when considering the actual design of a valve.

shortly, tubes. We shall, however, adhere to British convention, reserving the term tubes for cathode-ray tubes.

5.2. Practical Forms of Valve

The whole assembly is enclosed in an evacuated bulb, usually of glass. The air is evacuated, partly to prevent the cathode from oxidation (burning) but also because any gas at normal pressure is an insulator and so would prevent the electrons from travelling from the cathode to the anode.

Occasionally a small amount of gas at a greatly reduced pressure is introduced. In these circumstances the electrons will occasionally meet an atom of gas and the force of the collision may produce ionization as explained in Section 1.6. These positively-charged ions will be attracted to the cathode, where they will deliver up their charge, the result being that the current flowing across the gap will be increased (since a positive charge flowing in one direction is equivalent to a negative charge flowing in the other).

This augmented current, however, is obviously variable, depending on the amount of ionization which has taken place, and moreover the relatively heavy ions are liable to damage the cathode. Hence except for certain types of high-current rectifier, the modern valve is made "hard," i.e. every possible particle of gas is exhausted from the bulb. A given voltage on the anode will then always produce a definite emission current, and its behaviour may be reproduced by a similar valve operating under identical circuit conditions.

In the manufacture of the modern valve, therefore, the free gas is first exhausted by very efficient pumps; it is found, however, that the glass and the metal of the anode (and any other electrodes) hold between their own molecules a certain number of molecules of gas, and these will ultimately come forth and the valve will become soft. To get rid of this "occluded" gas, therefore, the bulb is warmed and the electrodes are made red hot by electrical means. This expels all the particles of gas that are likely to escape in the normal operation of the valve.

Cleaning up the residual gas is further assisted by introducing into the valve some active agent known as a *getter*. Magnesium is one such agent, a small piece of metallic magnesium being attached to the electrodes. When the valve is made hot the magnesium volatilizes and combines chemically with any residual gas which is left. The products of such combustion are drawn off in pumps, while the remaining magnesium is deposited over part of the glass of the bulb. Other forms of "getter" are employed in special circumstances.

The Diode

A simple valve containing a cathode and an anode as just described is called a *diode* and is illustrated diagrammatically in Fig. 5.1 in which there is an L.T. battery for heating the filament and an H.T. battery to apply the positive potential to the anode.

The figure also shows the characteristics of the diode, i.e. the manner in which the anode current varies as the H.T. or L.T. voltages are varied. Several interesting points will be noted.

(*a*) The relationship between anode voltage and anode current is not linear, the current increasing more rapidly as the voltage is increased. Actually the relationship obeys what is called the *three-halves law*, i.e. the current $i = E_a^{3/2}$.

(*b*) At a certain point the anode current ceases to obey this law and reaches a constant value so that further increase in anode volts

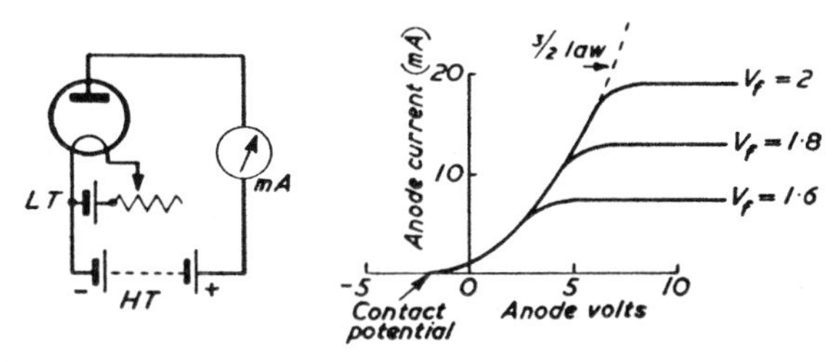

FIG. 5.1. SIMPLE DIODE CIRCUIT AND (IDEALIZED) CHARACTERISTICS

produces no increase in current. This is called *saturation* and indicates that all the electrons emitted from the cathode have been drawn to the anode. It will be noted that the saturation current increases with the filament current as one would expect, since the increase in filament current will produce a higher temperature and hence a greater supply of electrons.

Normally a diode (or any other form of valve) is so designed as to provide far more electron emission than is required, in which case the anode current follows the three-halves law throughout the range of working conditions. Such a valve is said to be *space-charge limited*.

Contact Potential

The third point of interest in the curves of Fig. 5.1 is the fact that there is still a small current even when the H.T. battery voltage is zero. This is because, as explained in Section 1.8, there is a small naturally-existing difference of potential between dissimilar elements. In order to reduce the current to zero it is necessary to apply a small negative potential to the anode in order to offset this *contact potential* as it is called. The contact potential is quite small, of the order of one volt, and is therefore only of importance around the region of zero current.

The Triode

The usefulness of the thermionic valve, however, can be considerably increased by the addition of further electrodes. The simplest such modification consists in the inclusion of a third electrode in the form of a spiral or wire-mesh grid between the cathode and the anode. Such a valve is called a *triode*.

The presence of the grid, of itself, has little influence on the electron stream, because of the small area which it presents, but if it is connected to a suitable source of potential it can either accelerate or decelerate the electron emission from the cathode, without itself taking any appreciable current. If the grid is negative with respect to the cathode it will tend to repel the electrons and prevent them from reaching the anode, even though the anode is at a positive potential, and a sufficient negative grid potential will cut off the anode current completely.

If the grid potential becomes positive, however, it begins to draw some small current itself. This is known as grid current, and while normally it is still only a small fraction of the anode current, it is usually to be avoided because it means that the grid begins to take power. As long as the grid is kept negative it exercises its control with a negligible consumption of power.

(Certain secondary effects arise at high frequencies but these will be discussed in later chapters.)

Characteristics

Let us now consider the characteristics of a triode. The two most important are:

1. The anode-current/anode-voltage curve, which shows the variation of the emission as the anode voltage is varied, with various fixed negative potentials on the grid.

2. The anode-current/grid-voltage curve, which shows the variation of emission with the voltage on the grid, at various fixed values of anode potential.

Fig. 5.2 shows these characteristics for a typical small valve. It will be seen that, as the anode voltage is increased, the anode current rises with increasing rapidity. Negative voltage on the grid delays the process so that more h.t. voltage has to be applied. The individual curves obey an approximate three-halves law, as with a diode, but the slope gradually decreases as the grid voltage is made more negative.

The second series of curves shows the variation of anode current with grid voltage. It will be observed that, as the grid voltage is made more negative, the anode current is ultimately reduced to zero.

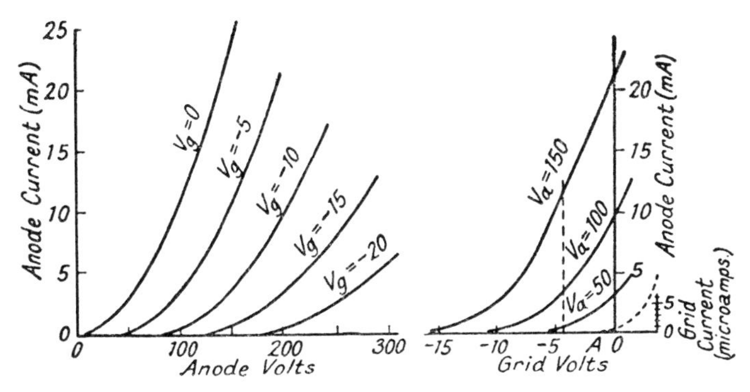

FIG. 5.2. CHARACTERISTICS OF SMALL TRIODE

The valve would be used, as discussed in the next section, with a small negative voltage on the grid, e.g. $4\frac{1}{2}$ volts with the curves shown. This is known as *grid bias*, and it will be seen that variations of grid potential around this bias point produce proportional variations of anode current.

As already explained, the valve is not normally used with positive voltage on the grid. The grid current in any case is only a few microamperes since most of the electrons shoot through the spaces in the grid to the anode, but the point *A* at which the grid current starts to flow is important for certain uses to which the valve can be put, and will be referred to later.

Valve Parameters

The performance of a valve can be assessed without reference to the full characteristics by the specification of certain parameters. The first of these is the *anode a.c. resistance*. This is the effective resistance of the valve to varying current and is obtained by observing the change in anode current which is produced by a given *small* change in anode voltage.

It is actually the slope of the valve characteristic for the particular conditions under consideration, so that the a.c. resistance

$$r_a = dv_a/di_a$$

The slope must be measured under conditions corresponding to the working conditions or, for purposes of comparison, at some accepted standard which is usually $v_a = 100$ and $v_g = 0$.

A second characteristic figure which we require to know is the relative effect of the grid voltage on the anode current. To determine this we observe the change in anode current produced by a given *small* change in grid voltage and divide the former by the latter. As we are here dividing a current by a voltage the result is not a resistance but a conductance, and this is known as the *mutual conductance* of the valve, usually written as $g_m = di_a/dv_g$.

A final parameter which is useful is the *amplification factor*. This is the change in anode voltage produced by a given change in grid voltage, and it will be clear that this figure is obtained by multiplying both the previous parameters together. Hence

$$\text{Amplification factor} = \mu = dv_a/dv_g = g_m r_a$$

With modern valves the a.c. resistance may range from something in the neighbourhood of 1,000 ohms for a power output valve up to several hundred thousand ohms for the tetrodes and pentodes discussed shortly.

Mutual conductance is usually specified in milliamperes per volt (mA/V)*, i.e., the change of anode current in mA divided by the change in grid voltage. This figure has improved considerably over the years. Whereas formerly it used to be less than 1 it is now possible to make valves having mutual conductances of over 40.

The amplification factor ranges from 2 or 3 for power valves up to 1,000 or more for tetrodes.

Cathodes

The earliest forms of valve utilized a filament of fine tungsten wire which was raised to a white heat by the passage of current. Modern valves use different cathode materials, there being two principal types. In the first type a small percentage of thorium is mixed with the tungsten. Under suitable conditions of heat treatment during manufacture this thorium forms a very thin layer on the outside of the filament and, having a lower work function than tungsten, gives the necessary electron emission at a dull yellow heat, requiring much less filament current, and having, incidentally, a

* Since the *siemens* is the unit of conductance, 1mA/V = 1mS.

much longer life, because the filament does not tend to crystallize and is thus not so brittle.

The other form of cathode, which is now more common, consists of a wire of platinum, nickel, or tungsten, on which is deposited a coating of what are known as "rare earths" such as oxides of barium, strontium, etc. These have an even lower work function so that it is only necessary to heat the filament to dull red to obtain the required emission.

Directly-heated cathodes are used for many purposes, mainly in valves required to deliver appreciable power, or small valves to be run off batteries. For the majority of uses, however, the heater current is obtained from the electric light mains, and for this special types of cathode have been developed.

In these valves the cathode is not a simple filament of wire, but is a small tube which is coated with electron-emitting oxides. Through the centre of this tube runs the heater (Fig. 5.3), which is supplied with current from the mains. The heat generated by this filament causes the cathode to warm up, and this in its turn gives off electrons.

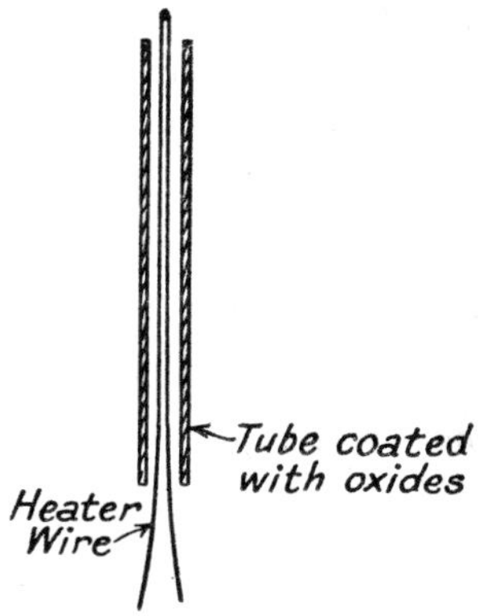

Fig. 5.3. Illustrating Construction of Indirectly Heated Valve

Owing to the relatively large mass of the cathode, it possesses appreciable heat inertia so that its temperature is not affected by the comparatively rapid fluctuations in the supply. Such a valve, therefore, may be operated direct from alternating current, the voltage being transformed down to a suitable small value for the purpose. The most common heater supply is 6·3 volts with heater current of 0·3 or 0·6 amps.

The valve behaves in exactly the same way as a battery-heated type, except that the return circuit connections are made to the cathode, which is independent of and insulated from the heater.

Tetrodes

To obtain a high amplification factor it is necessary for the grid wires to be relatively close together. Such a grid has a greater effect on the electrons passing through it but it also increases the space-charge effect. In a *tetrode* a *screen grid* is interposed between the anode and the *control grid*. This is connected to a positive potential, usually about one-third of the anode potential, and this attracts electrons from the filament, so overcoming the repelling effect of the control grid.

By the time the electrons reach this second grid they are travelling so fast that they shoot right through it and reach the anode without difficulty, while the electron flow can be controlled by varying the voltage on the inner grid in just the same manner as with a triode. This provides good mutual conductance with a high amplification factor.

An even more important advantage arises from the avoidance of capacitance coupling between the anode and grid of the valve which permits some of the (amplified) energy in the anode circuit to be fed back into the input (grid) circuit and so causes instability. This can be avoided by interposing an earthed screen between anode and grid, which is effectively provided by this second grid. The subject is discussed more fully in Chapters 7 and 9.

Secondary Emission

The anode-current/grid-voltage characteristic of a tetrode valve is similar to that of a triode except that it usually has a much steeper slope. The anode-current/anode-voltage characteristic, however, is different, due to an important secondary effect. As the voltage on the anode increases from zero, leaving the voltage on the screen at its normal value, the anode current at first increases normally. A point is reached, however, where the electrons shooting through the screen acquire sufficient velocity to displace some of the electrons in the material of the anode, producing what is called *secondary emission*. The electrons emitted by the anode are attracted back to the screen because its potential is higher than that of the anode. The total anode current ceases to increase and actually falls, so that we have the curious result that the anode current *decreases* while the anode voltage is increasing. It is indeed, possible for the secondary emission to exceed the emission from the cathode so that over a small range of anode voltage the anode current actually becomes negative as shown in one of the curves in Fig. 5.4.

As the anode voltage is increased further, it becomes comparable with the screen voltage and there is no longer sufficient difference of potential for the screen to capture the secondary-emission electrons. The anode current then rises rapidly, but when the anode voltage is about 4/3 times the screen voltage the current begins to limit and a form of saturation sets in.

This is because the total current drawn from the cathode is determined mainly by the voltage on the screen, which is nearer to the cathode and has the major influence on the electric field surrounding the cathode, which is what primarily determines the current. This screen voltage is constant so that beyond the "knee" the tetrode becomes very nearly a constant-current device.

Unlike the control grid, the screen itself draws appreciable current, to an extent depending upon the relative values of anode and screen voltage. In fact, the sum of the anode and screen currents is constant so that the variation of screen current with anode voltage is as shown dotted in Fig. 5.4.

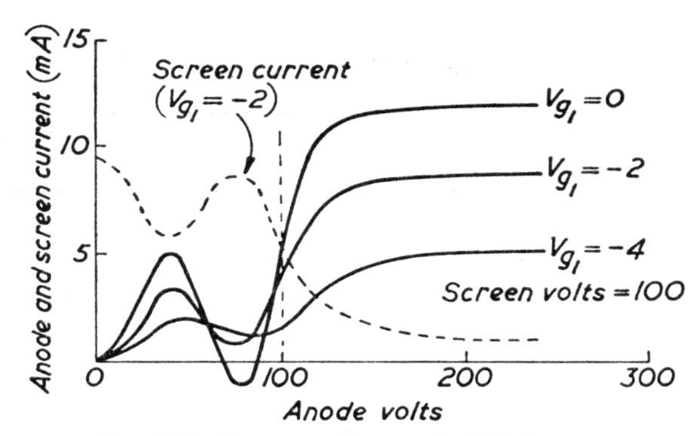

FIG. 5.4. TETRODE CHARACTERISTICS

The total current is controlled by the grid voltage exactly as it is in a triode. Fig. 5.4 thus shows a family of curves for different values of grid voltage.

Pentodes

The presence of the secondary-emission kink in a tetrode limits its usefulness because it makes an appreciable part of the characteristic unsuitable for normal use. This limitation, however, can be overcome by the introduction of another grid between the screen and the anode. This third, *suppressor grid* is in the form of a wide spiral, normally connected to cathode, and serves to introduce a small negative field close to the anode. This has very little influence on the rapidly moving electrons passing from the screen to the anode, but it serves to repel the secondary electrons emitted from the anode and thus prevents them from reaching the screen, even when the screen is at a higher potential than the anode.

The effect of this suppressor grid is to remove the kink in the characteristic completely so that a *pentode* characteristic is of the form shown in Fig. 5.5. This will be seen to retain all the desirable characteristics of the tetrode with the further advantage that there is no unusable portion of the characteristics until the anode voltage falls to quite low values.

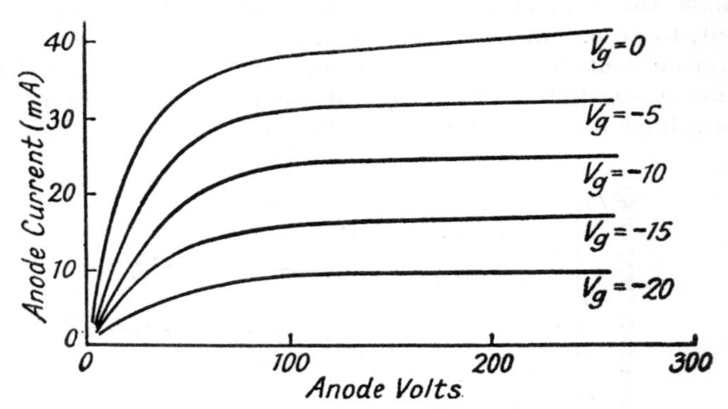

FIG. 5.5. TYPICAL PENTODE CHARACTERISTICS

It is possible to obtain a similar suppression of the negative emission by a suitable choice of the relative distances of the screen and anode from the cathode but the anode diameter has to be larger than normal. Such valves are known as *critical-distance tetrodes*.

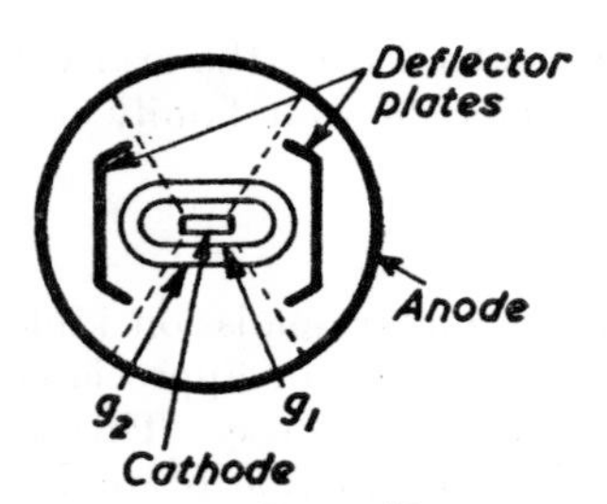

FIG. 5.6. BEAM TETRODE

A variant of this arrangement is the *beam tetrode*, which is frequently used for power output valves. Here the anode is located at the critical distance with deflector plates at zero potential between the screen and anode as shown in Fig. 5.6. They produce a negative electric field which has the twofold effect of concentrating the main electron stream into a beam, and at the same time inhibiting the emission of secondary electrons.

Suppressor Control

It should be noted that the suppressor grid may be used for additional control purposes, although this grid is normally connected to the cathode, and in many valves it is connected internally, so that it is not available for use. If the suppressor grid is externally available, however, it may be found that some control of the anode current can be obtained by altering the voltage on the suppressor grid. This is sometimes convenient, particularly where two controls are required. The normal signal may be applied to the usual control grid while some subsidiary signal may be applied to the suppressor grid. A particularly useful application of this technique is where a valve is to be brought in and out of use quickly without disturbing its other functions. Applying a large negative voltage to the suppressor grid may completely cut the valve off and prevent it from functioning at all, while restoration of the normal suppressor grid voltage will bring it back into its full operating condition practically instantaneously.

Variable-mu Valves

For use in amplifiers where it is desired to vary the amplification by suitable adjustment of the operating conditions, there is a type of valve called a variable-mu valve. This, as its name implies, has

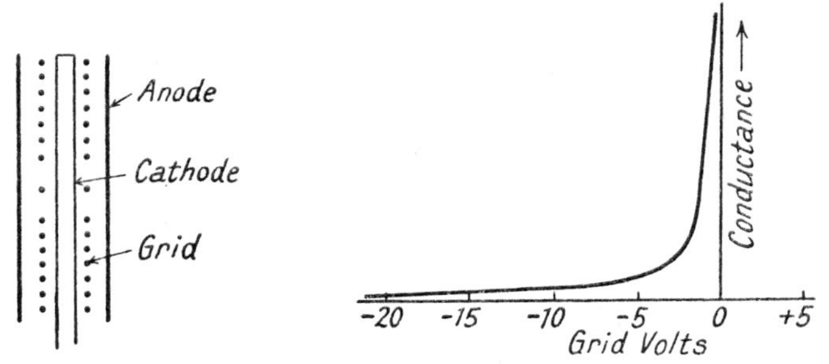

Fig. 5.7. Illustrating Cut-away Portion in Centre of Grid to Give Variable-mu Action and the Type of Characteristic Obtained

a variable amplification factor, which is achieved by winding the control grid with a non-uniform spiral. This means that the control exercised by the grid varies over the length of the cathode so that when the close-mesh portion has cut off the emission there is still current available from the open-mesh portion. It is found that if the spacing of the turns on the grid wires is logarithmically graded an exponential variation of anode current with grid voltage is produced. A similar effect may be obtained rather more cheaply by omitting some of the grid wires as shown in Fig. 5.7.

With such a valve the mutual conductance (and hence the amplification) varies smoothly with the grid bias so that a gradual control is obtained instead of a sudden cut-off.

Mixer Valves

In superheterodyne receivers (and certain other applications) there is a requirement for a valve which provides an anode current change proportional to the product of two control voltages. This

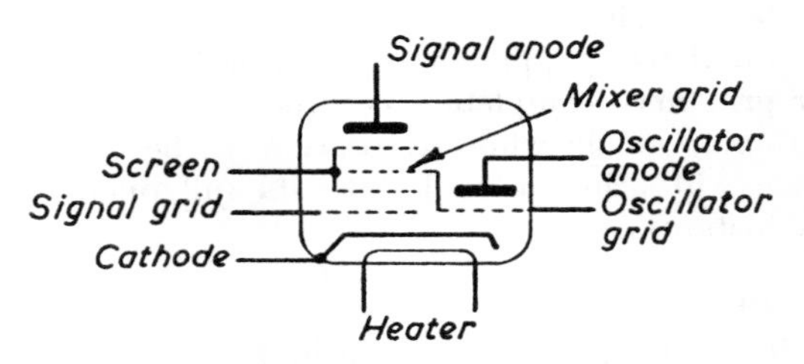

FIG. 5.8. DIAGRAM OF TRIODE-HEXODE

Here two valves are included in the same envelope, a triode for generating the local oscillation and a hexode for providing the mixing.

can be done in various ways, the most usual being by means of a special valve known as a *hexode*. This is similar to a tetrode in construction but the screen comprises two concentric grids, with a separate mixer grid in between them, as indicated in Fig. 5.8. This mixer grid exercises a control on the anode current in a similar manner to the control grid, but the presence of the additional screen around it reduces interaction between the circuits.

Typical Construction

In a practical valve the various electrodes are supported on stems mounted in a glass foot, while there is usually a support of mica or other insulating material at the top of the assembly in order to keep it rigid. Any movement of the electrodes relative to each other will influence the electron stream and therefore give rise to spurious signals. This effect is known as *microphony* because the valve is in fact acting as a microphone in that it will translate mechanical vibrations into electrical e.m.f.s.

The glass foot is then sealed into a bulb and a small tube is provided either in the bulb or in the foot, which can be connected to the pumps so that the air can be evacuated. This tube is then sealed off when the vacuum is complete, and the remaining gas is then removed by raising the temperature of the electrodes by some suitable means so that the getter fires as explained on page 232.

The connections to the various electrodes are then brought out to a suitable base which plugs into a socket so that the valve may be removed and replaced as and when necessary. With many modern receiving valves the base is made part of the foot itself, a

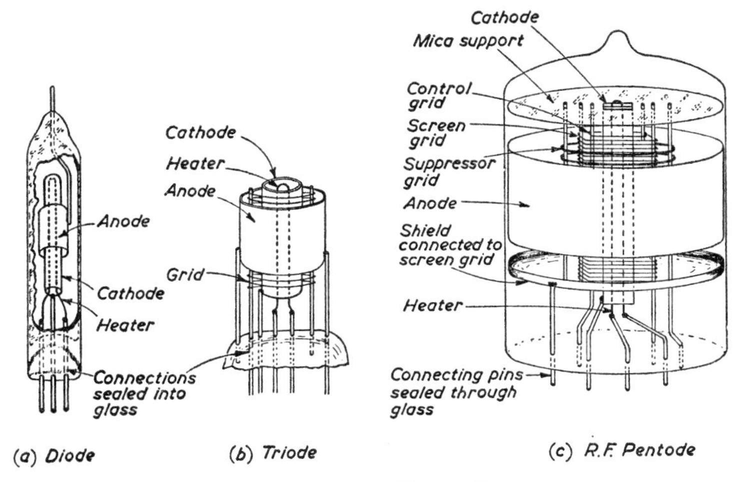

Fig. 5.9. Representative Valve Constructions

series of short pins being located on the glass of the foot at the correct spacing, while for very small constructions the connections to the electrodes are brought out on thin semi-flexible wires which are actually soldered directly into the circuit.

Fig. 5.9 illustrates a variety of modern valve constructions.

Multiple valves are often used incorporating two or more assemblies round a common cathode. Typical examples are the *double-diode triode*, or the *triode-hexode* shown in Fig. 5.8.

5.3. Basic Valve Usage

The manner in which valves are used in practical circuits is discussed in Part II, but it is appropriate here to review briefly the basic principles involved. Valves are used in three main ways: as amplifiers, rectifiers and oscillators.

Use of Valve as Amplifier

We have seen that the variation of potential on the control grid of a valve will produce corresponding variations in the anode current. To enable this variation of current to be utilized it is necessary to include in the anode circuit a suitable load, across which the variations in current will develop useful voltage, or in which useful power can be produced.

Fig. 5.10 shows a simple triode circuit in which an impedance Z has been included in the anode circuit. If a signal e.m.f. e_g is applied between grid and cathode there will be a variation in anode current

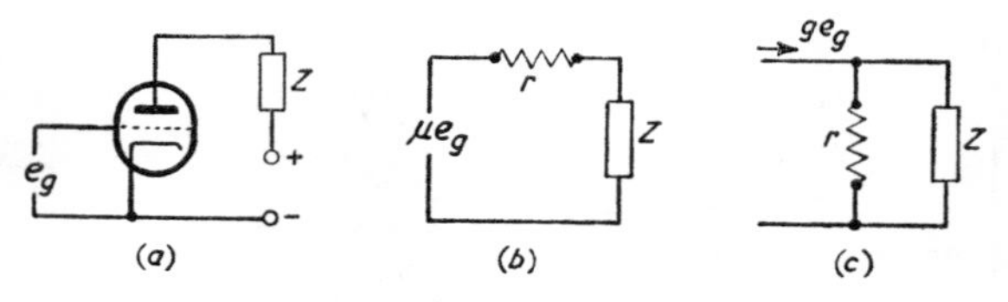

FIG. 5.10. TRIODE AMPLIFIER

equal to ge_g, where g is the mutual conductance of the valve; alternatively we can say that this variation of anode current is that which would be produced by an e.m.f. μe_g operating in the anode circuit, where μ is the amplification factor. We can therefore draw an equivalent circuit using these relationships, which can be done in two ways according to whether one uses the voltage or the current as the primary variable.

Fig. 5.10 (*b*) shows the voltage form. The effective anode voltage μe_g will pass current through a circuit consisting of the internal resistance of the valve (r_a) and the load impedance (Z) in series. Then clearly

$$i_a = \mu e_g/(r_a + Z)$$

$$\text{The output voltage} \quad V = Zi_a = \frac{Z}{r_a + Z} \cdot \mu e_g$$

and the amplification $\quad A = \mu Z/(r_a + Z)$

(Note that if Z is not a pure resistance the term $(r_a + Z)$ is the vector sum.)

Fig. 5.10 (*c*) shows the alternative (current) form. The anode current $i_a = g_m e_g$ flows through r_a and Z in parallel, and the voltage

V is thus given by i_a multiplied by the impedance of r_a and Z in parallel, so that

$$V = \frac{r_a Z}{r_a + Z} \cdot g_m e_g$$

and $$A = g_m r_a Z/(r_a + Z) = \mu Z/(r_a + Z)$$

since $g_m r_a = \mu$. Thus either approach gives the same answer.

In a practical circuit, as will be discussed in later chapters, there are other factors to be allowed for, notably the capacitance between the various electrodes in the valve, but their effect may be assessed by suitable modification of the equivalent circuit. Either form may be used depending on which is more convenient. The same type of equivalent circuit would be used for a tetrode or pentode because the additional electrodes are at a steady potential and do not affect the basic operation of the valve.

It will be noted that the amplification, or *stage gain*, depends on the ratio of the external load impedance to the total impedance, so that when possible the load impedance is made high compared with the valve resistance. This is practicable with triodes where r_a is relatively low, but with tetrodes or pentodes r_a is very high and, in general, Z is only a small fraction of r_a. On the other hand μ is also high which more than compensates for this apparent inefficiency. In fact where r_a is large compared with Z the expression for stage gain above becomes simply $A = g_m Z$.

Phase Relations

Because of the varying anode current the anode potential will also vary. In fact it is clear from the circuit of Fig. 5.1(a) that the anode voltage will be the steady h.t. voltage plus or minus the voltage developed across the load Z. If the grid voltage becomes more positive, the anode current will increase, the voltage drop $i_a Z$ will increase, and the anode voltage will fall, while if the grid goes negative, the anode current will fall and the anode voltage will rise.

Hence the anode voltage variations will be in the opposite direction to the grid voltage variations, being in fact 180° out of phase, if the anode load is a pure resistance. If the anode load is reactive the anode voltage will be $(180 + \phi)°$ out of phase, where ϕ is the phase angle of the load and may be anything between $\pm 90°$.

Choice of Operating Conditions

It will be clear that, in use, the valve does not operate over the static characteristics of Figs. **5**.2 which are taken with zero anode load, because every change in anode current causes a corresponding variation in anode voltage. To assess the performance under working conditions it is necessary to superpose on the appropriate static characteristics a *load line* representing the load impedance.

Consider the characteristics of a triode as shown in Fig. **5.11**. The standing condition with zero signal applied to the grid is at O,

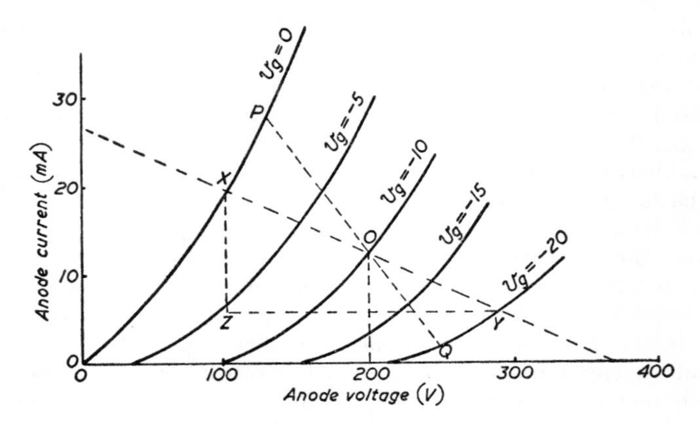

Fig. 5.11. Showing Use of Load Line

with $v_a = 200$ and $v_g = -10$, at which point $i_a = 13$ mA. Suppose the grid voltage is now reduced to zero. Then the current will lie somewhere on the $v_g = 0$ curve. At the same time the anode voltage will have fallen because of the increased voltage drop produced by the increased anode current and we have to find a condition which satisfies both requirements. To do this we draw through the point O a line having a slope proportional to $-R$, where R is the anode load. The line XOY is such a line representing a load of 13,500 ohms for it will be seen that if it is produced it cuts the axes at 375 volts and 27·5 mA respectively, corresponding to $R = 375/0{\cdot}0275 = 13{,}500$.

This line cuts the $v_g = 0$ curve at X so that the anode volts will fall to 100 and the current will rise to 20 mA. Similarly if the grid voltage is increased to -20 the valve will adjust its operating condition to the point Y where $v_a = 295$ and $i_a = 6$ mA.

The peak anode voltage swing is thus 195 volts which has been produced by a grid swing of 20 volts, giving a stage gain of 9·75. It will be noted that the positive anode swing is only 95 volts as against 100 for the negative swing so that there is a small amount of distortion, though this would be quite acceptable for most requirements. If, however, we use a load line such as *POQ* corresponding to a load of 4,500 ohms the anode swing is +50 and −75, a quite marked discrepancy, which would not be acceptable.

The higher the load, the less the slope of the load line and with a triode it is found that, in general, the higher the load the less the distortion. With tetrodes and pentodes, however, the reverse is the case, as is explained in Chapter 10. The design of practical amplifiers thus involves a choice of operating conditions which gives the best approximation to the particular requirements.

It will be noted that the anode voltage can rise appreciably above the standing value. If the anode load is a high resistance, as is often the case, the h.t. supply must be high enough to allow for this. Thus in the example just given the standing current is 13 mA which with 13,500 ohms anode load will cause a voltage drop of 175 volts so that the h.t. supply would need to be 375 volts, which is, in fact, the point where the line *XOY* cuts the axis. If the h.t. voltage were less, say 250 volts, the conditions would need to be recalculated and the input signal would have to be limited.

This limitation only applies with a circuit having a resistive anode load. If the stage is transformer-coupled there is only a negligible steady voltage drop and the standing anode voltage is practically the full h.t. supply voltage. The voltage on the (inductive) transformer primary winding, however, can be either positive or negative according to the change of current through it, so that the actual anode voltage can swing up above the steady h.t. value.

It will be noted that in this example the operations take place around a mean grid voltage of −10 V. The circuit has to be arranged to provide this standing (no-signal) bias by some convenient method, as is described in Part II.

Load Ellipse

We have assumed a straight line for the load characteristic. This is only true when the load is a pure resistance. With a load containing reactance, there will be phase displacement which will cause the load characteristic to become an ellipse. This may run the valve into undesirable conditions during certain parts of its swing, but the effect is considered more fully in Chapter 10 (Fig. 10.17).

Use of Valve as Rectifier

As said earlier, a simple diode valve will only conduct when its anode is positive to the cathode. This *rectifying* action can be utilized in various ways. Consider the circuit of Fig. 5.12 (*a*). An alternating e.m.f. applied to the input will produce pulses of current on the positive half-cycles which will charge the capacitor C. In the absence of any load R, the capacitor voltage will build up to the peak value of the applied a.c.

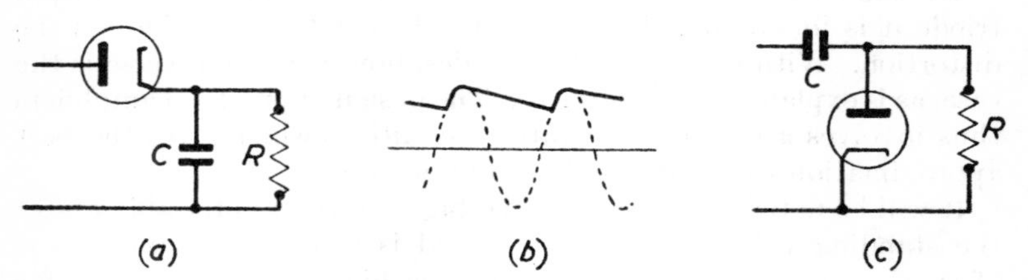

FIG. 5.12. DIODE WITH CAPACITOR

If a load is present it will partially discharge the capacitor during (rather more than) the negative half-cycle, but the charge will be replenished by a pulse of current through the diode during that portion of the positive half-cycle when the instantaneous value of the applied e.m.f. exceeds the capacitor voltage. The output voltage will thus vary as shown in Fig. 5.12 (*b*).

Such a circuit (or its solid-state equivalent) may be used in the detector stage of a receiver, as explained in Chapter 9.

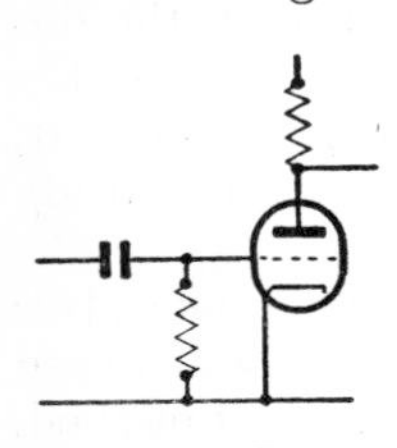

FIG. 5.13. LEAKY-GRID RECTIFIER

Fig. 5.12 (*c*) shows a modification of the circuit in which the diode is connected across the resistor. The action is similar: when the signal e.m.f. is positive, current will flow through the capacitor and the diode (which will have a much lower resistance than the resistor and will thus virtually short-circuit it), while during the negative half-cycle the diode will not conduct and the charge will leak away through R. Note that with this arrangement the polarity of the charge on the capacitor is such as to make negative the side connected to the anode.

Fig. 5.13 shows what is known as a *leaky-grid detector*. Here a triode is used with a capacitor in the grid lead and a resistor between grid and cathode. Thus the grid acts like the diode anode in Fig. 5.12 (*c*) and acquires a negative potential. This will cause the anode current to fall and so develop an amplified signal across the anode resistor. The arrangement is a convenient combination of rectifier and amplifier but is now mainly of academic interest.

Anode Rectification

If a valve is biased at a point towards the bottom of its $i_a v_g$ characteristic (cf. Fig. 5.2), rectification will be produced by virtue of the curvature of the characteristic, since a positive grid swing will produce a greater change of anode current than a corresponding negative swing. A valve in this condition is called an *anode-bend detector*, but this usage is virtually obsolete.

However, if the grid swing is small the mean anode current is very nearly proportional to the square of the grid-voltage swing. This square-law relationship is useful for certain applications, one advantage being that it is not primarily dependent on the exact shape of the valve characteristic and can therefore be used for measurement.*

A similar action can be obtained with semiconductor diodes which are more usually employed today.

Valve Oscillators

If a circuit is set in oscillation the currents will ultimately die away due to the losses in the circuit. If at each oscillation, however, a small e.m.f. could be introduced, sufficient to bring the current up to its original value, the circuit would continue to oscillate indefinitely. Owing to its amplifying properties, a valve can readily be arranged to supply this e.m.f. and so can be made to generate continuous oscillations.

Fig. 5.14 shows the simplest way of doing this. L_1C is an oscillating circuit and due to the oscillations therein an alternating e.m.f. will be impressed on the grid of the valve. This e.m.f. will cause magnified oscillations to take place in the anode circuit, which will flow through the coil L_2. If L_2 is suitably coupled to L_1, an e.m.f. will be induced in L_1 which will be in the same direction

* This may appear strange in view of the fact that the emission characteristic of a valve obeys a three-halves law. However, the characteristic may be represented mathematically in the form

$$i_a = ge + g'e^2/2! + g''e^3/3! + \dots$$

where $e = (e_g + e_a/\mu)$

$g = di_a/de_g$

and g', g'', etc, are the derivatives dg/de_g, $d^2g/de_g{}^2$, etc.

We are interested in the mean value of i_a when e_g varies about its steady value. Provided the valve is not biased to cut-off and the signal is never large enough to reduce the anode current to zero, then the mean value of the first and third terms does not change but the second term represents an increase in i_a (since e^2 is always positive) proportional to e^2.

as the e.m.f. already existing there, and will therefore restore the currents to their original value, as described above, and the circuit L_1C will oscillate indefinitely. The coil L_2 must obviously be coupled in the right direction or the e.m.f. induced will be opposite to the e.m.f. existing in L_1 and will stop the oscillation instead of maintaining it; also the coupling must be greater than a certain critical value, or else the energy fed back will not be sufficient to make up the losses in the oscillating circuit.

A useful rule may be cited in this connection. If the coil L_2 is wound in a given direction from anode to H.T. +, the coil L_1 must

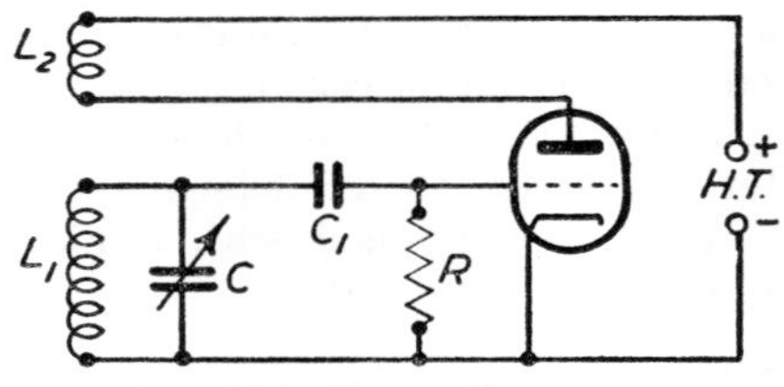

FIG. 5.14. VALVE OSCILLATOR

be in the same direction from cathode to grid if it is desired to sustain the oscillation.

It will be noted that a capacitor C_1 has been included in the lead to the grid. By an action similar to that of the leaky grid detector this capacitor will acquire a charge which will make the grid negative. This will reduce the conductance of the valve and limit the anode current. The valve will thus adjust itself to a suitable operating point on its characteristic such that there is just sufficient feedback to sustain the oscillation.

Various modifications of this arrangement are possible. The tuning capacitance may be connected across the anode coil instead of the grid coil, if preferred, while it is possible for both grid and anode circuits to be tuned. In such a case the coupling required to maintain oscillation is very small and it will usually be found that the internal capacitance between anode and grid of the valve is sufficient to set up and sustain an oscillation.

This internal feedback, in fact, introduces problems in the design of r.f. amplifiers, and precautions have to be taken to prevent unwanted oscillation. The action is discussed in detail in Section 9.2, where it is shown that the feedback is positive when the anode circuit is inductive. Hence if both grid and anode circuits are tuned, oscillation will occur at a frequency slightly above the resonant point, at which the anode circuit exhibits an inductive reactance.

Another circuit, often used because of its simplicity, is shown in Fig. 5.15 (*a*) in which a direct coupling is employed. The oscillatory circuit is connected directly across the grid and anode of the valve. The cathode is connected to an intermediate point on the coil, at or near the centre. The potential of the anode end of the tuned circuit is clearly 180° out of phase, relative to the cathode, with the grid end. But the anode voltage variations produced by the e.m.f. applied to the grid are also 180° out of phase so that they reinforce

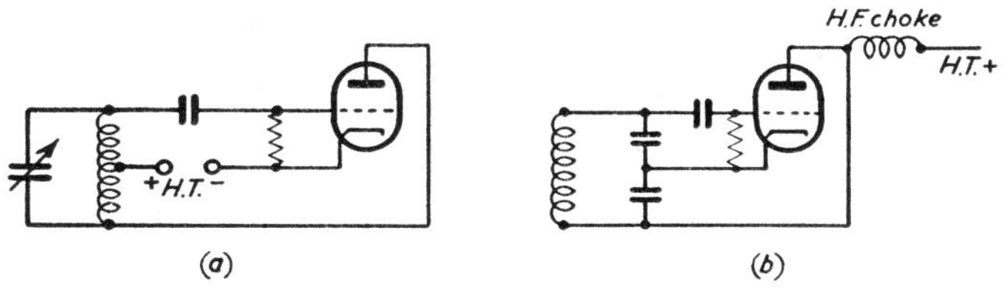

FIG. 5.15. HARTLEY AND COLPITTS OSCILLATORS

the tuned circuit e.m.f. and oscillation results. This circuit is known as the Hartley circuit, named after its originator.

Fig. 5.15 (*b*) shows a modification known as a Colpitts oscillator in which the tapping is taken on the capacitance.

There are other forms of oscillator, some of which do not use conventional tuned circuits. The subject is discussed in more detail in Chapter 12.

5.4. Gas-filled Tubes

It has already been pointed out that in an ordinary valve all gas is removed. For certain applications, however, a small amount of inert gas is deliberately introduced. Consider a triode in which a small amount of gas is present. If the grid is at a certain potential, depending upon the anode voltage, a current will begin to flow. The electrons in the cathode–anode space will encounter molecules of gas which they will ionize, releasing further electrons which in their turn will ionize further molecules and so there will be a cumulative build-up of current producing the avalanche effect mentioned in Section 1.6.

This means that once the current has started to flow it will no longer be under the control of the grid potential but will actually rise to a very high value, limited only by the resistance of the circuit, and moreover this current will continue to flow as long as there is voltage on the anode. It can only be stopped by reducing the anode potential substantially to zero (actually to a value less than the ionization potential of the gas). The electrons in the gas will then recombine so that after a sufficient time interval (which is normally about one millisecond) the gas is again non-conducting and increasing the anode potential will not produce any current until the critical voltage is reached.

The ratio of anode/grid voltage at the "firing" point is called the *control ratio.* Thus with a control ratio of 20 and an anode voltage of 100, no current will flow if the negative grid voltage is greater than 5 volts. Conversely if the grid is set to —5 volts no current will flow until the anode voltage exceeds 100.

It will be seen that such a valve acts as a highly sensitive relay. It passes no current as long as the grid voltage is more negative than the critical value but as soon as this value is reached a large current flows which can only be stopped by cutting off the anode voltage. There are many circuits in which this sort of behaviour is convenient, as will be seen in later chapters. Valves of this kind are known as gas-filled triodes, or *thyratrons.* The gas used is usually neon or argon, though sometimes mercury vapour is used. One difficulty with this class of valve is that the heavy ions which drift towards the cathode tend to destroy the emitting surface of the cathode and therefore precautions have to be taken, both in the manufacture and in the conditions of use, to minimize the damage caused in this way.

This type of valve is now obsolete, having been superseded by a range of much more convenient semiconductor devices known as *thyristors* which are discussed in Chapter 6.

High-Current Rectifiers

The presence of gas within a valve permits a considerably increased current to flow. Hence extensive use has been made of diodes containing a suitable gas filling—usually mercury vapour—as rectifiers in power supply circuits such as are discussed in Chapter 11.

Such valves will pass currents from a few amperes in small sizes to hundreds of amperes in larger sizes. The main disadvantage is that the heavy ions can produce destructive bombardment of the cathodes. This can be minimized by special constructions, but it is necessary to delay the application of the anode voltage for a suitable time—of the order of 30 seconds—to permit stable internal conditions to become established.

Such valves are again obsolescent, since the requirements can be more conveniently met with solid-state devices.

Cold-cathode Tubes

A special type of gas-filled device is the *cold-cathode tube*, which does not require a heated cathode. This contains an inert gas at low pressure but the conditions of operation are rather different from an ordinary valve. Current flows, in fact, as a result of a glow discharge as explained in Chapter 1.6, and hence differs from emission current in two respects:

(*a*) Current will not flow until the voltage on the anode exceeds a certain value.

(*b*) The current is not unidirectional but can flow with either positive or negative anode potential.

The simplest form of the device is the familiar neon tube commonly employed as an indicator lamp. This will glow when the applied e.m.f. (in either direction) exceeds the striking voltage. which is of the order of 80 to 130 volts according to the construction, It may be operated from the standard a.c. supply if a suitable series resistor is incorporated to limit the current once the tube has struck.

By making the cathode of much smaller surface area than the anode the conductance can be made greater when the anode is positive, and the effect may be augmented by coating the cathode with suitable emissive material, but the tube will still conduct with the anode negative and any appreciable current in this reverse direction may damage the cathode surface. With this type of tube, therefore, it is necessary to take precautions to prevent the anode from becoming negative in use.

Stabilizer Tubes

The cold-cathode diode is frequently used as a stabilizing device because of the constant voltage drop which it displays. A typical characteristic is shown in Fig. 5.16. The tube will not conduct appreciably until the anode voltage is sufficient to produce ionization by collision, when the current will rise rapidly to the saturation value. Once the tube has struck, however, the current can be maintained with a somewhat lower anode potential, and moreover, as will be seen from Fig. 5.16 (*b*), there is a region where a very small change in potential occurs over a wide range of current.

In practice such a tube would be used in a circuit such as Fig. 5.16 (*a*). Provided the circuit conditions are such that when the tube is not conducting the voltage across AB exceeds the critical value, the tube will strike. The additional current will cause an additional voltage drop on R_1 which will reduce V_{AB} to some point in the region XX in Fig. 5.20 (*b*). The total current will now be divided between R_2 and the tube, and any variation in the load current through R_2 will be compensated by a corresponding change in the current in the tube. But since the voltage drop across the tube is substantially constant irrespective of the current (within the range

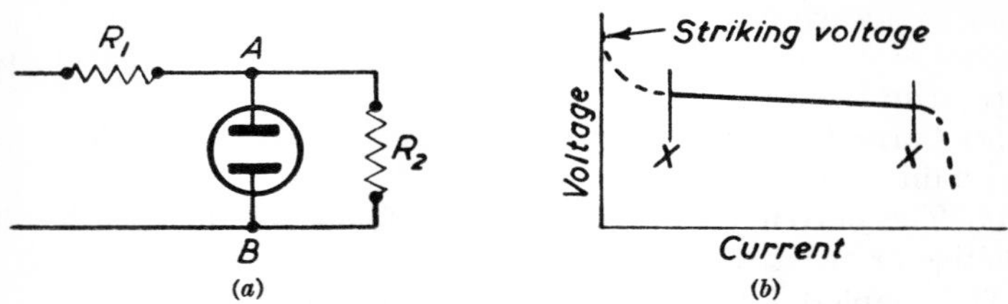

FIG. 5.16. ILLUSTRATING USE OF STABILIZER TUBE

XX) the voltage across R_2 is held constant. Such tubes are therefore usually known as *stabilizer tubes.*

TRIGGER TUBES

If a third electrode is introduced between anode and cathode the point at which the tube becomes conducting may be controlled by varying the potential of this "trigger" electrode, so providing a relay action similar to that of a thyratron. Once started, the current continues to flow until the anode voltage falls below the ionization potential.

Current will flow when the potentials of the trigger and anode exceed a certain value *in either direction*, but to avoid damage to the cathode, current should only be allowed to flow when both anode and trigger electrodes are positive to the cathode.

INDICATOR TUBES

With suitable modification a cold-cathode tube can be arranged to display a series of numbers or characters. A typical device uses a series of cathodes fashioned in the shape of the number or character required. These are assembled in a stack about 1 mm apart, so arranged as to produce the minimum optical interference, so that when any cathode is energized its character is clearly displayed.

The cathodes are normally held at such a potential relative to the (common) anode that no discharge occurs, but by reducing the potential of any given cathode to zero a discharge takes place to this cathode causing it to glow clearly. As a variant the indication can be provided by a series of (seven) co-planar segments. By activating the appropriate segments a stylized figure or letter is displayed. (This type of tube is also made with a heated cathode, the segments being coated with fluorescent material which glows green when actuated.)

An alternative arrangement uses a series of cathodes in a single matrix, such as 7 rows of 5. Then by earthing the appropriate cathodes a suitable representation of the required figure or letter can be produced. This type of tube, of course, requires more elaborate control circuitry.

The use of such tubes is discussed further in Chapter 15.

An earlier form of tube which combines the functions of display and counting is known as a *dekatron*. This carries a ring of 10 cathodes arranged around a central anode. By arranging a series of "guide" electrodes between the cathodes the discharge can be arranged to step from one cathode to the next every time an input pulse is supplied. This device is also discussed in Chapter 15, but developments in integrated circuitry have made it more convenient to perform any counting operations in the control circuits, and to display the results on a simple indicator tube.

5.5. Photocells

As mentioned in Chapter 1 certain materials can absorb the energy in light waves, causing some of the electrons to jump to orbits of higher energy level.

Under suitable conditions electrons may actually leave the material in a similar manner to thermionic emission. Whether this happens or not depends on the wavelength of the light and the work function of the material. Electrons will, in fact, be emitted provided the wavelength is less than $1{\cdot}24/\phi\ \mu m$, where ϕ is the work function.

Only a small range of materials have values of ϕ such that this *threshold wavelength* is within the visible spectrum, the most common being cæsium for which $\phi = 1{\cdot}8$ so that $\lambda = 0{\cdot}685\ \mu m$ which is in the red region of the spectrum.

All the alkali metals, however, have low work functions and so exhibit photo-electric effects. For best results the photo-sensitive material should be in a very thin layer and the usual practice is to deposit a film of cæsium on silver-oxide or antimony, but various other combinations are possible. The wavelength at which the maximum photo-emission is obtained depends on the materials used. There is always a pronounced peak in the wavelength/sensitivity curve.

Vacuum Cells

To utilize the emission it is necessary to provide a suitable anode at a positive potential, as in a valve. A practical photocell thus consists of an evacuated glass bulb, containing a photo-sensitive cathode either in the form of a plate or deposited directly on the inner surface of the glass, while a simple anode in the form of a

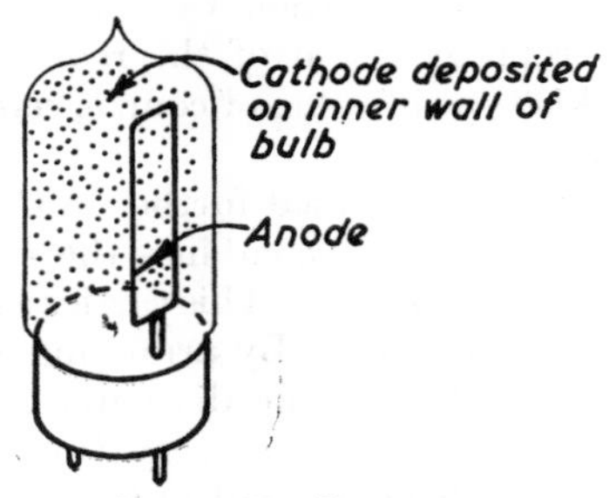

FIG. 5.17. PHOTOCELL

straight wire or loop is located a suitable distance away as in Fig. 5.17. The anode is made in this simple form to avoid obstructing the light falling on the cathode.

The characteristics of a vacuum photocell are shown in Fig. 5.18. With a given illumination the current gradually increases as the anode potential is increased until all the emitted electrons are captured and saturation occurs. As the illumination is increased similar curves are obtained of progressively higher saturation level.

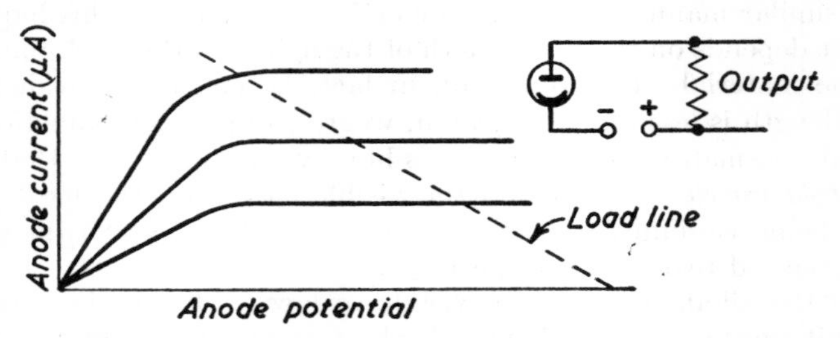

FIG. 5.18. CHARACTERISTICS OF VACUUM PHOTOCELL

The resulting characteristics are similar to those of a pentode with illumination taking the place of grid voltage, and if a high resistance is connected in the anode circuit, as shown, an e.m.f. will be developed across the resistor proportional to the illumination. The extent of this e.m.f. can readily be assessed by drawing a load line through the operating point as shown in Fig. 5.18.

It should be noted that the currents obtained are of the order of microamperes so that load resistances of several megohms are customarily used.

Gas-filled Cells

The anode current for a given illumination may be increased by including a small amount of gas in the cell. The primary photoelectric electrons then produce ionization by collision which results in an increased current.

This process, however, must be strictly limited. When the anode

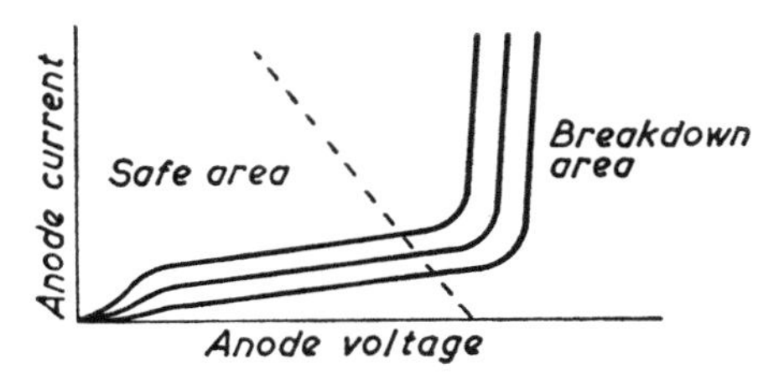

FIG. 5.19. CHARACTERISTICS OF GAS-FILLED PHOTOCELL

voltage reaches a certain value dependent on the illumination, avalanche effect occurs and a continuous discharge takes place which will destroy the cathode if it is allowed to persist. Moreover, even in the safe region the current for a given condition is not precise, and when the illumination is removed the gas takes an appreciable fraction of a millisecond to deionize. Gas-filled cells, therefore, are mainly used as low-frequency relays or occasionally for sound reproduction, under suitably controlled conditions. They are not suitable for measurement. A typical characteristic is shown in Fig. 5.19.

Electron Multipliers

The sensitivity of a vacuum photocell may be increased some thousands of times by utilizing secondary emission. A series of anodes is used at progressively higher potentials, these anodes being coated with a material which has a low work function and is thus relatively easily persuaded to emit secondary electrons as a result of bombardment by primary electrons. Thus each anode in turn acts as a fresh cathode giving out three or four times as many electrons as it receives.

Further details of this technique will be found in Chapter 16.

Photo-conducting and Barrier-layer Cells

There is a further class of photocell which depends for its action on the properties of semiconductors. For example, selenium changes its resistance under the influence of light and can thus be used as a photocell, while if a thin film of a noble metal is deposited on the surface of a semiconductor a transfer of electrons takes place in the presence of light across the barrier between the two materials which actually generates an e.m.f. These devices are a particular application of semiconductor phenomena and are discussed in the next chapter. Because of their smaller size and more rugged construction they have largely replaced the thermionic type of photocell.

6

Semiconductors

6.1. Semiconductor Physics

Reference has been made in Chapter 1 to materials known as semiconductors, which behave partly as insulators and partly as conductors. Many semiconducting materials are known, the most widely used being germanium and silicon, though there are various compounds such as copper-oxide and indium arsenide which exhibit semiconducting properties.

Intrinsic Semiconductors

Broadly speaking, semiconductors are crystalline substances, having their atoms arranged in an orderly pattern known as a lattice. Germanium and silicon both occur in Group 4 of the periodic table, having four electrons in the outer or valence shell. This results in a crystal having a diamond structure with each atom equidistant from four neighbouring atoms and each valence electron being shared between the parent atom and one of the neighbouring atoms. Two electrons, one from each neighbouring atom, thus form what is known as a covalent bond. We see, therefore, that a perfect crystal of germanium is an insulator because there are no available current carriers. In practice, even at ordinary temperatures, a few electrons acquire sufficient thermal energy to break their bonds, thus becoming available as current carriers. Therefore there is a small degree of conductivity and the material is known as an *intrinsic semiconductor*. This conduction clearly increases with temperature and is one of the main limitations to the use of semiconductors.

Impurity Semiconductors

Conducting properties are also conferred by the presence of certain impurities. Consider the effect of the addition to pure germanium of a very small quantity of atomic phosphorus, antimony, or arsenic; these atoms are pentavalent, having five electrons in their outer or valence shells and are comparable in size with germanium atoms. Such an atom can hence occupy a position normally filled by a germanium atom, and owing to the relatively small number of impurity atoms, will normally be completely surrounded by germanium atoms. The four nearest germanium atoms then form covalent bonds with four of the five valence electrons of the impurity, but the fifth electron is held by no such binding force and becomes available for conduction in the same way as a free electron in an intrinsic semiconductor. The material thus becomes an *impurity* semiconductor, and since the current carriers are negative, it is called an *n-type* semiconductor. The impurity atoms are called *donors* since they donate free electrons.

Impurity semiconductors can also be formed by the introduction of small quantities of a trivalent impurity such as boron, gallium, or indium. In this case there are only three valence electrons of the impurity to form covalent bonds with the germanium, so that one of the electrons of a neighbouring atom is left without a partner. This vacancy where an electron would normally be in a perfect lattice is called a positive *hole*: when an electron moves to, say, the left to fill this hole, it leaves a corresponding vacancy behind and the hole appears to move to the right. It is convenient to regard this moving hole as a current carrier and the resulting material is called a *p-type* semiconductor. The impurity atoms are called *acceptors* since they accept electrons from the germanium atoms.

Now that the concept of positive holes has been introduced it may be mentioned that in intrinsic semiconductors every time a valence bond is broken thermally not only is an electron released for conduction but a positive hole is also created, so that positive and negative current carriers exist in equal numbers. Thus in an impurity semiconductors there will always be a small concentration of carriers of opposite sign to those produced by the impurity action. These carriers are known as *minority carriers* and those produced by the impurity as *majority carriers.*

Energy Levels

To appreciate the behaviour of semiconductors fully, however, it is desirable to consider the phenomenon of electrical conduction in greater detail, particularly from the standpoint of energy levels. Quantum theory postulates that electrons in matter can exist only at specified levels of potential energy corresponding to the distance of their orbits from the nucleus. Since it takes energy to move an electron from an inner orbit to one that is more remote from the nucleus the electrons of an atom will prefer the inner orbits: it is, however, a fundamental law of nature that no more than two electrons in an atom may occupy the same energy level. Therefore as further electrons are accommodated in an atom they have to occupy orbits of progressively higher energy level.

It is also found that when two similar atoms are brought close together there is an interaction or coupling between the orbits of their electrons that causes each individual energy to be split into two slightly different levels. Consequently in a solid, where the concentration of atoms is high, interaction between atoms produces a large number of energy levels so close together as to form a band that can be regarded as essentially continuous.

Conduction of electricity is possible only when there is present an energy band that contains some electrons but not the theoretical

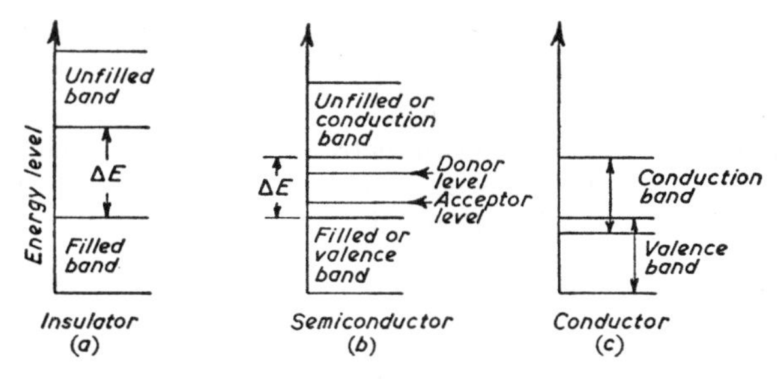

FIG. 6.1. ENERGY BANDS IN VARIOUS TYPES OF MATERIAL

maximum number that it is permitted to accommodate. A completely filled band means that there are no free electrons whereas a band that is partially filled denotes that free electrons are present. Since the electrons always fill the lower energy bands first the conductivity of a solid is determined by the situation existing in the highest level energy band that contains electrons and by the relationship of this band to the next possible band. The various situations that can exist are shown in Fig. 6.1.

In insulators the lower band is filled while the upper one is empty and separated from the lower by a gap of several electron-volts: the probability of an electron gaining enough energy at ordinary temperatures to jump the gap to the higher level is very small and thus the conductivity is very low.

In semiconductor-type material the gap between the energy bands is much smaller, 0·78 eV for germanium and 1·12 eV for silicon, so that even at ordinary temperatures some electrons can gain enough thermal energy to reach the upper energy band; conduction then takes place due to the fact that the upper energy band now has some electrons in it and also because the lower energy band is no longer completely filled. The addition of impurity atoms to the semiconductor provides new permissible energy levels, just below the conduction band with an n-type impurity and just above the valence band for a p-type impurity. In consequence the gap is reduced to a small fraction of an electron-volt, so that much smaller increases in energy permit electrons to cross the relative gaps, allowing conduction due to their presence in the conduction band or their absence from the valence band.

In good conductors the valence and conduction bands overlap, allowing free movement of electrons and high conductivity.

6.2. Semiconductor Diodes

The essential feature of semiconductors is that, since their current carriers are predominantly electrons *or* holes, they exhibit substantially unidirectional conductivity.

The simplest form of semiconductor device is the diode, which is basically a unidirectional conductor like the thermionic diode and has many applications in communications engineering.

The Junction Diode

Consider a piece of semiconducting material in which there are both p-type and n-type regions. The boundary between such regions is called a *p–n junction*, and it is upon the electrical properties of this junction that modern semiconductor electronics is based. Such a semiconductor is called a *junction diode* and may be constructed by one of two principal techniques. The junction may either be formed by the introduction of the appropriate impurities at different times during the growth of the crystal or by fusing a small quantity of p-type impurity into n-type material or vice versa. The two types of junction so formed are known as "grown" and "alloyed" junctions and have substantially the same electrical properties.

Various refinements of these processes are adopted to meet specific requirements, as detailed in Appendix 4, but they all basically depend on the provision of a boundary between p- and n-type materials.

Consider the electrical properties of such a junction. Since the majority carriers (positive holes in p-type material and electrons in n-type material) are free to wander at random, they tend to diffuse across the boundary between the two regions. As a result the n-region acquires a positive charge and the p-region a negative charge, and an equilibrium is reached. These charges are confined to the boundary, forming an ionized *depletion layer*; the existence of this ionized space-charge region gives asymmetrical conducting properties to the junction and the transfer of an electron all the way across the junction requires the expenditure of a certain amount of energy which is termed the barrier energy or barrier height.

If a potential is applied across a junction so as to make the p-type side positive and the n-type side negative, positive holes will move

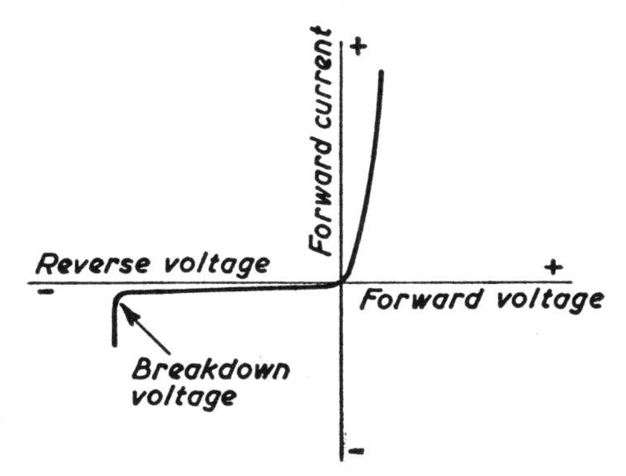

FIG. 6.2. CHARACTERISTICS OF A TYPICAL JUNCTION DIODE
The reverse current is greatly magnified

across the junction from p to n and negative electrons from n to p. If, however, the potential is applied the other way round there will theoretically be no current flow because there are no available current carriers of the right polarity; a perfect diode would then be formed by the junction. In practice there is a small reverse current due to the presence of minority carriers, i.e. electrons in p-type material or holes in n-type material, these minority carriers being formed by the same thermal action that produces carriers in intrinsic semiconductors.

The basic form of characteristic is shown in Fig. 6.2. The forward current does not begin to flow to an appreciable extent until the applied voltage exceeds a small amount known as the forward voltage drop, which is of the order of 0·15 V for germanium and 0·6V for silicon. Thereafter it rises rapidly in an approximately exponential manner (i.e. $di/dv = k(v - v_f)$ approximately, where v_f is the forward voltage drop or "hop-off").

In fact, with zero volts across the diode there is a very small reverse leakage current i_0. The characteristic can thus be expressed in the form $i = i_0\ (e^{kv} - 1)$, where $v = V - v_f$. When the applied voltage $V = 0$, $i = (1 - 1) = 0$. Below this value i is (slightly) negative, while above it i increases rapidly, as in Fig. 6.2.

The reverse current is due to the presence of minority carriers in the material and is thus very much smaller than the forward current. As the reverse voltage is increased the reverse current reaches a saturation level and thereafter only increases slowly. Ultimately, however, a point is reached where some of the covalent bonds are broken, causing the release of additional carriers and a cumulative (avalanche) effect occurs resulting in a rapid increase in current. This causes the temperature of the junction to rise, which increases the current further, leading to a condition known as *thermal runaway*, which destroys the material.

It may be noted that this condition can also arise if the forward current is excessive, so that with any semiconductor device suitable arrangements have to be made to limit the temperature of the junction, partly by providing adequate heat conduction and partly by limiting the current. For small power applications it is sufficient to limit the current, but for larger powers it is necessary to mount the material in a relatively massive housing which can be bolted to a metal heat sink having a large radiating area. With such precautions quite small semiconductor elements will handle currents of several hundred amperes.

This is a much better performance than can be obtained with intrinsic semiconductors like copper-oxide or selenium, or thermionic devices, which are now obsolescent. Germanium diodes will withstand a reverse voltage of several hundred volts with junction temperatures up to 70 to 90°C. Silicon types, though having a slightly greater forward voltage drop, will withstand reverse voltages of the order of 1,000 and can be used at temperatures up to 150°C.

Zener Diodes

Special forms of silicon diode can be constructed in which the avalanche effect occurs at voltages well below the normal breakdown voltage. These use a heavily doped p–n junction with a very small (normal) reverse current. At a certain voltage, usually within the range 5 to 15 volts, an avalanche effect occurs producing a very sharp rise in current, but because of the low voltage and leakage current there is little temperature rise and thermal runaway does not occur (provided that the permissible dissipation is not exceeded). If the voltage is reduced, the current falls to its original small value, so that the action is reversible. This is known as the *Zener* effect and the point where the abrupt transition occurs, called the Zener voltage, is precise and consistent, so that it may be used as a reference voltage.

Zener diodes are frequently used in stabilized supply circuits as described in Chapter 11 (Figs. 11.14 and 11.15).

The Point-contact Diode

An alternative form of diode is the *point-contact diode.* This consists of a springy pointed wire in contact with the surface of a piece of impurity semiconductor material. The material of the contact is usually a tungsten wire, while the semiconductor used is either silicon or germanium.

The point-contact diode can be regarded as a special case of the junction diode. The forming process used in the manufacture, in which carefully controlled pulses of current are passed through the rectifier, is thought to produce, for instance, a small p-region directly under a metal point in contact with an n-type semiconductor. Thus a very-small-area p-n junction is formed under the metal point.

The semiconductor used can be either p- or n-type. P-type silicon point-contact diodes are extensively used as detectors and mixers at microwave frequencies, where their higher forward resistance more readily matches that of the associated circuit. Because of the small contact area, the capacitance of these diodes is less than 1 pF.

Diode Capacitance

In the forward direction the diode impedance is purely resistive. A reverse bias across the junction, however, forms an ionized depletion layer so that the device holds a small charge and behaves

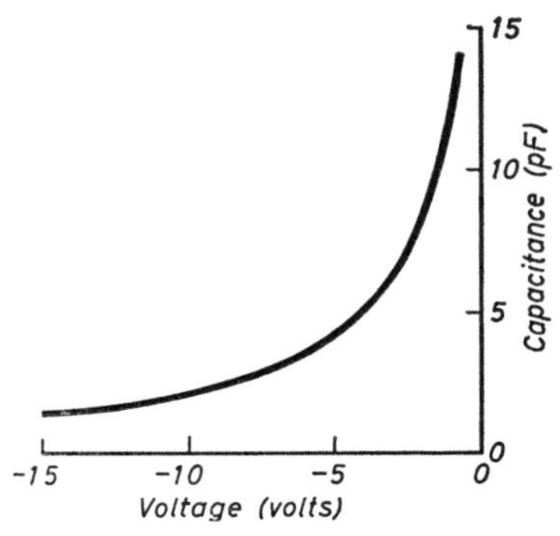

FIG. 6.3. TYPICAL CAPACITANCE CURVE FOR A REVERSE-BIASED P–N DIODE

as a small capacitor shunted by a high resistance. The value of this capacitance is an inverse function of the bias, being of the form $C = kV^{-1/n}$ where n is between 2 and 3. Fig. 6.3 shows a typical characteristic.

This effect can be used to provide a controllable variation of frequency in an oscillator for f.m. or a.f.c. circuits.

Applications of Diodes

Semiconductor diodes are normally used for the purposes of rectification and detection, as with the now obsolete thermionic diode. They have, however, the important advantage that by suitable choice of material (as explained in Appendix 4) they can be made to exhibit certain special properties. These will be discussed more fully in their appropriate context, it being sufficient here to note some of the more important of these special forms.

Varactor Diode

This is a modification of the normal diode which follows a specific capacitance-variation law, and is also designed to have a low internal loss at high frequencies. In one form the capacitance is given by $C = kV^{-1/2}$. Then $Q = kV^{1/2}$. If the diode current is $i = I \sin \omega t$, then $Q = \int i\,dt = (I/\omega) \cos \omega t = kV^{1/2}$. Hence $V = (I/k\omega)^2 \cos^2 \omega t$, so that the device has a square-law characteristic, which has various uses. In particular, since $\cos^2 \omega t = \frac{1}{2}(1 + \cos 2\omega t)$, it can be used as an efficient frequency doubler.

Tunnel Diode

This uses a very thin wafer of material with a high concentration of impurity. With this construction the reverse-bias region does not exhibit the usual high resistance, while the forward region possesses a negative-slope region as shown in Fig. 6.4 (a). It was discovered

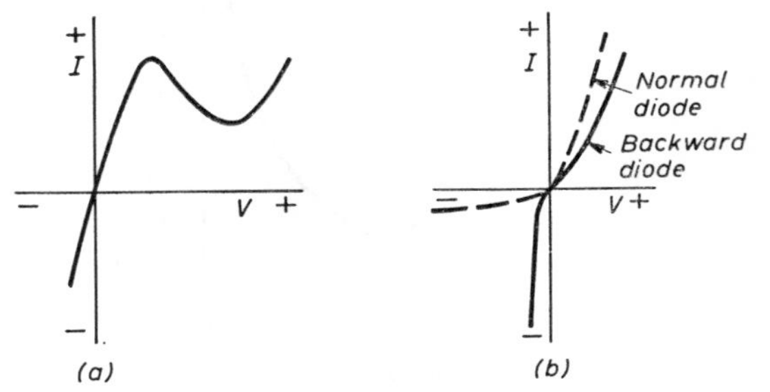

FIG. 6.4. CHARACTERISTICS OF SEMICONDUCTOR DIODES
(a) Tunnel diode (b) Backward diode

by Esaki in 1958 and has obvious uses in generating dynatron-type oscillations, for which it can be used at frequencies up to several GHz.

Gunn Diode

This is named after J. B. Gunn, who discovered that, by passing current through n-type gallium arsenide at high field strength, electrons can transfer from states of high to low mobility. This produces a decrease in average electron velocity with field strength, which is equivalent to a negative resistance characteristic and hence can be used to generate oscillations. The frequency is dependent on the thickness of the layer and is in the range 1–100 GHz.

Backward Diode

By arranging the Zener point of a diode to coincide with the origin, the reverse-bias region has a higher conductance than the forward region, as shown in Fig. 6.4 (*b*). Such a device has certain special applications, notably as a microwave mixer.

Switching Diodes

Computer applications involve circuits which are required to change rapidly from an "off" to an "on" state. With a normal diode (or transistor) there is a small time lag but with suitable construction this can be reduced to less than 1 nanosec (10^{-9} sec). This type of usage is discussed further in Chapter 18.

P–I–N Diode

This is a high-resistance diode which is used as a controllable variable resistor rather than a rectifier. It is described more fully in Chapter 13 (page 609).

6.3. The Transistor

The usefulness of a semiconductor element would clearly be greatly increased if the conductivity could be made responsive to some control electrode, analogous to the grid in a thermionic valve. Such a device is known as a *transistor*.

The junction triode transistor is a logical development from the junction diode and consists essentially of two junction diodes arranged physically so that the forward current of one influences the reverse current of the other. Such an arrangement is shown in Fig. 6.5, which will be seen to be a sandwich of p–n–p material. The portions E and B constitute one junction diode biased in the forward direction and the portions B and C constitute another diode biased in the reverse direction.

Consider first the behaviour of the mobile charges at the junction EB. The majority carriers in the portion E, which is called the *emitter*, will be swept across the junction EB into the region B, which is called the *base*, where they become, in effect, minority carriers. If the base region were thick or of high conductivity these minority carriers would quickly combine with the majority carriers already present in the region; when, however, the base is made only a few thousandths of an inch thick and of lower conductivity than the emitter, the minority carriers diffuse through the base to the junction BC. Here they come under the influence of the strong electric field due to the reverse bias across the junction and are swept across into the portion C, which is hence called the *collector*.

Current Transfer Ratio

Under these conditions nearly all the holes emitted into the base will reach the collector, so that a change in emitter current will

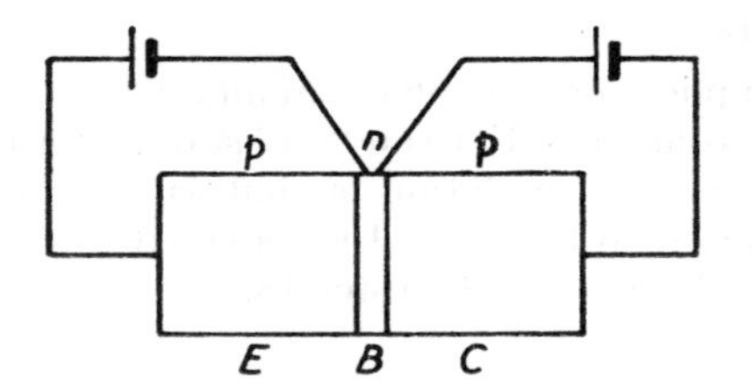

FIG. 6.5. THE P-N-P JUNCTION TRANSISTOR

cause nearly as great a change in collector current. The ratio of these two changes is called the *current transfer ratio*, α, and is typically just less than unity for a junction transistor. We can now see that a transistor has amplifying properties since it can be regarded as two diodes back-to-back, the connections being such that the emitter–base circuit is of low resistance while the base–collector circuit is of high resistance; thus, since the current gain is only just less than unity, the power gain will be considerable.

This factor α is a fundamental property of the transistor and is a measure of the current transfer through the device. It is sometimes loosely, but incorrectly, called the current gain, but in fact, as will be seen shortly, the transistor is not necessarily used in the manner shown in Fig. 6.5. In particular, a transistor is often used with the input applied to the *base*, in which case the current gain has to be expressed differently. Any such modifications, however, are merely adaptations of the basic parameter α which remains the fundamental and most important property of the transistor.

The arrangement of Fig. 6.5 constitutes what is called a *p–n–p transistor*. A similar action can be obtained with a sandwich of p-type material between two n-type materials. This is known as an *n–p–n transistor* and behaves in a similar manner, except that the battery voltages have to be reversed. In this case the majority carriers are electrons which have a somewhat better mobility than holes, and therefore provide a better high-frequency performance.

Because the operation involves both types of charge carrier, devices of this type are known as *bipolar* transistors, as distinct from the field-effect types described in Section 6.6. which only involve one type of carrier.

Practical Forms of Circuit

There are various ways of using a transistor in practice. Fig. 6.6 illustrates the arrangement corresponding to that of Fig. 6.5. where

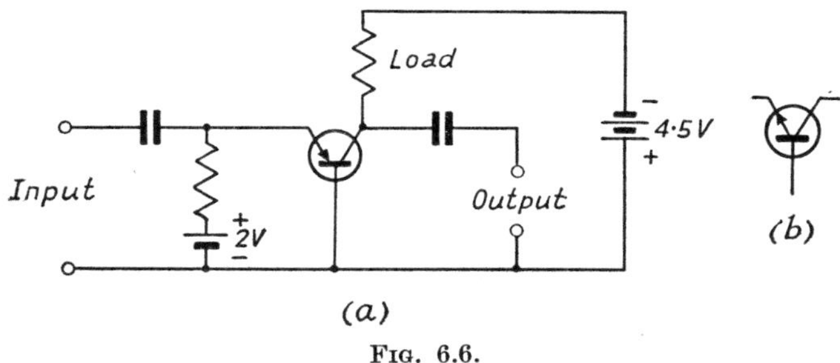

FIG. 6.6.

(*a*) Common-base circuit with p-n-p transistor
(*b*) Graphical symbol for an n-p-n transistor

the input is applied to the emitter with the output taken from a suitable load in the collector circuit. The base is thus common to both input and output so that the arrangement is known as the *common-base* or grounded-base mode. The conventional symbol for the transistor will be noted, the base being denoted by a short (thick) line with the collector and emitter drawn from it as shown. The emitter is distinguished by an arrow which, with a p–n–p transistor is directed towards the base. For an n–p–n transistor the arrow is directed away from the base as in Fig. 6.6 (*b*), the circuit being otherwise unaltered except that the battery voltages would be reversed.

The characteristics of a typical transistor connected in this manner are shown in Fig. 6.7. They will be seen to be similar to those for a pentode valve. In particular it will be noted that:

(*a*) The collector current i_c is almost independent of the collector voltage over the whole working range down to zero. This means that the transistor can be operated (for small signals) on very low battery voltages.

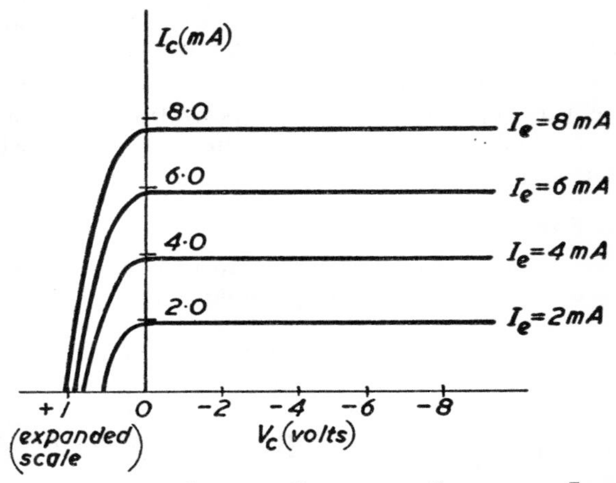

FIG. 6.7. COLLECTOR CURRENT CURVES FOR GERMANIUM JUNCTION P–N–P TRANSISTOR IN COMMON-BASE CONNECTION

(*b*) The almost horizontal slope of the characteristics indicates a very high (a.c.) collector resistance. In practice the effective resistance is modified by the circuit conditions, but it is still very large compared with the input resistance, which is the reason for the amplifying action, as said earlier.

(*c*) The factor which mainly influences i_c is the emitter current i_e; the two currents are, in fact, nearly the same, since $i_c = \alpha i_e$. Hence whereas in a pentode valve the anode current is controlled by the grid voltage, the equivalent control in a transistor is exercised by the emitter *current*. For a proper understanding of transistor behaviour, in fact, it is essential to regard it as a current-operated device.

These characteristics (and those of Fig. 6.10) relate to a p-n-p transistor. For an n–p–n type they would be similar but the values of V_c would be positive.

Since $i_c = \alpha i_e$, the current transfer ratio in this mode is α, which typically has a value of the order of 0·98.

Current Relationships

The currents in the different elements are shown in Fig. 6.8. The (a.c.) emitter current i_e divides between the collector and the base, the greater part ($= \alpha i_e$) flowing in the collector, the remainder (i.e. the small proportion lost due to recombination) flowing in the base. Hence $i_e = i_b + i_c$. This is a fundamental relationship which applies to any transistor.

It is not a static relationship because of the presence of leakage currents. In particular there is a current due to the minority carriers in the collector, which is called the *collector leakage current*, I_{co}. The d.c. collector current I_c is thus $\alpha I_e + I_{co}$. The current I_{co} also

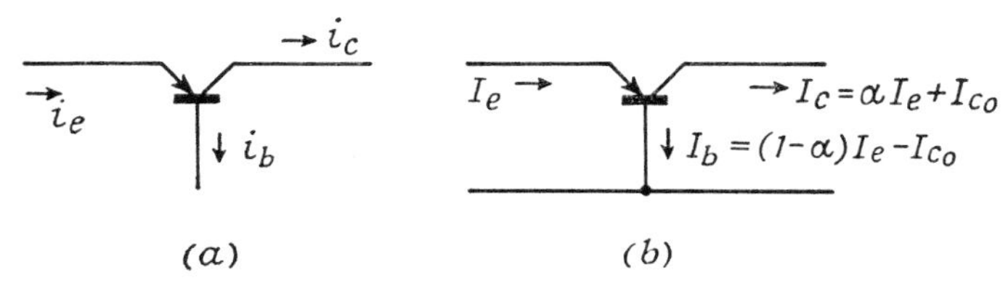

FIG. 6.8. CURRENT RELATIONSHIPS IN COMMON-BASE MODE

flows through the base in opposition to that due to the true transistor action, as shown in Fig. 6.8 (*b*), so that the base current

$$I_b = (1 - \alpha)I_e - I_{co}$$

The difference is insignificant unless $(1 - \alpha)I_e$ is comparable with I_{co}. Hence the directions of the currents shown in Fig. 6.8 (*a*) apply for d.c. conditions unless I_e is very small, when I_b may be reversed.

In practice the direction of the d.c. current I_b is only important in its effect on the biasing arrangements. It is frequently convenient to develop a bias voltage by inserting a resistor in the base connection, which will normally develop a bias voltage in such a direction as to reduce the effective emitter–base potential.

The collector leakage current with the common-base mode of Fig. 6.6 is very small, being of the order of 5 μA for a small-signal germanium transistor and virtually negligible with silicon types.

It may be noted that, in relation to the collector voltage, the collector carries a reverse current. Hence in some diagrams i_c and i_b are shown in the opposite direction to those in Fig. 6.8. This is a matter of convenience, depending upon the voltage to which the currents are referred.

Common-emitter Mode

The form of circuit shown in Fig. 6.6 has certain disadvantages in practice and for many purposes the alternative arrangement shown in Fig. 6.9 is used. Here the input is taken to the *base*, with the

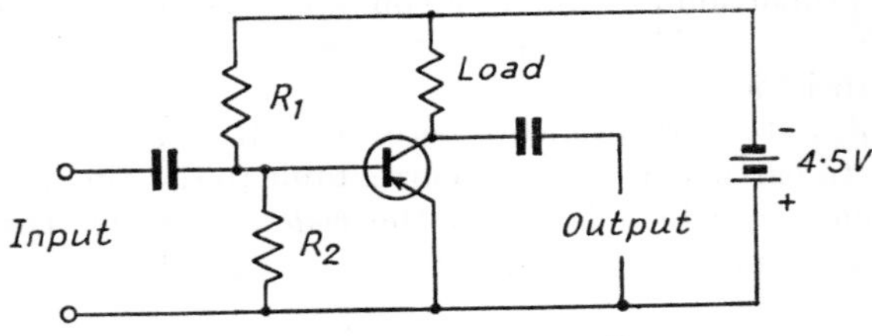

FIG. 6.9. COMMON-EMITTER CIRCUIT

emitter earthed, the arrangement being known as the *common-emitter* mode. With this circuit only one battery is required, the voltage divider R_1R_2 being adjusted to provide a suitable base–emitter potential.

Here, although the basic action is the same, the controlling factor is now the *base* current $i_b(= i_e - i_c)$; and since i_c is nearly equal to i_e, i_b is quite small—usually between 2 and 3 per cent of i_e. With this arrangement the current transfer ratio is no longer $di_c/di_e(= \alpha)$, but is di_c/di_b. Now since $i_b = i_e - i_c$ we can write

$$di_b/di_c = di_e/di_c - di_c/di_c = \frac{1}{\alpha} - 1$$

Hence the current transfer ratio $di_c/di_b = \alpha/(1 - \alpha)$.

It is customary to use the symbol α' or β for this modified factor applicable to the common-emitter mode. If $\alpha = 0{\cdot}98$, $\alpha' = 49$, which means that a transistor is some 50 times more sensitive to changes in base current than to changes in emitter current.

The characteristics of a transistor in common-emitter mode are shown in Fig. 6.10. It will be seen that they are similar to those for the common-base circuit, but the slope of the "horizontal" portion is increased, indicating a lower output resistance; on the other hand, the smaller values of control current (i_b) indicate a much higher input resistance. Finally, the knee of the curves now appears at a small negative potential, but this is still only a fraction of a volt.

Collector Leakage Current

The collector leakage current is the same as in the common-base mode, but its influence on the performance *expressed in terms of base*

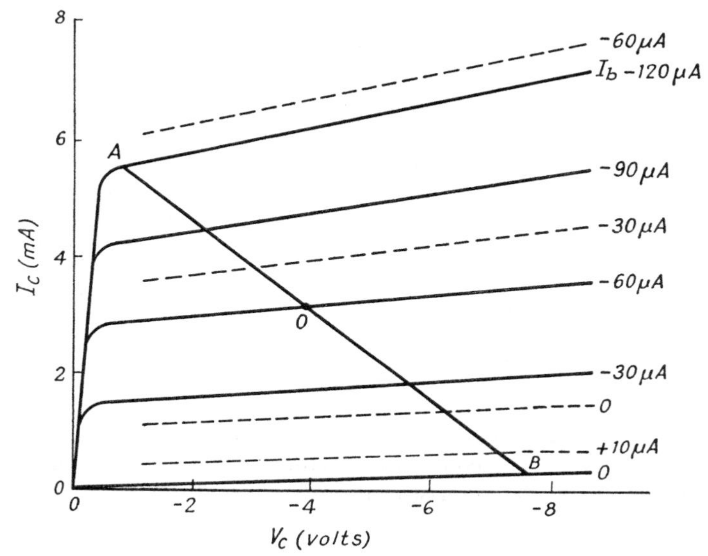

FIG. 6.10. CHARACTERISTICS OF A TYPICAL GERMANIUM P–N–P TRANSISTOR IN A COMMON-EMITTER CONNECTION

Solid curves 25°C Dotted curves 45°C

With silicon types the variation of current with temperature is considerably less

current is considerably greater. The fundamental relationship is $I_c = I_{co} + \alpha I_e$.

But $I_e = I_c + I_b$, from which the expression may be written

$$I_c = I_{co}/(1 - \alpha) + \alpha' I_b$$

In terms of the base current therefore the collector leakage $I'_{co} = I_{co}/(1 - \alpha)$, which is some 50 times greater than I_{co}. Its effect can be seen in the characteristics of Fig. 6.10 in which there is a small collector current even when the base is open-circuited ($i_b = 0$). This is I'_{co} and will be seen to increase proportionally with V_c.

The expression above may be written in the form

$$I_b = (I_c - I'_{co})/\alpha',$$

whence it will be seen that if I_c (the true transistor current) is less than I'_{co}, the base current will become negative, tending to a value I_{co}/α when $I_c = 0$.

The effect of collector leakage is much less marked with silicon transistors, with which, as said earlier, I_{co} is usually very small.

Working Point

In both common-base and common-emitter modes the steady (d.c.) potentials on the electrodes need to be suitably chosen to accommodate the necessary current swings. Thus in Fig. 6.10 a suitable working point would be O, the line AOB representing a resistive load of 1250 ohms in the collector circuit of Fig. 6.9, with a steady base current of 60 μA. This brings the working value of collector voltage to 3·9 volts, with a swing from 0·7 to 7·5 volts, so that a 9 volt battery would suffice. This would, in fact, be the maximum permissible swing. The point A coincides with the knee of the curves so that any increase in i_b (along OA) produces no further increase in i_c and the transistor is said to be "bottomed." Similarly at B the collector current is (nearly) zero and the transistor is said to be "cut off."

Input Characteristics

As said earlier, the controlling factor in a transistor is the input current, and the characteristics in Figs. 6.7 and 6.10 are plotted for various values of i_e and i_b respectively. The voltage required to produce these currents, however, is not directly proportional to the current, which increases rapidly beyond the hop-off point in an exponential manner. The emitter–base junction is, in fact, a forward-biased diode, so that with the common-base connection the input current is of the form $I = I_o(e^{kv} - 1)$, as discussed earlier (page 264).

With the common-emitter connection the value of I_o is increased by reason of the collector leakage current I_{co}, which in fact becomes the predominant factor, so that with $V_b = 0, I_b = -I_{co}/\alpha$, as shown in Fig. 6.11.

A transistor will thus only provide a linear output, i.e. an output proportional to the input, when used under *current-drive* conditions, in which the input signal is arranged to produce appropriate variations in input *current*. It is nevertheless often required to operate under *voltage-drive* conditions, with the input providing variations of input *voltage*. In these circumstances, due allowance has to be made for the variation of input impedance, as is discussed in Section 6.4.

Effect of Temperature

The basic mechanism of semiconduction is essentially dependent on temperature, as was explained in Section 6.1. Hence the performance of a transistor is considerably affected by the temperature of the junction both in respect of the ambient temperature and the local heating resulting from the current in the device itself. An important part of transistor design is thus concerned with the limitation of temperature rise, particularly in large-signal (output) stages.

The current transfer ratio is only slightly affected, increasing by about 0·5 per cent per degree C, but the leakage currents I_{co} and I'_{co} are considerably increased as the temperature rises, as one would expect since the collector leakage is due to the intrinsic minority carriers in the material. For example, with a small-signal germanium transistor, I'_{co} at 45°C could be some eight times its value at 25°C. This affects the input characteristic, as shown dotted

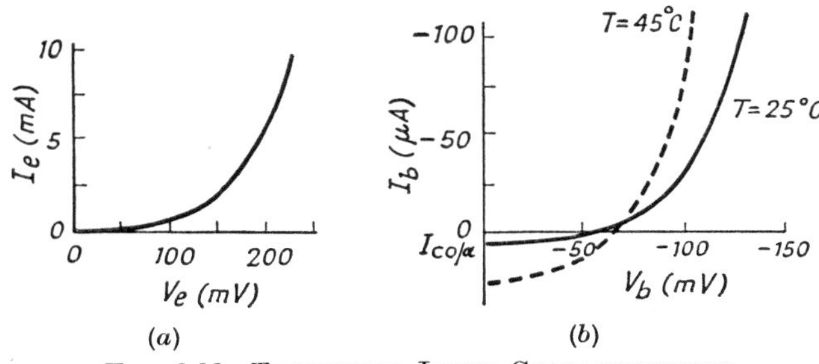

FIG. 6.11. TRANSISTOR INPUT CHARACTERISTICS
(a) Common Base (b) Common Emitter

in Fig. 6.11. It also affects the output characteristics of Fig. 6.10 where it has the effect of raising all the characteristics, as shown by the dotted lines, with consequent alteration of the working point and limitation of the available swing.

This restricts the use of germanium transistors to conditions of reasonable, and not widely varying, temperature. At temperatures approaching 100°C the leakage current may entirely swamp the transistor action, and for safety the maximum temperature *of the junction* should not exceed 80°C.

Silicon transistors are much better in this respect and can be safely used with junction temperatures as high as 150°C. The only operational difference with silicon types is the slightly greater "hop-off", so that current does not flow until the base voltage exceeds between 0·6 and 0·7 V. This has negligible influence on the performance as an amplifier, but may have some significance in switching circuits.

Effect of Frequency

A more detailed discussion of the various circuit arrangements appears in the following sections, including a third (*common-collector*) mode which is sometimes used. It will be evident, however, that whatever the mode, the performance is fundamentally dependent on the basic parameter $\alpha = di_c/di_e$. (Note that this is *not* the ratio of the standing currents I_c/I_e.) This is not independent of frequency, but falls off as the frequency is increased. The value at low frequencies (approaching zero) may be determined from the static characteristics, or by direct measurement, and is usually given the symbol α_0.

As the frequency is increased, various internal effects begin to appear which introduce limitations. In the first place, the diffusion of the current carriers across the base is a relatively slow process so that at high frequencies only a proportion of the carriers reach the collector before the polarity of the applied e.m.f. reverses. As said earlier, n–p–n transistors, in which the carriers are electrons (which have greater mobility), are appreciably better than the p–n–p type at high frequencies, but they are still subject to limitations.

In addition to this transit-time effect, the performance is limited by the internal capacitances, particularly that between the collector and the base, and these various internal factors combine to produce a progressive reduction in the current gain as the frequency is increased. The frequency at which the current transfer ratio falls to $1/\sqrt{2}$ times its zero-frequency value (i.e. 3 dB down) is called the *alpha cut-off frequency* and is given the symbol f_α in common-base mode, or $f_{\alpha'}$ in common emitter mode.

The alpha cut-off, however, is not easy to measure, and its use has certain other disadvantages in practice, so that it has been largely replaced by an alternative parameter f_1 (or f_T), which is discussed in Section 6.5 (page 288).

Alloy-diffusion Transistors

As just said, the high-frequency limitations in a transistor are due mainly to the relatively slow diffusion of holes (or electrons) through the base material. The effect may be minimized by making the base region very thin, and then providing an accelerating or "drift" field across the junction. Thus in a p–n–p transistor the p-type emitter pellet would be doped with both p- and n-type additives. The assembly is heated to a high temperature for a carefully-controlled time during which the n-type additive penetrates the base more deeply than the p-type and so forms a graded layer which, in use, accelerates the holes towards the collector. This is known as the *alloy-diffusion* process.

A modification of the alloy-diffusion process is what is called an *epitaxial* construction. The normal transistor has a high collector resistance and high reverse potential, but this necessarily limits the current which can be passed. To overcome this, the main portion of the collector is made from low-resistance material, on which a thin layer of high-resistance material is deposited. Base and emitter regions are then formed in the epitaxial layer by the conventional alloy-diffusion process. This maintains the advantages of high reverse potential (and low leakage) with a lower effective collector resistance, permitting greater current without saturation. By providing a thin silicon-oxide layer over the epitaxial layer the leakage may be still further reduced. This is known as planar epitaxial construction.

Further details of these and other forms of construction are given in Appendix 4.

Avalanche Effect

The collector of a transistor is a reversed-biased diode so that if the voltage across it becomes excessive it will run into the avalanche region as explained on page 264. The characteristics in this region are illustrated in Fig. 6.12 and will be seen to display certain

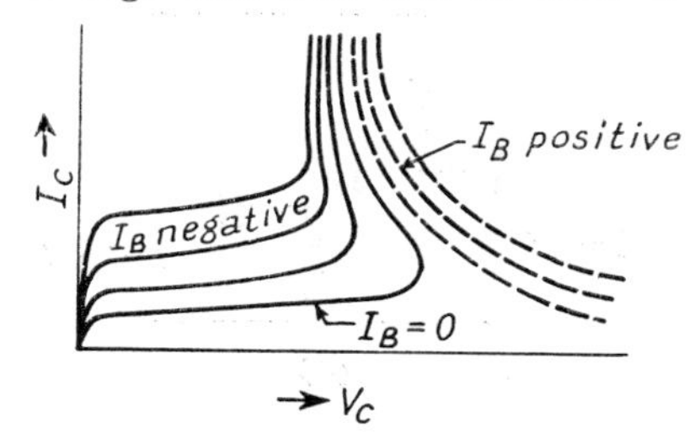

FIG. 6.12. ILLUSTRATING AVALANCHE EFFECT

interesting effects. The onset of the effect depends upon the base current, but once the action has started the base loses control until a substantial *positive* current is injected into the base. The characteristics, in fact, turn back on themselves, as shown.

Provided that precautions are taken to limit the current so that thermal runaway does not occur, this effect can be used to produce large controlled current pulses, as discussed in Chapter 18.

6.4. Transistor Performance

There are three basic circuit configurations for a transistor, depending on the relative connections of the electrodes. These are illustrated in Fig. 6.13, and may be summarized as follows:

(*a*) *Common-base mode.* Here the input is fed to the emitter with the output taken from the collector, with the base common to both as shown in Fig. 6.13 (*a*). The arrangement is characterized by a very low input impedance and high output impedance.

(*b*) *Common-emitter mode.* Here the input is fed to the base, with the emitter earthed (Fig. 6.13 (*b*)). This has a substantially higher input impedance and is therefore much more frequently adopted, but its high-frequency performance is not as good as with the common-base mode.

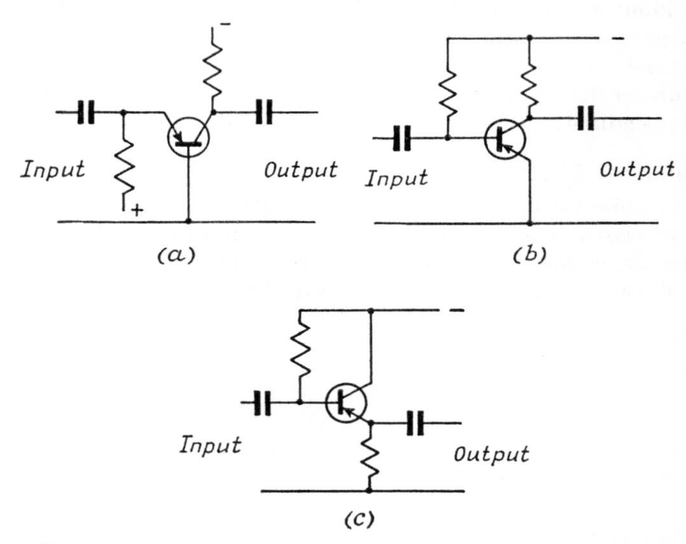

FIG. 6.13. ILLUSTRATING THE THREE BASIC TRANSISTOR MODES

(*c*) *Common-collector mode.* Here the input is applied to the base with the output taken from the emitter as shown in Fig. 6.13 (*c*). This is an arrangement similar to the cathode-follower in valve circuits, and is sometimes called an emitter-follower. It has a high input impedance and a low output impedance and is thus useful as an impedance changer, with a (voltage) gain of slightly less than unity. It is also useful as a current amplifier.

The performance of a transistor in any of these modes can be assessed by constructing equivalent circuits, of varying complexity according to the accuracy of the representation required. It is convenient to obtain a general appraisal of the performance of the three configurations by the use of relatively simple equivalent circuits, after which certain aspects can be considered in greater detail by more sophisticated analysis.

Common-base Mode

Fig. 6.14 shows the equivalent circuits for the common-base connection of Fig. 6.13 (*a*). The resistors r_e, r_b and r_c represent the effective a.c. resistances of the emitter, base and collector respectively.

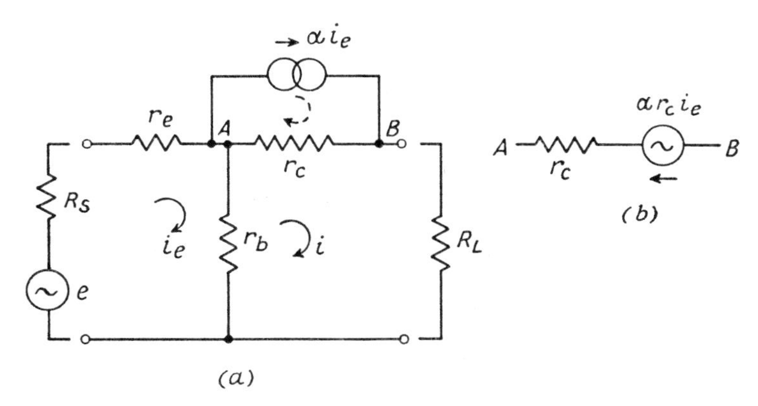

FIG. 6.14. EQUIVALENT CIRCUIT FOR TRANSISTOR IN COMMON-BASE MODE

R_S and R_L are the source and load impedances respectively. The source e.m.f. e will produce an input current i_e as shown, and the greater part of this will be transferred to the collector, where it is represented by an infinite-impedance current generator αi_e. This current will be distributed through the total network across the points AB, part being dissipated internally and part appearing usefully in the load R_L.

Alternatively the current generator αi_e may be replaced by a zero-impedance voltage generator $\alpha r_c i_e$ in series with r_c, as shown in Fig. 6.14 (*b*), but as many of the later equivalent circuits involve the current-generator concept it is preferable to use this from the outset.

From Kirchhoff's laws we can write

$$e = (R_S + r_e)i_e + (i_e - i)r_b \tag{1}$$

$$0 = iR_L + r_c(i - \alpha i_e) + r_b(i - i_e) \tag{2}$$

From these expressions we can deduce the input and output impedances, and the stage gain.

INPUT IMPEDANCE

From (2), $$i(R_L + r_b + r_c) = i_e(r_b + \alpha r_e) \quad (3)$$

So that $$e = \left[(R_S + r_e + r_b) - \frac{r_b(r_b + \alpha r_c)}{R_L + r_b + r_c}\right] i_e$$

The input impedance $Z_i = e/i_e - R_S$

It is customary to add a second suffix, b, e or c to denote the relevant mode. Using this nomenclature and substituting the values in eqn. (3) we can write

$$Z_{ib} = r_e + r_b\left[1 - \frac{r_b + \alpha r_c}{r_b + r_c + R_L}\right]$$

$$= r_e + r_b\left[1 - \frac{\alpha}{1 + R_L/r_c}\right] \quad \text{if } r_b \ll r_c$$

If $R_L = \infty$, $Z_{ib} = r_e + r_b$

If $R_L = 0$, $Z_{ib} = r_e + r_b(1 - \alpha)$

In practice $r_b(1 - \alpha)$ is of the same order as r_e and since r_b is several times r_e it is clear that a considerable variation in Z_{ib} (between 10 and 20 to 1) may be produced by changes in R_L.

OUTPUT IMPEDANCE

To deduce the output impedance a slightly indirect approach is required. The output current, i, may be regarded as being produced by some internal e.m.f. v operating into $Z_o + R_L$. This internal e.m.f. will bear some relationship to the input e.m.f., e, so that we can write $v = f(e)$ and hence $f(e) = i(Z_o + R_L)$.

If now we can obtain from the two Kirchhoff relationships an expression in this form, it will disclose the required value of Z_o (and $f(e)$ if required).

From (1) above

$$e = (R_S + r_e + r_b)i_e - ir_b$$

From (3), i_e may be written in terms of i, whence

$$e = \left[(R_S + r_e + r_b)\,\frac{R_L + r_b + r_c}{r_b + \alpha r_c} - r_b\right] i \quad (4)$$

This may be re-written as

$$\frac{r_b + \alpha r_c}{R_S + r_e + r_b} \cdot e = \left[R_L + r_b + r_c - \frac{(r_b + \alpha r_c)r_b}{R_S + r_e + r_b}\right] i$$

This is in the required form $f(e) = (R_L + Z_o)i$

so that
$$Z_{ob} = r_b + r_c - \frac{(r_b + \alpha r_c)r_b}{R_S + r_e + r_b}$$

$$= r_c \left[1 - \frac{\alpha r_b}{r_e + r_b + R_S}\right] \text{if } r_b \ll r_c$$

If $R_S = \infty$, $Z_{ob} = r_c$

$R_S = 0$, $Z_{ob} = r_c(1 - \alpha)$ very nearly,

so that again variations in R_S can produce large changes in Z_{ob}.

STAGE GAIN

If R_L is small compared with r_c the voltage gain $A_v = iR_L/e$, which from (4) above is

$$A_{vb} = \frac{(r_b + \alpha r_c)R_L}{(R_S + r_e + r_b)(R_L + r_b + r_c) - r_b(r_b + \alpha r_c)}$$

$$\simeq \frac{\alpha R_L}{R_S + r_e + r_b(1 - \alpha)} \text{ since } r_b \ll r_c$$

This expression is positive, indicating that the output is in phase with the input, which is a characteristic of the common-base mode.

It is useful to evaluate these expressions with typical values. For a typical small-signal transistor $r_e = 12{\cdot}5\ \Omega$, $r_b = 700\ \Omega$ and $r_c = 1{\cdot}25\ \text{M}\Omega$.

If $R_S = 25\ \Omega$ and $R_L = 2\ \text{k}\Omega$, then

$$Z_{ib} = 26{\cdot}5\ \Omega,\ Z_{ob} = 87\ \text{k}\Omega \text{ and } A_{vb} = 38$$

(The value of R_L may seem low, but it must be remembered that the output will probably be fed into a second transistor which will have a low input resistance and a value of $R_L = 2\ \text{k}\Omega$ is not unrepresentative.)

Common-emitter Mode

The equivalent circuit for the common-emitter connection (Fig. 6.13 (*b*)) is shown in Fig. 6.15. This will be seen to be similar to Fig. 6.14 with r_e and r_b interchanged. However, since the primary variable is now the *base* current, i_b, the current generator should be expressed in terms of i_b instead of i_e. Now, $i_b = i_e - i_c = i_e(1 - \alpha)$, so that

$$i_e = i_b/(1 - \alpha) \text{ and } \alpha i_e = \frac{\alpha}{1 - \alpha} \cdot i_b$$

In terms of base current, therefore, the current transfer ratio. is $\alpha' = \alpha/(1 - \alpha)$. Hence the current generator is shown as $\alpha' i_b$ and

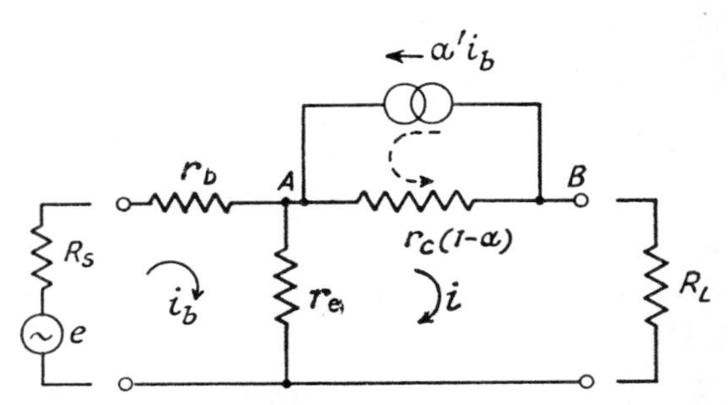

FIG. 6.15. EQUIVALENT CIRCUIT FOR TRANSISTOR IN COMMON-EMITTER MODE

of reversed polarity (because i_b is in the opposite direction to i_e). (The symbol β is often used for α'.)

The effective value of r_c also needs to be modified. Writing r'_c for the modified value, then if the e.m.f. across the points AB is to be the same, $\alpha i_e r_c = \alpha' i_b r'_c$, whence $r'_c = r_c \,.\, \alpha/\alpha' = r_c(1 - \alpha)$.

The Kirchhoff equations are

$$e = i_b(R_S + r_b + r_e) - i r_e \tag{5}$$

$$0 = iR_L + r'_c(i + \alpha' i_b) + r_e(i - i_b) \tag{6}$$

where $\quad r'_c = r_c(1 - \alpha)$ and $\alpha' = \alpha/(1 - \alpha)$

INPUT IMPEDANCE

From (6), $\quad i(R_L + r'_c + r_e) = -i_b(\alpha' r'_c - r_e) \quad (7)$

whence, by similar reasoning to that for the common-base mode,

$$Z_{ie} = r_e + r_b + \frac{r_e(\alpha r_c - r_e)}{R_L + r_e + r_c(1 - \alpha)} \text{ (since } \alpha' r'_c = \alpha r_c)$$

$$\simeq r_e + r_b + \frac{\alpha r_c r_e}{R_L + r_c(1 - \alpha)}$$

since $r_e \ll r_c$ and is usually small compared with $r_c(1 - \alpha)$.

If $$R_L = \infty,\ Z_{ie} = r_e + r_b$$

If $$R_L = 0,\ Z_{ie} = r_b + \alpha' r_e$$

In practice, r_b is several times r_e, so that the extreme variation of Z_{ie} is usually only about 2 to 1.

Output Impedance

By the same procedure as for the common-base calculation it is found that

$$Z_{oe} = r_e + r_c(1 - \alpha) + \frac{r_e(\alpha r_c - r_e)}{R_S + r_b + r_e}$$

$$\simeq r_c(1 - \alpha) + \frac{\alpha r_e r_c}{R_S + r_b + r_e} \quad \text{since } r_e \ll r_c$$

If $$R_S = \infty,\ Z_{oe} = r_c(1 - \alpha)$$

If $$R_S = 0,\ Z_{oe} = r_c(1 - \alpha) + \alpha r_e r_c/(r_b + r_e)$$

In practice again, the second term is of the same order as the first so that the maximum variation is about 2:1. Thus the common emitter connection has the advantage that the input and output resistances are not seriously affected by the load or source resistances.

Stage Gain

$$A_{ve} = -\frac{(\alpha r_c - r_e)R_L}{(R_L + r_e + r'_c)(R_S + r_b + r_e) + r_e(\alpha r_c - r_e)}$$

In practice r_e is small compared with r_b and r_c, and R_L is small compared with r'_c. The expression can then be simplified to

$$A_{ve} \simeq -\alpha' R_L/[(R_S + r_b) + \alpha' r_e]$$

This is similar to that for the common-base mode, but depends on α' rather than α. The negative sign indicates that there is a phase reversal with the common-emitter mode.

Again using typical values (which will be the same as for the common-base mode, but $r'_c = r_c(1 - \alpha) = 25\ \text{k}\Omega$) and assuming $R_S = 1{\cdot}25\ \text{k}\Omega$ with $R_L = 2\ \text{k}\Omega$ as before, we find that

$$Z_{ie} = 1{\cdot}27\ \text{k}\Omega,\ Z_{oe} = 25\ \text{k}\Omega \text{ and } A_{ve} = 38$$

Hence the common-emitter mode provides a much higher input resistance, with a lower output resistance, so that it is preferred for many applications. The gain is appreciably affected by R_S. Making $R_S = 0$ in the above example increases A_{ve} to 74, but this is not a preferred condition as is explained shortly.

Common-collector Mode

The equivalent circuit for the common-collector mode is shown in Fig. 6.16. Since the primary variable is again i_b the current generator is $\alpha' i_b$ and the collector resistance is modified to $r_c(1-\alpha)$.

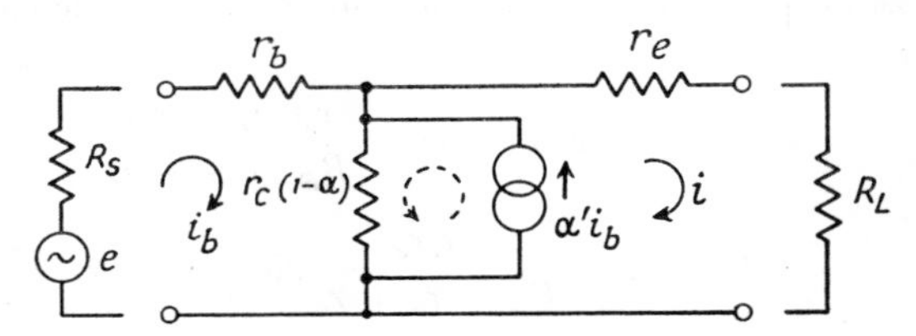

FIG. 6.16. EQUIVALENT CIRCUIT FOR TRANSISTOR IN COMMON-COLLECTOR MODE

The current transfer ratio is i/i_b. But here $i = i_e$, so that

$$i/i_b = i_e/i_b = (i_c + i_b)/i_b = i_c/i_b + 1$$

But $$i_c/i_b = \alpha/(1 - \alpha)$$

Hence the current transfer ratio $\alpha'' = 1/(1 - \alpha) = 1 + \alpha'$

From the Kirchhoff equations, as before, we find that

$$Z_{ic} = r_b + \frac{(r_e + R_L)r_c}{r_e + R_L + r_c(1 - \alpha)}$$

In practice R_L is small compared with $r_c(1 - \alpha)$ and r_e is small compared with R_L. The input impedance then becomes

$$Z_{ic} = r_b + R_L/(1 - \alpha) \simeq \alpha'' R_L$$

which with $R_L = 1000\ \Omega$ would be approximately 50 kΩ.

Similarly the output impedance

$$Z_{oc} = r_e + \frac{(1 - \alpha)(R_S + r_b)}{1 + (R_S + r_b)/r_c}$$

If $R_S \to \infty$ (current drive) $Z_{oc} \simeq r_c(1 - \alpha)$

$R_S = 0$ (voltage drive) $Z_{oc} = r_e + r_b(1 - \alpha)$

R_S is small but finite $Z_{oc} = r_e + (r_b + R_S)(1 - \alpha)$

The maximum stage gain is unity and in a practical circuit, assuming $R_L \gg r_e$ and r_b, is

$$A_{vc} = \frac{R_L}{R_L(1 + r_b/r_c) + (1 - \alpha)} \simeq \frac{R_L}{R_L + (1 - \alpha)}$$

This connection is thus principally used for impedance changing, having a high input resistance and a low output resistance.

On the other hand, the current transfer ratio, α'', is high so that the mode is useful where pure current gain is required, with no voltage gain.

Current and Voltage Drive

The evaluation of the foregoing expressions is clearly dependent on the values or r_e, r_b and r_c. These, however, are not constant, but depend upon the operating conditions, as is discussed in Section 6.5. In particular, the emitter resistance is inversely proportional to the emitter current. Now, in the common-base mode, and to a somewhat less extent in the common-emittor mode, the input impedance is largely dependent on r_e and is therefore by no means constant. A signal e.m.f. applied to the input of a transistor will thus encounter a varying impedance, so that the current will not be proportional to the signal e.m.f.

Since a transistor is basically a current amplifier it should be operated under conditions where the controlling factor is the input current, which is known as *current drive*, and the circuits should be arranged accordingly. However, where *voltage drive* conditions exist, as is frequently the case, the source resistance, R_S, should be large compared with Z_{in} so that the input current is not seriously affected by variations in Z_{in}. This is essential with a common-base circuit. With a common-emitter circuit the input impedance is less dependent on r_e and satisfactory operation can usually be obtained if R_S is approximately equal to Z_{in}.

Mutual Conductance

The collector current is αi_e, where i_e is the current in the emitter. This may be expressed in terms of the input voltage in the form $i_c = g_m e$, where $g_m = di_c/dv_e$, and e is the voltage across the emitter. Since in many circuits the emitter voltage is not equal to the input voltage v_i, the expression may be written $i_c = G_m v_i$, where G_m is (calculably) somewhat less than g_m. (See page 292.)

This parameter is called the *mutual conductance* (or forward transconductance) of the transistor, and is analogous to the corresponding parameter in valve technique, though of considerably different value being of the order of 40 mA/V.

Power Gain

Since the emitter and collector currents of a transistor are nearly equal, while the output impedance is many times greater than the input impedance, there is a substantial power gain. The action is illustrated by the simple equivalent circuit of Fig. 6.17, in which the

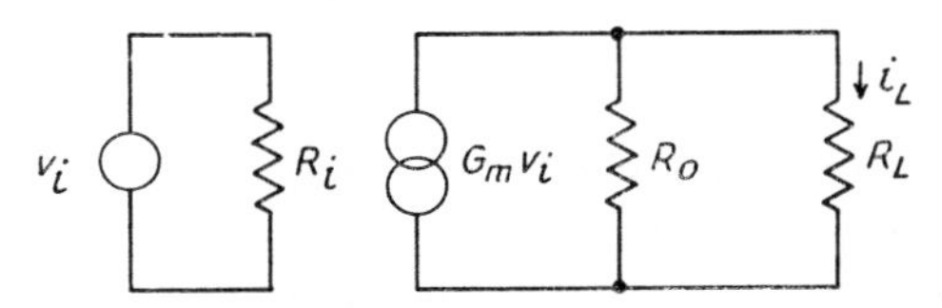

FIG. 6.17. EQUIVALENT CIRCUIT ILLUSTRATING POWER GAIN

generator supplies an input power $= v_i^2/R_i$, which appears in the output circuit as a current generator $G_m v_i$ operating into R_o and R_L in parallel. This current divides in the inverse ratio of R_o and R_L, the proportion actually flowing in R_L being $R_o/(R_o + R_L)$. Hence the current $i_L = G_m v_i R_o/(R_o + R_L)$ and the *useful* output power

$$P_o = i_L^2 R_L = G_m^2 v_i^2 \left(\frac{R_o}{R_o + R_L}\right)^2 R_L$$

The input power is v_i^2/R_i, so that the power gain

$$A_p = P_o R_i / v_i^2 = G_m^2 R_i R_L \left(\frac{R_o}{R_o + R_L}\right)^2$$

This is a maximum when $R_L = R_o$, in which case

$$(A_p)_{max} = \tfrac{1}{4} G_m^2 R_i R_L$$

With practical values this is of the order of 10,000 (40 dB). Whether it can be fully utilized depends on the nature of the load R_L. This is discussed further in Chapters 9 and 10.

6.5. Transistor Parameters

Design calculations in transistor circuits are necessarily based on the parameters r_e, r_b and r_c (or certain four-terminal forms derived therefrom, as described later in this section). However, the values of these parameters are not constant, but are considerably influenced by the operating conditions.

There are three main factors which influence the performance, namely the operating current, the frequency and the internal feed-back path.

Effect of Operating Current

The emitter resistance is the forward resistance of the emitter–base diode. It can be shown that for any type of transistor, irrespective of size or type, this has a value

$$r_{e0} = kT/eI_e$$

where k = Boltzman's constant ($= 1{\cdot}380 \times 10^{-23}$)
e = Charge on an electron ($= 1{\cdot}602 \times 10^{-19}$)
T = Temperature (K = °C + 273)
I_e = Emitter current

At a temperature of 20°C and a current of 1 mA this works out at 25 ohms, which is a useful figure to remember. Hence r_{e0} is inversely proportional to the emitter current but increases with the temperature of the junction (though to a secondary extent because T is in kelvins. Thus a 55°C rise in temperature only produces a 20 per cent rise in T).

The additional suffix 0 indicates that this is the zero-frequency or d.c. value. Moreover, since $I_c = I_e$ very nearly, one can say that $r_{e0} \simeq 25/I_0$, where I_0 is the mean working current in milliamperes. This is obviously quite low, and the a.c. value is even less, as will be seen later.

The collector resistance r_{c0} is also inversely proportional to I_e, but its value depends on the materials and construction of the transistor, being generally in the range 50 kΩ to 5 MΩ.

The base resistance r_{b0} is not appreciably dependent on I_e, having a roughly constant value determined by the construction, ranging from about 500 Ω for a small alloy transistor to 5 Ω or less for a large power type.

The current transfer ratio α_0 is not greatly affected by I_e. It varies by about 5 per cent above or below the value at $I_e = 1$ mA, increasing at first as I_e increases, but falling off afterwards.

The mutual conductance $g_{m0} = di_c/dv_e = \alpha_0/r_{e0}$. It is thus not constant, but is proportional to I_e. Hence for consistent operation it is necessary to stabilize the operating current, as explained in Section 6.6.

Variation of α with Frequency

The parameters r_e, r_b and r_c are not affected by frequency, as such, though their effective value is modified by the internal feedback paths, as discussed shortly. The current transfer ratio, α, however, is markedly frequency-dependent. This was mentioned in Section 6.3 (page 276) where it was shown that, because of the finite time required for the current carriers to diffuse through the base, the effective value of α decreases as the frequency is raised. The incidence of this effect depends upon the construction and the connection mode, and for any given transistor and mode there is a critical frequency known as the *alpha cut-off*, at which the current transfer ratio (α, α' or α'' according to the mode) falls to $1/\sqrt{2}$ times its zero-frequency value (i.e. 3 dB down).

The alpha cut-off, however, is not easy to measure, and its use has certain other disadvantages in practice, so that it has been largely replaced by an alternative parameter f_1. Basically the variation in current gain is given to a close approximation by the expression

$$\alpha = \alpha_0 \, e^{-j\phi h/(1+jh)}$$

where α_0 is the zero-frequency current transfer ratio, $h = f/f_\alpha$ and ϕ is a constant depending on the construction of the transistor. This is a complex relationship having real and quadrature components and the frequency f_1 is that at which the real part of the expression $= 0{\cdot}5$.

This point corresponds very nearly with the frequency, known as the *transition frequency*, f_T, at which β $(= \alpha/(1 - \alpha))$ falls to unity. This is more easily measured than f_α, and indicates, in fact, the highest frequency at which the transistor can be used as an amplifier (in common-emitter mode).

At frequencies not too greatly different from the transition frequency, β can be regarded as being proportional to $1/f_T$. Thus at a frequency $0{\cdot}5 f_T$, $\beta = 2$. The parameter f_T is therefore sometimes called the *gain-bandwidth* product.

Internal Feedback

The third factor affecting the circuit parameters is the influence of the internal capacitances and feedback paths within the transistor itself. To appreciate this slightly more detailed equivalent circuits are required.

As said earlier, the current carriers from the emitter do not all reach the collector. In a p–n–p transistor some of the holes from the emitter, in passing through the base will combine with free electrons in the base material, but because the base is very thin the majority will be swept on until they come under the influence of the collector field. In this region any free electrons will have been drawn into the collector by the action of the field so that no recombination is possible and the holes continue to move into the collector material. This barrier region is called the *depletion layer*, and its width is obviously dependent on the collector voltage.

The base can thus be represented by a transmission line having series elements representing the forward resistance and shunt elements containing resistance representing the effect of recombination, shunted by capacitances representing the emitter–base capacitances, as shown in Fig. 6.18. At the end of the line, the holes

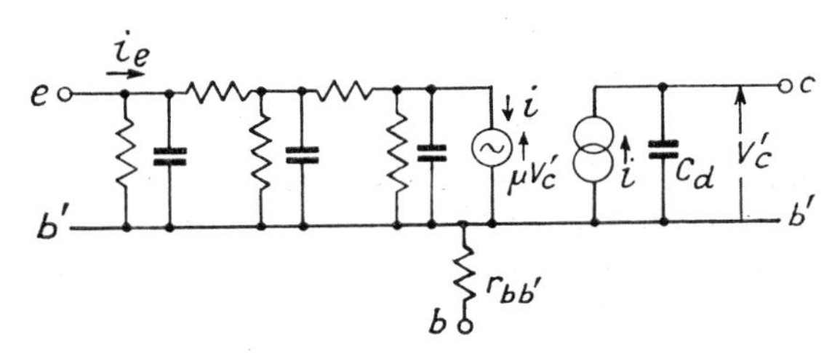

FIG. 6.18. MORE COMPLETE EQUIVALENT CIRCUIT

reaching the depletion layer are swept into the collector so that the line is effectively short-circuited. These holes, however, appear in the collector as a current so that a separate (infinite impedance) current generator is inserted in the collector portion to represent the current, shunted by the capacitor C_d to represent the capacitance of the depletion layer.

It was said, however, that the width of the depletion layer depends on the collector voltage. There is, in fact, a small voltage feedback from the collector to the emitter and this is represented by a zero-impedance generator in series with the short-circuit across the end of the line. The e.m.f. of this generator is $\mu V'_c$, where μ is the voltage feedback factor (having a value of the order of 1/2000).*

Finally, in a practical transistor it is not possible to make contact with the active portion of the base material so that a resistance $r_{bb'}$ has to be included to represent the resistance between the active part of the base and the actual base connection. (This is usually small, of the order of 75 ohms).

Practical Forms of Equivalent Circuit

While the equivalent circuit of Fig. 6.18 illustrates the basic factors involved, it is not in a convenient form for practical calculations.

* This factor must not be confused with the voltage amplification factor $\mu(= de_a/de_g)$ used in valve technique.

However, various more amenable circuits can be derived from it by suitable simplifications.

One such transformation is shown in Fig. 6.19 (*a*), where the various impedances have been combined into a T network comparable with that of Fig. 6.14. The internal feedback generator $\mu V'_c$ has been eliminated as such, its effect being allowed for by suitable modifications to the parameters, the values of which now become

$$r_e = r_{e0} - (1 - \alpha_0)\mu r_{c0}$$
$$r_b = \mu r_{c0}$$
$$r_c = r_{c0}(1 - \mu) \simeq r_{c0}$$
$$\alpha = (\alpha_0 - \mu)/(1 - \mu) \simeq \alpha_0$$

The significant modifications are thus that r_e is appreciably less than r_{e0} and that an additional resistance $r_{bb'}$ has been included in the base, while both input and output impedances are modified by the capacitances $c_{b'e}$ and $C_{b'c}$.

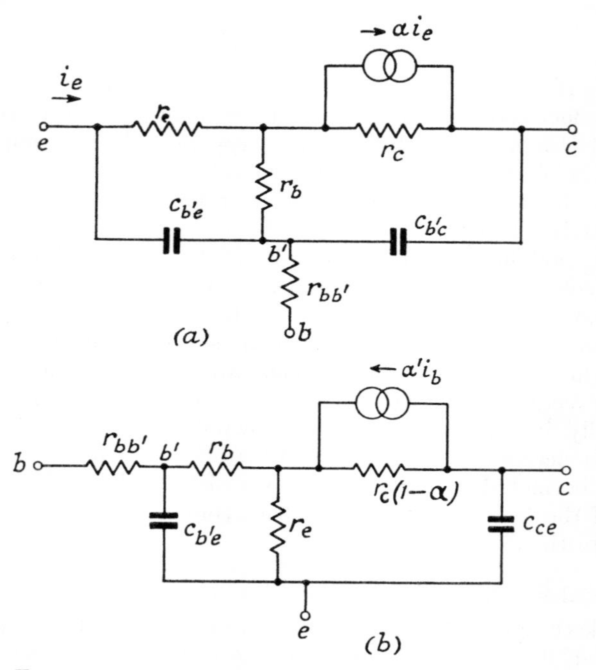

FIG. 6.19. PRACTICAL FORMS OF EQUIVALENT CIRCUIT
(*a*) Common-base mode
(*b*) Common-emitter mode

Fig. 6.19 (*b*) shows the circuit rearranged for the common-emitter mode. The values of the elements are unchanged except that, since the current generator is shown as αi_b, the effective value of r_c is modified to $r_c(1 - \alpha)$ as explained earlier (c.f. Fig. 6.15).

Equivalent-π Circuits

An alternative transformation makes use of the four-terminal, or π, network shown in Fig. 6.20. This form is often more amenable to calculation, particularly with common-emitter circuits, and is

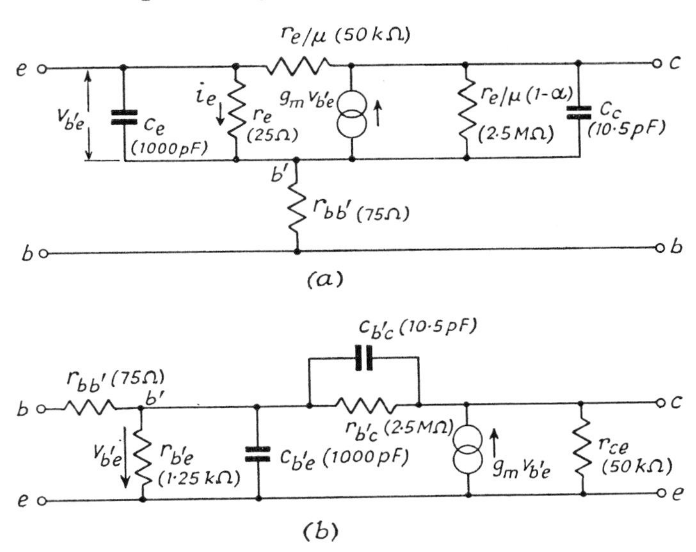

FIG. 6.20. EQUIVALENT-π CIRCUITS

(*a*) Common-base mode
(*b*) Common-emitter mode

extensively employed in the practical treatments discussed in Part II.

The relevant transformations are

$$r_{b'e} = r_{e0}/(1 - \alpha)$$
$$r_{ce} = r_{e0}/\mu$$
$$r_{b'c} = r_{e0}/(1 - \alpha)\mu$$

With this form of circuit it is more convenient to represent the current generator in terms of the input e.m.f., i.e. as $g_m V_{b'e}$, where g_m is the forward transconductance. It will be noted, however, that the effective input $V_{b'e}$ is less than the applied signal by reason of the voltage drop on $r_{bb'}$. In consequence, as was said earlier, the effective transconductance is slightly reduced, the modified value being

$$G_m = g_m Z_{b'e}/(r_{bb'} + Z_{b'e})$$
$$= g_m r_{b'e}/(r_{bb'} + r_{b'e} + j\omega C_{b'e} r_{bb'} r_{b'e})$$

As a first approximation the capacitance term may be neglected, so that $G_m = g_m/(1 + r_{bb'}/r_{b'e})$, but as ω increases the capacitance term becomes increasingly important. Using the values shown in Fig. 6.20 (*b*), with $g_m = 39$ mA/V, and assuming $f = 500$ kHz, the value of G_m becomes 35·8 mA/V. (The approximate calculation makes $G_m = 36·7$ which may be close enough for practical purposes.)

It may be noted that in these circuits the effect of the internal voltage feedback (the generator $\mu V'_c$ in Fig. 6.18) is represented by an appropriate impedance between the collector and the emitter or base. Typical values for a small-signal transistor are shown in brackets against the various elements.

Four Terminal Parameters

The T parameters employed so far are useful in providing a physical appreciation of the behaviour of a transistor. In practice, however, the calculations can be simplified if the transistor is regarded simply as a four-terminal network, as shown in Fig. 6.21.

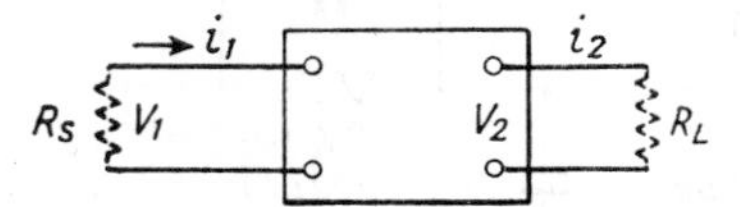

FIG. 6.21. FOUR-TERMINAL (BLACK-BOX) NETWORK

For example, two parameters of immediate practical interest are the input and output impedances, which are v_1/i_1, and v_2/i_2 respectively. These impedances can be evaluated from an equivalent circuit of the form of Fig. 6.20, but more usually they are specified by the transistor manufacturer.

Hybrid Parameters

For many purposes the relevant information can conveniently be obtained from four parameters, as follows:

h_i = Input *impedance* (dv_1/di_1) with $R_L = 0$ (to a.c.)

h_f = Forward (current) transfer characteristic (di_2/di_1) with $R_L = 0$

h_o = Output *conductance* (di_2/dv_2) with $R_S = \infty$

h_r = Reverse (voltage) transfer characteristic (dv_1/dv_2) with $R_S = \infty$

The first pair are evaluated with the output short-circuited (to a.c.) and the second pair with the input open-circuited. It will be seen that whereas h_i and h_o have the dimensions of resistance or conductance, h_f and h_r are ratios. Hence they constitute a *hybrid* system (which is why the symbol h is used).

The values of these parameters depend on the circuit configuration. Hence it is customary to add a further subscript indicating the circuit to which they apply. Thus h_{fb} is the current transfer ratio with common-base connection ($= -\alpha$), while h_{fe} is the equivalent parameter for the common-emitter mode ($= \alpha'$ or β).

They are also dependent to some extent on the operating conditions, which are usually specified, and also on the frequency, so that unless otherwise stated they should be regarded as relating to the low-frequency (sometimes called static) conditions.

Table 6.1 shows the relationships between these parameters and the relevant T parameters.

TABLE 6.1. h-PARAMETERS

$h_{ib} = r_e + (1 - \alpha)r_b$	$h_{ie} = r_b + \beta r_e$	$h_{ic} = r_b + r_e/(1 - \alpha)$
$h_{ob} = 1/r_c$	$h_{oe} = 1/r_c(1 - \alpha)$	$h_{oc} = 1/[r_e + r_c(1 - \alpha)]$
$h_{fb} = -\alpha$	$h_{fe} = \beta$	$h_{fc} = -(1 + \beta)$
$h_{rb} = r_b/r_c$	$h_{rc} = (1 + \beta)r_e/r_c$	$h_{rc} = r_c/[r_c + r_e(1 + \beta)]$
	$\beta = \alpha/(1 - \alpha) = \alpha'$	

The T-parameters can be derived from the h-parameters by the following conversions

$$r_e = h_{ib} - (1 - \alpha)h_{rb}/h_{ob}$$
$$r_b = h_{rb}/h_{ob}$$
$$r_c = 1/h_{ob}$$

y and z Parameters

Parameters can also be derived from the four-terminal network having the dimensions of impedance (z-parameters) or admittance (y-parameters) the latter being the more useful in practice. Typical y-parameters are shown in Table 6.2.

As with h parameters, a second suffix is added to indicate the relevant mode.

Y-parameters are frequently used in r.f. calculations, since they are easily measurable. Thus by short-circuiting the a.c. output of a common-emitter transistor with a large capacitor, so that $v_o = 0$, the input admittance y_{ie} is simply i_i/v_i.

TABLE 6.2. y-PARAMETERS

Symbol	Parameter	Condition
y_i	Input admittance	Output short-circuited
g_i	Input conductance	
c_i	Input capacitance	
ϕ_i	Phase angle of input admittance	
y_o	Output admittance	Input short-circuited
g_o	Output conductance	
c_o	Output capacitance	
ϕ_o	Phase angle of output admittance	
y_f	Transfer admittance	Output short-circuited
g_f	Transfer conductance	
c_f	Transfer capacitance	
ϕ_f	Phase angle of transfer admittance	
y_r	Feedback admittance	Input short-circuited
g_r	Feedback conductance	
c_r	Feedback capacitance	
ϕ_r	Phase angle of feedback admittance	

All these parameters relate to the basic four-terminal network of Fig. 6.21 and, as said, are more convenient than the T parameters which are rarely quoted as such.

Limiting Values

A consideration of paramount importance in transistor circuitry is that the rated maximum voltages and currents *must not be exceeded*. Whereas with a valve overloads may do no permanent damage, even a momentary excess with a transistor will destroy it instantly, and irreparably.

The manufacturers therefore specify a series of absolute maximum ratings, and the circuit conditions have to be arranged so that these limiting values are not exceeded under any conditions. The principal voltage limits specified are

V_{CBO} = Collector–base voltage with emitter open-circuited
V_{CEO} = Collector–emitter voltage with base open-circuited
V_{EBO} = Emitter–base voltage with collector open-circuited

The last of these is particularly important because it is, in general, only a few volts even though the rated collector voltage may be 50 or more. It is thus essential, when choosing a transistor for a particular application, to ensure that under the worst possible conditions the limiting values are not exceeded.

6.6. Practical Transistor Circuits

The design of transistor circuitry is considered in detail in the appropriate chapters in Part II, but there are certain fundamental requirements which can usefully be summarized here. A significant feature of semiconductor devices is the allowance which has to be made for variations in the parameters. With the common-emitter mode, which is by far the most usual, these variations can be of major importance.

It was shown earlier that with this mode the stage gain is approximately βR_L. But since $\beta = \alpha/(1 - \alpha)$ it only requires a small variation in α to produce considerable change in β. In practice the production spread can be as much as 3:1, so that with a transistor type having a nominal h_{fe} (β) of 80, the actual value may lie between 40 and 120. The circuit must be designed to give the required gain with a low-limit sample, and still be stable with a high-limit sample.

D. C. Stabilization

A further factor of importance is that the performance of a transistor stage is substantially affected by variations in the collector current. A small-signal amplifier is designed to work in a Class A condition about a mean working point as shown in Fig. 6.22 (*a*).

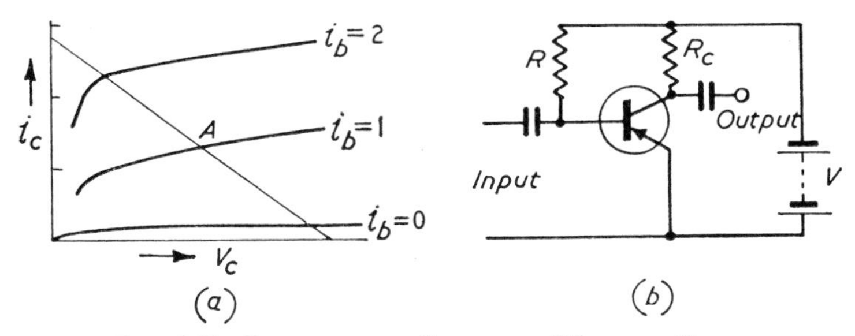

FIG. 6.22. ILLUSTRATING CHOICE OF WORKING POINT

Any change in collector current will not only shift the working point but will also modify the input and output impedances, and the gain.

A suitable working point could be established by feeding the base from the supply through a resistor, as in Fig. 6.22 (*b*), the value being chosen to provide the required conditions. The base current, however, is influenced by any collector leakage current which may be flowing, being in fact $I_c/\alpha' - I_{co}$, as was explained on page 273). The leakage current I_{co} increases with temperature, the variation with germanium transistors being considerable, while β is also affected by temperature *and* the actual value of I_c.

Hence the working point can change considerably, and a more stable arrangement is shown in Fig. 6.23 in which the base is fed through a potential divider R_1, R_2 with a resistor R_3 in the emitter circuit. Any increase in collector current decreases the base–emitter potential, so reducing I_b and hence I_c. R_3 also produces a feedback at signal frequencies, which may be obviated if desired by connecting a capacitor across it as shown.

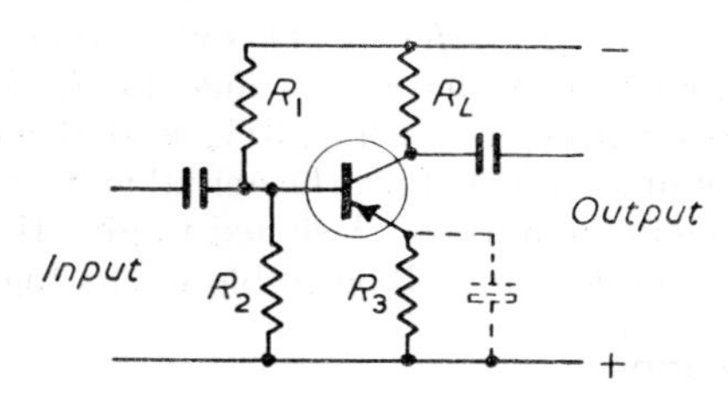

FIG. 6.23. CIRCUIT PROVIDING BETTER STABILIZATION

An alternative circuit is shown in Fig. 6.24. Here the base potential is derived from the collector. Any increase in collector current will cause V_c (and hence V_b) to fall, tending to hold I_c constant. The circuit will again produce negative signal feedback which may be avoided by decoupling the feedback resistor as at (*b*).

Neither circuit will compensate for changes in the supply voltage, which, with a battery supply, may fall considerably below the nominal value. A measure of compensation is possible by making

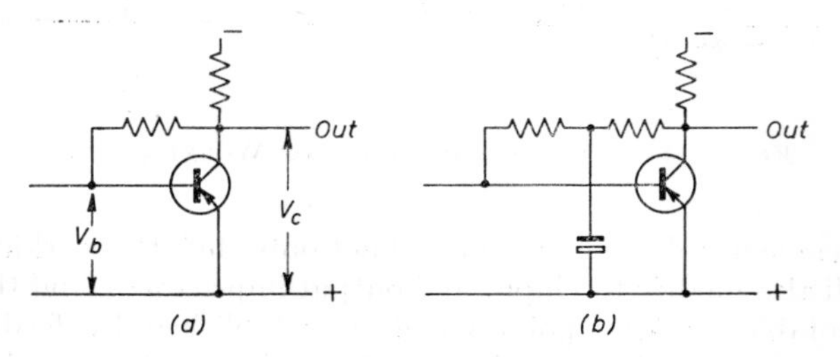

FIG. 6.24. ALTERNATIVE METHODS OF D.C. STABILIZATION

R_2 in Fig. 6.23 non-linear so that its value increases as the current falls, thereby holding V_b constant. A convenient method is to connect a diode across, or in series with, R_2 and this arrangement is often used to stabilize battery receivers. If this diode is of similar material to that of the transistor, and is mounted close to it, it can also compensate for variations due to temperature changes.

The output resistor R_L may be replaced by a transformer, as shown in Fig. 6.25. The effective a.c. collector load is then n^2R, where R is the secondary load and n is the *step-down* ratio. R could be the input impedance of a second stage, or a loudspeaker; in either case it would have a low value so that to provide a reasonable collector load (of 10 kΩ or more) a step-down ratio would be needed, and this is usual with transistor circuits, as is discussed in Chapters 9 and 10.

In this case the collector voltage will be virtually that of the battery since there will only be a small voltage drop on the trans-

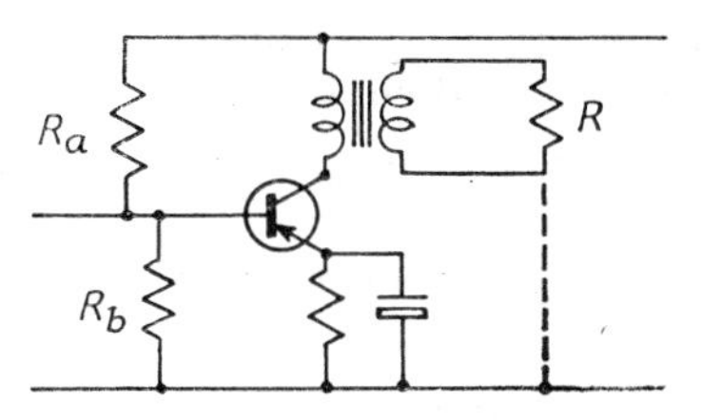

FIG. 6.25. TRANSFORMER-COUPLED TRANSISTOR CIRCUIT

former primary, and the voltage divider R_aR_b will have to be differently proportioned not only to provide the required working point, but to maintain adequate thermal stability.

Thermal Runaway

This question of thermal stability is an important factor in the design of transistor circuits. The current in a transistor necessarily dissipates a certain amount of power, mainly in the collector portion owing to its higher resistance.

The collector current is the true current I_c plus the leakage current I_{co}. As the collector current increases, the dissipation, and hence the internal temperature, begins to rise and this causes a rapid increase in I_{co}, providing a further increase in temperature. This results in a cumulative effect known as *thermal runaway* which can destroy the transistor.

The effect may be limited in two ways. If the (d.c.) collector load is appreciable, the increased current will cause an increased voltage drop so that the collector voltage will fall and the transistor will ultimately "bottom." It can be shown that if the collector voltage is never more than half the supply voltage this limiting action will be sufficient to prevent thermal runaway. This is known as the "half-supply-voltage" principle, and is generally used in small-signal stages.

Alternatively (or in addition) the transistor may be provided with a heat sink, being mounted in good physical contact with a structure of relatively large surface area and good thermal conductivity so that the heat is dissipated externally and the temperature rise is thereby limited.

The system, in fact, forms a thermal amplifier, and if the loop gain is greater than unity, the conditions are unstable and thermal runaway will occur. Now, the power dissipation in the collector $P_c = I_c V_{ce}$, where V_{ce} is the collector–emitter voltage $= V - I_c R$, where V is the supply (battery) voltage and R is the total d.c. resistance in the collector and emitter circuits. Hence

$$P_c = I_c(V - I_c R) = VI_c - I_c^2 R$$

If the collector current now increases by a small amount δI_c the dissipation will increase to

$$P_c + \delta P_c = V(I_c + \delta I_c) - (I_c + \delta I_c)^2 R$$

By subtraction, and neglecting δI_c^2,

$$\delta P_c = \delta I_c(V - 2I_c R)$$

This is clearly zero if $I_c R = \frac{1}{2}V$, which is the half-supply-voltage criterion mentioned above.

If $I_c R$ is less than $\frac{1}{2}V$, as it may well be in a transformer-coupled stage, it is necessary to assess the thermal loop gain, which is $\theta \delta P_c / \delta T$, where θ is the rise in junction temperature per unit collector dissipation. This must be less than unity, and for safety about 0·5.

The factor θ depends on I_c and on the thermal conductivity of the system so that it may be kept within bounds by limiting I_c or, if this is impracticable or undesirable, by providing an adequate heat sink to conduct the heat away and so limit the rise in temperature.

The problem is not serious in small-signal stages, but where large currents have to be handled, as in an output stage, the transistors are mounted in relatively massive metal casings which may be bolted direct to a chassis or heat sink. If electrical insulation is required a thin mica washer is interposed. This limits the thermal conduction since mica is a poor conductor of heat, but the overall dissipation is usually adequate. Fig. 6.26 illustrates a typical arrangement.

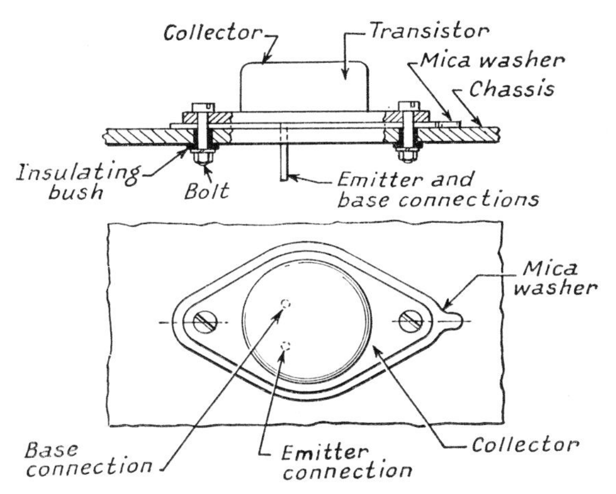

FIG. 6.26. CONSTRUCTION OF POWER TRANSISTOR

Large-signal Stages

Where large currents are required, as in an output stage, the same basic circuitry is used with appropriately larger transistors. Steps must be taken to avoid thermal runaway, as just described, while the input current must be correspondingly increased. This usually involves a transformer-coupled input stage of the form of Fig. 6.25, as is discussed more fully in Chapter 10.

As with valve circuits, the standing current may be reduced, with corresponding increase in efficiency, by using two transistors in a Class B connection, as shown in Fig. 6.27.

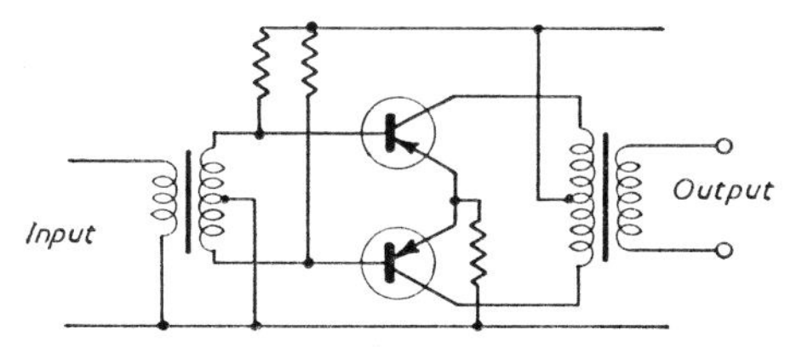

FIG. 6.27. CLASS B OUTPUT STAGE

Variable-gain Circuits

The gain of a transistor stage is normally controlled by replacing one of the circuit resistors with a variable voltage divider and thereby controlling the proportion of the signal which is transferred to the succeeding stage.

Occasions arise, however (as in a.g.c. circuits) where it is desired to control the gain electrically. This can be done by varying the potential of the base and thereby the standing base current. The $i_b v_b$ characteristic is approximately exponential in form, as shown in Fig. 6.28. Hence a given value of input signal (± 50 mV in the figure) will produce considerably different excursions of i_b, depending on the bias voltage V_b; and since the output in the collector is dependent upon the variation of base *current*, the gain can be controlled by altering the bias.

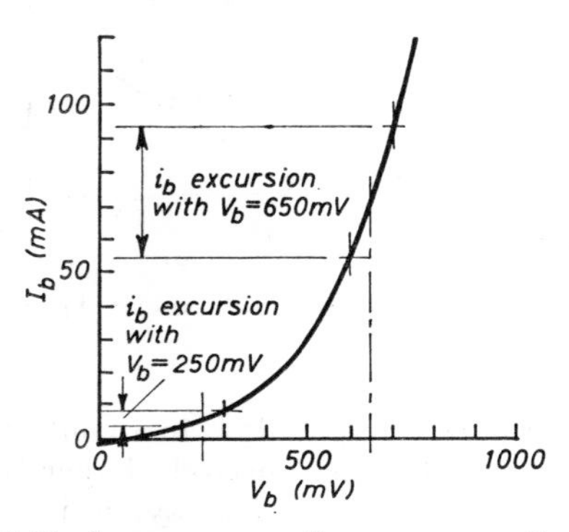

FIG. 6.28. ILLUSTRATING REVERSE-BIAS CONTROL

Provided that the input signal is small an exponential control of the gain can be obtained, similar to that with a vari-mu valve, but the range is limited to about 30 dB (31·6 times). This form of control is called *reverse-bias control*.

An alternative method applicable principally to r.f. circuits involves increasing the base potential. This increases the collector current, and if a (decoupled) series resistor is included in the collector circuit, as shown in Fig. 6.29, this increased current causes a reduction in the effective collector potential. As a result, the operating point on the characteristics moves towards the knee of the curves so that the output swing is limited. The conditions are illustrated in Fig. 6.29 (*b*) which shows some representative characteristics. If R_1 is 1 kΩ and the battery voltage is 9, the normal working point would be at O, with $I_c = 2$ mA and hence $V_c = 9 - 2 = 7$ V. The load line AOB then produces an output swing of 14 V. If now I_c is increased to 8 mA, V_c is reduced to 1 V, and the output swing A′O′B′ for the same load and i_b swing is only 3 V.

The output is obviously distorted, being in the form of pulses which become of increasingly shorter duration as I_c is increased,

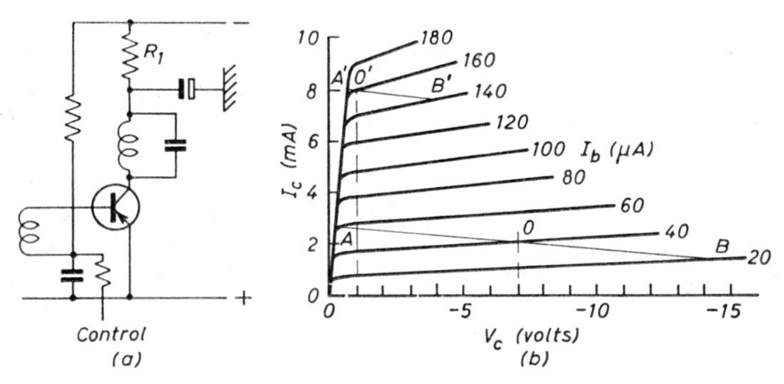

FIG. 6.29. ILLUSTRATING FORWARD-BIAS CONTROL

but when the collector circuit is tuned this is tolerable, being in fact an example of the Class C operation described on page 323. This is known as *forward-bias control* and can provide a range of up to 40 dB. It has the advantage that it will accept large input swings which is useful in a.g.c. circuits.

Common-base Amplifiers

The circuits shown so far have all utilized the common-emitter mode, which is by far the most convenient for the majority of applications. The alternative common-base configuration was discussed on page 279, where it was shown to exhibit a relatively high output impedance but a very low input impedance, of the order of 25 ohms. This is sometimes useful in circuits which have a very low source impedance, such as a moving-coil microphone. It is also used in wide-band r.f. amplifiers where its better h.f. performance is an advantage.

A practical disadvantage is that a separate supply of reverse polarity is required for the emitter bias, though this can sometimes be obviated by an arrangement such as in Fig. 6.24 (*a*). Here the base is held at a small negative potential (assuming a p–n–p transistor so that the emitter is positive to the base as required).

A feature of the common-base circuit is that the output is in phase with the input, whereas with the common-emitter mode there is a phase reversal, and this is sometimes useful.

Emitter Followers

The third (common-collector) mode, often called an emitter follower analogous to the cathode follower in valve technique, is frequently used as an impedance changer. As shown on page 285, this circuit has a gain of slightly less than unity, but has a relatively high input impedance. It may be used as a buffer stage between a high-impedance source such as a photocell and a common-emitter amplifier.

A practical circuit is shown in Fig. 6.30. The base potential is controlled by the resistance R_1 which may be calculated as follows. To obtain the maximum swing of the output signal, the volt drop across R_2 should be half the supply voltage. V_e is then $\frac{1}{2}V_{cc}$ and the

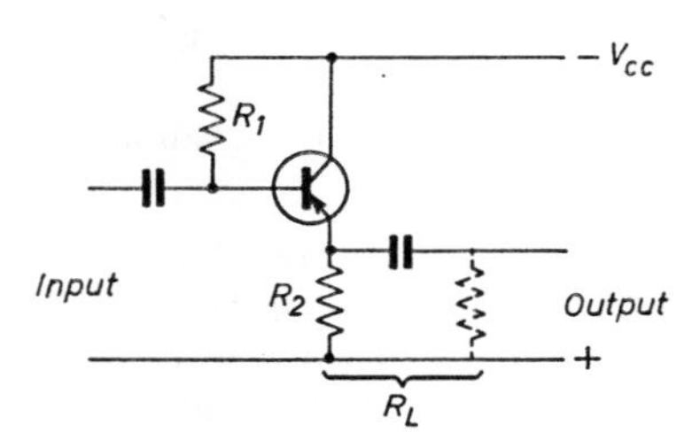

FIG. 6.30. EMITTER FOLLOWER

collector current $I_c = \frac{1}{2}V_{cc}/R_2$. The base potential must be $V_e + V_o$, where V_o is the hop-off voltage, approximately 0·15 V for germanium and 0·7 V for silicon. The base current is I_c/β, so that

$$R_1 = \beta(V_{cc} - V_b)/I_c$$

With $V_{cc} = 6\text{ V}$, $R_2 = 3\text{ k}\Omega$ and a silicon transistor having $\beta = 50$, R_1 would be 115 kΩ.

The load impedance R_L will be less than R_2 because of the impedance of the following stage, and the a.c. conditions must take this into account. The gain is not appreciably affected, but the input impedance, which is $r_b + R_L/(1 - \alpha)$, falls off considerably if R_L is low, as shown in Fig. 6.31. Hence it must not be assumed that the input impedance of an emitter follower is always high.

The emitter follower differs from a valve cathode-follower in one important respect, namely that the input impedance is not independent of the output. Hence under operating conditions some portion of the output signal is fed back to the input. This is not normally of great significance unless the input signal is appreciably dependent on the input impedance of the transistor. In general, it results in a reduction of gain, which becomes somewhat less than the expected value.

Darlington Circuit

Simple combinations of two transistors are often used to obtain a modified performance. One of the most popular is the Darlington pair, or super-alpha circuit, illustrated in Fig. 6.32. This is used in a variety of ways.

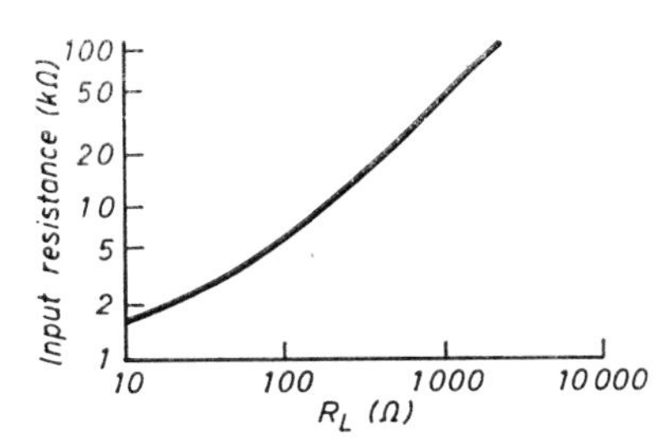

FIG. 6.31. TYPICAL VARIATION OF INPUT RESISTANCE OF AN EMITTER FOLLOWER ($\beta = 50$)

Fig. 6.32 (*a*) shows the arrangement as an impedance changer, which is of use where the load resistance is too low to be conveniently handled by a normal circuit. With the circuit shown, the input resistance of Tr_2 is approximately βR_L, and since this constitutes the emitter load of Tr_1, the input resistance of Tr_1 is $\beta^2 R_L$. Thus,

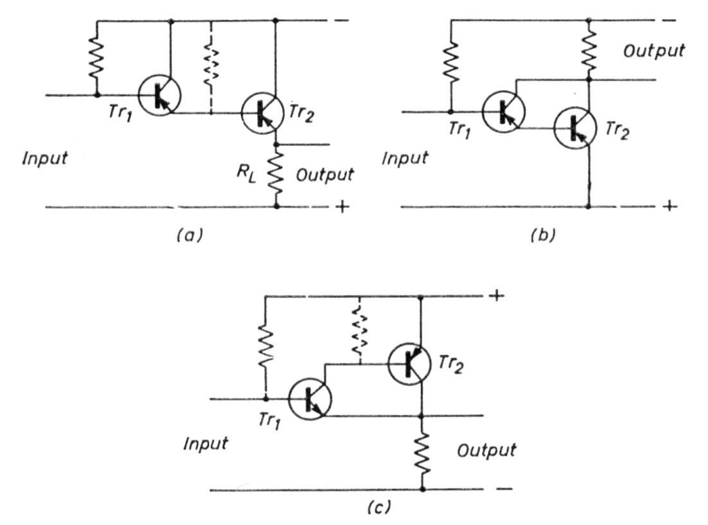

FIG. 6.32. DARLINGTON CIRCUITS

using transistors having $\beta = 70$, a load as low as $10\,\Omega$ would provide an input resistance of approximately $50\,k\Omega$.

If Tr_2 is connected as a common-emitter stage, as shown in Fig. 6.32 (*b*) the arrangement behaves as a single stage having a very high current gain $\beta_1\beta_2$. In this case R_L would be several kilohms, so that the emitter load of Tr_1 would be several hundred ohms, and the input resistance $(\beta_1 R_E)$ would still be reasonably high.

A disadvantage of the circuit of Fig. 6.32 (*a*) is that the hop-off voltages of the transistors are added, which with silicon transistors may be inconvenient. This may be overcome by using complementary transistors, as shown in Fig. 6.32 (*c*).

It should be noted that in arrangements (*a*) and (*c*) the collector current of Tr_1 is the *base* current of Tr_2. Hence if Tr_1 is to be working under suitable conditions to provide an adequate value of β. Tr_2 must have a considerably higher collector current than Tr_1, If this is not convenient, the collector current of Tr_1 can be suitably increased by including a resistor from the supply line as shown dotted.

Cascading

Various other forms of cascade connection are used, the transistor lending itself readily to such circuitry. One such arrangement is the *cascode* circuit shown in Fig. 6.33. Here Tr_1 is a normal common-emitter amplifier with its output fed to the emitter of a second

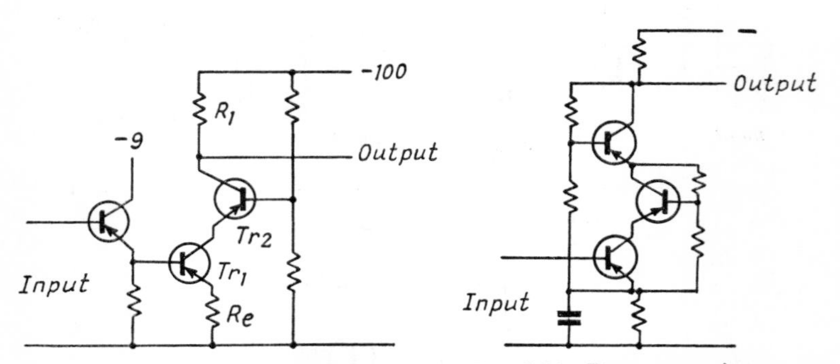

FIG. 6.33. CASCODE AMPLIFIER

FIG. 6.34. BEANSTALK AMPLIFIER

transistor Tr_2, which is thus operating in the common-base mode. It thus contributes nothing to the overall (current) gain, but permits the output voltage (and hence the output signal) to be approximately twice that which could be handled by either transistor alone.

Because Tr_1 is feeding into a very low load (the input impedance of Tr_2, which is approximately 25 Ω), the overall voltage gain is low, but for this reason the internal feedback through Tr_1 is negligible, and the circuit is sometimes used in r.f. amplifiers where stability is more important than gain.

The transistors must obviously be of a type having similar nominal collector currents since they both carry the same current. The process may be continued indefinitely, as in the *beanstalk* amplifier of Fig. 6.34. This is a useful method of extending the output voltage, but since transistors are now made to handle well over 100 volts, it is only occasionally used.

Transistor Oscillators

By arranging a circuit having a loop gain greater than unity a transistor will maintain continuous oscillations, as with a valve,

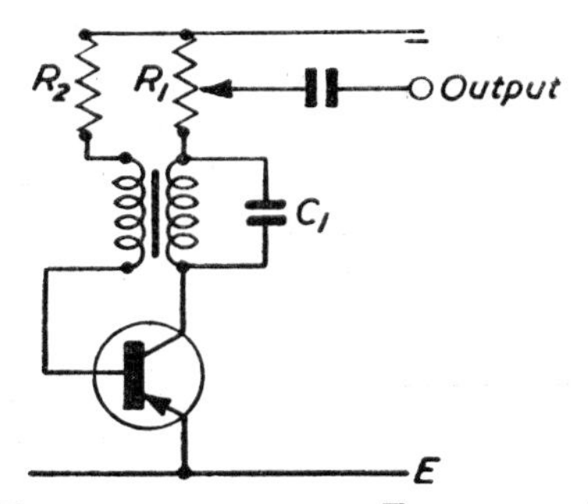

Fig. 6.35. Transformer-coupled Transistor Oscillator

and any of the customary circuits may be used. Fig. 6.35 shows a simple transformer-coupled oscillator. The transformer ratio is designed to match the high-impedance collector circuit into the low-impedance base, while the primary is tuned with the capacitance C_1 to obtain the frequency required. The output is taken from the voltage divider R_1 in the collector circuit, while R_2 is chosen to provide a suitable operating point.

Feedback and phase-shift oscillators can also be adapted for transistor operation, as is discussed further in Chapter 12.

Transistor Inverters

Transistor circuits can be devised to convert d.c. to a.c. power, and/or to replace rotary convertors or vibrators when a low-voltage d.c. supply is required to be converted to a higher voltage. Such arrangements are not only smaller and cheaper, but have a higher efficiency, while the reliability is greatly enhanced since no moving parts are involved. To achieve high efficiency in a transistor inverter it is necessary that the transistors are either bottomed or cut off; in each case power dissipation is low. During switching between these states the power dissipation will rise, so that the switching time should be kept as short as possible.

A transformer-coupled self-oscillating inverter is shown in Fig. 6.36. Consider that the circuit is in operation with Tr_1 conducting; the

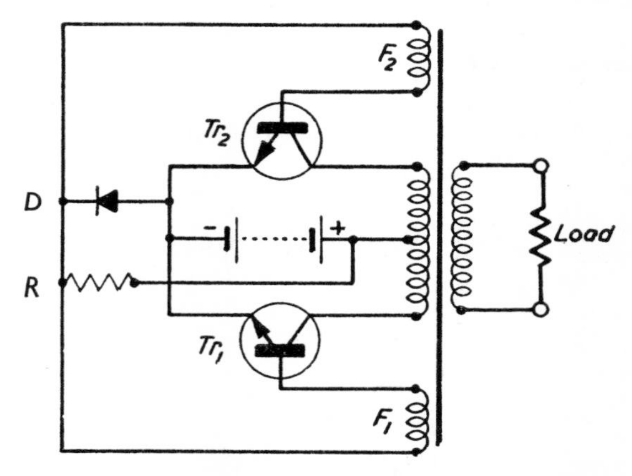

FIG. 6.36. PUSH-PULL SELF-OSCILLATING INVERTER

primary inductance L will at first be practically constant and the primary current will rise according to the relationship $\frac{di}{dt} = \frac{V}{L}$, which initially will be virtually linear. A constant voltage will be induced at the feedback windings F_1 and F_2. The polarity of these will be such as to maintain conduction in Tr_1 and to cut off Tr_2. When the core material begins to saturate, the primary inductance will fall and the collector current in Tr_1 will rise even faster until it reaches a value $\alpha' I_b$.

$\frac{di}{dt}$ then becomes zero and the base voltage will fall to zero. The collector current then begins to fall, reversing the polarities of the voltages at F_1 and F_2 so that Tr_2 will begin to conduct. This process will then continue as an oscillation.

The output waveform at the secondary will be substantially square and can be rectified and smoothed to produce a d.c. output. The addition of the resistor R and diode D shown in Fig. 6.36 is a useful way of starting oscillations. The diode will not conduct initially and a current will flow through the resistor to the bases of both transistors causing the transistor of higher α' to conduct first. When the action has started, base current for the conducting transistor can flow through D, but this will result in only a small voltage drop.

The Transistor as a Switch

Because transistors in general, and silicon types in particular, have an abrupt transition from the non-conducting (reverse-biased) state to the conducting state they lend themselves to a variety of switching applications. As with diodes, the transition is not instantaneous, but with suitable manufacturing technique the transition time can be reduced to less than one nanosecond (10^{-9} sec).

Very considerable use is, in fact, made of switching circuitry, ranging from the control of very large currents by means of special devices known as *thyristors*, to simple low-voltage logic switching, often employing multiple-emitter gates. The techniques warrant more than brief mention and are discussed extensively in Chapter 18.

6.7. Field-Effect Transistors

The essential feature of the bipolar transistor is that it is a current-operated device, as distinct from the valve, which is voltage operated. The conductance of semiconductor material, however, can be controlled by using an auxiliary electrode to vary the electric field within the material, so producing a device which is voltage controlled. As will be seen, the mechanism only involves one type of charge carrier, so that devices of this type are known as *unipolar* transistors.

Fig. 6.37 illustrates such an arrangement, which is called a *field-effect transistor* (f.e.t.). A slab of n-type material is provided with ohmic connections at the two ends so that a voltage applied across the slab will produce a current by the normal transfer of free electrons through the material. A pellet of p-type material is now introduced in the middle of the slab. This is negatively biased and so produces in its vicinity a *depletion area*, i.e. a region in which no charge carriers are available. As the potential of this electrode is increased, the depletion region spreads until it finally inhibits the passage of current through the slab.

Hence the current through the slab, called the *drain current* can be controlled by varying the gate voltage, and since the gate–slab junction forms a reverse-biased diode, the input resistance of the gate is very high—of the order of 10^4 MΩ, with a shunt capacitance of only a few picofarads. The three electrodes are called the *source*, *gate* and *drain*.

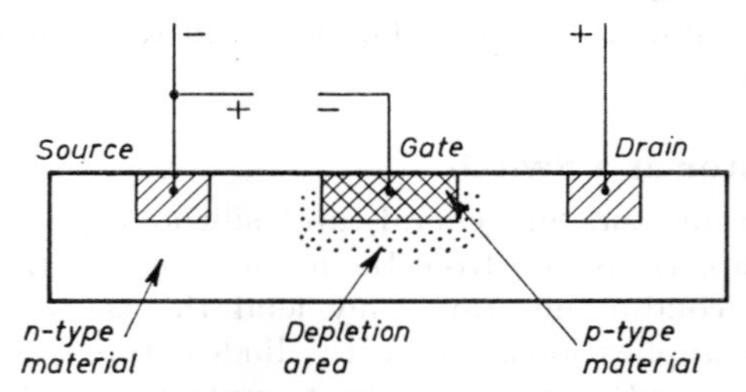

FIG. 6.37. BASIC STRUCTURE OF A JUNCTION-GATE FIELD-EFFECT TRANSISTOR (JUGFET)

Such a device is known as a *junction-gate* field-effect transistor, or *jugfet*, and has characteristics of the form shown in Fig. 6.38, which will be seen to be similar to those of a pentode. The gate voltage at which the drain current is reduced to zero is called the *pinch-off voltage*, and the device is normally operated with a standing gate bias of about $\frac{1}{2}V_p$.

The symbols for jugfets are shown in Fig. 6.41. With the n-channel type just described a negative gate voltage is required.

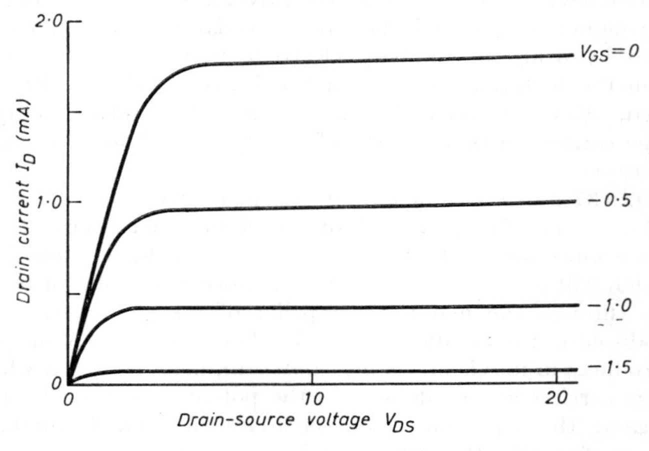

FIG. 6.38. CHARACTERISTICS OF *n*-CHANNEL JUGFET

Complementary, p-channel types are available, with n-type gate material, in which case a positive gate voltage is required. In the symbols the distinction is indicated by the direction of the arrow in the gate lead.

Insulated-gate F.E.T.s

The gate of a jugfet is not isolated from the rest of the circuit. However, as explained in Appendix 4, it is a simple matter with silicon material to form an insulating layer of silicon dioxide (or nitride) on the surface. This permits the construction of devices having insulated gates, as shown in Fig. 6.39. The construction is

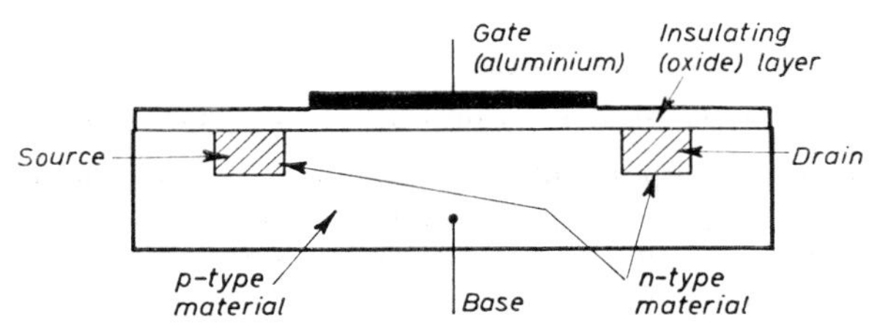

FIG. 6.39. STRUCTURE OF IGFET (ENHANCEMENT TYPE)

slightly modified, a slab of p-type material being used with n-channel regions under the source and drain connections, the whole being covered with a thin insulating film. An aluminium coating is then formed over the insulating layer between the n-regions, which forms with the base a small capacitor.

If the gate potential is zero it exercises no influence and the drain current is simply the (negligible) leakage current of one of the p–n junctions, but if the gate potential is positive to the base, electrons are attracted to the surface, so permitting the flow of a drain current which is controlled by the gate potential as with the junction type.

Such devices are known as *insulated-gate* f.e.t.s, *igfets* or *mosts* (metal-oxide-semiconductor transistors). With the arrangement described the drain current is virtually zero with $V_G = 0$, and increases as V_G becomes positive. It is therefore said to be working in the *enhancement* mode.

If the two separate n-type regions are replaced with a very thin continuous layer on top of the p-type base material, as in Fig. 6.40, drain current will flow even when $V_G = 0$, as with the jugfet, which is sometimes more convenient; but because this layer is very thin, the field between gate and base still controls the current, which is decreased by a negative bias and in the limit reduced to zero. The device is therefore said to be working in the *depletion* mode.

An important proviso with any type of igfet is that the gate must never be allowed to remain open-circuited. Under such conditions it acquires a static charge which, because of the negligible

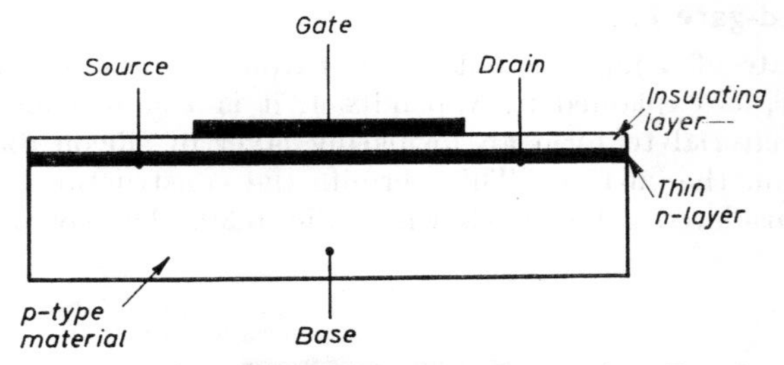

Fig. 6.40. Structure of Igfet (Depletion Type)

leakage, can build up a voltage sufficient to destroy the insulating layer. Hence they are supplied with the leads twisted together. Later forms, however, incorporate a gate-protection diode which obviates the difficulty.

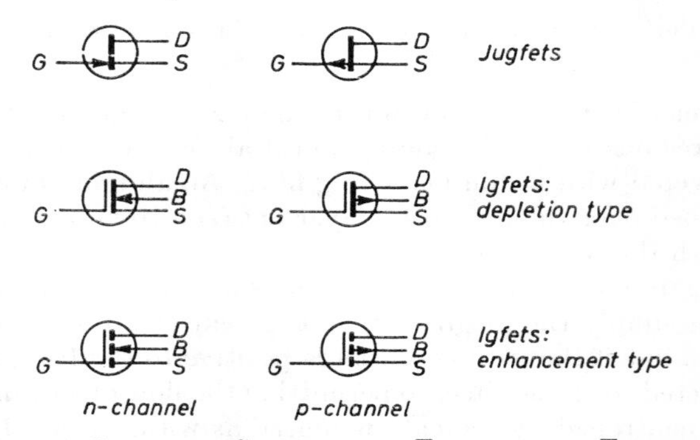

Fig. 6.41. Graphical Symbols for Field-effect Transistors

B. Substrate — *G*. Gate
D. Drain — *S*. Source

As before, the materials may be interchanged to provide complementary types, the graphical symbols being indicated in Fig. 6.41. The fact that with the enhancement type the drain current is zero when $V_G = 0$, so that the circuit is virtually open-circuited, is indicated by breaks in the symbol as shown.

Further details of the construction of igfets are given in Appendix 4.

F.E.T. Parameters

The relevant parameters for f.e.t.s are customarily expressed in four-terminal form, usually in terms of admittances (y-parameters) which is most convenient for design. These are

y_i = Input admittance with output short-circuited to a.c.
y_f = Forward transfer admittance with output short-circuited to a.c.
y_o = Output admittance with input short-circuited to a.c.
y_r = Reverse (feedback) admittance with input short-circuited to a.c.

The mutual conductance g_m is the real part of y_f. The reactive part only becomes significant at high frequencies so that for many purposes y_f may be considered equal to g_m.

These parameters are not constant but depend on the circuit conditions. Thus since they include the internal capacitances, they are frequency-dependent, though the effect is usually small below about 50 MHz.

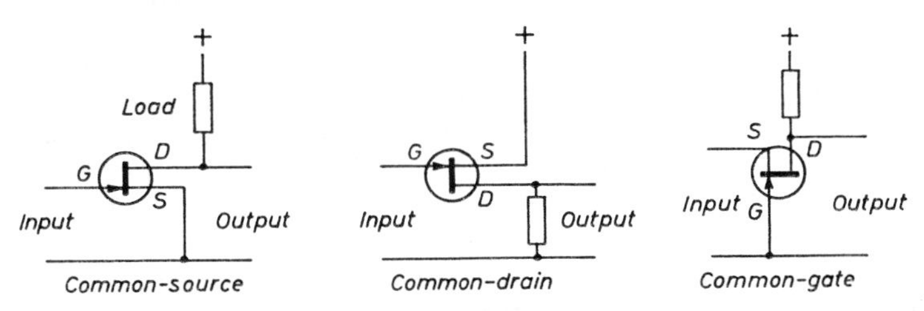

FIG. 6.42. F.E.T. CONFIGURATIONS

They are also dependent on the circuit mode, there being three possible configurations as shown in Fig. 6.42, and the suffixes s, d and g are added to denote the common-source, common-drain and common-gate modes respectively. For example, y_{is} represents the common-source input admittance.

F.E.T. Circuits

A typical f.e.t. has values of $r_i = 10^{12}$, $r_o = 100\ \text{k}\Omega$ and $y_{fs} = 2{\cdot}5\,\text{mA/V}$. These are comparable with the parameters of an r.f. pentode, so that the appropriate circuitry tends towards valve techniques. Fig. 6.43 shows a simple circuit in the common-source mode. The voltage gain would be $g_m Z r_o/(r_o + Z)$; but whereas with a valve r_o is many times Z, so that the gain simplifies to $g_m Z$, this is not necessarily the case with an f.e.t.

A further difference is that the value of y_{fs} ($\simeq g_m$) is not constant, as with a valve, but depends on the drain current I_d as shown in Fig. 6.44. It also falls off with frequency, as already mentioned. Subject to these reservations, however, the performance is similar to that of a valve.

The common-drain connection is, in effect a cathode follower while the common-gate mode is equivalent to the grounded-grid

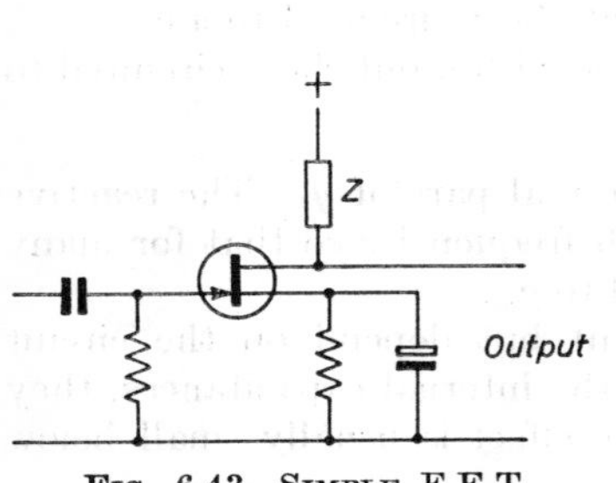

FIG. 6.43. SIMPLE F.E.T. CIRCUIT

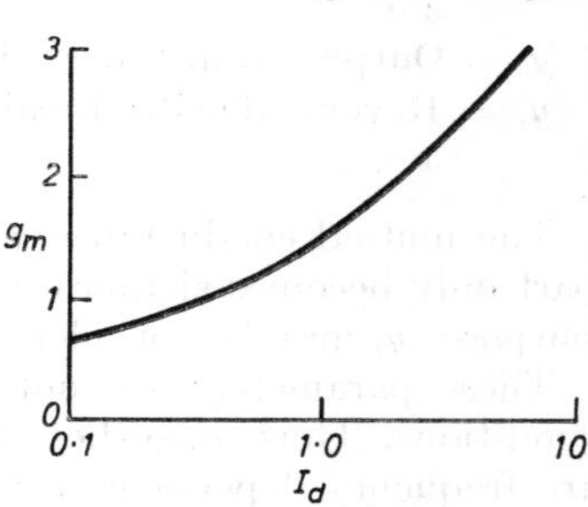

FIG. 6.44. TYPICAL VARIATION OF g_m WITH DRAIN CURRENT

valve circuit discussed on page 589, and performance calculations may be made accordingly, using the appropriate f.e.t. parameters.

As with the grounded-grid valve, the input impedance of the common-gate mode is low, being approximately Z/g_m, but this can be tolerated in v.h.f. applications where the impedance of the feed lines is low in any case, and the arrangement has the advantage that the feedback through the internal capacitance is negligible. Typical circuits are discussed in Chapter 13.

The common-source connection provides a phase reversal, as with a valve or common-emitter transistor, whereas with the other two modes the output and input are in phase.

6.8. Photosensitive Semiconductors

We have seen that the conductivity of a semiconductor is affected by heat. Certain semiconductor materials are also responsive to energy in the visible or infra-red range of the spectrum, and this forms the basis for the use of semiconductors as photo-sensitive devices.

There are three main forms of photo-sensitive semiconductor, namely:

(*a*) Variable-resistance devices,
(*b*) Barrier-layer cells,
(*c*) Photo-transistors.

Selenium Cells

The most common variable-resistance device is the *selenium cell.* Selenium is an intrinsic semiconductor and hence its conductivity depends on the existence of the relatively small number of electrons which acquire sufficient energy to break their covalent bonds. The presence of light waves produces an increase in the number of these

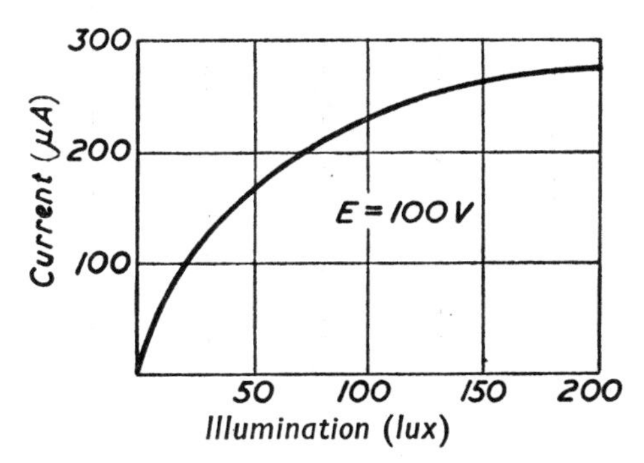

Fig. 6.45. Response Curve of Selenium Cell

free electrons, causing the resistance of the material to decrease. Hence if a suitably constructed selenium cell is connected across a source of e.m.f. the current which flows will vary with the illumination. The response is not linear, being of the form shown in Fig. 6.45.

Selenium cells are made up by depositing a thin layer of a suitable selenium compound (about 0·25 mm thick) onto a ground-glass plate, which has previously been provided with two intermeshed grids of a suitable inert metal to serve as connections. The cell is then annealed to convert the selenium to its pure metallic form and suitably mounted.

Barrier-layer Cells

The second type of cell, known as the *barrier-layer cell,* is formed by depositing a very thin layer of noble metal, such as gold, onto the surface of a semiconductor. The deposited layer is so thin as to be transparent and it is found that with such an arrangement the energy acquired by the electrons under the impact of the light appears as an e.m.f. across the boundary between the noble metal and the semiconductor.

This type of cell thus generates its own e.m.f. and requires no external supply, which makes it very convenient for light meters and similar devices. The cell is essentially a current generator and hence works best into a low-resistance load. As the load is increased the output falls off as indicated in Fig. 6.46.

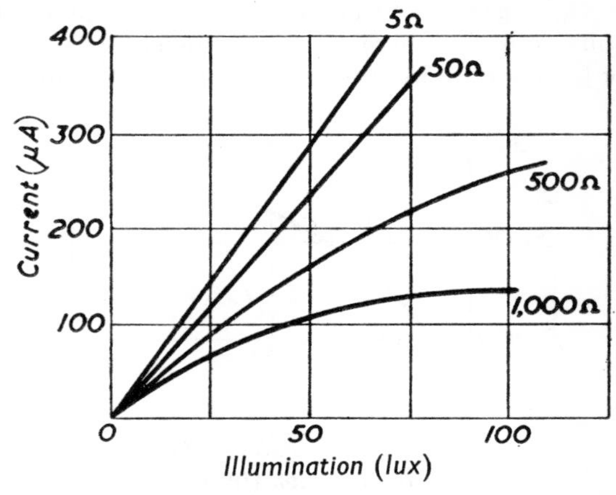

FIG. 6.46. RESPONSE OF BARRIER-LAYER CELL WITH DIFFERENT LOAD RESISTANCES

PHOTO-TRANSISTORS

The third type of semiconductor photocell is a development of normal transistor technique. If a p–n diode is biased in the reverse direction under conditions of darkness, little current will flow owing to a scarcity of carriers of suitable polarity; if such a junction is now illuminated, hole-electron pairs will be liberated and more current will flow owing to the increased minority carrier concentration. Such a device is known as a *photo-diode*, the small current in the absence of illumination being called the *dark current*. If, however, the illuminated junction forms part of a transistor, substantial current amplification is obtained.

Such a device is called a *photo-transistor* and may be used with simple circuits of the type illustrated in Fig. 6.47. The base is open circuited so that the collector current is simply I'_{co}. If the base is now illuminated a current I_{ph} will appear in the base which will be multiplied α' times by the normal (common-emitter) amplification. A ratio of light/dark current of several hundred may be obtained with a photo-transistor, the actual ratio depending on I'_{co}, which should therefore be kept as low as possible.

The characteristics of a photo-transistor are in fact similar to those of Fig. 6.10 with I_b expressed in terms of the incident light. Hence the output is not greatly affected by the collector voltage as long as this is beyond the knee.

The dark current can sometimes be reduced by connecting the base to earth through a suitable resistor, as shown dotted, thereby improving the light/dark ratio, but this is not usually necessary.

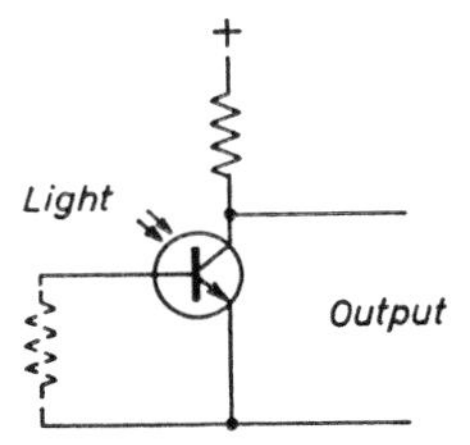

FIG. 6.47. PHOTO-TRANSISTOR CIRCUIT

Photocell circuitry is usually d.c. coupled. Hence if the output is fed to an amplifying stage, this must have a high input resistance; otherwise, it will draw current through the collector resistance, so that the dark current is no longer simply I'_{co}, but some larger value, and the light/dark ratio will be reduced. It is best therefore to use

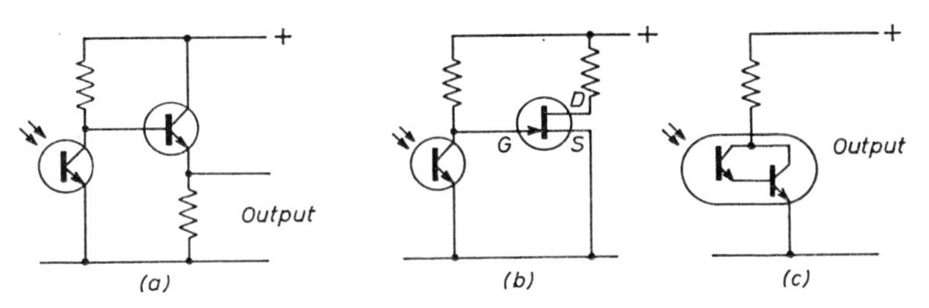

FIG. 6.48. PHOTOCELL AMPLIFIER CIRCUITS

an emitter follower or an f.e.t for the first amplifier stage as in Fig. 6.48. Some photo-transistors incorporate a Darlington amplifier as part of the (monolithic) construction, as in Fig. 6.48 (*c*).

Photometric Units

The incident illumination is measured in *lux*, the significance of which is discussed, with other photometric units, in Appendix 2.

Part II

Practical Applications

7

The Radio Transmitter

THE radio transmitter consists essentially of three parts:

(*a*) A means of producing an oscillating current of the required frequency and magnitude.

(*b*) A method of modulating the current so that the required intelligence can be communicated.

(*c*) An aerial system wherein these currents produce the electromagnetic waves required.

The present chapter is concerned with items (*a*) and (*b*), item (*c*) being dealt with in Chapter 8.

For medium and high powers, valve techniques predominate, though for small powers transistor operation is more efficient. Since the requirements are basically the same it is convenient to discuss valve circuitry first. The modifications required for transistor operation can then be discussed at appropriate points.

7.1. Valve Transmitter Circuits

The essentials of a transmitter are shown in Fig. 7.1 which will be seen to comprise a tuned circuit consisting of an aerial which, at medium and long waves, behaves substantially as a capacitance, tuned with the inductance L_1 (called the aerial tuning inductance

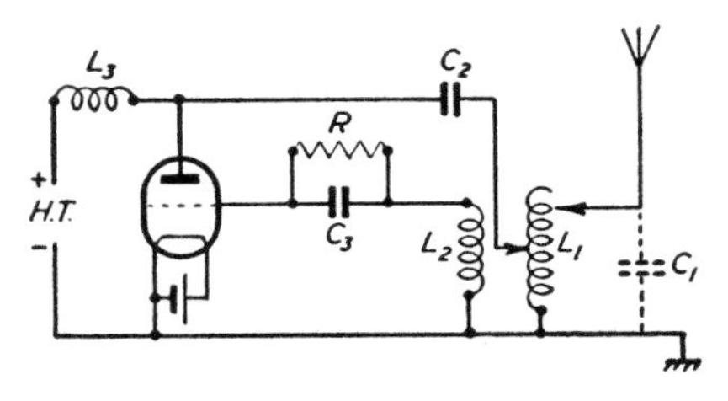

FIG. 7.1. SIMPLE TRANSMITTER CIRCUIT

or A.T.I.). This is effectively in the anode circuit of a valve while the coil L_2 is coupled to L_1 in such a manner as to supply the grid with e.m.f. of the right phase and magnitude to sustain continuous oscillation.

The H.T. supply to the anode is not connected in series with the tuned circuit, because this is connected to earth. It is fed therefore through the inductance L_3 which is called a *radio-frequency choke*. The inductance of L_3 is such that its reactance is large compared with the tuned-circuit impedance. The varying anode current is thus separated into two components, a steady current which passes through L_3 and an alternating component which passes through the capacitor C_2 into the tuned circuit L_1C_1.

Oscillator Operating Conditions

It will be noted that the anode connection to the tuned circuit is made variable. This is to permit the effective impedance to be matched to the valve as explained later.

The network C_3R in the grid circuit is for the purpose of automatically adjusting the bias. When the grid swings positive, grid current flows which charges the capacitor C_3 and thus establishes a negative bias. During the next half cycle this charge leaks away through the resistance R, and the conditions adjust themselves so that the pulse of grid current each cycle is just sufficient to make up the charge lost through the resistor. This action automatically provides the optimum bias, which builds up to a point where the oscillation is a maximum. Any further increase in bias then reduces the grid swing so that the pulse of grid current is no longer sufficient to maintain the charge and the grid bias falls slightly to restore the optimum condition.

The time constant of the circuit must not be too large or the bias will not be able to vary sufficiently rapidly. Suppose the grid bias increases beyond the optimum point. The amplitude of the oscillation will begin to decrease at a rate depending on the decrement of the tuned circuit, but if the charge on the capacitor does not fall at the same rate the oscillation will actually cease for a short period until the bias has fallen sufficiently.

Under these conditions, the oscillation will be continually starting and stopping, giving rise to what is known as "grid tick" or "squegging." The frequency with which the oscillations stop and start may vary from one every few seconds to several hundreds a second. The remedy is to reduce the time constant of the grid circuit as already explained.

Drive Oscillator System

The circuit of Fig. 7.1, however, is only suitable for a very simple low-power transmitter and various modifications would be introduced in a practical transmitter. In particular, except in the simplest cases, the required oscillation is generated by a low-power oscillator, handling only a few watts of oscillating power. It is a comparatively easy matter to start and stop such an oscillation, or to modulate it with speech or music for radio telephony. A further advantage is that variations in the constants of the aerial and associated circuits do not affect the frequency of the oscillation, but only have a minor effect on its amplitude. This is most important, particularly with short waves.

The output from this oscillator is then fed to an amplifier which supplies the power to the aerial. When considerable power output is required several stages of amplification are employed, using valves capable of dissipating more power at each successive stage. Banks of such valves in parallel are used, and the final bank may consist of a number of water-cooled valves in parallel.

High-efficiency Amplifiers; Class A, B, and C Operation

It is important that these amplifying stages shall convert as much as possible of the input power into oscillating energy. A valve can be used in three main conditions, as follows:

1. *Class A*. The valve is biased so that the anode current never falls to zero and the mean value remains unaltered whether there is a signal on the grid or not.
2. *Class B*. The bias is adjusted to cut-off so that anode current only flows during the positive half-cycles of signal.
3. *Class C*. The valve is biased beyond cut-off so that the anode current pulses occupy less than half a cycle.

Class A Operation

In the Class A condition the anode current swings between nearly zero and twice the normal steady value as shown in Fig. 7.2. If the mean value is I_0, the peak current swing $I_{\text{max}} = 2I_0$ (very nearly) and the r.m.s. current is $I_{\text{max}}/2\sqrt{2}$.

The anode voltage swing will depend upon the effective anode impedance. If this were such as to cause the anode voltage to fall to zero at the instant of maximum anode current, the anode voltage swing would be $\pm V$, where V is the steady (H.T.) value, and the r.m.s. swing would be $V/\sqrt{2}$. The alternating power is then

$$[V/\sqrt{2}][I_{\text{max}}/2\sqrt{2}] = \tfrac{1}{4}VI_{\text{max}}$$

The d.c. power supplied to the valve is $VI_0 = \frac{1}{2}VI_{max}$, so that the maximum efficiency with Class A operation is 50 per cent. In

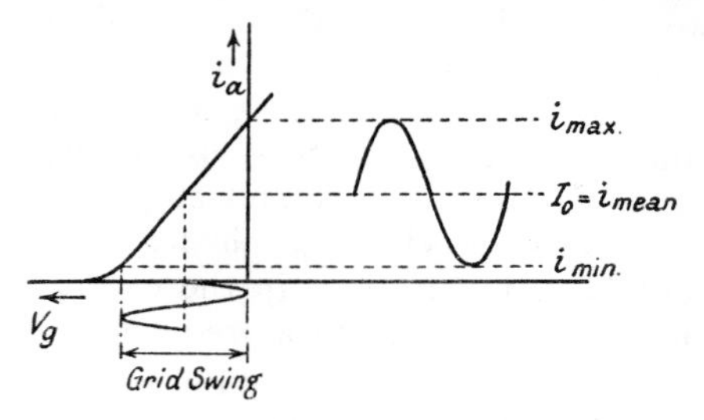

FIG. 7.2. VOLTAGE AND CURRENT RELATIONS WITH CLASS A OPERATION

practice neither the anode voltage nor the anode current is reduced to zero so that the voltage and current swings are less than the theoretical maxima and the efficiency rarely exceeds 40 per cent.

CLASS B OPERATION

Greater efficiency is obtainable by operating in a Class B condition. Here the valve is biased to cut-off, as shown in Fig. 7.3. Over the positive half cycle of grid swing the anode current varies in proportion; over the negative half cycle no anode current will flow

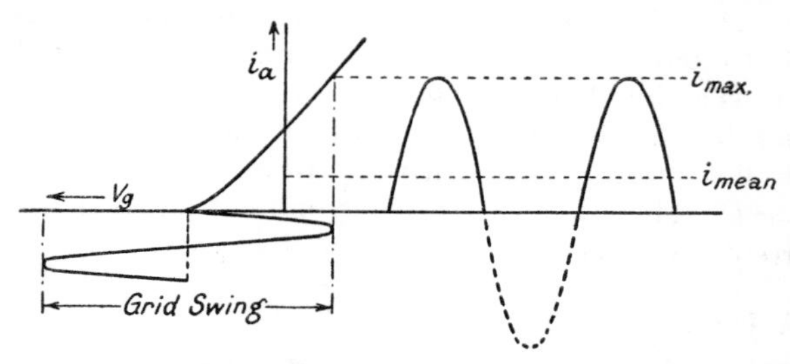

FIG. 7.3. WITH CLASS B OPERATION THE VALVE IS BIASED TO CUT-OFF

but if the anode circuit is tuned the resonant action will maintain the oscillation so that the anode voltage will continue to vary sinusoidally.

Thus, although anode current only flows every half-cycle, the anode voltage will be a faithful copy of the grid voltage. Note that the grid is shown as running appreciably positive on the positive peak. This is normal with power output and transmitting valves, which are designed to permit short excursions into the region of grid current, thereby considerably increasing the peak anode current.

The anode current with Class B working is thus a series of half sine waves. The average value of a sine wave is $2/\pi$ times the maximum value, but since we are only using every alternate half wave the average value in a Class B circuit is I_{max}/π. As before, the alternating power is $\frac{1}{4}VI_{max}$, but the d.c. power is now VI_{max}/π. Hence the theoretical efficiency is $\pi/4 = 78{\cdot}5$ per cent. Again this is not fully realizable in practice, but efficiencies of 60 to 70 per cent are attainable.

CLASS C OPERATION

Still better efficiencies are obtainable by increasing the grid bias beyond the cut-off point as in Fig. 7.4. Under these conditions, sometimes called *flick impulsing*, anode current flows in short

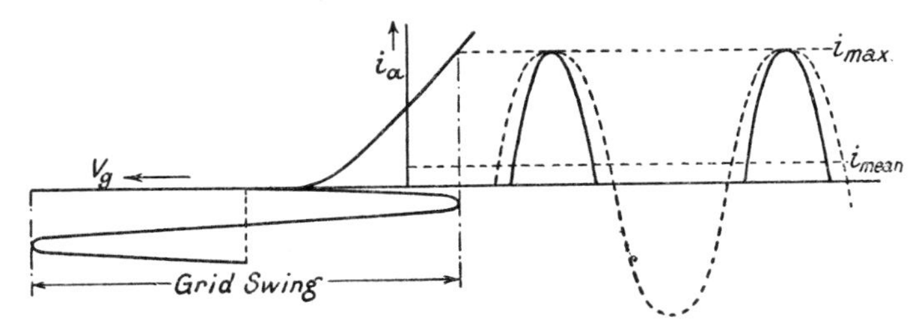

FIG. 7.4. CLASS C OPERATING CONDITIONS

pulses of a high peak value, and efficiencies of 80 to 85 per cent are obtained. The anode voltage is no longer directly proportional to the grid voltage, so that the circuit can only be used for maintaining or amplifying a steady oscillation. Where a modulated wave is being handled Class B working must be used.

Angle of Flow

The proportion of the cycle during which the anode is conducting is called the *angle of flow*. If the valve conducts for a complete half-cycle (Class B condition) θ is 180°. For Class C conditions θ is usually around 100° to 150°.

The mean anode current can then be assessed in terms of θ. With Class B working ($\theta = 180°$) we have seen that the mean anode current is I_{max}/π. When θ is less than 180° the current flows for a smaller proportion of the total time so that the mean value is less, and in fact, to a reasonable approximation, $I_0/I_{max} = \theta/180\pi$.

Amplifier Operating Conditions

The adjustment of a Class B or a Class C transmitter depends upon the matching of the external impedance in the anode circuit to the valve, but the conditions are appreciably different from those of a Class A circuit. In general, the optimum impedance is lower than that for maximum power output under Class A conditions, which require a load impedance approximately equal to that of the valve.

Fig. 7.5 illustrates a typical series of characteristics of a transmitting valve. Since we are concerned with obtaining the largest

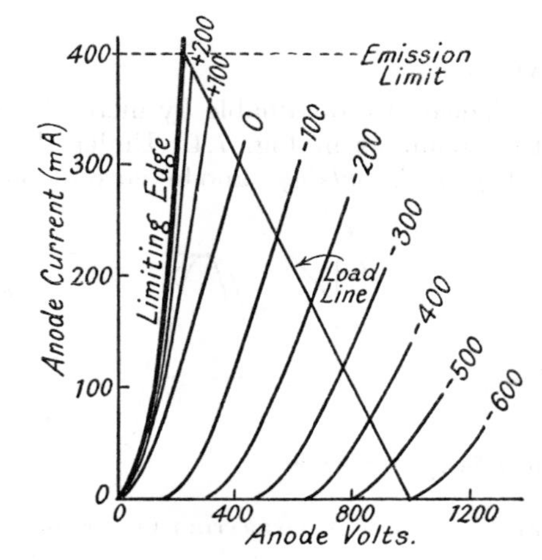

FIG. 7.5. CHARACTERISTICS OF TYPICAL LOW-POWER TRANSMITTING VALVE

possible anode current, the grid is allowed to go positive, but if the grid voltage becomes comparable with that on the anode, most of the current will go to the grid. Hence beyond this point, little further increase in anode current results, and the thick line on the characteristic in Fig. 7.5 is known as the *limiting edge*.

We can draw a load line from the operating voltage V at such an angle that it corresponds with the load in the anode circuit. The line shown, for instance, represents 2,000 ohms, since a change of 400 volts corresponds to a change of 200 milliamperes. The point where this load line cuts the limiting edge represents the maximum peak anode swing.

Now, it will be clear that the lower we make the load (and hence the more vertical the load line) the greater the peak anode current obtained. There is, however, a limit to this, determined by the maximum safe emission of the valve. This is considerably higher than the safe steady anode current, for two reasons. Firstly, this peak anode current occurs at the point when the anode voltage is at its lowest, and secondly, it is only momentary. Nevertheless, a limit, known as the *emission limit*, does exist, and the load must be such that it cuts the limiting edge of the characteristic at or below the emission limit.

It will be noted that the anode voltage is still not reduced to zero, and, in general, the anode voltage swing is about 80 per cent of the H.T. voltage. The anode current depends upon the design of the valve. Although the maximum anode current occurs at periods of low anode voltage, there is nevertheless an appreciable quantity of power to be dissipated at the anode. This energy is dispersed in the form of heat, and the methods adopted to radiate this heat effectively are discussed later in the chapter. Transmitting valves are usually designed to withstand a peak emission of six to ten times the average value, but the overriding consideration is the dissipation, as discussed in the next paragraph.

Performance Calculations

Having determined the optimum load we can assess the operating conditions more closely, but we need first to know the signal-frequency component of the anode current. As we have seen, the anode current is a series of pulses followed by an idle period of zero current. This asymmetrical waveform can be represented by

(*a*) A d.c. component, I_0.

(*b*) An a.c. component at the fundamental signal frequency, having a maximum value I_s.

(*c*) A series of smaller components at harmonics of the fundamental frequency.

It is the first two components in which we are primarily interested. In a Class B amplifier I_s is approximately equal to half the peak anode current I_{max}, but in a Class C amplifier the ratio is less, since to maintain the same value of I_s with a shorter pulse of anode current the peak current will necessarily be greater. The ratio I_{max}/I_s is clearly dependent on the angle of flow, though the relationship is not quite linear. The actual ratio depends upon circuit conditions but is given to a fair approximation by the following table:

θ	80	100	120	140	160	180
I_{max}/I_s	3·75	3·0	2·7	2·4	2·2	2·0

Given this ratio we can then estimate the efficiency. The anode voltage swings from the H.T. value E_o down to a value E_{min} as shown in Fig 7.5. Hence the peak signal voltage, E_s, is $E_o - E_{\min}$. The r.m.s. power will be $(E_s/\sqrt{2})(I_s/\sqrt{2})$, so that

$$\text{Output power} = P_s = \tfrac{1}{2}(E_o - E_{\min})I_s$$

The input power P_o is E_oI_o, where $I_o = (\theta/180\pi)I_{\max}$, as explained on page 323, whence the operating efficiency P_s/P_o is readily calculable. The difference between P_o and P_s is the power lost in the valve, which, as said earlier, is dissipated as heat, and this must not exceed the permitted rating for the valve. If it does, the operating conditions must be suitably amended (e.g. by limiting the peak current).

The final calculation required is that of the grid bias. This must be such that the peak input signal is just sufficient to swing the anode current up to the limiting value. This may be determined from the characteristics, or it may be estimated with sufficient accuracy for practical purposes from the expression

$$E_{bias} = E_o/\mu + E_g \cos \tfrac{1}{2}\theta/(1 - \cos \tfrac{1}{2}\theta)$$

where E_g is the peak positive grid swing; if this is not known, it can be taken as 80 per cent of the minimum anode voltage.

Anode Tap

Having found the correct working load, which will be seen from Fig. 7.5 to be $(E_o - E_{\min})/I_s$, it is necessary to "match" the tuned

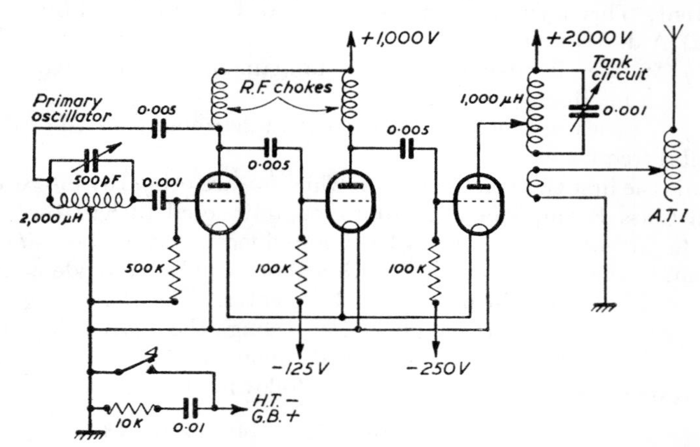

FIG. 7.6. CIRCUIT OF LOW-POWER TRANSMITTER FOR 150–500 kHz

Keying is effected by breaking the earthy side of the H.T. and G.B. supplies. The network across the key is to absorb the energy when the current is suddenly broken, and so eliminate sparking.

circuit, i.e. to adjust its effective impedance to the required value. This is done by tapping the anode connection across a suitable portion of the inductance. Then, as was shown in Section 2.4, $Z_1' = k^2 Z_2/n^2$, where k is the coupling factor and $n = \sqrt{(L_2/L_1)}$. Since $Z_2 = L_2/CR$, this can be written $Z_1' = k^2 L_1/CR$.

In practice a rough preliminary calculation is made to determine the approximate tap position, and the final result obtained by trial. Fig. 7.6 shows the circuit of a simple low-power telegraphy transmitter utilizing these principles. The master oscillator is followed by two triode amplifiers, the first of which is not tuned but uses a simple inductance for its anode load, while the second incorporates a tuned anode circuit with a matching tap as just described. Keying is achieved by making or breaking the negative H.T. connection, with a resistance-capacitance network across the key to suppress sparking.

Push-Pull Output

Where the required output is more than can be provided by a single valve, several valves may be used in parallel. In such cases it is often convenient to arrange the valves in a *push-pull* circuit. The valves are arranged in pairs with a centre-tapped input circuit. During the first half-cycle of the input the first valve operates in the normal way, while the second is driven negative and hence is inoperative. During the next half-cycle the conditions are reversed, so that each valve supplies power to the load in turn. The advantage of the method is that the peak current demand *from the supply* is only that of one valve instead of twice as much, as would be required if the valves were in parallel.

The valves feed a common load through a transformer or tap which reflects into each the same load as it would require if operating alone. The arrangement is used considerably in audio-frequency circuits and is more fully discussed in Chapter 10. Fig. 7.16 illustrates a push-pull output stage.

It should be noted that when, as is usual, an output stage is driven slightly into the positive grid region, the grid current which flows on the peaks imposes a load on the preceding stage which has therefore to be designed to supply a (small) power. Such stages are known as *driver stages*.

Energy Transfer

It will be seen that the aerial is inductively coupled to the tuned anode circuit of the final amplifier stage, the effective resistance of which is then $R_1 + (M^2\omega^2/Z^2)R_2$ where M is the mutual inductance and R_2 is the resistance of the aerial circuit, including the *radiation resistance*, which is the apparent resistance due to the power expended in radiation as explained in the next chapter. Since the secondary is tuned $Z_2 = R_2$ so that $R_1' = R_1 + M^2\omega^2/R_2$.

The coupling between the primary and aerial circuits is subject to the limitations discussed in Chapter 2. If the circuit were self-oscillating the coupling would be limited to the critical coupling $M^2\omega^2 = R_1R_2$, at which condition 71 per cent of the primary power is transferred to the aerial. Tighter coupling produces double tuning points and the circuit is liable to change spasmodically from one frequency to the other.

With a driven circuit this limitation does not apply. Increasing the coupling will cause the aerial resonance curve to become double-humped, but the frequency of the oscillation is unaffected since this is entirely determined by the master oscillator. Some broadening of the top of the aerial tuning curve is, in fact, an advantage in handling the sidebands of a modulated transmission. Hence tighter coupling is permissible, with $M^2\omega^2$ equal to 2 or 3 times R_1R_2, which results in an energy transfer of 85 to 90 per cent.

Neutralizing

It is essential to ensure that the r.f. amplifying stages do not, themselves, oscillate or even exhibit any tendency to do so. It is therefore important to avoid any feedback between the output and input whether round the whole system or in the individual stages. Accidental coupling through mutual inductance or stray capacitance can be avoided by proper attention to layout, but there remains the coupling via the internal capacitance between anode and grid of the valves themselves.

This is discussed in Chapter 9 (page 413), where it is shown that the effect may be overcome by arranging to neutralize the energy fed back through the valve by an equal and opposite amount of energy introduced into the circuit in the opposite direction. This could be done in Fig. 7.6 by connecting a small capacitor from the anode of the final valve to the grid of the *preceding* stage. By suitable adjustment of this capacitance any internal feedback may be counteracted.

A different, and more symmetrical, arrangement is adopted in the circuits of Figs. 7.15 and 7.16, where it will be seen that the tuned circuits in the intermediate stages are centre-tapped, one end feeding the grid and the other connected to the anode through a small neutralizing capacitor which is adjusted to obtain stable operation (which will occur when the capacitance is of the same order as the internal valve capacitance).

Screened Valves

The alternative is to use a tetrode or pentode instead of a triode and this is frequently done in the early stages of a power amplifier though triodes are still used in many instances for the final stages.

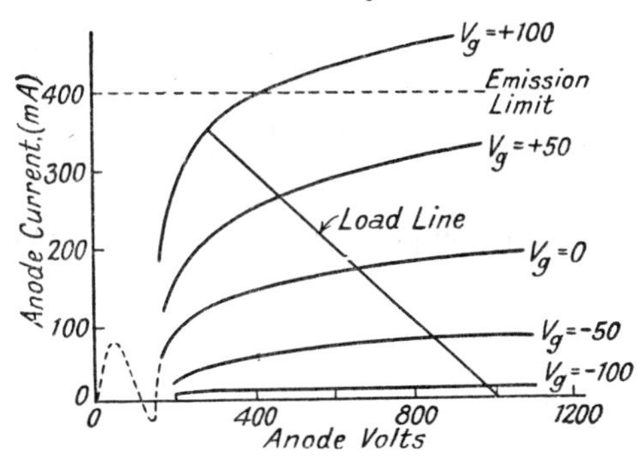

FIG. 7.7. CHARACTERISTICS OF POSITIVE-DRIVE TETRODE

The characteristics of a tetrode power amplifier intended for positive drive are shown in Fig. 7.7. The limiting effect with positive grid volts is not apparent here, and in general a tetrode or pentode has a greater efficiency. This is discussed more fully in Chapter 10.

7.2. Frequency Stability

Modern conditions require that the frequency of transmitters shall be both accurate and constant. Any variation of frequency will cause the modulation sidebands to encroach upon transmissions occupying the adjacent frequency channel, so causing interference. Hence it is important to minimize these variations as much as possible. The coils and capacitors of the master oscillating circuit should be minimally dependent on temperature and other atmospheric conditions.

The next requirement is to minimize the variations caused by changes in the valve impedances since these are connected across part of the tuned circuit, and influence the frequency to a small extent.

It can be shown that stable operation is obtained if either the anode resistance or the grid resistance of the valve can be maintained constant. One method of maintaining stability, therefore, is to use the highest value of grid leak which can be employed without intermittent operation (squegging). Under these conditions, the grid resistance of the valve is approximately half that of the grid leak, and therefore tends to remain constant.

A circuit with a high Q is an advantage, because a change of frequency then produces a rapid change of phase, and this exercises a correcting influence on the frequency.

Crystal Control

The more usual method of obtaining the requisite stability, however, is to use some form of electro-mechanical oscillator as the primary source of frequency. The most commonly used device is the quartz crystal, which makes use of the piezo-electric effect mentioned in Chapter 1. If a thin slice of quartz is subjected to strain, either tension or compression, an e.m.f. is developed across it.

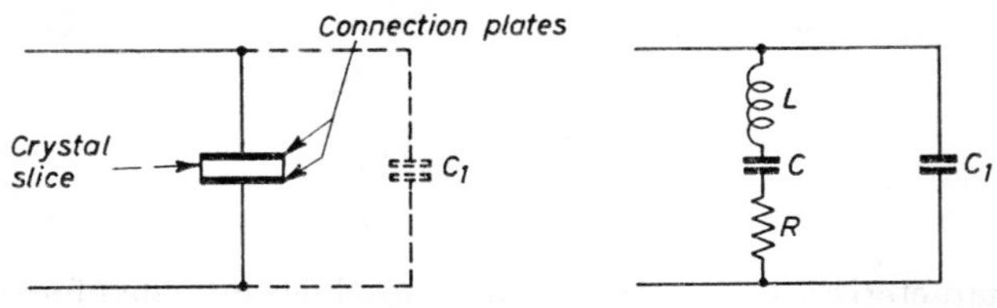

FIG. 7.8. EQUIVALENT CIRCUIT OF QUARTZ CRYSTAL

Conversely an alternating e.m.f. applied across the crystal will cause it to expand or contract. The effect is very small except at the frequency of mechanical resonance of the crystal slice, when it becomes appreciable.

The resonant frequency is high. It depends on the dimensions of the crystal slice and is generally in the megahertz region, though by special constructions resonant frequencies as low as 200 Hz are possible. It will be evident that such devices can be used as a source of extremely constant frequency. A quartz crystal can be represented in electrical terms by the equivalent circuit of Fig. 7.8, in which the inductance L simulates the mechanical inertia, while the elasticity is represented by the capacitance C, with the resistance R representing the molecular friction.

It will be seen that there are two resonant conditions. One is the series resonance between L and C, at which the impedance is simply R, the loss due to molecular friction. The other is a parallel-resonant condition determined by C_1, at which the effective impedance is very high. In practice these two frequencies are quite close, differing only by about 1 per cent. The actual operative mode depends on the circuit. In either case the resonance is extremely sharp, corresponding to a Q of 20,000 or more.

Fig. 7.9 (*a*) shows a crystal-controlled valve oscillator. This is effectively a Colpitts circuit (Fig. 5.15) in which the tuning capacitances are the anode–cathode and grid–cathode capacitances of the valve. The crystal operates at a frequency just sufficiently below resonance to permit it to exhibit an inductive reactance. Only approximate tuning of the anode circuit is therefore required, but

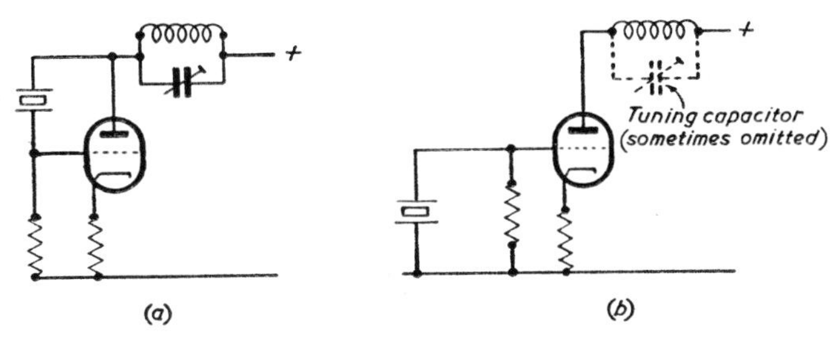

Fig. 7.9. Two Forms of Crystal Oscillator Circuit

the circuit has the disadvantage that, if the crystal has several possible resonances (close together) as explained in the next section, it will select the mode having the highest Q.

An alternative circuit is shown in Fig. 7.9 (b) which is effectively a tuned-anode tuned-grid arrangement. Oscillation is maintained

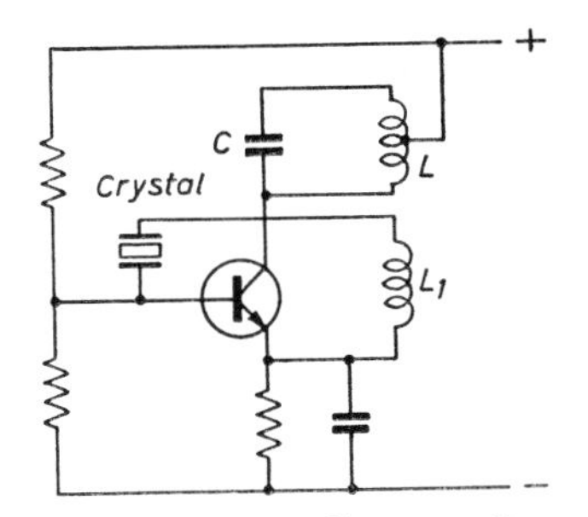

Fig. 7.10. Transistor Crystal Oscillator

here by feedback through the anode–grid capacitance of the valve, which is only positive when, and to the extent that, the anode circuit is inductive (as explained on page 415). Hence, by tuning the anode circuit to a frequency slightly higher than the desired parallel-resonant frequency of the crystal, precise operation is obtained. If the crystal has only one mode the anode tuning may be omitted.

The problem of frequency stability is even more important with transistor oscillators since the parameters of solid-state devices are markedly temperature-dependent. Fig. 7.10 shows a crystal-controlled transistor circuit. This contains a tuned circuit in the collector which is maintained in oscillation by feedback to the base via the coil L_1. The crystal, however, presents a high impedance to any but its (series) resonant frequency, so that the feedback is only operative at this point, resulting in a precisely-determined frequency of oscillation.

Crystal Modes

The crystal slice has to be cut in a particular way since the piezo-electric and elastic constants are dependent on the orientation of

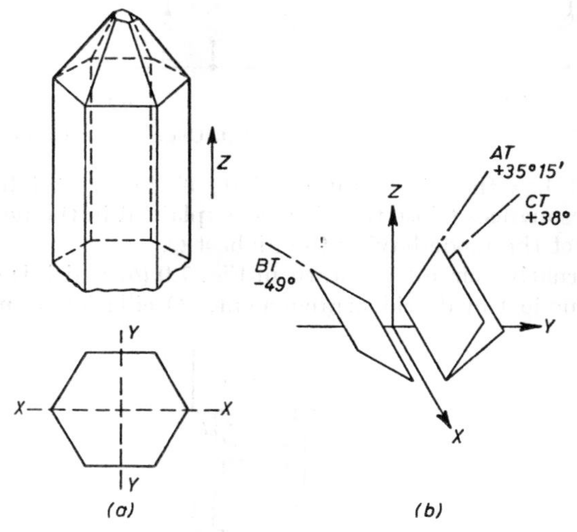

FIG. 7.11. ILLUSTRATING QUARTZ CRYSTAL SLICES

the slice within the crystal. Quartz crystals are hexagonal in cross-section as shown in Fig. 7.11 (*a*). The axis joining two opposite corners is called the *X*-axis, the axis at right angles to this the *Y*-axis, and the axis along the length of the crystal the *Z*-axis.

The simplest mode of oscillation which can be produced in quartz is longitudinal, and for this application, slices are normally cut from the crystal with sides nearly parallel to the *Y*-axis. This is known as an *X*-cut crystal since its face is normal to the *X*-axis. The resonant frequency with this cut is approximately 2,730/*t* hertz, where *t* is the thickness of the slice in millimetres. In fact, optimum temperature characteristics are obtained from crystal slices cut with the length at an angle of $+5°$ from the *Y*-axis, and these slices are known as $+5°$ X-cut bars. Such crystals are used for oscillators in the range 40–200 kHz. An alternative X-cut at $-18°$ to the Y-axis is frequently used in crystal filters because it produces a very pure frequency spectrum.

Oscillations at lower frequencies can be obtained by operating X-cut crystals in the thickness-flexure mode. Here the crystal is held by four wires mounted at the nodal points, which occur at a distance 0·224 times the length from each end of the crystal. Such crystals are used in oscillators down to 2 kHz, while even lower frequencies can be obtained by adopting a bimorph construction in which two thin plates of crystal are bonded together. These bimorph or duplex crystals operated in the thickness-flexure mode can be used in oscillators at frequencies as low as 200 Hz.

Higher-frequency oscillators can be made using X-cut crystals in the thickness-longitudinal mode, but owing to the poor temperature coefficient of X-cut plates, they have largely been replaced for this purpose by rotated Y-cut crystals operating in one of the shear modes in which squares in the crystal are distorted into rhombi. These crystals are cut from planes which are rotated by various angles from the Z-axis, and some of the more commonly used cuts are shown diagrammatically in Fig. 7.11 (*b*). The angles of cut are chosen to give the best combination of temperature coefficient, activity and harmonic distortion.

Oscillators in the range 150–950 kHz normally use $CT(+38°)$, $DT(-52°)$ or $ET(+66°)$ cut crystals operating in the face-shear mode, while the $AT(+35°\ 15')$ and $BT(-49°)$ cuts are used for higher frequencies. AT-cut crystals operating in the thickness-shear mode are made to resonate at fundamental frequencies up to 20 MHz, while BT-cut crystals can be made to produce fundamental resonances at even higher frequencies. However, for frequencies above 20 MHz *overtone* crystals are generally used, as described later.

The temperature coefficient of frequency is very dependent on the type and angle of cut, but typical AT-cut crystals can have a normal tolerance of ± 25 parts in 10^6 from $-20°C$ to $+70°$, and this can be reduced to ± 10 parts in 10^6 by careful cutting. Most other crystals have a parabolic frequency-temperature characteristic with the zero point normally between 30 and 50°C but giving a rather worse tolerance that the AT-cut over a wide frequency range.

Where very close tolerance is required the crystal is housed in a temperature-controlled oven. These ovens are very often double, the inner oven being maintained at a slightly higher temperature than the outer, and the outer being kept at a temperature higher than the highest ambient temperature ever likely to be encountered in the particular climate. The cut of the crystal is then arranged so that the zero-temperature-coefficient point of the parabola occurs as near as possible to the oven temperature.

Crystal Mounting

The crystal slice has to be ground to a high precision, since any unevenness can result in spurious modes of vibration relatively close together, and the crystal may prefer to operate in one of these spurious modes in preference to the intended one. Any grease or scratches on the surface of the slice can also cause erratic behaviour.

Hence it is usual to form the electrodes by depositing a suitable metal film on the face of the slice, from which connections are taken at a nodal point, the assembly being then suitably encapsulated. In many cases the crystal is supported by these connecting wires within a glass envelope, which may be evacuated to reduce molecular friction and so increase the Q.

The parallel-resonant frequency is affected to a minor extent by the circuit capacitance. This is sometimes utilized to "pull" the frequency to the exact value required.

Frequency Multiplication

At frequencies above about 25 MHz the quartz slice becomes so thin as to be too fragile for convenience. It is therefore customary at the higher frequencies to generate a frequency which is a submultiple of the value required. This frequency is then doubled one

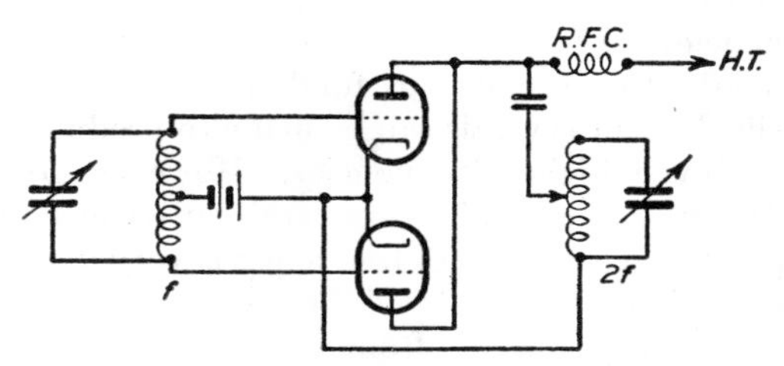

FIG. 7.12. FREQUENCY DOUBLING CIRCUIT

or more times by suitable circuits. The most usual is a variant of the push-pull arrangement, as shown in Fig. 7.12. The input circuit is tuned to the fundamental frequency and the anode circuit to twice this frequency. Since one of the anodes receives an impulse every half-cycle, which occurs at twice the fundamental frequency, it will be seen that the anode circuit is satisfactorily energized. The valves, of course, are biased to cut off.

Several stages of frequency doubling are often used to provide four or eight times the initial frequency. Alternatively, the circuit can be arranged to provide an output frequency several times the input frequency, and this is often done in f.m. transmitters as explained on page 354.

Overtone Crystals

An alternative arrangement for generating high-frequency oscillations is to use an *overtone* crystal. So far we have considered the crystal as vibrating in its fundamental mode, but it will be appreciated that it can also vibrate in a harmonic mode, though with decreased amplitude. By suitable choice of dimensions and mounting, quartz plates can be designed to operate in one of these harmonic modes, so avoiding the necessity for subsequent frequency multiplication.

It is important to note the difference between operating a crystal in its fundamental mode and electrically selecting a harmonic, and actually driving the crystal to vibrate at a mechanical overtone which is never exactly integrally related to the fundamental because of the necessarily imperfect mechanical structure. Oscillators are constructed using third, fifth and seventh overtone AT (or BT) cut crystals to resonate at frequencies between 15 and 200 MHz.

Electro-mechanical Oscillators

Crystal oscillators are not normally used below about 50 kHz. Below this frequency the crystals become unwieldy and fragile, though crystals are made for frequencies as low as 200 Hz. It is more usual for such frequencies, however, to divide down from a higher frequency as explained in Chapter 10.

Alternatively, for these lower frequencies use may be made of tuning forks or magnetostriction oscillators. A tuning fork constitutes a mechanical resonant circuit and can be maintained in a state of oscillation by a valve in similar fashion to an electrical tuned circuit. The fork is mounted on a rigid base with two coils located near the ends of the tines, as shown in Fig. 7.13. The vibration induces e.m.f. in the grid coil which causes amplified currents to appear in the anode coil and if these are in the right phase the vibration will be sustained.

The frequency is thus dependent almost entirely on the mechanical constants of the fork and is only influenced to a minor degree by the maintaining circuit. The primary frequency-determining factors are the length of the tines and the elasticity of the material, both of which vary with temperature. By the use of special alloy steels, however, this variation can be limited to about +10 parts per million per degree Celsius.

Still greater accuracy is attainable by the use of "slice" forks made up of a sandwich of two materials having positive and negative temperature coefficients, so proportioned as to produce virtually

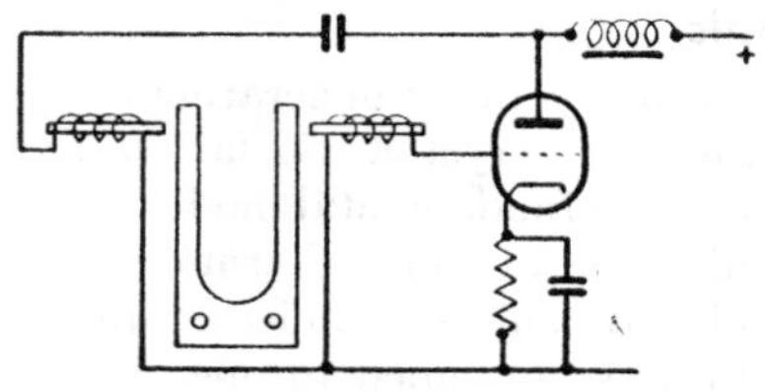

FIG. 7.13. CIRCUIT OF VALVE-MAINTAINED TUNING FORK

zero overall coefficient. With such forks the influence of the maintaining circuitry becomes appreciable, but with suitable precautions the overall stability can be within a few parts per million, which is comparable with a quartz crystal.

By using a transistor instead of a valve for the maintaining circuit, as is now more usual, complete units can be produced only a few cubic centimetres in volume.

An alternative form of electro-mechanical oscillator makes use of the effect known as *magnetostriction*. Certain materials, such as cobalt-steel or nickel, show an appreciable change in length when subjected to a magnetic field. A rod of the material, housed in a long solenoid carrying a.c., will thus vibrate longitudinally at the frequency of the a.c. (provided a d.c. polarizing current is also applied to prevent the magnetic field from reversing. Otherwise it will vibrate at twice the applied frequency, since the magnetostriction effect is independent of the direction of the magnetism).

An oscillator can be constructed by coupling two coils to the rod, each extending from near the centre to the end, and connecting these in the anode and grid circuit of a valve. A small current in the anode coil will set the rod vibrating, and this will induce voltages in the grid coil, which, if correctly phased, will provide amplified currents in the anode coil and hence build up the oscillation. The rod will vibrate as a half-wave resonator, so that the centre point is a node and may be rigidly clamped.

The frequency of vibration is determined almost entirely by the rod, irrespective of the circuit conditions, and is given by v/l, where v is the velocity of sound in the material and l the length of the rod. Frequencies ranging from one to several hundred kHz can be produced by this means, the upper limit occurring when the skin effect in the material prevents the magnetic field from penetrating to an adequate depth. The stability is comparable with that obtained with a tuning fork.

7.3. Modulation

We have now to consider the methods used to *modulate* the signal, i.e. to vary its character in such a way as to communicate the intelligence required. With a telegraphy transmitter this is done by starting and stopping the oscillation at suitable intervals to provide signals in terms of the Morse code or other system.

This may be done easily enough with a driven transmitter by simply rendering one of the amplifier valves inoperative in the key-up position. With a self-oscillating transmitter the oscillation itself must be started and stopped, though it is usually more convenient not to stop the oscillation completely but to reduce the amplitude to a small value in the key-up position. Alternatively the frequency may be varied by a small amount, so providing "marking" and "spacing" waves of different frequency.

With a telephony transmitter the amplitude or frequency of the oscillation has to be varied in conformity with the audio-frequency signal. The fundamental aspects of modulation were discussed in Section 3.2. We are concerned here with the methods adopted to produce the required modulation.

Choke Modulation

With amplitude modulation the most usual and convenient method is what is known as *choke modulation*. The anode supply to the modulating amplifier is obtained through a low-frequency choke, which also supplies the anode of the modulator valve as shown in Fig. 7.14. Variations of the grid voltage of the modulator cause

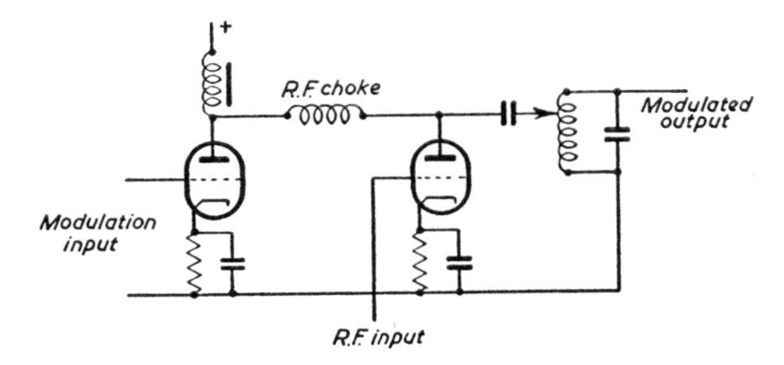

FIG. 7.14. CHOKE-MODULATION CIRCUIT

variations in the anode voltage, and since the anode of the modulating amplifier is tied to the anode of the modulator, this voltage varies in the same manner and so modulates the amplitude of the high-frequency oscillations as required.

As explained in Section 3.2, with 100 per cent modulation the instantaneous maximum value of the oscillations is twice the mean value of the carrier. Hence, the instantaneous power radiated is four times the carrier power. This has to be supplied by the modulator, which must therefore be considerably larger than the oscillator valve itself. Alternatively several modulator valves may be used in parallel, and in practice this is usually more convenient.

The r.f. amplifier must be operated under such conditions that the oscillating current is directly proportional to the anode voltage. A Class B amplifier or oscillator under suitable conditions may fulfil these requirements, but it is equally possible that it will not do so and this point must be verified. Moreover, to obtain true fidelity of modulation it is necessary for the impedance of the modulating choke to be substantially constant over the full range of modulation frequencies.

Since the frequency of the oscillations generated by a valve is dependent to some extent on the constants of the valve, and since these tend to vary with the anode voltage, it follows that an ordinary self-oscillating valve modulated as described will vary in frequency as well as amplitude during the process. This is undesirable, particularly at high frequencies, where a small percentage variation in the frequency may produce a larger change in the received signal than the whole of the modulation. Modulation, therefore, is seldom employed on the actual oscillating valve. The oscillator is isolated by passing the signal through a buffer amplifier before reaching the modulation circuit, as shown in Figs 7.15 and 7.16.

Fig. 7.15 shows the circuit of a medium-power telephony transmitter. A crystal oscillator is followed by a buffer amplifier which feeds the modulator stage. The modulated signal is then passed through a further amplifier which drives the output valve. The power stages following the modulation cannot be Class C operated because such a system is not linear, and it is essential that the amplitude of the oscillations in the anode circuit shall be strictly proportional to the excitation applied to the grid. Hence for the final stages Class B amplification is used.

Fig. 7.16 shows a simplified circuit of a 50-kW transmitter. A crystal oscillator feeds a buffer amplifier and then a 50-W amplifier on which the modulation is provided. The modulated output is then amplified by two neutralized 250-W valves, which then feed

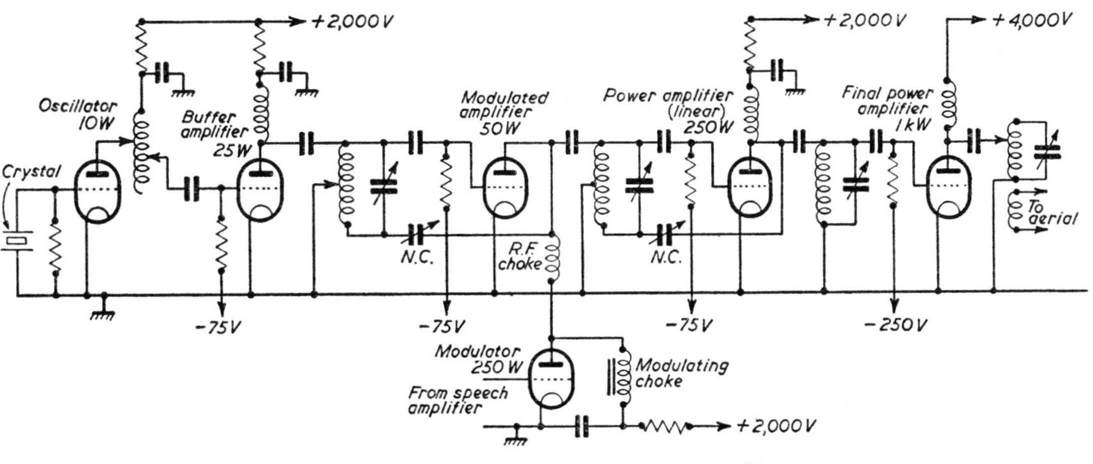

FIG. 7.15. TYPICAL MEDIUM-POWER TELEPHONY TRANSMITTER
Pentode valves are often used in the early stages to avoid neutralizing.

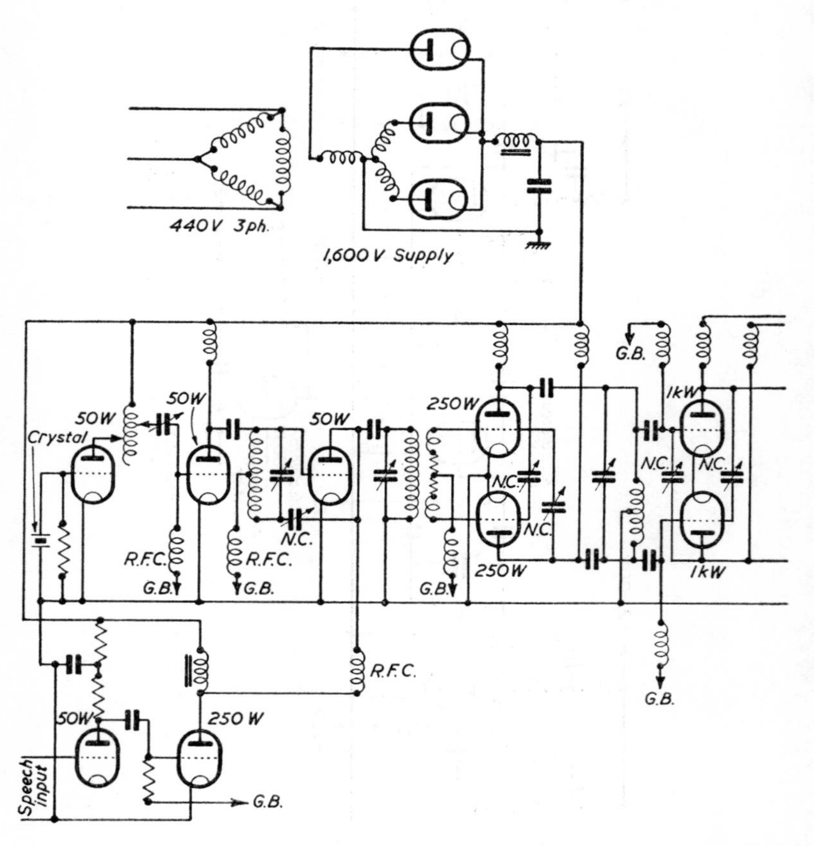

FIG. 7.16. SIMPLIFIED CIRCUIT

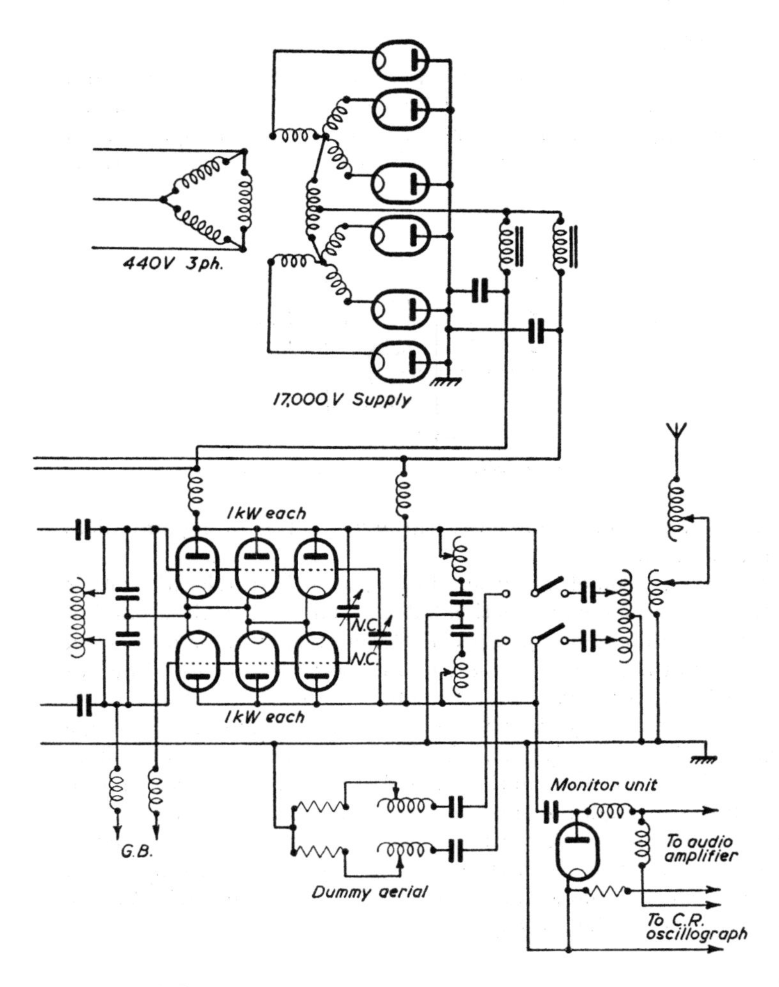

OF A 50-kW TRANSMITTER

two water-cooled valves and then finally feed six water-cooled valves in parallel push-pull.

Balanced Modulators

Reference has already been made in Section 3.2 to single-sideband operation. This may be achieved by passing the signals from the modulator stage through a band-pass filter which suppresses all frequencies except one sideband and some or all of the carrier.

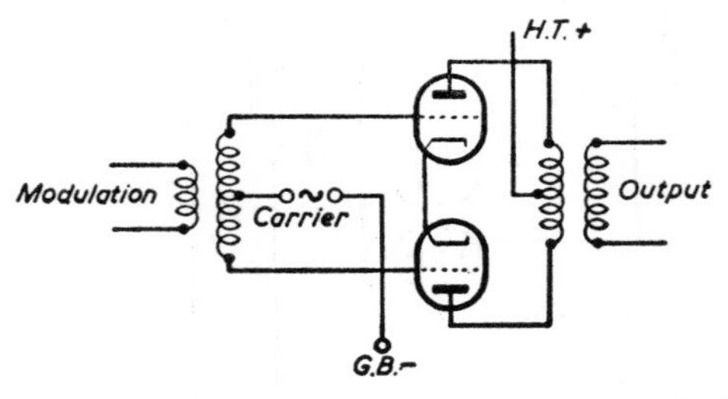

FIG. 7.17. CARSON BALANCED MODULATOR CIRCUIT

The modified signal is then passed through the remaining amplifiers in the normal way.

It is, however, sometimes desired to remove the carrier altogether (although as was shown in Section 3.2 this is not, by itself, a practicable form of transmission). Special forms of modulator have been devised to do this, one of the best-known being the balanced modulator due to Carson, which is illustrated in Fig. 7.17. A push-pull arrangement is employed and the carrier and modulating voltage are introduced separately. The former is introduced so as to affect the grids of the two valves in the same phase, whereas the modulation is introduced in push-pull manner, so that the grids of the valves are in opposite phase. A centre-tapped output transformer is used, with the result that the carrier voltage does not appear in the secondary, whereas the modulation combines with the carrier to give sidebands in the ordinary way.

Ring Modulator

An alternative form of modulator which suppresses both the carrier and the modulation frequency, leaving only the sidebands, is the ring modulator shown in Fig. 7.18. The carrier is impressed across the points P, Q and flows through the two halves of the transformer windings in opposition. The positive half cycles flow through the rectifiers A, B and the negative half cycles through C, D. While the rectifiers A, B are working, C and D will be subject to a negative voltage and therefore will remain non-conducting even if further voltages (e.g. modulation voltages) are impressed across them, provided these additional voltages do not exceed the carrier voltage. Similarly when the rectifiers C, D are operative A and B are cut off.

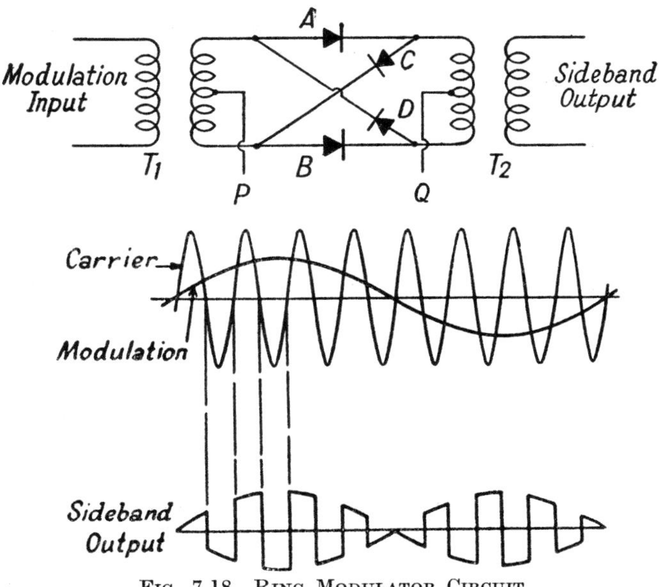

FIG. 7.18. RING MODULATOR CIRCUIT

No carrier voltage appears in the output due to the symmetry of the arrangement.

The modulation is impressed across T_1 and current flows alternately through A, B and C, D producing side frequencies in T_2 by combination with the carrier. These side tones do not cancel out so that the desired sidebands appear in the output while the carrier is suppressed.

The arrangement, in fact, is a switch of which the connections are changed over by the carrier, provided that the carrier voltage exceeds the modulation voltage (as it always does in practice).

Centre-tapped resistances may be used in place of T_1 and T_2.

Single-Sideband Operation

Most modern high-power transmitters, particularly in the H.F. band, use single-sideband (s.s.b.) technique. Only one sideband is transmitted, together with a small "pilot" carrier (since complete suppression of the carrier is not a practicable arrangement). This has the advantages of avoiding the waste of power which results when the carrier is transmitted at full strength, and also reducing the overall bandwidth which increases the signal/noise ratio.

The process is usually carried out in three stages, as indicated diagrammatically in Fig. 7.19. The speech signal is first applied to a balanced modulator with a carrier frequency of 100 kHz. The

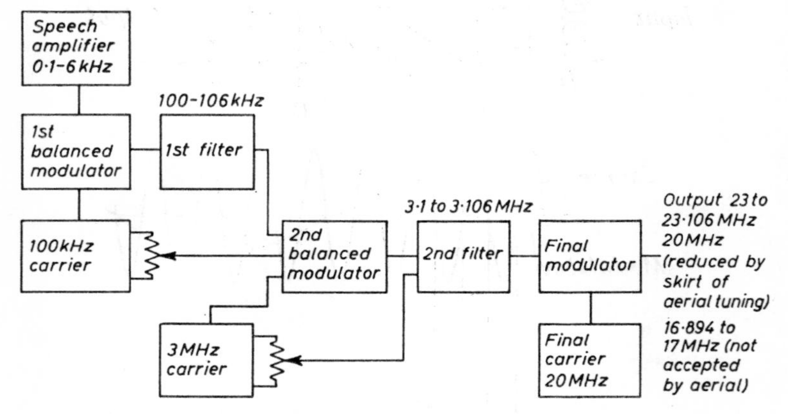

FIG. 7.19. SINGLE-SIDEBAND OPERATION

output, containing the two sidebands and a vestigial amount of carrier, is then passed through a band-pass filter which selects one sideband only. Assuming an initial modulation band of 6 kHz, and choosing the upper sideband, this filter will have a pass-band of 100 to 106 kHz.

A small amount of carrier is reintroduced and the combined output is then applied to a second balanced modulator with a carrier frequency of 3 MHz, the output from which is then passed through a second filter accepting 3·1 to 3·106 MHz. This output is then used to modulate a final carrier at a frequency determined by the actual transmission frequency chosen. If this is, say, 20 MHz the output frequency bands will be 23 to 23·106 and 16·894 to 17 MHz. These are sufficiently widely separated for the normal aerial tuning to select the upper band and suppress the lower one, resulting in a transmission containing a single sideband with a vestigial amount of carrier.

At the receiving end the carrier is restored to its proper amplitude either by the use of circuits which accentuate the carrier frequency in relation to the modulation frequencies, or by using the carrier component to synchronize a local oscillator. In the latter case it is only necessary to vary the frequency of the local oscillator around the correct nominal value until intelligible speech is obtained, at which point the local oscillator will lock onto the pilot carrier. (See Chapter 12.)

Vestigial-sideband Transmission

A modification of the single-sideband system which is more suitable for broadcast transmissions is what is known as vestigial-sideband operation. Here the full carrier is transmitted with one full sideband but only a small portion of the other. This avoids the necessity to reintroduce or reinforce the carrier at the receiving end, while still providing appreciable reduction in the overall bandwidth. This form of transmission is standard practice for television transmissions, as is explained in Chapter 16 (cf. Fig. 16.9).

Quiescent Carrier Systems

An alternative method of saving power which is more applicable to medium and small power transmitters is what is known as *suppressed carrier operation*. It is found that with a normal two-way telephone conversation, speech is only transmitted on either channel for about 13 per cent of the total time, the remaining time being taken up by pauses or listening to the reply. Hence considerable saving of power results if the carrier is only transmitted during actual speech.

This can easily be arranged by using the speech signal to control the carrier, which is transmitted at full amplitude as long as speech is present, but which is cut off when the speech ceases. The control circuitry is provided with a slight delay to prolong the carrier for a short period after the cessation of the speech to avoid clipping between syllables or during legitimate pauses.

The system has the further advantage of reducing noise. If the carrier is still being radiated during reception periods it can induce currents in nearby structures, particularly on board ship, which can re-radiate spurious noise into the receiver.

Pulse Modulation

Reference was made in Chapter 3 to the pulse modulation technique which has certain advantages in commercial or military communication links. This involves the transmission of a succession of pulses of which the amplitude, width or separation is varied to communicate the information required. This is basically a particular form of amplitude modulation, but if the pulses are to be clearly defined, the circuitry has to be able to handle much higher frequencies than the normal speech range.

The pulse duration is very brief, usually of the order of 1 μsec, and if a sharp edge is to be preserved the carrier must be able to change from zero to full amplitude in something like one-tenth of this time. The requirements for the satisfactory handling of such waveforms are discussed in Chapter 17, which also describes the methods used to provide variable width or separation of the pulses. It is shown there that the bandwidth required is approximately $0{\cdot}4/t$, where t is the time of rise. If this is taken as 0·1 μsec the bandwidth is 4 MHz.

Because of the brief duration of the pulses, the carrier only radiates full power for a fraction of the time, and the idle period may be occupied with a succession of suitably delayed additional pulses. This is known as multiplexing, and permits 8 or 10 channels to be handled simultaneously.

A further advantage of pulse modulation is the improvement in signal/noise ratio. In general, any modulating signal has superposed on it a proportion of background noise which interferes with the clarity of the reception, as discussed on page 418. If the amplitude of the pulses is appreciably greater than the noise level, the tops may be squared off by a limiter, thus virtually eliminating the noise.

Inversion and Scrambling

Privacy is one of the principal requirements of a commercial telephone service, and the fact that ordinary broadcast telephone signals could be easily picked up militated at first against the use of radio telephony. The difficulty has largely been overcome with modern short-wave circuits which employ beam transmission and reception, while to make doubly sure the speech is often inverted.

For this purpose the speech frequencies before modulation are combined with a fairly high frequency, producing sidebands in the ordinary course of events. The lower sideband only is chosen, so that the lowest speech frequency is now the highest modulation frequency. These frequencies are now heterodyned with another frequency lower than the lowest frequency in the band. The difference tone is used, all other frequencies being filtered out, and we are left with a modulation much the same as the original as regards range, but completely inverted.

Such a modulation is quite unintelligible to the ordinary listener, and privacy is thus assured. It is translated at the receiving end by going through the reverse processes, in the requisite order.

As a typical example we may take a frequency range of 100–3,000 Hz. This would be arranged to modulate an oscillator of, say, 50 kHz. The lower sideband, ranging from 47,000 to 49,900 Hz, would be selected, and subsequently made to heterodyne another frequency to reduce the actual order of frequency. This is often done in two stages. For example, the first stage would use an oscillator of 42,000 Hz giving difference frequencies of 5,000 to 7,900 Hz, the other frequencies being suppressed, and this if necessary could again be heterodyned with an oscillator of 4,900 Hz which would give the original scale of frequencies ranging from 100 to 3,000 Hz, but completely inverted.

Another method is to transpose certain bands of frequency, which is done by selecting the required bands with filters and heterodyning them to different degrees so that when recombined they occur in a different order.

In practice both methods are employed either separately or together and speech broken up in this general manner is said to be *scrambled*.

Valves for Transmitters

Transmitting valves are similar to those employed for receiving purposes, except that they are larger in size and are designed to dissipate heat very readily from the anode and to a smaller extent from the grid. The valves are designed to give a high peak emission, the ratio of peak emission to mean anode current being of the order of 10 to 1.

In order to increase the peak dissipation at the anode, materials are used which will run at a higher temperature. Molybdenum is one such material, and anodes made with this will run at a cherry red. Another method of increasing radiation is to blacken the anode, since a black body radiates heat better than a bright one. Still another method is to use carbon for the anode.

Heat dissipation is by direct radiation from the anode to the glass and thence by ordinary convection. Since a large bulb has to be employed, in order that the glass shall not soften, this ultimately sets a limit to the size of the valve. Forced draught cooling is sometimes employed.

The next step was in the direction of silica valves, since this material will run at a higher temperature than glass. A silica envelope, however, is very expensive, and valves of this type are usually made demountable, the construction being such that if the filament burns out the base of the valve can be removed, a new filament fitted, the base relined up and welded to the envelope, and the whole re-exhausted.

The most successful technique is to use cooled-anode valves. Here the anode itself is made the outer container of the valve. The grid and filament assemblies are mounted on a glass foot as usual, and this is sealed into the anode. For medium powers this is sufficient in itself. The anode is usually formed with large cooling fins to increase the radiation.

For still larger powers the anode itself is immersed in a water jacket through which a supply of cooling water circulates to dissipate the heat. Since the anode is at a potential of several kV, the water circuit has to be suitably insulated, and this is usually done by connecting the supply to the anode through a long length of rubber hose suitably coiled up. Valves of this type are in daily use handling powers of 25 and 50 kilowatts each, and banks of such valves handling radio-frequency outputs of 500 kilowatts or more are employed in modern technique.

Certain specialized developments are necessitated by short-wave working, but these are discussed more fully in Chapter 13.

7.4. Terminal Arrangements

The transmitting apparatus itself is usually housed in cubicles of steel frame construction, so arranged that access to the "live" parts cannot be obtained without disconnecting the supply. This is an important proviso in view of the high voltages involved, and considerable care has to be taken to make the operation safe.

In many instances, as in the transmitters of the B.B.C., for example, each stage in the chain is duplicated, including the valves, and by throwing over a switch the current may be routed through the alternative stage. This enables continuity of service to be maintained. Both channels are tuned up and the valves kept alight, so that a simple change-over of connections is all that is required.

The main controls are on the transmitter, but controls of input, modulation, etc., are effected from a remote control desk, on which are duplicated the principal circuit meters and a monitor of the output, so that the control engineer can see the situation at a glance.

Land Line Connection

A special circuit is necessary to link a radio channel to a land line. The radio channel is a four-wire system having a pair of leads for the transmitted speech and another pair for the received signals. These are combined in a special transformer known as a *hybrid coil*, which enables both signals to be carried over the same pair of wires as is done with ordinary telephony.

Fig. 7.20 illustrates the system. Speech from the subscriber induces voltages in windings A and C. The former passes to the transmitting repeater (which is merely an amplifier) and this in turn modulates the transmitter. The voltage in winding C can produce no effect, because it is in the output of the receiving repeater

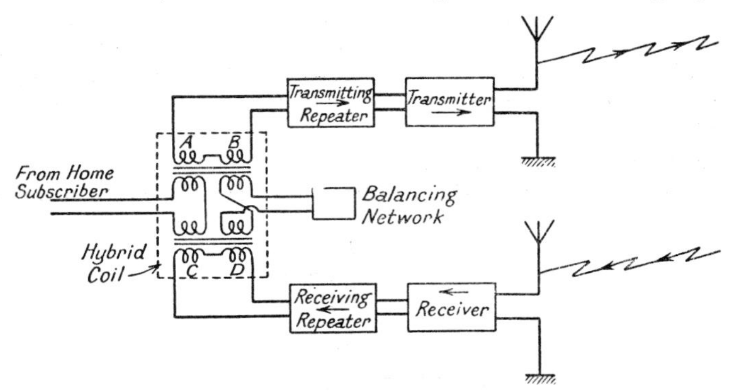

Fig. 7.20. Illustrating Connection of Radio to Land Line

and cannot force its way back owing to the unidirectional characteristic of any valve amplifier.

Speech picked up on the receiver is amplified by the receiving repeater and appears in windings C and D. Currents are induced in the land line, where they travel on to the subscriber, and also in the balancing network, which is an artificial line having the same characteristics as the actual line. These currents also induce voltages in A and B, but as one of the transformers is reversed the voltages are in opposition and cancel out. Thus, the received speech is prevented from operating the transmitter.

Singing

Any interaction between the transmitter and receiver will give rise to a continuous oscillation or *sing*. There are two forms encountered in practice, one being a half sing when energy is fed back from the local transmitter to the local receiver, while the other is an all-round sing from the local transmitter to the distant receiver and back through the distant transmitter to the local receiver.

The first difficulty can be overcome by adequate design of the local receiver. Given sufficient wavelength separation and a selective circuit, no trouble should be experienced.

The all-round sing depends on accuracy of the hybrid balance. This, of course, involves careful adjustment of the balancing network which is connected across one side of the hybrid coil to be equivalent to the land line. Generally speaking, singing is more difficult to prevent with high-gain receivers, as one might expect.

Echo Suppressors

As a general rule, therefore, suppressors are included in the circuit, so arranged that, as soon as the local subscriber commences to speak, the receiver is cut off. If the distant subscriber speaks, the receiver takes command, provided the home subscriber is not speaking.

This may be done with voice-operated relays. The speech frequency is rectified and the resulting d.c. is applied to a suitable

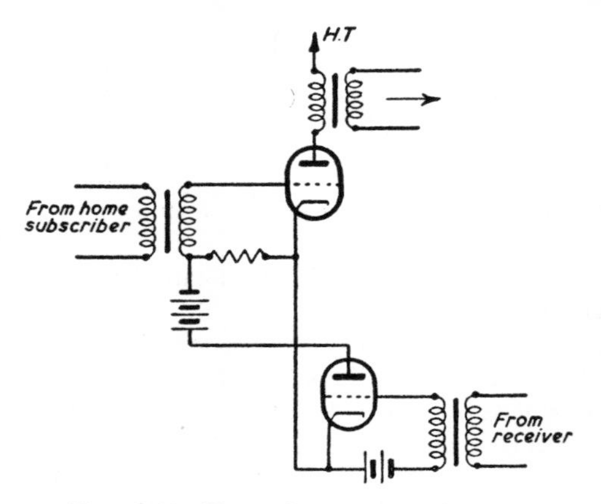

FIG. 7.21. ECHO SUPPRESSOR CIRCUIT

relay which changes over the connections. Alternatively electrical methods may be used, one such system being shown in Fig. 7.21. In the grid return of one of the amplifier valves is a high resistance which is arranged to form the anode load of a second valve operating as an anode-bend rectifier. This normally carries no current, but when speech voltages are applied to its grid a voltage is developed across the anode load which applies a large negative bias to the first valve and so renders it inoperative.

Volume Compression and Expansion

The power radiated by a transmitter, and hence the field strength at the receiver, is continually varying in accordance with the modulation. Moreover a large proportion of normal speech sounds are of low intensity, and it is only occasionally that the level rises to full volume, so that the available power is only in use for a small portion of the total time.

An immediate improvement would result if the speech waves could be compressed so that the difference between normal and peak values was reduced. If this compression were in the ratio of 2:1 only, the average level would be doubled and the power radiated would increase four times.

Commercial long-distance channels, therefore, utilize a device called a *compandor*, which is a portmanteau word denoting a compressor (at the transmitter) and an expander (at the receiver). Both devices are similar but act in opposite directions.

They comprise an audio-frequency amplifier of which the gain can be controlled by the application of a suitable voltage. For example, two variable-mu valves in push-pull would provide a gain proportional to the grid bias, the use of two valves being preferable to cancel out any second harmonic distortion which might arise from the non-linear characteristic. A hexode can also be arranged to provide variable a.f. gain by applying speech input to the modulator grid and the control voltage to the signal grid.

Control voltage is obtained from an amplifier followed by a detector, which develops a d.c. voltage proportional to the instantaneous speech level, and by applying this to the non-linear amplifier the modulation may be compressed or expanded (according to the direction in which the control voltage is applied) to any desired extent within reason.

In practice a compression of about 2:1 suffices, and at the receiver the modulation is expanded in the same ratio to restore it to normal.

7.5. F.M. Transmitters

With a frequency-modulated transmission the procedure is basically the same as with amplitude modulation except for the method of modulation. There are certain differences due to the fact that frequency modulation is generally only employed in transmissions operating in and above the v.h.f. band, but the special techniques used at such frequencies apply both to a.m. and f.m. transmissions and are discussed in Chapter 13.

The main special requirement in a f.m. transmitter is the ability to vary the carrier frequency in accordance with the modulation. One method of accomplishing this is to connect a *reactance valve* across the tuned circuit of the oscillator, as shown in Fig. 7.22. The anode of the reactance valve will then be subjected to the r.f. voltage

e_a. This will pass a current through the network CR, and if $1/\omega C$ is much larger than R this current will be very nearly $j\omega Ce_a$ and the voltage at the grid will be $j\omega CRe_a$.

This will produce an anode current $j\omega CRge_a$, where g is the mutual conductance of the valve; but this current will be e_a/Z, where Z is the effective impedance of the valve. Hence $Z = 1/j\omega CRg$. In other words the valve "looks like" a small capacitance CRg, so

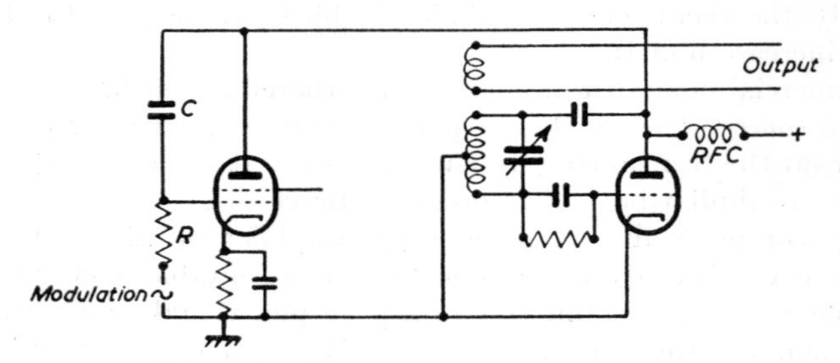

FIG. 7.22. REACTANCE-VALVE MODULATOR

that if we vary g we shall change the effective capacitance and so alter the tuning of the oscillator.

We can vary g by altering the potential of any of the grids of the reactance valve, whichever is the most convenient, and by suitable choice of valve and components we can obtain the requisite variation of frequency. The arrangement will also work if C and R are interchanged, in which case the valve looks like a variable inductance.

Such an arrangement clearly provides the modulation required, for the change of oscillator frequency is proportional to the input to the control valve (i.e. the instantaneous amplitude of the speech or music), while the rate at which the frequency changes is determined by the frequency of the modulation.

Phase Modulation

The arrangement of Fig. 7.22, however, has one serious disadvantage, namely that the oscillator must be of a type in which the frequency is determined by the tuned circuit values. But we have seen that for modern conditions this is inadequate and the main frequency-determining element has to be a quartz crystal or equivalent device.

Fortunately this difficulty can be surmounted by using phase modulation which, as was pointed out in Section 3.2, is an equivalent process. A phase-modulated wave is also frequency-modulated, and vice versa.

Now, it was also shown in Section 3.2 that, if the carrier in an amplitude-modulated wave is 90° out of phase, the modulation does not produce any appreciable change of amplitude but only causes the phase of the carrier to swing around its mean position. By making use of this effect, therefore, we can modulate in a separate stage, deriving the primary frequency from an independent oscillator, which can be crystal-controlled if desired.

Fig. 7.23 shows one way of doing this, due to E. H. Armstrong, who was the "father" of frequency modulation. The initial carrier frequency is passed through a balanced modulator which removes the carrier and leaves only the sidebands. These are then combined with a fresh carrier derived from the original oscillator but shifted by 90°. The result is a phase-modulated wave, which is then passed

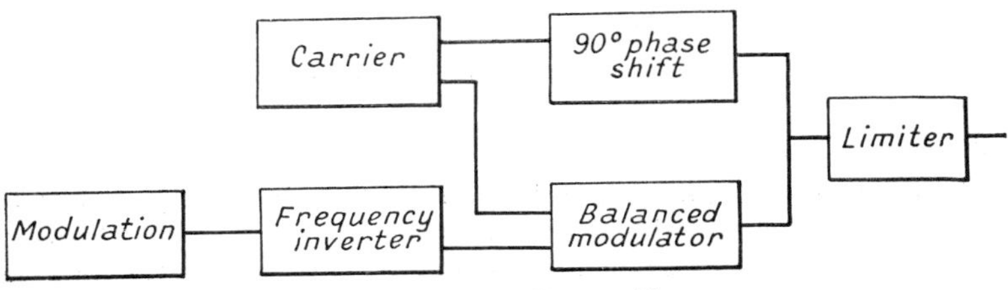

FIG. 7.23. ARMSTRONG PHASE MODULATOR

through a limiter, which is an amplifier so designed that with any input exceeding a certain value the output is constant. This removes any small changes of amplitude which may have been introduced during the modulation process.

With phase modulation, however, as explained in Section 3.2, the frequency deviation is proportional to the amplitude and *frequency* of the signal, whereas with frequency modulation the frequency deviation is independent of frequency. Hence if a frequency-modulated output is required from the circuit of Fig. 7.23 the modulating signal must be passed through a frequency-inverting network. A simple RC series circuit will do this if $R \gg 1/\omega C$. The voltage across the capacitor is then inversely proportional to ω.

A rather simpler form of phase modulator is shown in Fig. 7.24. The initial carrier, which can be supplied from a crystal-controlled oscillator, is applied to the grid of a valve having a low-Q tuned circuit in its anode. Voltage will be developed across this circuit by direct coupling through the grid–anode capacitance. This capacitance in series with the dynamic resistance L/CR will constitute a phase-advance network so that the voltage, E_1, across the tuned circuit will suffer a 90° phase shift. An e.m.f., E_2, will be developed across the tuned circuit, however, by the normal amplifying action. This further e.m.f. will be in phase with the input and can be controlled by the modulation, and it will be seen from Fig. 7.24 (*b*) that variation in the amplitude of this in-phase e.m.f. will have the effect of swinging the phase of the resultant signal across the anode load.

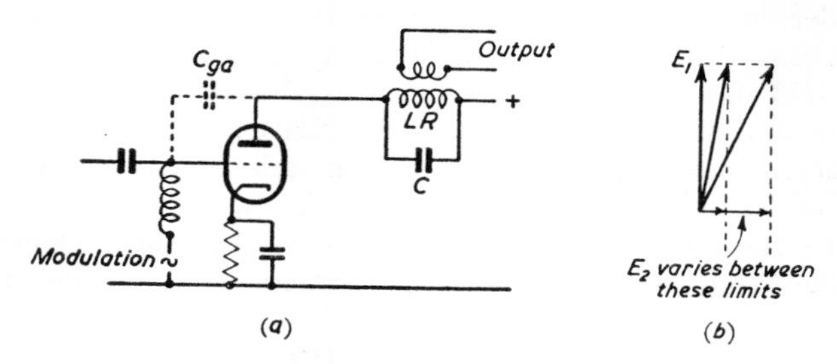

FIG. 7.24. SIMPLE PHASE MODULATOR

Frequency Multiplication

A limitation with phase modulation is that the frequency deviation produced is very small—of the order of 25 Hz as against 75 kHz usually employed for frequency modulation. This disadvantage is overcome by generating the initial oscillation at a relatively low frequency of, say, 200 kHz and then passing this through frequency multipliers which multiply both the carrier and the deviation.

A frequency multiplier is simply a tuned r.f. amplifier in which the anode circuit is tuned to a frequency which is a harmonic of the input signal. The arrangement is thus a Class C amplifier of the type shown in Fig. 7.4, but the anode current pulse only occurs once every n cycles, where n is the multiplication ratio.

Current pulses of the form of Fig. 7.4 contain appreciable quantities of low-order harmonics, and by suitable adjustment of the operating conditions the particular harmonic required can be emphasized. Since the amplitude of the harmonic falls off with increasing values of n the process can only be used with relatively small multiplications, from 2 to 5 times being usual.

It may be that the multiplication required to obtain the full deviation is such that the final carrier frequency is too high. Thus to bring a deviation of ± 25 Hz to ± 75 kHz a multiplication of 3,000 is necessary which, with an initial carrier of 200 kHz, would give a final carrier frequency of 600 MHz.

This difficulty is overcome by introducing, at some suitable point in the process, a heterodyne stage. For example, the initial carrier of 200 kHz $\pm$ 25 Hz could be mixed with an oscillation of, say,

130 kHz, which would produced a beat frequency of 70 kHz ± 25 Hz, resulting in a final carrier of only 210 MHz. The heterodyne process, in fact, reduces the carrier frequency but *not* the deviation.

Pre-emphasis

The advantage of frequency modulation is that the receiver, which is designed to respond only to changes in carrier frequency, will disregard interference produced by atmospheric disturbances and/or circuit noise which causes amplitude variations. However, as explained in Section 9.4, certain secondary effects occur, as a result of which interference does enter the system to an extent which increases with the modulation frequency. To counteract this it is usual to accentuate the upper modulation frequencies at the transmitter by including a *pre-emphasis* network. This is simply a network which increases the amplitude of the modulation-frequency components above about 1 kHz to an increasing extent. A similar, but inverse, *de-emphasis* network is then included in the receiver after the detector, which restores the modulation to its original value but with considerably reduced interference.

This pre-emphasis network should not be confused with the "distorter" network in Armstrong modulation, which is a network giving linear inversion of the output with frequency over the whole range.

All these functions can be performed equally well by equivalent solid-state circuitry, as explained in the next section.

7.6. Transistor Transmitters

For small powers and portable units, transmitters may be constructed using transistors in preference to valves. This not only provides an economy in size and weight, but permits a greater overall efficiency by reason of the saving of the power required to heat the valve cathodes.

The design of a transistor transmitter is basically the same as for valve circuitry, involving a primary oscillator, a modulation amplifier and an output stage. Each of these has to fulfil the various requirements already discussed, the only difference being that the circuitry has to be modified to accommodate the different matching necessitated by the low-impedance current-operated behaviour of the transistor. This was outlined in Chapter 6 and is considered more specifically in Chapters 9, 10 and 12.

Fig. 7.25 shows a typical circuit for a 4-watt medium-frequency a.m. transmitter. It will be seen to comprise a modulated oscillator feeding an output stage.

The primary oscillator produces oscillations in the circuit L_1C_1

by feedback from collector to base of Tr_1, via the coupling coil L_2. The operating conditions and matching requirements are determined in accordance with the principles discussed in Chapter 12. If a high stability is required, this may be obtained by including a quartz crystal in the feedback circuit as shown within the dotted rectangle. This presents a high impedance to any but the (sharply-tuned) resonant frequency so that feedback only occurs at the crystal frequency.

This is followed by an r.f. modulation amplifier which is tuned to the oscillator frequency, as described in Chapter 9. The output of

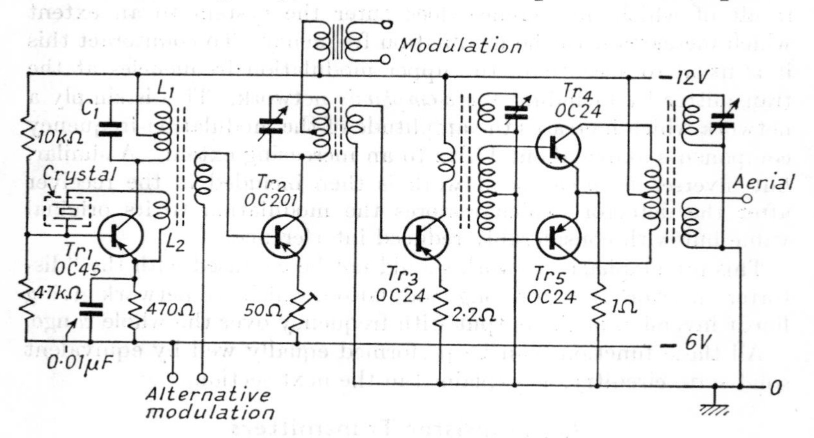

FIG. 7.25. LOW-POWER TRANSISTOR TRANSMITTER

this amplifier may be modulated by varying the effective supply voltage, as in the choke modulation circuit of Fig. 7.14. This can be done by introducing a transformer in the collector lead of Tr_2 as shown. This will provide true modulation since the collector voltage will vary above and below the mean value in accordance with the modulation signal.

To avoid undue d.c. voltage drop the secondary of the transformer must be of low resistance, but it must also have an adequate inductance to deal with the lowest modulation frequency. Moreover, since it is "looking into" a low impedance it must be fed from a modulating amplifier capable of delivering adequate power (four times the carrier power for full modulation, as was explained on page 338).

It is therefore simpler, for small modulation depths up to about 40 per cent, to modulate the signal by varying the *base* potential.

As explained in Chapter 6 (page 299) the gain of a transistor amplifier may be varied by such means, but except for small variations the output is subject to increasing distortion. The method also does not provide true modulation since variation of the base potential can only produce a reduction in output, so that the mean carrier power is reduced. However, as said, the method is satisfactory, and simpler, for small modulation depths.

Output Stage

The modulated output from Tr_2 is then fed to an output stage comprising a driver stage Tr_3 (described later) feeding two transistors in Class B push-pull. Since the full power output is only required

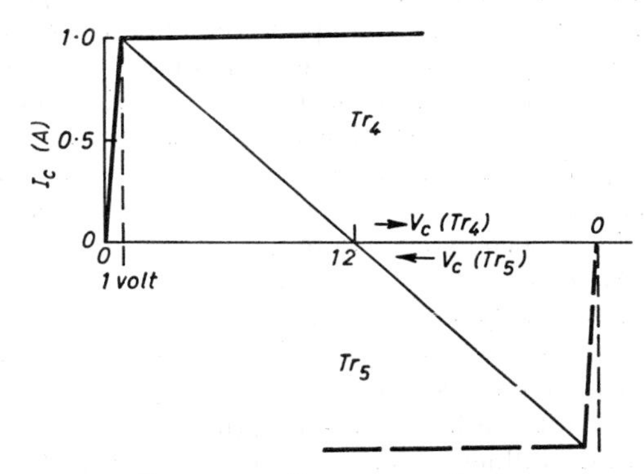

FIG. 7.26. IDEALIZED PUSH-PULL CHARACTERISTICS

with 100 per cent modulation whereas the average modulation depth is only about 30 per cent, considerable saving of power can be obtained if the supply current can be made proportional to the output.

This can be done, as explained on page 322, by biasing the transistors to cut off as shown in Fig. 7.26. Negative excursions of base current in Tr_4 then produce variations of collector current proportional to the input, as required for faithful modulation. On the positive half-cycle Tr_4 is inoperative, but because the collector circuit is tuned a sinusoidal output will be maintained over the complete cycle.

However, by providing a second transistor, Tr_5, of which the base is supplied with signals in anti-phase as shown, this transistor will be operative during the quiescent half-cycle of Tr_4 so that energy is supplied to the load over the complete cycle and the effective power output is doubled.

This form of Class B push-pull operation is widely used in a.f. amplifiers and is discussed in detail in Chapter 10 (page 501). A load line is calculated which provides voltage and current swings as shown in Fig. 7.26 ranging from cut-off to the knee of the characteristic. In an r.f. stage the load is determined by the effective collector impedance $Z_w = Q_w L\omega$, and this must be such that, with the circuit shown, the output transistors can swing over a range of 11 volts (allowing 1 V for the knee of the characteristics) with a peak current of 1 A. This provides a peak power output (from the pair) of 11 W. However, this is the condition for 100 per cent modulation, and the mean carrier power would be one-quarter of this, or 3·75 W, as explained on page 338.

The effective load for each transistor must be $11/1 = 11$ ohms. Now, as explained in the next chapter, the effective radiation resistance of a simple (quarter-wave) aerial is of the order of 40 ohms, so that to match this to the collectors a step-up ratio of approximately 2:1 will be required (for each half of the transformer). The effective Q of the aerial circuit will be low because of the radiation resistance, and a figure of 15 would be representative. The inductance of the secondary winding $= Z_w/Q_w\omega$, which for a frequency of 500 kHz would be 0·85 μH.

This is inconveniently small. Hence a tertiary winding is provided having an inductance of, say, 100 μH, tuned with an appropriate capacitance, which for 500 kHz would be approximately 1,000 pF. This tertiary circuit can have a high Q_0 (in excess of 200) which will have a negligible influence on the overall working Q so that the matching will not be affected.

Care must be taken to limit the operating temperature to within the prescribed limits. The output transistors are of the relatively massive construction illustrated in Fig. 6.26, and they must be mounted direct on to the chassis to provide an adequate heat sink. The 1 ohm resistor in the (common) emitter connection is a further precaution against thermal runaway.

Driver Stage

It remains to design a suitable input stage to drive the output pair. The design of such stages, again, is considered in Chapter 10. Briefly, in order to drive the output transistors up to a peak current of 1 A, base current swings of $1/\beta$ will be required. If β is taken as 150 the peak base currents will be of the order of 6 mA, which will require, with the transistor shown, a swing of approximately 5·5 V. Hence the peak drive power will be some 330 mW, and the stage must be designed to supply a power of this order.

Because the collector circuit is tuned, this stage could also be operated in Class B, but in view of the small peak current the saving would be small.

Applications of Transistor Transmitters

This typical design has only been discussed in broad terms to indicate the general approach. The individual circuitry would be based on the principles discussed in Chapters 9 and 10. It is evident

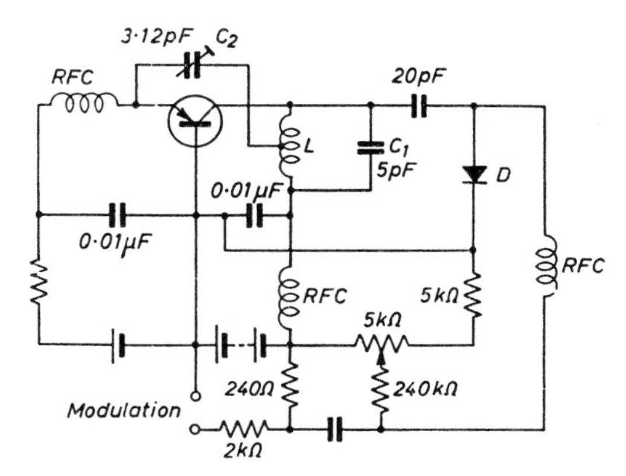

FIG. 7.27. TRANSISTOR FREQUENCY MODULATION

that since no cathode heating power is required the overall efficiency can be high, approaching 70 per cent.

For this reason, transistor circuits are often used as pre-amplifiers for high-power transmitters, the final output stages using valves. The applications are not limited to medium frequencies, and with appropriate modifications transistor techniques can be used well into the megahertz region, as is discussed in Chapter 13.

Solid-state techniques in fact can be used for most of the applications formerly performed with valves. For example, Fig. 7.27 shows a frequency-modulated oscillator using a varactor diode to perform the function of the reactance valve of Fig. 7.22. The oscillator uses a transistor in common-base mode, oscillation being produced by feedback from collector to emitter through the capacitor C_2. The frequency is controlled by C_1, in parallel with which is the variable-capacitance diode D, the capacitance of which is varied in accordance with the a.f. modulation, as was explained in Chapter 6.

8

Aerials and Feeders

HAVING generated radio-frequency oscillations in a radio transmitter, it is then necessary to cause these oscillations to radiate wireless waves as effectively as possible. To do this it is necessary to cause the current to flow in an extended structure known as an *aerial* or *antenna*.

Similarly at the receiving end a suitable structure is required to extract the maximum energy from the waves at that point.

Finally it is often required to locate the transmitter or the receiver at some distance from the aerial so that it is necessary to design feeders which will carry the currents from one point to another with as little loss as possible.

These three aspects of aerial systems will be considered separately.

8.1. The Transmitting Aerial

The concept of a wave as a ripple in a line of force has already been propounded in Chapter 3, and it will be clear that the larger we can make this ripple the stronger will be the wave produced.

Our aim, therefore, is to cause the electric current to flow between two points as far apart as possible, and for most transmissions these points are situated vertically apart. Hence, it becomes necessary to erect a mast or tower of some kind to support a vertical wire into which the current may be fed. The passage of the electrons up and down this wire will produce the desired radiation.

Natural Wavelength

Now, since a wire possesses inductance and capacitance of its own, there will obviously be some particular frequency at which the inductance and capacitance of the wire resonate so that the aerial itself constitutes an oscillating circuit. This occurs when the length of the wire is about 80 per cent of the wavelength of the radiation which would be generated by current of that particular frequency.

An aerial in such a condition is said to be operating at its *natural wavelength*.

The inductance and capacitance are not concentrated, but are more or less uniformly distributed over the whole length. Consequently, the current and voltage will not be uniform, but will vary from place to place along the wire. The current at the end of the

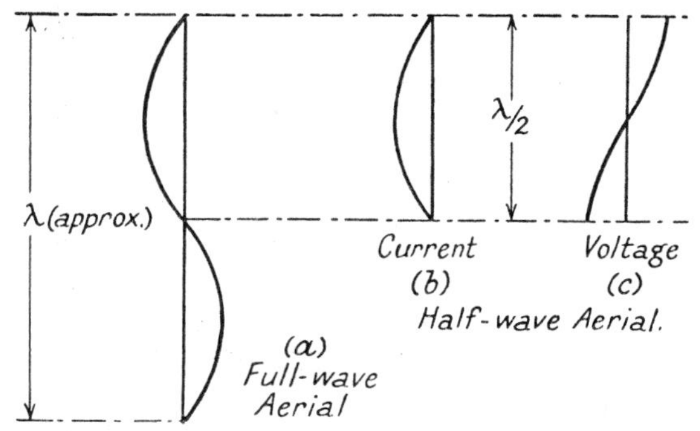

FIG. 8.1. DISTRIBUTION OF CURRENT AND VOLTAGE IN A WIRE OSCILLATING NATURALLY

wire must clearly be zero, because there is no other path for the current, and since the arrangement is symmetrical the current at the centre will also be zero, the distribution being sinusoidal in form as shown in Fig. 8.1 (*a*).

In such an aerial, however, the current is in the opposite direction in the two halves, so that the radiations would cancel out. If we want to radiate we must limit the length to a half wavelength as shown in Fig. 8.1 (*b*), where we have zero current at each end and a maximum in the middle.

Dipole Aerial

Such an aerial is called a *dipole*. It may be considered as being made up of a large number of small elements symmetrical about the centre of the wire, and we can consider each element as forming a capacitor with a corresponding element located at an equal distance from the centre.

The voltage will thus gradually increase as we go from the centre to the ends since the voltages on the individual elements will all add together. Similarly if the small "elemental" capacitors discharge and then recharge in reverse direction the currents will all add together, increasing from nothing at the ends to a maximum in the centre.

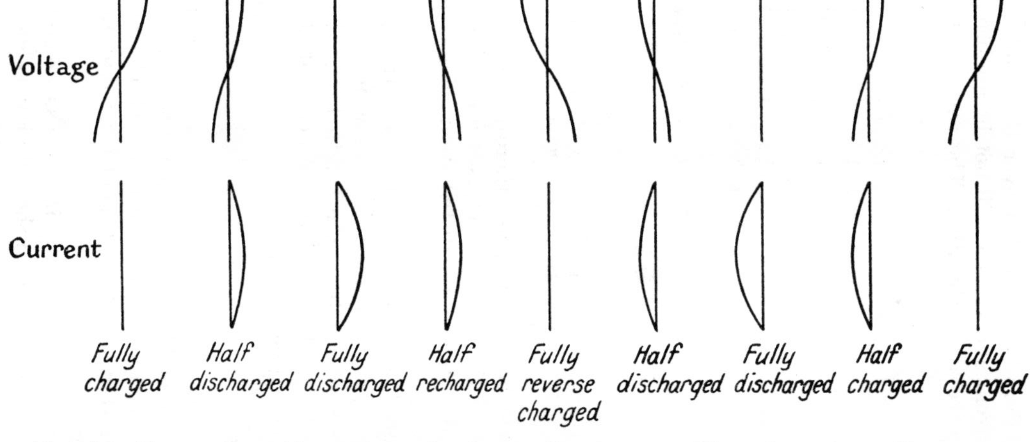

FIG. 8.2. VOLTAGE AND CURRENT AT SUCCESSIVE INTERVALS ON WIRE OSCILLATING NATURALLY

Thus the voltage and current distributions are of the form shown in Fig. 8.1 (*b*) and (*c*), and it is easily shown that the distribution along the wire is sinusoidal in form. It must be remembered that this is a *space* distribution along the wire quite distinct from the change in the value of the current or voltage with time, due to the fact that the circuit is oscillating. Fig. 8.2 shows the change in amplitude of the voltage and current at successive instants during the oscillation. It will be seen that the strength of the current at any point in the wire passes from a maximum through zero to a maximum in the opposite direction and back again, while at any instant the distribution of the current at different points of the wire increases from zero at the end to a maximum in the middle. The voltage distribution is similar in character but 90° out of phase both in time and space.

Practical Form of Aerial

Dipole aerials such as this are used to some extent where the wavelength is of the order of a few metres but are not very practicable for long waves several hundred metres long. The practical form of the aerial, indeed, depends very considerably on the wavelength, and we can divide aerials into two main classes:

(*a*) Earthed aerials, suitable for medium and long wavelengths.

(*b*) Free aerials, suitable for short waves (though sometimes these short-wave aerials also are earthed at one end).

Earthed Aerials

For long waves where the length of even a half wavelength would be impracticable it is convenient to connect the bottom end of the aerial to ground as shown in Fig. 8.3. We then find that the current in the aerial induces currents in the earth, which produce the effect of an image of the aerial in the ground. This enables the aerial length to be reduced by half, so that we have a quarter wavelength above the ground, the remainder of the aerial being supplied by the image in the earth itself.

This clearly gives a convenient arrangement, for it brings the point of zero voltage and maximum current actually at the ground. The aerial would be coupled through a coil or by other suitable means to the closed oscillating circuit of the transmitter.

Even so, it is only feasible to use an aerial oscillating naturally for wavelengths of a few hundred metres. Beyond this it is necessary to increase the inductance or capacitance. For example, if the aerial is provided with a horizontal top, as shown in Fig. 8.4 the capacitance of the upper portion to ground is increased and the resonant frequency can be attained with a height considerably less than $\lambda/4$. The capacitance is no longer uniformly distributed, being almost entirely due to the flat top and the current in the vertical

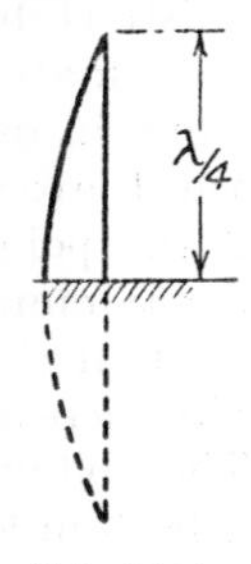

FIG. 8.3. QUARTER-WAVE AERIAL

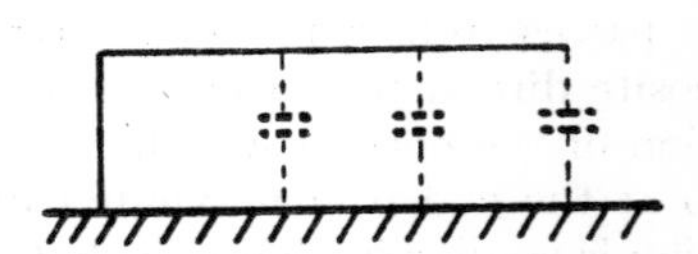

FIG. 8.4. FLAT-TOP AERIAL

portion is nearly uniform (which is an advantage, as is shown later when discussing effective height).

Alternatively the aerial may be loaded by including an inductance at the earth end, as illustrated in Fig. 8.5. This again reduces the length of aerial wire required to tune to the required frequency, but it has the disadvantage that the maximum current is flowing in the coil, which has poor radiating properties. In practice, therefore, a compromise between the two methods is used for medium and long waves, the aerial being provided with a flat top, usually from one to three times the actual height, and a loading inductance included at the base to tune the aerial to the required frequency.

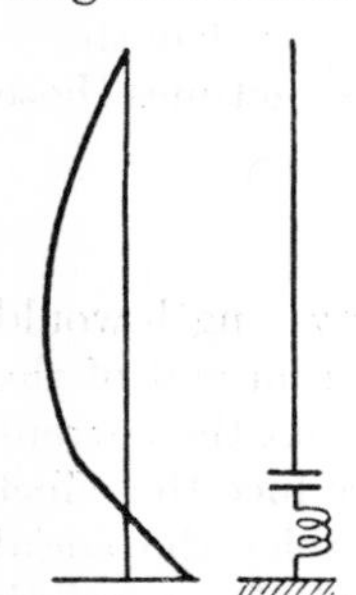

FIG. 8.5. ILLUSTRATING USE OF "SHORTENING" CAPACITOR

A type of aerial occasionally employed for the shorter medium waves, ranging from 150 to 250 metres, is shown in Fig. 8.5. In order to maintain a large effective height, masts 50–100 m high may have to be employed, and the length of the aerial in this case becomes greater than the quarter wavelength. It is then no longer practicable to feed in current at the current node, nor is the earth point at zero potential. It is necessary to insert a capacitance in the aerial circuit as well as an inductance, and to adjust these so that the total effective length of the aerial is three-quarters of a wavelength. The earth, then, again becomes a point of maximum current and zero voltage, as shown in the figure. Such a capacitor is usually called a "shortening" capacitor.

Directional Radiation

The radiation from a simple dipole or a quarter-wave aerial is uniform in all directions around the axis of the wire, but there are occasions when it is desirable to concentrate the radiation more strongly in one direction. Certain aerial arrangements have a natural tendency to do this. With a flat-topped aerial, for example, the radiation from the horizontal portion interacts with that from the vertical portion and strengthens the radiation in one direction while diminishing it in the other. An "L" aerial, such as that shown in Fig. 8.6, would thus tend to radiate more strongly upwards and away from the horizontal top, as indicated in the figure.

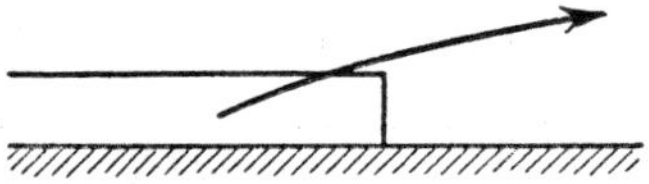

FIG. 8.6. AN "L" AERIAL IS PARTIALLY DIRECTIONAL

This is only a partial directivity, however, and effective directional characteristics can only be obtained by the use of aerial arrays, in which the currents are so disposed as to produce a concentration of the radiation in one direction and a cancellation in the other. Such arrangements are discussed more fully in Section 8.4 and Chapter 13.

Non-Fading Aerial

A special type of directional radiation is sometimes employed with aerials for broadcasting transmitters to minimize fading. Beyond the immediate neighbourhood of the station the transmission is accomplished almost entirely by the indirect ray reflected from the ionosphere, which is liable to fading and distortion as explained in Chapter 3. The broadcast engineer therefore only considers his station to be effective over the area served by the ground ray.

Unfortunately, before the ground ray has really died out, interference begins to creep in from high-angle radiation which leaves the transmitter in an upward direction and is sharply reflected by the ionosphere so that it returns to earth in a comparatively short distance (of the order of 150 km with ordinary broadcast wavelengths). This high-angle radiation suffers from fading and distortion in the same manner as the longer range indirect ray, and if it is comparable in strength with the ground ray at any given point serious fading will obviously be obtained.

Special forms of aerial can be devised which minimize this high-angle radiation and at the same time accentuate the horizontal radiation, both of which will increase the ratio of ground wave to sky wave at a given distance from the transmitter and consequently will increase the service area of the station.

Consider a simple vertical wire half a wavelength long. Each portion of the wire will radiate energy according to the current at that point. Consider a wave generated by the current at the base of the aerial and travelling in an upward direction. At the same instant a similar wave is being generated by current at the top. But the base wave will take half a period to travel to the top of the aerial, by which time the current at the top will be flowing in the opposite direction and will generate a wave that will cancel out the wave which arrives from the base.

Waves leaving from intermediate points on the aerial will not completely cancel out, but it will be clear that the upward radiation will be distinctly limited. Moreover, if we make the aerial still more than half a wavelength long the current in the top portion will begin to emit reverse radiation. On the other hand, the ground wave will suffer if the aerial is made much more than half a wavelength long because the radiations in the horizontal plane are in phase and hence the waves from the top portion will oppose those from the bottom portion due to the reversal of the current.

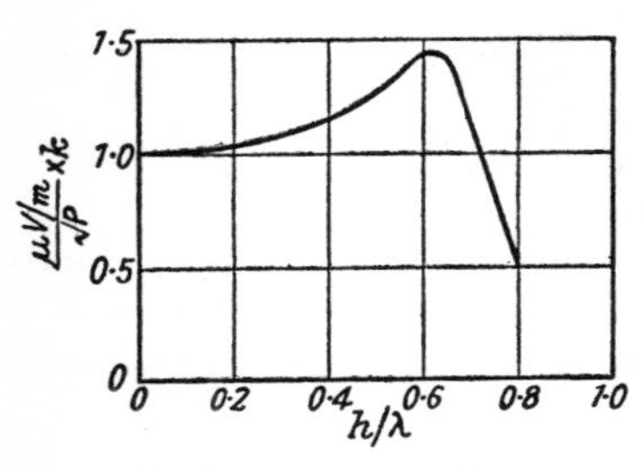

FIG. 8.7. ILLUSTRATING RELATIVE INCREASE IN HORIZONTAL RADIATION WITH HALF-WAVE AERIAL

Experiments show that an optimum condition is obtained when the length of the aerial is about $0{\cdot}6\lambda$. The usual procedure is to make the mast itself the aerial. It is built in a gradually increasing section up to about $0{\cdot}25\lambda$ after which the section decreases again. The ultimate section is often a simple vertical rod on top of the mast proper. The base of the mast has to be insulated because it is not at earth potential.

Fig. 8.7 shows the improvement in conditions with increasing height. The vertical axis is the horizontal field strength divided by the square root of the power input, which is the criterion which concerns the engineer whose object is to produce the maximum horizontal radiation for a given power. It will be seen that with a height of $0{\cdot}6\lambda$ there is an increase of nearly 50 per cent in the horizontal radiation. For further information refer to an article by Ballantine, entitled "High Quality Radio Broadcasting," *Proc. I.R.E.*, Vol. 22, page 616.

It will be appreciated that this technique does not eliminate fading but reduces its severity by increasing the range over which the ground wave predominates.

Effective Height

We saw in Chapter 3 that the field strength at a distance r from an earthed transmitting aerial is given by the expression

$$E_r = 377Ih/\lambda r \text{ volts/metre}$$

in which I is the aerial current, λ is the wavelength, and h is the effective height of the aerial. This is not the actual height, because the current in the aerial is not uniform. It is, in fact the height of an equivalent aerial in which the current is uniform throughout.

With a long-wave aerial having a large horizontal top portion the majority of the capacitance is provided by the flat top and the current in the down lead is nearly uniform. In such circumstances the effective height may be between 0·8 and 0·9 times the actual height. On the other hand, in a half-wave or quarter-wave aerial the current distribution is sinusoidal, in which case the effective height may be taken as the actual height multiplied by the average value of the current, which is $2/\pi$ or 0·64. Thirdly, if the aerial is loaded with inductance the current varies nearly uniformly from zero at the top to a maximum at the bottom so that the effective height is approximately 0·5 times the actual height. Hence the effective height varies for each individual aerial, lying somewhere between 0·5 and 0·8 times the actual height.

Radiation Resistance

The current in a transmitting aerial is given by V/Z, where V is the induced e.m.f. and Z is the impedance. Since the aerial is tuned, Z is simply the effective resistance R, but this will not be just the conductor resistance because power is absorbed in radiating the electric waves.

From the expression for the field strength it is possible, by integrating the values through the whole of the hemisphere surrounding the aerial, to arrive at an expression for the total power radiated, which comes to $160\pi^2h^2I^2/\lambda^2$, where I is the r.m.s. value of the aerial current, λ is the wavelength and h is the effective height. This power may be assumed to be dissipated in an imaginary *radiation resistance*, so that since the power is I^2R_r, we can say that

$$R_r = 160\pi^2h^2/\lambda^2 = 1{,}580h^2/\lambda^2$$

In a long-wave aerial this radiation resistance accounts for about 50 per cent of the total, because in general it is not practicable to make h/λ more than a small fraction. At shorter wavelengths, however, this limitation does not apply and practically all the resistance is radiation resistance, with consequent increase in efficiency. For example, the quarter-wave aerial of Fig. 8.3 has an effective height of $(2/\pi)(\lambda/4)$, and substituting this in the above expression gives $R_r = 40$ ohms, whereas conductor resistance and losses will only amount to a few ohms at most.

The Earth System

Efficient radiation with an earthed aerial depends to a large extent upon having a good earth connection, so that the ground may act as an efficient mirror as already described. For this purpose the transmitting site is usually located on ground which has a good conductivity, and the earth connection is made by burying a series of wires or pipes running from the transmitting building. These may be concentrated immediately under the aerial system or may run in all directions, the deciding factor being largely one of cost.

Earth losses are important from the viewpoint of overall efficiency, and attempts have been made to reduce them to negligible proportions by using a *counterpoise* or *earth screen.* This consists of a network of wires a few feet above the earth running under the aerial and for a considerable distance each side.

Experience shows that, just as a network of wires in the aerial system has an efficiency nearly as great as would be obtained by a solid sheet, so an earth screen of this nature constitutes an efficient mirror with less loss than would be obtained from a conventional earth connection.

Aerial Losses

The importance of earth resistance, and aerial losses generally, depends upon the wavelength. At long and medium waves where h/λ (and hence R_r) is small, the aerial losses can absorb an appreciable proportion of the available power and it is necessary to take into account not only the conductor loss (including the earth system) but also the dielectric loss caused by the presence of buildings, trees and other objects in the electric field of the aerial. These factors vary with frequency, so that for any given structure there is an optimum wavelength, and vice versa.

It is not necessary to discuss this in greater detail here, though it may be noted that with aerials operating on long and medium wavelengths it is customary to minimize the conductor resistance by using several wires in parallel, spaced round a ring to reduce skin effect. The down-lead in Fig 8.19 illustrates this type of construction.

8.2. The Receiving Aerial

The requirements for a receiving aerial are somewhat simpler than those for the transmitter. We have seen that the transmitting aerial radiates an electric field which travels outwards and gradually decreases in strength as the distance from the transmitter increases. This electric field is actually a potential gradient, as we saw in Chapter 1, so that if we set up a wire in the path of the wave an e.m.f. will be induced equal to the product of the field strength and the (effective) height of the wire.

The actual field strength, in practice, is of the order of millivolts, rather than volts, per metre so that the e.m.f. induced is quite small, but this may be amplified to any desired extent provided that the received signal is appreciably greater than the spurious e.m.f.s customarily known as background noise. Such background may arise from random e.m.f.s generated internally in the receiving equipment, as discussed in the next chapter, or may be produced by atmospheric disturbances or locally-generated interference. The table indicates the order of field strength encountered in practice.

Field Strength μV per m	*Remarks*
10,000 (10 mV)	Very good signal, well audible above local interference.
1,000 (1 mV)	Good signal requiring a measure of h.f. amplification, liable to interference but capable of good service.
100	Fair signal requiring good tuning circuits and high amplification. Will be subject to serious interference.
10	The smallest commercial signal. Will require full resources of selective amplification, directional aerials, etc. (But much weaker signals can be handled by parametric amplifiers as discussed in Chapter 19.)

It will be seen that as the field strength falls off the influence of the background noise becomes increasingly greater. Its effect may be minimized by tuning the aerial and the subsequent amplifying circuits to the frequency of the received signal, which also serves to discriminate between the required signal and the many unwanted signals, all of which are inducing e.m.f.s in the aerial simultaneously. Still further discrimination may be obtained by giving the receiving aerial directional properties, as is often done for commercial reception.

It is clear, therefore, that the design of the receiving aerial is closely associated with the subsequent receiving equipment. It also depends upon the wavelength of the signal. With long or medium waves it is neither necessary nor convenient to make the length of the aerial comparable with the wavelength, particularly as it is often required to receive signals over a wide range of wavelengths. At short waves, however, it is customary to use half-wave aerials or dipoles, often arranged in special directive arrays, but this is a specialized technique which is discussed in more detail in Chapter 13.

At long and medium wavelengths it is sufficient to use a vertical wire, sometimes with a horizontal top portion, a few metres in length. The e.m.f. induced will depend upon its effective height which, since the aerial length is only a small fraction of a wavelength, will be nearly the same as the actual height.

Aerial Tuning Circuits

Such an aerial will contain inductance and capacitance, but since it is operating well below its natural wavelength, additional inductance or capacitance, or both, will be required to tune it. However, it is more convenient to couple the aerial to a tuned secondary circuit as shown in Fig. 8.8. Since the aerial is not tuned, a tight

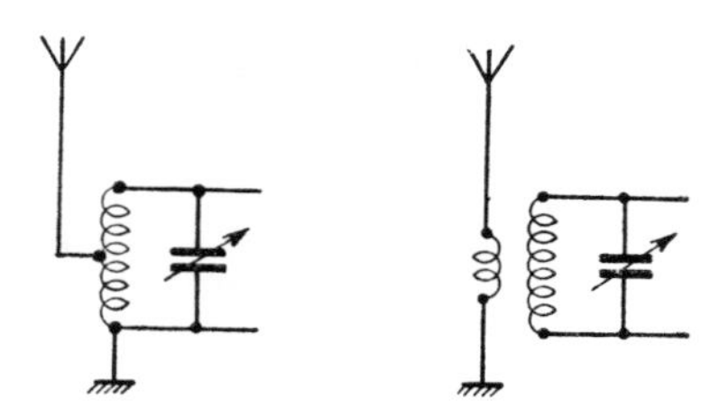

FIG. 8.8. TWO TYPES OF INDUCTIVE AERIAL COUPLING

coupling is permissible and it is usual to provide a small step-up ratio which not only increases the e.m.f. developed across the secondary but reduces the effective aerial constants reflected into the secondary and so permits a wider tuning range to be obtained with a given tuning capacitor.

The equivalent circuit is shown in Fig. 8.9. Since the secondary is tuned, $X_2 = 0$, and the equivalent primary resistance is $M^2\omega^2/R$, which is, in practice, large compared with the aerial resistance, which has therefore been neglected.

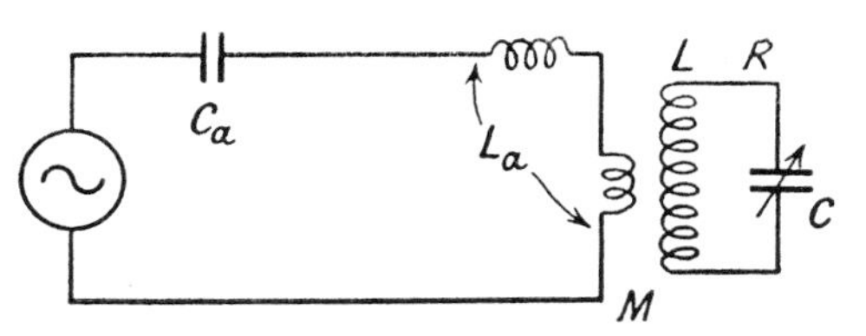

FIG. 8.9. EQUIVALENT AERIAL CIRCUIT

Then
$$i_1 = \frac{e}{M^2\omega^2/R + j(\omega L_a - 1/\omega C_a)}$$

where L_a includes the primary inductance of the transformer;

$$e_2 = M\omega i_1$$

and v_2, the voltage across the secondary coil, is given by

$$v_2 = M\omega \,.\, (L\omega/R)i_1 = (M/RC)i_1$$

since $\omega^2 = 1/LC$. Hence the effective step-up is

$$\frac{v_2}{e} = \frac{M/RC}{M^2\omega^2/R + j(\omega L_a - 1/\omega C_a)} = \frac{M/C}{M^2\omega^2 + jX_1R}$$

where X_1 is the reactance of the aerial $= \omega L_a - 1/\omega C_a$.

Maximum output is obtained when the two terms in the denominator are equal—i.e. $M^2\omega^2 = X_1R$. This usually gives a step-up of the order of five or ten to one.

The secondary e.m.f. is not constant but increases with frequency. A capacitance coupling would give the reverse effect, and aerial circuits are sometimes arranged to combine both forms of coupling in order to obtain a uniform transfer of energy over a range of frequency. Fig. 8.10 shows such a circuit. The inductance of L_1 is higher than usual, so that it would tune with the aerial capacitance just above the highest wavelength to be received.

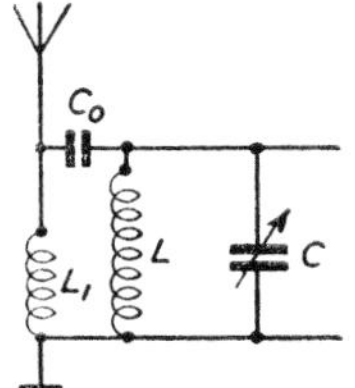

FIG. 8.10. CIRCUIT FOR OBTAINING UNIFORM AERIAL STEP-UP

It is loosely coupled to the secondary and the energy is transferred partly by inductive coupling and partly directly to the top of the secondary through C_o.

The direction of L_1 is such that these two couplings are in opposition. At low frequencies the energy transfer is nearly all inductive since C_o is small and offers a large reactance to the currents. As the frequency rises, more voltage is induced in L from L_1, but more current flows through C_o in opposition and so keeps the total secondary voltage fairly constant.

Transistor Aerial Circuits

In the foregoing expressions the secondary resistance R must include the reflected aerial resistance R_a/n^2, which is usually negligible, plus the equivalent resistance due to the input resistance of the amplifier (or detector) to which the circuit is connected. With valve circuits this is normally so high as to produce negligible loading, though with a diode detector some loading is present, as explained on p. 453.

This is no longer true, however, with transistor circuits, where the input resistance with bipolar transistors is quite low. It is

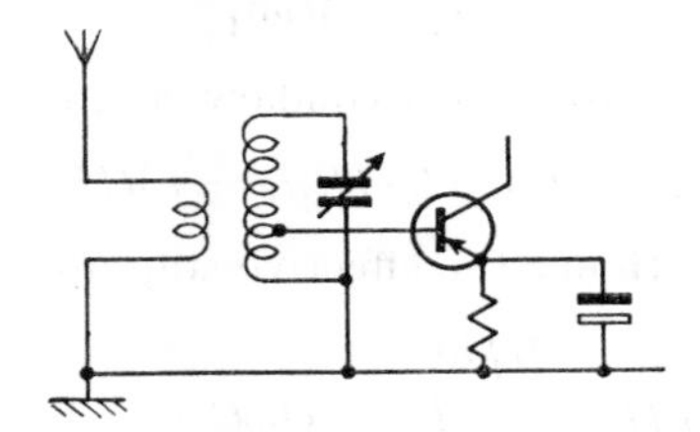

FIG. 8.11. TRANSISTOR AERIAL COUPLING

possible, of course, if a high input resistance is essential, to use a field-effect transistor, as discussed in Chapter 6, in which case the technique is similar to that for a valve, but with the bipolar transistor the input resistance is only a few thousand ohms. Now, a resistance P across a tuned circuit is equivalent (at resonance) to a series resistance of L/CP, which with typical values is several hundred ohms. This would reduce the Q of the circuit disastrously, so that the transistor must be fed from a tertiary winding (or an equivalent tapping) such that the reflected load Pn^2 does not reduce the circuit Q too much. The arrangement is illustrated in Fig. 8.11, and it may easily be shown that optimum results are obtained when the reflected load reduces the Q by one-half, i.e. $L/CPn^2 = R$.

The use of a step down in this manner will reduce the output voltage, but it must be remembered that the transistor is a current-operated device, so that the requirement is for maximum secondary *current*.

Ferrite Rod Aerials

It is more usual with transistor receivers to use a different type of aerial consisting of a short length of Ferrite rod $\frac{1}{2}$ to $1\frac{1}{2}$ cm in diameter. If this is aligned approximately in the direction of the wave, the magnetic component of the travelling wave is concentrated in the higher-permeability material of the rod, and e.m.f. is induced in a coil wound round the rod. This coil is designed to have an

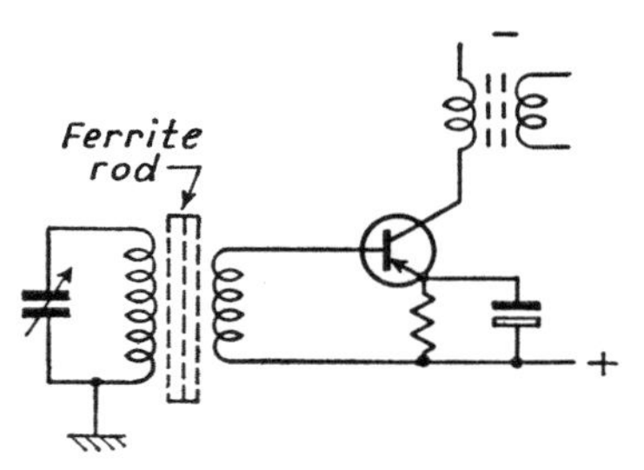

FIG. 8.12. FERRITE AERIAL

inductance which, with the appropriate variable capacitor, can be tuned over the frequency band required.

The output is taken from a secondary winding as shown in Fig. 8.12, the (step-down) ratio being chosen to provide an effective Q equal to half the unloaded Q, as mentioned above.

Ferrite rod aerials are directive, producing maximum signal when aligned with the direction of the wave, and so behave similarly to the frame aerials discussed in Section 8.4.

Band-pass Coupling

In order to obtain increased selectivity the aerial and the secondary are sometimes separately tuned. When this is done the system becomes a species of coupled circuit, and the coupling must be kept below a certain limiting value for satisfactory operation.

The critical coupling is obtained when the effective resistance introduced into the primary by the secondary circuit coupled to it is equal to the initial primary resistance, which occurs when $M\omega = \sqrt{(R_1R_2)}$ (see Chapter 2). Under these conditions the maximum current is obtained in the secondary and the maximum energy transferred from the primary.

With critical coupling the secondary resonance curve shows only one peak, while that of the primary shows two peaks very close together. As the coupling is increased the primary peaks move farther apart, and a double hump also appears in the secondary resonance curve.

Receiving circuits are sometimes operated in this condition, the peaks being 5 to 10 kHz apart so that the sidebands are received at a strength comparable with the carrier, whereas with the ordinary peaked resonance curve the frequencies 5 to 10 kHz off tune are appreciably attenuated, particularly with a sharply tuned circuit. This gives rise to a loss of the upper frequencies, or "top cut," which is minimized by band-pass tuning. Beyond the limit of the

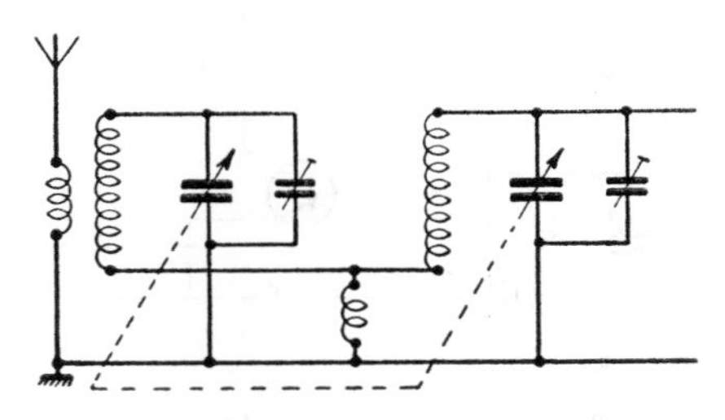

Fig. 8.13. Band-pass Aerial Circuit

peaks the current falls very sharply, giving a better selectivity than could be obtained with one circuit alone.

A typical circuit is shown in Fig. 8.13. The two circuits are tuned together with a two-gang capacitor, but to allow for the differing self-capacitance of the two circuits, including the effect of the aerial, small preset "trimmers" are connected in parallel with the main tuning capacitance as shown.

Anti-interference Devices

One of the principal troubles in modern reception, particularly in broadcast receiving, is that of local interference. Much modern electrical apparatus produces electrical disturbances. These take the form of very rapidly damped waves, which contain components covering a very wide range of frequencies, and will thus produce interference in any receiver in the vicinity, irrespective of the frequency to which it is tuned (just as the dropping of a weight near a piano will set all the strings vibrating).

The interference is transmitted in three ways:

(*a*) Direct radiation.
(*b*) Direct conduction.
(*c*) Mains radiation.

The first is self-explanatory. The interfering currents at the source generate waves which are radiated and picked up by the receiver. The range of such interference is limited to 10 or 20 metres, and this form of disturbance is most troublesome in industrial districts, though lifts and even household appliances such as refrigerators may be annoying.

Direct conduction is usually low-frequency in character and is only occasionally encountered in the case of sets operated from the electric mains. It usually arises from inadequate smoothing in the h.t. supply portion of the set or from bad layout or poor earthing. It is thus mainly a matter of design.

Ninety per cent of the interference is of class (*c*). The interference

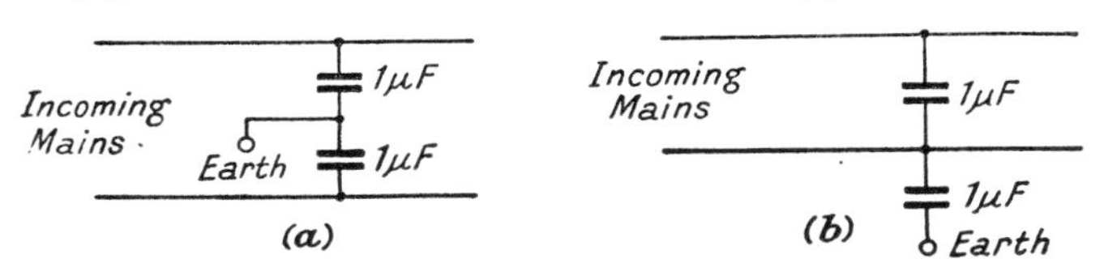

FIG. 8.14. SUPPRESSION CIRCUITS FOR FILTERING MAINS-BORNE INTERFERENCE

is conducted by the mains, acting as a radio transmission line, to the point where the receiver is located and is then re-radiated on to the receiving aerial. It may travel either symmetrically, going out on one main and returning on the other, or asymmetrically, going out on both mains together with the earth as a return. Both forms are encountered and the interference may be conducted in this way several miles from the source.

One remedy is to filter the mains at the point where they enter the building by connecting a capacitor to earth from each line as shown in Fig. 8.14. With the first arrangement the centre point is live, until it is earthed. The second arrangement is preferable as the earth terminal is always dead.

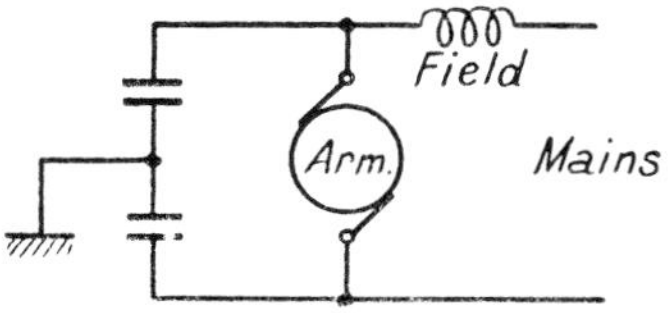

FIG. 8.15. SUPPRESSOR CIRCUIT FITTED TO MOTOR

The most satisfactory remedy, of course, is to fit suitable suppressors at the actual source of the interference. The sparking at the brushes of a motor, for example, can often be prevented from causing interference by connecting two capacitors across the brushes and earthing the centre point as shown in Fig. 8.15. The radio-frequency oscillations generated in the motor are then short-circuited to earth through the by-pass capacitors and prevented from travelling along the mains. In some cases it is necessary to insert radio-frequency chokes in the leads as well as the by-pass capacitors, while of course all these methods only prevent the radiation of the disturbance from the mains and do not affect the direct radiation from the offending machine. The only remedy against this latter trouble is to enclose the whole machine in a shielding box of metal sheet or gauze, and this is not always practicable.

The subject is one of some complexity and numerous special remedies are available for particular circuits. Electro-medical

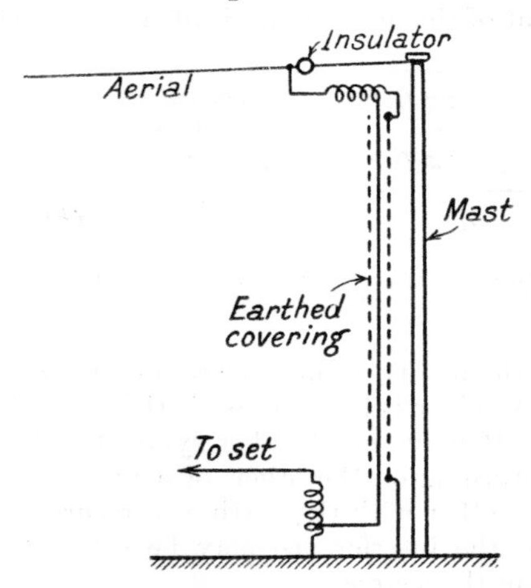

FIG. 8.16. SHIELDED LEAD-IN CABLE

apparatus is very troublesome, for it actually uses radio-frequency currents and is thus a small transmitter of very bad noise. It can only be silenced by enclosing the equipment in an earthed screened room and filtering all outgoing mains and telephone wires inside the screen.

Trams and trolley buses are also troublesome, and it is frequently the practice to fit special h.f. chokes in the trolley arm leads and also in severe cases to connect capacitors from the overhead lines to earth every hundred metres or so.

Shielded Lead-in (Fig. 8.16)

Apart from the elimination of the disturbance at the source some measure of relief can be obtained by using a shielded down-lead for the aerial. The aerial is provided with a short horizontal top and the down-lead itself is enclosed in metal braiding. The fairly local direct radiation from interfering plant is thus prevented from affecting the greater part of the aerial, and, although there is some loss in the effective signal strength received from the required signal, the overall result is a marked improvement.

To avoid undue attenuation of the signal due to the capacitance between the down-lead and the earthed covering it is necessary to include matching transformers at each end, as shown. The design of such transformers is discussed in the next section.

8.3. Feeders and Transmission Lines

In a simple transmitting system it is possible to lead the aerial straight into the transmitting building, and to situate the aerial tuning inductance close to the transmitter. This, however, is not always convenient. One may have several transmitters to be housed in the same building, and it is obviously better if the aerial systems can be kept well spaced from one another to avoid any possibility of interaction. Consequently, it is necessary to carry the energy from the transmitting building to the aerial through a transmission line,

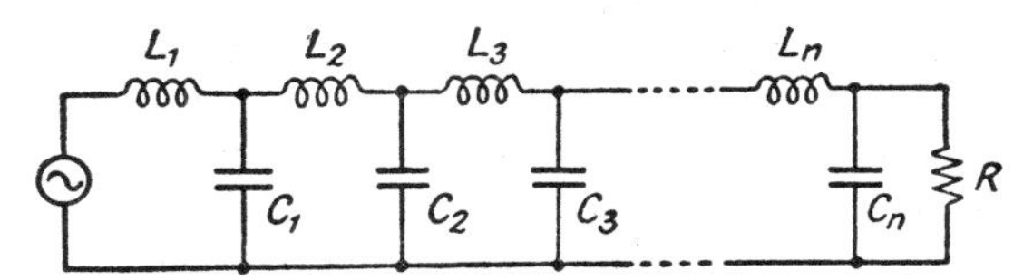

FIG. 8.17. ILLUSTRATING FEEDER AS A SERIES OF SECTIONS

in much the same way as power is carried at low frequencies for industrial purposes.

The development of this feeder technique received a considerable impetus following the introduction of short waves, for some type of feeder is essential in the modern short-wave transmitting station. We will, therefore, discuss briefly the technique of the transmission of energy through feeder wires.

Suppose we consider two wires of indefinite length connected at the sending end to a source of a.c. voltage. The wires will each have a small self-inductance which will be uniformly distributed along their lengths. There will also be a capacitance between them, likewise uniformly distributed along the wires. Let us break the feeder up into small sections and consider the inductance and capacitance concentrated in each section, so that the system looks like Fig. 8.17.

The voltage at the sending end will cause a current to flow into the first capacitor. This current will lag slightly behind the voltage, because of the inductance in the wires. The capacitor will build up a charge, again with a slight time lag, so that voltage will appear across C_1, similar to the input voltage but slightly delayed.

This voltage, in turn, will feed current into the next section, and C_2 will commence to charge up. This will feed current into C_3, and so the voltage will be transmitted along the feeder until we reach the far end. Here the feeder is terminated in an impedance, which we will assume to be a resistance (since we nearly always terminate a feeder in a tuned circuit which will have no reactance at resonance).

Reflection

What happens here depends upon the value of the resistance. Suppose it is infinitely high. The last capacitor is unable to discharge through the load and will therefore force current *back along the feeder*, and energy will commence to surge back to the transmitting end. The wave will, in fact, be *reflected*.

Similar reflection occurs if we short-circuit the far end, for here the last capacitor is not able to accept any charge, and the energy from the preceding section of the feeder has nowhere to go except back along the line, which it proceeds to do.

If the resistance is high but not infinite, it will accept some current from the last section, but not enough. The discharge of a capacitor through a resistance takes time, and if the resistance will not discharge the capacitor fast enough, the excess charge can only go back along the feeder. Thus, we have a partial absorption and partial reflection. If the resistance is too low the capacitor discharges too fast, and again we have partial reflection.

There is clearly one particular value of resistance which accepts energy just as fast as it is being supplied by the feeder, and with this critical value *no reflection occurs*. Energy is passed unhindered right along the feeder and is absorbed at the far end, with very little loss in transit.

Characteristic Impedance

What is this value of resistance? It can be deduced by finding the rate at which the capacitors charge and equating this to the rate at which the end section will discharge through the load. This involves calculus, and the treatment is beyond the scope of the present work, but the result is surprisingly simple.

The conditions are fulfilled if the resistance is made equal to $\sqrt{(L/C)}$, where L and C are the inductance and the capacitance of the line,* or of a given length of line which comes to the same thing,

* In a transmission line we have to consider two forms of loss—

1. Loss of voltage due to resistance and inductance.
2. Loss of current due to leakage and capacitance.

The shunt current is dependent on the capacitance, and the larger this capacitance the greater the loss. The leakage current across the insulation since the total inductance and capacitance each depend upon the length. This is called the *characteristic* or *surge impedance.* It is the impedance of the line itself when the far end is open, and if we terminate it with a resistance of equal value, energy will be accepted from the line without reflection, irrespective of the length of the line.

The inductance of a pair of parallel wires was shown in Chapter 1 to be

$$0{\cdot}2 \log_e (d/r)\ \mu\text{H per metre}$$

where d is the spacing of the wires and r the radius.

The capacitance between them is

$$\frac{10^{-3}}{36 \log_e(d/r)}\ \mu\text{F per metre}$$

Hence the characteristic impedance $\sqrt{(L/C)} = 276 \log_{10} (d/r)$ ohms. Over a wide range of practical values for d/r this gives an impedance of the order of 500 to 700 ohms, and an average value of 600 ohms will be found to give results of the right order.

For short-wave work tubular feeders are often used, one wire running inside the other. The principle is the same, but the characteristic impedance is usually about 70 ohms.

Matching the Line

A feeder, therefore, provides a very flexible arrangement, which we can use for any frequency. All that is necessary is to ensure that the termination is correct, and we do this by matching the line to the output by high-frequency transformers, or by tapping the feeder across the tuned circuit in the usual way. The impedance of a parallel circuit at resonance is L/CR. This is usually very high, but by tapping down the coil we can make the effective primary impedance what we wish.

Fig. 8.18 shows a termination using a feeder of this type. Both the input and the output are tapped across the respective tuned circuits, so that the maximum efficiency is obtained. The input is

may be expressed in terms of the *leakance*, which is the reciprocal of the insulation resistance. Hence, the larger the leakance the greater the current, which is a similar variation to that of the capacitance.

The complete expression, therefore, involves both R, the line resistance, and C, the leakance, and is of the form

$$Z = \sqrt{[(R + j\omega L)/(G + j\omega C)]}$$

At radio frequencies $L\omega$ is much greater than R and $C\omega$ is much greater than G, so that the expression simplifies to $\sqrt{(L/C)}$.

Those readers who desire a more mathematical treatment should refer to *High Frequency Alternating Currents*, by McIlwain and Brainerd.

matched to the feeder because, as is so often the case in communication work, the maximum efficiency is obtained when the internal and external impedances are equal. Hence we obtain maximum energy transfer into the feeder when it is suitably matched at the transmitting end.

The receiving end is matched to avoid reflection, which is another

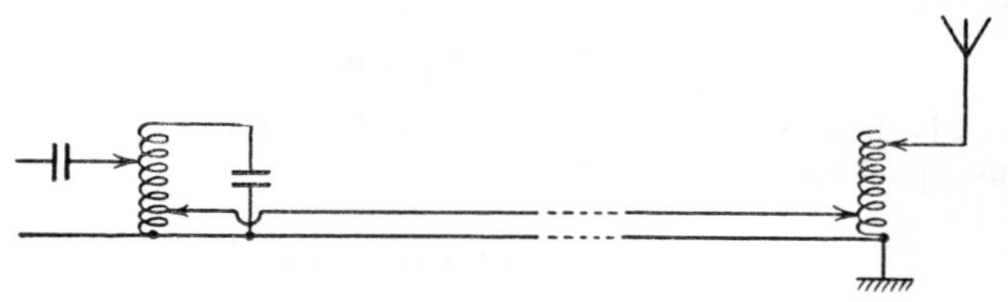

FIG. 8.18. MATCHING THE INPUT AND OUTPUT BY AUTO-TRANSFORMERS

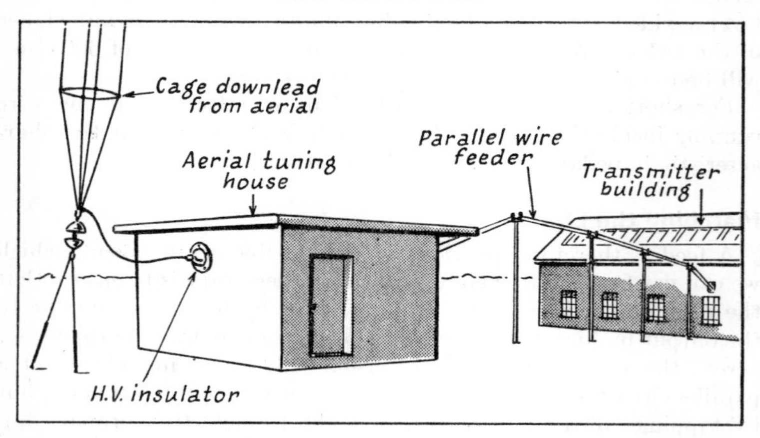

FIG. 8.19. FEEDER HOUSE WITH AERIAL DOWN-LEAD AT A TYPICAL HIGH-POWER BROADCASTING STATION

way of saying that the maximum energy transfer is obtained at this point as well.

It should be noted that the feeder only carries a current necessary to transmit the requisite power, whereas the oscillating current in the circuits at either end is much greater than this, due to the usual resonant action. The process is the same as in the valve oscillator or amplifier, where the anode feed current is only large enough to maintain the losses, while the current oscillating in the tuned "fly-wheel" circuit is much greater.

Fig. 8.19 illustrates a typical feeder arrangement. The building in the foreground houses the aerial tuning inductance, which has to handle a current of several hundred amperes, while the feeder from the main building only carries ten or twenty amperes and is thus quite light in construction.

In short-wave transmitters, it is often necessary to supply a number of aerials from the same feeder, and in this case special junction arrangements have to be made. The feeder is split into two, and each half into two again, and so on. At each point a matching transformer is arranged so that the impedance of the line viewed from either direction always appears to be the characteristic impedance $\sqrt{(L/C)}$.

Travelling Waves

It is important to have a clear understanding of the voltage and current distribution in a transmission line. If we apply an alternating e.m.f. to the input of a feeder, this voltage will be transmitted along the line by the process just described, but since, as we have seen, there is a slight time lag in the transmission, there will be a progressive delay along the length of the line.

At the distant end, therefore, the voltage will vary in accordance with the impressed e.m.f. at the input end, but will go through the cycle a fraction of a second later, having been transmitted along the line by a *travelling wave*. This wave travels with a finite velocity very nearly equal to that of light (though if there is appreciable loss in the line the velocity is reduced, as explained on page 389).

Because of this finite velocity of travel the voltage (and current) are not the same at all points in the line *at any given moment*. At any instant the voltage and current will be distributed sinusoidally along the wire (assuming sine-wave input) but, since the wave is travelling, every part of the line in turn goes through the complete cycle of the e.m.f. applied to the input.

Tuned Feeders; Standing Waves

Let us see what happens if the feeder is not correctly terminated. Consider a feeder open at the far end so that reflection occurs. The wave reaching the far end will be reflected and will thus travel back along the line towards the sending end. But this reflected wave will interfere with the forward wave so that the voltage at any intermediate point will be the (vector) sum of the e.m.f.s due to the forward wave and the reflected wave.

When this happens we find that there are places where the two waves cancel each other out, while in between they reinforce one another, so producing what is called a *standing wave*. The system is, in fact, nothing but an elongated oscillating circuit, rather like the simple aerial discussed in the previous section, but as there are two wires, the currents in which are equal and opposite at any given section, little or no radiation takes place.

The actual voltage at each successive maximum point will increase as we approach the sending end, because the attenuation of the forward wave is less, while the reflected wave is more attenuated (having had farther to travel). The variation is only appreciable, however, in lines where the resistance and leakance are comparable with the relevant reactances, as they may be in telephone lines because of the relatively low frequencies involved. At radio frequencies the attenuation is very small and the standing waves are of practically constant amplitude.

We can assess the effect mathematically relatively simply. Let us assume that at some point at a distance x from the receiving end the voltage due to the forward wave is $e_f = E \sin \omega t$. The reflected wave will have had to travel a small additional distance, actually equal to $2x$, which will take a time $\delta t = 2x/v$, where v is the wave velocity. Hence the e.m.f. at x due to the reflected wave (neglecting any reflection loss) is the same as would result from a forward wave which started δt earlier, so that $e_r = E \sin \omega(t - \delta t)$.

The resultant e.m.f. at x is thus

$$\begin{aligned} e_f + e_r &= E \sin \omega t + E \sin \omega(t - \delta t) \\ &= 2E \sin \tfrac{1}{2}[\omega t + \omega(t - \delta t)] \cos \tfrac{1}{2}[\omega t - \omega(t - \delta t)] \\ &= 2E \cos \tfrac{1}{2}\omega\delta t \sin \omega(t - \tfrac{1}{2}\delta t) \\ &= 2E \cos (\omega x/v) \sin \omega(t - x/v) \end{aligned}$$

This will be seen to represent a wave which is varying sinusoidally with time, but the amplitude of which is *varying sinusoidally along the length of the line*, being proportional to $2E \cos \omega x/v$. But $v = \lambda f$, so that the amplitude may be written

$$2E \cos 2\pi f x/\lambda f = 2E \cos 2\pi x/\lambda$$

This is zero when $2\pi x/\lambda$ is an odd multiple of $\pi/2$, i.e. when $x = \lambda/4, 3\lambda/4, \ldots$, and is a maximum (equal to $2E$) when $2\pi x/\lambda$ is an even multiple of $\pi/2$, i.e. when $x = \lambda/2, \lambda, 3\lambda/2, \ldots$

Thus, we have alternate maxima and minima of both current and voltage. The points of zero voltage are called *nodes* and are separated by half a wavelength. The current nodes occur at points of maximum voltage, called *antinodes*, midway between the voltage nodes.

Fig. 8.20 shows the distribution of voltage and current on an open-circuited feeder which is an even multiple of half a wavelength long.

Such a device is known as a *tuned feeder* or *Lecher wire*, and is sometimes used for measuring the wavelength of a transmitter. By finding the current nodes and measuring the distance between them, an accurate estimation of the wavelength can be arrived at.

The nodes are usually located by means of a shorting bar carrying a flashlamp or other indicating device which is moved along the feeder, maximum brilliance indicating a current antinode.

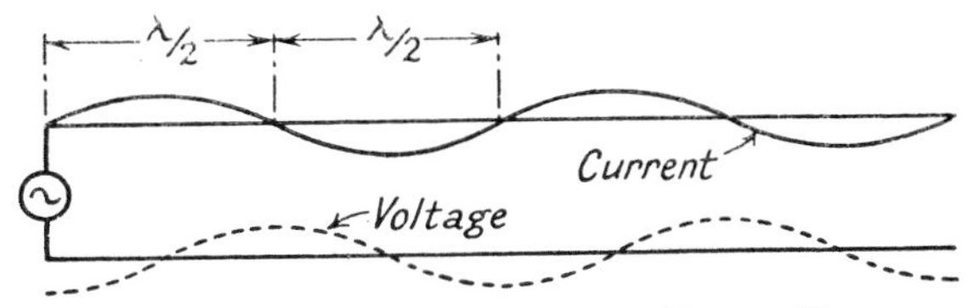

Fig. 8.20. Standing Waves on a Tuned Feeder

Volt-ampere Requirements

If the feeder length is not an exact multiple of $\lambda/2$ the situation is as if the generator were located at some intermediate point along the line. The voltage at the far end is still a maximum and the current at this point is clearly zero, as before. As we travel back towards the input end the voltage and current vary in the manner just described, giving nodes of either current or voltage every quarter wavelength until the generator is reached.

The voltage and current supplied by the generator are then determined by its position. If it were $\lambda/4$ to the right in Fig. 8.20, it would be supplying no voltage* and maximum current, and at any other point it would supply something less than the maximum. A feeder can thus provide a step-up action like an ordinary tuned circuit, the voltage at the far end being greater than the generator voltage, and this effect is sometimes utilized.

Tuned feeders may also be used to supply an aerial on short-wave systems, by merely extending the end of the feeder to form the aerial (*see* Fig. 8.21). The current and voltage relations in the system can now continue without any hindrance, and all that we have done is to increase the length of the aerial very considerably, but since the two wires of the feeder are running parallel, the feeder does not radiate as explained above, all the actual radiation coming from the aerial itself.

This form of tuned feeder, however, is only used where the run is short—not more than about one wavelength. As distinct from

* Assuming no resistance. Actually some voltage would be required to make up for losses.

the untuned transmission line, the resonant feeder has to carry the full oscillating current, so that the losses are greater and this restricts its usefulness to short runs.

The subject is further discussed in Chapter 13.

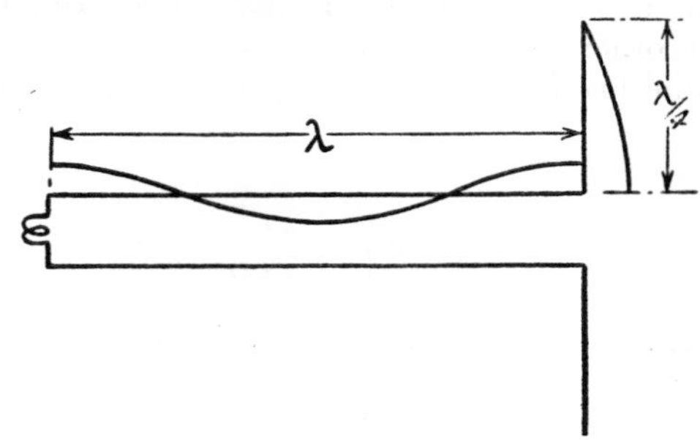

FIG. 8.21. TUNED FEEDER SUPPLYING A HALF-WAVE AERIAL

Input Impedance of Tuned Feeder

The input impedance of a tuned feeder clearly varies between zero at a voltage node and infinity at a current node. The actual impedance (neglecting resistance) can be shown to be

$R_0 \cot \omega\sqrt{LC}x$ for a feeder with the far end open, and

$R_0 \tan \omega\sqrt{LC}x$ when the far end is short-circuited,

R_0 being the characteristic impedance $= \sqrt{(L/C)}$, and x the distance from the commencement of the feeder.

As shown later, $\sqrt{(LC)} = 1/v$ where v is the wave velocity, which in an r.f. transmission line is substantially equal to c, the velocity of light. We can then rewrite the above expressions in the form

$$Z = R_0 \cot 2\pi x/\lambda = R_0 \cot 2\pi fx/c \quad \text{when} \quad Z_r = \infty$$
$$= R_0 \tan 2\pi x/\lambda = R_0 \tan 2\pi fx/c \quad \text{when} \quad Z_r = 0$$

Thus Z will be seen to vary from zero to infinity according to the familiar tangent law, the pattern repeating every wavelength. It may be noted also that $Z = Z_0$, the characteristic impedance, when $x = \lambda/8$.

Matching with a Feeder

A feeder may thus be used to match two different impedances, the arrangement depending upon conditions. One example is shown at (*a*) in Fig. 8.22 where an untuned transmission line is being used to feed a half-wave aerial.

Here the aerial is fed with voltage through a short length of tuned feeder made equal to an odd multiple of $\lambda/4$ and having its bottom end short-circuited.

The impedance of the feeder thus rises from zero at A to infinity at B, and the transmission line is tapped across at the point C at which the impedance, as determined by the expression above (or,

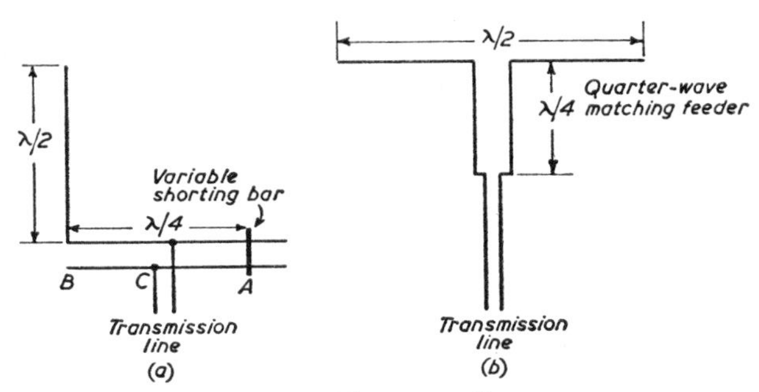

FIG. 8.22. MATCHING FEEDERS

more likely in practice, by trial and error) is equal to the characteristic impedance of the transmission line.

Another method is to use a quarter-wave feeder in series with the aerial as shown at (*b*). Under these conditions, it can be shown that no reflection occurs if the characteristic impedance of the matching line $Z_m = \sqrt{(Z_1 Z_2)}$, where Z_1 and Z_2 are the impedances of the aerial and transmission line respectively.

The same device may be used to match any two impedances provided they are not too widely different.

Reflection Coefficient

So far we have assumed that any voltage at the far end of a mismatched line is completely reflected, but this is only correct in the two extreme cases of open- or short-circuit termination. With intermediate values of Z_r some of the energy may be accepted by the termination so that the reflection is only partial. As we have seen, no reflection occurs if $Z_r = Z_0$, and it can be shown that for any other value of Z_r the ratio of the reflected e.m.f to the forward e.m.f. is given by

$$\rho = (Z_0 - Z_r)/(Z_0 + Z_r)$$

where Z_0 is the characteristic impedance of the line and Z_r the terminal impedance.

This ratio is called the *reflection coefficient*. (It should be noted that Z_0 and Z_r are *vector* quantities and must be expressed as such, i.e. in the form $Z_r = R_r + jX_r$.)

Standing-wave Ratio

When the reflection coefficient is unity the standing waves vary in amplitude from zero to $2E$, as in Fig. 8.20, but when ρ is less than unity the variation is obviously not so great and the performance of a feeder is sometimes assessed in terms of its standing-wave ratio (S.W.R.) or, more precisely, the voltage standing-wave ratio (V.S.W.R.), which is the ratio of maximum to minimum amplitudes of the standing waves.

This may be expressed in terms of the reflection coefficient as

$$\text{V.S.W.R.} = (1 + \rho)/(1 - \rho) = Z_0/Z_r$$

Losses in Feeders

We have neglected resistance throughout. In practice, resistance and other losses are inevitably present while radiation, though small, accounts for some loss. At radio frequencies, however, the expressions deduced are not appreciably affected with normal lines.

The presence of losses will cause the current to be less than the theoretical value, but even this is only serious on long tuned feeders for the reasons already stated. In a large commercial station the matter receives some consideration and the feeders are designed to give the best efficiency.

With an open-wire feeder, radiation loss decreases as the spacing of the wires is reduced, but resistance loss increases due to proximity effect, and a spacing which makes $Z_0 = 600$ ohms is about the best. With concentric feeders radiation loss is negligible and the problem is one of conductor loss, which is least if the radii of the external tube and central conductor have a ratio of 3·6:1. This is the order of spacing used in coaxial television cable and gives a value of Z_0 equal to about 75 ohms.

Balanced Feeders

If, in a two-wire feeder, one line is earthy, the other, being subject to the full variation of potential, may induce voltages in near-by wires and feeders giving rise to *crosstalk*. To avoid this a balanced arrangement is often used such that as the potential on one feeder increases that on the other decreases by an equal amount. This can be arranged, for example, by earthing the mid-point of the transformer winding to which the feeder is connected or by some similar form of symmetrical connection.

Transmission Equations

We have discussed the properties of transmission lines in qualitative terms. A quantitative analysis involves mathematical treatment beyond the scope of this book, but it is desirable to state briefly the main expressions dealing with wave propagation.

By considering the line as composed of a number of small elements each receiving energy from its predecessor and passing it on to its successor, we can derive an expression for the voltage at any distance x from the sending end. The complete expression is of the form

$$v = V_s e^{-\gamma x} + V_r e^{\gamma x}$$

Thus the voltage at a distance x is the sum of two voltages—

(1) $V_s e^{-\gamma x}$, called the *forward wave*, which decreases exponentially from its initial value V_s, and

(2) $V_r e^{\gamma x}$, called the *reflected wave*, which has a value V_r at the receiving end but decreases as it travels back towards the sending end, and hence *increases* with the distance x.

Assuming a correct termination, however, so that there is no reflection, the second term vanishes, so that for most practical cases we can write

$$v = V_s e^{-\gamma x}$$

where V_s is the voltage at the input or sending end and γ is a factor known as the *propagation coefficient*.

We can evaluate γ in terms of the line constants. The series impedance is $Z = R + j\omega L$. The shunt admittance is $Y = G + j\omega C$ (*see* footnote to page 378). It is then found that $\gamma = \sqrt{(ZY)}$. Note that since we are dealing with distributed constants the symbols R, G, L, and C represent quantities per unit length.

Now, $$\sqrt{(ZY)} = \sqrt{[(R + j\omega L)(G + j\omega C)]}$$

If we expand this and collect the resistive and reactive terms together we may write it in the form $\alpha + j\beta$, where α and β are somewhat complex terms involving the four quantities R, G, L, and C.

Hence we can write the expression for v, the voltage at a distance x from the sending end, in the form

$$v = V_s e^{-\gamma x} = V_s e^{-(\alpha + j\beta)x}$$
$$= V_s e^{-\alpha x} e^{-j\beta x}$$

Now, $e^{-j\beta x}$ can be written $(\cos \beta x - j \sin \beta x)$ so that

$$v = V_s e^{-\alpha x} (\cos \beta x - j \sin \beta x)$$

Thus the wave is subject to a gradual decay, of exponential form, determined by the term $e^{-\alpha x}$, accompanied by a progressive phase displacement determined by the second term involving β. Hence the factor α is called the *attenuation coefficient*, while β is known as the *phase-change coefficient*.

As already stated, the evaluation of α and β is cumbersome.* It is found that, provided $L\omega \gg R$,

$$\alpha = \tfrac{1}{2}[R\sqrt{(C/L)} + G\sqrt{(L/C)}]$$
$$= \tfrac{1}{2}[R/Z_0 + GZ_0]$$

The expression for β is clumsy, but it may conveniently be written in terms of α, in the form

$$\beta = \sqrt{[\alpha^2 - (RG - \omega^2 LC)]}$$

Clearly α is independent of frequency but β is not, so that the phase shift varies with the frequency. The actual phase shift at a distance x is $\tan^{-1} \beta x$.

Phase Displacement

Hence, even in a properly terminated line which does not exhibit standing waves of voltage or current, there is a progressive phase displacement along the line. If we consider a feeder having no losses, the voltage at a distance x becomes simply

$$E_x = E(\cos \beta x - j \sin \beta x)$$

where β is the phase-change coefficient.

Since $\cos A - j \sin A = A$, this means that the numerical value of the voltage is always the same, and equal to the voltage E, but the phase will be continually changing, making a complete rotation through 2π every time βx becomes a multiple of 2π. In other words there will be standing waves of *phase*, of wavelength $\lambda = 2\pi/\beta$.

This changing phase may or may not be of importance. In short-wave feeders where various aerials have to be supplied with current in the correct phase, the length of the feeder has to be exactly chosen. Otherwise it is immaterial.

* Because of the $\sqrt{}$ sign. The expression $\sqrt{(ZY)}$ can be written in the form $\sqrt{(p + jq)}$ quite easily. But α is *not* equal to p nor β to q as the reader may quickly verify for himself.

Velocity of Propagation

Since the phase changes by one complete cycle every time βx becomes a multiple of 2π, the wavelength $\lambda = 2\pi/\beta$. The *velocity of propagation* along the line can thus be deduced from the fundamental relation $v = \lambda \times f$, so that $v = 2\pi f/\beta = \omega/\beta$.

If R and G are sufficiently small to be negligible by comparison with ωL and ωC, so that the line may be considered as loss-free, α becomes zero and β reduces simply to $\omega\sqrt{(LC)}$; hence

$$v = 1/\sqrt{(LC)} \text{ metres/sec}$$

For a simple parallel-wire feeder the values of L and C (see page 378) are such that $1/\sqrt{(LC)} = 3 \times 10^8$ metres/sec, which is the velocity of electric waves in free space. Hence for a loss-free parallel-wire line $v = c$.

This is true of loss-free lines in general, but only as long as the above assumptions are justified: in telephone lines, for instance, where the frequencies involved are much lower, R and G are often not negligible by comparison with ωL and ωC, and the velocity of transmission is appreciably less than c.

Effect of Mismatching

As stated above, if the line is not terminated correctly the term $V_r e^{\gamma x}$ does not disappear and at any point the voltage is the sum of the two terms. It is important to realize that this is the *vector* sum because the forward voltage V_f and the reflected voltage V_r are not necessarily in phase. We have seen that the forward wave is subject to a progressive phase lag of β radians/metre; the reflected wave is subject to a similar phase lag travelling in the reverse direction.

We can represent the effect by two vectors V_f and V_r rotating at the same speed but in opposite directions (remembering that the rotation here represents travel along the line); and the relative effect is not altered if we consider the forward-wave vector as stationary with the reflected-wave vector rotating backwards at twice the speed. Then as we progress back along the line the combined voltage will vary rhythmically between $V_f + V_r$ and $V_f - V_r$. At intermediate points the voltage will be the vector sum depending upon the phase displacement between V_f and V_r.

We must not assume that the two vectors are necessarily in phase at the receiving end. In fact, they will only be so provided that both Z_0 and Z_1 are purely resistive. If the load is reactive, then V_f and V_r are displaced by an angle θ depending on the (vector) relationship between Z_1 and Z_0. The effect of this is that the standing waves do not start from the termination but build up as if they had started from a point beyond the end of the line, so that the first node is not $\lambda/4$ back along the line. The effect is illustrated in Fig. 8.23.

The actual voltage developed at the termination and the phase of the standing waves depend upon both the magnitude and the phase angle of the load, and the relationships are somewhat involved. An

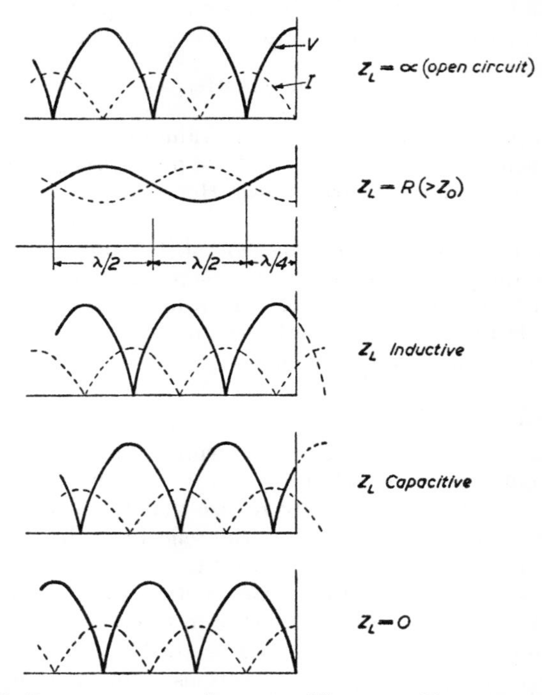

FIG. 8.23. DISTRIBUTION OF STANDING WAVES ON MISMATCHED FEEDER

appreciation of the conditions can be obtained from the vector diagrams of Fig. 8.24.

If V_L and I_L are the voltage and current in the load, the terminal conditions are

(a) $V_L = V_f + V_r$, and

(b) $I_L = I_f - I_r$ (the reflected current I_r being in opposition to V_r).

Now, $V_f = I_f Z_0$. Hence we can use the same vector to represent both these voltages, and thence produce a combined diagram as shown in Fig. 8.24, which displays the terminal conditions very clearly. It shows the relative magnitude and phase of V_L, V_f and V_r and also the phase angle between V_L and I_L.

The angle between V_f and V_r depends on the terminal conditions

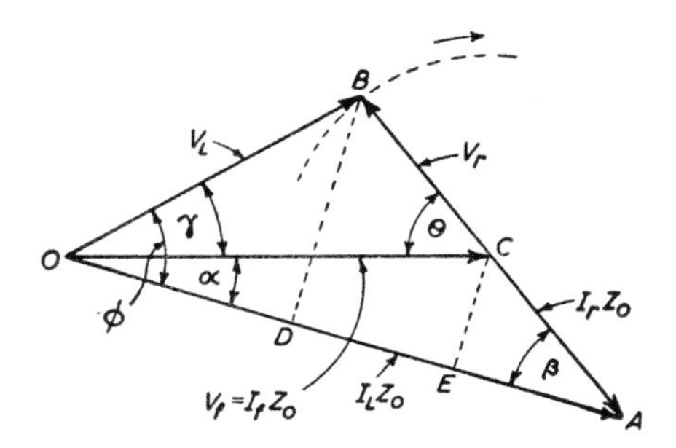

Fig. 8.24. Vector Diagram of Mismatched Termination

and we can deduce from this the position of the nodes in the standing waves. If we regard the vector V_r as rotating relative to V_f we shall obtain the fluctuation between $V_f - V_r$ and $V_f + V_r$ already mentioned. The rotation is clockwise because we are moving *back* along

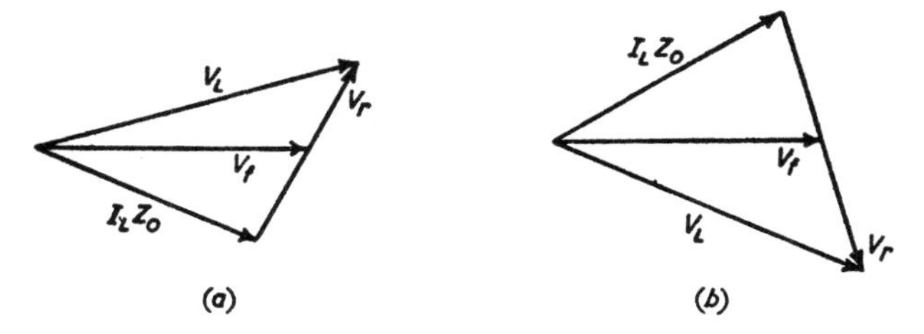

Fig. 8.25. Vector Diagrams for Different Loads
(a) Z_L inductive $> Z_0$; (b) Z_L capacitive

the line and a rotation through 2π represents a complete cycle, which as we have seen, occurs in a line length of $\lambda/2$.

Fig. 8.24 shows the conditions for an inductive load less than Z_0. Fig. 8.25 shows two other possible conditions for $Z_L > Z_0$, and for a capacitive load.

This graphical construction* provides a rapid and convenient

* An excellent discussion of this vector aspect of reflection is given in *Short Wave Wireless Communication* by A. W. Ladner and C. R. Stoner, 5th ed. (London, Chapman & Hall, 1950).

solution to terminal problems for which the corresponding mathematical expressions are often complex.

From the geometry of the figure we can write

$$V_f = \tfrac{1}{2}I(Z_0 + Z_L)$$
$$V_r = \tfrac{1}{2}I(Z_0 - Z_L)$$

whence the reflection coefficient

$$\rho = V_r/V_f = (Z_0 - Z_L)/(Z_0 + Z_L)$$

as previously stated.

When Z_0 is resistive (as it usually is in radio-frequency lines) this may be written

$$\rho = \frac{\sqrt{[(Z_0 - R_L)^2 + X_L{}^2]}}{\sqrt{[(Z_0 + R_L)^2 + X_L{}^2]}}$$

The expression for the phase angle between V_f and V_r is cumbersome but it can conveniently be expressed as

$$\theta = \alpha + \beta$$

where $\alpha = \text{arc sin } X_L/(Z_0 + Z_L)$,
and $\beta = \text{arc sin } X_L/(Z_0 - Z_L)$.

Conversely, if we know V_f, V_r and θ, the phase angle of the *load* is

$$\phi = \alpha + \gamma$$

where $\alpha = \text{arc tan } V_r \sin \theta/(V_f + V_r \cos \theta)$,

and $\gamma = \text{arc tan } V_r \sin \theta/(V_f - V_r \cos \theta)$.

As stated earlier, the voltage at a distance l from the termination depends upon the angle θ between V_f and V_r. Hence when θ is known, the voltage at a distance l from the receiving end is given by

$$V_l = V_f - V_r (\cos \psi + \text{j} \sin \psi)$$

where $\psi = (4\pi l/\lambda + \theta)$.

This is a maximum when $\psi = n\pi$ and a minimum when $\psi = 0$.

8.4. Directional Aerials

The simple aerials so far considered are not appreciably directional but receive signals equally well from any direction. It is often required, however, to use systems which are directional in character, which can be achieved by a combination of aerials so spaced as to produce an appreciable time lag between the voltage induced by the waves at the two ends of the system. By suitable phasing of the currents set up in the aerial this time lag can be caused to accept a signal coming from one direction and to reject one coming from the opposite direction.

Consider the effect of an electromagnetic wave travelling from right to left across the loop of wire $ABCD$ (see Fig. 8.26). It will reach the side AB first and will induce an e.m.f. therein. While travelling from A to D no effect occurs because the wave does not move *across* either of the wires AD or BC. On reaching DC, however, an e.m.f. will again be induced in the same direction as before. Since DC, however, constitutes the opposite side of the loop, this e.m.f. will oppose that induced in AB, and the two will tend to cancel out. Actually they will not quite do so because the e.m.f. in AB was set up a very short time before that in DC, and hence the two e.m.f.s will be slightly out of phase. There will be a small resultant e.m.f., therefore, due to the passage of the wave.

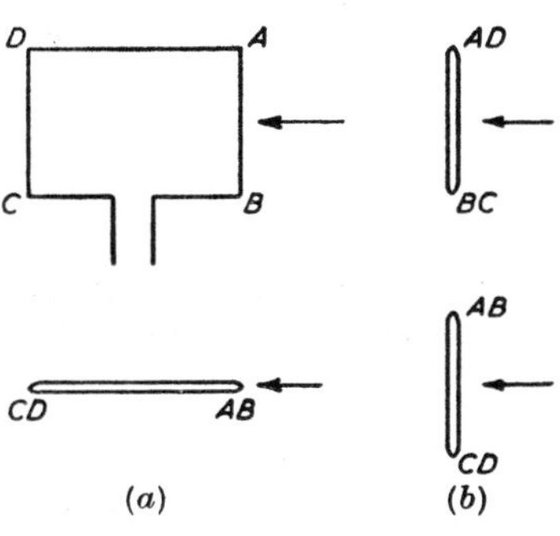

Fig. 8.26. Loop Aerial

Suppose now that the loop is turned through an angle of 90°, and takes the position shown at (b). Here the wave will affect the two sides AB and DC *simultaneously*. The e.m.f.s as before will be in opposition, but there is no phase difference between them in this position, and they will cancel out completely without leaving any resultant.

The e.m.f. induced, therefore, depends on the phase difference, due to the fact that the wave reaches one side of the loop before the other. Consequently, as the loop is rotated, the e.m.f. induced will gradually vary from a maximum (when the loop is pointing directly towards the transmitter) to zero (when the loop is at right angles). Further rotation will cause the e.m.f. to increase again to a maximum when the loop is again in line with the transmitter, while a second zero position will be obtained when the loop is again at right angles.

Polar Diagram

Consider a single-turn loop as in Fig. 8.27 at an angle θ with the direction of the incoming wave. The time δt which the wave will take to pass across the loop is $(d/c) \cos \theta$ where d is the distance across the loop and c is the velocity of the wave. The field strength of the wave will be $E \sin \omega t$ so that the resultant e.m.f. is

$$Eh[\sin \omega(t + \delta t) - \sin \omega t]$$

where h is the height of the loop.

If we put $t = 0$, this reduces to

$$\begin{aligned} e &= Eh \sin \omega \delta t \\ &\simeq Eh\omega\delta t \text{ (because } \omega\delta t \text{ is small)} \\ &= Eh(2\pi c/\lambda)(d/c) \cos \theta \\ &= 2\pi Eh(d/\lambda) \cos \theta \end{aligned}$$

Thus the loop e.m.f. is dependent on the dimensions relative to the wavelength and on the cosine of the angle θ. If we plot the e.m.f.

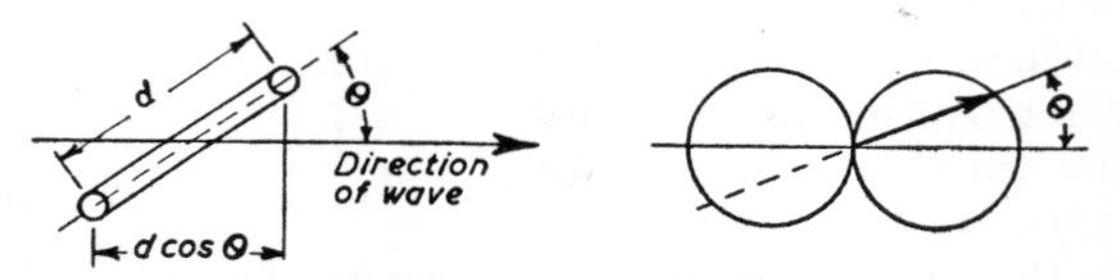

Fig. 8.27. Polar Diagram of Loop

in terms of θ we obtain what is called a *polar diagram*, as shown in Fig. 8.27, which for a simple loop is a "figure of eight" having two maxima when the loop is pointing directly towards the transmitter and two minima when it is at right angles to the direction of the oncoming wave.

Practical Forms of Loop

So far only one turn has been considered. Several turns may be wound, however, each of which will have e.m.f. induced in it in exactly the same way, and the total e.m.f. will be the sum of the e.m.f.s in the several turns. A loop with several turns is called a *frame aerial*. The actual number of turns which may be employed with a given area of frame is limited by the inductance, which increases rapidly as the turns are increased; the total inductance must obviously be such as to tune conveniently to the wavelength required.

Direction Finding

The directional properties of a frame may be employed to obtain the bearing of a given station. The station is tuned in the usual manner, and the frame is then rotated until no signals are heard; in this position the frame will be at right angles to the direction of the transmitting station. The sharpness of the bearing depends on obtaining a good zero. One of the chief sources of error is the tendency of the frame as a whole to behave like a simple vertical wire aerial. Signals received in this mode are independent of the direction of the transmitting station. The effect is normally small, but becomes appreciable when the e.m.f. induced in the frame, as such, is very small, i.e. near the zero point.

The effect will thus be to give a minimum signal only, instead of a sharp zero, when the frame is at right angles to the transmitter. To avoid, as far as possible, errors from this source, the frame leads and connections must be made as symmetrical as possible. As a further precaution the middle point of the frame itself is very often connected direct to earth.

Effect of Frame Width; Robinson System

Other errors arise due to the width of the frame itself. With such a frame, even in the zero position, the turns nearest the direction from which the signal is coming will be affected slightly before the remainder, giving a small phase difference which will still provide a small e.m.f. and so destroy the zero. If the frame is slightly rotated beyond the zero the small e.m.f. induced in the frame acting normally may cancel out that due to the "width effect" and provide a crisp, but quite false, zero.

In an endeavour to overcome this E. H. Robinson suggested the use of a main frame with an auxiliary frame at right angles. This additional frame is connected in series with the main frame through a switch which permits it to be connected in either direction. When the main frame is at zero there will still be a signal picked up on the auxiliary frame, and this will be many times greater than the small e.m.f. due to width effect.

Under these conditions, since the e.m.f. from the main frame is zero, the residual signal will be the same whichever way the auxiliary frame is connected, and the true zero is therefore located by the fact that no difference in signal is observed as the auxiliary frame is reversed. It was claimed that it was easier to match two small signals to equality than to determine a zero accurately, but the method is now mainly of academic interest.

Crossed-loop System

Another method of minimizing the difficulty is to increase the size of the frame so that it becomes a single turn or loop, and to rotate it electrically. This is known as the *crossed-loop* or *Bellini-Tosi* system and is illustrated in Fig. 8.28.

Two large (triangular) loops are erected at right angles to each other. The loops are coupled to the receiver by a special arrangement, having two fixed coils mounted at right angles to each other and a single rotating coil inside.

The fixed coils are connected one in series with each loop, and the moving coil is connected to the receiver. When the coupling coil is in the same plane as one of the fixed coils, it picks up the

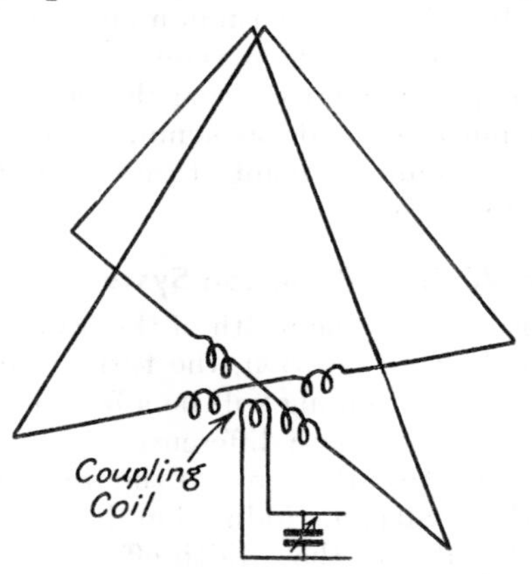

FIG. 8.28. BELLINI-TOSI AERIAL

e.m.f. from that loop, and none from the other, which is at right angles. The system, therefore, has the directional properties of the loop to which the coil is coupled. Similarly, if the receiver is coupled to the second loop, the system has the directional properties of this loop, i.e. receives from a direction at right angles. Hence rotating the coupling coil through 90° is equivalent to rotating the aerial system through the same angle.

In intermediate positions the coupling coil picks up some e.m.f. from each loop and intermediate directional effects are obtained, the rotation of the coupling coil producing the same effect as the rotation of the whole system. The arrangement is known as a *radio-goniometer*. (Fig. 8.29)

Signals comparable with those obtained on an ordinary aerial may be obtained with this system. It is also capable of being erected very symmetrically, so that the zeros obtained are quite sharp. The coupling between the aerials and the secondary is usually made tight so that the tuning of the secondary also tunes the aerials.

Night Effect

A phenomenon which considerably affects D.F. work is what is known as "night effect." It is found that by day the bearings obtained on given fixed stations are reasonably reliable (within 2°). At sunrise or sunset, however, the apparent bearing may vary considerably, often as much as 180° in a few seconds. After dark, conditions become steadier but are still unreliable. This effect was shown by Eckersley to be due to polarization of the waves reflected

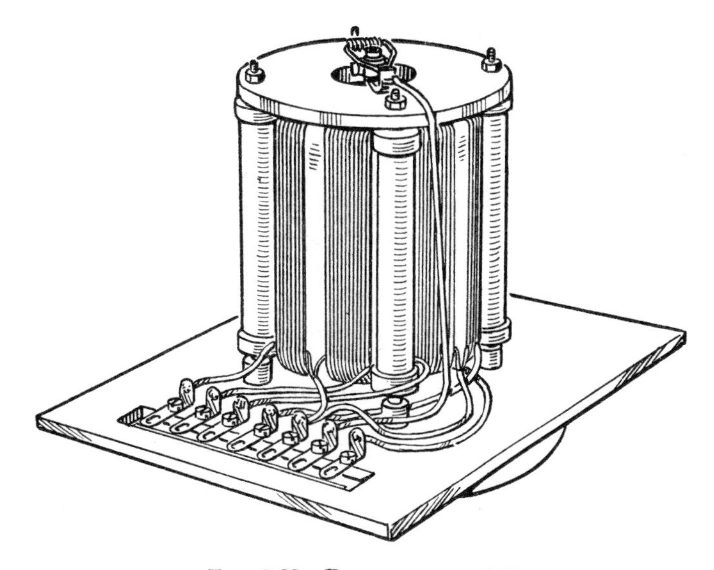

FIG. 8.29. RADIO-GONIOMETER

from the upper atmosphere, which is more or less permanently electrified as was shown in Chapter 3.

These reflected waves arrive twisted so that the electric fields arrive partly or wholly horizontal instead of vertical. Under such conditions, since the waves are coming downwards at an angle, the top of the frame is affected before the bottom and a strong signal is induced. As with width effect, this may be cancelled out by the true signal in some other and quite false position of the frame giving wildly inaccurate bearings.

Eckersley pointed out that the errors produced by this effect were worst when the frame was at right angles to the wave, but did not affect the direction of maximum reception.

Heart-shaped Balance

This led to the suggestion of a method known as the *heart-shaped balance*, in which the frame is used in its maximum position. Both a frame and an aerial are employed in this scheme. The polar diagram of a frame, it will be remembered, is a figure of eight, while that of an aerial is a circle (since it receives equally well in all directions). Now, although a frame has two maximum positions, the current in the frame must clearly flow in the opposite direction in the second position, since the side which received the wave last now receives it first. If the frame and aerial are thus coupled in opposition to a second circuit, the frame will help the aerial in one

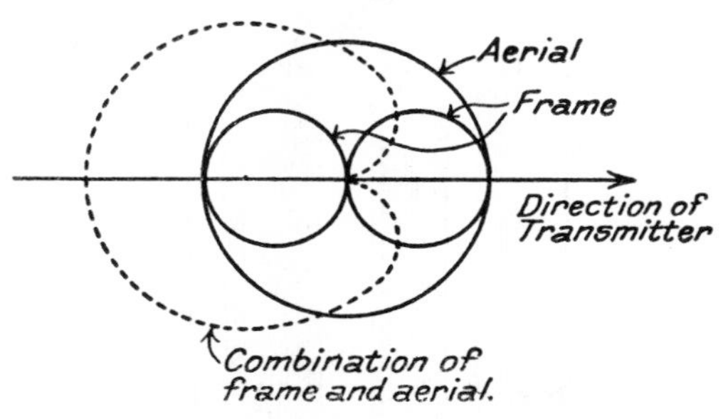

FIG. 8.30. ILLUSTRATING HEART-SHAPED BALANCE

position and will oppose it in the other. If the aerial current can be made equal to the frame current, the result will be that the effects of the aerial and frame cancel out in one position, while if the frame is turned through 180° the two will help and a maximum e.m.f. will result.

Intermediate positions will give values of e.m.f. in between the maximum and zero, the polar diagram for such an arrangement being shown in Fig. 8.30. It will be seen to be a heart-shaped figure, which explains the name given to the method. The zero will be seen to occur when the frame is pointing towards the transmitting station so that night effect is obviated.

Fig. 8.31 shows one method of producing this balance. The frame and aerial are untuned but are tightly coupled (in opposite directions) to the secondary circuit; the tuning of this circuit also tunes the frame and aerial owing to the tight coupling. The resistance is for the purpose of reducing the aerial current until it is equal to that of the frame (while it also serves to ensure that the currents in the aerial and frame are in phase, as unless they are, no balance can be obtained). The secondary circuit is loosely coupled to a third tuned circuit connected to the receiver. The middle point of the frame is earthed as previously explained to reduce the capacitance effect to earth.

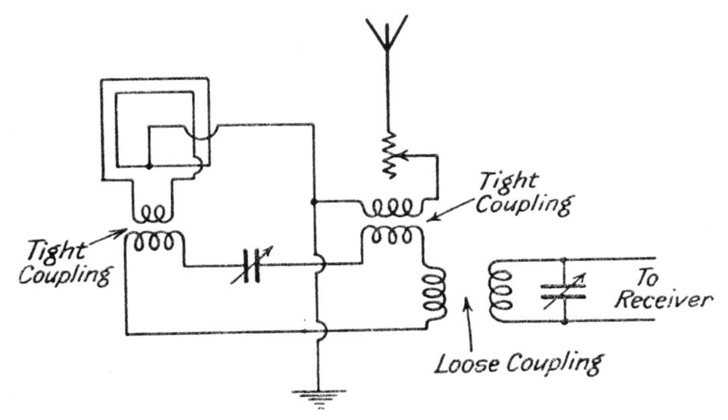

FIG. 8.31. CIRCUIT FOR PRODUCING HEART-SHAPED BALANCE

Sense Finding

This arrangement, however, is not completely successful, as under many conditions the minimum, although correct, is not sharp. The heart-shaped balance, however, is still employed to determine the *sense* of the bearing.

The ordinary frame is ambiguous in this respect, since it is only possible to determine the general direction of the signal and not on which side of the receiving station the transmitter lies. If the frame is rotated through 180 degrees a similar zero point will be obtained, but the heart-shaped balance is definite in that only one zero point is obtained (when the frame is pointing towards the station).

Screened Loops

It is not essential to use large loops and a radio-goniometer. With suitable precautions, quite a small frame is possible, and this may be rotated mechanically in the ordinary manner. The loop or frame is enclosed in a metal tube, which is, however, split at one point so that it does not form a completely closed loop. This, therefore, effectively screens the frame and prevents it from acting as an aerial, although it does not prevent it from picking up signals as a frame. By this means the vertical component is reduced to quite a small value and practicable direction finding is possible.

In order to obtain still greater accuracy, however, a small vertical aerial is fitted close to the frame, connected to the receiver through a coupling coil which is adjusted to provide a reverse e.m.f. sufficient to cancel the vertical component from the frame. The aerial is also employed in conjunction with the frame to obtain the heart-shaped balance for determining the sense of the bearing as already described.

Two such screened frame aerials are sometimes used, mounted at right angles, with a goniometer, instead of the large triangular loops illustrated in Fig. 8.28, to obtain a more compact arrangement.

Calibration

Any direction-finding apparatus set up on board ship or even on shore is usually calibrated by visual observation or some similar system, in order to determine the effect of local obstacles. Massive metal objects or wires in the vicinity can, and will, pick up energy from the wireless waves and re-radiate this energy, so that the wave passing the frame is not the pure wave from the transmitter but is distorted by the addition of these small extraneous radiations. The apparent direction of the wave will, therefore, be changed, and the only method of finding out the extent of this variation is by actual calibration of the apparatus on site.

The error is generally found to be of a "quadrantal" nature, being alternatively positive and negative each 90 degrees. The error, in fact, varies from a maximum in one direction to a maximum in the other direction rather in the manner of sine waves. There are also likely to be isolated objects introducing errors of their own, but, provided that the general superstructure around the equipment is not altered, the calibration, once obtained, is found to remain accurate.

Adcock System

The most satisfactory system of direction finding, however, is that suggested by Adcock. In essence this system consists of four vertical aerials, situated 90° apart, and arranged as shown diagrammatically in Fig. 8.32. It will be seen to be somewhat similar to the crossed-loop arrangement, but instead of using frame aerials, only vertical aerials are employed which are not responsive to other than normal waves.

Each aerial is a simple vertical wire. Two leads are taken from the mid-point of each to the radio-goniometer, and these leads are run close together, so that any voltage induced in the top one is immediately cancelled out by a corresponding voltage induced in the bottom one, and the net e.m.f. left in the aerial is simply that induced in the vertical portion.

The connections to the goniometer are such that the e.m.f. induced from one aerial is opposed to that from the diametrically opposite aerial, so that the effective e.m.f. is due simply to the spacing of the aerials, in exactly the same manner as in a loop. The same applies to the two other aerials at right angles, and by coupling the receiver to the field coils with a goniometer the orientation of the whole system may be varied.

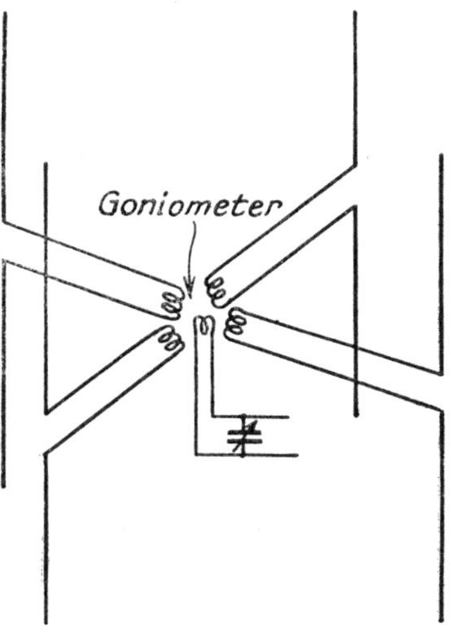

FIG. 8.32. AERIAL ARRANGEMENTS FOR ADCOCK SYSTEM

It is possible with this arrangement to receive bearings which are accurate to within plus or minus 1°, even at night, and practically all land D.F. stations now utilize this system.

A modification of the system which is frequently used for obtaining bearings from aircraft utilizes a cathode-ray tube. Signals are derived from two Adcock aerials at right angles but instead of coupling these through a goniometer they are applied to the X and Y plates of a cathode-ray tube. This produces a line of light on the tube face at an angle depending on the relative strength of the two signals, and hence the bearing can be read off directly and instantly. This is essential with an aircraft which can change its position by several miles in less than a minute.

Position Finding

A single bearing only gives the direction of the transmitting station. If a ship or aircraft requires its position it is necessary to obtain several bearings from a number of stations. If two stations some distance from one another are asked for bearings and these are then plotted on a map, the intersection of the two lines will indicate the position of the transmitter. This is called a "fix."

The stations asked for the bearing should be chosen so that the bearings intersect at an angle between 60 and 120 degrees. Otherwise the actual intersection is indefinite since a small error in the direction of one or both bearings may produce a large error in the apparent position.

If three bearings can be obtained the accuracy is greatly increased.

The three lines should all meet at a point. Actually they do not, due to inevitable small errors in the bearings, but they form a small triangle, often called a "cocked hat," and the true position is somewhere within this triangle.

Aerial Arrays

Apart from direction finding, it is often desired to provide a receiving aerial with directional properties. This can be achieved by suitable arrays of aerials which are spaced in relation to the wavelength to produce a reinforcement of the signals from one direction. Such arrays are frequently employed with short-wave reception, and are discussed in detail in Chapter 13.

The Beverage Aerial

A different form of directional aerial suitable for fixed-station work was developed by C. W. Beverage. It consists of a long horizontal wire a few feet off the ground pointing in the approximate direction from which the signal is coming. It operates by virtue of the fact that the lower portion of the vertical field in a wireless

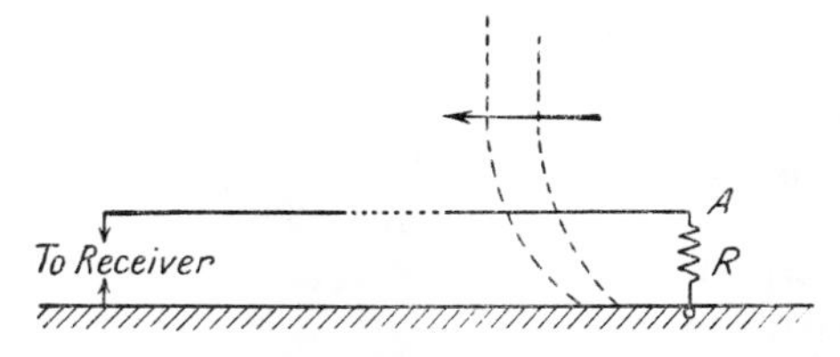

FIG. 8.33. SHOWING HOW WAVE DRAG IS UTILIZED IN THE BEVERAGE AERIAL

wave drags slightly due to the resistance of the earth, as shown in Fig. 8.33.

There is thus a horizontal component which induces voltage in the wire. The voltage in the first element will travel along the wire to the far end. Meanwhile the advancing wave is inducing further changes in the successive elements of the wire and the effect is cumulative so that if the wire is several wavelengths long a considerable signal builds up at the far end.

Signals from the reverse direction produce the same result but on arrival at the point A they are absorbed by the resistance R, which is made equal to the characteristic impedance of the line and thus absorbs without reflection. Signals from any other direction clearly cannot build up to the same extent.

Diversity Reception

Reference may be made to a particular form of receiving aerial which is intended to overcome fading. As explained on page 131, fading is produced by irregular reflection in the ionosphere. In particular, the plane of polarization is twisted so that a wave may arrive horizontally polarized, in which case it will have no effect on the ordinary vertical receiving aerial.

This variation in the plane of polarization may be spasmodic or it may be regular, arising in the latter case from a circularly polarized wave in which the plane of polarization is rotating continuously.

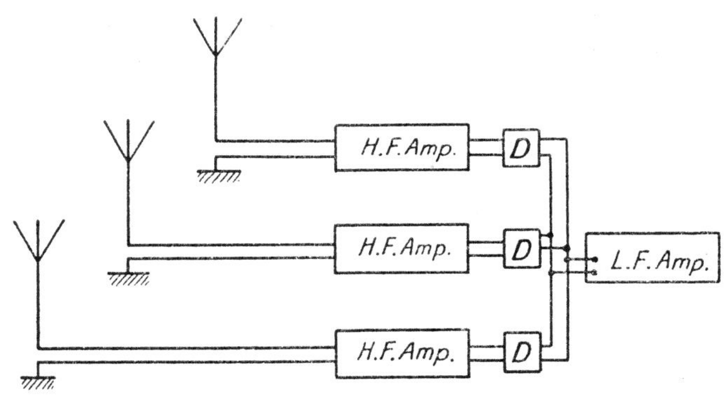

FIG. 8.34. ARRANGEMENTS FOR DIVERSITY RECEPTION

Attempts may be made to minimize fading by the use of automatic volume control, but this is not entirely satisfactory, for no amount of increased amplification can compensate for a complete fade-out, while, in addition, the increasing amplification always brings with it an increased background noise.

It is found, however, that the reception at different localities is not the same at the same instant. When the signal has faded to vanishing point in one locality it may be quite strong only a short distance away. Consequently, if two or more receiving aerials are erected a few wavelengths apart and each one individually tuned to the required signal, and the output of each mixed subsequent to rectification, we should obtain a reasonably uniform signal. When one aerial is receiving practically nothing there will be some signal in one at least of the others, and in practice such a combination usually provides satisfactory reception. The arrangement is illustrated in Fig. 8.34.

Certain difficulties are experienced in applying this system to telephone reception, because the modulation is not always in phase on the three receivers. To overcome this, square-law detectors are used, together with an automatic volume-control device following the detectors, which operates on all three receivers simultaneously. Since the square-law detector operates much more effectively on a strong signal than on a weak one, this arrangement ensures that the greater part of the output obtained from the system comes from the particular aerial which is receiving the best signals at that moment. The system is reasonably satisfactory in practice and is used to a considerable extent.

The aerials are spaced a few wavelengths apart and the signals are brought away by means of r.f. feeders of the type discussed in the previous section.

9

The Radio Receiver

THE radio receiver is concerned with the amplification of the signals picked up on the receiving aerial and, in particular, with the selection of the wanted signal in preference to unwanted signals, of whatever sort. This involves the use of the amplifying and tuning techniques already discussed in basic terms in earlier chapters.

Three separate functions are required, as illustrated in Fig. 9.1. These are:

(*a*) Amplification of the r.f. signals induced in the aerial to an extent sufficient to permit efficient demodulation. This function

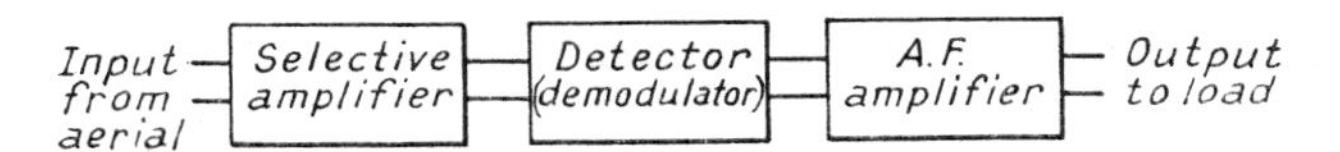

FIG. 9.1. ESSENTIALS OF A RADIO RECEIVER

also involves the selection of the required signal and the rejection of unwanted signals, so that tuned amplifiers are customarily employed.

(*b*) Demodulation of the r.f. signal so that the intelligence contained in the modulation can be extracted.

(*c*) Amplification of the audio-frequency signal to the required extent, and its application to a suitable sound-reproducing or recording device.

These functions will be considered separately. The development of transistor techniques has rendered the use of valves obsolete in many cases, but there are still occasions where valves are used. The basic principles are the same for both techniques so that, as far as possible, a fundamental approach has been adopted. It is then shown how these requirements are applied to the two different types of circuitry.

9.1. Radio-frequency Amplification

The simplest form of receiver requires only a tuned aerial circuit feeding a detector, as in Fig. 3.15. However, this is only suitable for the reception of powerful local signals where considerations of selectivity do not present any problem.

For the majority of requirements considerable r.f. amplification is necessary. This may be carried out at the frequency of the incoming signal, in which case both the aerial and the interstage couplings must be capable of being tuned to the signal. Such an arrangement is known as a *tuned-radio-frequency* (t.r.f.) amplifier and is convenient for certain requirements, particularly with commercial stations (including marine and aircraft services) which operate on a fixed frequency. For broadcast reception, however, where the required signals cover a band of frequencies, the need to readjust all the tuning circuits when selecting another station introduces

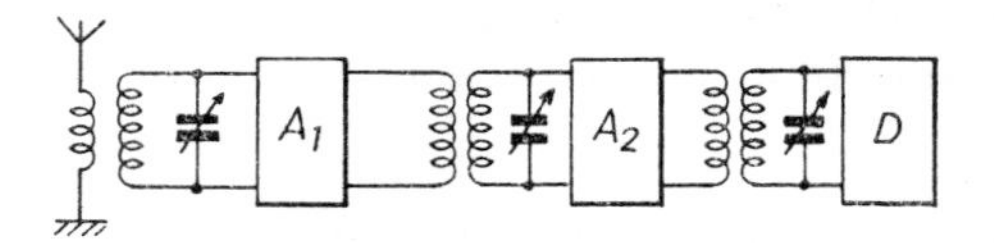

FIG. 9.2. TWO-STAGE R.F. RECEIVER

practical difficulties and it is customary to change the frequency of the incoming signal to an intermediate frequency for which a highly-selective fixed-tuned amplifier may be used. This is called a *superheterodyne* receiver, and while it is merely a special form of r.f. amplifier, it employs certain special techniques which are discussed in Section 9.4.

Fig. 9.2 is a diagrammatic representation of a two-stage r.f. amplifier. It comprises a tuned aerial circuit feeding an amplifying device A_1 which is coupled, in turn, to a second amplifier A_2, the output from which feeds the detector (demodulator) D. The design of the aerial circuit has already been discussed (page 370). We are concerned here with the amplifier stages which are required to provide maximum amplification of the desired signal with a maximum rejection of unwanted signals. These requirements are to some extent conflicting, because the amplifying device, whether it be valve or transistor, influences the performance of the tuned circuits, and the condition for maximum gain is not the same as for maximum selectivity, so that a practical compromise has to be chosen.

Basically, the device receives a certain input which is magnified by its internal action and develops at its output an amplified signal of which the form and magnitude depend on the relative external and internal impedances. As shown in Chapters 5 and 6, the basic stage gain can be written in the form $A_v = \mu Z/(r + Z)$ where Z is the external load, r is the internal resistance and μ is the amplification factor. When r is large compared with Z, as is usual with the valves or transistors used for r.f. amplification, the gain may be written in the form $A_v = gZ$, where g is the forward transconductance $= \delta i_{out}/\delta e_{in}$. Hence the stage gain is simply proportional to Z, and can theoretically be made very large. In practice, it is limited by internal (and external) feedbacks through which a small proportion of the output is transferred to the input and may cause continuous oscillation. This limitation is discussed later.

Variation of Gain with Frequency

If the circuit is tuned with a variable capacitor the impedance (L/CR), and hence the gain, will vary with frequency. With modern receivers the main amplification is provided at a fixed frequency by using the superheterodyne principle described later, but occasionally t.r.f. stages are required.

In such cases it is possible to obtain a uniform gain by using a mixed-coupled transformer, as in Fig. 9.3. The e.m.f. induced in the secondary is then partly inductive, which rises with frequency, and partly capacitive, which falls as the frequency increases, and by correct proportioning a constant voltage transfer can be arranged.

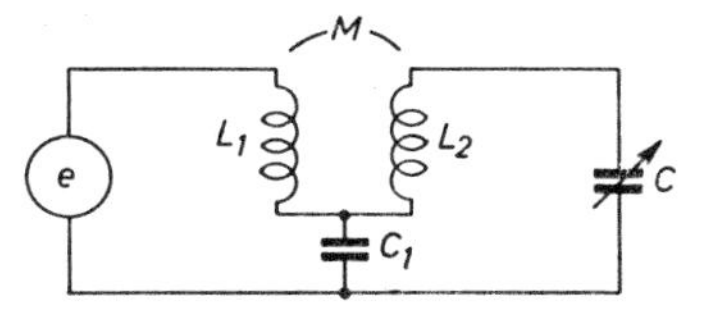

FIG. 9.3. MIXED-COUPLED TRANSFORMER

The capacitive energy transfer may be achieved in other ways, e.g. by a small top capacitance coupling.

Stability

It is important in any amplifying stage to avoid any unauthorized feedback of energy from the anode to the grid circuit. If this feedback is sufficient in magnitude and in the right direction, continuous oscillation will result, as in the case of a simple valve oscillator. Conversely, if the feedback is in the reverse direction the gain of the stage will be reduced.

There are two principal methods by which feedback takes place. One is due to direct magnetic or electrostatic coupling between portions of the circuit or the wiring. Currents are induced from one circuit to the other, and, owing to the very high amplification which can be obtained (a voltage gain of 100–150 is quite normal in a modern receiver), this feedback has only to be very slight in order to produce a marked effect.

Shielding

To minimize this interaction it is customary to enclose the tuning coils in metal shields or cans. These shielding boxes are made of copper or aluminium, and operate as follows. The magnetic fields generated by the coils induce eddy-currents in the material of the shielding. These eddy-currents in turn produce magnetic fields of their own which, by Lenz's law, are in opposition to the original field. Consequently, the field outside the can is negligibly small. The process obviously involves a loss of energy, since the eddy currents circulating in the material of the can must absorb power, but provided the screen is not too close to the coil the loss of energy is small. The magnetic field set up by the shield reduces the effective magnetic field within the coil, so that a coil has a lower inductance inside the can than out of it.

It is important that each circuit shall have its own screen. One common screen for all the circuits, even if it is partitioned off, is unsuitable, as circulating currents in the screen may induce voltages from one circuit to another and so defeat the whole object of the screening. For the same reason, any electrostatic screening, including the chassis, *must not be allowed to carry any r.f. currents*, otherwise coupling will be introduced between the circuits. The chassis should not be used as an earth line. A separate heavy-gauge busbar should be run between the appropriate points, this busbar being connected at one point only to the chassis, as close as possible to the low-potential end of the supply.

With coils or transformers housed in completely closed magnetic circuits, as is more usual with transistor circuitry, the field is suitably confined and there is negligible interaction. It is, however, still necessary to avoid feedback due to common impedances in the supply lines. This is discussed more fully on page 416.

A third source of instability is the internal feedback within the valve or transistor itself. This is discussed in Sections 9.2 and 9.3.

Selectivity

As said earlier, the design of the interstage couplings has to allow for the requirements of selectivity as well as gain.

The selectivity of a circuit is the measure of the frequency discrimination exercised by the tuning. It may be expressed in terms of the ratio of the voltage (or current) for a given deviation to that at resonance. This was shown in Chapter 2 (page 95) to be

$$I/I_0 = 1/(1 + j2Q\delta) = 1/\sqrt{(1 + 4Q^2\delta^2)}$$

where $\delta = (f - f_0)/f_0$ and $Q = L\omega_0/R$.

In this expression Q is the working, or "loaded," value allowing for the presence of the circuit impedances. These will, in general, be in the form of a resistance across the tuned circuit and will thus increase the effective resistance. If the shunt resistance is P, the effective series resistance is increased by L/CP.

Alternatively, if P_0 is the unloaded parallel impedance of the circuit ($= L/CR$), then $Q_w = Q_0P_0/(P_0 + P)$.

Bandwidth

From these expressions the relative response can be estimated for a given value of δ, and it will be noted that the attenuation is dependent upon $Q = (1/R)\sqrt{(L/C)}$. Hence, in general, the higher the L/C ratio the better the selectivity.

However, in practice the circuit requirements are determined by considerations not so much of selectivity as of *bandwidth*. As explained in Chapter 3, the information in a modulated wave is entirely contained in a series of side frequencies extending over a small range on either side of the carrier, so that if the information is to be properly received, the tuned circuits must maintain an adequate response over the full range of these sidebands.

The ratio δ above relates to the deviation on either side of the resonant frequency; the bandwidth is twice this deviation, so that $B = 2(f - f_0) = 2\delta f_0$. It is customarily accepted that adequate reception is obtained if the attenuation over the stipulated bandwidth is not more than 3 dB down ($I/I_0 = 1/\sqrt{2}$). This requires $1 + 4Q^2\delta^2 = 2$, whence $Q\delta = \frac{1}{2}$. Hence $\delta = 1/2Q$. The bandwidth $B = 2\delta f_0 = f_0/Q$, so that the required $Q = f_0/B$.

This is the criterion for a single circuit, but in practice several circuits are usually employed in cascade. The overall response is then $(I/I_0)^n$, where n is the number of stages. Hence, if the *overall* response is to be 3 dB down, the individual circuits must produce less attenuation, which involves a lower individual value of Q. It was shown on page 111 that, if the attenuation is expressed in dB, then for n stages the criterion is that for each individual stage the response at the specified bandwidth must be only $3/n$ dB down. With two stages this gives $B = 0{\cdot}65 f_r/Q$, while with three $B = 0{\cdot}5 f_r/Q$.

If the circuit is transformer-coupled it is necessary to calculate the equivalent single circuit using the normal coupled-circuit laws. In the case of two similar loose-coupled circuits (i.e. a transformer with both primary and secondary tuned) then with critical coupling and identical circuits $I/I_0 = 1/\sqrt{(1 + 4Q^4\delta^4)}$, as shown on page 110. For a single stage the 3 dB bandwidth is then $\sqrt{2}f_r/Q$. This again must be modified if more than one stage is used, the criteria for two and three stages being $1{\cdot}15f_r/Q$ and f_r/Q respectively.

It will be shown in the sections which follow that the design of an r.f. stage is determined from the Q value required. This must be chosen to meet the specified bandwidth requirements with the total number of stages involved.

9.2. R.F. Valve Amplifiers

The simplest form of valve-operated r.f. amplifier uses a tuned-anode circuit as shown in Fig. 9.4. The input is applied to the grid, while the anode is fed from the h.t. line through an inductance tuned with a (variable) capacitor. The output is passed to the succeeding stage through the isolating capacitor C_2.

The valve is biased to its correct operating point by means of the

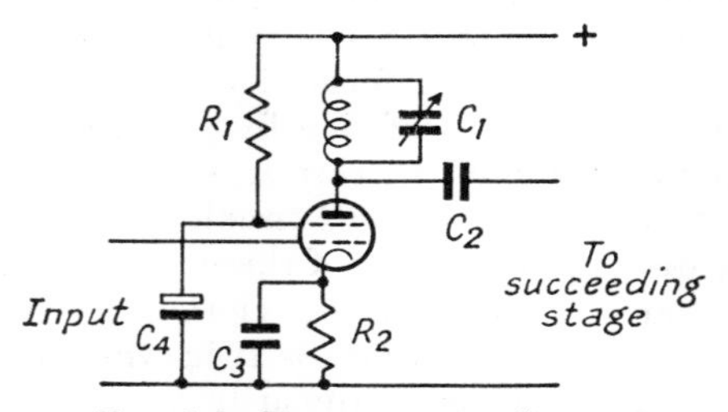

FIG. 9.4. TUNED-ANODE CIRCUIT

cathode resistor R_2 which is so chosen that the total cathode current $(i_a + i_{g2})$ develops across it a voltage equal to the bias required. The cathode is thus positive to the earth line, so that the grid is biased negative to the required extent. The cathode resistor is shunted by a small capacitor (of the order of $0{\cdot}1\ \mu$F) so that the signal-frequency component of the cathode current is by-passed and the only e.m.f. developed across R_2 is that due to the steady current. (If this is not done the signal currents will produce a negative feedback e.m.f. which will reduce the stage gain, as is discussed on page 479.)

The screen is fed from the h.t. line through a resistor R_1, the value being chosen to drop the voltage to the correct rated value; again, to eliminate the effect of any signal-frequency currents, the screen is by-passed to earth through a capacitor. In many instances the valve is designed for operation with the screen at full h.t. potential so that R_1 and C_4 are not required.

Such a circuit will give a gain $A_v = gZ$, where g is the mutual conductance of the valve. The value of the load impedance Z, however, must not be taken as that for the tuned circuit alone $(= L/CR)$ but must allow for the shunting action of the internal impedance of the valve. Neglecting any secondary effects, this is r_a and, as shown on page 101, this will increase the effective tuned circuit resistance by L/Cr_a. Hence the true gain is

$$gL/C(R + L/Cr_a) = gL/(CR + L/r_a),$$

and the working $Q\ (= Q_w) = \omega L/(R + L/Cr_a)$.

Alternatively it may be more convenient to express the relationships in terms of impedance. If the unloaded impedance $L/CR = Z_0$, then $Z_w = Z_0 r_a/(Z_0 + r_a)$.

The effective Q is

$$Q_w = Q_0 r_a/(Z_0 + r_a)$$

and the effective bandwidth is

$$B_w = f_r/Q_w = f_r(Z_o + r_a)/Q_o r_a$$

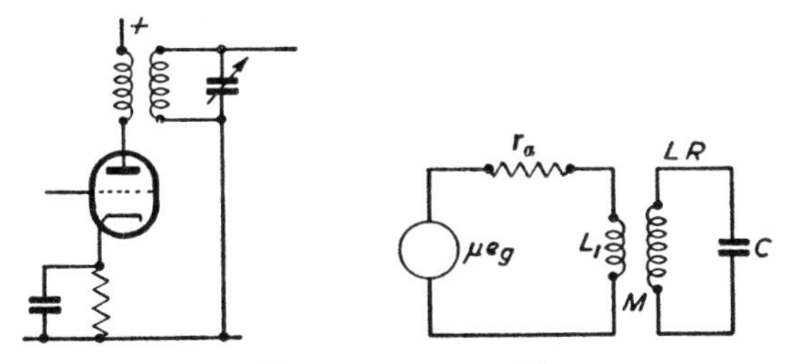

FIG. 9.5. SIMPLE HIGH-FREQUENCY TRANSFORMER AND EQUIVALENT CIRCUIT

R.F. Transformers

It will be seen that the shunting effect of the valve impedance reduces the effective Q of the tuned circuit. The effect is not normally of major importance because, as will be seen later, it is customary, in order to maintain adequate stability, to use pentodes in r.f. stages, in which case r_a is large compared with Z.

However, if r_a is comparable with, or appreciably smaller than Z, it may be better to use an r.f. transformer, as shown in Fig. 9.5. The effective primary impedance is then

$$r_a + M^2\omega^2/R$$

and the primary current

$$i_1 = e_g/(r_a + M^2\omega^2/R)$$

The voltage induced in the secondary is $M\omega i_1$, and the voltage across the secondary coil is $(L\omega/R)e_2$

$$= \frac{L\omega}{R} \cdot M\omega \cdot \frac{\mu e_g}{r_a + M^2\omega^2/R}$$

$$= \frac{ML\omega^2\mu}{Rr_a + M^2\omega^2} \cdot e_g$$

The first term of this expression represents the gain of the stage, and by differentiating this it can be shown that the maximum gain results when $M^2\omega^2 = Rr_a$. This can be written $r_a = M^2\omega^2/R = Z_1$. In other words, the internal and external impedances are equal.

In this condition, the effective Q of the secondary is reduced to one-half the unloaded Q, which is a useful criterion to remember.

The mutual inductance between two windings perfectly coupled without leakage is $\sqrt{(LL_1)}$, where L and L_1 are the inductances. In practice, M is less than this because the primary flux does not all link with the secondary, to allow for which we introduce a *coupling factor*, k, so that $M = k\sqrt{(LL_1)}$.

The criterion for maximum gain can thus be rewritten

$$Rr_a = M^2\omega^2 = k^2LL_1\omega^2 = k^2L_1/C$$

since $\omega^2 = 1/LC$.

Hence $$L_1 = CRr_a/k^2$$

Now, the step-up ratio t is approximately $\sqrt{(L/L_1)}$ and the optimum value of this is obtained by substituting the value of L_1 just obtained. Hence

$$\text{Optimum step-up} = k\sqrt{(L/CRr_a)} = k\sqrt{(Z/r_a)}$$

where Z is the dynamic impedance, L/CR, of the secondary circuit.

With a triode the optimum value of t is usually between 2 and 3. Modern equipment, however, almost invariably uses pentode valves in which r is anything from 0·5 megohm upwards. With such conditions t becomes fractional.

For example, if $r = 500{,}000$ ohms and the circuit has a dynamic impedance of 100,000 ohms (which is about the average for commercial receivers) the optimum ratio, with $k = 0{\cdot}8$, becomes 0·36, representing a step-*down* of 2·8 to 1. Assuming $g = 1\text{mA/V}$ ($\mu = 500$), this would give a stage gain of 108·5, as against 83·5 with a tuned anode circuit.

In practice, this refinement is rarely necessary because typical r.f. pentodes have a mutual conductance appreciably greater than unity, so that a tuned anode circuit will develop all the gain that the circuit can handle before instability occurs due to stray couplings; but where considerations of selectivity are important the use of an r.f. transformer may be desirable.

Neutralizing

Mention was made earlier to the possibility of undesirable feedback between output and input of an r.f. amplifier. In a valve circuit this can take place through the internal capacitance between anode and grid of the valve itself. The magnitude and direction of this internal feedback depends upon the nature of the anode impedance, as shown on page 415. When the anode circuit is nearly

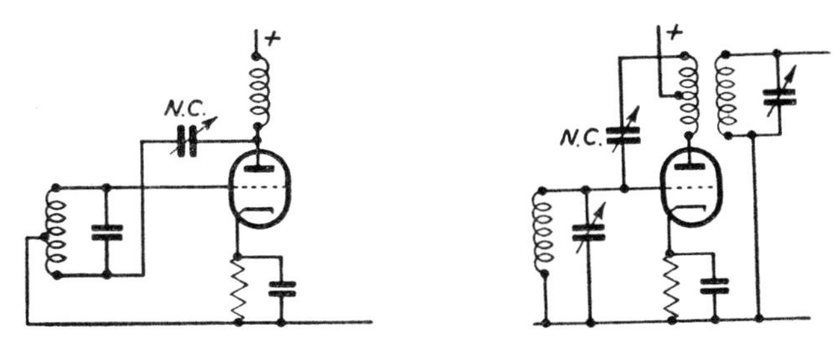

FIG. 9.6. TWO FORMS OF NEUTRALIZED CIRCUIT

in tune the feedback is positive and may easily be sufficient to produce continuous oscillation. In a triode valve this is a very serious difficulty, and it is overcome by the use of neutralizing circuits in which energy is fed from the output to the input, through a special circuit, in opposition to that which passes through the valve. If these circuits are symmetrically designed, the adjustment remains adequate over a wide range of frequency, such as would be covered by the normal tuning operation.

Fig. 9.6 shows two satisfactory forms of circuit. The first circuit, due to Rice, is very symmetrical, but suffers from the disadvantage that only half the full voltage is applied to the grid of the valve.

In the second, due to Hazeltine, a separate neutralizing winding is used, exactly similar to the primary winding, but connected in opposite phase. It is essential that the primary and neutralizing winding shall be very tightly coupled, and the usual practice is to wind one over the other.

Modifications of these circuits are used, usually employing one or other of these two basic forms. There is, however, a third method, known as coil neutralization, in which an inductance is connected between anode and grid. If the reactance of this inductance is equal to that of the anode–grid capacitance it will clearly provide a feed-back in anti-phase and so eliminate any tendency to oscillate. This cancellation only applies for one particular frequency so that the method is of limited application but is used in amplifiers operating at fixed frequencies such as in radio transmitters.

Screened Valves

The alternative remedy is to reduce the anode–grid capacitance to negligible proportions in the valve itself. This is done by introducing an earthed screening grid between anode and grid, as explained in Chapter 5, and the modern receiver uses tetrodes, or more usually pentodes, in the r.f. stages. Such valves have internal anode–grid capacitances of 0·01 pF or less. It must be remembered, however, that if feedback is to be avoided the stray capacitance *external to the valve* must be reduced to a similarly low value, which requires careful screening and attention to layout.

Feedback can also be eliminated by modifying the circuit to maintain the grid at earth potential with both anode and cathode at varying signal potentials. This *grounded-grid* technique is not convenient at normal frequencies but is occasionally used in short-wave receivers, and is discussed in Chapter 13.

Miller Effect

The feedback of energy from anode to grid of a valve through the internal capacitance was first analysed by John M. Miller,* who showed that the effective input impedance depended on the anode load.

Fig. 9.7 shows the equivalent circuit of a valve, this being a simple network except for the fact that there is the amplified voltage μe_g in the anode circuit. By ordinary a.c. theory it can be proved that

$$\text{Input resistance } R_g = -\frac{1/\omega C_{ga}}{A \sin \theta}, \text{ and}$$

$$\text{Input capacitance } C_g = C_{gc} + C_{ga}(1 + A \cos \theta)$$

* *Bureau of Standards Bulletin, No.* 351.

where A is the stage gain, C_{ga} and C_{gc} are the grid–anode and grid–cathode capacitances respectively, and θ is the angle by which the voltage on the load leads the a.c. anode voltage μe_g.

Let us examine these equations further. The input capacitance is obviously always greater than the "geometric" capacitance of the valve, C_{gc}, while the input resistance may be positive or negative according to the sign of the denominator $A \sin \theta$. If it is positive it means that extra damping is introduced into the circuit. If negative, the damping is reduced and self-oscillation may occur.

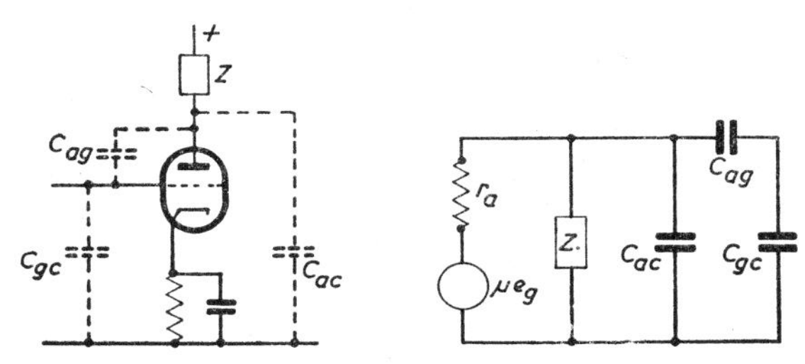

Fig. 9.7. Equivalent Circuit of Triode

If the anode load is inductive (at the frequency under consideration), θ is positive and hence the input resistance is negative. This is the cause of instability in h.f. amplifiers. Cos θ is fairly small and C_g is equal to C_{gc} plus two or three times C_{ga}.

If the anode load is resistive, θ is zero and the input resistance is infinite. This condition can only be obtained by a tuned anode circuit, and hence an r.f. stage with the circuits dead in tune is not unstable at that frequency (but it *is* at lower frequencies, which make the anode load inductive). C_g would be very high since $\cos \theta = 1$, but the condition is an impracticable one and need not be analysed further.

With a capacitive anode load, $\sin \theta$ is negative and the input resistance is positive, i.e. the damping is increased. Cos θ remains positive (for θ never exceeds $-\pi/2$) and C_g is thus always more than C_{gc}, the limiting value being $C_{gc} + C_{ga}$ when $\cos \theta$ is nearly zero.

Thus in an r.f. amplifier the effect is that as the anode circuit is tuned the input capacitance rises to a maximum and falls away again. This effect distorts the tuning of the input circuit and is known as "pulling."

The shunt resistance also varies. If the anode circuit is capacitive, $\sin \theta$ is negative and the grid resistance is positive, i.e. the input circuit is subjected to additional damping. At resonance, $\sin \theta = 0$ and the input resistance is infinite, so that there is no effect on the circuit other than the additional capacitance referred to above. If the anode circuit is inductive the input resistance becomes negative, causing a reduction in damping which, if sufficient, will cause continuous self-oscillation.

The two effects combine to produce a distortion of the tuning curve. At frequencies above resonance the additional damping reduces the response, while below resonance the response is maintained beyond the true resonance point by the negative input resistance. The result is a flat-topped resonance curve slightly displaced from the true resonant point. The effect is normally only serious with triode valves, though it may still be appreciable with pentodes because, though C_{ga} is of the order of 0·01 pF only, the stage gain can be quite large.

There is a second effect which occurs when the time of the oscillation is comparable with the time taken by the electrons to move between the electrodes in the valve. This results in a serious decrease in the effective input resistance. The effect was first described by W. R. Ferris,* but is only troublesome at frequencies above about 10 MHz. It is referred to in more detail in Chapter 13.

Multi-stage Amplifiers; Decoupling

For simple receivers a single stage may suffice, but usually several stages are used, either at the signal frequency or, more usually, at an intermediate frequency as in the superheterodyne technique described later. Since substantial overall gain may be developed it is essential to eliminate accidental feedback from output to input. The layout of the components must be such as to avoid stray coupling, whether inductive or capacitive, which could provide a feedback path, and it is sometimes necessary to run some leads in screened cable.

A more important source of feedback is the presence of common impedances. In Fig. 9.8, for example, the anode currents of both valves have to run through the common impedance Z (which is inherent in the circuit). The voltage developed across this impedance by the anode current of the second valve is thus automatically introduced into the anode circuit of the first valve, whence it will be transferred to the grid of the second valve, causing either an increase or decrease in amplification according to the phase. This difficulty is overcome by decoupling the various leads. A filter comprising an r.f. choke or resistance with a capacitor by-pass is inserted in each lead, as shown in Fig. 9.9. The r.f currents are thus

* Ferris, W. R. "Input resistance of vacuum tubes as ultra high frequency amplifiers." *Proc. Inst. Rad. Engrs.*, **24**, p. 82.

by-passed to earth without going through the battery or power-supply unit.

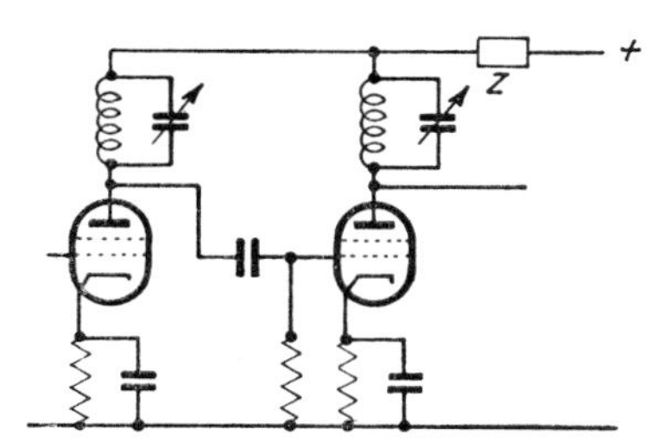

FIG. 9.8. THE COMMON IMPEDANCE Z IN THE HIGH-TENSION LEAD WILL CAUSE COUPLING BETWEEN THE VALVES

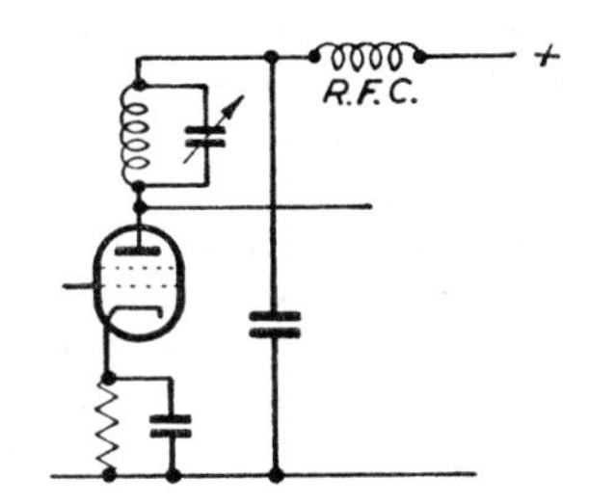

FIG. 9.9. HIGH-FREQUENCY DECOUPLING CIRCUIT

Residual Signal

The layout of an amplifier is also of vital importance from the point of view of selectivity. From the knowledge of the constants of the circuits it is possible to calculate the resonance curves (allowing for valve damping and similar factors) and arrive at an estimate of what is called the *adjacent channel selectivity*, which means the reduction in strength likely to be obtained on a signal relatively close in frequency to (only a few kHz away from) the wanted station.

A still more important requirement, however, is that the selecting action shall continue to be progressive, and even with good circuits this may not necessarily be the case. Once the relatively large resonant gain has been lost, i.e. at frequencies 20 kHz or more off resonance, the circuit gives a small but nearly equal response to any frequency within quite a wide range.

With a powerful local station close at hand, the residual voltage even well away from the tuning point may be quite large and it is necessary therefore to ensure that each circuit continues to exercise its full selecting action.

This is partly a matter of circuit design, for the use of a high L/C ratio ensures a fairly steep "skirt" to the resonance curve, but is equally a matter of layout. Suppose we have two circuits, separated by one or more amplifying valves, each containing a short length of common lead, perhaps in the earthy or low-potential side where it might be assumed unimportant, then two effects will occur.

The currents in the last stage will set up voltages due to the impedance of this common lead, which will be reintroduced via the first circuit into an earlier portion of the amplifier. This will produce feedback which may either increase the gain (and in the limit produce instability) or reduce the gain and cause the set to be less sensitive than it should be.

The second and even more pernicious effect is that residual signals from a powerful station, partly filtered by the first stage, have a direct entry into the later stage without having to go through the filtering action of the intervening stages, so that the *remote channel selectivity* suffers seriously.

Such matters require careful attention if an amplifier is to behave properly. A good test is to see whether the amplifier "cascades" as it should. If the gains of two adjacent stages are A and A_1 the gain of the two together should be AA_1. If it is not, then some common coupling is present and it should be located and removed.

Noise Level

In broadcast technique the number of stages is limited, mainly from considerations of cost. Where this does not apply, it is possible to increase the number of stages and to obtain thereby an improved performance, principally in the direction of increased selectivity. It is necessary to limit the amplification per stage because there is a maximum overall amplification beyond which useful results cannot be obtained.

This limit is set by the noise level in the amplifier, which manifests itself as an indeterminate background which is amplified at the same time as the signal, and it is clearly necessary that the signal-to-noise ratio shall be such as to enable the signal to be clearly distinguished. The cause of background noise is threefold.

1. *Atmospheric Disturbances or Interference.* Atmospherics are random waves which emanate from natural sources. They are in the form of damped waves of so short a duration as to be practically instantaneous, and they appear indiscriminately over the whole frequency spectrum.

The only satisfactory method of combating them is to use a combination of sharp selectivity and directional methods, or to use frequency modulation, in which case the required information is largely independent of the amplitude, as explained in Section 9.6.

Interference from electrical plant is a local disturbance and is more troublesome in broadcasting than in commercial practice. The commercial receiving station is located away from such sources and adequate suppression is fitted to any machinery which has to be in the vicinity.

With broadcast technique this is not always possible, and

additional devices have to be used at the receiver to minimize the disturbance.

2. *Shot Noise.* This is a noise arising from uneven emission of the electrons in a valve. If adequate emission is provided from the cathode, however, this is continually surrounded by a space charge which acts as a reservoir so that the effect is reduced considerably.

The equivalent grid noise due to shot effect can be shown to be given by

$$v_s^2 = 2keF\Delta i_a/g^2$$

where e is the charge on an electron,
i_a is the anode current,
g is the mutual conductance,
Δ is the frequency band of the receiver,
k is a constant,
and F is a factor dependent on the operating conditions.

With a temperature-limited cathode $F = 1$, but with a valve such as a triode the space charge, acting as a reservoir as already described, considerably reduces the noise, and F has a value of about 0·05.

With a pentode valve the presence of the screen overcomes the space charge to a large extent, while the random distribution of electrons between anode and screen further increases the noise. In other words, although the steady anode and screen currents bear a fixed relation for any given operating conditions, the instantaneous values are fluctuating slightly, and as a result of these various factors F increases to a figure of the order of 0·25 to 0·3.

A screened valve, therefore, produces five or six times the noise of a triode, while with a frequency changer the conditions are still worse, as explained on page 445.

The criterion of goodness, from a noise point of view, is determined, for a given class of valve, by the ratio g^2/i_a. The higher this can be made the less will be the noise.

3. *Thermal Agitation,* sometimes called *Johnson noise.* This arises from the movement of the electrons in the material of the conductors. Both this and the shot noise are only important in the first stage of the amplifier where they are followed by the full amplification of the remaining stages. In the case of thermal agitation, the impedance across the grid and cathode of the first valve develops a noise which is proportional to the effective resistance and to the absolute temperature. As is to be expected, the noise is random and is distributed over the whole frequency spectrum, so that

a receiver with a wide bandwidth inherently possesses more background noise than a sharply selective receiver, which is a further argument for restriction of the bandwidth to the minimum necessary to fulfil the particular requirements. The uniformly-distributed random noise is sometimes called "white" noise, by analogy with white light, which includes all the frequencies in the visible spectrum.

The magnitude of the voltage produced can be calculated from the formula

$$v_n^2 = 4kT \int_{f_1}^{f_2} R\, df$$

where k = Boltzmann's constant = $1{\cdot}380 \times 10^{-23}$ J/K,
T = absolute temperature = 273 + temp. in °C,
R = resistance component of input impedance, Ω,
and f_1 and f_2 are the frequency limits of the receiver Hz.

This is a general formula which assumes that the resistive component of input impedance is not constant but varies with frequency. In the more usual case the resistive component is constant, in which case the formula reduces simply to

$$v_n^2 = 4kTR(f_2 - f_1)$$

It should be noted that a tuned circuit behaves as a high resistance $= L/CR$ and that this dynamic resistance should be employed in the formula. The magnitude of the effect may be easily assessed. For example, a tuned circuit having an effective resistance of 0·25 megohm at a temperature of 300 K, followed by an amplifier with a 10-kilohertz bandwidth would develop 6·4 microvolts. To provide a satisfactory signal-to-noise ratio, therefore, we must have a signal of 50 or 60 microvolts, which accounts for the statement made on page 369, that the smallest commercial signal is of the order of 10 microvolts per metre.

Noise Factor

To distinguish between the noise generated by the valve circuitry and that due to the source impedance it is customary to specify the *noise factor* of the valve. This is defined as the ratio of the total noise power fed into the load to that due to the thermal noise in the source. Thus

$$N = \int_{f_1}^{f_2} v_{out}^2\, df \Big/ \int_{f_1}^{f_2} A^2 v_{in}^2\, df$$

where A is the stage gain at frequency f within the band $f_2 - f_1$.

The factor is often expressed in logarithmic units, in which case $N(\text{dB}) = 10 \log_{10} N$.

As said above, v_{in} is usually not frequency-dependent, and if v_{out} is also substantially independent of frequency over the band, the noise factor becomes simply

$$N = 10 \log_{10} v_{out}^2/4kTRA^2$$

Gain Control; Variable-mu Valves

The signal at the output end of the receiver, either in the form of audio-frequency output to a loudspeaker or telegraphic signals to a recorder, must obviously be capable of control. The necessity for keeping the voltage applied to the detector within certain limits has already been mentioned, and this leads to the need for some means of controlling the amplification of the r.f. stages.

A most satisfactory way of achieving this is by using what are known as *variable-mu* valves. These are provided with a specially-designed grid so that the slope decreases progressively as the grid bias is increased. Consequently, the effective amplification is under easy and simple control by merely altering the grid bias applied to the system.

Theoretically, the valve should obey an exponential law so that the rate of change of slope at any point is proportional to the slope itself at that point. This results in a valve having a rather low maximum conductance, and therefore a compromise is adopted, by omitting some of the wires in the centre of the grid. Then, when the major part of the electron emission is cut off by the negative voltage on the grid, there is still a "hole" through which some electrons can flow, and since the grid at this point is a wide-mesh one, the effective amplification factor is quite small. By suitable design a smooth variation of slope can be obtained.

Cross-Modulation

Any ordinary valve will, of course, give decreased amplification as the grid bias is increased, but the variation is not uniform, and this gives rise to a pernicious effect known as *cross-modulation.* Let us assume that we have an interfering signal several times as strong as the wanted signal. Normally we should pass the signals through still further tuned stages at each of which the ratio of wanted to unwanted signal would improve, so that in the end the interference would be eliminated.

In the early stages, however, the strength of the interfering signal may well be large enough to swing the valve over an extended portion of the characteristic. If over this portion the slope of the valve varies considerably, the effective gain of the stage will depend on the strength of the interfering signal so that the output of the wanted signal will be controlled by the modulation of the interfering signal in addition to its own legitimate modulation.*

This cross-modulation, once introduced, *cannot be removed by subsequent tuning*, because it contains a term at the frequency of the wanted signal. It must therefore be avoided at the start, firstly by the best practicable pre-selection in the tuned circuits prior to the first valve, and secondly by ensuring that the valve will handle a really large input without serious distortion. A variable-mu valve is so designed, particularly with the bias run back.

Other methods of radio-frequency gain control are sometimes used, such as variation of screen voltage, alteration of aerial coupling, etc. These, however, are of limited application and the majority of modern circuits use the variable-mu technique.

Aperiodic Amplifiers

In certain circumstances an r.f. amplifier is required to operate in an untuned condition. For this purpose an r.f. pentode may be used with a resistive anode load, the arrangement being simply a special example of the a.f. technique discussed in Chapter 10. The gain is gR (g being the mutual conductance and R the anode load) up to a frequency for which the stray circuit capacitance, which is in parallel with the load, becomes comparable with R.

In general, this requires a very low value of load but with modern valves appreciable gain is still possible. For example a capacitance of 13 pF at 6 MHz has a reactance of 2,000 ohms. If this is not to exercise appreciable shunting action the anode resistor would

* As previously mentioned on page 249, the valve characteristic can be represented by the expression

$$i = ge + g'e^2/2! + g''e^3/3! + \ldots$$

In a simple treatment, adequate for most practical purposes, it is sufficient to assume a linear relationship and ignore the subsequent terms, but in certain conditions these have to be taken into account.

If $e = E \sin \omega t$, then $e^2 = E^2 \sin^2 \omega t = \frac{1}{2}E^2(1 - \cos 2\omega t)$, so that the second term introduces a d.c. component and a second harmonic. The third term, however, involves $e^3 = E^3 \sin^3 \omega t = \frac{1}{4}E^3 (3 \sin \omega t - \sin 3\omega t)$, which thus contains not only a third harmonic but also a component of fundamental frequency which adds to that resulting from the first (linear) term.

If $e = E_1 \sin \omega_1 t + E_2 \sin \omega_2 t$, the third term contains fundamental components *at both frequencies*, thus introducing the cross-modulation. In a variable-mu valve this effect is minimized by avoiding sudden changes in the slope, the curve being approximately exponential in character so that the rate of change of slope is always proportional to the slope itself.

need to be limited to about 750 ohms which, with a valve having a slope of 7, would give a gain of 5·25.

Wide-band Amplifiers

An alternative requirement is to be found in r.f. amplifiers for television frequencies. Here it is essential that the gain shall be reasonably uniform over an appreciable deviation from the mean, e.g. ± 3 MHz at a mean frequency of 50 MHz. This requires a very broadly-tuned circuit, and the usual procedure is to make the anode load an inductance of such a value as to tune to the mean frequency with the stray circuit capacitance, and then to shunt this with a low resistance to broaden the tuning.

In such circumstances the damping of the parallel resistance P far outweighs any other resistance in the circuit and the network behaves like a parallel-tuned circuit having an equivalent resistance simply equal to P. Its effective series resistance is then L/CP and the Q of the circuit $L\omega/R = L\omega/(L/CP) = 2\pi CPf_0$, where f_0 is the resonant frequency.

Now, it was shown on page 96 that the relative response of a circuit when off tune is $1/\sqrt{(1 + 4Q^2\delta^2)}$, where $\delta = (f - f_0)/f_0$. The bandwidth of a network is usually assessed as the point where the response is 3 dB down, i.e. $V/V_0 = 1/\sqrt{2}$. For this to obtain, the term $4Q^2\delta^2$ must be unity, so that $Q = 1/2\delta$.

Now $\delta = (f - f_0)/f_0$ which is the relative deviation on each side of the tuning point. The total deviation is thus 2δ and the total *bandwidth*

$$\Delta = 2\delta . f_0 = f_0/Q = f_0/2\pi f_0 CP = 1/2\pi CP$$

From this the value of P may be calculated. For example, if the bandwidth required is 6 MHz and C is 25 pF,

$$P = 1/2\pi C\Delta = 1{,}070 \text{ ohms}$$

As already explained, the impedance at the mid-frequency is simply P, so that the gain is gP. With an EF91 valve having $g = 7{\cdot}6$ the gain would be 8·1.

9.3. R.F. Transistor Amplifiers

In a valve amplifier the design is based on considerations of voltage gain. A transistor, on the other hand, is a low-impedance current-operated device so that the basic requirement is a matter of obtaining the optimum *power* transfer. Hence the general principles outlined in Section 9.1 have to be applied in a somewhat different manner.

In particular, the use of simple tuned circuits is no longer practicable in an r.f. stage, but must be replaced by r.f. transformers to obtain proper matching, and since fairly high ratios (between 5 and 10 to 1) are involved these transformers are usually housed in Ferrite cores of the type described in Chapter 4. With such cores the coupling factor is very nearly unity, which simplifies the calculations.

Fig. 9.10 illustrates a basic r.f. stage. The transistor Tr_1 is

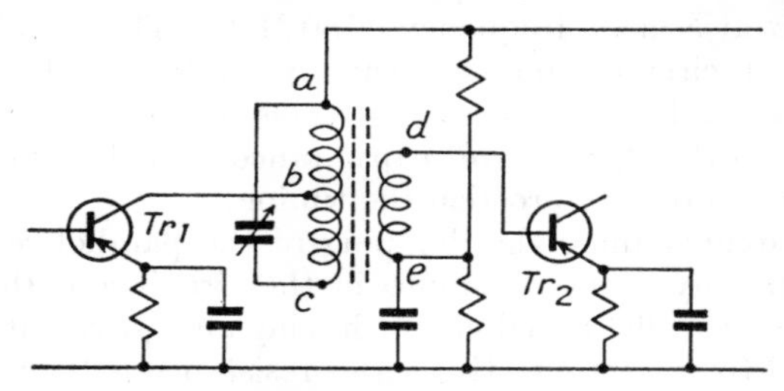

FIG. 9.10. SIMPLE TRANSISTOR R.F. STAGE

coupled to Tr_2 through a transformer of a suitable ratio to match the relatively high output impedance of Tr_1 into the low input impedance of Tr_2. Since this requires an appreciable step-down ratio it is more convenient to tune the primary winding; even so, the primary inductance calculated from purely matching considerations is usually inconveniently low, requiring a large tuning capacitance. Hence the primary inductance is increased to a more

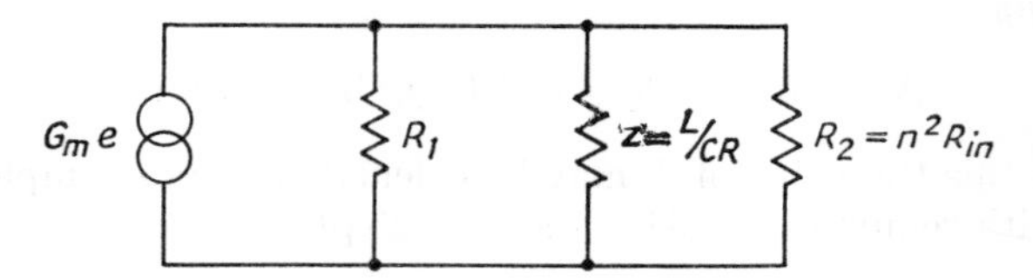

FIG. 9.11. SIMPLIFIED EQUIVALENT CIRCUIT OF FIG. 9.10

convenient value, the collector being connected to an appropriate tap to maintain the matching.

A simplified equivalent circuit is shown in Fig. 9.11. The transistor Tr_1 is represented as a current generator $G_m e$ developing power in the network $R_1 Z R_2$. G_m is the effective transconductance di_{out}/de_{in}, the significance of which is discussed later. R_1 is the output impedance of the transistor, while R_2 is the reflected impedance of the input impedance of the following stage. R_2 is thus $n^2 R_{in}$ where n is the step-down ratio of the transformer. Z is the dynamic impedance of the tuned circuit $= L/CR$. For the present these can be assumed to be all resistive, the capacitive components of R_1 and R_2 constituting part of the tuned-circuit capacitance.

The optimum conditions are obtained when the external and internal loads are equal, in which case half the total power is dissipated internally and half is available externally. In this case, however, the external load is made up of R_2 and Z in parallel so that a proportion of the available power is dissipated in the tuned circuit impedance Z, and the power actually available externally is appreciably less than half the total. If there were no coil loss ($Z = \infty$) the optimum condition would result when $R_1 = R_2$. In a practical case the tuned circuit absorbs some of the power, but the condition for maximum power transfer (i.e. maximum power in R_2) is still that $R_1 = R_2$. This may be proved as follows:

The working impedance $Z_w = Q_w X$, where $X = L\omega$.

Similarly $Z = Q_o X$. But Z_w is R_1, Z and R_2 in parallel;

$$\text{hence } Z_w = Q_w X = \frac{R_2[R_1 Q_0 X/(R_1 + Q_0 X)]}{R_2 + R_1 Q_0 X/(R_1 + Q_0 X)} \qquad (1)$$

This may be rewritten in terms of Q_0 and Q_w as

$$Z_w = R_1 R_2 (Q_0 - Q_w)/(R_1 + R_2) Q_0 \qquad (2)$$

The power in R_2 is $V^2/R_2 = (IZ_w)^2/R_2$

$$\text{whence } P = \frac{I^2}{R_2}\left[\frac{R_1 R_2 (Q_0 - Q_w)}{(R_1 + R_2) Q_0}\right]^2 \qquad (3)$$

This is a maximum when $dP/dR_2 = 0$, which occurs when $R_2 = R_1$.

This is an important result, for it would seem at first sight that the optimum result would be obtained by matching the effective load (R_2 and Z in parallel) to the resistance R_1; but in fact the requirement is for maximum power to be transferred to Tr_2 (i.e. maximum power in R_2, which is obtained when $R_2 = R_1$).

Insertion Loss

From these expressions we can deduce the loss introduced by the presence of the load Z due to the tuned circuit. If Z is infinite, and $R_1 = R_2 = R$, the voltage across the network $= \frac{1}{2}IR$. The working voltage is IZ_w, which (from eqn. (2)) is $\frac{1}{2}IR(Q_0 - Q_w)/Q_0$.

The power is proportional to V^2, so that

$$\frac{\text{Actual power output}}{\text{Output with perfect coil}} = \left(\frac{Q_0 - Q_w}{Q_0}\right)^2$$

In logarithmic form this may be written

$$20 \log_{10} (Q_0 - Q_w)/Q_0 = 20 \log_{10} (1 - Q_w/Q_0) \text{ dB} \tag{4}$$

This is sometimes called the *insertion loss*, though strictly speaking the *loss* is the inverse of this, namely

$$20 \log_{10} (Q_0/(Q_0 - Q_w) \text{ dB} \tag{5}$$

which is actually a more convenient form. Thus if $Q_w = \frac{1}{2}Q_0$, expression (4) becomes

$$20 \log_{10} 0{\cdot}5 = 20(\bar{1}{\cdot}6990) = 20(-0{\cdot}3010) = -6 \text{ dB}$$

the minus sign denoting a loss. The inverted expression (5) gives the loss directly as $20 \log_{10} 2 = 6$ dB.

Effective Gain

The current generated in the transistor divides between the internal and external loads R_1 and R_L as shown in Fig. 9.12. The proportion in R_L is $i_L = iR_1/(R_1 + R_L) = G_m e R_1/(R_1 + R_L)$ and the external power is thus $i_L{}^2 R_L = G_m{}^2 e^2 R_L \{R_1/(R_1 + R_L)\}^2$.

The load R_L, however, is composed of R_2 in parallel with the tuned circuit impedance Z, so that the current will divide further,

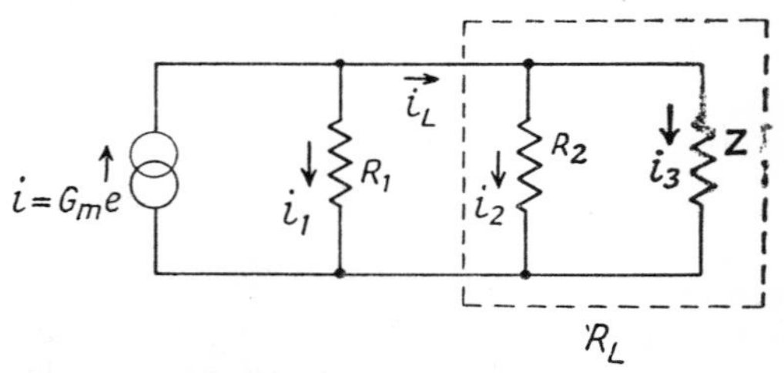

FIG. 9.12. ILLUSTRATING DIVISION OF CURRENT

and the proportion of the total external power which is actually developed in R_2 is $Z/(Z + R_2)$. Hence the power in R_2 is

$$P_2 = G_m{}^2 e^2 R_L \left(\frac{R_1}{R_1 + R_L}\right)^2 \left(\frac{Z}{Z + R_2}\right)$$

The input power is e^2/R_{in}, so that the power gain

$$A_p = G_m{}^2 R_L R_{in} \left(\frac{R_1}{R_1 + R_L}\right)^2 \left(\frac{Z}{Z + R_2}\right)$$

Knowing the circuit values, the gain can be evaluated from these expressions. In practice, however, it is simpler to work out the maximum gain and then calculate the insertion loss from eqn. (5).

The maximum gain is obtained when Z is infinite. R_L then $= R_2$, and $A_p(\max) = \frac{1}{4}G_m R_1 R_2$. The actual gain is then

$$A_p = \tfrac{1}{4}G_m R_1 R_2[(Q_0 - Q_w)/Q_0]^2 \qquad (6)$$

The voltage gain is the square root of the power gain. If the gains are expressed in dB, $A_v = \frac{1}{2}A_p$.

Design Procedure

The first step in the design is to determine Q_w. This is determined by the bandwidth requirements. It was shown on page 110 that the bandwidth of a single tuned circuit $B = f/Q$, where f is the resonant frequency. Hence $Q_w = f/B$.

The value of Q_0 is arbitrary. There is no point in making it too large since this will only increase the insertion loss, and would involve needless expense. It is customary to make Z equal to the combined impedance of R_1 and R_2 in parallel, $= \frac{1}{2}R_1$ if $R_1 = R_2$. This makes $Q_w = \frac{1}{2}Q_0$, giving an insertion loss of 6 dB.

The working impedance $Z_w = L\omega Q_w$, whence $L = Z_w/\omega Q_w$. This determines the value of inductance required. If $R_1 = R_2$, and $Z = \frac{1}{2}R_1$ as just stated, $Z_w = \frac{1}{4}R_1$. The inductance is then

$$L = R_1/4\omega Q_w$$

The capacitance required is then $1/L\omega^2$. This is usually inconveniently large, but can be translated into practical terms by an auto-transformer arrangement as described below.

It will be useful to work out the values for a practical example. A typical r.f. transistor would have $G_m = 35$ mA/V, h_{ie} $(= R_2) =$ 800 Ω and h_{oe} $(= R_1) = 29$ kΩ.

The step-down ratio to make $R_2 = R_1$ will then be

$$n = \sqrt{(29{,}000/800)} = 6$$

The maximum gain is

$$(A_p)_{max} = \tfrac{1}{4} \times 35^2 \,.\, 10^{-6} \times 29 \,.\, 10^3 \times 0{\cdot}8 \,.\, 10^3 = 7100 = 38{\cdot}5 \text{ dB}$$

The actual gain will be 6 dB down on this, namely 32·5 dB = 1780.

The inductance is $R_1/4\omega Q_w$. Assuming a bandwidth of 9 kHz and a resonant frequency of 500 kHz, $Q_w = 56$, so that

$$L = 29.10^3/(4 \,.\, 2\pi \,.\, 500 \,.\, 10^3 \,.\, 56) = 41{\cdot}5\ \mu\text{H}$$

The capacitance $C = 1/L\omega^2$ then becomes 2450 pF, which is too high, a figure of 250 pF being more practicable. The output capacitance of Tr_1 would be about 40 pF while the reflected capacitance of

Tr_2 input would be about 25 pF (as is discussed shortly). These two strays will thus contribute 65 pF to the total, leaving 2385 pF to be supplied as the reflected capacitance of 250 pF across the full primary. Hence the tap ratio is $\sqrt{(2385/250)} = 3{\cdot}1$, and the total inductance is $41{\cdot}5 \times 3{\cdot}1^2 = 400\ \mu$H.

As said earlier, these windings will be housed in a Ferrite core assembly. Such cores are very consistent in production and the makers specify the number of turns per microhenry. A typical core would provide an inductance of $N^2/50\ \mu$H, where N is the number of turns, so that for 400 μH the number of turns required is 140. The wire used must be such that the resistance of the coil (at 500 kHz) is 11 ohms to provide $Q_0 = 112$. The coil would be tapped at $140/3{\cdot}1 = 45$ turns, while the secondary winding would be $45/6 = 7\frac{1}{2}$ turns.

The wire used must be such that the effective resistance (including core losses) $= L/CZ$. For the example quoted this would be $41{\cdot}5 \,.\, 10^{-6}/(2450 \,.\, 10^{-12} \times 14{\cdot}5 \,.\, 10^3) = 1{\cdot}17$ ohms, for which it would be necessary to use stranded wire.

Effective Transconductance

The gain calculations involve the transconductance G_m. As explained in Chapter 6 (page 292), this is somewhat less than the intrinsic transconductance g_m by reason of the finite resistance $r_{bb'}$ in the connection to the base.

In practice the difference is not great, G_m being of the order of $0{\cdot}95g_m$.

Internal Feedback

The performance of the simple circuit of Fig. 9.11 is modified in practice because of internal feedback within the transistor. This arises from the inevitable small capacitance between collector and base, as shown in Fig. 9.13 (*a*), as a result of which a small fraction

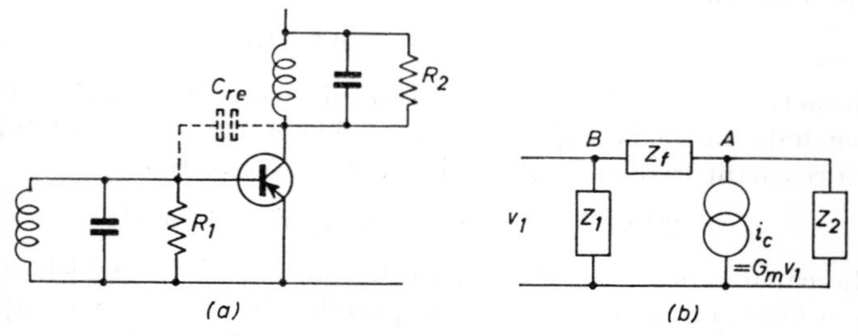

FIG. 9.13. ILLUSTRATING INTERNAL FEEDBACK

of the output voltage is transferred back to the input. This produces a slight modification of the values of R_1 and R_2, but the more important effect is that, if the magnitude and phase of the feedback are such as to increase the overall gain, the circuit may oscillate continuously.

The equivalent circuit is shown in Fig. 9.13 (*b*), where Z_1 and Z_2 represent the input and output impedance and Z_f is the impedance of the feedback path. This is actually a small capacitance shunted by a high resistance, but for practical purposes the parallel resistance may be ignored so that $Z_f = 1/\omega C_{re}$. If Z_1 and Z_2 are resistive, the voltage at A is in phase with i_c, and since $1/\omega C_{re}$ is much greater than Z_1, the feedback current i_f, and hence the voltage at B, will lead by 90° and will not affect the signal.

If the loads are tuned circuits the impedances will only be resistive at the resonant frequency. Below resonance the voltage at A will lead on i_c so that i_f leads by more than 90°; the voltage at B will also lead on i_f so that the combined effect produces a component at B in phase with the input, and if this is greater than v_i oscillation will result.

Above resonance the feedback current will lead by less than 90°, thus having a component in opposition to the input, resulting in negative feedback. In either case the maximum feedback occurs at the frequencies for which the reactance and (parallel) resistance are equal, producing a phase angle of 45° (90° total).

It may be that even at this frequency the loop gain is less than unity so that oscillation does not occur. The feedback can, nevertheless, produce appreciable distortion of the tuning, since below resonance the gain will be increased while above resonance it will be decreased, producing a lop-sided resonance curve with a peak at a frequency slightly below the true resonant point.

Stability Factor

In Fig. 9.13 (*b*) the voltage at A is $G_m v_i Z'$, where Z' is the impedance Z_2 shunted by the feedback path $Z_f + Z_1$. However, Z_f is much higher than Z_2 so that Z' may be taken as equal to Z_2. The proportion of this voltage developed at B is $Z_1/(Z_f + Z_1)$, but since Z_f is much greater than Z_1 this can be simplified to Z_1/Z_f. The voltage at B is thus $G_m v_i Z_2 Z_1/Z_f$. For stability this must not exceed v_i. Hence, writing $1/\omega C_{re}$ for Z_f the criterion for stability is

$$\omega C_{re} G_m Z_1 Z_2 < 1$$

If the input and output circuits are tuned, maximum feedback will occur at the frequency for which the phase angle (of each circuit) is 45°. This occurs when $Z_1 = R_1/\sqrt{2}$ and $Z_2 = R_2/\sqrt{2}$. Hence the stability criterion can be written

$$\omega C_{re} G_m R_1 R_2 < 2$$

This is the minimum requirement. In practice allowance must be made for tolerances in the parameters—known as the production spread—so that the criterion is taken as $2/K$, where K is a factor of safety called the stability factor, which should be at least 2 and is usually 4 or 5.

Thus with a stability factor of 4, $\omega C_{re} G_m R_1 R_2$ must be less than 0·5. With early transistors C_{re} was of the order of 10 pF, but with a modern r.f. transistor the capacitance is only some 1 to 1·5 pF, and special v.h.f. types have C_{re} as low as 0·15 pF. Using the values in the example worked previously, with $C_{re} = 1{\cdot}5$ pF the criterion works out at 0·385, which would be acceptable with a stability factor of 4; but any appreciable increase in frequency (e.g. to 1 or 1·5 MHz) would introduce the risk of instability.

Neutralizing

A circuit which is inherently stable is said to be stability limited, but such circuits in general do not make the best use of the transistor. To enable the full gain to be developed it is necessary to counteract

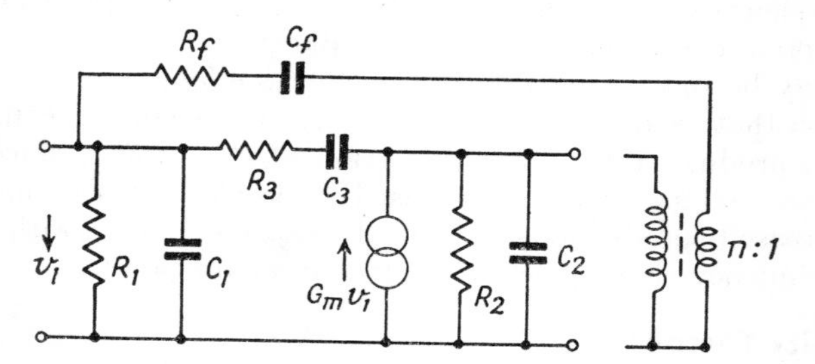

FIG. 9.14. MODIFIED FORM OF FIG. 9.13 WITH ADDITION OF NEUTRALIZING CIRCUIT

the effect of the internal feedback by some suitable external circuit which provides a compensating feedback in anti-phase in a manner similar to that adopted with valve techniques.

For complete compensation both capacitive and resistive components of the feedback should be neutralized. Z_f is actually C_{re} shunted by a high resistance of the order of megohms. This can be replaced by ordinary transformations to an equivalent series form as has been done in Fig. 9.14. C_3 is then slightly less than C_{re}, while R_3 is only of the order of 10 kΩ. The internal feedback can then be neutralized by an external circuit taken from the secondary of the coupling transformer (which must be connected to provide the required anti-phase). Since the transformer will normally be arranged with a step-down ratio the neutralizing circuit impedance must be correspondingly reduced to equate the internal and external feedback, so that $C_f = nC_3$ and $R_f = R_3/n$.

Such a circuit is said to be *unilateralized.* It has the advantage that there is no transfer of energy between output and input *in either direction,* which means that the input and output impedances

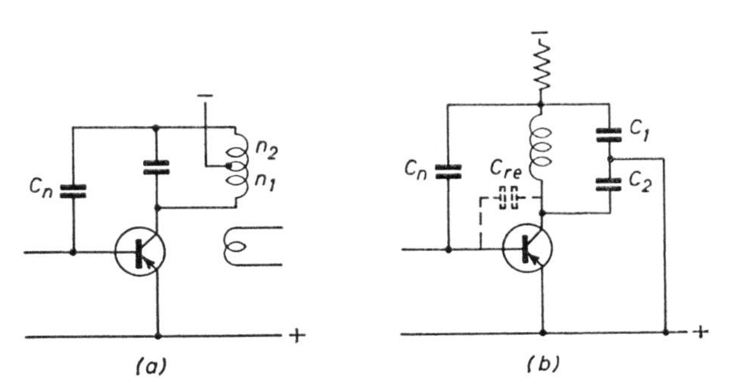

FIG. 9.15. TWO FORMS OF NEUTRALIZING CIRCUIT

are not affected by the internal feedback. However, improvements in technique have rendered this complete compensation of secondary importance, and for the majority of circuits the resistance R_f is omitted, adequate neutralization being effected merely by a correct choice of C_f.

Various alternative arrangements are possible. Thus, by rearranging Fig. 9.10 to obtain the correct anti-phase, the neutralizing capacitor may be fed from the primary, as shown in Fig. 9.15 (*a*). C_n must then be $C_{re}n_1/n_2$. Still another arrangement is the bridge circuit of Fig. 9.15 (*b*), in which $C_n = C_{re}C_1/C_2$.

Noise Level

Transistor amplifiers suffer from the same limitations as valve amplifiers in respect of background "noise." This was discussed on page 418 and was shown to arise partly from small irregularities in the electron flow (shot noise) and partly from thermal agitation of the electrons in the circuit impedances. The impedances involved in transistor circuits are much lower than with valves but the currents are correspondingly greater, so that the input noise is much the same. The low input impedance of a transistor, however, introduces appreciable thermal noise within the transistor itself (mainly in $r_{bb'}$) in addition to the shot noise which is inevitably present.

Transistor noise appears to have two principal constituents, namely intrinsic noise and excess noise. Intrinsic noise arises from the normal transistor action and depends mainly upon the emitter current. Up to about 1 mA the noise is roughly constant but then begins to increase proportionally. This is mainly due to a partition effect (similar to that in tetrode or pentode valves) arising from the

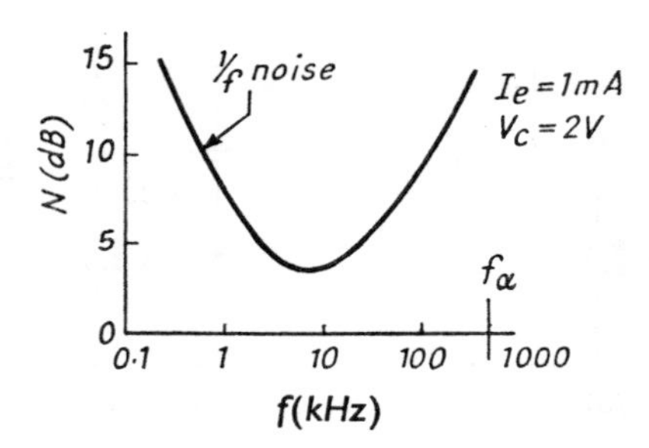

Fig. 9.16. Variation of Noise Factor with Frequency

division of the current between base and collector. This second factor also increases with the collector voltage so that to keep the noise level low I_e and V_c should be kept small. It is also desirable to keep r_b and α' small, but these are factors in the design of the transistor.

The noise is directly proportional to (absolute) temperature, and to the bandwidth of the circuit. It is also dependent on the source impedance, falling to a broad minimum when $R_s = R_{in}$ and thereafter rising again.

The "excess noise" component is an additional factor which appears to be connected with leakage current. It is most important at low frequencies and in fact decreases in a linear manner as the frequency increases; it is sometimes called the $1/f$ component for this reason.

As the frequency continues to increase, α becomes complex and there is an increase in the partition effect so that the intrinsic noise begins to increase rapidly. Fig. 9.16 shows the type of variation encountered.

Noise Factor

Since the total noise is produced partly by the (external) input impedance and partly by the transistor it is desirable to separate the two by specifying the *noise factor* of the circuit. This is defined as the ratio of the total noise power fed into the load of the transistor to that portion which is due to the thermal noise of the source impedance. Hence

$$N = 10 \log_{10}\left[\int_{f_1}^{f_2} v_o^2\, df \Big/ \int_{f_1}^{f_2} A^2 v_n^2\, df\right] \text{ dB}$$

where v_o is the r.m.s. noise output voltage, v_n is the r.m.s. noise due to the source and A is the voltage gain. If A, and/or v_o and v_n, are frequency dependent they must be expressed in terms of f.

If the factors are not greatly dependent on frequency then (using the expression for v_n^2 on page 420) the noise factor is simply

$$N = 10 \log_{10} v_0^2/4kTRA^2$$

where k = Boltzmann's constant
T = absolute temperature
R = source impedance
v_0 = r.m.s. noise output voltage
A = voltage gain.

This may be expressed in another way by defining the noise factor as

$$N = S_{in}/S_{out}$$

where S_{in} = signal/noise ratio of the source, and S_{out} = signal/noise ratio of the output. If S is expressed in dB, then $N = S_{in} - S_{out}$.

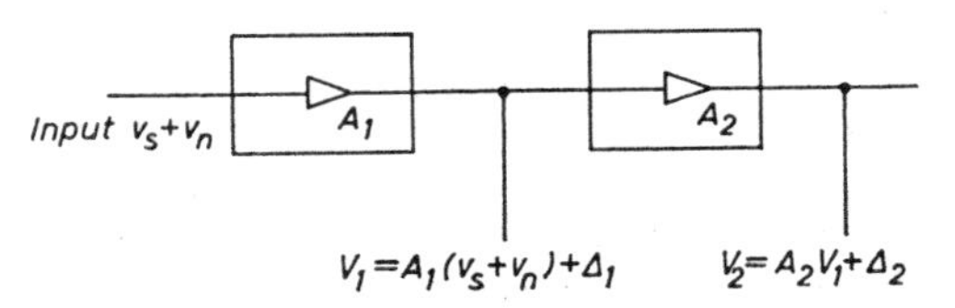

Fig. 9.17. Illustrating Noise Level in Cascade Amplifier

Thus if S_{in} is, say, 50 dB, the additional noise generated in the stage will increase the relative noise level, so that S_{out} is less than S_{in}, becoming perhaps 46 dB. The noise factor for the stage is then 4 dB.

It is the first stage of an amplifier which is of major importance. Consider the two-stage amplifier illustrated diagrammatically in Fig. 9.17. Let the input signal be v_s and the input noise be v_n, so that $S_o = v_s/v_n$. Assuming that the signal and noise voltages are amplified equally (which is not strictly correct) the respective outputs from the first stage will be A_1v_s and $A_1v_n + \Delta_1$, where Δ_1 is the additional noise generated internally in the stage.

The signal/noise ratio at this point will then be

$$S_1 = A_1v_s/(A_1v_n + \Delta_1) = v_s/(v_n + \Delta_1/A_1)$$

which is clearly less than S_o.

The signal output from the second stage will be $A_1A_2v_s$, while the noise output will be

$$V_{2n} = A_2(A_1v_n + \Delta_1) + \Delta_2$$
$$= A_1A_2(v_n + \Delta_1/A_1 + \Delta_2/A_1A_2)$$

whence $$S_2 = v_s/(v_n + \Delta_1/A_1 + \Delta_2/A_1A_2)$$

This is very little different from S_1 because Δ_2 is divided by A_1 *and* A_2. In other words the internal noise of stages subsequent to the first contributes very little to the overall signal/noise ratio.

Hence it is the first stage which is of significance, and here it will be observed that the operative factor is Δ/A, so that to maintain a high signal/noise ratio the gain of the first stage should be high, while Δ should be minimized by using a small value of I_e (less than 1 mA), with a low value of source impedance ($= R_{in}$ for optimum performance).

9.4. The Superheterodyne Receiver

Amplification of the incoming r.f. signals as such has certain limitations, particularly at the higher frequencies, and it is more usual to arrange for the amplification to be effected at an intermediate frequency by using the superheterodyne principle. The procedure is to convert the incoming signal to a different (lower) frequency by mixing it with a suitable local oscillation. A modulated wave is then produced having a frequency equal to the difference between the incoming oscillation and the local oscillation. This difference frequency is then accepted by the use of tuned circuits, and is suitably amplified until the voltage is sufficient for the particular detector in use.

The advantage of this procedure is that the tuned circuits employed are all fixed-tuned (except the aerial and oscillator). Hence, high-efficiency circuits can be used giving high gain and improved selectivity without the troubles attendant on the use of ganged r.f. stages.

In particular, the amplification in a t.r.f. stage varies with the setting of the tuning capacitor since the dynamic impedance L/CR

becomes less as C is increased. This difficulty is obviated in a superheterodyne receiver, which gives uniform gain *and selectivity*, an important advantage.

The essential components of a superheterodyne receiver (commonly known as a superhet) are shown in Fig. 9.18. The incoming signal is first passed through an r.f. stage, which is usually a simple aerial tuning circuit but may include an r.f. amplifier. The signal is then passed to a mixer stage in which the carrier frequency is translated into a chosen (lower) intermediate frequency. It is then passed through an i.f. amplifier tuned to this new frequency, followed

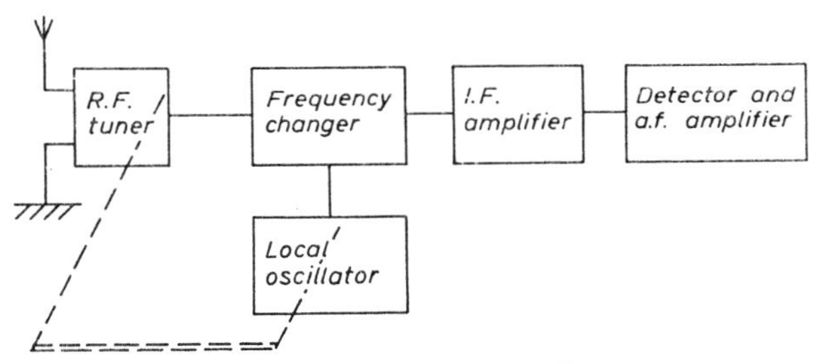

FIG. 9.18. SUPERHETERODYNE RECEIVER

by a detector and audio-frequency output stage. The local oscillator in the mixer is ganged with the r.f. tuner in such a manner as to provide a constant frequency difference between the incoming signal and the local oscillation, thus providing the required fixed intermediate frequency.

It is evident that the mixer, or frequency changer, is a vital part of the system. Before discussing this in detail, however, it is convenient to review briefly the requirements of the i.f. stages since to some extent the technique is the same for both circuits.

For ordinary broadcast reception an i.f. of 465 or 470 kHz is customarily adopted and the details which follow will be mainly concerned with frequencies of this order. For frequency-modulated transmissions in the v.h.f. band an i.f. of 10·7 MHz is usual. The basic principle remains the same but the circuitry has to be adapted to the higher frequency, as is discussed in Section 9.6. Still higher frequencies are involved in television receivers, which require a video bandwidth of 6 MHz, to accommodate which an i.f. of the order of 40 MHz is necessary as is discussed in Chapter 16.

Valve I.F. Amplifiers

A valve-operated i.f. stage follows normal r.f. practice except that because of the fixed frequency it is practicable to use band-pass tuning, so improving the selectivity. The coupling transformers thus have both windings tuned, with critical coupling between them. The coils are usually housed in the same screening can, tuned by small pre-set capacitors, or by adjustable ferrite slugs, as illustrated in Fig. 9.19.

The stage gain will be $gZ_1/(r_a + Z_1)$, where g is the mutual conductance of the valve, and r_a the internal resistance. With a normal r.f. pentode r_a is large compared with Z_1 so that the gain is simply gZ_1. However, Z_1 is the *effective* impedance of the primary,

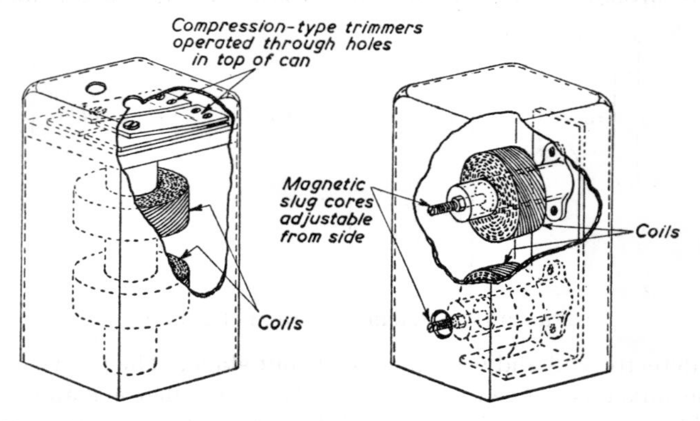

FIG. 9.19. TYPICAL VALVE I.F. TRANSFORMER CONSTRUCTIONS

allowing for the transferred secondary impedance, which with critical coupling is half that of either circuit alone.

The design procedure is thus first to determine the overall Q_w necessary to meet the bandwidth requirements, as discussed earlier. The individual circuits must then be designed to have an unloaded Q_0 equal to twice this value, and since $Q = L\omega/R$ this enables a value for L to be determined such that with the particular construction proposed the resistance (at the i.f.) is of the right order. With critical coupling, assuming identical circuits, $M\omega = R$, so that the coil separation can then be arranged to provide the required mutual inductance $M = R/\omega$, and finally the capacitance can be evaluated from $C = 1/L\omega^2$.

Z_0 is then L/CR $(= Q_0L\omega)$, and $Z_w = \frac{1}{2}Z_0$, from which the gain can be calculated.

For the final stage, which feeds the detector, a slight modification is required, because the detector load is not negligible, as is explained in Section 9.5. This means that the Q of the secondary is no longer equal to that of the primary, in which case $k_{crit} = 1/\sqrt{(Q_1Q_2)}$, which is slightly higher than before. Both circuits would be designed to have the same Q_0 as before, but a modified value of Q_2 would be evaluated to allow for the load across it, and this would be used in calculating k_{crit}. With this modification the effective primary impedance will still be $\frac{1}{2}Z_0$.

Transistor I.F. Amplifiers

Similar procedures can be adopted with transistor i.f. amplifiers, but in the majority of popular receivers only a single-tuned circuit is used, with an additional stage or stages if the selectivity is insufficient. The design of a single-tuned i.f. stage is thus similar to that of the r.f. stage previously discussed.

As with a valve i.f., the final stage feeding the detector requires modified treatment, but with a single-tuned circuit this simply involves different matching, as is discussed in detail in Section 9.5 (page 454).

For more sophisticated receivers double-tuned stages are employed, but since the coils are normally housed in closed ferrite cores the necessary coupling has to be provided externally, by the use of a suitable tapping or coupling coil, or if this is inconvenient, by an equivalent capacitance coupling as shown in Fig. 2.29 (page 112). An example of a double-tuned i.f. stage is given in the section on f.m. receivers (page 470).

Stagger Tuning

While a single circuit may comply with the bandwidth requirements, this may not provide adequate selectivity as was discussed on page 417. It is customary therefore to use several i.f. stages, when if the circuits are properly cascaded the off-tune response is $(s_x)^n$ times that of a single circuit, where s_x is the relative response of a single circuit x kHz off tune, and n is the number of stages.

If this is not adequate, the successive tuned circuits may be adjusted to resonate at frequencies slightly displaced from the mid-frequency f_0. This is called *stagger tuning* and has the effect of providing an approximation to a band-pass circuit. It can be shown that, with two circuits each mistuned by $\delta/2$ on either side of f_0, the response curve is the same as for a coupled circuit with a coupling factor $k = \delta/f_0$, as illustrated in Fig. 9.20.

It is customary to apply stagger tuning to the first two stages, the third being tuned to f_0 (but with a modified Q_w to allow for the detector load as explained above).

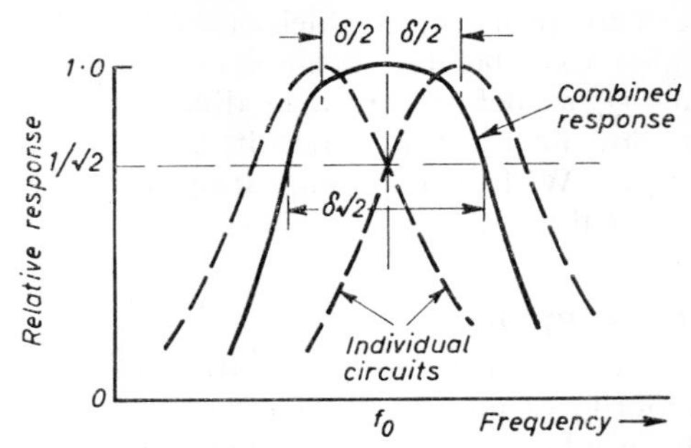

FIG. 9.20. ILLUSTRATING STAGGER TUNING

Frequency Changers

The input to the i.f. amplifier is derived from the frequency changer, which is the device through which the incoming r.f. carrier frequency is converted to the chosen intermediate frequency. This involves the process of combining, or mixing, the r.f. signal with a local oscillation. The circuitry is necessarily somewhat different according to whether valves or transistors are used, and we shall therefore first discuss the basic requirements and then show how these are achieved with the two types of circuit. In either case a local oscillator is required, for which any convenient circuit may be used, but it is important to ensure that the waveform is as pure as possible.

Harmonics are to be avoided because they will heterodyne with stations other than the wanted one and introduce audible whistles, particularly if they happen to coincide with the frequency of a powerful local station or a small multiple thereof. It is desirable, therefore, to use a good oscillator circuit having a high Q.

Theory of Mixing

There are two methods by which two oscillations can be mixed. The first is to add them together by the heterodyning process described earlier. The second is to modulate one with the other, which produces a similar result but has certain advantages.

Consider first the additive process. The sum of two sine waves $\sin \omega_o t$ and $\sin \omega_s t$ is

$$2 \sin \tfrac{1}{2}(\omega_o + \omega_s)t \cos \tfrac{1}{2}(\omega_o - \omega_s)t$$

which is a sine wave of the mean frequency modulated at what appears to be *half* the difference frequency. This simple expression,

Conversion Conductance

In frequency changers a term analogous to the customary mutual conductance is usually quoted. This is the change in anode current *at beat frequency* divided by the change in grid voltage at signal frequency, and it is known as the *conversion conductance*.

The gain of a frequency-changer stage may be determined in the same way as an ordinary r.f. pentode, being equal to cZ, where c is the conversion conductance and Z is the anode impedance at the intermediate frequency. This, of course, assumes that the anode impedance is small compared with the internal resistance of the valve, which is justifiable since the valve is made with characteristics similar to those of an ordinary r.f. pentode.

It will be noted that in the expression for electronic mixing on page 440 the output contains a factor of $\frac{1}{2}$. This indicates that the conversion conductance is only half the equivalent mutual conductance, and in practice, if a frequency changer is supplied with i.f. on the signal grid, the gain will be twice as great as that which is obtained when the valve is operating as a frequency changer.

Transistor Frequency Changers

The simplest form of transistor frequency changer uses a self-oscillating mixer circuit such as is shown in Fig. 9.23. The signal is

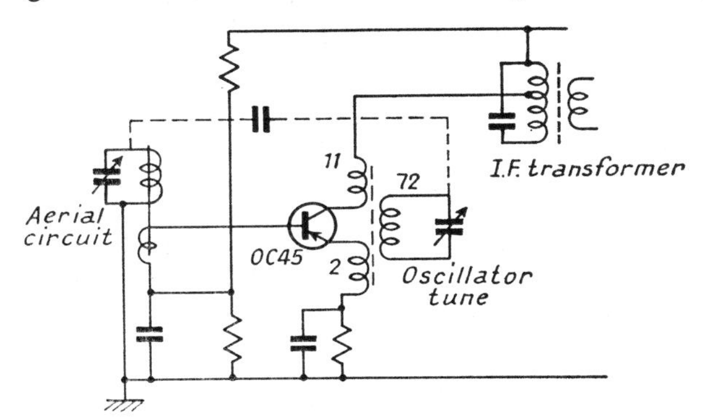

FIG. 9.23. SELF-OSCILLATING TRANSISTOR MIXER

applied to the base, while the collector is coupled back to the emitter through a secondary circuit tuned to the oscillator frequency. The emitter will thus carry currents at both signal and oscillator frequencies so that an additive mixing is obtained, which is utilized by including in the collector a transformer tuned to the difference (i.f.) frequency.

The oscillator tuned circuit is designed to cover the frequency band required (as explained on page 447) and has a high Q, which permits the oscillation to be maintained with only a loose coupling to the transistor, thereby minimizing circuit damping. The coupling windings themselves are so proportioned as to match the collector/emitter impedances.

Oscillator design is considered more fully in Section 12.1 and it will suffice here to note the values in a typical frequency changer. If the relevant impedances Z_o and Z_i are 30 kΩ and 1,000 Ω, the ratio of collector/emitter windings will be $n = \sqrt{(30/1)} = 5{\cdot}5$. The combined loading of the collector and emitter is then that of Z_o and n^2Z_i in parallel $= \frac{1}{2}Z_o$. In this case, however, it is not desired to match this to the tuned circuit. The requirement is that the transistor loading shall only have a small influence on the tuned circuit, so that a substantial step-down ratio is needed.

By the usual circuit laws it can be shown that this ratio is given by

$$n_1^2 = 2L\omega Q_oQ_w/Z_o(Q_o - Q_w)$$

If the i.f. is 470 kHz, and the lowest signal frequency is 550 kHz, f_{osc} will be 1,020 kHz. If the tuning capacitance is 140 pF, L will be 175 μH. Taking $Q_w = 100$ and $Q_o = 120$, $n_1 = 6{\cdot}6$. The actual turns depend upon the core used. With a typical core an inductance of 175 μH would be obtained with 72 turns, in which case the primaries will be 11 and 2 turns respectively, as shown.

Input Transformer

The input impedance to the signal (into which the aerial transformer must be matched) can be estimated from the expression for Z_i (common-emitter mode) with $R_L = 0$, because the collector load will be tuned to the i.f. and will have a greatly reduced impedance (approaching zero) at signal frequency. With this reservation the design of the input transformer is in accordance with the procedure already discussed in Section 8.2.

Output Transformer

The output is fed to the i.f. stages through a coupling transformer matched to the transistor as in a normal r.f. stage. In a frequency changer, however, it is not possible to offset the internal feedback by unilateralization because of the range of frequency involved, so that the effective output impedance of the transistor varies widely. Hence a compromise has to be adopted, and it is customary to make the coupling transformer the same as in the succeeding i.f. stages. In practice this provides a relatively low effective load in the collector circuit which masks the variations in Z_o.

Effect of Stray Feedback

The oscillatory voltage at the collector in Fig. 9.23 will produce a current through the internal feedback path in the transistor which will tend to reduce the amplitude of the oscillation. This internal feedback is usually small and may be neglected, but there is a further and more serious feedback path through the stray capacitance between the oscillator and aerial circuits, as shown dotted in Fig. 9.23. The voltage on the tuned circuit is some 6 times that on the collector and the feedback *current* reaching the base is still further increased by the amount of the step-down ratio of the aerial transformer. Since this is of the order of 10:1 the total feedback is some 60 or more times as serious as that in the transistor itself. The internal capacitance is of the order of 2 pF so that a stray capacitance as low as 1 pF can produce 30 times as much feedback as that due to the transistor alone.

The effect of this feedback depends upon the phase of the oscillator windings and may thus cause either an increase or decrease in the

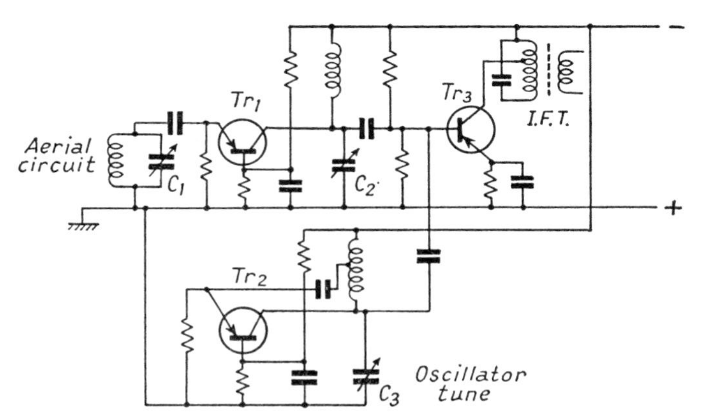

FIG. 9.24. FREQUENCY CHANGER CIRCUIT WITH SEPARATE OSCILLATOR

amplitude of the oscillation. If the increase is large the waveform is distorted and *squegging* may result.* The stray feedback increases with frequency and will ultimately cause the oscillator to pull into step with the signal so that no i.f. is produced.

* Squegging is an unstable condition in which a rapid increase in amplitude shifts the operating conditions so that the transistor is momentarily paralysed. The circuit thus produces bursts of oscillation at a repetition rate (usually audible) depending on the time constants. (The action is discussed further in Chapter 12.)

This "pulling," which is discussed in Chapter 12, can only be avoided by reducing the stray capacitance. Since the aerial and oscillator tuning capacitors are usually ganged there is appreciable capacitance between them which must be minimized by providing an earthed screen between the sections. Even so, the pulling may still be present at high frequencies. In such circumstances the oscillator and mixer must be kept separate. Such a circuit, suitable for frequencies of the order of 100 MHz, is shown in Fig. 9.24. The signal is amplified in a buffer stage before being fed to the frequency changer, which is also fed through a small capacitance from an entirely separate oscillator. It will be noted that in this circuit the r.f. and oscillator stages use the common-base mode, which is essential to obtain a sufficiently high cut-off frequency. C_1, C_2 and C_3 would be ganged.

Conversion Conductance

The gain of an r.f. stage is approximately $g_m Z$, where Z is the load and g_m is the effective mutual conductance $= di_c/dv_1$ (*see* Fig. 9.25). The gain of a mixer stage (at the intermediate frequency)

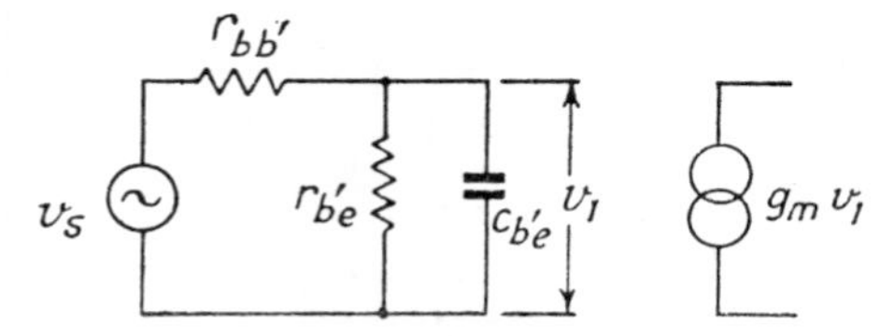

FIG. 9.25. SIMPLIFIED FREQUENCY CHANGER EQUIVALENT CIRCUIT

may be similarly assessed if the value of g_m is $(di_c)_{i.f.}/(dv_1)_{r.f.}$. This modified value of g_m is called the *conversion conductance*, g_c.

In a transistor the value of g_c is not a definite proportion of g_m (as in a valve) but depends on the frequency. Thus in the simplified equivalent circuit of Fig. 9.25 the collector current is $g_m v_1$ and $v_1 = v_s Z/(r_{bb'} + Z)$, where Z is the impedance of $r_{b'e}$ and $C_{b'e}$ in parallel $= r_{b'e}/(1 + j\omega C_{b'e} r_{b'e})$. If the collector current is to be expressed as $G_c v_s (= g_m v_i)$,

$$G_c = g_m Z/(r_{bb'} + Z)$$
$$= g_m/[r_{bb'}/r_{b'e} + 1 + j\omega C_{b'e} r_{bb'}]$$

At low frequencies the reactive term in the denominator is negligible, so that the expression becomes $g_m/(1 + r_{bb'}/r_{b'e})$, which is the same as the expression for the normal transconductance G_m

(*see* page 292). As the frequency increases, however, this term becomes appreciable and the expression is better written in the form

$$G_c \simeq g_m/(1 + j\omega C_{b'e} r_{bb'})$$

This is frequency-dependent for two reasons. Firstly the value of $g_m \simeq \alpha/r_{b'e}$ falls off with frequency, being 3 dB down at the cut-off frequency f_α. In addition, there is a further falling off due to the increasing effect of the input capacitance $C_{b'e}$.

It must be remembered that ω is 2π times the *oscillator* frequency, which is higher than the signal frequency, and a simple calculation with representative values shows that if $f_{osc} = 1{\cdot}5$ MHz, G_c could be 3 dB down on g_m.

Noise in Frequency Changers

A frequency changer is subject to increased shot noise, particularly in a valve. Referring again to the basic expression $v_s^2 = 2keF\Delta i_a/g^2$ quoted on page 419, it will be clear that the factor i_a/g^2 is increased because the value of g is approximately halved, while the factor F is found to be of the order 0·5 to 0·6.

Up to a point the designer still aims to keep g^2/i_a high, but this involves long cathodes and consequent increase in inter-electrode capacitances, which causes increased "pulling" between the circuits, and often a valve is deliberately designed to have more shot noise in order to permit improvement in these other respects.

In a transistor frequency changer the noise factor is somewhat increased by reason of the increased emitter current, but is not greatly different from that for normal amplifier operation.

However, if the stage is fed with an amplified signal from one or more r.f. stages, the signal-to-noise ratio of the frequency changer becomes of minor importance, the Johnson noise on the input to the r.f. stage becoming the predominating factor.

Undesired Response

Probably the greatest defect of the superheterodyne system is its liability to produce false signals. This arises from the large number of combinations which will affect the frequency changer. The most obvious form of false response is that known as *image* or *second channel* interference.

The local oscillator is adjusted to differ from the wanted signal by the intermediate frequency and is usually arranged higher than the signal frequency for convenience of ganging. It could, however, equally well be lower, and it follows, therefore, that for any setting of the oscillator frequency there will be two signal frequencies, one lower and the other higher, each of which will produce the required intermediate frequency. Only one of these is the wanted signal.

Hence, to avoid interference the pre-selecting circuit, i.e. the tuning circuit prior to the frequency changer, must be capable of discriminating between the wanted signal and another signal separated by twice the intermediate frequency, *even though this second signal may be many hundred times stronger.* With low intermediate frequencies this is not always easy. The criterion is actually the relative values of the intermediate frequency and the signal frequency, and as the signal frequency becomes higher it becomes desirable to increase the intermediate frequency correspondingly.

Broadcast receivers usually use at least two tuned circuits prior to the frequency changer, often associated with a radio-frequency amplifying stage to improve the signal-to-noise ratio. This is adequate to avoid second channel interference on medium and long waves but is not sufficient for short waves where the interference is present, and each station will be found to show two tuning points.

Commercial receivers often use two frequency changers. The first i.f. is relatively high to eliminate second-channel interference, followed by a second frequency changer which converts the signal to a lower frequency at which the main amplification can be more easily obtained.

Whistles

An even more annoying form of interference is that which gives rise to whistles at certain parts of the tuning scale. These whistles arise from interaction between the oscillator or one of its harmonics and some received oscillation which differs by the i.f. plus or minus some small amount. This oscillation will be accepted by the i.f. amplifier, and since it is also handling the carrier of the wanted station (converted to intermediate frequency), the two oscillations will beat and produce an audible whistle.

Another form of interference is obtained if two strong local stations differ in frequency by an amount equal to the intermediate frequency chosen. These will beat to produce an oscillation of the order of the i.f., which in turn will beat with the carrier of the station being received and will produce an audible beat.

Ideally, of course, no signal should reach the frequency changer at any frequency other than that required, but in practice the pre-selection is by no means perfect, and strong local stations will invariably give rise to these beats. The only remedy is to improve the quality of the tuning prior to the frequency changer so that the residual voltage which still comes through the chain when the circuit is off tune is very small. With receivers working on a fixed frequency, or within a restricted band, it is possible to choose the intermediate frequency so that the beats which would be produced are outside the audible range. A simple calculation of frequency differences, taking into account not only fundamentals but harmonics up to the fifth, will soon show what interferences are likely to occur.

With a broadcast receiver, owing to the wide range and the different conditions under which it has to be used so that the "local" stations are different at different parts of the country, it is not practicable to make an ideal choice and adequate pre-selection is the only remedy.

Another form of whistle arises from harmonics of the intermediate frequency. If reception is attempted at a frequency corresponding to a harmonic of the i.f. and if any coupling exists between the second detector and the input of the set, then whistles can arise since the second detector is bound to produce some measure of harmonics of the intermediate frequency.

In any case very careful screening and layout are essential on the stages prior to the mixer. Otherwise, signals will find their way to the signal input of the frequency changer

(*a*) by direct pick-up on the signal frequency circuit of the frequency changer (or the oscillator circuit);

(*b*) by leakage from the aerial circuit to the frequency changer via stray capacitances or conductive paths which will short-circuit the selecting action of the tuned circuits.

The choice of earthing points is important, for a relatively short length of common earth lead will form a direct coupling from beginning to end of the chain and seriously reduce the selectivity, as explained on page 417.

Ganging in Superheterodyne Receivers

The usual practice is to tune all the signal-frequency circuits, including the oscillator, together. The ganging of the band-pass and/or r.f. stages is simple and is effected by the methods described on page 374.

The oscillator circuit, however, is more troublesome owing to the somewhat different frequency range to be covered. For example, using an intermediate frequency of 450 kHz the medium-wave signal frequency range will be from, say, 550 to 1,500 kHz, while the oscillator must range from 1,000 to 1,950 kHz. Reducing the inductance of the oscillator coil will be necessary, but we are still left with the fact that the requisite range of the capacitor will be too great.

Thus to cover a range of 550 to 1,500 kHz a capacitance range of $(1{,}500/550)^2 = 7{\cdot}45$ will be required. If we make the oscillator inductance such that it will tune with the minimum capacitance to 1,950 kHz, it will give 715 kHz at the maximum, whereas we require 1,000 kHz. Clearly the capacitance range must be restricted.

This can be done by including a series or "padding" capacitor in the oscillator section of the gang capacitor which will reduce the effective maximum capacitance to a value sufficient to tune the particular oscillator inductance to 1,000 kHz. Such a capacitance will be fairly large and will thus have little effect on the minimum, but we have no guarantee that the capacitance will be correct at intermediate points on the tuning scale.

An alternative method is to add a large "trimming" capacitor in parallel with the oscillator section. This again restricts the capacitance range, and a value can be so chosen as to give correct capacitance at top and bottom of the scale (though with a different oscillator inductance).

As before, the intermediate "tracking" may be incorrect and actually both methods produce errors in opposite directions. Hence, in practice, a combination of the two is used, a third point being chosen in the middle of the scale at which it is arbitrarily assumed that the tracking is correct. We then have three "spot-on" frequencies and three variables—the oscillator inductance, the padder and the trimmer, and we can find definite values for each which will fulfil the three conditions. Elsewhere the tracking is still a little out, but the method can be made to give very satisfactory results.*

The padding and trimming capacitances must, of course, be altered for each wave range and are usually incorporated in the coil system and changed over by the switching which alters the coils. One arrangement is shown in Fig. 9.26.

* For the three points chosen, at the top, middle and bottom of the scale, we note

(*a*) The signal frequency to which the circuit is tuned,
(*b*) The capacitance of the (oscillator) capacitor, C_o.

The oscillator frequency at each of the three points must be the signal frequency plus the intermediate frequency in use, so that we can calculate the LC product for the three frequencies.

The effective capacitance at each point is given by

$$C = C_t + C_oC_p/(C_o + C_p)$$

where C_t and C_p are the trimmer and padder capacitances respectively, while C_o is known. Hence three equations can be written down in terms of L, C_t and C_p.

Alternatively, specially-shaped vanes may be used on the oscillator section of the capacitor, which is made to have a smaller maximum capacitance than the other sections, but such a capacitor

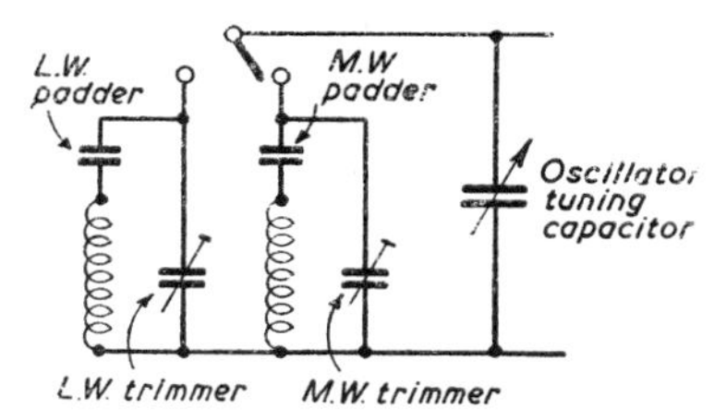

FIG. 9.26. TRACKING CIRCUIT FOR SUPERHETERODYNE OSCILLATOR

can only be used with a specified i.f. and on one waveband. For any other wave range, padders and trimmers must still be used.

Frequency Drift

A most important practical consideration is that of frequency drift. As the parts of the receiver warm up, the oscillator frequency tends to change and with bad design this drift is serious, necessitating constant re-tuning. Moreover the drift may continue almost unabated for some hours.

This is due to change in the relative permittivity of the insulation of the tuning capacitor. Bakelized linen, for example, shows a change of +0·2 per cent per degree C. Ebonite and high grade synthetic compounds are subject to changes of the order of +0·02 per cent.

Ordinary ceramic materials have a positive temperature coefficient between 0·01 and 0·02 per cent, but certain materials incorporating oxides of titanium have a negative coefficient of 0·06 to 0·07 per cent. By allocating a suitable proportion of the circuit capacitance to this negative-coefficient material the positive coefficient not only of the remaining capacitance but also of the inductance in the circuit may be offset, so that the circuit is substantially free from frequency drift.

An advantage of transistor operation is that the internally generated heat is usually negligible.

Flutter

As explained in the next section, it is customary to arrange that the gain of the r.f. amplifier is controlled by the strength of the signal, the technique being known as *automatic gain* (or *volume*) *control*.

If this control voltage causes a change in the oscillator frequency an effect known as *frequency flutter* arises. If the frequency drifts the detector voltage falls, causing alteration of the bias on the frequency changer (via the a.g.c.) which pulls the frequency back. The detector voltage rises again, removing the a.g.c. and the original state is resumed. The detector voltage is thus in a continual state of fluctuation causing a disagreeable fluttering of the speech or music.

9.5. The Detector Stage

A very important link in any radio receiver is the detector. This should normally obey a linear law—i.e. the rectified (a.f.) output should be proportional to the r.f. input. This is particularly important for telephony reception, not only on the score of freedom from distortion (so that the a.f. output may follow faithfully any variations in the r.f. input due to modulation) but also from considerations of selectivity.

As explained in Chapter 3, a linear detector produces an output corresponding to the variations in amplitude of the modulation envelope. If a second (weaker) signal is present, the modulation envelope is subject to a superposed fluctuation determined mainly by the difference between the two carrier frequencies. If this is above the audible limit the form of the modulation envelope is not seriously modified, so that the detector only responds to the modulation of the desired (stronger) signal. It is found in practice that, if the desired carrier has a mean amplitude more than twice that of the unwanted carrier, the detector ignores the interfering modulation completely, and the tuning circuits prior to the detector must therefore be such that this requirement is met.

This rejection does not occur if the carrier frequencies differ by an amount within the audible range. In this case an audible whistle appears in the detector output (together with the desired modulation). If the frequencies are very close the interaction appears as a sub-audible flutter and/or an interference between the wanted and unwanted modulations which is known as "sideband splash." Such effects cannot be eliminated by normal tuning methods. Where the expense is justified they may be combated by rejector circuits tuned to the interfering station and/or the use of directional aerials.

Diode Detectors

The simplest form of detector is the diode, which is practically linear and has a high rectification efficiency. Fig. 9.27 (*a*) shows a simple diode circuit. The r.f. voltage across the tuned circuit will produce a current through the diode which will charge capacitor C_1

to a voltage approximately equal to the peak value of the applied e.m.f. During the remainder of the cycle this charge will leak away through the resistor R_1 until the next peak, when the diode will again conduct and the charge will be restored.

Fig. 9.27 (*b*) shows an alternative form of circuit in which the diode and the capacitor are interchanged.

In either case the e.m.f. across R_1 is equal to the peak amplitude of the r.f. input, so that if the signal is modulated the e.m.f. across R_1 follows the modulation. The time-constant C_1R_1 must be such that the charge on the capacitor can vary at least as rapidly as the highest modulation frequency.

Assuming 80 per cent modulation this requires the capacitor to

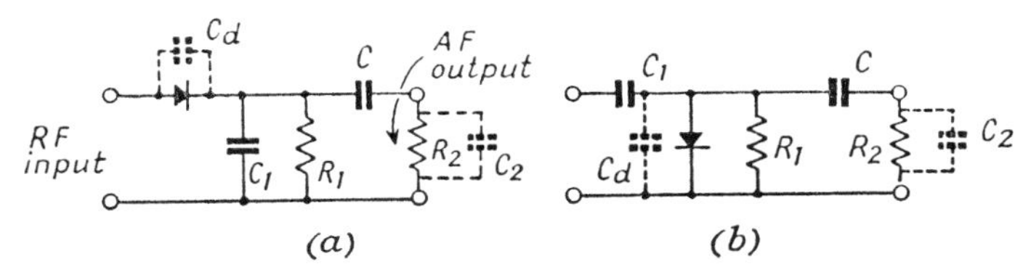

FIG. 9.27. SIMPLE DIODE DETECTOR

discharge to 20 per cent of its peak voltage in a time of one half-cycle, which requires a CR product of approximately 2 (cf. Fig. 1.7). Hence C_1R_1 should not be greater than $1/4f$. The value of R_1 must not be too low or it will impose too much load on the input, while C_1 must be large compared with C_d to ensure that the voltage drop on C_d is negligible. Typical values with a valve circuit would be 100 pF and 0·25 MΩ, which would handle up to 10 kHz. With the lower-impedance transistor circuitry, values of 0·01 μF and 5 kΩ would be suitable.

The a.f. output across R_1 will be fed to the a.f. amplifier (or sometimes direct to the output stage), and this is usually done through an isolating capacitor C to permit the necessary bias to be applied independently. In many cases R_1 is a voltage divider, as in Fig. 9.31, to provide a convenient control of the output.

It is clear, however, that these output arrangements modify the behaviour of the circuit. The e.m.f. across R_1 is a steady d.c. voltage with a superposed alternating component due to the modulation, and the circuit presents different impedances to the two components. To the steady d.c., the impedance is simply R_1. The alternating component, however, has a parallel path through C and R_2 with C_2 in parallel.

The alternating component, therefore, will produce relatively more current than the d.c. component, so that distortion is liable to be introduced. The maximum modulation which can be handled without distortion can be arrived at as follows.

Let the mean value of the capacitor voltage be e. Then the modulation component will be $me \sin pt$ where m is the depth of modulation and p is the modulation frequency. If Z is the effective impedance of the circuit to a.c., the modulation current will be $(me/Z) \sin pt$, and the peak value of this will be me/Z.

Now, this peak value cannot exceed the d.c. current. Hence we can write $me/Z = e/R_1$, whence $m = Z/R_1$.

At low frequencies the capacitances exercise a negligible effect so that Z is simply $R_1R_2/(R_1 + R_2)$. The maximum modulation depth is thus $m = R_2/(R_1 + R_2)$.

At higher frequencies Z is less than the value just quoted so that the permissible modulation falls off still more. By making R_2 three or four times R_1 we can approach full modulation, but some distortion is unavoidable. Special networks have been suggested to obviate the difficulty, but the usual practice is to make R_2 large and admit a small distortion on the peaks.

It is worth noting that up to the critical value the efficiency of rectification is very high because, although the a.c. impedance is lower than R_1, the current is greater, and in fact the a.c. modulation voltage is the same as it would be if the circuit only contained the one resistance R_1. The rectification efficiency is then $R_1/(R_1 + r)$, where r is the diode resistance, this being the ratio of the a.c. voltage across R_1 to the modulation voltage across the whole circuit.

The operating conditions of a diode may be represented by characteristics of the form of Fig. 9.28. The application of a signal to the circuits of Fig. 9.27 will charge the capacitor C_1 to a voltage nearly equal to the peak value of the applied e.m.f. This has the effect of shifting the diode characteristic to the left so that a family of curves can be drawn for different values of input signal, as shown.

With a signal of varying amplitude the mean operating point would be as at P, corresponding to a mean (negative) potential V. Excursions on either side of this will then cause the operating point to shift on to a corresponding characteristic. The point P will be determined by drawing a load line OA having a slope corresponding to the d.c. load R_1. The a.c. load, however, will be less than this as just explained, corresponding to a load line such as XPY, which cuts the zero axis before the full modulation is developed.

A possible remedy is to apply a small positive bias to the diode as shown at Q, which is sometimes done in high-fidelity equipment, but more usually the circuitry is arranged so that the a.c. load is not seriously lower than the d.c. load, the small distortion present being accepted.

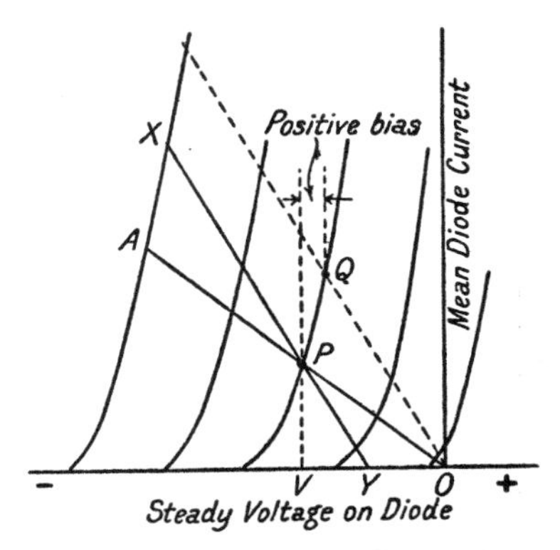

FIG. 9.28. DYNAMIC DIODE CHARACTERISTICS

Input Resistance

The input resistance of a detector has an important bearing on the design of the circuit. Since it is only a unidirectional conductor, special methods have to be adopted to estimate the input resistance. A simple method is to assume that the diode only conducts on the peaks of the signal. The power absorbed is then $\hat{E}i_{mean}$, where $\hat{E}$ is the peak voltage and i_{mean} the average diode current. Now,

$$i_{mean} = \beta\hat{E}/R_1$$

where R_1 is the diode load resistance and β is the rectification efficiency. Thus

$$\text{Power absorbed} = \beta\hat{E}^2/R_1$$

But $\hat{E}^2 = 2E^2{}_{r.m.s.}$ so that

$$\text{Power} = \frac{E^2}{R_1/2\beta}$$

Hence the effective input resistance $= R_1/2\beta$.

β is usually about 0·7 to 0·8 so that the effective input resistance is approximately $\frac{2}{3}R_1$.

Design of Input Circuit

This finite impedance must be allowed for in the design of the input circuit. The presence of a shunt resistance R_p across a tuned circuit reduces the Q in the ratio $Q_w/Q_o = R_p/(Q_oL\omega + R_p)$. In the case of a valve circuit with a diode resistance of 0·25 MΩ, R_p would be about $\frac{2}{3}R = 160$ kΩ. Assuming $Q_o = 150$, Q_w would be 49, resulting in an appreciable loss of gain and selectivity. Some improvement can be obtained by tapping the diode across part of the coil, in which case R_p is replaced by R_pt^2, where t is the step-down ratio. This appreciably improves Q_w and, for small values of t, gives a slight improvement in overall gain.

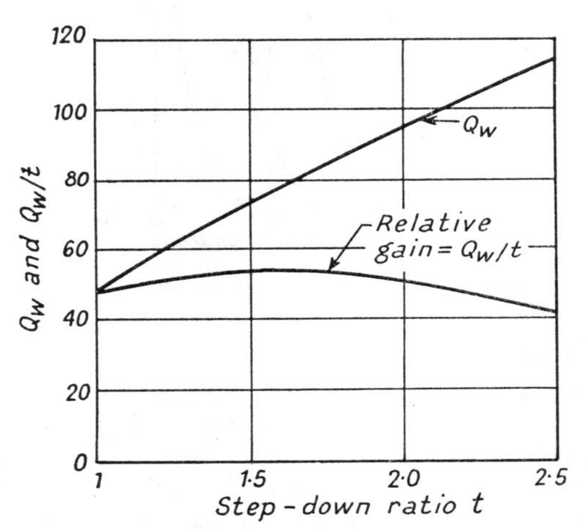

Fig. 9.29. Effect of Detector Tap

$Q_0 = 150$ $L = 750\ \mu$H $\omega = 3{\cdot}10^6$

Typical results with an i.f. secondary having a Q of 150 are shown n Fig. 9.29. The gain is assumed to be proportional to Q_w/t, and the rectification efficiency β is taken as 0·75. It will be seen that a step-down of 1·5:1 gives a 10 per cent increase in gain and 50 per cent increase in Q_w.

Transistor Detector Stage

With transistor operation there is an additional consideration, namely the necessity to develop an adequate voltage across the diode, which for reasonable efficiency requires a signal input of at least 3 V. The requirement here is for maximum power in the diode circuit, which is not the same as the criterion for maximum power gain.

Consider a transistor having an output resistance h_{oe} of 32 kΩ operating from a 9 V supply with a standing current of 2·5 mA. The maximum collector voltage swing, allowing for a small emitter bias, would be ± 8 V. If the load is matched for maximum gain the current swing will be $8/32{,}000 = 0{\cdot}25$ mA. Assuming a diode load of 5 kΩ the effective impedance will be approximately two-thirds of this, as just explained, i.e. 3 kΩ. The transformer ratio will then be $\sqrt{(32/3)} = 3{\cdot}27$, so that a primary current swing of 0·25 mA will produce a diode current swing of $0{\cdot}25 \times 3{\cdot}27 = 0{\cdot}82$ mA. With a diode efficiency of 70 per cent this will develop an output voltage of $0{\cdot}82 \times 10^{-3} \times 3{\cdot}10^{3} \times 0{\cdot}7 = 1{\cdot}72$ V (peak), which is inadequate.

The permissible current swing in the transistor, however, is at least 1 mA, to obtain which the effective load should be 8 kΩ. This would require a transformer ratio of $\sqrt{(8/3)} = 1{\cdot}63$. The output would then be $1{\cdot}63 \times 10^{-3} \times 3{\cdot}10^{3} \times 0{\cdot}7 = 3{\cdot}43$ V, which would be satisfactory. Hence the transformer ratio for a detector stage is generally only about half that for an amplifying stage.

The a.c. diode load will be somewhat less than 3 kΩ because of the shunting effect of the succeeding a.f. stage. Hence it is desirable to follow the detector with a transistor having a relatively high input resistance. Values of 5 to 10 kΩ are commonly available.

Square-law Detectors

With a linear detector of the type so far considered (Fig. 9.27) the form of the diode characteristic is irrelevant. The diode conducts when the applied e.m.f. exceeds the voltage on the capacitor C_1,

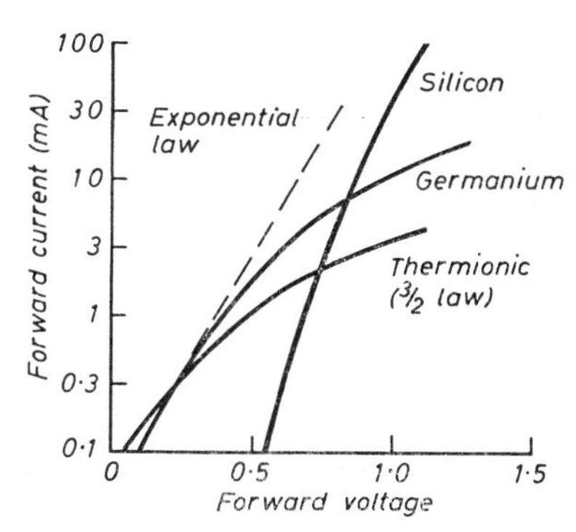

FIG. 9.30. COMPARISON OF TYPICAL DIODE CHARACTERISTICS

so that, provided it can supply sufficient current, the capacitor voltage is maintained at the instantaneous peak value of the carrier, and so follows the modulation. Since a typical receiving diode will pass a current in excess of 50 mA there is no difficulty in meeting this requirement, the only reservation being that with the silicon diodes commonly employed there is a hop-off voltage of the order of 0·6 V, as shown in Fig. 9.30, so that the diode will not conduct appreciably until this voltage is exceed.

However, if the capacitor is omitted, and the subsequent circuit is arranged to respond to the actual diode current, the output can be arranged to be proportional to the square of the input voltage, and this possibility is utilized for certain special applications. One such usage is as a mixer at high frequencies in the microwave region as discussed in Chapter 13. At these frequencies the conventional form of frequency changer is inadequate but, as was shown on page 439, if the signal and local oscillator e.m.f.s are applied together to the input of a square-law detector, the output contains a term proportional to the product of the two inputs at the difference frequency, thus providing the required i.f. signal.

It was shown on page 249 that with small inputs a thermionic diode provides a square-law output, despite the fact that the characteristic obeys a nominal three-halves law. A semiconductor diode obeys an approximately exponential law of the form $i = ae^{bv}$, but by similar reasoning it can be shown that for small inputs the output follows a square law.

Automatic Gain Control

The range of signal strengths to be handled by a practical receiver varies widely. Moreover the signal from any particular (distant)

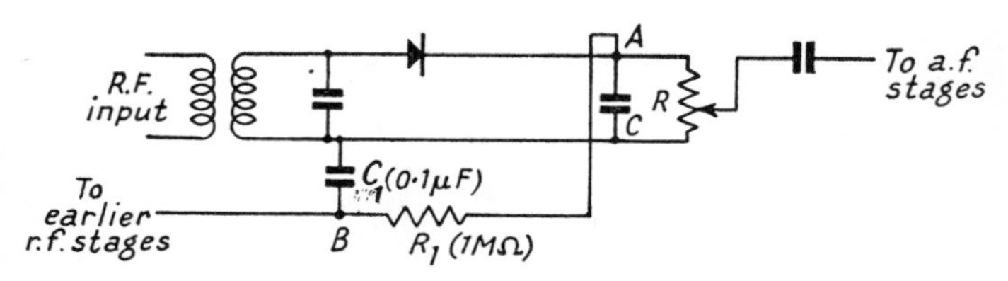

FIG. 9.31. BASIS OF AUTOMATIC GAIN CONTROL

station is liable to considerable variation due to fading. Hence it is desirable to control the r.f. gain in inverse proportion to the signal. This condition can be partially achieved by utilizing the d.c. component of the detector output to control the gain of the r.f. stages, as in Fig. 9.31.

The d.c. voltage across R is proportional and nearly equal to the peak value of the r.f. signal input. The time constant CR is chosen so that this voltage can follow the variations in amplitude due to the modulation, but if the voltage at A is fed through a second network R_1C_1 which has a long time constant, the voltage at B will be simply proportional to the peak r.f. signal and will ignore the modulation, and so provide the necessary control. The network R_1C_1 also effectively suppresses any r.f. component which would otherwise produce r.f. feedback and hence possible instability.

It is important that the time constant of these a.g.c. filter circuits should not be too large. Otherwise there is an appreciable time lag between the variations in the carrier and the correcting action. Sudden variations in strength, such as are obtained when tuning in, may momentarily paralyse the receiver because a large voltage is built up on the capacitors which takes an appreciable time to leak away.

The direction of the control voltage depends upon the connection of the diode. In a valve receiver a negative control voltage is required which can be used to increase the bias on one or more of the r.f. amplifying valves and so reduce the gain. By using variable-mu valves a smooth control of the gain can be obtained without distortion, as explained in Chapter 5, over a range of several hundred to one.

With transistor circuitry the diode may be arranged to provide either positive or negative control voltage according to the requirements. As was explained in Chapter 6 (page 300) the gain of a transistor can be varied either by reducing the base–emitter voltage or by increasing the collector current. Control which operates by reducing the base voltage is known as *reverse a.g.c.*, the alternative methods being called *forward a.g.c.*

The polarity depends upon the type of transistor. Thus for a p–n–p type reverse a.g.c. requires a positive control voltage, while an n–p–n type will require a negative control voltage.

Delayed A.G.C.

In some cases it is desirable to arrange that the automatic gain control does not come into operation until a certain minimum strength of signal has been reached. This is accomplished by using two diodes, one of which provides the signal frequency voltages and the other, which is tied in parallel with the first, provides the a.g.c. voltage.

This second diode, however, is provided with a small permanent negative voltage so that rectification does not take place until the peak value of the applied signal has exceeded this delay voltage. Fig. 9.32 shows a circuit of this type. The operation is otherwise the same as already described.

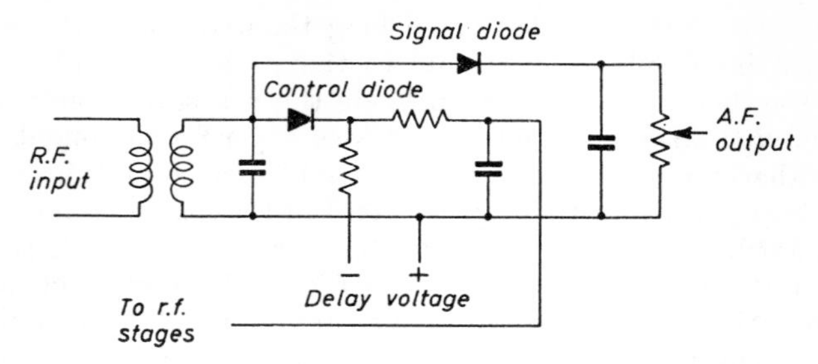

FIG. 9.32. CIRCUIT FOR PROVIDING DELAYED A.G.C.

Control Ratio

The aim of the designer is to obtain a good *control ratio*, i.e. a large control voltage for a small change in signal strength. To some extent, delayed a.g.c. automatically provides an improvement, as the curves in Fig. 9.33 show.

The full curve represents the output voltage in terms of input, without a.g.c. If plain a.g.c. is used, this becomes operative at

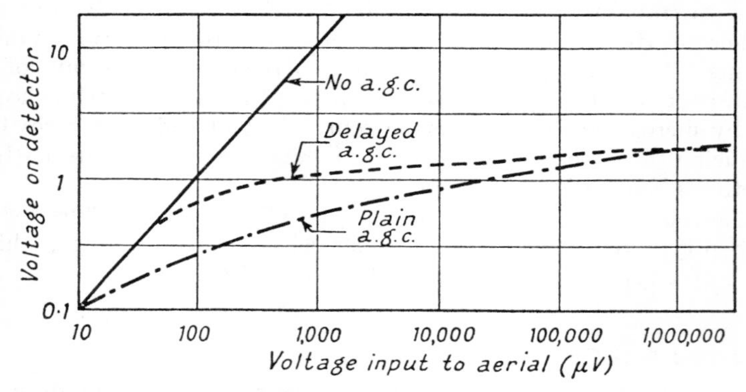

FIG. 9.33. SHOWING THE EFFECT OF AUTOMATIC GAIN CONTROL ON RECEIVER OUTPUT

once and cannot therefore be allowed to be too fierce in its operation or the output would never reach its normal value. The chain dotted curve shows a typical performance without delay. It will be noted that the gain on weak signals is appreciably reduced.

The third curve shows delayed a.g.c. which does not operate until the output is nearly up to the full amount and then comes into play with full action, giving an almost level output and a much better control ratio.

It may be that the voltage at which the detector delivers full output is insufficient to give a sharp enough control. Thus with a delayed system requiring 5 volts radio frequency to load the detector, we could use, say, 4 volts delay. This gives only 1·4 volts to provide a.g.c. which would be quite inadequate.

In commercial practice a separate i.f. amplifier is used for the a.g.c., supplied from some point subsequent to the frequency changer. This amplifier is *not controlled by the a.g.c.*, so that the full increase of signal is allowed to develop a.g.c. voltage. With the normal system, of course, as already explained, the action of the a.g.c. automatically cuts down the source of the a.g.c. voltage itself, so that it is impossible to obtain a truly level characteristic.

Quiet A.G.C.

As a development of this system the circuit is sometimes arranged to give no audible output until the carrier reaches a predetermined value. One method is to over-bias the output stage and allow the

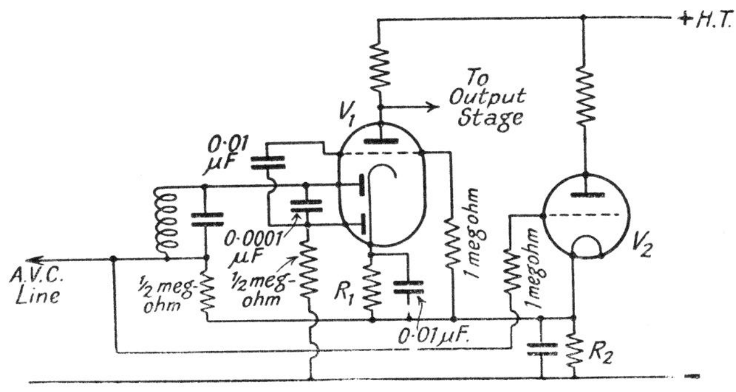

Fig. 9.34. Muting Circuit for Interstation Noise Suppression

a.g.c. circuit to "knock off" the bias when the carrier voltage exceeds the delay voltage. The process is known as *muting*.

Another method is to put negative bias on the signal diode so that it normally does not operate. The a.g.c. diode, however, is left operating normally with a small delay if necessary. On the arrival of a signal sufficiently strong to operate the a.g.c., the signal diode is released and the circuit functions normally.

The release of the signal diode (or the a.f. stages) may be accomplished by suitable circuit arrangements of the normal network, but it is more usual to employ a separate muting diode for the purpose. Fig. 9.34 shows such a circuit. V_1 is a double-diode-triode, the first diode being used for a.g.c. and therefore made negative to the cathode by the drop in the resistor R_1, which provides bias for the triode section of the valve in the usual way.

The second (signal) diode is coupled through a small capacitance to isolate it from the first as regards d.c. potential, but the leak in this case is returned to earth so that it is held negative to the cathode of V_1 by the voltage drops on R_1 and R_2. V_2 is a high-current high-slope valve. As soon as the a.g.c. diode operates, negative potential is applied to the grid of V_2. The anode current rapidly decreases, knocking off the bias on the signal diode, which thus comes into operation.

Tuning Indicators

In a receiver fitted with automatic gain control, the strength of the signal does not vary appreciably as the station is tuned in. This is because the normal increase in strength as the tuning point is approached is counteracted by the automatic gain control coming into operation. It is, therefore, possible for an inexpert user to adjust the receiver quite incorrectly, to avoid which tuning indicators are sometimes fitted. As the bias on the r.f. valves is run back the anode current decreases, and at the point where the carrier is a maximum (i.e. at the tuning point) the anode current reaches its lowest value. Hence, a meter in the anode circuit can be used to indicate the exact tuning point.

Alternatively a magic eye, or cathode-ray tuning indicator, may be used in which the height or spread of a fluorescent glow is controlled by the carrier voltage and so indicates the correct tuning point.

Transistor A.G.C. Circuits

The principles of a.g.c. are equally applicable to transistor circuitry. The gain of a transistor r.f. stage cannot be controlled with the same smoothness as is possible with a variable-mu valve, but a satisfactory measure of control is obtainable by altering the base potential.

As said earlier, the control may be arranged to decrease the base–emitter potential. This is called reverse a.g.c. and will provide a control range of some 30 dB (31·6 times), but has the disadvantage of reducing the signal-handling capacity of the stage. Fig. 9.35 illustrates a suitable circuit.

Alternatively, the base potential may be increased. This increases the collector current, and if the collector circuit includes a suitable (decoupled) resistor the collector voltage is reduced, which again reduces the gain as explained in Chapter 6 (page 300). This is called forward a.g.c. and has the advantage that the signal-handling capacity is increased, while the range of control is rather greater.

Either method may be used according to circumstances; sometimes both are used together, operating on different stages.

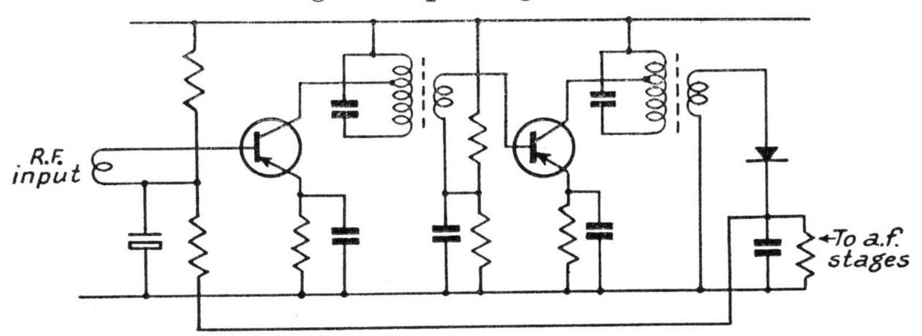

FIG. 9.35. TRANSISTOR AMPLIFIER WITH A.G.C.

Damping Diode

The control ratio may be substantially increased by the introduction of a damping diode across the first i.f. transformer, as shown in Fig. 9.36. A separate winding is used with a damping

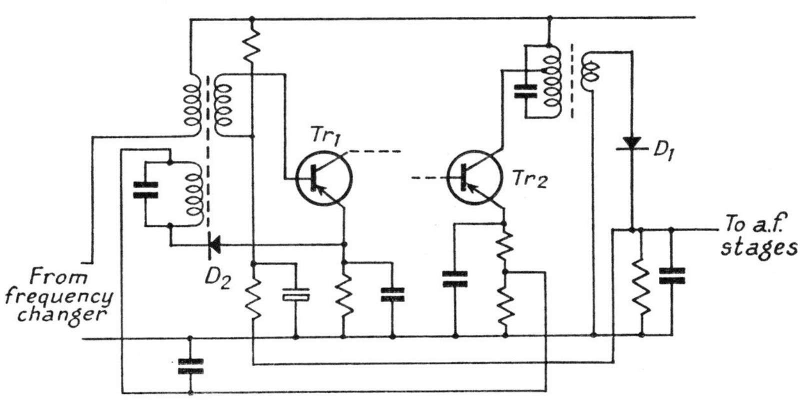

FIG. 9.36. A.G.C. CIRCUIT WITH DAMPING DIODE

diode D_2 across it. The connections are such that the d.c. potential across this diode is the difference between the emitter bias on Tr_1 and a suitable portion of that on Tr_2, which in the absence of a signal provides a reverse bias on D_2 so that it is non-conducting and does not damp the tuned winding. When a signal is present, however, the a.g.c. voltage provided by D_1 reduces the base (and emitter) voltage on Tr_1 until at the predetermined level the diode D_2 begins to conduct and damps the tuned winding to an increasing extent.

The arrangement thus provides a form of delayed a.g.c., and has the advantage that the effective load of the first i.f. stage is reduced, which enables it to handle larger inputs. Fig. 9.37 illustrates the effect of the a.g.c. with and without the damping diode.

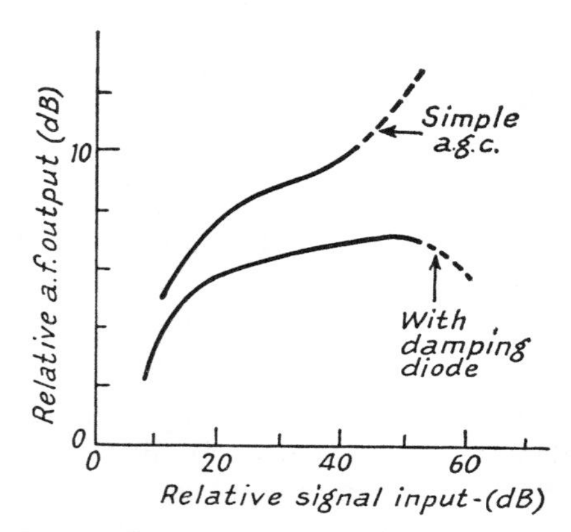

FIG. 9.37. ILLUSTRATING EFFECT OF DAMPING DIODE

Various more sophisticated forms of control have been devised providing amplified a.g.c. and muting, as in valve technique. These are beyond the scope of the present discussion, but reference may be made to a system providing a linear control over a range of 90 dB which was described by Willis and Richardson *Proc. I.E.E.*, **106B**, Supplements 15–18, p. 780 (1959).

Automatic Frequency Control

A development of some importance is the application of a.g.c. to control the *tuning* of a receiver. The system is principally applied to superheterodyne receivers where the fine control of the tuning is a function of one circuit—the oscillator. If the frequency of this circuit can be automatically adjusted to give correct tuning the effects of slightly incorrect setting and, more important, frequency drift, can be largely counteracted. The method is known as *a.f.c.* (*automatic frequency control*) or *a.t.c.* (*automatic tuning control*).

This result is accomplished in the first place by using a discriminator circuit as shown on the left of Fig. 9.38. Two i.f. transformers are used, sharply tuned to peak a few kHz above and below the true tuning point. If the oscillator setting is correct so that the intermediate frequency generated lies accurately in between these two tunes, both T_1 and T_2 will develop the same voltage, each being partially off tune. The currents in the diode loads R_1 and R_2 will be equal and opposite and will offset one another, and no "control" voltage will be developed.

If the oscillator now drifts causing the i.f. to rise, the higher tuned transformer, say T_1, will develop more voltage, while T_2 will develop less. Consequently R_1 will develop more voltage than R_2

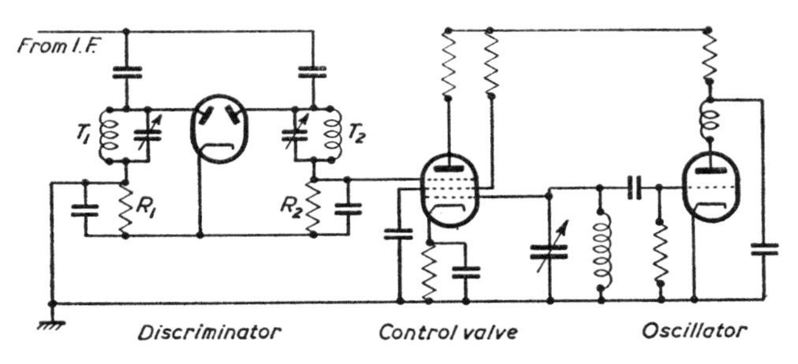

FIG. 9.38. DISCRIMINATOR CIRCUIT FOR OBTAINING AUTOMATIC FREQUENCY CONTROL

and there will be a resultant voltage applied to the control valve. If the oscillator drifts in the opposite direction, a similar but opposite control voltage will be produced.

This control voltage may be utilized in various ways. The most usual is to employ the Miller effect of a valve, which depends on the effective amplification. In Fig. 9.38 the control is applied to the suppressor grid which is an effective method of altering the gain. A positive control voltage will cause the gain to increase, while a negative voltage will cause a decrease. The effective grid–cathode capacitance of the valve will change accordingly, and as this is in parallel with the oscillator tuning circuit the oscillator frequency will be modified in such a direction as to counteract the drift. The effective gain of the control valve must be arranged, by suitable choice of anode load, to produce this result.

Alternatively, an inverted Miller effect may be used to vary the effective anode–cathode reactance of a valve, as in the frequency-modulated circuits discussed on page 352.

An equivalent arrangement using transistor circuitry is shown in Fig. 9.39. It uses a slightly different form of discriminator (actually of the Foster–Seeley type described in the next section), but the principle is the same, namely that any deviation in frequency causes a variation in the d.c. output above or below the mean value. This output is applied to a varactor diode connected across the oscillator circuit of the mixer. Varactor diodes have the property, as explained in Chapter 6, of changing their capacitance in inverse

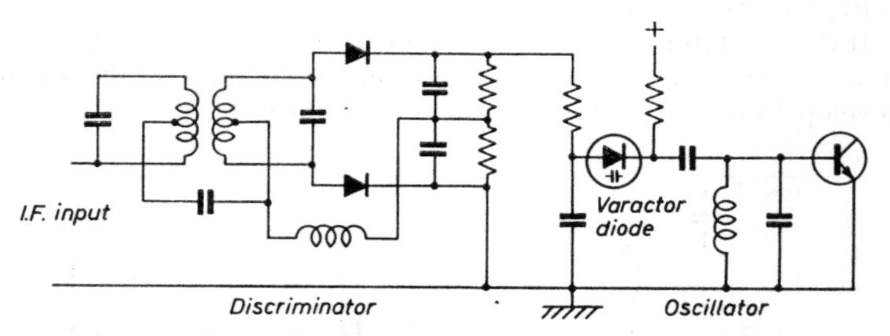

FIG. 9.39. TRANSISTOR A.F.C. CIRCUIT

proportion to the voltage across them. Hence any deviation in frequency can be arranged to produce a correcting change in oscillator frequency.

9.6. F.M. Receivers

For the reception of frequency-modulated signals the technique prior to the detector is the same as for amplitude modulation. At the detector stage, however, it is necessary to convert the variations of frequency of the carrier into corresponding variations in amplitude. One obvious way of doing this is to mistune the last i.f. stage slightly, so that the output is dependent on the frequency.

This is not a practicable arrangement, however, for it requires critical setting and it is more usual to employ a discriminator circuit similar to that used for the automatic frequency control described earlier (Fig. 9.38). At the same time it is customary to pass the signals through a limiter stage to remove any spurious amplitude variations which may have been introduced.

Limiters

This limiter is actually an important part of the process. The intelligence in a frequency-modulated signal is contained entirely in the variation of frequency. Atmospheric and local interference, on the other hand, is almost entirely amplitude-modulated so that the received signal will be a mixture of the two. If this signal is passed through a limiting stage which gives a constant output irrespective of the input (above a certain minimum level), any amplitude modulation (and hence the interference) will be removed leaving the frequency modulation unaffected.

Such a limiter is easily provided by a short-base r.f. pentode operated with low anode and screen voltages. Under such conditions anode-current saturation occurs as the grid potential approaches zero, while if the grid runs appreciably negative the anode current is cut off. A corresponding transistor circuit would be designed to handle only a small input, any increase in signal causing bottoming and cut-off on the respective peaks. It is a simple matter to arrange the conditions such that any signal greater than a given minimum produces no increase in output. The waveform of the carrier is, of course, distorted, becoming flattened, which introduces certain secondary effects, discussed later.

Discriminators

The output from this limiter stage is then coupled to a double secondary circuit, as shown in Fig. 9.40. These secondary circuits are tuned to frequencies slightly above and below the intermediate

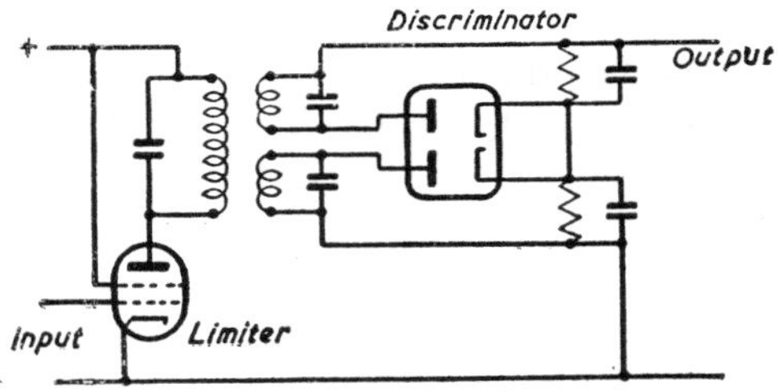

FIG. 9.40. DOUBLE-TUNED DISCRIMINATOR

frequency and the voltage developed across each is rectified with a diode. The rectified outputs are connected in opposition as shown in Fig. 9.41, which will be seen to produce a substantially linear

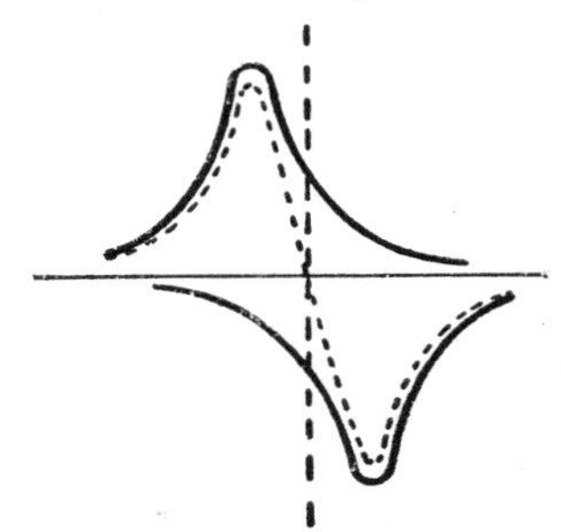

FIG. 9.41. DISCRIMINATOR CURVE

variation of amplitude with change of frequency over the operative range.

A somewhat simpler, and actually more efficient, arrangement is the phase-difference discriminator shown in Fig. 9.42. Here a single centre-tapped secondary is used, but the centre point is also fed with signal from the primary, through a capacitor.

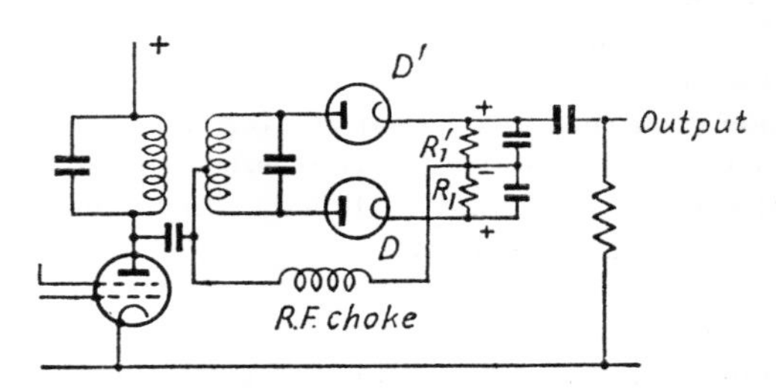

Fig. 9.42. Phase-difference Discriminator

[This circuit is often called the Foster–Seeley discriminator, from its originators D. E. Foster and S. W. Seeley, *Proc. Inst. Radio Engrs.* **25**, p. 289.]

If the current in the primary is i_1, the voltage across the primary will be $j\omega L i_1$. The secondary current i_2 is $j\omega M i_1/Z_2$ and the voltage across the secondary is $j\omega L_2 i_2 = (\omega^2 M L_2/Z_2)i_1$. Now, if the secondary is tuned, Z_2 is resistive, so that the secondary voltage is 90° out of phase with the primary. Hence the voltages across the

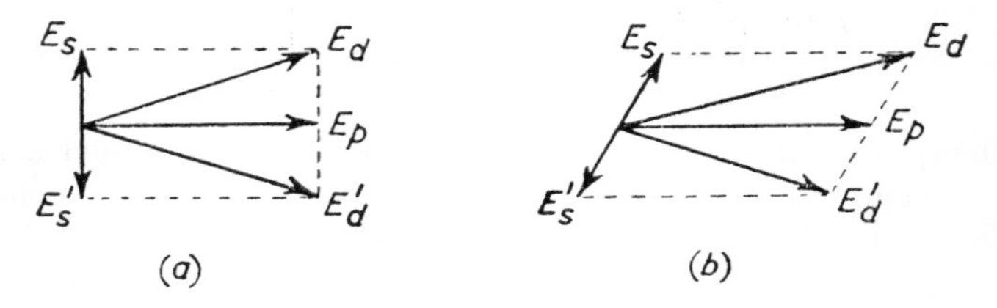

Fig. 9.43. Vector Diagram of Fig. 9.42 Circuit

diodes are the vector sum of the half-secondary voltage plus the primary voltage in quadrature, as shown in Fig. 9.43 (*a*). Because of the symmetry of the system E_d and E'_d are (numerically) equal so that the d.c. voltages developed across R_1 and R'_1 cancel out.

At any other frequency, however, Z_2 is not resistive, so that the secondary voltage will not be exactly 90° out of phase but will be as shown in Fig. 9.43 (*b*). The diode voltages are then not equal and there is a resultant signal. In fact, as the frequency deviates from the mid-frequency to which the circuits are tuned, the d.c. output varies in an exactly similar manner to that of the double-tuned discriminator as in Fig. 9.41.

This form of discriminator has a higher output than the double-tuned type. Because of its symmetrical form, it tends to ignore amplitude variations, but in practice this is insufficient and it is still necessary to precede it with a limiter.

Ratio Detector

A modified form of circuit which is somewhat simpler to produce, and provides much better elimination of amplitude modulation, is the ratio detector illustrated in Fig. 9.44. This will be seen to be similar

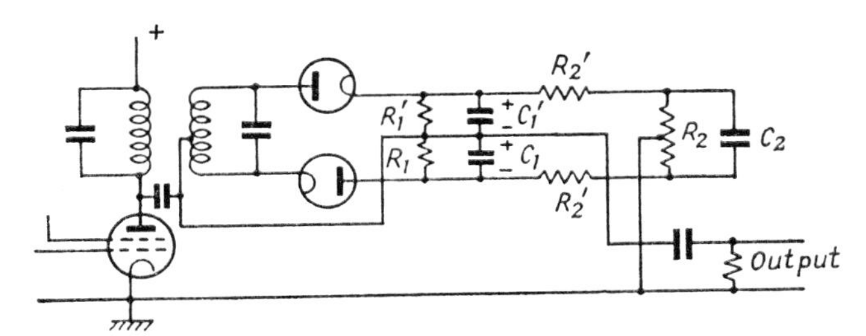

FIG. 9.44. RATIO DETECTOR

to that of Fig. 9.42, but has one of the diodes reversed. The diode load network is shunted by the network C_2R_2 (via the resistors R'_2) and the centre point of R_2 is earthed. The arrangement thus forms a bridge network giving zero output when the voltages across R_1 and R'_1 are equal, but a difference output when these voltages differ, by reason of the frequency deviation.

The difference output is only half that which is obtained with the arrangement of Fig. 9.42, but this is compensated by other advantages. In particular the network $R'_2CR'_2$ provides a reservoir action which tends to maintain the voltage across R_2 constant irrespective of changes in the input signal due to unwanted amplitude modulation, so that the a.m. rejection is improved.

This action is assisted by the fact that an increase in input causes the diodes to draw more current which reduces the Q of the secondary. This causes a small phase shift which has the effect of reducing the difference voltage, and by choosing R'_2 suitably the variation due to change in amplitude can be substantially eliminated. A reduction in input causes an increase in Q which results in a corresponding reverse correction.

Since R_2 provides a d.c. discharge path for the capacitors $C_1C'_1$ the resistors $R_1R'_1$ may be omitted in a practical circuit.

Fig. 9.45 shows a transistorized version of a ratio detector. By making the time constant of the load, RC, of the order of 0·1 sec

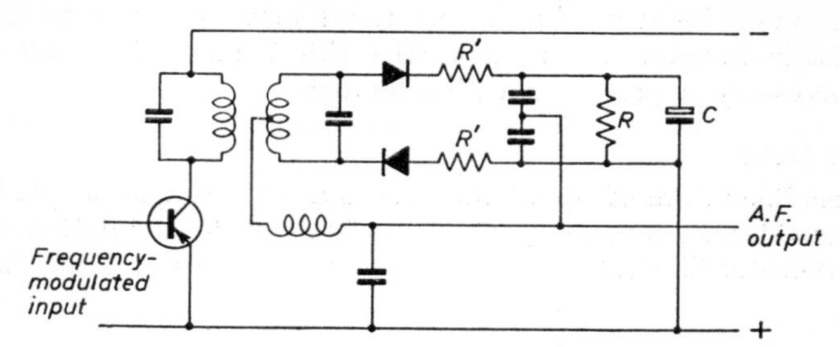

FIG. 9.45. TRANSISTORIZED RATIO DETECTOR

the circuit will virtually disregard amplitude variations in the incoming signal.

Because of its inherent stability and high a.m. rejection this type of detector is now almost universally adopted.

Locked Oscillator Discriminator

A detector circuit which operates on a different principle is the arrangement known as the locked-oscillator discriminator illustrated in Fig. 9.46. This is based on the fact that if two oscillators are

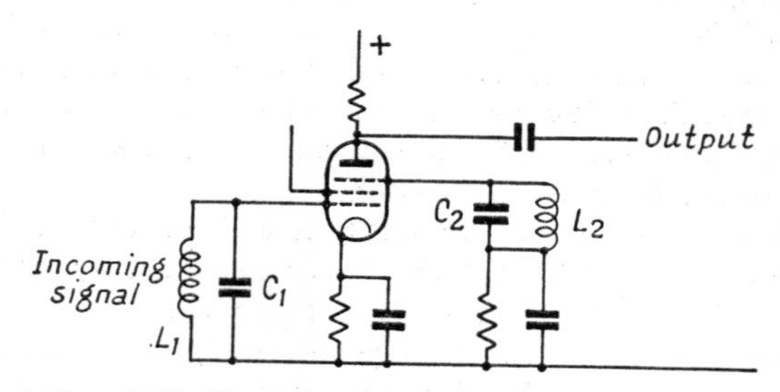

FIG. 9.46. LOCKED-OSCILLATOR DISCRIMINATOR

operating at nearly identical frequencies the weaker oscillator will tend to synchronize with the stronger (the "pulling" effect referred to in Chapter 12).

In the circuit of Fig. 9.46 the signal grid is supplied from the final i.f. stage while the suppressor grid is connected to a second circuit tuned to the mean frequency of the incoming frequency-modulated signal. The incoming signal modulates the electron stream in the valve, and the suppressor, being in this stream, acquires a charge which varies accordingly. The suppressor current ($= dq/dt$) will thus vary in quadrature with the input and so provides a negative reactance effect which maintains the circuit L_2C_2 in oscillation.

If the frequency of the incoming signal, however, deviates from that to which L_2C_2 is tuned the phase angle of the suppressor current is no longer exactly 90° but becomes greater or less depending on the direction of the deviation, and since the anode current is controlled by the vector sum of the grid and suppressor voltages there will be a variation in the anode current which is directly proportional to the frequency deviation.

Since two oscillators will only lock over a limited range the circuit is only effective for small deviations, but it copes adequately with the deviations in a practical f.m. transmission.*

Pre-emphasis

The use of a limiter to remove any amplitude modulation in the signal does not result in the elimination of all interference. For one thing the capture effect already referred to in the previous section (page 450) still applies, so that if the interference is greater than the signal the true signal is swamped.

Even when the interference is relatively small, however, it cannot be entirely eliminated by limiting because, although the amplitude variations are levelled out, the interference will remain as a phase modulation of the frequency-modulated signal. This is not important in itself but it can introduce a certain amount of spurious frequency modulation as a by-product to an extent which depends upon the strength of the interference and its frequency separation from the wanted signal. There is, in fact, a direct relation between the spurious frequency deviation produced by the interference and the frequency separation. This means that the higher the (audio) frequency of the noise the greater the amplitude at which it is reproduced.

The effect is aggravated because in a normal modulation the amplitude of the higher-frequency components tends to be less than at the lower frequencies. To counteract this effect, therefore, it is customary to accentuate the high frequencies in the modulation at the transmitter. The modulation is passed through a pre-emphasis network, as previously mentioned in discussing f.m. transmitters

* For further details refer to "A Locked Oscillator Quadrature Grid F.M. Detector" by J. Avins and T. Brady, *R.C.A. Review*, **16**, p. 648.

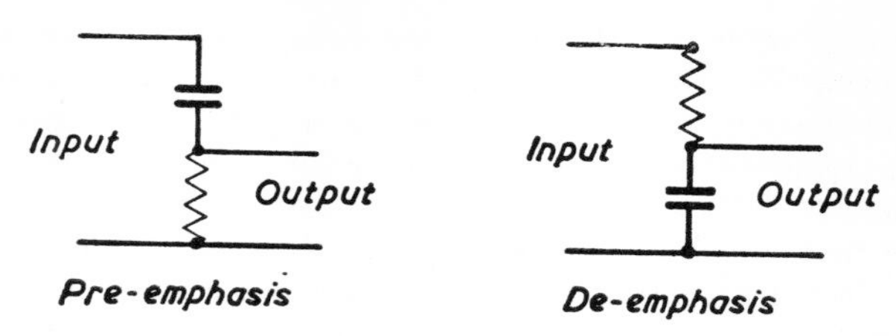

FIG. 9.47. FREQUENCY-CORRECTING NETWORKS

(page 355), while at the receiver a corresponding de-emphasis network is included, usually immediately following the discriminator.

These networks are simple RC circuits as shown in Fig. 9.47. They are usually designed to have a time constant of the order of 100 μsec, which begins to operate at about 1,000 Hz and produces an accentuation (or attenuation) of 20 dB at about 20 kHz.

I.F. Amplifier

The bandwidth of a frequency-modulated transmission is much greater than with a.m., being approximately $2(\Delta + p)$, where Δ is the frequency deviation and p is the highest modulation frequency. The usual deviation is 75 kHz, so that, if p is 15 kHz, $B = 180$ kHz, and in practice a figure of 200 kHz is usually allowed. This clearly demands a much higher intermediate frequency than the 470 kHz used with a.m. transmissions, and for certain practical reasons a frequency of 10·7 MHz has become standard.

To obtain the necessary selectivity three i.f. stages are usually employed, using double-tuned transformers to preserve the frequency response over the required bandwidth. With critical coupling this requires the circuit Q to be $f_0/B = 10{\cdot}7/0{\cdot}2 = 53$ (cf. page 111).

The circuits, however, cannot be matched to the transistors in the normal way because of the greater internal feedback arising from the higher frequency involved. It was shown earlier (page 429) that with a stability factor of 4, $\omega C_{re} G_m R_1 R_2 = 0{\cdot}5$, so that $R_1 R_2 = 1/2\omega C_{re} G_m$. The feedback can be minimized by using transistors having a specially low internal capacitance. If C_{re} is taken as 0·2 pF, then with $G_m = 35$ mA/V,

$$R_1 R_2 = 1/(4\pi \times 10{\cdot}7 \times 10^6 \times 0{\cdot}2 \times 10^{-12} \times 35 \times 10^{-3}) \simeq 10^6$$

This is very much less than the product $R_{out} R_{in}$ for the transistor itself, so that normal matching is not possible. However, it was shown on page 455 that, to obtain maximum output from a typical transistor, R_1 can be 8 kΩ. The stability requirement can then be met by making $R_2 = 125\ \Omega$.

To provide a load $Z_w = 8$ kΩ the inductance required will be $Z_w/\omega Q_w = 8.10^3/(2\pi \times 10{\cdot}7 \times 10^6 \times 53) \simeq 2{\cdot}25\ \mu$H, and C $(= Q_w/\omega Z_w) \simeq 100$ pF.

The secondary step-down ratio will be $\sqrt{(8{,}000/125)} = 8$. However, since an inductance of the order of 2 μH would only require a

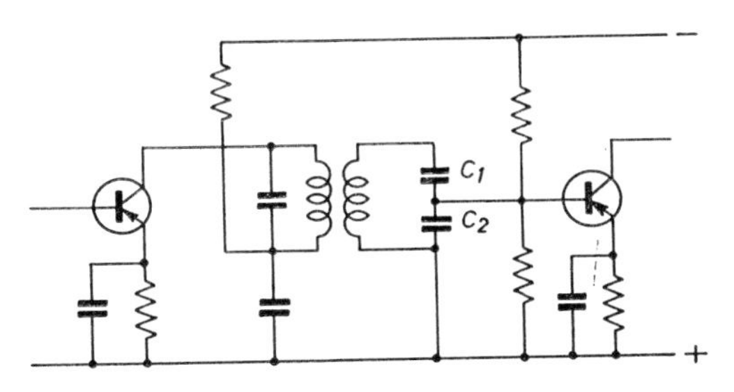

FIG. 9.48. 10·7MHz I.F. AMPLIFIER STAGE

few turns, it is difficult to arrange the correct tapping point, and it is more convenient to use a capacitive tapping as shown in Fig. 9.48.

If n is the step-down ratio, $C_2 = nC_1$, while C_1 and C_2 in series must equal the tuning capacitance C. C_1 must thus be $C(1 + n)/n$, which with the values above makes $C_1 = 112{\cdot}5$ pF and $C_2 = 900$ pF.

10

Audio-frequency Technique

AUDIO-FREQUENCY technique is concerned with the handling of relatively low-frequency signals which do not fall within the radio-frequency range already discussed. Strictly speaking, audio-frequency signals are those which fall within the range of audible sound (approximately 16 to 16,000 Hz), but the techniques are applicable, with appropriate modifications, to signals in the sub-audio range down to zero frequency (d.c.) and the supersonic range up to 100 kHz or more, while at the upper end of the spectrum are the video frequencies up to some 6 MHz involved in television practice, for which a blend of a.f. and r.f. technique is used.

The main difference between a.f. and r.f. circuits is that, whereas r.f. amplifiers are concerned with a single (adjustable) frequency, a.f. amplifiers are usually required to handle equally effectively all the frequencies within the required spectrum. The early stages are concerned with voltage amplification, which is approximately gZ where g is the effective transconductance of the device and Z is the load, so that the design is concerned with the provision of a load which is either constant or varies in a specified manner over the range of frequency to be handled. The later stages are concerned with the delivery of power to the reproducing device in use, which requires a slightly different approach, but is still subject to the same basic requirements.

The interpretation of the requirements depends on the type of amplifying device. A valve, being voltage-operated, requires a different approach to a bipolar transistor, which is current-operated, so that the two systems are best discussed separately, with suitable cross-references to factors which are common to both.

10.1. Feedback

There is, however, one important aspect of a.f. technique which applies to any type of amplifier, namely *feedback*. This may be accidental or deliberate, and results in the transfer of some part of the output energy to the input, either around an individual stage or the amplifier as a whole. The result depends on the relative phase of the feedback and may vary between positive feedback which increases the gain (and may result in oscillation), and negative feedback which will reduce the effective gain.

Accidental feedback may occur through stray coupling (usually capacitive) between output and input wiring or components so that attention to layout is important. It can also occur as a result of e.m.f. developed across any impedances in the supply lines which are common to several stages. This was referred to in the previous Chapter (page 416) and may be avoided by the provision of a bypass capacitor to earth having a negligible impedance at the signal frequency.

Feedback may also occur through the internal capacitance of the stage, but this is not usually troublesome at audio frequencies. The effect has been discussed in relation to r.f. circuits in Chapter 9, and it is a simple matter to determine whether it is likely to be significant in any particular a.f. circuit.

Basic Feedback Conditions

Let us deduce a generalized expression for an amplifier with feedback, as shown in Fig. 10.1. If the amplifier gain is A and the signal

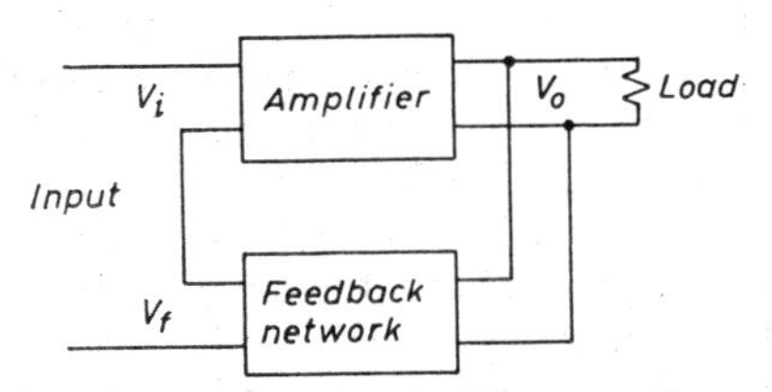

Fig. 10.1. Amplifier with Voltage Feedback

input is V_i, the normal output V_o will be AV_i. If now a proportion β of the output is fed back to the input, the effective input voltage becomes $V_i + \beta V_o$.

The output voltage will be modified to V_o', so we can write $V_o' = A(V_i + \beta V_o')$ so that $AV_i = (1 - \beta A)V_o'$. Hence the modified gain is

$$A' = V_o'/V_i = A/(1 - \beta A)$$

This is the fundamental feedback relationship. The factor β is called the *reverse transfer ratio*, and βA is called the *feedback factor*. Their values depend upon the circuit configuration and are, in general, frequency dependent.

If β is positive (so that the feedback voltage is in phase with V_i) the overall gain is increased, and if $(1 - \beta A)$ is greater than unity, oscillation will result. Such a condition is thus to be avoided unless deliberate oscillation is required.

Negative Feedback

If β is negative, conditions are reversed. The overall gain is reduced, but tends to a limit $1/\beta$, which is *independent of A*. In fact,

$$A' = A/(1 + \beta A) = 1/\beta \quad \text{if} \quad \beta A \gg 1$$

This means that the overall performance is virtually independent of the amplifier parameters, providing considerably improved stability. The arrangement also reduces distortion since if there is initially a harmonic component V_h in the output, this will be reduced by the feedback by the same factor $(1 + \beta A)$. These advantages, of course, are obtained at the expense of overall gain since to obtain a given output the input signal has to be appropriately increased.

In practice the implementation of the technique involves a number of subsidiary factors. The feedback may be a suitable proportion of the output voltage (*voltage feedback*); or it may be made proportional to the output current (*current feedback*). It may be applied either in series with the input signal (*series feedback*) or in parallel therewith (*shunt feedback*).

Finally, the actual feedback voltage appearing at the input is determined by the relative circuit impedances, so that the feedback factor βA has to be assessed individually. Both A and β are subject to phase change with varying frequency, and if the overall change is more than 180° the sign of the term βA will be reversed and positive feedback will result, producing a tendency to instability.

By plotting the vector of βA against phase angle in what is called a *Nyquist diagram*, the magnitude and phase of the feedback factor may be determined over any desired frequency range, and the stability conditions established.

The feedback network is often made deliberately frequency-dependent to provide a specific form of amplifier response.

Effect of Negative Feedback on Circuit Impedances

Voltage feedback tends to maintain the output voltage constant, which is equivalent to reducing the effective internal impedance. In fact, if R_o is the normal output impedance, the modified impedance R_o' becomes $R_o/(1 + \beta A)$.

Current feedback, on the other hand, tends to maintain the output current constant, which is equivalent to an increase in output resistance. In this case $R_o' = R_o(1 + \beta A)$.

There is a similar influence on the input impedance depending on whether series or shunt feedback is employed. Where the feedback (of either voltage or current form) is introduced in series with the signal, the effective input impedance becomes $R_i' = R_i(1 + \beta A)$.

With shunt feedback the input resistance is reduced becoming $R_i' = R_i/(1 + \beta A)$.

The practical applications of feedback will be seen in the sections which follow, in which valve and transistor operation will be considered separately.

10.2. A.F. Valve Amplifiers

The simplest form of a.f. amplifier is the resistance-coupled stage illustrated in Fig. 10.2. Here the load is a resistance R which is constructed (usually of carbon composition) so that it is substantially non-inductive over the frequency range involved. Its effective value, however, is modified at high frequencies by the stray circuit capacitance represented by C_o, while at low frequencies the performance is affected by the network CR_1 through which the signal is transferred to the succeeding stage.

Consider first the gain at a medium frequency such that the

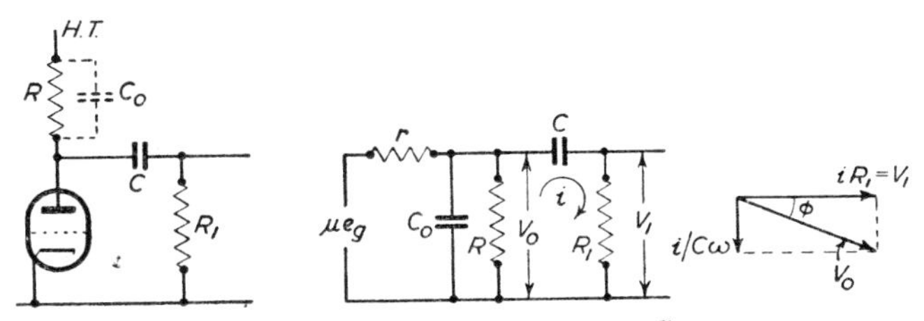

FIG. 10.2. RESISTANCE-COUPLED STAGE

reactance of C_o is negligibly high by comparison with R and the reactance of C negligibly low compared with R_1.

The voltage μe_g passes current through the anode resistance r in series with R and R_1 in parallel. The voltage transferred to the next stage is that across R_1, so that the gain is $R_p\mu/(R_p + r)$, where

$$R_p = RR_1/(R + R_1)$$
$$\simeq R \quad \text{if} \quad R_1 \gg R$$

At low frequencies the reactance of C is not negligible and the voltage V_1 is thus only a part of the voltage developed, V_o. If i is the current through CR_1, the proportion of V_o developed across R_1 is

$$\frac{R_1}{(R_1 + 1/jC\omega)} = \frac{R_1}{\sqrt{(R_1^2 + 1/C^2\omega^2)}}$$

It is easy to calculate the relative values of C and R_1 to make V_1/V_o a given percentage, and a criterion often quoted is for the product CR_1 to be 0·0065 which makes $V_1/V_o = 0{\cdot}9$ at 50 Hz. (It will be clear that the fundamental requirement is that the reactance of C shall be low compared with R_1 and so we can either increase C or R_1. Thus it is the product of the two which is important. An alternative criterion is $R_1C\omega = 2$, which provides 90 per cent of the full gain at frequency $\omega/2\pi$.)

Phase Angle

This criterion is by no means adequate for high-quality amplification, particularly where several stages are in use. The loss in each stage is cumulative, so that with three stages V_1/V_o would only be 0·73.

A still more serious defect, however, is the change of phase which occurs, for the current I, and hence the voltage V_1, is by no means in phase with V_o but leads by an angle $\tan^{-1}(1/RC\omega)$, as will be clear from Fig. 10.2. If $CR = 0{\cdot}0065$ this angle is 25·8 degrees, and three stages would give a lead of 77·4 degrees. There are many circuits in which correct phase relationship is of vital importance. In such circumstances the product CR must be increased to reduce the phase angle to the required value.

High-frequency Loss

At the upper frequencies the capacitance C_o becomes troublesome and shunts the resistance R. The amplification is then reduced by the lowering of the effective impedance. R and C_o in parallel provide an impedance $Z = R/(1 + jRC_o\omega)$ and the gain is $\mu Z/(r + Z)$ taking due account of the vectorial nature of Z. Under given conditions of C_o and ω, the value of R necessary to maintain 90 per cent of the full amplification is given by $RC_0\omega = \frac{1}{2}$. (The higher ω or C_o the less R must be, and this soon limits the effectiveness of the circuit.) C_o includes the anode-cathode capacitance of the valve in question, the stray circuit capacitances, *and the input capacitance of the succeeding valve.* This latter capacitance may entirely swamp the previous ones, for the input capacitance of a valve is not merely its static or geometric capacitance but is increased by the Miller effect.

Miller Effect; Use of Screened Valves

From the formula quoted in the previous chapter the input capacitance $C_g = C_{gc} + C_{ga}(1 + A\cos\theta)$. Assuming an effective gain, A, of 30 and $\cos\theta = 0{\cdot}8$, with C_{gc} and $C_{ga} = 7$ pF each, we have $C_g = 7 + 7(1 + 24) = 182$ pF, which would have a reactance at 10 kHz of only 87,500 ohms.

Hence, if effective amplification of the upper frequencies is to be maintained, R must be kept low, which means a limited gain per stage. The remedy is to use pentode valves which have grid–anode capacitances of 0·01 pF or less. Miller effect with such valves is negligible, though the static capacitance C_{gc} still remains, plus the anode–cathode capacitance of the preceding valve and the stray capacitances.

This permits anode loads of 50,000 to 100,000 ohms to be used without undue loss of high-frequency response, and since with a pentode the stage gain is gR, gains of several hundred are easily obtained with modern valves.

High-frequency Phase Displacement

Phase shift occurs at high frequency as well as at low frequencies for a similar reason. The presence of C_o across R results in a lag, as will be seen from Fig. 10.3. The current through R is in phase with V_o. The total current through R and C_o in parallel, however, leads on V_o as shown, and this is the current supplied by the valve. Hence V_o leads on μe_g by an angle $\tan^{-1} RC_o\omega$, giving a phase shift which increases with R, C_o and ω.

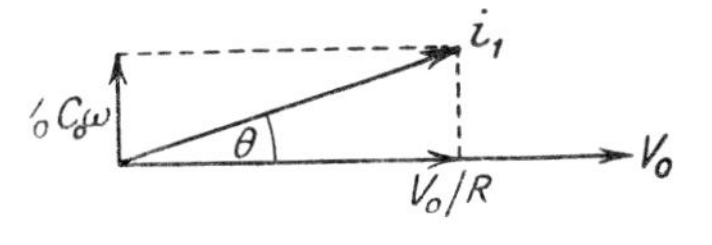

FIG. 10.3. VECTOR DIAGRAM OF FIG. 10.2 CIRCUIT AT HIGH FREQUENCIES

Once again the effect is cumulative and if the phase shift at the end of an n-stage amplifier is not to exceed α degrees, the phase shift per stage, θ, must not exceed α/n. This angle θ is the phase shift in the Miller effect formulae already quoted.

It is possible to maintain the upper frequencies up to a point by including a choke in series with the resistance as shown in Fig. 10.4. This tunes with the stray capacitance and so maintains a high effective anode impedance at the upper frequencies. The effective impedance of the combination can be calculated from the usual circuit laws (see page 97), from which it can be shown that if $L = C_oR^2$ the response rises to $\sqrt{2}$ times the normal at the frequency for which $\omega^2 L_o C = 1$ and then falls off sharply. If $L = 0{\cdot}5\, C_oR^2$ the response is practically level up to the "resonant" point.

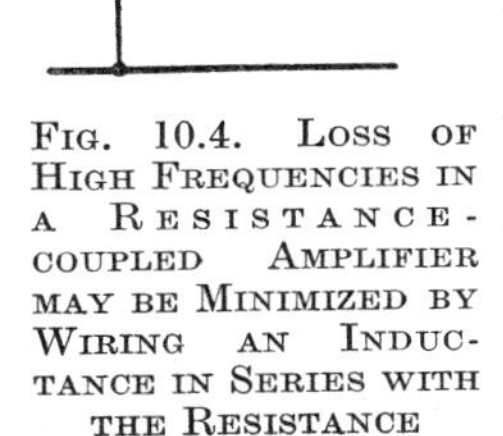

FIG. 10.4. LOSS OF HIGH FREQUENCIES IN A RESISTANCE-COUPLED AMPLIFIER MAY BE MINIMIZED BY WIRING AN INDUCTANCE IN SERIES WITH THE RESISTANCE

Direct Coupling

Where very low frequencies are to be handled, a direct-coupled amplifier may be used. Here the anode of the first valve is connected either directly or through a resistance to the grid of the next valve. In the former connection, to avoid the heavy positive grid voltage which would otherwise result, the cathode of the second valve must also be raised to a slightly more positive potential. This method is of limited application since it involves an increase in the overall h.t. voltage. An alternative arrangement is shown in Fig. 10.5.

Here the voltage developed at the anode of V_1 is applied across resistances R_1 and R_2 in series, the bottom end being connected to a point of negative potential. By suitable choice of R_1 and R_2, therefore, the potential of the point A may be made slightly negative with respect to the cathode of V_2.

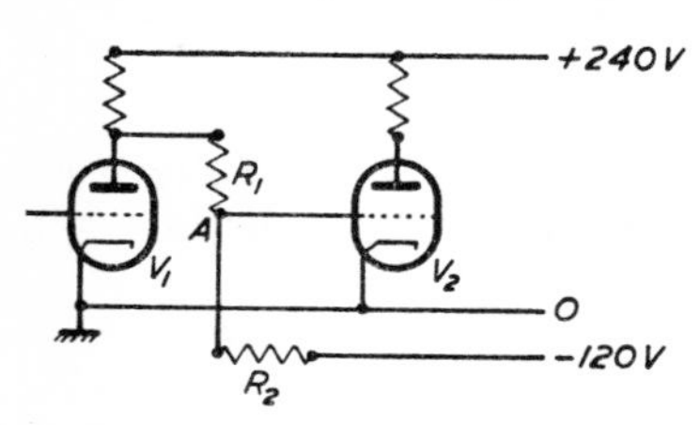

FIG. 10.5. DIRECT-COUPLED AMPLIFIER

The signal voltage is divided in the ratio $R_2/(R_1 + R_2)$ so that there is appreciable loss of gain, but it is usually easy to obtain more gain from the valve than is actually needed. The method has the advantage that all the cathodes may be at the same potential (and all earthy).

A direct-coupled amplifier will respond to frequencies down to zero (d.c.), but it suffers from the same disadvantages as the ordinary resistance-coupled amplifier at the high frequencies. It is also liable to drift, since small changes of current through the network may upset the somewhat delicate balance of the voltages. Other forms of circuit are discussed in Chapter 17.

Multi-stage Amplifiers

Several stages may be used in cascade to produce increased overall gain. The individual stages are designed as already described but it is important to ensure that

(*a*) The overall phase shift is tolerable. Normally this will be kept small, though sometimes a deliberate phase shift is permitted, as in the phase-shift oscillator described in Chapter 12.

(*b*) Stray feedback is avoided, particularly that arising from common impedance coupling in the h.t. line. (See page 416.)

Fig. 10.6 shows a two-stage amplifier in which deliberate negative feedback has been introduced. Common-impedance coupling has been eliminated by the decoupling circuit in the anode circuit of the first stage. A measure of individual negative feedback has been provided in each stage by the omission of bypass capacitors across the cathode resistors, while a further feedback is provided from output to input. When the grid of V_1 goes positive the voltage at the anode of V_2 will also go positive, and some part of this e.m.f.

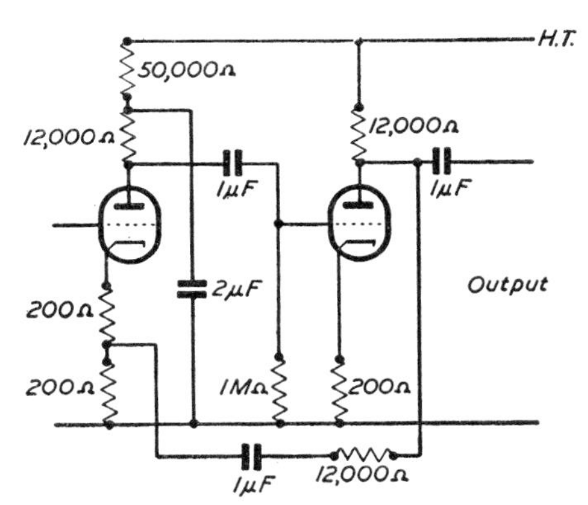

FIG. 10.6. TWO-STAGE AMPLIFIER WITH NEGATIVE FEEDBACK

is transferred to the *cathode* of V_1 which will thus move in opposition to the grid–cathode signal input. At medium frequencies $\beta = 200/12{,}200$ so that the gain $1/\beta$ will be 61.

Cathode Feedback

The negative feedback which results from the omission of the usual by-pass capacitor can be utilized in various ways. Consider the circuit of Fig. 10.7 (*a*). If i_a is the a.c. component of the anode current, we can write

$$i_a = -\mu e_g/[r_a + (R + R_c)]; \quad v_R = i_a R; \quad v_{in} = e_g - i_a R_c$$

$$\text{Stage gain} = i_a R/(e_g - i_a R_c) = R/[(e_g/i_a) - R_c]$$

Substituting the expression above for i_a, this reduces to

$$A = \mu R/[r_a + R + R_c(1 + \mu)]$$

This is the same as the normal expression except that R in the denominator is increased by $R_c(1 + \mu)$, with consequent reduction in gain (and, possibly, improvement in linearity).

The cathode resistor may be increased beyond the value required for bias, the grid leak being tapped across a portion only, as in Fig. 10.7 (*b*). Under such conditions, apart from the reduction in gain due to feedback, there is a further loss of output because the anode resistance is only a fraction of the total external load, and an appreciable fraction of the output voltage appears across the cathode resistor. This effect is sometimes utilized to obtain two output

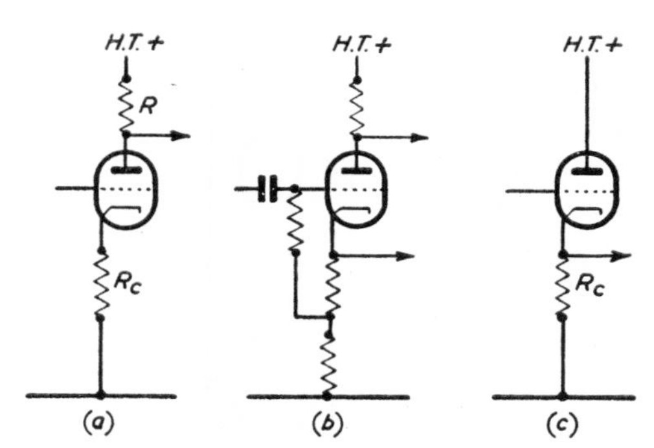

FIG. 10.7. DEVELOPMENT OF THE CATHODE-FOLLOWER CIRCUIT

voltages in opposition—one from the anode and the other from the cathode.

It will be clear that an increased anode current will cause an increased voltage drop on both anode and cathode resistors. But since the anode is already negative *with respect to earth* (h.t. +) it will become still more negative. The cathode is positive to earth and will increase its positive potential, so that anode and cathode (signal) potentials move in opposite directions.

The voltage across the cathode resistor is $i_a R_c$ so that the gain at the cathode becomes

$$A = \mu R_c/[r_a + R + R_c(1 + \mu)]$$

Cathode-follower Circuit

If we reduce the anode resistance to zero, as in Fig. 10.7 (*c*), this reduces to

$$A = \mu R_c/[r_a + R_c(1 + \mu)]$$

and in the limit, if $R_c(1 + \mu)$ is much greater than r_a we have simply $A = \mu/(\mu + 1)$.

Such a circuit is called a *cathode follower*. It will be seen that the maximum gain is just short of unity and may not be more than a fraction if $R_c(1 + \mu)$ is of the same order as r_a. The arrangement, in fact, operates not as an amplifier but as an impedance changer, which is often of great convenience.

For example, a high-resistance voltage divider cannot be used to feed the grid of a valve if the frequency is high because the parallel reactance of the valve input and stray capacitance will be small. This largely short-circuits the bottom portion of the voltage divider so that it ceases to provide adequate control. If, however, the input is applied to the grid of a cathode follower, the cathode resistance may be in the form at a voltage divider of quite low resistance such that the capacitance of the succeeding stage does not affect the operation. The input impedance to the cathode follower, on the other hand, is very high being simply the grid–cathode capacitance (including strays). Miller effect is not present.

The circuit is also often used as a buffer stage to prevent one circuit from interacting with another.

The expression for gain may be rewritten in the form

$$A = \frac{\mu}{\mu + 1} \cdot \frac{R_c}{r_a/(\mu + 1) + R_c}$$

Thus the effective amplification factor is $\mu/(\mu + 1)$, and the effective internal impedance is $r_a/(\mu + 1)$ which, when $\mu \gg 1$, becomes simply $r_a/\mu = 1/g_m$.

A.F. Transformers

Transformer coupling is used in the simpler types of circuit, usually following triode valves. Amplification of the bass frequencies is here dependent on maintaining a high effective primary impedance relative to the valve resistance. This involves using a transformer with a high inductance obtained either by using a large iron circuit (so that the saturation introduced by the heavy anode current shall not be serious) of a high-permeability material with a parallel-feed circuit.

The upper frequencies with a transformer are entirely limited by the self-capacitance in the circuit, including the effective grid–cathode capacitance of the succeeding valve. By proper design a resonant effect may again be called into play to assist in maintaining a good response curve.

The ordinary transformer possesses appreciable leakage inductance, the magnetic field of the primary being incompletely linked with the secondary. This may be represented by showing a perfect transformer in series with a small leakage inductance to represent that portion of the magnetic flux which does not link with the secondary.

By ordinary transformer laws the quantities may all be referred to the primary, in which case we can represent the transformer itself by a simple choke.

The equivalent circuit of a transformer, therefore, is as shown in Fig. 10.8, and it will be seen that there are two possibilities of

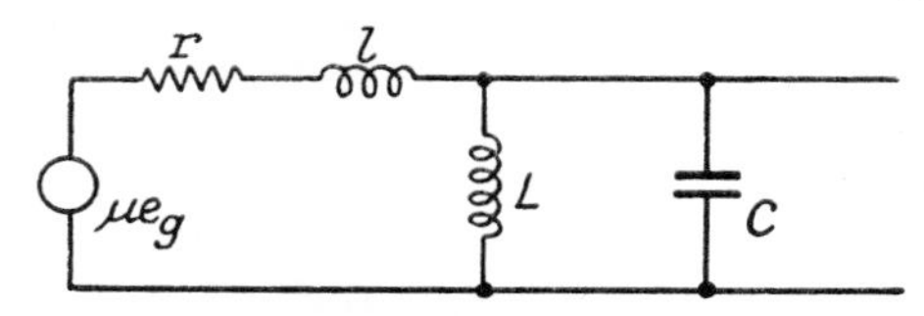

FIG. 10.8. EQUIVALENT CIRCUIT OF AUDIO-FREQUENCY TRANSFORMER

resonance. The first of these occurs between the main primary inductance L and the self-capacitance C in the circuit, and is usually arranged to fall between 50 and 100 Hz in order to maintain good bass response. The second is between the leakage inductance l and the self-capacitance C which occurs at a high frequency, under which conditions the primary inductance acts as an infinite impedance and may be ignored. By suitable design of the leakage

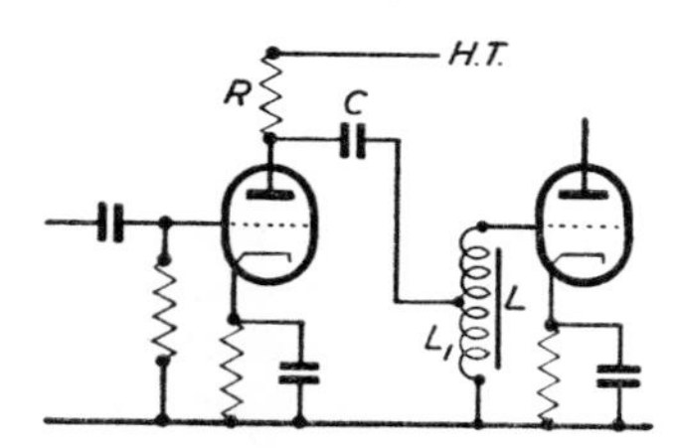

FIG. 10.9. AUTO-TRANSFORMER ARRANGEMENT

inductance this second resonance can be made to occur between 5,000 and 10,000 Hz and thus to maintain the amplification of the upper frequencies.

The presence of the steady d.c. component of the anode current reduces the effective inductance, so that a relatively large iron circuit is required. To avoid this a parallel-feed circuit may be used as shown in Fig. 10.9, the primary being conveniently (but not necessarily) in the form of a tapping on the secondary. This permits the use of small and compact transformers on high-permeability cores.

The capacitance C can be arranged to resonate with L_1 at around

100 Hz, and from the ordinary laws of resonance it will be clear that the voltage across L_1 can exceed the input voltage supplied by the valve, giving a definite rise in the amplification. Fig. 10.10 illustrates the effect.

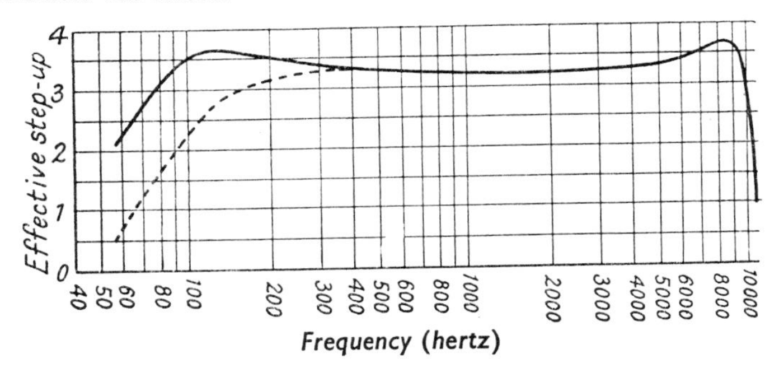

FIG. 10.10. ILLUSTRATING THE EFFECT OF WINDING RESONANCES

The Output Stage

The choice of the valves to be used in the amplifier depends on the voltages to be handled. It is usual to work back from the output stage which is, first of all, designed to work into the optimum load for maximum undistorted output.

The methods of determining this vary with the circumstances and the degree of accuracy required. The most general method is to adopt the graphical tactics outlined in Chapter 5, and it will be as well to review the process briefly.

Fig. 10.11 shows typical characteristics for a triode. The working point O is chosen in this case corresponding to 200 volts on the anode and $12\frac{1}{2}$ volts grid bias. The variations of anode voltage and current may be determined by drawing a line through the point O having a slope corresponding to the effective resistance of the load, and for absolutely distortionless working the points where this *load line* cuts the limiting characteristics should correspond to equal excursions of anode voltage and current from the mean value.

In Fig. 10.11, for example, the mean bias is $12\frac{1}{2}$ volts, so that the limits of grid swing will be 0 and 25, assuming the valve to be fully loaded, and the distances OA and OA' should be equal if the load represented by AA' is correct.

Actually they are far from equal, showing that considerable distortion is occurring. This particular distortion, wherein one half of the wave is flattened and the other peaked, is called *second harmonic distortion* because the form of wave is the same as would result if a pure wave were mixed with a (smaller) wave of twice the frequency as shown in Fig. 10.12.

It will be clear from Fig. 10.11 that a non-linear operation such as this will be accompanied by an increase in the mean anode current (because the current swing OA' is greater than OA). To

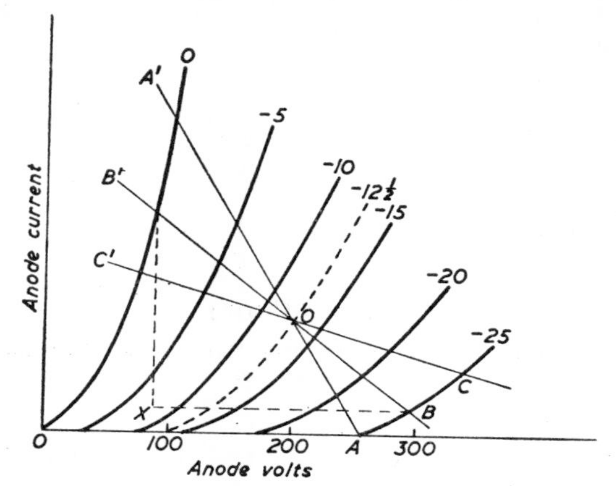

FIG. 10.11. ILLUSTRATING CHOICE OF OPTIMUM LOAD FOR A TRIODE

The figures on the curves refer to the grid voltage.

allow for this the second harmonic in Fig. 10.12 has been shown displaced above the zero line for the fundamental.

Now let V be the maximum amplitude of the fundamental and dV that of the harmonic. It will be seen that the positive peak of the combined wave is $V + 2dV$ and that of the negative wave is $V - 2dV$. Hence the ratio of the positive to negative peaks is $(V + 2dV)/(V - 2dV)$, i.e. OA'/OA in Fig. 10.11. If $d = 0{\cdot}05$ (5 per cent distortion) this is $1{\cdot}1/0{\cdot}9 = 1{\cdot}22$. Hence if OA'/OA is

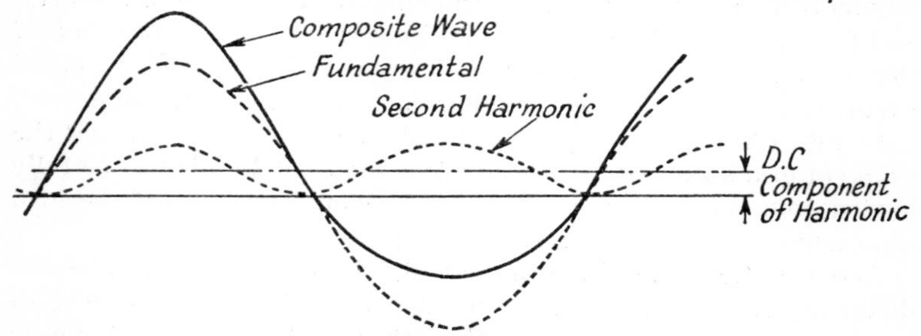

FIG. 10.12. ILLUSTRATING FLATTENING OF WAVE DUE TO SECOND HARMONIC

not greater than 1·22 we shall not introduce more than 5 per cent distortion and this criterion is often used by circuit designers.

The load line $A'OA$ does not fulfil this requirement, but $B'OB$ does. It will be noted that the anode current is no longer reduced to zero at each swing, so that the valve is operating less efficiently, while the power output (which is $\frac{1}{8}XB \,.\, XB'$) is also less than before.* As we reduce the slope of the load line the power output falls off still further while the distortion at first becomes less and then begins to increase again, due to the tendency of the characteristics to become flatter with high grid bias.

A steep load line corresponds to a low load resistance. Hence, we find with a triode that as we increase the load the power output rises to a maximum and then falls off again, while the distortion falls to a minimum and then begins to rise again. The optimum distortion condition does not coincide with that for maximum power output.

Output Tetrodes and Pentodes

A similar though slightly modified procedure is adopted with pentodes, and critical-distance or beam tetrodes. The latter valves

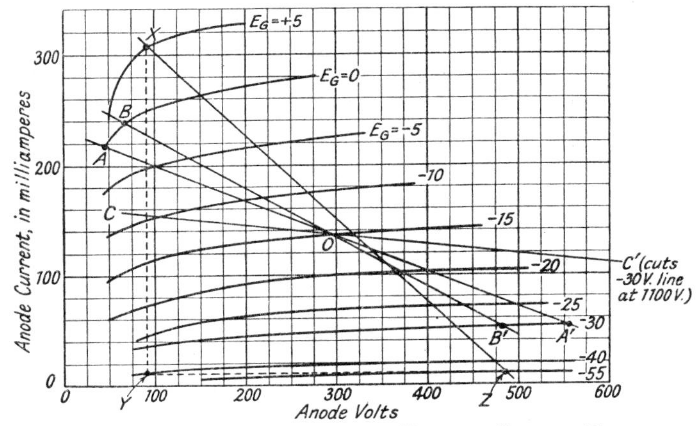

FIG. 10.13. CHARACTERISTICS OF A TYPICAL OUTPUT TETRODE

have the anode located farther from the screen and so disposed that the secondary-emission kink is removed. At the same time the screen current is appreciably reduced by aligning the screen and control grid, so that a more efficient valve results.

* XB is the peak voltage swing $= 2\sqrt{2}$ times the r.m.s. value. Similarly XB' is $2\sqrt{2}$ times the r.m.s. current. Hence the power is $XB \,.\, XB'/(2\sqrt{2})^2 = \frac{1}{8}XB \,.\, XB'$.

In use, the valves behave similarly to pentodes, and a typical set of characteristics is shown in Fig. 10.13. The line AOA' represents a condition giving no second harmonic distortion at all, for $AO = OA'$. This, however, is not the best condition, for there is a marked third-harmonic component which can be detected by noting the increments in anode current for equal increments of grid voltage. These are shown in the table below, which shows the change in current per **5** volts grid change.

TABLE OF ANODE CURRENTS IN FIG. 10.13

Grid Bias	*Load Line AA'*		*Load Line BB'*	
	I_a	*Increments*	I_a	*Increments*
−30	55		52	
		15		18
−25	70		70	
		35		30
−20	105		100	
		30		35
−15	135		135	
		30		35
−10	165		170	
		35		35
− 5	200		205	
		15		35
0	215		240	

It will be seen then working from the middle (−**15 V**) we have, on each side, changes of 30, 35, and 15 mA respectively for successive increments of 5 volts. This will give a wave with a marked third harmonic, as shown in Fig. 10.14. Note that with third harmonic the wave is symmetrical. This is true of any odd harmonic.

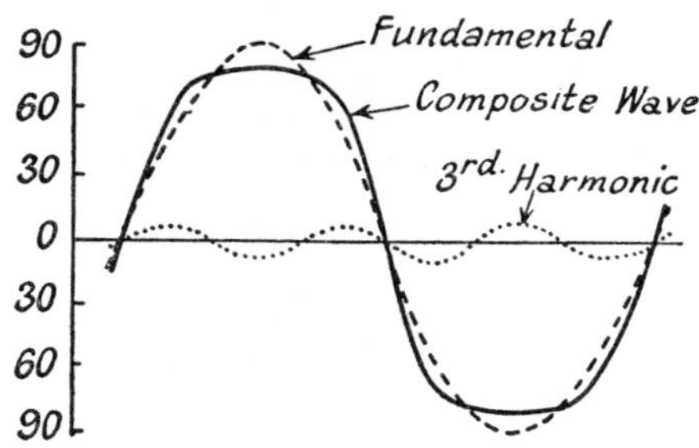

FIG. 10.14. ILLUSTRATING INFLUENCE OF THIRD HARMONIC ON WAVEFORM

On the other hand, if we reduce the load slightly to that shown by BOB' we introduce about 5 per cent second harmonic, for $OB = 1{\cdot}26\,OB'$, but over most of the line equal increments of grid voltage give equal anode current changes, so that we have very little third harmonic.

This behaviour is characteristic of pentodes and tetrodes. As we increase the load the second-harmonic content falls to zero and then rises again (this time due to a flattening of the top of the curve instead of the bottom). The third harmonic behaves similarly but

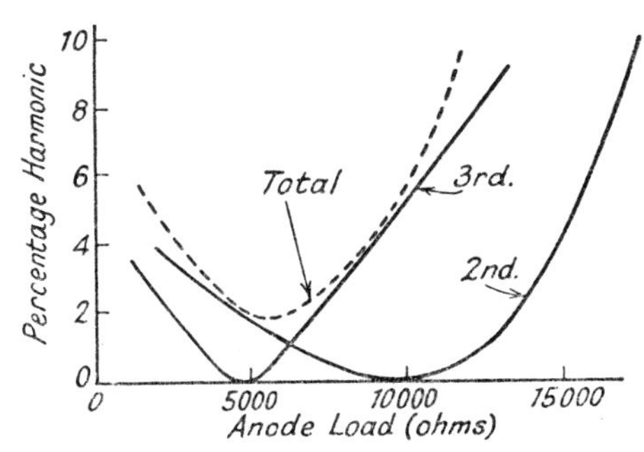

FIG. 10.15. ILLUSTRATING THE EFFECT OF LOAD ON HARMONICS

falls to zero much earlier and we usually find that at the point of zero second harmonic the third has risen alarmingly. The best condition is between the two as shown in Fig. 10.15.

It is difficult to locate the best load from the characteristics but curves similar to Fig. 10.15 are usually available from the makers.

As an approximation the following procedure may be used. First determine

α = current swing for a positive grid swing of 0·71 times the peak grid swing.

β = current swing for a negative grid swing of the same value.

Then evaluate three parameters as under

$$A = \frac{\text{positive peak current}}{\text{negative peak current}}$$

$$B = 1 + \alpha/\beta$$

$$C = 1{\cdot}41\beta/\text{negative peak current}$$

Then

$$\text{Percentage 2nd harmonic} = \frac{A - 1}{1 + A + BC} \times 100$$

$$\text{Percentage 3rd harmonic} = \frac{1 + A - BC}{1 + A + BC} \times 100$$

Impedance Limiting

It should be noted that tetrodes or pentodes should not be run with high loads. The line COC' (Fig. 10.13) represents such a load, and shows that if the full $\pm$ 15 volts grid swing is applied the anode voltage will swing to a voltage of over one thousand, which may easily cause a breakdown of the insulation.

A safety load of several times the normal load should therefore be connected permanently across the output or else a safety spark gap incorporated.

The average loudspeaker impedance, while fairly constant over the middle registers, rises rapidly in the upper frequencies. This causes the load line to swing round, giving rise to the same trouble if a large high (audio) frequency input is applied. Even if no damage results, very bad distortion will be produced. To avoid this, a limiting load is often connected in parallel, sometimes with a capacitor in series, the capacitance being chosen such that its reactance is high at normal frequencies, but falls rapidly above a few thousand hertz thereby bringing the limiting resistance into play.

Choice of Valves

Having decided upon the valve to obtain the required output one knows the grid input required. If sufficient voltage can be obtained direct from the detector this is all that is necessary. Valves are made to-day which will give several watts out with only a few volts peak signal input, which can easily be obtained from a diode detector. Some commercial receivers use this system.

In general, however, one a.f. amplifying stage is desirable. This may be the detector itself if an amplifying detector is used. If not, a first a.f. valve must be employed capable of providing this voltage. Due to the coupling impedance in the anode circuit the anode voltage will, in general, be less than the h.t. voltage, particularly if resistance coupling is used, and the valve must be capable, with the actual effective anode voltage, of providing without distortion an anode swing equal to the voltage required. This involves as a corollary that to provide this anode swing the grid swing required on the input to the valve must be within the limits of the grid bias used.

It must be remembered that, in a design of this type, allowance has to be made for the peak condition, i.e. the strongest possible signal likely to be handled under full modulation.

Class B Operation

The ordinary low-frequency amplifier operates as a Class A amplifier, being biased approximately to the middle of the straight portion of the characteristic. This involves a large steady anode

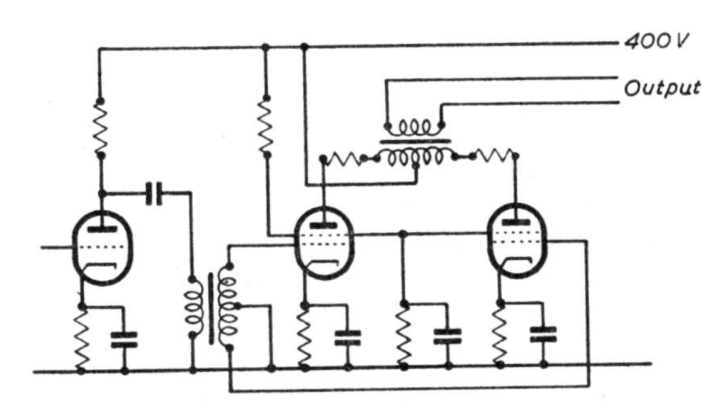

Fig. 10.16. Two Tetrodes Arranged for Class AB2 Operation (Positive Drive)

current which is the same whether the signal is large or small. Particularly with battery-operated receivers, this involves a waste of current, since the periods when anything like full modulation is in operation are only a small fraction of the total time.

This has led to the introduction of Class B amplifiers for audio-frequency purposes. As explained in the chapter on transmitters, this form of operation consists in biasing the valves to such a point that the anode current is nearly zero. Positive half-cycles of grid voltage then cause the anode current to swing over the operative portion of the characteristic, while the negative half-cycles produce no appreciable effect.

In a transmitter the anode voltage maintains itself over this portion of the cycle by virtue of the tuned circuit in the anode. With an audio-frequency amplifier this is not permissible because we are dealing with a variety of frequencies and any undue resonance cannot be tolerated. It is necessary, therefore, to use two valves in push-pull, one of which handles one half-cycle and the other the next. The outputs are combined in the normal manner. A typical circuit is shown in Fig. 10.16.

For Class B operation the valves are often designed to run into the positive grid region. Such a valve absorbs power on its input and has therefore to be fed from a small power valve operating through a *driver transformer* which is designed in the same way as an output transformer to feed into the load provided by the grid current drive.

The nomenclature with push-pull output stages is similar to that for transmitting valves, as discussed on page 321, except that, because a linear output is required, Class C working is not used. The various arrangements are thus as follows

Class A: Valve biased to its mean current, which does not vary with the signal.

Class B: Valve biased to cut-off so that $i_a = 0$ with no signal and peaks to a maximum on positive grid swings. (This is rarely used in a.f. amplifiers.)

Class AB1: Valve biased to a small standing current, between 5 and 25 per cent of the peak current.

Class AB2: As Class AB1 but with the grid running positive on the peaks.

Fig. 10.16 shows two tetrodes arranged to operate in this Class AB2 (positive drive) condition. The output of the two valves in normal Class A push-pull is 14·5 watts with 9 per cent total distortion. Under Class AB2 conditions 60 watts output may be obtained with only 3 per cent distortion. It is important to keep the screen voltage constant, but a separate power unit should *not* be used, for if the main h.t. supply to the anodes fails the screens will draw excessive current and the valves will be damaged.

The operating condition of each valve is shown at XZ in Fig. 10.13. The valve normally operates at 25 volts bias and runs up to +5 volts. As the valve runs towards zero its partner runs negative. The total swing, therefore, is from -25 to $+5$ on one valve and from -25 to -55 on the other. The peak power output is thus represented by the triangle XYZ. YX is the peak current swing and YZ the peak voltage swing so that the r.m.s. power is $\frac{1}{2}YX.YZ$.

This is the power supplied for the first half-cycle of input, and as we have just seen, this power is provided almost entirely by one valve of the pair. During the next half-cycle of input the second valve becomes operative and supplies a similar amount of power. Hence the power $\frac{1}{2}YX \times YZ$ is the total power output, being derived from each valve in turn. With the conditions shown $P = \frac{1}{2}.300.396.10^{-3} = 59{\cdot}4$ watts.

It will be seen that the arrangement makes a more effective use of the valve for, whereas the total current swing is not greatly different from that which could be obtained in a Class A condition, the voltage swing is practically twice as great.

If it is desired to reduce the standing current, the bias may be increased, in which case the operation approaches a true Class B condition. This is frequently done with transistor circuitry, as discussed on page 501.

Output Transformer

The output transformer must be designed so that it reflects the correct load into the valves. With the example just given, the load line ZX corresponds to $396/0{\cdot}3 = 1{,}320$ ohms. If the secondary load is 600 ohms this will require the secondary turns to be $\sqrt{(600/1{,}320)} = 0{\cdot}675$ times the *half* primary. Hence the *overall* step-down will be $2(1/0{\cdot}675) = 2{\cdot}96{:}1$.

This is equivalent to saying that the effective anode–anode load for the two valves should be $600 \,.\, 2{\cdot}96^2 = 5{,}280$ ohms, and optimum loads are customarily expressed in this form. It will be seen that the anode–anode load is not twice the figure for one valve alone but *four* times.

The performance of a Class B pair can be expressed in terms of the voltage and current swings without actually plotting a load line (though this is necessary to assess the probable distortion).

The peak current swing (YX) is very nearly the same as the peak current $\hat{I}$. If E_0 is the h.t. voltage and E_s is the peak voltage swing (YZ) then

$$\begin{aligned}
\text{Power output (two valves)} &= \tfrac{1}{2}E_s\hat{I} \\
\text{Mean anode current} &= \hat{I}/\pi \quad \text{(per valve)} \\
&= 2\hat{I}/\pi \quad \text{(total)} \\
\text{Input power} &= 2E_0\hat{I}/\pi \\
\text{Efficiency} &= \tfrac{1}{4}\pi E_s/E_0 \\
\text{Primary load resistance } (a\text{-}a) &= 4E_s/\hat{I}
\end{aligned}$$

The difference between the input and output power is dissipated in the valves as heat, and the operating conditions must be chosen so that this dissipation does not exceed the safe limit for the valves in use.

Maximum Power Output

The requirement that the internal dissipation shall not exceed the safe limit applies to any valve, though it is of greatest importance in output stages. The manufacturers state the maximum permissible dissipation and the working conditions must be chosen so that this limit is not exceeded *at any part of the operating cycle.*

The internal dissipation is the product of the instantaneous anode voltage and current, and it is possible to plot a curve of the characteristics showing the maximum permitted anode current for varying anode voltages as shown in Fig. 10.17 for a maximum dissipation of 10 watts. If the standing bias is to be —10 V, a load of 4250 ohms

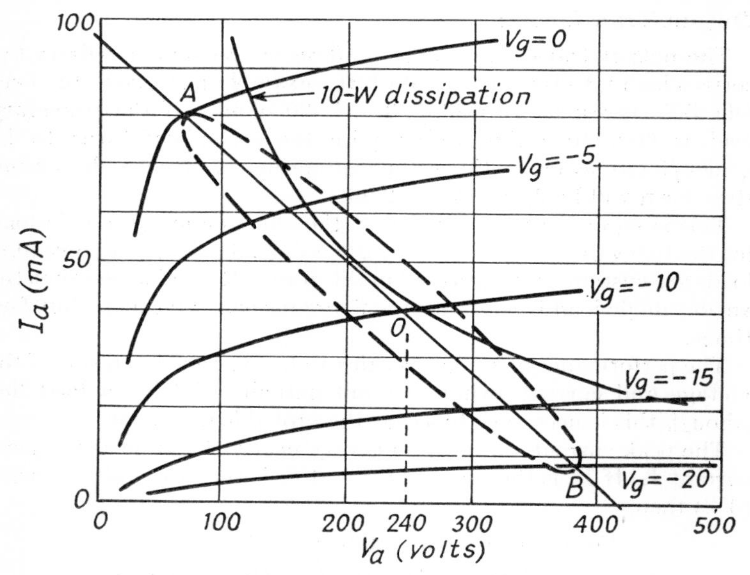

FIG. 10.17. ILLUSTRATING DISSIPATION LIMIT AND LOAD ELLIPSE

must be used to comply with the 5 per cent distortion criterion ($AO = 1{\cdot}22\ OB$) and still lie within the 10 W dissipation boundary.

It will be seen that this gives a working point of 240 V and 40 mA which is only 9·6 W; but as V_a falls the load line becomes tangential to the 10 W dissipation line. Hence it is not correct to assume that the standing conditions represent the maximum dissipation.

These conditions, moreover, only apply for a resistive load, and while this is usual there are cases where the load is partially reactive. In such circumstances the load line becomes an ellipse and the conditions are entirely changed. Let us assume, for example, that the load is inductive and that at the working frequency $L\omega = \frac{1}{4}R$. The current will then lag 14·5° behind the voltage. At the point O

where the a.c. component of the voltage swing is zero the current will be $-\frac{1}{4}I_{max}$ and will not go through zero (i.e. the standing current of 40 mA) until the voltage swing is 45 V (i.e. an anode voltage of 240 − 45 = 195 V). By plotting the instantaneous values of (a.c.) voltage and current over the cycle it will be found that the load line is an ellipse as shown dotted in Fig. 10.17.

It will be seen that this exceeds the maximum dissipation over a considerable portion of the cycle and to avoid over-loading, the working point would need to be moved appreciably to the left (lower V_a and I_a) while the effective slope would need to be reduced to avoid distortion. Hence the presence of the inductance would result in a considerable reduction in the output.

The conditions are slightly modified when two valves are being used as a Class B pair. Here each valve is only in operation for half the complete cycle so that it is permissible to use a load line which crosses the dissipation line for part of the swing. Theoretically this should permit the use of a dissipation line corresponding to twice the normal rating, but in practice a more conservative rating is usually necessary. The question is considered further in the discussion of transistor output stages on page 501.

10.3. A.F. Transistor Amplifiers

The considerations which govern audio-frequency transistor circuitry are not basically different from those discussed in the previous

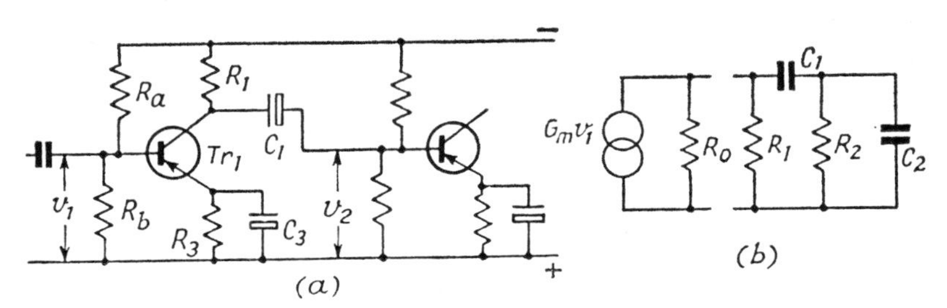

FIG. 10.18. TRANSISTOR R-C AMPLIFIER

section, but since the bipolar transistor is a current-operated device the impedances involved are very much lower.

Thus Fig. 10.18 illustrates an R-C coupled transistor amplifier using the common-emitter mode which, as said earlier, is the most suitable for low and medium frequencies*. The equivalent circuit

* Throughout this section p–n–p transistors are shown, but the circuitry is equally applicable to n–p–n types, using a positive supply voltage.

is shown in Fig. 10.18 (*b*), from which it will be seen that the load in the collector circuit of Tr_1 is made up of the network $R_1C_1R_2C_2$, where R_2 and C_2 are the input resistance and capacitance of the following stage. At medium frequencies the reactances of C_1 and C_2 may be neglected so that the effective load $Z = R_1R_2/(R_1 + R_2)$ and the stage gain is $G_mZR_o/(Z + R_o)$. The resistance R_2, however, is not an external component as is the case in a valve circuit (c.f. Fig. 10.1), but is the input resistance of the succeeding transistor Tr_2. This is of the order of 1000 ohms so that there is little advantage in making R_1 much larger than this. Actually R_1 would be several thousand ohms (e.g. 4·7 kΩ) in order to provide correct operating conditions, as explained later, but the gain is determined principally by R_2.

This is not such a serious limitation as might at first appear. Taking $G_m = 35$ mA/V, $R_2 = 1000\ \Omega$ and $R_o = 30$ kΩ the stage gain would be 34.

At low frequencies the reactance of C_1 will begin to become appreciable, so that v_2 is less than the full amount, the proportion actually applied to Tr_2 being $R_2/(R_2 + 1/j\omega C_1)$. There is, however, a compensating effect. R_2 is normally small compared with R_1 so that the effective load is determined mainly by R_2. But if the reactance of C_1 becomes appreciable this shunting action is reduced so that more gain is obtained from the transistor and this partially offsets the attenuation of the C_1R_2 network.

If we assume that the effective load Z is small compared with R_o (as it is in practice) the stage gain is approximately G_mZ. Now, Z is the impedance of R_1 in parallel with C_1R_2, so that

$$\frac{v_2}{v_1} = G_m \frac{R_1\,(R_2 + jX)}{R_1 + R_2 + jX} \cdot \frac{R_2}{R_2 + jX}$$

$$= G_mR_1R_2/(R_1 + R_2 + jX)$$

where $$X = 1/\omega C_1$$

When the reactance of C_1 is negligible, this reduces to

$$G_mR_1R_2/(R_1 + R_2)$$

Hence the ratio of the output at low frequencies to that at mid-frequencies becomes

$$(R_1 + R_2)/(R_1 + R_2 + jX)$$

If $R_1 = kR_2$ and $X = nR_2$, the expression may be written in the form

$$(1 + k)/\sqrt{[(1 + k)^2 + n^2]}$$

from which the relative bass cut may be assessed at any frequency. If $k = 5$, the loss will be 10 per cent when $n = 2{\cdot}8$; i.e. $R_2C_1\omega = 0{\cdot}36$. This is an interesting result which shows that the bass response is maintained at appreciably lower frequencies than with a valve (for which 10 per cent cut results when $R_2C_1\omega = 2$). With the values previously assumed ($R_1 = 5\ \text{k}\Omega$, $R_2 = 1\ \text{k}\Omega$) a capacitor of $1{\cdot}15\ \mu$F would maintain 90 per cent response at 50 Hz.

At the high-frequency end of the scale the reactance of C_1 is negligible but C_2 begins to shunt R_2. The effective gain thus becomes $G_mR_p/(1 + R_pC_2\omega)$, where $R_p = R_1R_2/(R_1 + R_2)$, which is 90 per cent of the mid-frequency gain when $RpC_2\omega = \frac{1}{2}$. With typical values this occurs at a frequency approaching 1 MHz. However, the performance deteriorates much earlier than this due to the falling off of the mutual conductance G_m. This is not the intrinsic mutual conductance g_m but is a derived parameter which decreases rapidly with frequency.

It was shown in Chapter 6 that the voltage gain in common-emitter mode (neglecting the source resistance R_s) is $\alpha' Z/(r_b + \alpha' r_e)$. If this is written as G_mZ, then $G_m = \alpha'/(r_b + a'r_e)$, which is roughly proportional to α'. This falls to unity at the cut-off frequency f_1, which for a small-signal a.f. transistor is of the order of 500 kHz, so that using the approximate 6 dB per octave rule one would expect the performance to begin to deteriorate at about 15 kHz.

With the common-base mode $G_m = g_m = \alpha/r_e$, and since α does not fall off as rapidly as α' (being 3 dB down at f_α = approximately $1{\cdot}2f_1$ for alloy-diffused transistors and $2f_1$ for drift-field types), the high-frequency response is much better. Nevertheless, because of its other advantages, the common-emitter mode is usually preferred.

Choice of Working Point

The operating conditions must be chosen to provide the required current swing, which can be done by drawing a load line across the characteristics as explained in Chapter 6 (Fig. 6.10).

This determines V_c and I_c, whence the collector resistance $R_1 = (V_{battery} - V_c)/I_c$. Note that this is *not* the load resistance, and will normally be about 5 times R_2. The base voltage may then be set by suitable proportioning of R_a and R_b (Fig. 10.18) to give the appropriate base current, and while this can be done with the emitter grounded a more stable condition is obtained with a resistor R_3 in the emitter circuit as was explained in Section 6.5, where it was also shown that to avoid the danger of thermal runaway the working point should be such that the voltage drop on R_1 is not less than half the battery voltage (the half-supply voltage principle).

R_3 must be by-passed with a capacitor C_3 to avoid signal-frequency feedback, the value being chosen so that $1/C_3\omega$ is not more than about one-tenth of R_3.

If several stages are being used, they may need to be decoupled to avoid common-impedance feedback (see p. 416). In determining the working point, the resistance R_1 must include the decoupling resistor.

Transformer Coupling

Fig. 10.19 illustrates a transformer-coupled transistor stage. The primary is in the collector circuit, while the secondary is connected to the base of Tr_2 with the earthy end taken to the tap on the

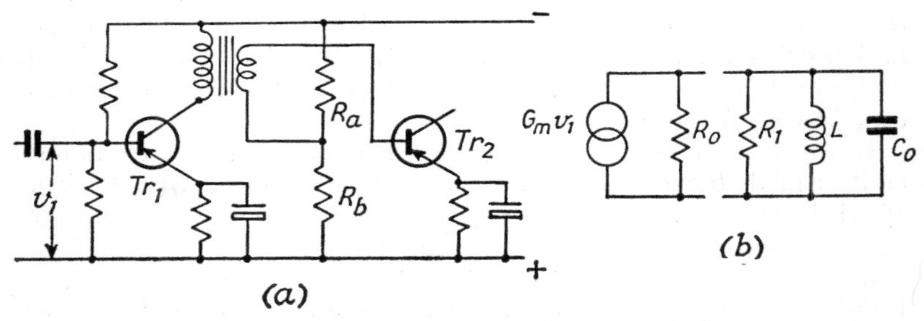

FIG. 10.19. TRANSFORMER-COUPLED TRANSISTOR AMPLIFIER

voltage divider R_aR_b, this slight rearrangement being necessary to prevent the negligible resistance of the secondary from short-circuiting R_b.

Such a circuit would be designed on a power-transfer basis. To obtain the maximum (external) power gain the equivalent primary impedance Z_p must be equal to the output impedance R_o of Tr_1. This can be achieved at mid-frequencies by adopting a (step-down) transformation ratio of $\sqrt{(R_o/R_i)}$, where R_i is the input impedance of Tr_2.

This equivalent primary impedance R_1 will be shunted by the primary inductance and the stray capacitance C_o (including the reflected input capacitance of Tr_2), as shown in the equivalent circuit of Fig. 10.19 (*b*). At low frequencies the effect of C_o can be ignored so that the current G_mv_1 will divide through R_o, R_1 and L. The portion of interest is that through R_1, which is in the ratio,

$$(1/R_1)/(1/R_o + 1/R_1 + 1/j\omega L)$$

Hence

$$i_{R1} = G_mv_1/\sqrt{[(1 + R_1/R_0)^2 + R_1{}^2/\omega^2L^2]}$$

and the voltage across the secondary, $V_2 = i_{R1}R_1/n$. When $\omega L \gg R_1$ (i.e. at medium frequencies) the second term is negligible and the voltage gain is

$$v_2/v_1 = G_m R_1 R_o / n(R_o + R_1)$$

If $R_1 = R_o = 30\ \text{k}\Omega$, $G_m = 35$ mA/V and $n = 5{\cdot}5$, the stage gain is 95.

As the frequency is reduced so that ωL becomes comparable with R_1, the current through R_1 (and hence the gain) falls. If $R_1 = R_o$ the gain falls to 90 per cent of its mid-frequency value when $\omega L = R_1$, and to $1/\sqrt{2} = 71$ per cent with $\omega L = \frac{1}{2}R_1$. The primary inductance may thus be calculated in terms of the permissible bass cut. With the values quoted, and assuming 10 per cent drop at 50 Hz, $L = 96$ H.

At high frequencies the effect of C_o becomes important. This is made up of the output capacitance of Tr_1 including circuit strays, plus the reflected input capacitance of $Tr_2 = C_{in}/n^2$. In a practical circuit this would be of the order of $30 + 1000/5{\cdot}5^2 = 63$ pF, which would resonate with an inductance of 96 H at a frequency of 12,800 Hz. Hence there would be a slight resonant rise in the output similar to that illustrated in Fig. 10.10 (though not in this case due to leakage inductance) followed by a rapid cut-off.

Transformer coupling offers little advantages for voltage amplification because two R-C stages in cascade will provide greater amplification at less cost. When current or power gain is required, however, as in a driver stage, transformer coupling is essential.

Output Stages

For the output stage the transistor is designed to handle large currents (of the order of amps rather than milliamps). Since this involves appreciable internal power dissipation a more massive construction is adopted as described in Section 6.5 (Fig. 6.26) and the transistor is mounted on the chassis or some form of heat sink to conduct the heat and prevent an undue rise in temperature.

The circuitry is similar to that already described with due allowance for the larger currents involved. A Class A output stage is illustrated in Fig. 10.20, using the common-emitter mode, which is the most usual. Since the peak collector current may be 1 A or more, the peak input current may be of the order of 20–30 mA, which requires a transformer coupling from the preceding (driver) stage, and to avoid undue variation of the working point the bottom portion of the bias voltage divider, $R_a R_b$, must be by-passed with a suitable capacitor.

The conditions are stabilized, as usual, by a resistor in the emitter circuit, but because of the large current this resistance is only of the order of 1 ohm or less. To by-pass this at signal frequencies would require quite impractical values of capacitance so that it is customary to omit the capacitor and accept the resulting negative feedback. This is no disadvantage in an output stage and in fact

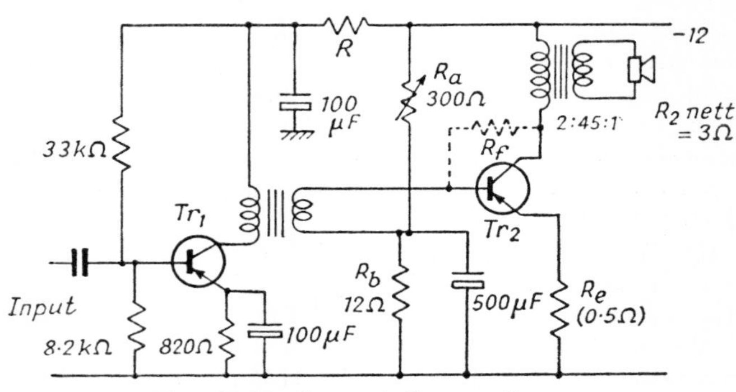

Fig. 10.20. Class A Output Stage

additional negative feedback is often provided, as shown by R_f in the figure.

The output load can be determined from the characteristics as in Fig. 10.21, which represent a typical power transistor. A supply voltage of 12 V has been assumed, which allowing for the voltage drop on the output transformer and the emitter resistor would leave

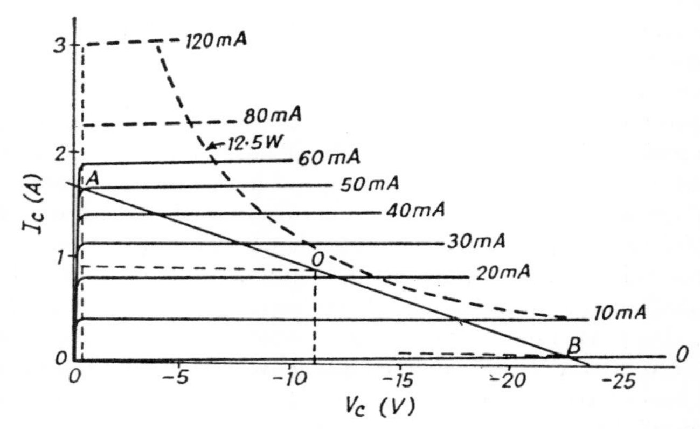

Fig. 10.21. Characteristics of Typical Output Transistor

The figures against the curves refer to I_b.

approximately 11 V for V_{ce}. To allow a margin of safety a standing current of 0·9 A could be chosen, giving a dissipation of 9·9 W, and a load of 14 ohms about this working point would provide a reasonably low distortion. ($OB = 1{\cdot}1\,OA$, giving less than 3 per cent second harmonic.)

The peak voltage swing is $22{\cdot}5 - 0{\cdot}5 = 22$ V while the peak current swing is $1{\cdot}65 - 0{\cdot}1 = 1{\cdot}55$ A. These are the overall peak swings, so that the r.m.s. power is

$$(V_s/2\sqrt{2})(I_s/2\sqrt{2}) = V_sI_s/8 = 22 \times 1{\cdot}55/8 = 4{\cdot}25 \text{ W}.$$

Part of the load is made up of the primary resistance of the transformer and the emitter resistance, which have been assumed to be 0·5 ohm each. The reflected secondary impedance should thus be $14 - 1 = 13$ ohms, and the power output is $4{\cdot}25(13/14) = 4$ W, giving an overall efficiency of $4/9{\cdot}9 = 40{\cdot}5$ per cent. (The maximum efficiency in Class A is 50 per cent, which would be obtained if $V_s = 2V_B$ and $I_s = 2I_o$, giving a power of $4V_BI_o/8 = \frac{1}{2}V_BI_o$. This is clearly not attainable in practice because of the knee on the characteristics.)

If the secondary load is 3 ohms net (after deducting the secondary resistance which should only be a small fraction of an ohm) the output transformer ratio is

$$\sqrt{(13/3)} = 2{\cdot}08$$

It remains to design the preceding driver stage which must be sufficient to swing the base current of Tr_2 over the required range. From the input characteristic (Fig. 10.22) it will be seen that the mean current of 25 mA requires a base voltage of 480 mV, so that to reduce i_b to zero a peak swing of this order is required. To this must be added voltage drop on the emitter resistor R_e plus the secondary resistance of the driver transformer. Allowing 1 ohm for these two together an additional 50 mV peak will be required, making 530 mV peak swing.

The supply voltage to Tr_1 will be less than 12 because of the drop on the decoupling resistor R. If we assume a net figure of 9 V we can safely assume a peak swing of 8 V in the collector of Tr_1, which will require a transformer ratio of $8/0{\cdot}53 = 15$. The primary current swing will then be $50/15 = 3{\cdot}3$ mA, which could be provided by a normal small-signal transistor with a standing current of 2 mA, requiring an input of the order of 5 mV depending on the effective G_m of Tr_1.

R_a would be adjusted to give a standing current of 0·9 A in Tr_2, the corresponding base voltage divider of Tr_1 being chosen to give a standing current of 2 mA.

Because of the non-linear input characteristic the positive swing of i_b from 25 to 50 mA only requires an increase of 120 mV in voltage; but by this very fact the input impedance (and hence the effective load on Tr_1) is correspondingly reduced and the drive voltage is reduced accordingly. Tr_1 thus operates with a peak *current* swing

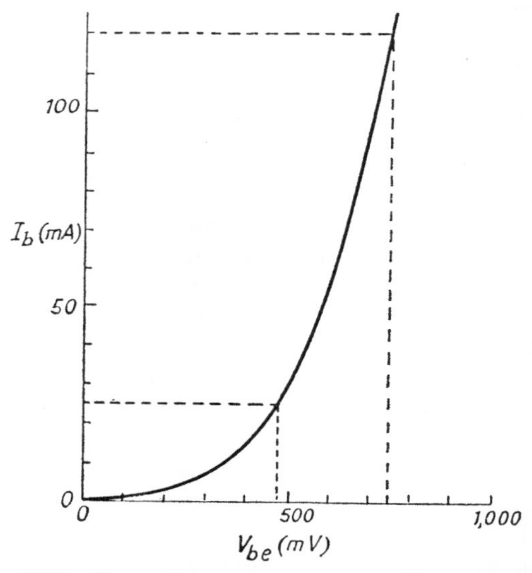

FIG. 10.22. INPUT CHARACTERISTIC FOR TRANSISTOR OF FIG. 10.21

of 1·65 mA in both directions, but the voltage developed automatically adjusts itself to the changing load. Nevertheless, a small increase in input voltage can produce a relatively large increase in i_b, and this may cause a disproportionate increase in i_c. This is known as overdriving and could produce thermal runaway if the heat dissipation were only just adequate for normal operation. It is desirable therefore to arrange the phasing of the driver transformer so that an increase in i_c of Tr_2 is produced by a decrease in i_c in Tr_1. Excess input will then drive Tr_1 to cut-off, and prevent overdriving Tr_2.

The non-linearity will also be responsible for introducing a certain amount of distortion. This is partially offset by the negative feedback produced by the un-bypassed emitter resistance R_3, but it may be still further reduced by applying additional negative feedback.

This may be done around the output stage alone by a resistor between collector and base as shown dotted as R_f in Fig. 10.20, or by feedback round the whole circuit from the secondary of the output transformer to the base of Tr_1.

If this is done the drive conditions must be suitably modified. Thus if R_f in Fig. 10.20 is such as to produce a negative feedback of 3 dB the drive voltage must be increased to $530\sqrt{2} = 750$ mV. The driver transformer ratio then becomes 10·7 and the current swing required from Tr_1 is 4·5 mA, which requires a slightly higher standing current of around 2·5 mA.

An approximate design may be derived without reference to the characteristics. If the total swing of the collector current in the output stage is I_c, the required base-current swing is I_c/α'. The transformer ratio may then be calculated to meet the requirements of the driver stage. Thus in the example above, I_c is 1·7 A, while α' is 35. Assuming a total driver current swing of 3·3 mA as before, the required ratio is $(1{,}700/35)/3{\cdot}3 = 14{\cdot}8$.

General Precautions

It is important in an output stage to ensure that the arrangements for heat conduction are adequate so that no undue rise in temperature can occur. Among other things, a rise in temperature causes a rapid rise in collector leakage current I_{co}, which can cause a further rise, leading ultimately to thermal runaway. The circuit should therefore be designed so that overdriving and/or changes in bias conditions do not lead to undue peak currents, and temperature-sensitive circuit elements are sometimes introduced (e.g. across R_b in Fig. 10.20) to limit the gain if the temperature rises.

The other major precaution is concerned with the maximum collector voltage. The load must be such that the peak negative voltage swing does not exceed the rated limits. In particular the load should never be disconnected when the circuit is in operation. This would make the load line horizontal requiring a (theoretically) infinite voltage swing, which would destroy the transistor.

Class B Stages

The disadvantages of a Class A stage are the limited efficiency (50 per cent maximum) and the high standing current (approximately $\frac{1}{2}\hat{I}_c$). More efficient performance can be obtained by using two transistors in push-pull and operating them in Class B conditions. This kind of operation was discussed on page 489 in relation to valve circuits and the same general principles apply.

In essence the basis of the arrangement is that the two transistors are biased nearly to cut-off. Input is supplied to the pair symmetrically as shown in Fig. 10.23. On the first half-wave Tr_1 is driven up to its peak current, while Tr_2 remains inoperative. On the next half-wave, conditions are reversed so that each transistor delivers one-half of the power in turn. The behaviour is illustrated by the

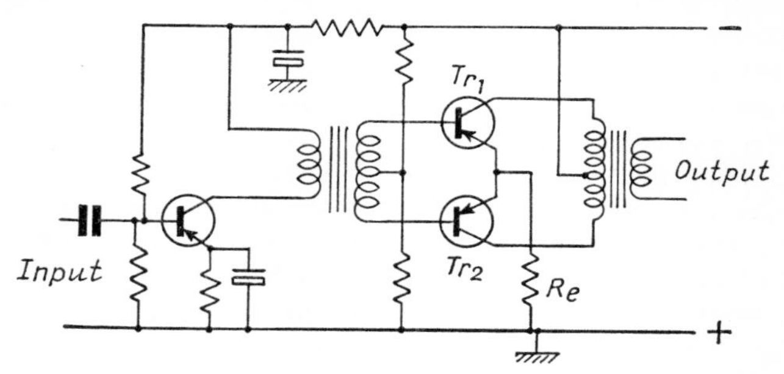

FIG. 10.23. PUSH-PULL OUTPUT STAGE

load line AOB in Fig. 10.24 in which the full-line characteristics relate to Tr_1 and the dotted curves to Tr_2.

The operating point O is the battery voltage but the base voltage (on each transistor) is such that the collector current is zero. Then Tr_1 supplies the power for the first half-wave, indicated by the

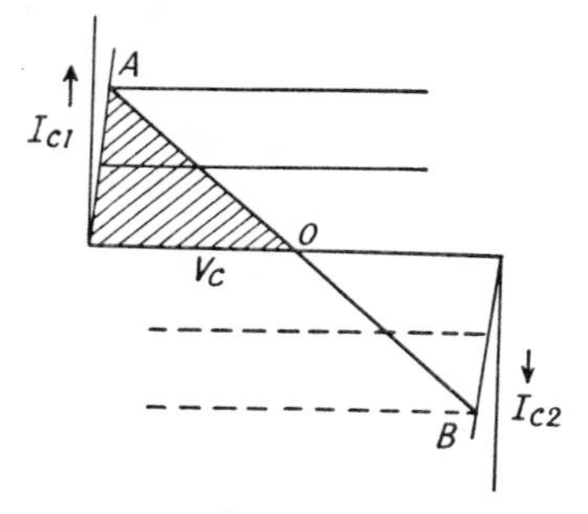

FIG. 10.24. ILLUSTRATING ACTION OF CLASS B OUTPUT STAGE

shaded area, while the second half is supplied by Tr_2, the two being combined in the centre-tapped output transformer of Fig. 10.23.

In practice true Class B operation such as this is not used. While two transistors (of the same type) may have similar characteristics over the normal range of operation (and are often selected as matched pairs) this does not hold good for very small values of i_b so that appreciable mismatching can occur around the point O, which produces what is known as *crossover distortion*. To avoid this the transistors are operated in Class AB1 conditions, i.e. with a small standing current about 5 per cent of $\hat{I}_c$. This results in a smooth transition from one to the other.

Class B working permits greater utilization of the transistor, with an appreciably lower load and larger peak currents. Fig. 10.25 illustrates the use of Class B operation with the characteristics of Fig. 10.21. The line AOB represents a load of 4 ohms through a working point of 12 V, 0·2 A. (Because of the smaller standing current the voltage drop on the primary and emitter resistances can be ignored.) This load causes the transistor to peak up to 3 A, so that from $i_b = 0$ to $i_b = 120$ mA the current swing I_s is 2·9 A, while the voltage swing $V_s = 12{\cdot}5 - 1 = 11{\cdot}5$ V. With two transistors in push-pull the combined output is

$$(V_s/\sqrt{2})(I_B/\sqrt{2}) = \tfrac{1}{2}V_sI_s = 16{\cdot}5 \text{ W}$$

which is over four times that obtainable from a single transistor in Class A.

The d.c. power is the product of the battery voltage V_o and the mean d.c. current. This is not the standing current I_o, but the mean of the sinusoidal swing from zero to $I_p = 2I_p/\pi$ (for the complete cycle). Hence d.c. power $= 2V_oI_p/\pi$ and the efficiency is $\tfrac{1}{2}V_sI_s/(2V_oI_p/\pi) = \dfrac{1}{4\pi}\cdot\dfrac{V_s}{V_o}\cdot\dfrac{I_s}{I_p}$. This is a maximum of 78·5 per cent if $V_p = V_o$ and $I_s = I_p$, but this is not attainable in practice as will be clear from Fig. 10.25. With the values shown $V_s/V_o = 11{\cdot}5/12$ and $I_s/I_p = 2{\cdot}9/3$ so that the efficiency is 74 per cent.

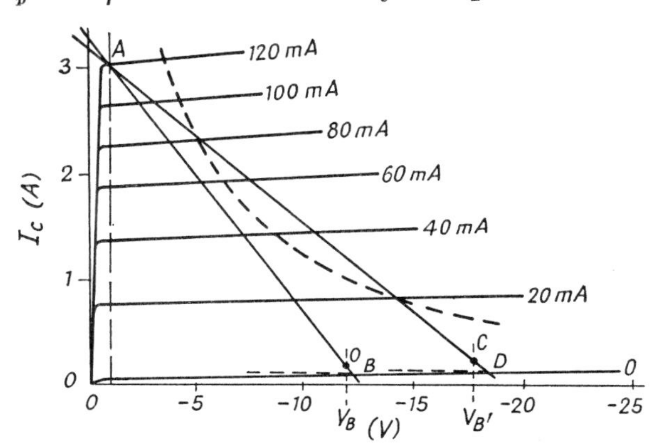

FIG. 10.25. CHARACTERISTICS OF FIG. 10.21 TRANSISTOR IN CLASS B

As said, the mean current is not I_o but $2I_p/\pi$. Hence in the quiescent state with no input a current I_o will be drawn by both transistors, making $2I_o$ total. When signal is applied this will rise to a maximum of $2I_p/\pi$ (drawn by each transistor in turn) at full drive. With the example chosen the quiescent current is 0·4 A rising to 1·92 A at full drive.

The load of 4 ohms will include R_e and the resistance of half the transformer primary. If $R_e = 0{\cdot}5\ \Omega$ and $\frac{1}{2}R_p = 0{\cdot}25\ \Omega$ the equivalent secondary load $= 3{\cdot}25\ \Omega$, which requires a ratio of $\sqrt{(3{\cdot}25/3)} = 1{\cdot}07$ per half, or 2·14 overall (assuming a 3 Ω secondary load as before).

The driver transformer is designed in the same manner as for a Class A stage. Each half of the secondary in turn has to supply a current of i_b peak. With the load AOB this requires 120 mA, which from Fig. 10.22 requires $V_{be} = 750$ mV. The mean current of 60 mA would require 650 mV, so that allowing as before for the drop on R_e and the secondary resistance a swing of some 700 mV will be required.

Because of the symmetrical nature of the output stage less decoupling of the driver stage is required, and may often be omitted altogether. If we assume that a peak voltage swing of 10 V is possible the transformer ratio *per half* will be $10/0{\cdot}7 = 14:1$. The current swing will then be $120/14 = 8{\cdot}5$ mA so that the driver will need to operate with a standing current of the order of 10 mA. This will require a somewhat larger transistor than for a Class A stage.

Dissipation

The load line AOB in Fig. 10.25 is well within the dissipation limit, and the dissipation is still further reduced by the fact that the transistor is only in operation for half the cycle. Over the half-cycle the mean voltage swing is V_s/π, and the mean current I_p/π. Hence the mean power dissipation is $V_sI_p/\pi^2 \simeq V_sI_p/10$ which with the load AOB is only 3·3 W.

However this is not the total loss. There is appreciable dissipation in the transistor due to the presence of the standing current I_o during the idle half-cycle. This is approximately

$$\tfrac{1}{2}V_BI_o = \tfrac{1}{2} \times 12 \times 0{\cdot}2 = 1{\cdot}2\ \text{W}$$

for the load line AOB. This is an appreciable fraction of the total and any increase in I_o due to a change in the bias conditions will increase the loss still further.

The collector leakage current I_{co} produces further losses. This is normally only a few per cent of $\hat{I}_c$, but it increases rapidly with temperature. Finally there is the disspation due to the base current, which will be V_b (mean) $\times$ $(2/\pi)\hat{I}_b$, but this is normally negligible.

Hence the total dissipation is approximately

$$0{\cdot}1 V_s I_p + \tfrac{1}{2} V_B I_o + V_B I_{co}$$

If we assume $I_{co} = 0{\cdot}1$ A, the total dissipation for the load AOB $= 0{\cdot}1 \times 11{\cdot}5 \times 3 + \frac{1}{2}(12 \times 0{\cdot}2) + 12 \times 0{\cdot}1 = 5{\cdot}7$ W.

Ambient Temperature

This is well below the permissible limit of $12\frac{1}{2}$ W for the particular transistor. However this limit is the absolute maximum and it is necessary to examine the permissible ambient temperature conditions to ensure the stability of the design. This depends on the efficiency of the arrangements, including the heat sink, for dissipating the heat. The relationship between power dissipation and junction temperature may be expressed in the form

$$P_d = (T_j - T_a)/(\theta_h + \theta_1 + \theta_t)$$

where T_j = maximum temperature of junction (°C)
T_a = ambient temperature (°C)
θ_h = heat sink coefficient = rise of temperature of the heat sink in °C per watt
θ_1 = transfer coefficient depending on the thermal conductivity between the heat sink and the transistor
θ_t = transistor thermal coefficient, which is a measure of the rise in junction temperature relative to the transistor casing.

θ_h and θ_1 are within the control of the designer, but θ_t is a function of the transistor itself. It is made as low as possible by the transistor manufacturer.

Typical values for a power transistor are $T_j = 90$°C and $\theta_t =$ 4°C/W. With a good heat sink (e.g. with the transistor mounted on the chassis) $\theta_h = 2{\cdot}2$°C/W, while θ_1 can vary between 0·2°C/W with the transistor in direct contact with the chassis to 0·5°C/W if a mica insulating washer is used.

Applying these values to the load AOB in Fig. 10.25, the maximum ambient temperature

$$\begin{aligned} T_a &= T_j - W(\theta_h + \theta_1 + \theta_t) \\ &= 90 - 5{\cdot}7(2{\cdot}2 + 0{\cdot}5 + 4) = 90 - 38{\cdot}5 = 51{\cdot}5\text{°C} \end{aligned}$$

Hence this arrangement would be inherently safe.

High-output Operation

It is evident, in fact, that the transistors of Fig. 10.25 would handle a greater power output if the battery voltage were increased. A possible arrangement is shown by the load line ACD. Here $V_B = 18$ and $I_o = 0{\cdot}25$ A.

$$\hat{I}_s = 3 - 0{\cdot}15 = 2{\cdot}85 \quad \text{while} \quad \hat{V}_s = 19 - 1 = 18$$

giving $P = \frac{1}{2} \times 2{\cdot}85 \times 18 = 26{\cdot}5$ W, at an efficiency of 75 per cent. It will be seen that this load line crosses the dissipation line over an appreciable part of its length but this is permissible because the transistor is only operating over half the cycle.

The dissipation is

$$0{\cdot}1(18 \times 3) + \tfrac{1}{2}(18 \times 0{\cdot}25) + 18 \times 0{\cdot}1 = 8{\cdot}45 \text{ W}$$

Applying this to the temperature equation above we find that the maximum ambient temperature is 33°C, which is a safe enough margin for normal requirements.

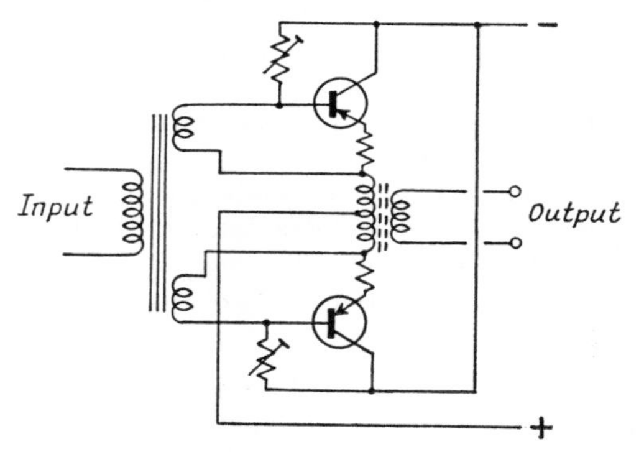

Fig. 10.26. Modified Form of Push-pull Circuit

Precautions are still necessary to avoid temperature rise (which can cause a rapid increase in I_{co}) which might be produced by overdriving or by a change in the bias arrangements causing an increase in I_o, which has a material influence on the dissipation as shown earlier. As a precaution, the emitter resistor R_e may be wound with copper or other wire having a positive temperature coefficient (as against the zero t.c. materials usually employed). Temperature-sensitive elements may also be incorporated in the biasing circuits to reduce the gain if the temperature rises abnormally.

There are many modifications of the basic push-pull circuit. Fig. 10.26 shows an arrangement in which the collectors are at battery potential, which permits the transistors to be mounted direct on the chassis, an arrangement which is not only simpler but provides slightly better thermal conductivity.

Fig. 10.27 shows a so-called transformerless Class B output stage. By using a centre-tapped supply the output transformer may be eliminated but a driver transformer is still required. The loud-speaker impedance must be designed to provide the requisite load.

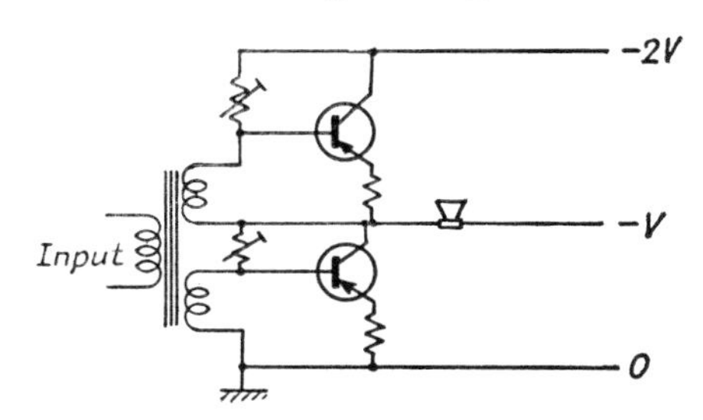

Fig. 10.27. Asymmetrical Class B Stage

Feedback arrangements may be incorporated in Class B stages in the same manner as in Class A amplifiers.

Because even a relatively small transistor will handle currents of the order of amperes the required power output can usually be obtained with collector voltages of the order of 50 V. Transistors are available, however, for operation at voltages well in excess of 100 V if required.

Complementary Operation

As said in Chapter 6, transistors are available having similar characteristics with either p–n–p or n–p–n construction. By using

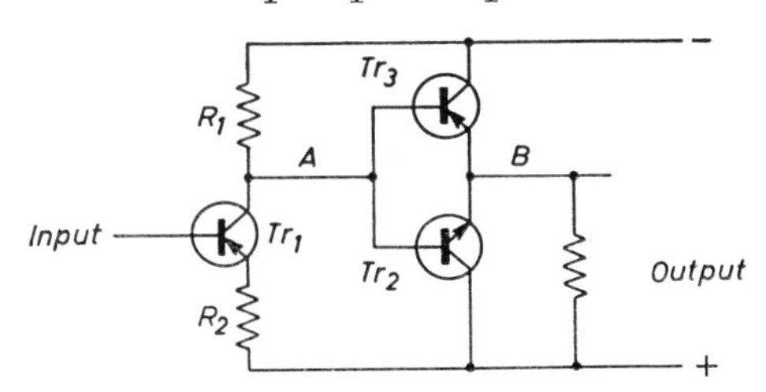

Fig. 10.28. Complementary Output Pair

such complementary pairs (preferably matched) the need for an input transformer can be obviated. Thus in Fig. 10.28, Tr_2 and Tr_3 are a complementary pair. If the voltage at A (derived from the input transistor Tr_1) goes positive, Tr_2 conducts while Tr_3 is cut off. Hence the potential of the point B moves towards the positive rail, while when A goes negative, B moves towards the negative rail. The resistors R_1 and R_2 must be arranged so that with no signal the point A stands midway between the positive and negative rails.

This arrangement is suitable for small outputs.

15 Watt Amplifier

For larger outputs matched pairs are not readily available. In such cases a normal push-pull output would be used with a symmetrical input derived from a complementary input pair, as in Fig. 10.29, which is a typical circuit incorporating many of the

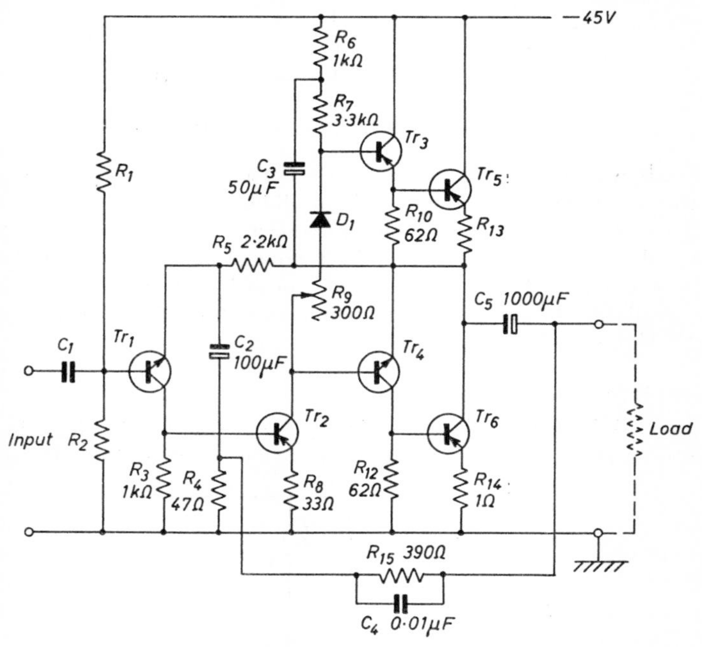

FIG. 10.29. TOBEY AND DINSDALE AMPLIFIER

features discussed in the preceding pages. This arrangement, which is reproduced by courtesy of the Editor of *Wireless World*, is known as the Tobey and Dinsdale circuit and is widely used in high-fidelity and public-address equipment.*

* Tobey, R., and Dinsdale, J., "Transistor audio power amplifiers," *Wireless World*, **67**, p. 565.

A feature of the circuit is that it is direct-coupled throughout, so that there is no need for any iron-cored components despite the fact that it will deliver a power output in excess of 15 W. Tr_1 is an n–p–n signal-frequency amplifier which feeds the base of the p–n–p transistor Tr_2. However, it also serves as a d.c. stabilizer to maintain the mid-point of the output stage accurately equal to half the supply voltage, which is essential for maximum undistorted output. The base of Tr_1 is held at a steady potential by the voltage divider R_1R_2, while its emitter is fed from the point P through R_5 (decoupled by C_2). Any variation in the quiescent potential of the point P is amplified by Tr_1 and the subsequent stages, providing a feedback which holds P constant.

The output of Tr_2 is applied to the bases of the complementary pair Tr_3, Tr_4. Tr_3 is an emitter follower, while Tr_4 is a common-emitter stage. The current gain of both stages is approximately α', so that, by making R_{10} and R_{12} equal and employing transistors having equal values of α', the output transistors Tr_5 and Tr_6 receive equal inputs in anti-phase. R_9 is adjusted to provide a small (symmetrical) unbalance of Tr_3, Tr_4, which sets the standing current of Tr_5 and Tr_6 to a small value sufficient to avoid crossover distortion, the diode D_1 serving to compensate for variations in the supply voltage as discussed in Chapter 6. It will be noted that the decoupling capacitance C_3 is taken to the point P, which is the zero-signal point of the output network.

Finally, signal-frequency feedback is provided from the output to the emitter of Tr_1, via R_{15}. The effective value of R_4 is modified by the impedance of Tr_1, but with the values shown the effective feedback factor is approximately 1/9, so that the overall gain is 9. With a supply voltage of 45 V the peak output voltage is 22·5 V, so that the peak input required is approximately 2·5 V. The capacitor C_4 limits the feedback at high frequencies (outside the pass band), which, because of phase shift within the amplifier, might become positive and so cause instability.

The r.m.s. power output is $(V_{cc}/2\sqrt{2})^2/R_{load}$, which with $V_{cc} = 45$ V and $R_{load} = 15\,\Omega$ is 17 W, and with the values shown the total harmonic distortion can be less than 1 per cent.

10.4. Sound Recording

Sound recording may broadly be divided into three sections according to the medium employed. Mechanical groove recording on cylinder or disc was the first method used, followed later by sound-on-film recording as used in cinema technique. A third method now in considerable use employs magnetic wire or tape on which the signal variations are imposed and subsequently reproduced by suitable magnetic pick-up heads.

The techniques of all three methods are highly specialized, and it is only possible here to discuss the basic principles. A more detailed exposition will be found in *Radio and Electronics*, Vol. 2, Chapter 7 (Pitman).

Disc Recording

Recording of sound on disc is carried out by cutting a spiral groove in the disc and then modulating the contours of the groove in accordance with the recorded sound. The modulation may be carried out in the vertical plane ("hill and dale") or in the horizontal plane (lateral recording); the latter system is now almost universally employed.

The recording may be cut directly on a lacquer-coated disc for direct playback, but in the gramophone record industry the original is made of wax from which a metal stamper is produced and used as a press tool to make a large quantity of copies in a plastic material. Early recordings used shellac discs rotating at 78 r.p.m., but these have now been superseded by vinyl discs running at 45 or 33⅓ r.p.m. Vinyl has the advantage of greatly reduced signal/noise ratio being virtually free from background noise (scratch), and it is also flexible so that the discs are much more durable than the early shellac discs which were very brittle.

The pick-up used to translate the mechanical motion of the needle into an electrical waveform is generally of the electromagnetic (moving iron), dynamic (moving coil), or piezo-electric crystal type. The principles of operation are fundamentally those of the corresponding types of microphone, the main consideration being lightness of weight to minimize wear on the record, and relative freedom of movement in a lateral direction with stiffness in the vertical direction. With the early 78 r.p.m. discs the needle or stylus was usually of steel (or sometimes fibre to minimise background noise). In the 45 and 33⅓ r.p.m. discs, known as E.P. (extended playing) and L.P. (long-playing) recordings, the grooves are narrower, and it is usual to use sapphire-tipped needles.

A typical piezo-electric pick-up construction is illustrated in Fig. 10.30. The output from such pick-ups is of the order of 0·5 to 2 volts so that pre-amplifiers are not required. However, the impedance is predominantly capacitive so that any load across the pick-up reduces the response at the lower frequencies, and a suitable correcting network may be needed as explained in Chapter 14.

Moving-iron types of pick-up are appreciably less sensitive and generally require an additional amplifying stage. They are predominantly inductive so that any resistance across the device reduces the high-frequency response.

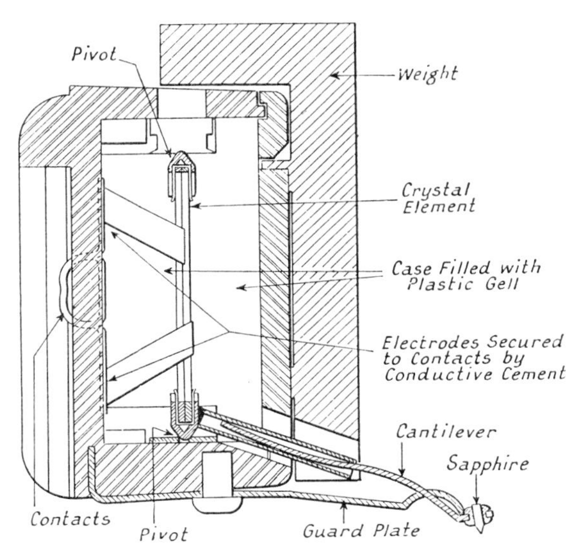

FIG. 10.30. SECTION THROUGH ACOS GP. 20 PICK-UP

Stereophonic Recording

Normal sounds have a directional quality because of the small phase differences between the arrival of the air waves at the two ears. This quality is not reproduced with a normal recording, but can be introduced by *stereophonic* techniques.

Two loudspeakers are used, fed independently with signals from two similarly spaced microphones at the recording studio. These are recorded on the disc by cutting the modulation on one side of the groove only. The two sides of the groove can thus carry entirely separate signals.

A special pick-up is used, having two styli set at 90°, each being 45° to the direction of the groove. The excursions of each stylus are controlled only by the modulation of its appropriate side of the groove, and do not come into contact with the opposite side at all. The pick-up is lowered into the groove in the normal way, and then delivers two separate outputs which are fed into separate amplifiers and loudspeakers, thus reproducing the stereophonic quality of the original.

Alternatively, the two channels may be recorded on film or magnetic tape, or provided by two separate radio transmissions.

Sound on Film

Just as in disc recording the groove may be modulated in the lateral or vertical direction so a photographic track may be modulated by either the variable area or variable density method. In the density method the track image is of constant width and its density is modulated to conform to the signal voltages, while in the area system the track is of constant density but occupies only a portion of the available track area. The modulation is then displaced laterally so that the image on the film corresponds to an oscillogram of the signal waveform.

In the reproduction of both systems the sound track is illuminated evenly from a fine slit at right angles to the direction of motion of the film. The movement of the film thus modulates the light beam and the light modulations are translated to electrical energy by a photoelectric cell. With correct adjustment the systems are capable of yielding equally good results and both are in general use.

The sound track is usually located some distance ahead of the picture frame to which it refers. This is necessary because the picture is jerked past the gate, frame by frame, whilst the sound track must be run smoothly past the slit. Consequently the two actions must take place some distance removed from each other.

Magnetic Tape Recording

In magnetic tape systems the signal is caused to produce variations in the magnetism of a thin iron wire or tape. The signal is applied to a recording head of toroidal form with a small air gap, as shown in Fig. 10.31. The tape or wire is drawn across this air gap so that the magnetic field is completed through a small longitudinal portion of the tape. Thus at any instant the small element of tape across the air gap becomes magnetized, and if the material has suitable remanence the tape as it moves past the recording head becomes a series of very small bar magnets having magnetizations proportional to the instantaneous signals.

The signal may be reproduced by drawing the tape at the same steady speed over the poles of a similar reproducing head; voltages are then induced in this coil which are proportional to the signal waveform.

The material used for the magnetic tape can be made either by coating a base with magnetic material or by impregnation. Steel wire can be used, but is less satisfactory. A typical coated tape would consist of a base of cellulose acetate coated with iron oxide, while an impregnated tape would consist of equal proportions of vinyl plastic and magnetic material.

In a typical tape recorder the tape is passed over an erasing head before reaching the recording head. The erasing head is fed with

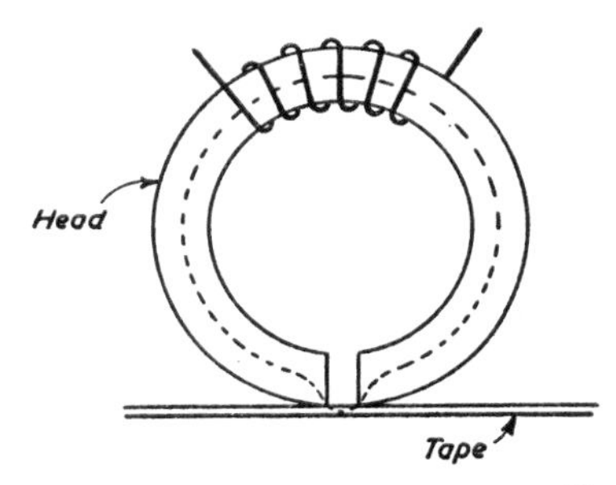

Fig. 10.31. Magnetic Recording Head

sufficient current at about 60 kHz to carry each element of the tape to saturation repeatedly in each direction. The result is that the magnetic domains are then aligned in a completely random manner and the result is equivalent to demagnetization. In fact this method is not completely efficient but the residual is usually 60 to 80 dB below the level of the peak signal.

In the basic recording system described above the residual magnetism after the tape leaves the recording head would be far from being proportional to the signal current owing to the pronounced curvature of the B–H curve (see Chapter 4). This difficulty is overcome by using h.f. bias at about 60 kHz on the recording head. The signal is applied as modulation on the bias so that for any value of signal the tape is driven through several complete cycles of magnetization by the bias signal and the residual magnetism follows the signal modulation quite closely. The same h.f. oscillator is used for the bias as is used for erasing, though at a suitably reduced amplitude.

The most notable advantage of magnetic recording is that the signal can easily be erased and the tape used over again many times. The recording can be reproduced many times without loss of quality and tape can be cut and stuck together again very simply during editing. Wire recorders can be pocket sized and are very useful for field work in broadcasting and as portable dictaphones; the technical quality is substantially inferior to tape but this can be tolerated for these applications.

The commercial disadvantage of magnetic tape recording as opposed to disc recording is the comparative difficulty of obtaining a large number of copies of a magnetic record.

For special purposes tape recorders can be designed to accommodate a number of separate tracks side by side on the same tape. One obvious application is the stereophonic recording mentioned earlier, the signals from the two microphones being recorded side by side on the one tape, and reproduced by suitably located pick-up heads.

The upper-frequency response with magnetic tape recording is determined by the width of the gap in the recording (or play-back) head, and the speed at which the tape moves over the head. The gap cannot be made too small or the magnetic flux would prefer to flow through the gap rather than the tape. A further problem is that, if the elements on the tape are too close together, they tend to demagnetize adjacent elements, while there is at high frequencies a certain skin effect which prevents the flux from penetrating the tape adequately.

In practice the minimum gap width is of the order of 0·25 to 0·5 mils and since two elements are involved in a complete cycle the maximum number of cycles is between 1000 and 2000 per inch. The standard tape speed is $7\frac{1}{2}$ in/sec, so that allowing for the effects just mentioned the maximum frequency is of the order of 10 kHz. For high fidelity a higher speed (e.g. 15 in/sec) may be used. For speech or medium-quality music a speed of $3\frac{3}{4}$ in/sec is often used. Considerable care has to be taken to maintain the tape speed constant, for obvious reasons.

Since the e.m.f. is determined by the rate of change of flux it is clear that with a constant tape speed the output will be proportional to the frequency of the recording. This is true up to about 1 kHz, after which the effects mentioned above cause a progressive limitation and the output begins to fall rapidly. Hence again there is usually a need to incorporate suitable equalizing networks.

Video Tape

Although the frequencies involved in a television modulation run into the megahertz region, as explained in Chapter 16, the number of picture elements in any one line is only of the order of 1000. Hence it is possible to arrange to record television signals on a tape one or two inches wide and arrange that the recording and reproducing heads are moved *across* the tape, at right angles to the normal direction of travel.

The head must be arranged to traverse the tape laterally in one line time, i.e. 0·64 millisec with a 625-line transmission, but by mounting a series of heads on a rotating arm this can be achieved with practicable speeds of rotation. The axial travel of the tape is then determined by the frame speed required, and by varying this, slow-motion or still pictures can be reproduced as required.

11

Power Supply Circuits

COMMUNICATIONS equipment normally requires a supply of d.c. power. In portable equipment this can be obtained from batteries but for larger units it is customary to derive the power from the a.c. supply mains. The current must be rectified to convert it to unidirectional form of the required polarity, and the pulsations are then filtered out by suitable smoothing circuits.

Basic Rectifier Circuits

Fig. 11.1 illustrates four simple types of circuit. In the first circuit the mains supply voltage is transformed up or down as

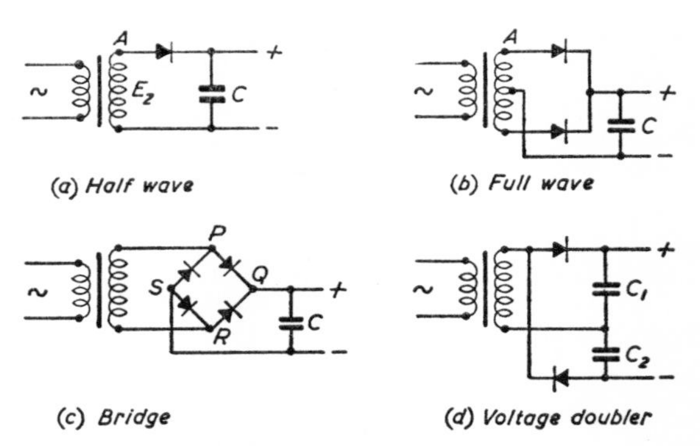

FIG. 11.1. FOUR MAIN TYPES OF RECTIFIER CIRCUIT

required and the voltage on the secondary is applied through a rectifier to a capacitor C. This capacitor receives a pulse of current every time the point A becomes positive so that a charge is built up which is then dissipated through the load across the circuit. Conditions adjust themselves until the current drawn from the transformer in the form of pulses is just sufficient to replenish the loss of charge during the idle portion of the cycle.

Fig. 11.1 (*b*) shows a full-wave circuit in which the secondary of the transformer is centre-tapped and two rectifiers are used, connected one from each extremity. During the half-cycle which makes the point A positive the top rectifier conducts, while during the succeeding half-cycle the bottom rectifier conducts. The current is fed into a single capacitor C so that the charge is replenished twice as quickly, which results in a higher output and less variation in the voltage.

Fig. 11.1 (*c*) shows a bridge rectifier, an arrangement which avoids the use of a centre-tapped transformer by employing four rectifiers in a bridge formation. The current flows first through PQ, through the capacitor C and back through SR. On the succeeding half-wave it flows via RQ, C, and SP. The performance of this class of circuit is similar to that of Fig. 11.1 (*b*). It has the advantage that only a single secondary winding is required and, as will be seen later, the transformer losses are reduced.

Fig. 11.1 (*d*) shows a voltage doubler circuit. This is an arrangement employed where it is desirable to restrict the voltage on the transformer secondary, or where, for some reason, there is only a limited voltage available. It gives approximately twice the voltage output as is obtained from Fig. 11.1 (*a*). Current flows through the top rectifier, and back through the capacitor C_1. The next half-cycle flows through the capacitor C_2 and back through the bottom rectifier. Each of these pulses charges the appropriate capacitor, so that the total voltage applied across the load is the sum of the two, which is thus twice the normal voltage. Since C_1 and C_2 are in series the effective capacitance is halved, in consequence of which, as will be seen later, the regulation may suffer. This class of circuit, then, is mainly used for the supply of high voltage at low current.

Regulation; Ripple Voltage

The current drawn from the capacitors during the "idle" portions of the cycle will cause the voltage to fall, and although this is replenished during the active portion it will be clear that the mean voltage will fall as the current drain increases, while the fluctuation of voltage about the mean value will become worse.

These are the two aspects of a rectifier system in which we are mostly interested. The variation in the mean output voltage with current is called the *regulation* which is defined as the variation of voltage from no load to full load expressed as a percentage of the full-load voltage. The fluctuation of voltage about the mean value is known as the *ripple* voltage, and both of these characteristics can be calculated from the circuit values.

Let us, at the outset, confine our attention to the first two circuits. It should be made clear that the rectifiers used may be of any type, either valve or semiconductor. Symbolic rectifiers have been illustrated to avoid any complication of the circuit by the introduction of heater windings on the transformers, etc. Considering Fig. 11.1 (*a*) it will be clear that, after the first few cycles, there will be a voltage on the capacitor C, of which the instantaneous value is fluctuating slightly, falling as current is taken from the capacitor by the load during the idle period of the cycle and rising

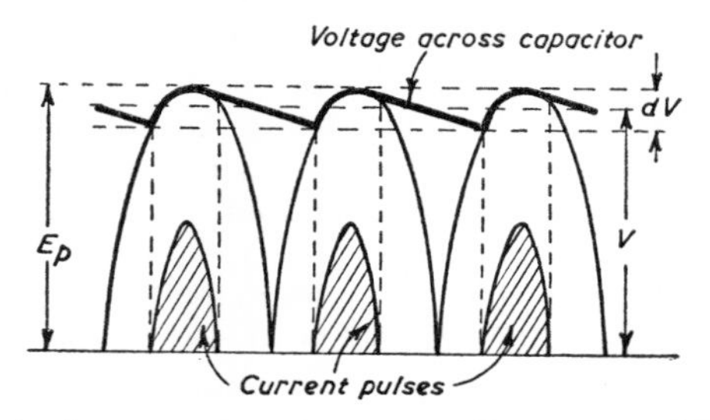

Fig. 11.2. Illustrating Action of Reservoir Capacitor

sharply during the period over which the charge is replenished again through the rectifier.

With a half-wave circuit there is one such pulse of current every cycle, while with a full-wave circuit there are two such pulses, one in each half-cycle as illustrated in Fig. 11.2, where the alternate half-cycles have been shown reversed (as, in effect, they are by the second rectifier) so that they all appear on one side of the zero line.

Now, the rectifier will not commence to conduct until the voltage on the secondary of the transformer exceeds the voltage on the capacitor. This does not happen at the beginning of the positive half cycle but is delayed until some point towards the top of the wave as shown in Fig. 11.2. At the point where the transformer voltage overtakes the steady voltage on the capacitor a large pulse of current flows through the rectifier causing the capacitor voltage to increase. When the input voltage falls below the capacitor voltage the rectifier ceases to conduct and the load current is then supplied from the charge stored in the capacitor which causes a steady fall in voltage over the idle period as shown.

The voltage on the capacitor will thus increase briefly and then fall away steadily, resulting in a mean voltage output V with a fluctuation above and below the mean value. The total value of this fluctuation dV may be estimated by assuming that the rectifier is only conducting for a small fraction of a cycle. Then if f is the number of times per second that the capacitor is charged, the loss of charge dQ due to the load current I is I/f (very nearly).

This charge must be replaced by the pulse of current through the rectifier, which will raise the voltage on the capacitor by the small amount dV. But since $V = Q/C$ we can write

$$dV = dQ/C = I/fC = V/RfC$$

where R is the load resistance. The peak value of the ripple is half this value, or $V/2RfC$.

(The actual fluctuation will be slightly less than this because the capacitor is not discharging for a full half-cycle but only about 90 per cent of this time, the balance being occupied by the charging period. The expressions, however, are close enough for practical purposes.)

The output voltage $V = E_p - V/2RfC$, which simplifies to

$$E_p/[1 + 1/(2RfC)] = E_p[1 - 1/(2RfC)] \text{ very nearly}$$

It will be clear that on no load (R infinite) the second term $1/2RfC$ becomes zero, so that the output voltage is simply E_p ($= \sqrt{2}$ times the r.m.s. input voltage), and the ripple becomes zero.

The regulation is

$$100(V/2RfC)/V = 50Rf/C \text{ per cent}$$

With a half-wave rectifier and a 50 Hz supply, f is 50. With a full-wave or bridge rectifier it is 100. With a three-phase rectifier such as we shall discuss later the value may be 150 or 300. This is the frequency of the ripple, and it is convenient to ignore the fact that the ripple voltage is not a true sine wave but to consider it as if it were a small sinusoidal voltage of frequency f superposed on the steady voltage V. We can then calculate suitable smoothing circuits to follow the reservoir capacitor C, and if these circuits are effective enough to remove the fundamental components they will be still more effective in removing the harmonic components so that this assumption of a sinusoidal ripple is quite justifiable.

Size and Rating of Reservoir Capacitor

It will be seen from the expressions just developed that the larger the reservoir capacitor the nearer does the output voltage approach the peak a.c. The result depends on the product CR. If this is small the factor $1/2RfC$ is large and V is considerably less than E_p, but as we increase CR by increasing C or R or both, V rises rapidly at first and then progressively more slowly as shown in Fig. 11.3.

It will be noted that if R is infinite (no load) $V = E_p$. Hence the capacitor must be designed to withstand the voltage E_p continuously.

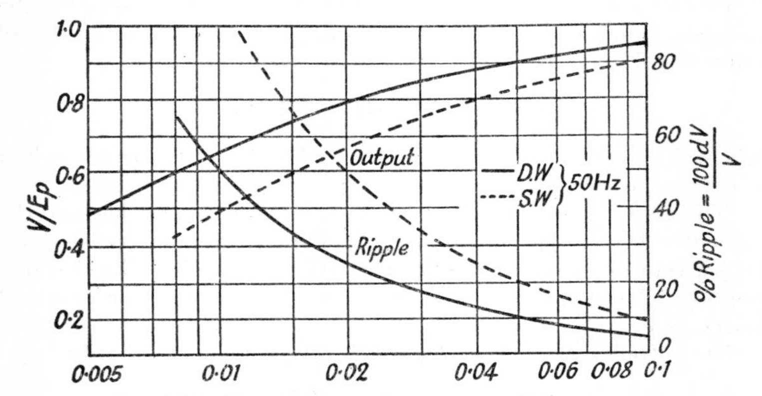

FIG. 11.3. VARIATION OF OUTPUT AND RIPPLE WITH PRODUCT CR FOR A 50 Hz SUPPLY

It is indeed desirable to allow a small factor of safety, for, as we shall see later, leakage inductance in the transformer may provide a resonance which will cause V to exceed E_p.

For normal loads the product fCR should be not less than 2. Thus with a load of 5,000 ohms and $f = 100$ (50 Hz double wave), $C = 4\ \mu$F. With smaller currents, such that $R = 50{,}000$ ohms, say, a capacitance of $\frac{1}{2}\ \mu$F would suffice.

Rectifier Current

In assessing the behaviour of the circuits it was assumed that the rectifier was only conducting for a brief time each cycle. This is an over simplification because the pulse of current from the rectifier, as shown in Fig. 11.2, must last long enough for the capacitor to be replenished. If the conducting period is $1/n$ of the cycle, the mean rectifier current must be n times the load current, and the peak value will be $2/\pi$ times this value (assuming sinusoidal pulses).

The rectifier has thus to supply pulses of current considerably greater than the output current, and this has to be allowed for in the design. The ratio n depends on the *conduction angle*, i.e. the proportion of the cycle during which the rectifier is conducting, and this in turn is dependent on the product fCR. The exact calculations are complex, but the curves of Fig. 11.4 shown the conduction angle θ in terms of this product. From this the peak rectifier current can be estimated approximately as

$$I_{rp} = 1{\cdot}5I(180 - \theta)/\theta \text{ for a full-wave circuit}$$
$$= 1{\cdot}5I(360 - \theta)/\theta \text{ for a single-wave circuit}$$

where I is the output-current.

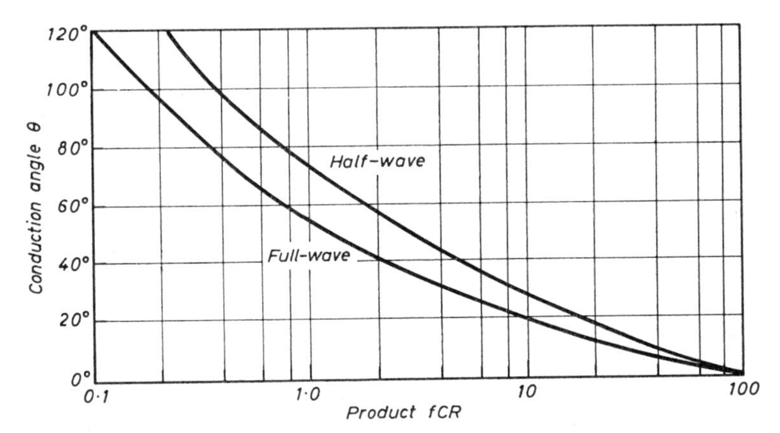

FIG. 11.4. VARIATION OF CONDUCTION ANGLE IN TERMS OF PRODUCT f_sCR

f_s = supply frequency

Thus with a full-wave circuit delivering 0·1 A at 250 V, $R = 2500\,\Omega$. If $C = 8\,\mu$F and f_s is 50 Hz, $f_sCR = 1$. Then $\theta = 55°$ so that $I_{rp} = 1{\cdot}5 \times 0{\cdot}1 \times (125/55) = 0{\cdot}34$ A. However, if $C = 64\,\mu$F, $I_{rp} = 2{\cdot}7$ A, which might overload the rectifier.

With the same values but with single-wave rectification the peak rectifier currents would be 0·57 A and 4·6 A, which would certainly not be tolerable.

Smoothing Circuits

Even if $fCR = 2$ there will still be 25 per cent ripple. This, of course, is far too great to be tolerated in a source of h.t. supply to a valve amplifier. We can reduce this by increasing C, but this is not economical. For one thing, C must not be made too large because this would increase I_{rp}.

Secondly, doubling C will only halve the ripple, whereas if we use the additional capacitance with a series inductance to form a filter, as shown in Fig. 11.5, we can obtain much more effective smoothing. The overall ripple voltage applied across the input to the filter is dV. This will cause a current $dI = dV/(L_1\omega - 1/C_1\omega)$, neglecting the resistance of the choke. The voltage across the output capacitor is $dI/C_1\omega = dV/(L_1C_1\omega^2 - 1)$, so that the ripple is attenuated by the amount indicated by the denominator, $L_1C_1\omega^2 - 1$, where $\omega = 2\pi$ times the *ripple* frequency.

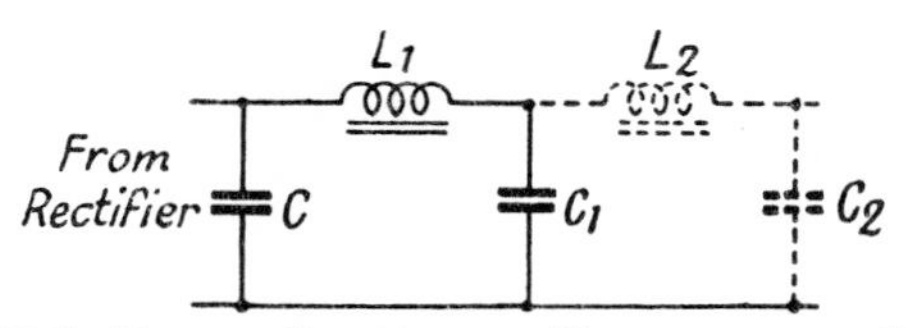

FIG. 11.5. FILTER CIRCUIT FOR REDUCING THE RIPPLE

If a second section is added this produces a further attenuation of $(L_2C_2\omega^2 - 1)$. Such double filters are occasionally required but usually one section suffices.

Design of Rectifier System

In practice we desire to arrange our system to deliver a certain d.c. voltage at a certain current. The first step is to decide how much voltage drop we shall experience on our system. The smoothing choke or chokes will have some d.c. resistance, and this must either be measured or estimated and due allowance made for the voltage drop. It is also necessary to allow some additional voltage drop for the loss in the rectifier, and for the resistance of the transformer winding.

The latter effect can be allowed for by designing the transformer to deliver the voltage required at the specified load current. The rectifier drop depends on the type of rectifier. With a semiconductor rectifier it is very small.

A typical calculation will indicate the procedure. Let us assume we wish to develop a voltage of 300 volts at a current of 60 mA. The load resistance R is thus 5,000 ohms. With a full-wave 50 Hz circuit $f = 100$. If $fCR = 2$, $C = 4\ \mu\text{F}$ and $dV = 1/2RfC$ then becomes 0·25.

If the smoothing choke has a resistance of 200 ohms and we assume a transformer resistance also equal to 200 ohms the combined drop will be $400 \times 0{\cdot}06 = 24$ volts, so that the input voltage required is 324 volts.

Thus $E_p = V(1 + 1/2RfC) = 324(1{\cdot}25) = 405$

whence $E_{r.m.s.} = 286$ (on load)

The ripple will be 25 per cent. If we follow this with a smoothing circuit consisting of a 30 H choke and an 8 μF smoothing capacitor, the ripple will be reduced 94 times, leaving 0·265 per cent or $\pm 0{\cdot}8$ volts ripple, which is negligible for all normal requirements.

Voltage-Doubler Circuits

The action of the voltage-doubler circuit is a little more complex than the simple circuits so far considered, because the pulses of current first cause an increase in the voltage on the top capacitor

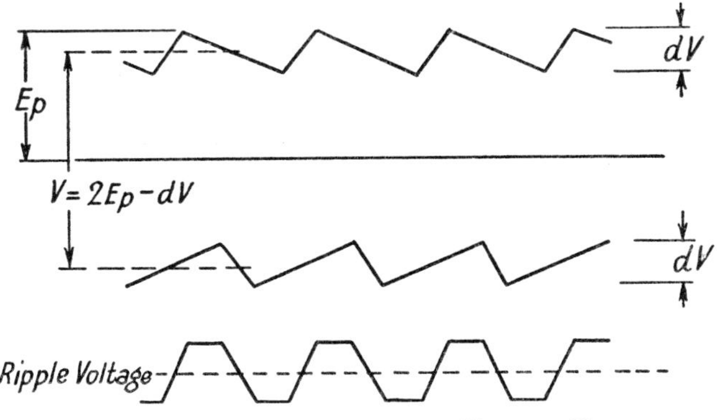

FIG. 11.6. ILLUSTRATING ACTION OF VOLTAGE-DOUBLER CIRCUIT

and then half a cycle later an increase *in the opposite direction* of the voltage on the bottom capacitor. The two capacitors swing away from the mid-point as shown in Fig. 11.6. This will be seen to produce a curious shape of ripple, and while this has a frequency equal to twice the supply frequency and thus behaves similarly to a full-wave rectifier, the regulation of the circuit is equivalent to that of a half-wave rectifier. This can readily be understood because only half the total capacitance is replenished every half-cycle, which is equivalent to replenishing the total capacitance every cycle.

It will be clear from the diagram that as a first approximation the output voltage is

$$V = 2E_p - dV$$
$$= 2E_p - V/RfC$$

whence $$V = \frac{2E_p}{(1 + 1/Rf_sC)}$$

In the above expressions

V is the total output voltage,

E_p is the peak transformer voltage,

C is the capacitance of *each* of the voltage-doubling capacitors (assumed equal), and

f_s is the *supply frequency* (usually 50 Hz).

Choke-Input Circuits

As the load current on the system increases, the peak current taken by the rectifier in charging up the reservoir capacitor becomes so large that there is a risk of damage to the rectifier. A second

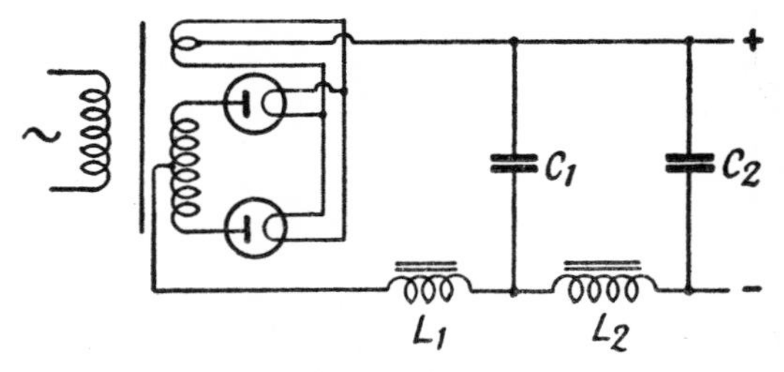

FIG. 11.7. CHOKE INPUT CIRCUIT

disadvantage of the type of circuit so far discussed is that the process of replenishing the charge in a sudden pulse over a small fraction of the cycle necessarily makes the output voltage very dependent upon the load. If we reduce the load resistance so that more current is drawn from the reservoir capacitor the voltage will fall so that we have a steadily dropping voltage characteristic.

The first factor is the more serious, particularly with high currents, and in these circumstances it becomes necessary to limit the rectifier current to a value more commensurate with the load current.

This can be achieved by introducing a choke before the first capacitor in the system, such an arrangement being known as a *choke-input filter*. The arrangement is shown in Fig. 11.7 for a full-wave rectifier. A choke, being a device which tends to maintain a constant current, will clearly not permit sudden pulses of current as shown in Fig. 11.2, but will endeavour to maintain a constant current flow.

Under such conditions the capacitor C_1 has little effect on the voltage developed, the d.c. voltage being determined by the transformer secondary voltage E_p less the voltage drop in the rectifier and the choke. If mercury vapour or semiconductor rectifiers are used the voltage drop in the rectifier is constant so that we are only left with the voltage drop on the choke and the regulation of the transformer, which involves not only its winding resistance but also the leakage inductance as explained in Chapter 4. It is, however, practicable to keep both these voltage drops small so that we are

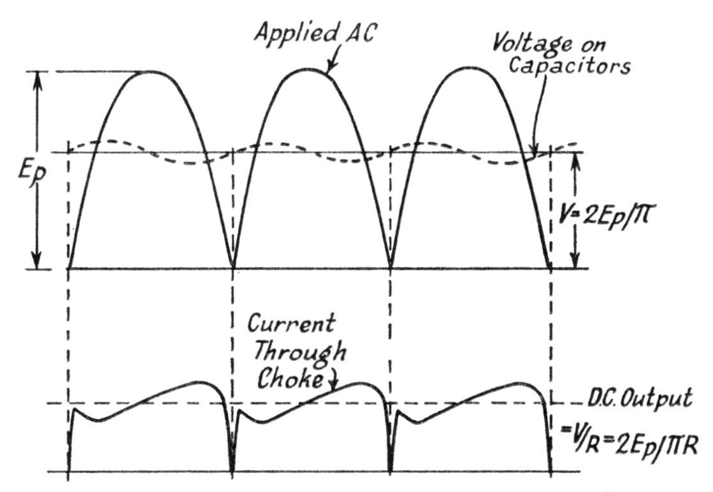

FIG. 11.8. VOLTAGE AND CURRENT RELATIONSHIP IN CHOKE-INPUT CIRCUIT

able to provide that the voltage delivered at the output terminals is tolerably constant irrespective of the load current.

Actually this good regulation cannot be maintained right down to no load, as we shall see, but over the majority of the current range from about one-tenth full load to full load the change in voltage can be made to be less than **5** per cent.

Fig. 11.8 shows the form of the current through the rectifier, from which it will be seen that the rectifier current is equal to the load current with a double-frequency ripple superposed, and if we increase the drain on the circuit, so that the load current increases, the rectifier current increases in similar proportion.

There is no longer any question of the rectifier not conducting until the a.c. voltage has exceeded a certain amount, except just at the very beginning of the cycle. This being so the d.c. voltage output is the mean value of the a.c. from the transformer secondary. We know that the mean value of a rectified sine wave is $2E_p/\pi$ whereas the r.m.s. value is $E_p/\sqrt{2}$.

Hence the ratio of the r.m.s. voltage on the transformer to the d.c. output is $\pi/2\sqrt{2} = 1{\cdot}11$ (which is, of course, the form factor of a sine wave). Conversely the d.c. output is 0·9 times the r.m.s. a.c. input. The actual output is reduced by reason of the voltage drops on the choke and the rectifier; the rectifier drop is about 15V for a mercury rectifier but only about 1V with a semiconductor type.

Value of Series Inductance

The difference in voltage between the applied a.c. from the transformer and the steady d.c. output is absorbed mainly by the choke. The choke will, in fact, carry a ripple current, and it will be clear from Fig. 11.8 that the limiting condition for the maintenance of a continuous current flow through the rectifier is that the peak ripple is less than the steady d.c. current.

Now, the form of the ripple voltage delivered by the rectifier is a series of rectified half-sine waves. This may be represented by a mean value $2E_p/\pi$ plus a series of even harmonics, the equation being

$$e = \frac{2E_p}{\pi}\left(1 - \frac{2}{3}\cos 2\omega t - \frac{2}{15}\cos 4\omega t - \frac{2}{35}\cos 6\,\omega t - \ldots\right)$$

We are only concerned with the second-harmonic term since the effect of the choke will be magnified on the higher harmonics, so that if we make it large enough to deal with the lowest-frequency ripple it will be more than adequate for the higher-frequency terms, which are in any case much smaller.

The peak amplitude of this second-harmonic term is $4E_p/3\pi$ so that the ripple current will be this voltage divided by the impedance of the circuit Z. In general the reactance of C will be low compared with the load R, so that we can say that $Z = L\omega - 1/C\omega$ (ω here being the value appropriate to the *ripple* frequency, i.e. 100 Hz with a 50 Hz full-wave rectifier).

The peak ripple current is thus $4E_p/3\pi(L\omega - 1/C\omega)$, and this must be less than the d.c., which is $2E_p/\pi R$ (the mean rectified output divided by the load).

Equating these two we find that $L\omega = 2R/3 + 1/C\omega$. If $C\omega$ is large so that $1/C\omega$ is negligible, this reduces to $L\omega/R = 0{\cdot}67$, a criterion which is often quoted. It may, however, be uneconomical to make C as large as would be necessary to justify this assumption, but it will be clear that if C is reduced, L must be increased.

A very able exposition of the requirements was given by C. R. Dunham (*J. Instn. Elect. Engrs.*, **75**, p. 278, 1943), in which an economical value of $1/C\omega = R/6$ is suggested. This results in a criterion $L\omega/R = 0{\cdot}84$.

Swinging Chokes

These values of L are the *minimum* values necessary to maintain a continuous current through the rectifier, and the actual value should be 20 to 30 per cent higher to allow a margin of safety. It will also be noted that the value of L depends on the load. As the load decreases (increasing R) the inductance must rise in direct proportion. Chokes can be constructed to comply with this condition over a range of current and are known as *swinging chokes*. They will not maintain their increase below about 1/10th full load, which is the reason for the statement previously made that good regulation cannot be maintained right down to no load, but it is always practicable to place a permanent load across the circuit drawing 10 to 20 per cent of the full-load current so that the system always delivers more than the critical current.

It should also be noted that, if the load falls below the value for which the choke has been designed so that the current is no longer continuous, the arrangement takes on the characteristics of a capacitor-input circuit and the voltage rises sharply, reaching, in the limit, the full value E_p at no load.

Peak Currents

With the criteria quoted above the peak current through the rectifier is limited to twice the mean current, a considerable improvement over the capacitor-input system.

It may, however, be desirable to work with a still lower ratio of peak/mean current. Rectifiers are usually rated in terms of peak current, and the permitted ratio of peak/mean may be less than 2, particularly with a small rectifier designed for high-efficiency working.

If such is the case the value of L must be increased in due proportion. Thus, if the ripple is halved, the ratio of peak to mean becomes 1·5 instead of 2. Thus, we should achieve the required result by making $L\omega/R = 1{\cdot}68$. It should be noted that increasing C will not suffice because $1/C\omega$ is already small and if C is made infinite it can only reduce the ripple in the ratio $0{\cdot}67/0{\cdot}84 = 0{\cdot}8$.

A final precaution is that the values of C and L should not be such as to resonate with the ripple. This can always be checked, but it will be found that with the values suggested there is no danger of this. If L is under one henry, however, the leakage inductance of the transformer (which is in series with L) may increase the value to such a point that resonance occurs.

Smoothing

It is still necessary to smooth the remaining ripple, for though the choke and capacitor constitute a filter the initial ripple may be much larger than with the shunt-capacitor input circuit. The peak ripple is $4E_p/3\pi$, and since the mean value of the d.c. is $2E_p/\pi$ the ratio of the two is $2/3 = 0{\cdot}67$, which is about three times as great as with a typical capacitor circuit.

The circuit is therefore followed, as a rule, by a second filter, the attenuation being calculated as already explained.

A typical design will illustrate all these points. Let us assume our requirements to be a supply of 1,250 volts at a current varying between 0·05 and 0·25 amps. Assuming a double filter with 50 volts drop on each choke and a further 15 volts in the rectifier, our input voltage = 1,365. Hence the r.m.s. transformer voltage (on load) must be $1{,}365 \times 1{\cdot}11 = 1{,}515$.

$$R \text{ is } 1{,}250/0{\cdot}25 = 5{,}000 \text{ ohms}$$

whence $$L\omega = 0{\cdot}84 \times 5{,}000 = 4{,}200$$
$$f = 100$$

so that $$L = 6{\cdot}7 \text{ H, say } 7 \text{ H}$$

This is at full load. At 50 mA the inductance must be 5 times as great, i.e. 35 H, so that the choke would be specified as 7/35 H at 250/50 mA.

$$C = 6/\omega R = 6/(628 \times 5{,}000) = 1{\cdot}93\ \mu\text{F, say } 2\ \mu\text{F}$$

$$\text{Initial ripple} = 1{,}365 \times 0{\cdot}67 = 915 \text{ volts}$$

Ripple across first capacitor

$$\begin{aligned} 915/L_1C_1\omega^2 &= 915/(7 \times 2 \times 10^{-6} \times 4\pi^2 \times 10^4) \\ &= 915/5{\cdot}5 = 166 \end{aligned}$$

If we wish to reduce our ripple to 1 per cent = 12·5 volts we require a further attenuation of 13·3 times which requires an LC product of 36 so that a 6 H choke and a 6 μF capacitor would suffice.

Insulation

Because of the high output voltage the chokes would preferably be placed in the negative lead as shown in Fig. 11.7. This avoids the need to insulate for the d.c. potential, but it is still necessary to insulate for the peak ripple in the first choke, i.e. $0{\cdot}67V$, so that for safety the choke should be insulated for a peak working voltage of V and tested at twice this voltage. If for any reason it is inconvenient to use the chokes in the negative lead the chokes must be insulated for a peak voltage of $2V$.

The first capacitor will have to withstand the d.c. voltage V plus the peak ripple across it, which we have seen to be 10 to 20 per cent of the total input ripple. Under normal conditions, therefore, it is sufficient to design C_1 for a working voltage of $1{\cdot}5V$ and C_2 for about $1{\cdot}1V$. It should be remembered, however, that if the load on the circuit is removed for any reason the voltage will rise to the full peak value of the a.c. input, on both capacitors. Sound design thus requires both capacitors to be designed to withstand a peak voltage E_p, and to be tested at twice this figure.

With capacitor input circuits, all capacitors must be designed to withstand the peak voltage E_p and should preferably be rated for twice the normal working voltage to provide a margin of safety. It should always be remembered that the a.c. mains may vary ± 10 per cent in voltage.

Any heater windings must, of course, be similarly insulated.

Peak Inverse Voltage

The rectifiers have to withstand an even greater strain, which is at its maximum during the non-conducting portion of the cycle and is thus known as the *peak inverse voltage.*

Consider Fig. 11.1 (*a*). On no load the capacitor C will charge to E_p. During the negative half-cycle the point A will be negative so that the transformer voltage will add to the capacitor voltage. The peak inverse across the rectifier is thus $2E_p$.

The same applies to each of the rectifiers in Fig. 11.1 (*b*) and 11.1 (*d*). In the bridge circuit of (*c*) the capacitor voltage does not add to the transformer voltage so that the inverse peak is simply E_p.

Two limits are usually specified, namely V_{RRM} = maximum permissible repetitive peak-peak voltage, V_{RSM} = maximum permissible non-repetitive peak voltage.

Three-phase Rectifier Circuits

Where three-phase supply is available it is possible to improve the conversion of a.c. to d.c. by using all three phases. If we charge a capacitor three times in every period our ripple frequency becomes 150 Hz, while if we can arrange a three-phase full-wave arrangement the ripple frequency becomes 300 Hz.

Fig. 11.9 shows a three-phase half-wave rectifier arrangement. The transformer is connected with its secondary in star and a rectifier is connected from the outer of each of the limbs. The cathodes of all three rectifiers are common and are taken to the input circuit. A capacitor input circuit may be used for small powers, but since three-phase working is usually employed with

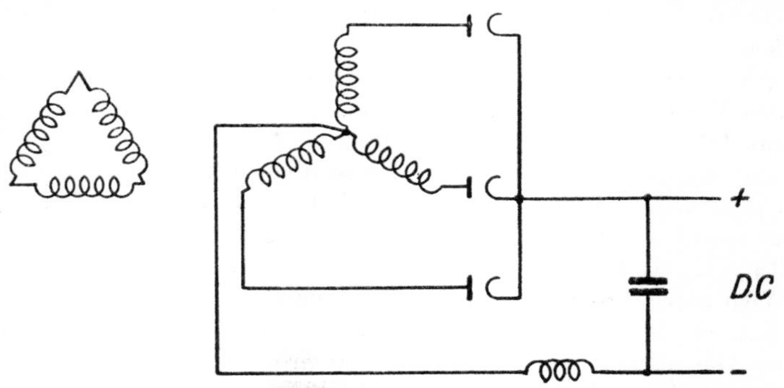

FIG. 11.9. THREE-PHASE HALF-WAVE RECTIFIER CIRCUIT

higher powers one usually finds a choke input circuit. For the reasons already mentioned the choke is connected in the neutral point so that it is at low potential.

The construction of three-phase transformers was described in Chapter 4. For small powers, three separate single-phase transformers may be used, but this arrangement is not used for higher powers because with any half-wave transformer the transformer secondary has to carry a d.c. component. The waveform of the secondary is, in fact, the rectified half-wave already mentioned, and the d.c. component may produce saturation of the iron. With a three-phase transformer in which the windings are arranged on three limbs of a special type of stamping this saturation is avoided.

The design of the filter is modified from that for a single-phase circuit because of the overlapping of the voltages as shown in Fig. 11.10. The effective voltage thus never falls to zero so that the ripple is considerably reduced.

Actually the mean value of a three-phase sequence of the type shown in Fig. 11.10 (*a*) is 0·83 E_p and the peak ripple is 0·3 E_p. This ripple has a frequency of three times the fundamental frequency, i.e. 150 Hz for a normal 50 Hz supply.

Using these figures the choke and first capacitor may be calculated by the same method as before, using $f = 150$, and the second choke and capacitance constituting the ripple filter can be calculated in a similar manner. Owing to the higher frequency of the ripple and smaller initial ripple content it is possible that a second filter may

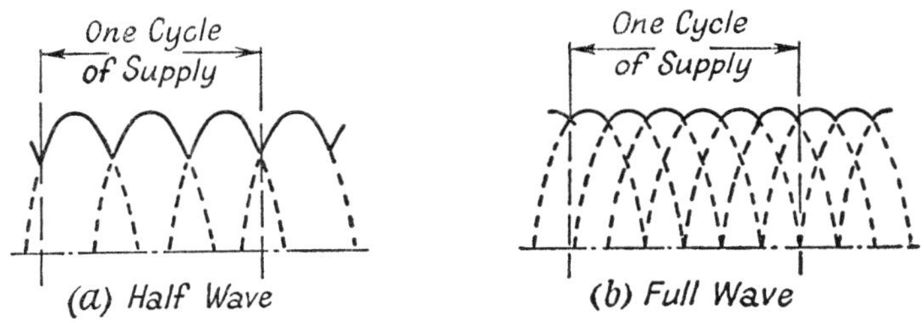

FIG. 11.10. ILLUSTRATING REDUCED RIPPLE WITH THREE-PHASE SUPPLY

not be required. If it is necessary, its inductance and capacitance will be appreciably smaller than with a single-phase circuit.

Fig. 11.11 shows a three-phase full-wave rectifier. With this arrangement the successive half cycles of voltage across the load

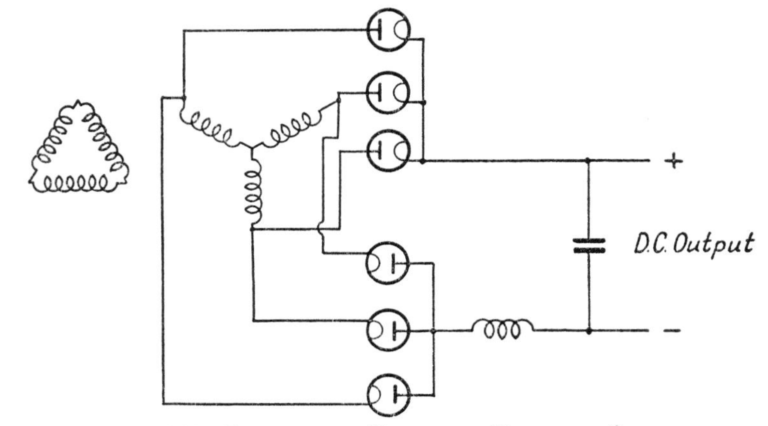

FIG. 11.11. THREE-PHASE FULL-WAVE RECTIFIER CIRCUIT

follow one another every 60 degrees, i.e. six times in a complete cycle. The result of this, as will be seen from Fig. 11.10 (*b*), is that the ripple voltage is extremely small. The voltage 30 degrees on each side of the peak of the wave is 0·87 times the peak value, so that we have only dropped 13 per cent. The total value of the ripple is thus only 6½ per cent and the mean value of the voltage is $0{\cdot}93E_p$. Consequently, with such a circuit, smoothing is the least of the problems and it is only necessary to design the choke to maintain a continuous current through the rectifier. The same laws as before apply, but in this case f is 300.

It will be seen that with this circuit two legs of the transformer are always in series, each feeding through its appropriate rectifier, so that the voltage per leg is only half that which is required in a half-wave circuit. (It must be remembered that we are dealing with peak values so that the phase difference between the voltages on the three legs does not enter into the calculations here.)

Rectifier Stacks

The construction of power supply circuits has been appreciably simplified by the introduction of semiconductor rectifiers. With valve rectifiers there is considerable inconvenience, and loss of power,

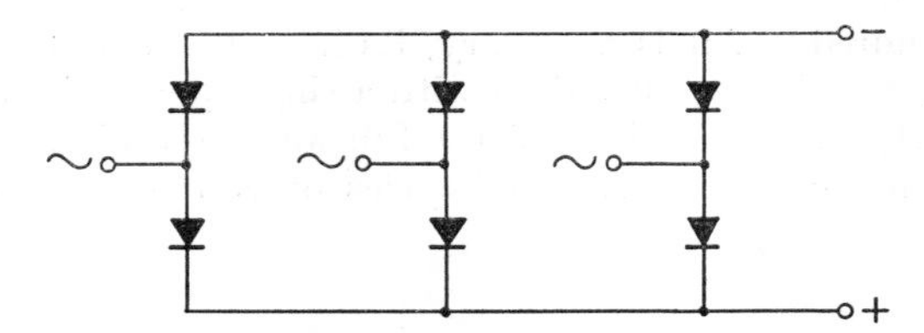

FIG. 11.12. THREE-PHASE BRIDGE RECTIFIER

arising from the necessity to provide heating for the cathodes of the rectifiers. This, in fact, virtually rules out the use of the bridge rectifier even for simple circuits.

This problem does not arise with semiconductor rectifiers, which can conveniently be assembled in bridge-connected stacks. Such units are available in a wide range of sizes, from low-voltage stacks delivering 20 V 1 A up to heavy-duty stacks delivering several hundred amperes at peak voltages of 1000 or more. They are highly efficient, having negligible forward voltage drop, and there is, of course, no cathode heating power required.

They are very convenient for three-phase circuits, since a three-phase bridge-rectifier stack can be assembled in a circuit such as that shown in Fig. 11.12 without the need for cathode heating windings.

For high-voltage low-current requirements such as the h.t. supplies for cathode-ray or X-ray tubes, half-wave stacks are made up comprising the necessary number of rectifiers in series, using the circuits of Fig. 11.1 (a) or 11.1 (c), often without subsequent smoothing since with the small current demand the ripple developed on the reservoir capacitor is usually not objectionable. Such stacks are available for voltages of 30 to 50 kV, at currents up to 50 mA.

Transformer Current

So far we have not discussed the current to be supplied by the transformer. The design of transformers is dealt with in Chapter 4, but it is relevant to consider here the requirements in a transformer feeding a rectifier.

The choke-input type of circuit is the simpler and will be considered first. As we have seen from Fig. 11.8, the current through the rectifier with a single-phase circuit is equal to the d.c. (plus the ripple). This is in the nature of a square-topped wave, which is different in its heating effect from a sine wave.

The average value of the secondary current in either limb is $I/2$. We can easily calculate the r.m.s. value; the current squared is I^2 for half the time and zero for the remainder, so that the mean squared value is $I^2/2$ and the square root of this (which is the r.m.s. value) is $I/\sqrt{2}$. Hence the effective current in each limb of the secondary winding is $I/\sqrt{2}$ and the transformer should be designed accordingly.

Similar calculations can be made for three-phase circuits. With a three-phase half-wave circuit, for example the current is I for one-third of the time so that the effective r.m.s. current is $I/\sqrt{3}$.

For a bridge or a three-phase full-wave circuit the current is symmetrical, being a square-topped wave of peak value I, alternately positive and negative. The r.m.s. value of such a wave is I, as the reader may easily verify for himself.

With capacitor input circuits the calculations are not so simple. As we have seen, the current in such a circuit is in the form of large peaks of short duration. Since the heating effect is proportional to the square of the current, such currents cause considerable heating. As a general guide the transformer secondary should be designed to supply (continuously) an r.m.s. current of 1·5 to 2 times the d.c. load current for a full-wave circuit, the ratio becoming larger with smaller d.c. current. For a half-wave or a voltage-doubler circuit the ratio should be between 2 and 3.

Table 11.1 summarizes the information.

Stabilized Power Supplies

The output from a normal power supply is not constant but varies with the load, due to the internal resistance, and is also dependent on the primary source (e.g. the a.c. mains). For many

TABLE 11.1 VOLTAGE AND CURRENT RELATIONS IN RECTIFIER CIRCUITS

Type of circuit	*Half-wave capacitor input*	*Full-wave capacitor input (centre-tapped or bridge)*	*Voltage doubler*	*Full-wave centre-tapped choke input*	*Bridge choke input*	*Three-phase half-wave*	*Three-phase double-Y*	*Three-phase full-wave*
R.M.S. transformer voltage (per limb)	$0{\cdot}71V(1+1/2RfC)$	$0{\cdot}71V(1+1/2RfC)$	$0{\cdot}35V(1+1/RfC)$	$1{\cdot}11V$	$1{\cdot}11V$	$0{\cdot}86V$	$0{\cdot}86V$	$0{\cdot}43V$
Peak ripple voltage	$V/2RfC$	$V/2RfC$	V/RfC	$0{\cdot}67V$	$0{\cdot}67V$	$0{\cdot}25V$	$0{\cdot}06V$	$0{\cdot}06V$
Ripple frequency	s	$2s$	$2s$	$2s$	$2s$	$3s$	$6s$	$6s$
R.M.S. transformer current (per limb)	$2I$ to $3I$	$1{\cdot}5I$ to $2I$	$3I$	$I/\sqrt{2}$	I	$I/\sqrt{3}$	$I/\sqrt{3}$	I

V is the d.c. voltage output
I is the d.c. load current
f is the *ripple* frequency
s is the supply frequency

purposes it is desirable to avoid such variations so that various methods are used to stabilize the output voltage.

This may be done very easily with the low voltages required for transistor circuits by using a Zener diode as in Fig. 11.13. The characteristic of a Zener diode is that the voltage across it is substantially independent of the current as explained on page 264.

If i_d is the diode current and i_L is the load current, the voltage drop on R will be $R(i_d + i_L)$.

Now $V_d = \text{constant} = V_{in} - R(i_d + i_L)$. Hence if i_L and/or V_{in} vary, i_d adjusts itself so that this equality is maintained. If $i_L = 0$ all the current flows through the diode and R must be chosen so that, with the maximum value of V_{in}, i_d does not exceed the permitted limit. (This is determined by the safe dissipation $V_d i_d$

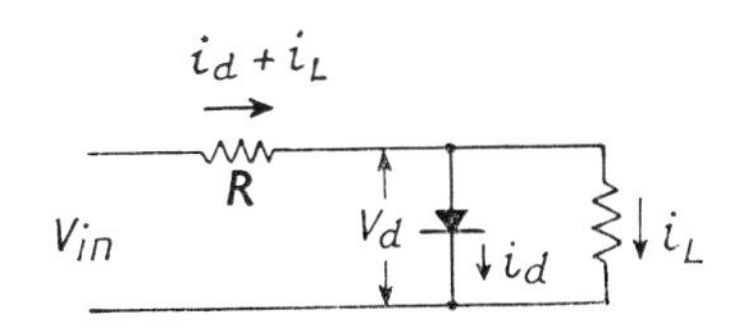

FIG. 11.13. ZENER-DIODE STABILIZER

beyond which thermal runaway is likely to occur.) As i_L is increased i_d will decrease by the same amount until it ultimately falls to zero. Beyond this point the voltage drop on R will become greater than $V_{in} - V_d$ so that the output voltage falls below the Zener voltage and is no longer stabilized.

Zener diodes are produced to operate at various voltages, usually in the range 2–40 volts, and in varying sizes according to the dissipation required.

For valve circuits a similar action may be obtained with the stabilizer tubes mentioned in Chapter 5 (page 253). Such tubes normally have working voltages in the range 80 to 150 though special tubes for higher voltages are made. Their range of stabilization is not as great as with Zener diodes, but the higher operating voltage is more convenient for valve circuits.

Both these arrangements are only suitable for small currents. Where large currents (and/or voltages) have to be handled a modified technique is adopted in which a Zener diode or stabilizer tube is used as a reference voltage. Thus in Fig. 11.14 the cathode of V_1 is held at a fixed potential above earth by the stabilizer tube S, while the grid potential is a definite fraction k of the output voltage, and is adjusted by the voltage divider P to make V_1 grid slightly negative to the cathode. Any change in output voltage develops an error signal between grid and cathode which is amplified and used to control the grid potential of V_2, causing the voltage drop to vary in such a manner as to restore the output voltage to normal.

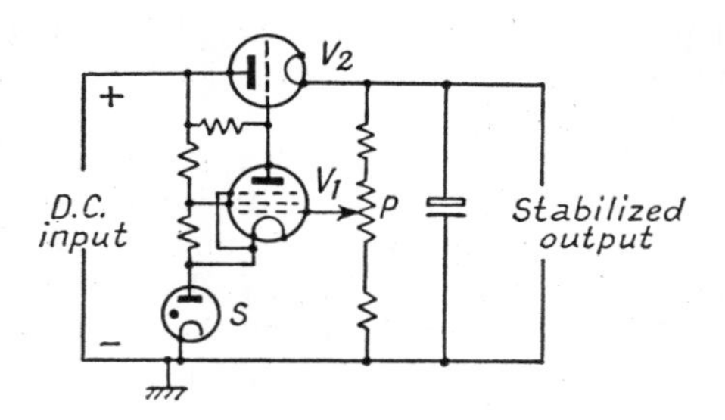

FIG. 11.14. STABILIZED SUPPLY USING REFERENCE TUBE

Fig. 11.15 shows a similar circuit using a transistor. A bridge rectifier provides an unregulated output about 20 per cent higher (on load) than the final output required. This is applied through the resistor R_2 to the Zener diode Z. This supplies a constant potential to the base of Tr_1, which is an emitter follower, and hence provides a constant output irrespective of minor changes in the load.

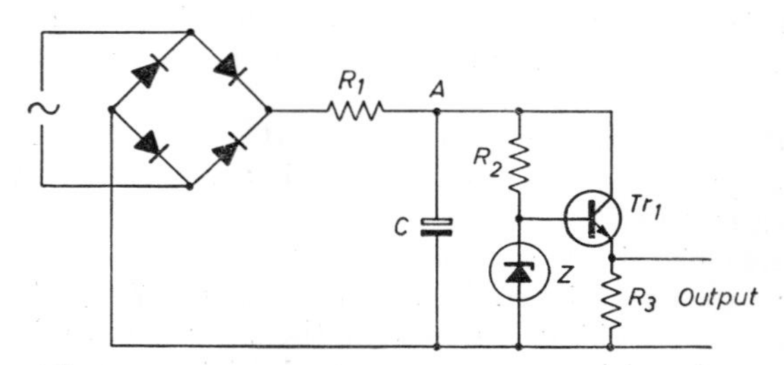

FIG. 11.15. SIMPLE TRANSISTOR VOLTAGE REGULATOR

The resistor R_2 will carry the Zener current plus the base current of Tr_1 and must be so proportioned that the diode current is within the permissible maximum. The transistor Tr_1 must be capable of supplying the full output current with a suitable margin of safety. It is desirable to include some current-limiting device in the collector circuit to protect the transistor in the event of a short-circuit on the output.

Transistor circuitry, however, has further possibilities. By including an amplifier between the reference diode and the output transistor, any error can be magnified so that a much closer control is possible. Thus in Fig. 11.16 the voltage on the reference diode Z is compared with a suitable proportion of the output. Any error is magnified by Tr_1, the output of which is applied to Tr_2 and Tr_3

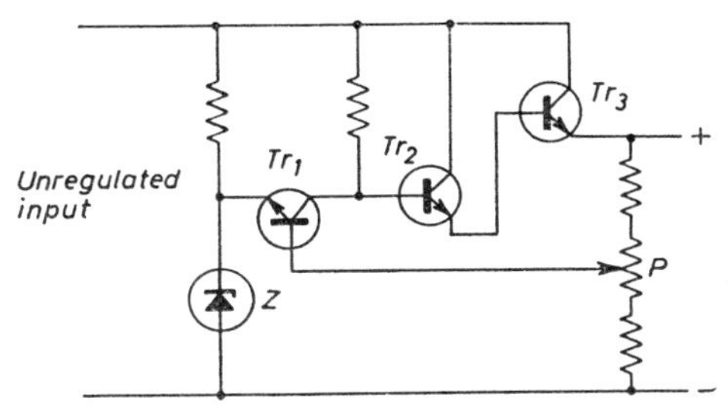

FIG. 11.16. REGULATOR WITH AMPLIFIED ERROR CONTROL

constituting a Darlington amplifier, the second stage of which controls the output current as in Fig. 11.15. This circuit has the further advantage that the output voltage does not have to equal the Zener voltage, but can be set by the voltage divider P to any desired value, while still being held stabilized.

Even more sophisticated circuitry is available with the integrated-circuit techniques discussed in Chapter 18, providing very compact units which hold the output steady within a fraction of 1 per cent against transient surges.

High-voltage C.R.T. Supplies

Cathode-ray tubes for oscilloscopes or T.V. sets require a supply of high voltage (several kV) but only a small current. For such low-power supplies it is sometimes convenient to derive the initial

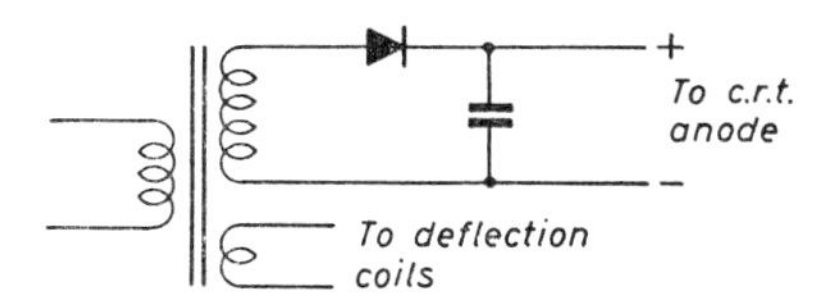

FIG. 11.17. E.H.T. SUPPLY CIRCUIT OPERATING FROM LINE-SCAN FLYBACK

power from an oscillator operating at a relatively high frequency. This reduces the size and cost of the transformer since the volume of iron is inversely proportional to the frequency.

Another method, used in T.V. sets, is to utilize the energy dissipated in the collapse of the line scan. The sawtooth current wave is fed to the scanning coils through a transformer which carries an additional secondary winding as in Fig. 11.17. The rapid change of flux during the flyback period creates a high-voltage pulse which can be fed to a rectifier and filter to supply the e.h.t. voltage to the tube.

12

Oscillator Circuits

ANY type of amplifier may be arranged to produce continuous oscillations by feeding a portion of the output back to the input in correct phase. Then, provided that the feedback is sufficient, continuous oscillation will occur at a frequency determined by the internal amplifier constants.

Let the output be V, and let a proportion β of this output be fed back to the input. Then, if the system is to oscillate, the output must be this input βV multiplied by the amplifier gain $A \sin \theta$, where θ is the internal phase shift. Hence $V = \beta V A \sin \theta$, so that $\beta A \sin \theta = 1$. The term $\beta A \sin \theta$ is called the *loop gain*, and the condition for oscillation is thus that the loop gain shall be positive and equal to (or greater than) unity. If it is greater than unity the amplitude of the oscillation will increase until internal limiting factors in the amplifier, such as increased losses and/or overloading, reduce the loop gain to unity.

Oscillators may be of three types, namely

(*a*) TUNED OSCILLATORS. Here the amplifier circuits are tuned so that the gain is only developed at a particular frequency.

(*b*) PHASE-SHIFT OSCILLATORS. Here the amplifier gain is largely independent of frequency but appreciable phase shift is permitted and oscillation will occur at the frequency for which $\theta = 180°$.

(*c*) RELAXATION OSCILLATORS. These are circuits in which a rhythmic change of conditions is produced by the charge or discharge of a capacitor.

These three types will now be considered in turn. It may be noted, however, that the conditions for oscillation may be satisfied accidentally, so that a circuit which is intended to serve as a normal amplifier may develop unwanted oscillations if stray (or common impedance) couplings produce a feedback giving a loop gain greater than unity at some frequency.

Tuned Oscillators

A single-stage amplifier is sufficient for a tuned oscillator since it is easy to arrange that the feedback is correctly phased. Fig. 12.1 illustrates a typical valve oscillator having a tuned circuit in the anode. The coupling coil L_1 feeds back an e.m.f. to the grid, in the correct phase, and oscillation builds up until the peaks of the grid swing run into grid current. This develops a charge on C_1 which runs back the bias and thus reduces the gain of the valve until a stable condition is attained (with the loop gain unity).

The time constant C_1R_1 should be sufficient to hold the charge

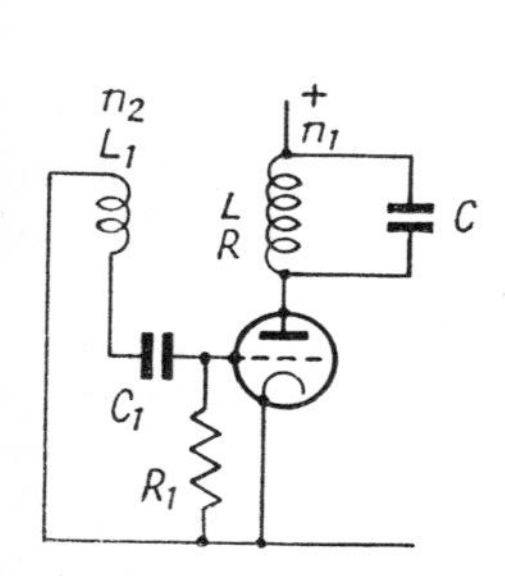

FIG. 12.1. VALVE TUNED OSCILLATOR CIRCUIT

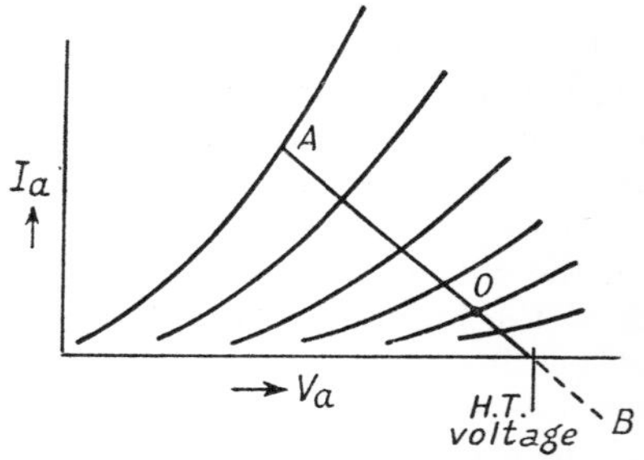

FIG. 12.2. LOAD LINE FOR OSCILLATOR CIRCUIT OF FIG. 12.1

on C_1 over a complete cycle but must not be too long or the grid bias may run back so far as to stop the oscillation momentarily. It will then not restart until the charge has leaked away, producing an intermittent starting and stopping, known as *squegging*. For the same reason the feedback coupling should not be excessive. The critical feedback to maintain oscillation is obtained when

$$n_2/n_1 = k/\mu$$

where k is the coupling factor. The coupling turns n_2 should be somewhat greater than this to provide a margin of safety but should not be more than about twice the critical value.

The output voltage may be determined by drawing a load line L/CR on the characteristics as shown in Fig. 12.2. Since the anode circuit is tuned the voltage can swing above the h.t. point but the overall swing is limited because of the reduction in μ at the lower values of i_a which will reduce the loop gain below unity. As just said, the working point is self-adjusting and in practice will adjust itself to a position approximately mid-way between Class A and

Class B conditions, as shown, giving an output about 50 per cent greater than would be obtained in Class A.

A valve oscillator may be arranged with a tuned grid circuit, as in Fig. 5.14. The action is similar but the amplitude of the tuned-circuit e.m.f. is obviously smaller, and less readily calculable.

A convenient modification of the circuit of Fig. 12.1 is to feed the supply to the mid-point of the tuned winding. The remote end is then in antiphase with the anode and so will provide the necessary feedback. This is known as a Hartley circuit, as shown in Fig. 5.15 (*a*), the equivalent arrangement using a capacitive tap, known as the Colpitts circuit, being shown in Fig. 5.15 (*b*); both forms are adaptable to transistor operation as described later.

The feedback ratio is unity so that the valve (or transistor) is driven well beyond cut-off, and so operates in a Class C condition, but provided the time constants are suitably chosen to avoid squegging the arrangement is convenient and is often used.

Transistor Oscillators

A similar technique may be used with transistors, a simple circuit being shown in Fig. 12.3. The tuned circuit is in the collector circuit and feedback to the base is provided in the correct phase by

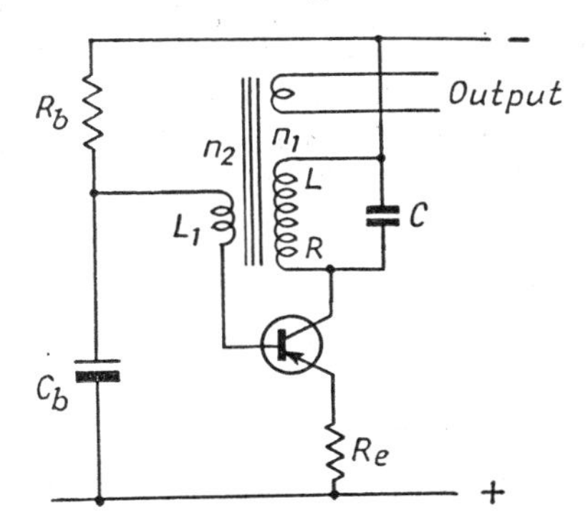

FIG. 12.3. SIMPLE TRANSISTOR TUNED OSCILLATOR

the secondary coil L_1. The loop gain is proportional to i_e (and hence to i_c) so that the circuit automatically adjusts itself to a value of i_c which is just sufficient to maintain a loop gain of unity. The collector signal current is thus a series of pulses, the circuit operating in a condition intermediate between Class A and Class B.

It is not usually necessary to plot a load line. An arbitrary value is assumed for the mean collector current (which would be of the same order as for an amplifier) and the base resistor, R_b, chosen accordingly. If now the transistor is assumed to be operating in a Class B condition the peak current will swing up to twice the mean, while the peak voltage swing will equal V_B very nearly (*see* Fig. 12.4). Hence the load will be V_B/I_{pk} approximately ($= 3$ kΩ in Fig. 12.4).

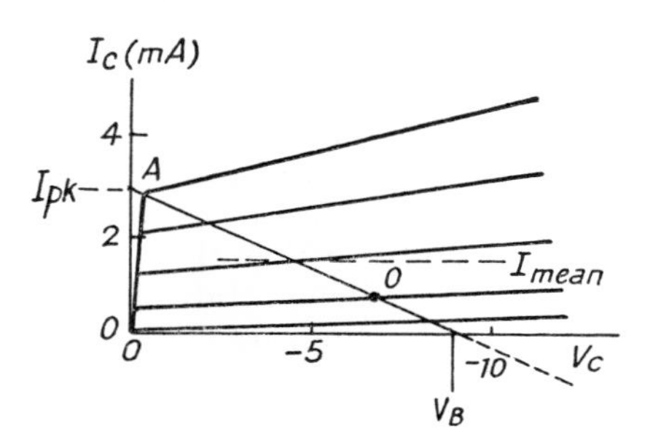

FIG. 12.4. TRANSISTOR OSCILLATOR CONDITIONS

Actually, as just said, the transistor will run back so that the working point will be something like O in Fig. 12.4, giving a peak voltage swing of OA, which in practice would be about 75 per cent of V_B. The loop gain is $g_m R_L/n \simeq R_L/nr_e$, where n is the transformer step-down ratio, from which an approximate value of n can be estimated, to provide a loop gain of unity.

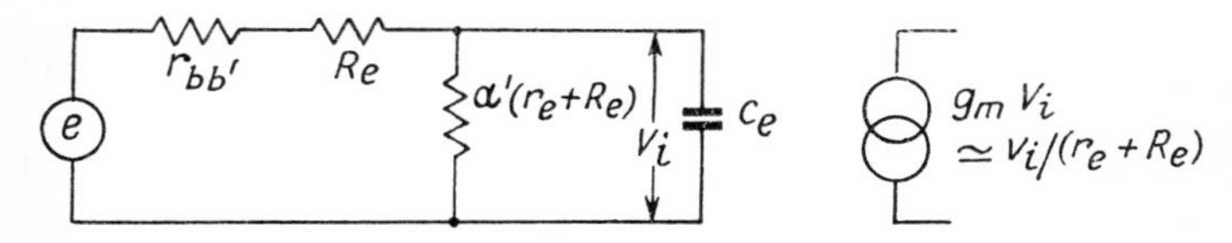

FIG. 12.5. EQUIVALENT CIRCUIT OF FIG. 12.3

For a more exact calculation the effect of r_{bb}' must be allowed for, as shown in the equivalent circuit of Fig. 12.5. The effect may be minimized by including a suitable (un-bypassed) resistor R_e in the emitter circuit as shown in Fig. 12.3. The loop gain then becomes $R_L/n(r_e + R_e)$. By making R_e of the same order as r_e the loop gain is halved, but the stability is improved. A correspondingly reduced value of n will be required to maintain a loop gain of unity.

At high frequencies the loop gain is further reduced by the

emitter capacitance C_e in the ratio $X/[(r_{bb}' + R_e) + jX]$, where $X = 1/\omega C_e$ so that n needs to be still further reduced.

Distortion

The oscillator waveform with the circuit of Fig. 12.3 is subject to some distortion because the transistor is bottoming on the peaks of the swing. This may be undesirable, particularly in a mixer where (as stated in Chapter 9) a pure waveform is desirable. Distortion may be minimized in two ways. One is to operate the transistor in

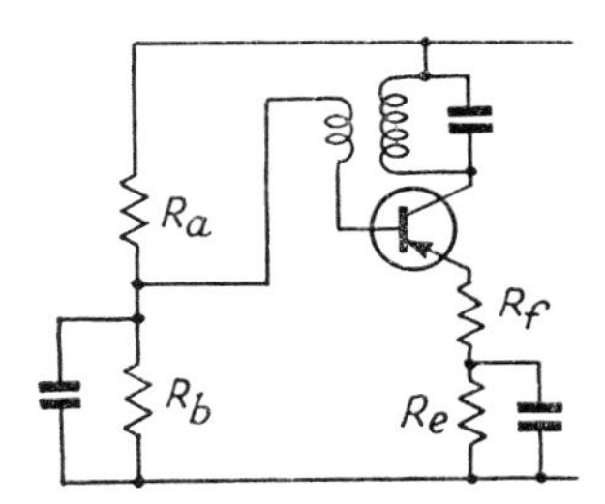

FIG. 12.6. FIXED BIAS TYPE OF OSCILLATOR

Class A conditions by controlling the base bias with a voltage divider, as shown in Fig. 12.6 and limiting the output with a feedback resistor R_f in the emitter circuit. R_a and R_b must be low so that variations of i_b do not affect the bias.

Alternatively, or in addition, the tuned circuit may be a separate circuit, of high Q, coupled to the transistor by two secondaries, as in the mixer circuit of Fig. 9.23.

Stability

The frequency of an oscillator is obviously affected by the impedances of the maintaining circuitry. With valve oscillators the effect is small and can be minimized by suitable precautions, as described on page 329. With transistor oscillators, however, the circuit capacitances are not only larger but vary appreciably with temperature and changes in operating conditions, so that if stability is important special precautions have to be taken (or a crystal-controlled circuit employed).

A form of circuit which can be used with advantage is shown in Fig. 12.7. This is known as the Clapp, or Gouriet, oscillator. It is basically a Colpitts circuit with the addition of a capacitor C_3 in series with L_1. If C_1 and C_2 are large compared with C_3, the frequency is determined mainly by L_1C_3, so that variations in C_1 and/or C_2 arising from changes in the transistor capacitances have little effect on the frequency.

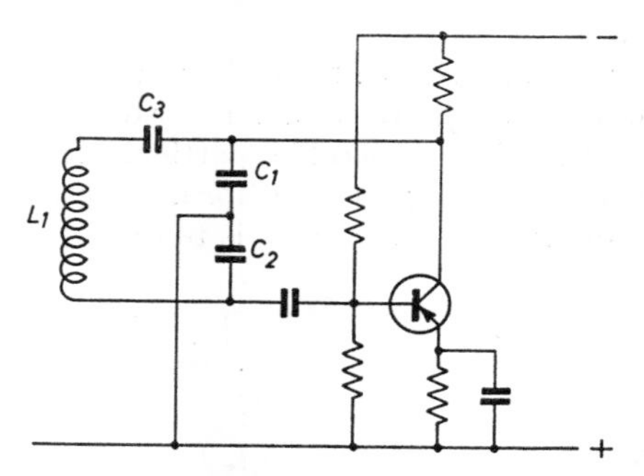

Fig. 12.7. Gouriet Oscillator

Negative-slope Oscillators

An alternative form of *LC* oscillator, sometimes called a *dynatron*, utilizes the negative slope which is exhibited, under suitable conditions, by certain types of valve or semiconductor. In these regions an increase in voltage produces a *decrease* in current, so that the effective (a.c.) resistance is negative.

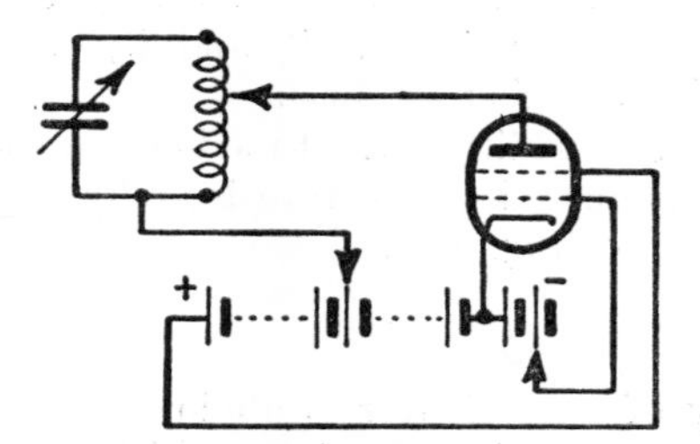

Fig. 12.8. Dynatron Oscillator

It was shown in Chapter 2 that a resistance P across a tuned circuit modifies the effective series resistance by an amount L/CP. Hence if P is negative, the effective resistance becomes $R - L/CP$, which in the limit can fall to zero, or even become sufficiently negative to offset any losses introduced by a load coupled to the circuit. In such circumstances the circuit will oscillate continuously, and this possibility is utilized for certain requirements.

Fig. 12.8 shows a dynatron circuit using a tetrode with the anode potential less than that of the screen. Then, as was shown in

Chapter 5 (Fig. 5.4), there is a small negative-slope region over the range $V_a = 0{\cdot}4$ to $0{\cdot}8\,V_s$ approximately. The actual (negative) slope can be varied by altering the grid bias, and if it is sufficient oscillations will be generated in the *LC* circuit. The amplitude will be automatically limited to the relatively small range over which the

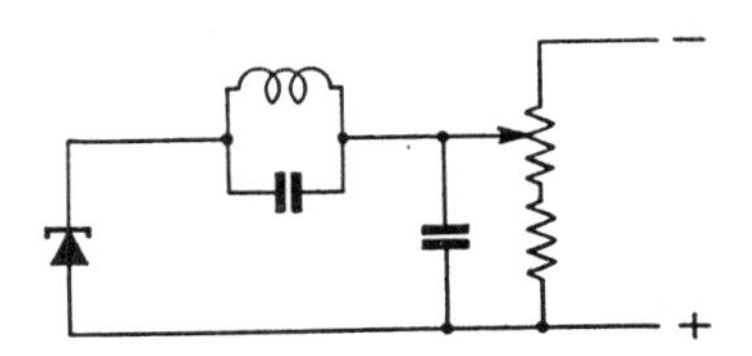

FIG. 12.9. TUNNEL-DIODE OSCILLATOR

negative slope effect operates, but the arrangement is sometimes convenient. In practice, a pentode would be used, with the screen and suppressor joined, and it is necessary to choose a valve which does exhibit the required negative slope to an adequate extent.

As said in Chapter 6, certain semiconductor devices will provide a similar action, notably the *tunnel diode* which was shown (Fig. 6.4) to have a negative slope over a small range of applied voltage. Fig. 12.9 illustrates such a circuit. There is no means of varying the negative slope, so that the tuned circuit must be designed so that Z_w $(= Q_w/C\omega)$ is equal to the negative slope of the diode used.

The output is limited, but the circuit has the merit of simplicity and will function at frequencies as high as a few gigahertz.

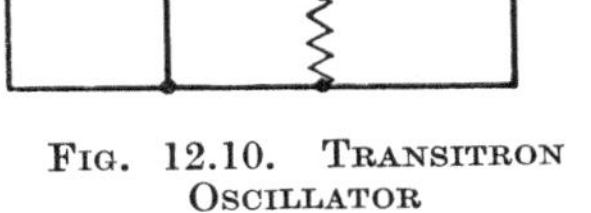

FIG. 12.10. TRANSITRON OSCILLATOR

A form of negative-slope oscillator often used with valve circuitry is the Transitron, illustrated in Fig. 12.10. This utilizes the fact that, over a limited range, positive increments of voltage on the suppressor grid of a pentode will produce proportional increases in anode current, and hence *decreases* in screen current, since $I_a + I_s$ is approximately constant. Hence the screen will exhibit a negative slope which can be used to sustain oscillation as shown.

Synchronization of Oscillators

It is found that if two oscillators are operating at frequencies not greatly different then, provided that there is some coupling between them, their frequencies are modified in such a manner as to reduce the difference. If the frequency difference is quite small the weaker oscillation will pull into step with the stronger so that they both operate at the same frequency. The effect is known as *frequency pulling* and may be a hindrance or a help.

In a superhet circuit, for example, as the signal frequency approaches that of the local oscillator serious pulling may occur if there is appreciable stray coupling between the circuits. In severe cases, particularly with transistor circuits, the two oscillations may pull into step so that there is no frequency conversion.

On the other hand, an oscillator may be deliberately synchronized with a control signal, as is often done in the reception of single-sideband transmissions where the carrier is re-introduced in the receiver by a local oscillator which is synchronized with the vestigial carrier signal. It is only necessary to adjust the local oscillator approximately to the required frequency when it will lock into synchronism with the carrier.

The effect is due to non-linearity in the maintaining system. If two signals $E_1 \sin \omega_1 t$ and $E_2 \sin \omega_2 t$ are applied to a non-linear system the third-order component contains terms at the fundamental frequencies of *both* inputs, as was explained in Chapter 9 when discussing cross-modulation (page 421). The feedback e.m.f. will thus be at a blend of these two frequencies, and if the external (synchronizing) signal is sufficiently strong it will take control provided that the frequency difference is not too great.

Phase-shift Oscillators

As said earlier, oscillators can be designed in which the frequency-controlling element is a phase-shifting network, instead of an *LC* circuit. The basic principle is that the feedback is normally negative so that no appreciable gain is produced. At a particular frequency, however, the phase is reversed by a suitable *RC* network, resulting in positive feedback and hence oscillation at this particular frequency.

Fig. 12.11 illustrates such an oscillator. In a common-emitter transistor the collector voltage is 180° out of phase with the input so that a connection from collector to base will produce negative feedback. In Fig. 12.8, however, this connection takes the form of a *CR* ladder network, the relative values of *C* and *R* being so chosen as to produce a 60° phase shift in each section. Since there are three sections the final output has a 180° phase shift and hence is in phase with the input so that if the loop gain is greater than unity, oscillation will take place at this particular frequency.

A 60° phase shift is produced in a single *CR* network when

$1/C\omega = 2R$, so that $f = 1/4\pi CR$. In a series of stages the reactance of the shunt arm is modified by the succeeding stage, so that in a 3-section network of identical stages the frequency is actually given by $f = 1/(2\pi CR\sqrt{6})$. The behaviour is modified, however, by the transistor impedances and in practice the expression $f = 1/4\pi CR$ gives a better approximation to the actual result.

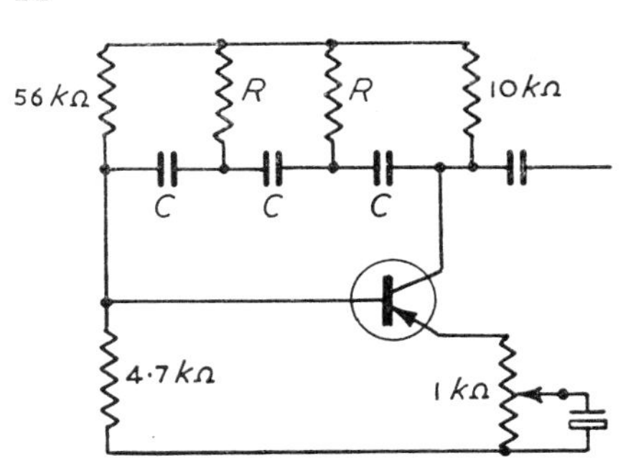

FIG. 12.11. PHASE-SHIFT OSCILLATOR

It is, of course, equally practicable to replace the transistor with a valve, in which case the influence of the (much higher) valve impedances is negligible and a highly stable oscillator results, having a frequency $f = 1/(2\pi CR\sqrt{6})$, assuming three identical stages.

Wien-bridge Oscillators

An alternative form of RC oscillator uses the reactive elements of a Wien bridge as the frequency-determining network. Such a bridge is illustrated in Fig. 12.12 and can be shown by the usual circuit laws to balance when $\omega CR = 1$. This means that the voltage

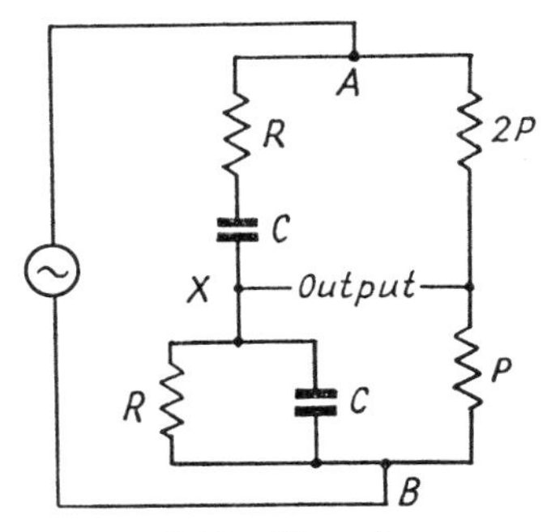

FIG. 12.12. WIEN BRIDGE

at X is in phase with (and one-third of) V_{AB}. Hence if such a network is introduced into the feedback path of an amplifier the system will oscillate at this frequency provided the loop gain is greater than unity.

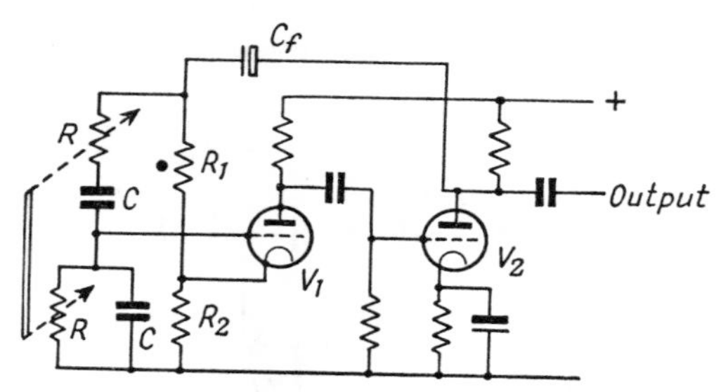

FIG. 12.13. WIEN BRIDGE VALVE OSCILLATOR

This is a more convenient form of oscillator if the frequency is required to be variable since a 2-gang variable resistor can be used to vary the bridge frequency over a convenient range. Fig. 12.13 shows a valve oscillator using this technique. Two stages are required in order to provide the correct phase of feedback through C_f. The resistor R_1 is a thermistor which adjusts itself approximately to the

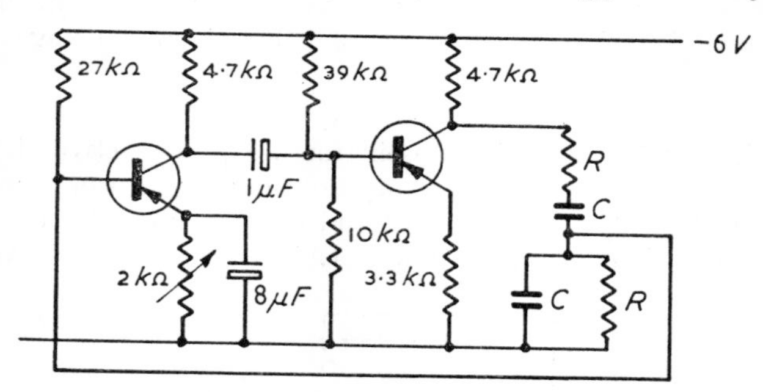

FIG. 12.14. WIEN BRIDGE TRANSISTOR OSCILLATOR

balance condition ($= 2R_2$), but not exactly so that a small out-of-balance voltage is developed sufficient to maintain the oscillation.

This circuit can be made extremely stable and is virtually independent of the valves. If the bridge components are suitably chosen, and appropriately temperature compensated, a frequency stability of 1 part in 10^5 is attainable.

With transistor operation the circuit elements can provide the resistive arms of the bridge, as in Fig. 12.14, which shows a transistor version of the arrangement. However, if this circuit is to cover a range of frequency the amplitude will vary appreciably; a more sophisticated circuit is shown in Fig. 12.15.

This circuit also uses a two-stage amplifier but an n–p–n transistor is used in the second stage so that the amplifier can be d.c. coupled with consequent saving in components and elimination of unwanted phase shift in the coupling capacitance. An emitter follower (Tr_3) gives a low-impedance output point from which feedback is taken via a thermistor (T) to a tapping on the emitter resistor of Tr_1. This feedback stabilizes the amplitude of oscillation so that the

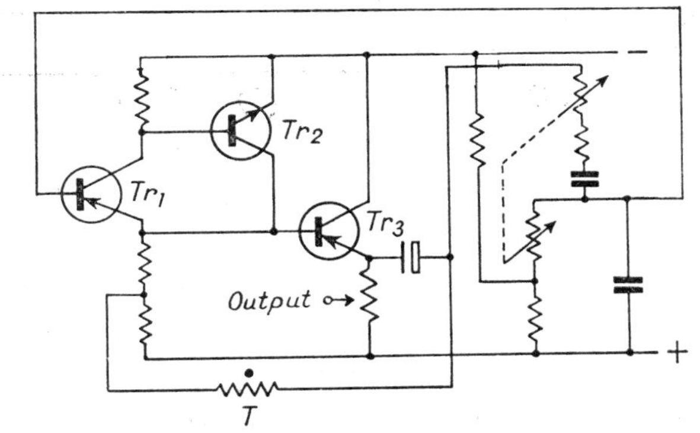

FIG. 12.15. IMPROVED FORM OF TRANSISTOR WIEN-BRIDGE OSCILLATOR

output is essentially independent of small changes in supply voltage or ambient temperature.

Transistor feedback oscillators are not as stable as the corresponding valve types because the frequency-determining networks are influenced by the transistor impedances, which are subject to variation due to changes of temperature and/or supply voltage.

Relaxation Oscillators

The oscillators so far considered produce a sinusoidal waveform, but there is a third class of oscillator involving an abrupt rhythmical transition from one state to another. These are called *relaxation oscillators* and are usually controlled by the time constant of a suitable CR network.

The simplest example of this type of circuit is the time-base circuit illustrated in Fig. 12.16. Here a capacitor C is charged through a resistor R from a battery. Across the capacitor is a neon tube. Such tubes will not conduct until the voltage across them reaches a certain level V_F; they will then "fire" and continue to conduct until the voltage falls below some lower value V_E (usually about two-thirds of the firing level). When the circuit is switched on the capacitor voltage rises exponentially until it reaches the firing level of the neon tube; this then conducts and discharges the capacitor until the voltage has fallen below the extinction level.

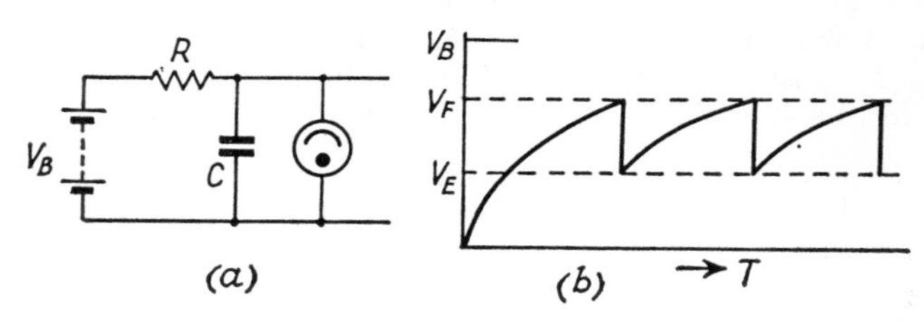

FIG. 12.16. NEON-TUBE OSCILLATOR

The tube then becomes non-conducting and the capacitor begins to recharge. There is thus a periodic charge followed by a sudden discharge as shown in Fig. 12.16 (*b*), the repetition rate being dependent upon the relative values of V_B, V_F and V_E, and the time-constant CR.

In practice a more sophisticated form of discharge circuit is usually adopted in which V_F is controlled and V_E is virtually zero,

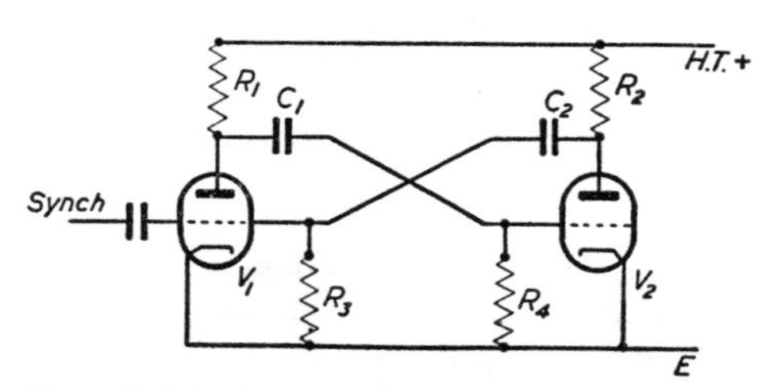

FIG. 12.17. ANODE-COUPLED MULTIVIBRATOR

thus obtaining a greater range of amplitude. Arrangements can also be made to provide constant-current charging of the capacitor, so producing a linear voltage rise instead of the exponential form of Fig. 12.16 (*b*). These modifications are discussed more fully under the heading of "Sawtooth Generators" in Chapters 15 and 17.

Another form of relaxation oscillator is the multivibrator illustrated in Fig. 12.17. This is a circuit which has two possible states, one with V_2 conducting and V_1 cut off, and the other with the conditions reversed. Assume V_1 is cut off and V_2 is "on." A positive pulse applied to V_1 grid will cause this valve to conduct. This will produce a negative pulse at the anode which will be applied through C_1 to V_2 grid and cut V_2 off. The charge on C_1 will leak away through R_4 until after a time V_2 again begins to conduct. This will produce a negative pulse at the anode which will cut V_1 off. Thus there is a rhythmic change of conditions between the two valves producing a square-wave oscillation.

The circuit is self-starting because of the inherent small difference between the g_m of the two valves, one of which will thus take more current and be rapidly driven to the fully-on condition. The circuit has many variants, including transistor arrangements, which are more fully discussed in Chapter 17. As shown there (page 707) the frequency of the oscillation $f = 1/2T$, where T is the switching period $\simeq 0{\cdot}5\,CR$, for a symmetrical circuit in which $C_1 = C_2 = C$ and $R_1 = R_2 = R$.

Blocking Oscillators

Another form of relaxation oscillator makes use of the "squegging" effect mentioned earlier. If the circuit of Fig. 12.1 is deliberately overcoupled the amplitude of the oscillation increases so rapidly that the capacitor C_1 builds up a large negative charge which cuts the valve off. The oscillation therefore ceases until this charge has leaked away through R_1 when it recommences and the whole process is repeated.

The circuit thus produces a series of short pulses of current at a rate determined by the time-constant C_1R_1; it is discussed in more detail in Chapter 17.

13

Short-wave Technique

A CONSIDERABLE proportion of modern radio transmissions utilizes what may be broadly termed *short waves,* which includes the wide range of categories listed in Table 3.1 (page 134). Such transmissions follow the basic principles already discussed, but the higher frequencies involved introduce certain special requirements, which will now be considered in greater detail.

The essential differences in short-wave transmission are threefold:

1. The propagation is by space waves instead of ground waves.
2. Because of the shorter wavelength highly directional aerial systems can be used.
3. Because of the higher frequency special techniques are required to generate and detect the signals.

13.1. Propagation

It was shown in Chapter 3 that, beyond the purely local (ground wave) range, transmission is possible over considerable distances by reflection from the ionosphere. This facility extends over a range of wavelengths down to about 10 metres, below which no reflection occurs and communication is limited to a roughly line-of-sight path.

Ionospheric Transmission

The mechanism of reflection was discussed in Chapter 3, where it was shown that the reflection depends on the angle at which the wave enters the ionosphere as illustrated in Fig. 13.1. If the angle is too steep the reflection is incomplete and the wave does not return, so that there is a *skip distance* within which no reception is possible. On the other hand, a wave which has reached the ground may be reflected upwards again and so arrive at a further distance by a second reflection from the ionosphere, as shown at B. If at this point another wave, having left the transmitter at a different angle, arrives by a single hop as shown dotted, the two waves may interfere, causing unreliable reception. It may thus be necessary to restrict the initial radiation by a directive aerial system to select either a single or double hop, but not both.

The situation is complicated by the fact that the reflection from the ionosphere is not uniform but varies with the time of day and season. As explained in Chapter 3 there are two main reflecting layers at different heights, known as the E and F layers. The behaviour of these is to some extent predictable, but it is often

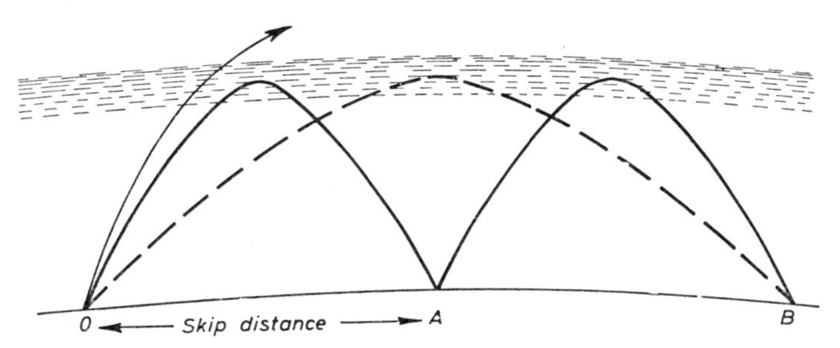

Fig. 13.1. Illustrating Reflection from the Ionosphere

found that the E layer contains random belts of ionization known as *sporadic E* which produce erratic and unpredictable reflections.

Hence the choice of wavelength and optimum angle of radiation are subject to a variety of factors, some predictable on the basis of practical experience, but others containing a considerable measure of uncertainty. In any case there is, under given conditions, a maximum usable frequency (M.U.F.) beyond which the waves are not reflected.

Moreover, any long-range transmission may easily pass through alternate regions of daylight and darkness in which the ionization conditions are widely different, so that it is rarely possible to find any single wavelength which will serve throughout the 24 hours. Very briefly, the conditions may be summarized as follows.

All Daylight

Daylight is the condition around midday (which is not affected by cloud cover). It extends for a greater period on each side of noon as one goes nearer the equator. The maximum distance which can ever be in the all-daylight zone is approximate 10,000 km, and for this coverage a wavelength of 14–15 metres would be used.

Shorter waves than this would have a skip distance exceeding the 10,000 km and in any case would be getting dangerously near the critical zone, where reflection from the ionosphere is incomplete.

The attenuation during daylight is greater than at any other time, but by the use of waves which produce the minimum number of reflections the losses are kept as low as possible.

Twilight Zone

This zone includes early morning, late afternoon, and the early hours of darkness. The attenuation is much less on all wavelengths in this zone, and particularly so for waves below 20 metres. The twilight zone extends over much greater distance than the daylight zone, and once again we use the shortest convenient wave. For maximum range—which may embrace the full 20,000 km of the earth's semi-circumference—we use waves between 15 and 18 metres, while for shorter distances longer waves up to about 40 metres are employed.

If two stations are situated so that the path of the wave is in the twilight zone all the time, the transmission conditions are exceedingly good and signals may be received which have travelled beyond the receiving point and right round the earth. This produces a faint "echo" signal one-seventh of a second later, this being the time a wave takes to encircle the earth, and these additional signals are known as $\frac{1}{7}$-second echoes.

Darkness Zone

As the darkness becomes more intense, the ionization in the upper atmosphere becomes less and hence the bending of the wave is not so great. The shorter waves are therefore barred from our use—not from any consideration of attenuation or distortion, but simply because they will never come back. This leads to a gradual increase in the wavelength used as darkness falls, and within the ordinary darkness zone wavelengths of 20–60 metres are customary. As the darkness becomes very intense, it is necessary to use still higher wavelengths, and, in fact, the critical wavelength below which the bending is insufficient may be as high as 30 metres.

Sky Wave Transmission

At wavelengths below about 10 metres there is no true reflection from the ionosphere and the waves do not return. Transmission therefore takes place along the direct path between transmitter and receiver, though as explained in Chapter 3 the range is somewhat greater than the purely optical distance because of refraction by the atmosphere. With very short wavelengths an effect known as scattering is also produced. *Tropospheric scatter* is produced by irregularities in the atmosphere which cause the waves to deviate from their normal path in a haphazard manner and a small portion is returned to earth.

A similar action is produced much higher up by irregularities in the ionosphere. This effect, known as *forward scatter*, can result in weak signals being received over distances of several thousand miles at frequencies of 50 MHz or more, well beyond the normal M.U.F. The signals are weak and subject to rapid variation but are dependable.

The techniques involved are clearly entirely different for the two types of propagation and will therefore be considered separately. For the purpose of this distinction we shall use the terms "short waves" to refer to ionospheric transmissions, and "ultra-short waves" as referring to sky-wave transmissions.

13.2. Short-wave Aerials

The basic form of aerial employed for both transmission and reception of short waves is the dipole or half-wave aerial previously discussed in Chapter 8. Such an aerial may be connected to the

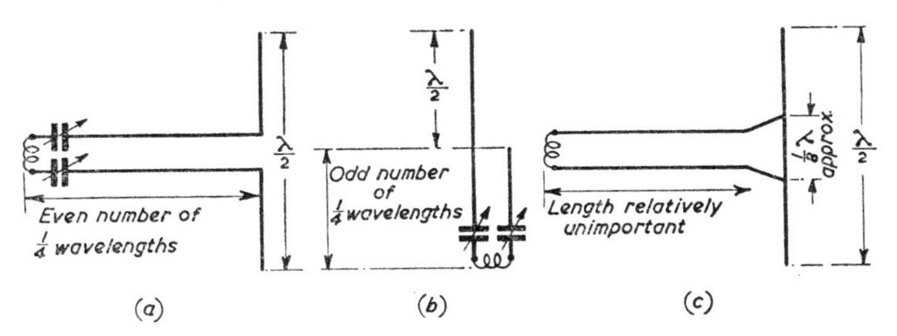

FIG. 13.2. METHODS OF FEEDING A HALF-WAVE AERIAL

transmitter (or receiver) with a short length of feeder, such as is shown in Fig. 13.2. With the arrangements at (*a*) or (*b*) the feeder is tuned, so that standing waves are produced as explained in Chapter 8 (page 381). With the correct number of quarter-wave-lengths there is then a current or voltage node at the actual feed point. In practice the feeders are made slightly over-length and then tuned by phasing capacitors which are adjusted to provide maximum current.

Fig. 13.2 (*c*) shows an untuned feeder through which the energy is transferred by a travelling wave. This requires a correct termination if reflection is to be avoided. As shown on page 379, a parallel-wire feeder has an impedance of approximately 600 ohms, which may be matched by tapping across two points at the centre of the aerial approximately $\lambda/8$ apart. If a coaxial feeder is used, the impedance is of the order of 70 ohms, which will require tapping points correspondingly closer (about $\lambda/20$).

The difficulty about a simple dipole aerial, however, is that the effective height is limited. The maximum is $\lambda/2$, but the current is

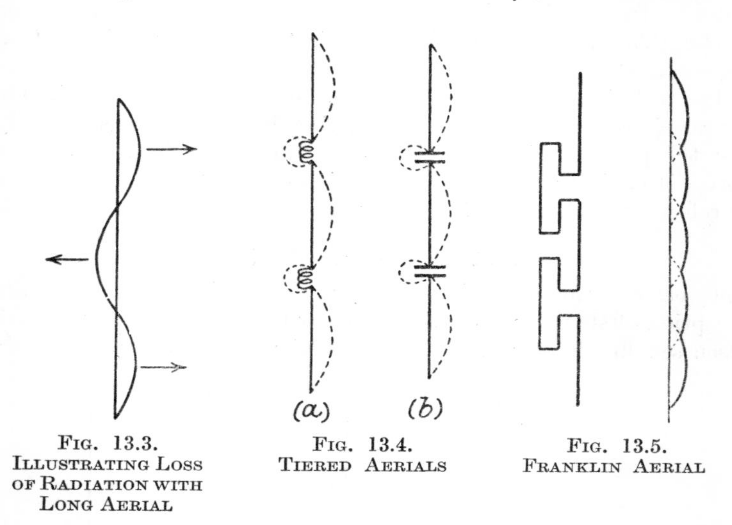

FIG. 13.3. ILLUSTRATING LOSS OF RADIATION WITH LONG AERIAL

FIG. 13.4. TIERED AERIALS

FIG. 13.5. FRANKLIN AERIAL

not uniformly distributed over the wire, so that the equivalent height is only about 80 per cent of the actual height.

No improvement is obtained by using a longer aerial. Consider a wire $1\frac{1}{2}$ wavelengths long, as shown in Fig. 13.3. Here radiation would be obtained from the top and bottom portions, and reverse radiation from the middle portion, which would cancel things out and leave the resulting radiation much the same as before. On the other hand, if we could reverse the direction of the current in the centre portion, then we should have all three half-wave aerials adding up to give a combined radiation. This can be accomplished by inserting phasing impedances either in the form of inductances as shown in Fig. 13.4 (*a*) or capacitances as at 13.4 (*b*). Such an aerial is known as a *tiered aerial*.

Still better results can be obtained by arranging to fold the wire backwards and forwards on itself, so that the only effective parts are the quarter-wavelength portions in the middle of each half-wavelength section, in which the current is a maximum and fairly uniform. Such an aerial is shown in Fig. 13.5, and was developed by C. S. Franklin of the Marconi Co.

Reflectors

Such arrangements, however, provide little directional effect, whereas because of the short wavelengths involved it is a simple matter to use groups of aerials to concentrate the radiation in one direction and so obtain greatly improved efficiency. The simplest method of doing this is to provide a reflector behind the aerial.

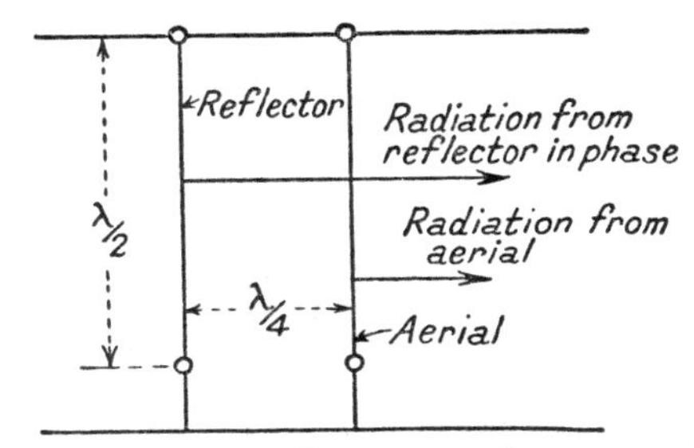

Fig. 13.6. Simple Reflector Arrangement

This need only to be a single wire of suitable size and separation, as in Fig. 13.6, which shows a half-wave aerial with a reflector wire, also half a wavelength long, spaced a quarter of a wavelength away. The aerial is supplied with current and induces e.m.f. in the reflector wire. This e.m.f. will be lagging by 90° in accordance with the usual a.c. laws, and since the reflector (being $\lambda/2$ long) is tuned, the current will be in phase with the e.m.f. This induced current will also radiate waves which will combine with those radiated by the aerial proper.

Consider the wave radiated from the reflector from left to right (i.e. in the direction of the aerial). Its phase is 90° behind the aerial radiation, but in its travel to the aerial a time equivalent to a 90° phase shift elapses, and the aerial current (and hence the radiation) is now 90° behind its original value, so that the two radiations are in phase. Thus the radiation from the reflector assists that from the aerial.

In the opposite direction, however, the reverse effect will happen, and the two radiations will cancel out. Thus we obtain directional transmission, the radiated field strength being distributed somewhat as indicated in Fig. 13.7. This is called a *polar diagram*, and represents the field strength in different directions, varying from a maximum in the forward direction, and falling to zero in the backward direction.

It will be clear that the phase relationships are only correct provided that the reflector is $\lambda/2$ long and spaced $\lambda/4$ behind the

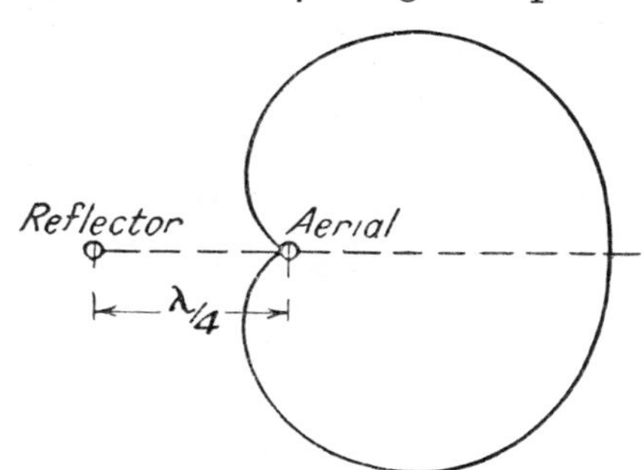

FIG. 13.7. POLAR DIAGRAM OF RADIATION WITH REFLECTOR AS IN FIG. 13.6

aerial. On very short waves, however, it is often convenient to use modified arrangements with closer spacing and reflectors which are not exactly $\lambda/2$ long. These arrangements are discussed on page 567.

Aerial Arrays

It will be noted, however, that there is still considerable radiation at the sides, whereas it would be much more satisfactory if we could concentrate the radiation almost entirely in the direction we require. This can be done by using several aerials side by side, each provided with its own reflector. Let us consider, first, the effect of radiation from two aerials without reflectors.

When the aerials are quite close together they act as one, but as the separation between them increases it is evident that there must be a phase difference between the radiations from the two aerials in certain directions. This phase difference will be most evident along a line joining the two, and we can draw polar diagrams showing the distribution of the field strength at different angles.

Fig. 13.8 (*a*) shows the diagram for two aerials spaced a quarter of a wavelength apart, from which it will be seen that there is a definite increase in the radiation on a line at right angles to the two aerials, and a reduction in the direction along the line joining the two.

The difference between this polar diagram and that shown in Fig. 13.7 should be noted quite clearly. The former diagram was for two aerials (or an aerial with reflector), in which the currents were 90° out of phase. In the present instance the two aerials are supplied with currents that are *in phase*, which produces an entirely different effect.

Now, as we increase the spacing between the aerials, the concentration of energy becomes more marked up to half a wavelength,

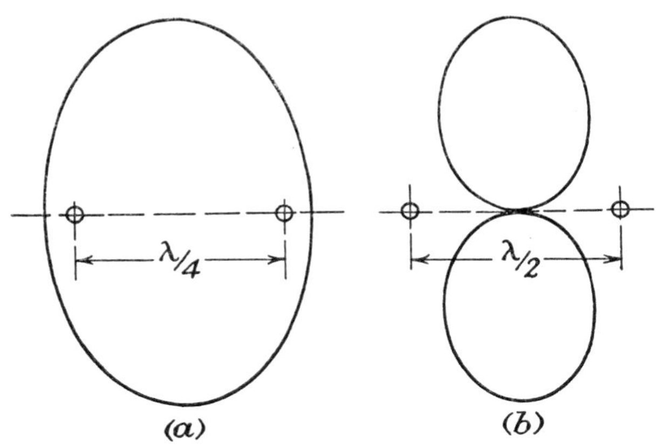

FIG. 13.8. RADIATION FROM TWO AERIALS IN PHASE

when we obtain a diagram of the form shown in Fig. 13.8 (*b*). Beyond this the effects become complex, producing diagrams having "rabbits' ears" and altering the general direction of concentration. A typical example is shown in Fig. 13.9 for two aerials spaced two wavelengths apart. More usually, therefore, the aerials are spaced half a wavelength apart, but by using several such aerials instead of

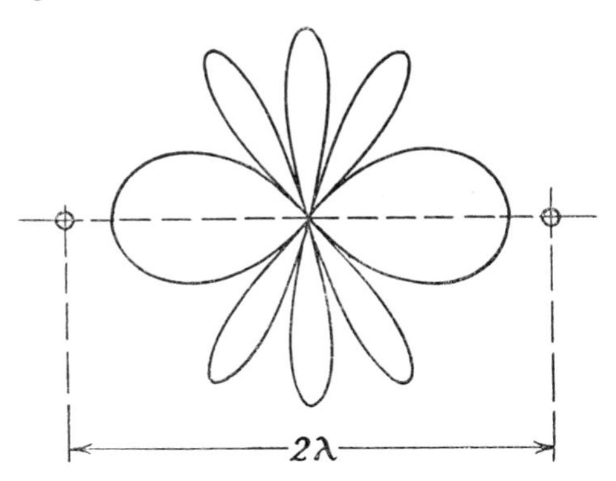

FIG. 13.9. TOO LARGE SPACING DESTROYS THE CONCENTRATION

only two, the concentration becomes more and more intense. Fig. 13.10 shows a polar diagram for four aerials which will be seen to concentrate the radiation within quite a narrow beam, plus two

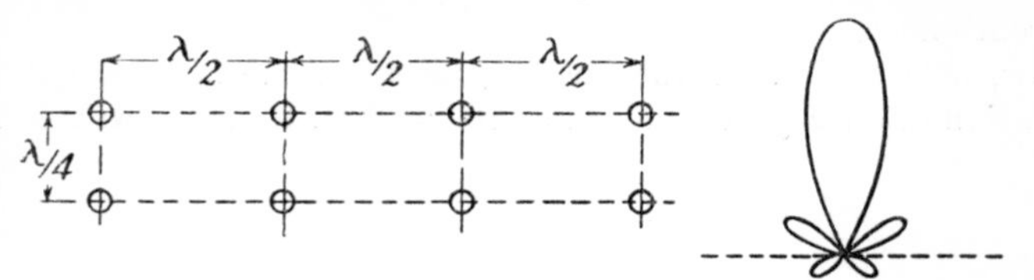

Fig. 13.10. Simple Array with Radiation Diagram

small subsidiary loops.* The radiation, however, is bidirectional, but we can make it unidirectional by providing reflectors behind each aerial when the backward radiation is cancelled practically completely.

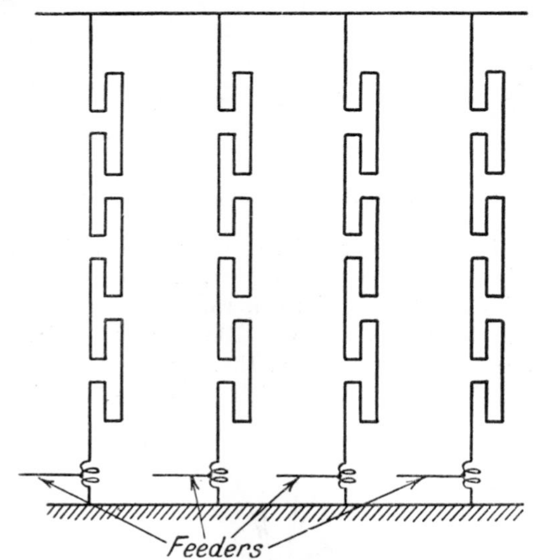

Fig. 13.11. Marconi Aerial Array

Still greater concentration of energy can be obtained by using more aerials in line, so that the modern short-wave transmitting aerial consists of a curtain of wires and reflectors broadside on to the direction of radiation. Fig. 13.11 shows a Marconi array which

* These polar diagrams can be built up from standard simple diagrams. For example, the four-aerial case is equivalent to two groups of two aerials one wavelength apart. By combining the diagram for two simple aerials λ apart with one for two aerials $\lambda/2$ apart, we obtain the diagram of Fig. 13.10.

comprises a series of tiered aerials separated by half a wavelength, while in the rear is a curtain of reflectors at a distance which is sometimes a quarter of a wavelength and sometimes three-quarters. Actually it is found that better results are obtained by having twice as many reflectors as radiators, so that the reflectors are spaced a quarter of a wavelength, and each one comprises two or three parallel wires a few centimetres apart. This added complexity is introduced not so much for improving the polar diagram as to give less critical tuning. An actual installation is illustrated in Fig. 13.24 (facing page 573).

Aperture

The total width of the curtain at right angles to the direction of radiation is usually called the *aperture*, and may be anything from two to eight wavelengths. The higher the aperture, the greater the concentration, but, of course, the greater the expense. The aerials are slung from triatics suspended between self-supporting towers, very elaborate cross-bracing being used to obtain rigidity of the structure, and thereby prevent relative movement between the aerials and the reflectors as far as possible. Counterweights are provided at the bottoms of both aerials and reflectors to keep them vertical.

Feeding the Aerials

It is necessary that all the aerials shall be supplied with current in the same phase. This is done by means of feeders all radiating from the transmitter, as illustrated diagrammatically in Fig. 13.12.

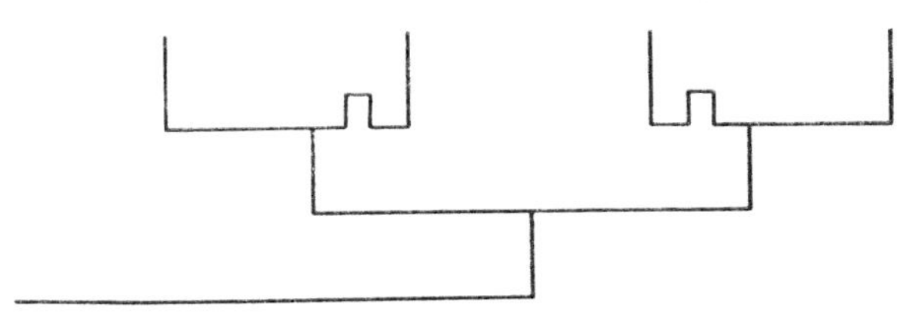

Fig. 13.12. Arrangement of Feeders to Secure Correct Phase

There is a main feeder line from the transmitter to a point in the centre of the array. From this point a number of subsidiary feeders radiate, each one being further subdivided, so that each aerial is ultimately supplied with current. The lengths of all feeders are made the same, being artificially zigzagged if necessary to provide the extra length, so that the voltages at the ends of the feeders are all in phase.

The junctions of the various feeder points have to be made through transformers to maintain the correct impedance, as was explained in Chapter 8.

Swinging the Beam

One advantage of this type of array is that the direction of radiation can be varied within small angles by altering the phase of the current at different ends of the curtain. If, for example, one end is supplied with current before the other, the interference between the radiations will obviously be modified, and actually it will bias the beam to one side. It is essential, of course, that the variation of phase shall be gradual throughout the aerial, and this is accomplished by adjusting the lengths of the feeders so that there is a gradually increasing phase lag in the aerial currents from one side of the array to the other.

The alteration in the length of the feeders is obtained by inserting extra sections of the required lengths, these being bent back on one another so that the two ends come together, and for this reason they are often known as *trombones*.

Standing Wave Arrays

The array just described uses entirely separate aerials fed in the correct phase, but alternative forms have been devised, using more or less continuous lengths of wire so constructed that standing waves are produced which add up to give the necessary radiation in

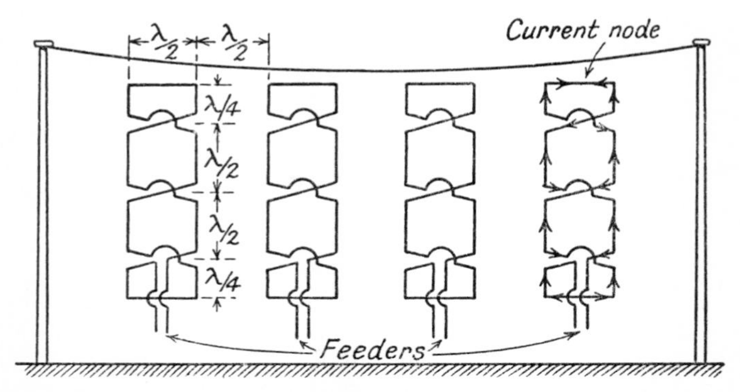

Fig. 13.13. "Sterba" Aerial Array for Short-wave Transmission

the forward direction. One of these is the *Sterba array* used by the International Telegraph and Telephone Corporation of America. This array consists of a number of groups arranged as shown in Fig. 13.13, wherein the successive half wavelengths are alternately vertical and horizontal.

The vertical components thus carry currents in phase, while the horizontal components cancel out. Each unit replaces two individual aerials of the Marconi array just discussed, and the complete array

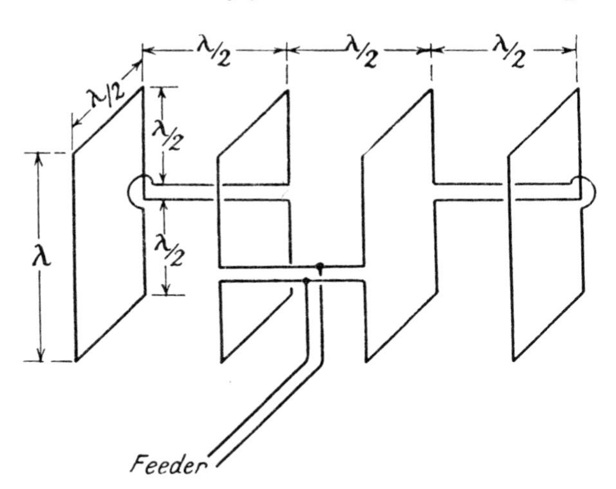

Fig. 13.14. T.W. Array

comprises a number of these units spaced half a wavelength apart, each fed by its own feeder. As before, the various units must all be in phase with one another. A similar construction spaced a quarter of a wavelength to the rear provides the reflector.

An extension of this idea is to be found in the *T.W. array* of the British Post Office, named after the designer, T. Walmsley. In this, the whole aerial is continuous, but is folded back on itself in such a way that the vertical components of the current add up, while the horizontal components cancel. This particular aerial is peculiar in that it has a depth of half a wavelength in the direction of transmission, the currents to the back sections being 180° out of phase with those in the front, and therefore producing additive radiation. The system only requires one feeder introduced at the central point, so that the necessity for accurate phasing of the various sections is obviated.

The radiation is bidirectional, so that the complete aerial still requires a reflector, which is provided by a simple curtain of wires a quarter of a wavelength behind the rear section, and hence three-quarters of a wavelength behind the front section.

Long Wire Radiators

There are other forms of aerial which will concentrate the energy in one direction, apart from the broadside reflector types so far considered.

We saw in Chapter 8 that a vertical wire several wavelengths long concentrates the radiation in an upward direction, and actually,

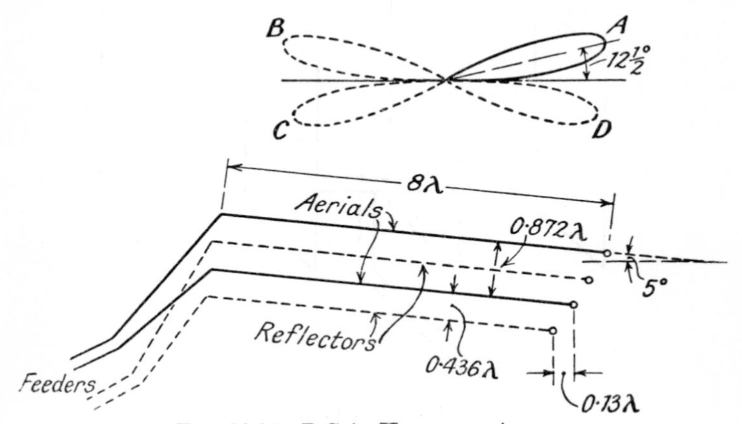

FIG. 13.15. R.C.A. HARMONIC AERIAL

as we increase the length, the effective radiation makes a decreasing angle with the wire itself, tending in the limit to radiate along the direction of the wire. This effect is used by the Radio Corporation of America in their *horizontal harmonic array* illustrated in Fig. 13.15.

This consists of a series of wires eight wavelengths long, arranged one above the other. The top wire by itself radiates mainly in a direction $17\frac{1}{2}°$ off the axis. This occurs all round the wire, giving a double conical sheath of radiation. The wires are arranged at an angle of 5° with the ground, giving an upward radiation at an angle of $12\frac{1}{2}°$, which is found to be the most favourable for long-distance communication.

A similar wire, a little less than a wavelength below the main wire, is provided with current in opposition to the main current. This wire is actually staggered slightly, and the vertical and horizontal displacements are such that radiation in the directions B and D are cancelled out. Finally, two reflector wires are provided with current 90° behind the current in the main wires, which cancels out radiation in the C direction and increases it in the A direction.

It should also be noted that the arrangement of two main radiating wires one above the other, carrying currents in opposition, largely cancels out any radiation in directions at right angles to the plane of the array, since in such directions the currents are exactly in phase opposition and the vertical spacing between the two wires does not introduce any phase angle, so that effective cancellation

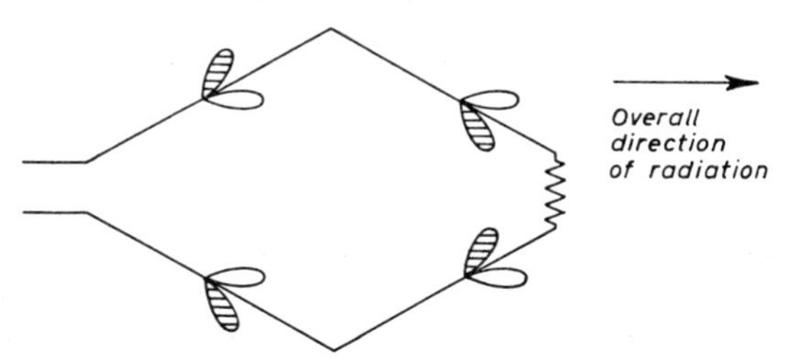

FIG. 13.16. RHOMBIC AERIAL

results. The whole aerial therefore directs the more or less concentrated beam in an upward angle of some 12½°, restricted to a few degrees on either side of the vertical plane containing the array.

End-fire Arrays

A single long wire terminated in its characteristic impedance will carry a travelling wave so that the successive elements of the wire radiate maximum energy in turn. There will thus be a phase difference between the radiation from any given element and those which precede it, as a result of which there is an overall directional effect which increases with the length of the wire. Such an arrangement is called an *end-fire array*. This produces a conical belt of radiation rather similar to that with a harmonic aerial.

The effect may be enhanced by using four such wires in a diamond formation, as in the *rhombic aerial* shown in Fig. 13.16. Each leg has a radiation pattern as just described (only the principal lobes being shown for clarity), and it is clear that by correct choice of the angles of the rhombus the shaded radiation lobes will cancel out while the others will combined to produce a marked directional effect.

Because the system is in effect a non-resonant transmission line, the legs do not have to be an integral multiple of the wavelength, but need only be relatively long to obtain a satisfactory end-fire performance. Hence the rhombic aerial can be used over a range of wavelength, and by tilting the whole assembly the radiation can be beamed in both lateral and vertical planes.

Receiving Aerials

For the reception of short waves various arrangements are used from simple dipoles to complete arrays similar to those at the

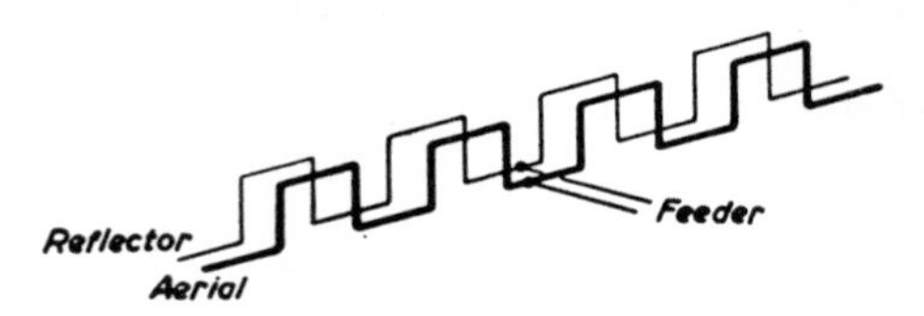

FIG. 13.17. ZIGZAG ARRAY

transmitter. There are, however, certain other forms of aerial which provide sufficiently directive reception with greater simplicity. One such arrangement is the Beverage aerial already discussed in Chapter 8.

Another form of wave aerial is the *zigzag array*, which consists of a long wire several wavelengths long bent into a series of zigzags of

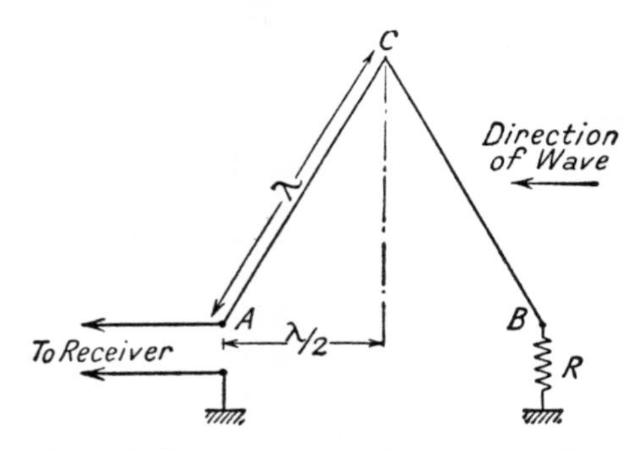

FIG. 13.18. "V" AERIAL FOR SHORT-WAVE RECEPTION

a quarter of a wavelength side, Fig. 13.17. Standing waves are built up which will have a cumulative reception from all the vertical sections, and the arrangement receives from the direction at right angles to its length. A similar zigzag a quarter wavelength behind the first acts as a reflector and provides unidirectional reception.

Another form of aerial which has the merit of simplicity is the *Inverted-V* type shown in Fig. 13.18. The sides of the V are one wavelength long and the action is as follows. Consider a vertical wire terminated at the earth end by its characteristic impedance. Voltages will be induced in each part of the wire by an advancing wave, and these voltages in travelling down to the base will all arrive in slightly different phase. It can be shown that, if the wire is made one wavelength long, the various component voltages arriving at the base all cancel out, leaving no voltage at all. If, however, we incline the wire into the direction of the wave and make the top half a wavelength ahead, the voltages at the top of the aerial arrive sufficiently in advance of those at the base to compensate for the extra distance they have to travel, and instead of zero voltage we obtain a maximum.

In practical form the aerial is usually completed with a further wire running from the top to the ground a further half wavelength ahead, this wire being terminated in a resistance as shown. This modification improves the cumulative action, while any wave arriving from the opposite direction builds up the voltage which is absorbed by the resistor R.

Horizontally-polarized Aerials

Because of the twisting of the plane of polarization which accompanies reflection at the ionosphere, short waves often arrive with as much horizontal as vertical polarization. Horizontally-polarized

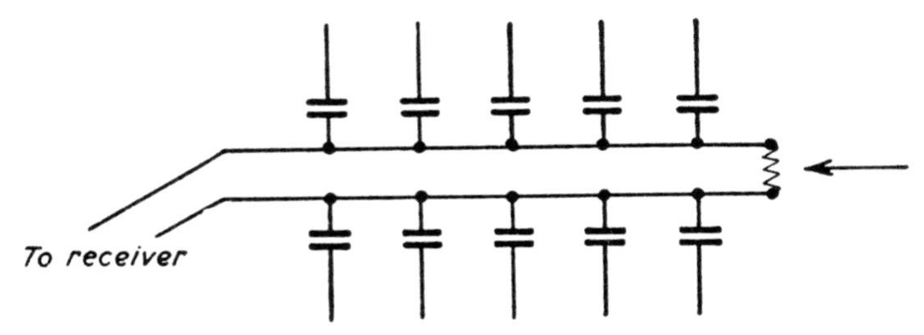

FIG. 13.19. R.C.A. FISHBONE AERIAL

receiving aerials are thus sometimes used, which have the advantage that they can conveniently be made several wavelengths wide.

One of these is the *R.C.A. fishbone array* shown in Fig. 13.19. It is rather similar to the Beverage aerial in operation. A series of horizontal half-wave aerials is erected, spaced by a small fraction of a wavelength, usually about $\lambda/6$, and connected to the transmission line. The voltages are induced in each aerial in turn and communicated to the transmission line. As with the Beverage aerial, we obtain a travelling wave which is reinforced by the voltages from the successive aerials as the wave travels along the aerial, while from other directions the building-up process does not occur. As before, the system is made unidirectional by connecting an absorbing resistance across the front end.

The rhombic aerial of Fig. 13.16 may also be adapted for the reception of horizontally-polarized waves.

Diversity Reception

The aerials we have discussed so far have been designed to make the greatest possible use of the signal available, but, as we have seen, short-wave signals suffer considerably during their reflection from the upper atmosphere. In particular, due to rotation of the plane of polarization, the signal received on any one aerial may completely vanish for short periods lasting from a fraction of a second up to several seconds.

Any normal variation in strength can be overcome by the use of automatic gain control on the receivers, but such systems work on the principle of reducing the amplification when the signal is strong, so that the overall result is no louder than when the signal is weak. No form of automatic control of this type can operate if there is no signal present, so that it is useless against a complete fade-out.

It is found, however, that when an aerial in one position is suffering from a fade, another aerial a few wavelengths away may be receiving quite satisfactorily; this is quite understandable, since a difference of a few wavelengths in the distance travelled by the wave may enable the plane of polarization to rotate from horizontal to vertical.

This has led to the introduction of what is called *diversity reception*, in which several aerials are erected a few wavelengths apart. They are all connected by correctly terminated transmission lines to the receiving room, where the signals are mixed. It is, however, impracticable to mix them straight away, for the phase of the modulation on the different aerials is not necessarily the same, and therefore each aerial is fed into its own radio-frequency amplifier, up to and including the detector stage, and then the outputs from all the detectors are mixed and passed through a common audio-frequency amplifier.

By making the detectors obey a square law characteristic, we automatically ensure that the greater part of the output comes from the receiver which is handling the greatest signal at that moment, and this automatic shifting of the load from one aerial to another operates quite well in practice. The inevitable distortion which accompanies fading, often due to selective fading, is partly reduced by diversity reception, though often not entirely overcome.

Musa System

Because of the rotating plane of polarization in a reflected wave the arrival angle which gives the best field strength is continually changing, and it is therefore necessary to make the vertical directivity of the receiving aerial system sufficiently broad to cover a range of angles.

To obviate this necessity, Friis and Feldman of the Bell Telephone Laboratories developed a system known as a Multiple Unit Steerable Antenna or *musa*.

The musa arrangement employs a series of fixed rhombic (horizontal diamond) aerials arranged in line and each fed to a phase shifter. The outputs of the phase shifters are combined and fed into a receiver. By operation of the phase shifters, which are all ganged together, it is possible to vary the vertical directivity of this array. Moreover, the beam is very much sharper than with a simple arrangement, so that, if the angle can be correctly chosen to coincide with the incoming signal, a marked improvement in signal-to-noise ratio will result.

Actually, the output from the six aerials is fed into three separate phase shifters operating in parallel. The first two select two vertical angles separated by a few degrees, and the outputs from these two channels are subsequently combined. This, therefore, provides a diversity reception, an audio-frequency delay being introduced into the channel receiving at the lower angle, the value of this delay being adjusted until the two audio signals are in phase. The audio-frequency phase is checked by means of a cathode-ray monitor tube, which produces the customary phase ellipse. Correct adjustment causes this ellipse to resolve itself into a single inclined line, while if the delay is incorrect, a maze of irregular circles and ellipses is seen.

It remains to ascertain which is the correct angle at any given time, and this is done by the third channel on which the phase shifter is continually rotated by a motor. The horizontal deflecting plates of a second cathode-ray tube are also linked with this motor drive in such a way that the spot moves horizontally across the screen once for each complete revolution. The vertical plates are operated from the output of a receiver so that at some part of the travel across the screen the spot will deflect sharply in a vertical direction, and the position of this peak will indicate the optimum angle of signal arrival at that time. The phase shifters of the other two units are therefore set at this angle. As conditions change, the position of the peak on the monitor tube varies, and the phase shifters are altered in conformity. Since the change of angle is relatively slow, this is a comparatively simple matter. A skeleton diagram of the arrangement is shown in Fig. 13.20.

Complex arrangements like this are justified in a commercial system by the increased serviceability obtained.

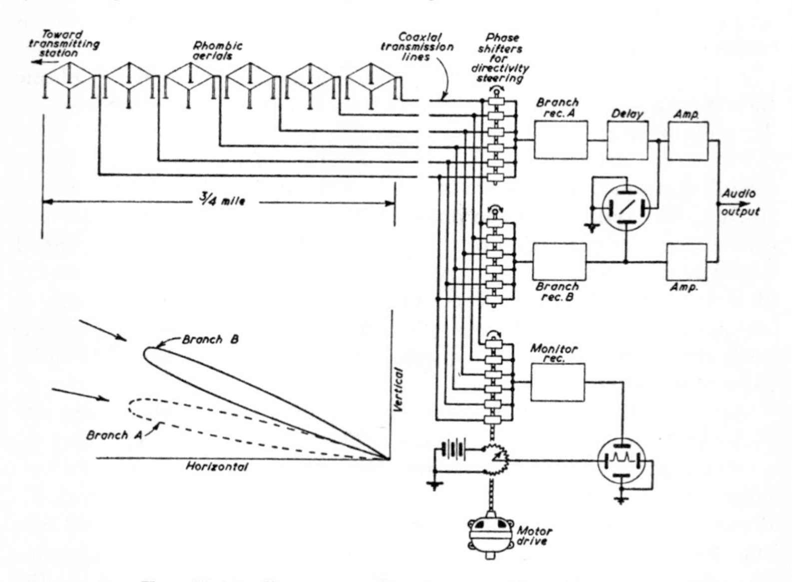

FIG. 13.20. SCHEMATIC DIAGRAM OF MUSA ARRAY

13.3. Ultra-short-wave Aerials

For the transmission and reception of waves below 10 metres elaborate arrays are no longer necessary. As we have seen, the propagation is limited to the quasi-optical path and satisfactory directivity is obtainable by using simple reflecting technique. The arrangements consist of a normal dipole accompanied by one or more *parasitic* aerials which are not directly energized but radiate as a result of currents which are induced from the main dipole, as explained on page 557 (Fig. 13.6).

It is found, however, that the directivity is increased if the spacing of the parasitic aerials is less than $\lambda/4$. In such circumstances the radiation from the parasitic aerial is not exactly in phase with (or in opposition to) that from the aerial proper, but the correct phasing can be obtained by altering the length of the parasitic aerial and thus altering the tuning, so that the current is no longer in phase with the induced e.m.f. but lags or leads by a suitable amount. In general, if the parasite is resonant at a lower frequency than that of the aerial (i.e. longer than $\lambda/2$) the radiation in the direction of the parasite is reduced, while if it is designed to resonate at a higher frequency, the radiation is increased.

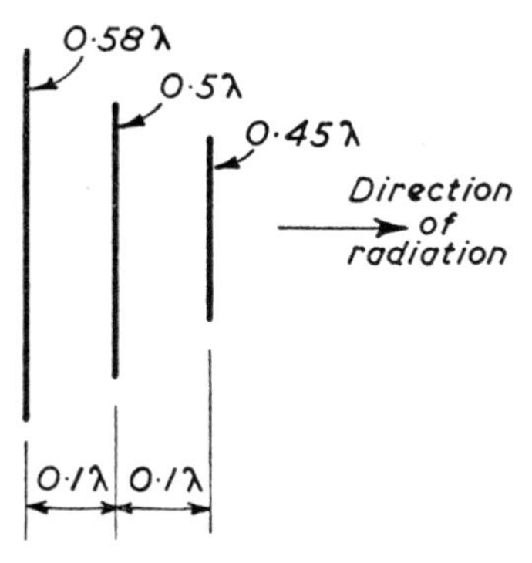

Fig. 13.21. Yagi Aerial

Simple arrays can thus be constructed having a parasite longer than $\lambda/2$ behind the dipole, acting as a reflector, and one or more shorter parasites in front acting as *directors*. A typical arrangement of this type is the *Yagi aerial* shown in Fig. 13.21 which is a reflector $0{\cdot}58\lambda$ long and a director $0{\cdot}45\lambda$ long, both spaced $0{\cdot}1\lambda$ from the dipole, which gives a directive gain of 6 times. Still sharper directivity can be obtained by using a series of directors.

Microwave Aerials

With still shorter waves having wavelengths of a few centimetres or less it becomes possible to use solid reflecting systems similar to optical techniques. This has one important advantage, for the arrays used for normal short-wave operation are only suitable for the particular wavelength for which they are designed, whereas a solid reflector operates equally well on any wavelength reasonably short compared with the dimensions of the reflector. An aperture (i.e. an effective diameter) of some 8 to 10 wavelengths is sufficient. The reflector need not be solid and is often constructed of suitably spaced wires forming a parabolic mesh.

A parabolic reflector will concentrate the rays falling on it into a parallel beam, exactly as in optics, provided the radiator is arranged at the focal point. The reflector, however, only operates on the waves radiated *back* from the aerial to the reflector. The forward radiation still spreads out in all directions and is thus largely wasted. To overcome this, a spherical mirror may be used in front of the aerial. This reflects the waves back through the focal point, as shown in Fig. 13.22 (*a*). Thus OAB is a backward wave reflected into a parallel beam in the normal way, whereas the forward wave OC is reflected back to O, on to D, and then out to E.

Even a simple plane copper sheet, however, will produce substantial reflection. A plane sheet $0{\cdot}1\lambda$ behind a dipole will provide a gain of about 8 times, while a corner reflector of the form shown in

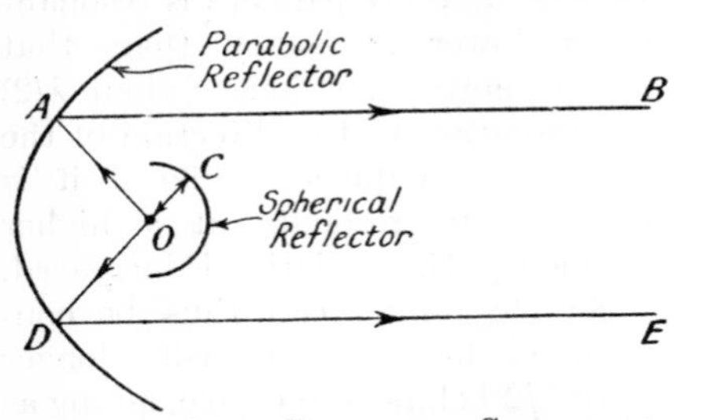

Fig. 13.22 (*a*). Reflector System for Microwaves

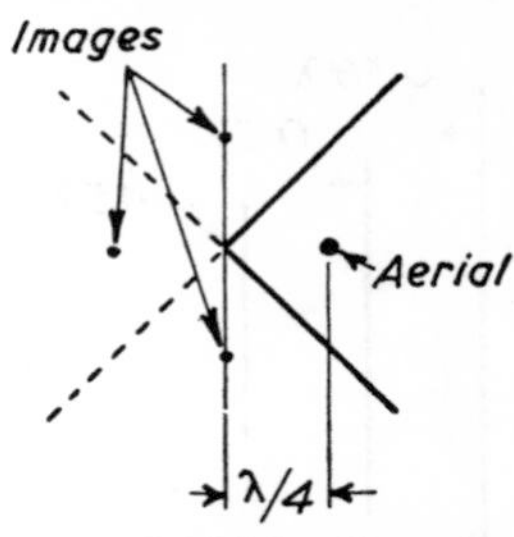

Fig. 13.22 (*b*). Corner Reflector

Fig. 13.22 (*b*) gives a gain of 16 with $\theta = 90°$ and 28 with $\theta = 60°$, where θ is the angle between the sides of the reflector.

Fig. 13.23 illustrates a microwave aerial system.

Slot Aerials

The whole technique of microwave handling is, indeed, quite different from that used for longer wavelengths and can only be referred to briefly here. Radiators can be employed in the form of horns, paraboloids, "cheeses," and many other variants. Moreover the manner in which the currents are led to the radiator are quite different, hollow tubes known as *waveguides* being used instead of the more usual feeders, as we shall see shortly.

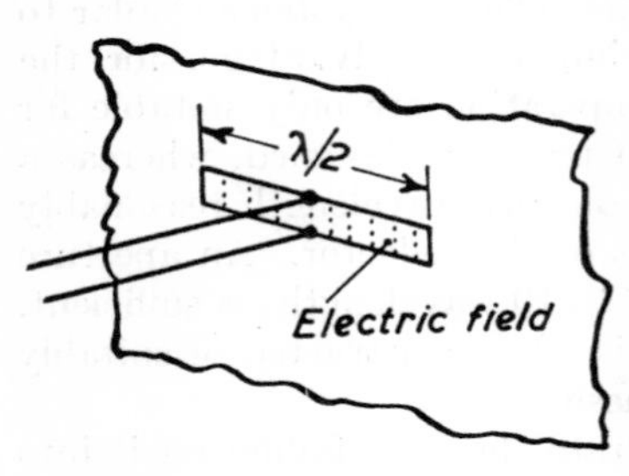

Fig. 13.25. Slot Aerial

Reference may be made, however, to a form of aerial which is frequently used, known as a *slot aerial*.

Consider a plane sheet with a slot cut in it as shown in Fig. 13.25. If this is fed with e.m.f. from a feeder as shown, there will be an electric field across the gap, and if the slot is $\lambda/2$ long it can be shown that the distribution of this field is sinusoidal. The arrangement thus behaves just like a dipole aerial except that the field is at right angles to the aerial. Thus the slot of Fig. 13.25 would radiate vertically polarized waves whereas its equivalent dipole would radiate horizontally polarized waves.

FIG. 13.23. STEERABLE DIRECTIONAL AERIAL AT GOONHILLY DOWNS RADIO STATION

(By courtesy of the Post Office)

Fig. 13.24. Aerial Array at Baldock Radio Station

(By courtesy of the Post Office)

13.4. Waveguides

As mentioned above, conventional forms of feeder are not employed for handling microwaves. Before the close of the last century Lord Rayleigh had suggested that electric waves could be propagated through metal tubes without the usual go-and-return conductor mechanism. He even pointed out that it was possible for waves to be transmitted through a cylinder consisting entirely of dielectric without any conductor at all. For over thirty years this possibility was never exploited, but in 1936 several investigators, notably Southworth and Barrow, operating independently, described experiments in the transmission of microwaves through hollow conductors in the form of cylindrical or rectangular tubes.

In order to envisage this it is necessary to abandon the idea of current flowing along one conductor and returning along another. and to replace it by the conception of electric or magnetic fields This is not so difficult for the radio engineer, who is already accustomed to visualize the propagation of energy through space in the form of wireless waves.

Fig. 13.26 shows how this can be accomplished. At the left-hand side of the figure we have a dipole aerial and the oscillation of electrons to and fro in this aerial generates closed loops of electric field which radiate outwards from the centre of the dipole. Now, we know that, if we enclose an electric field within a metal box, eddy currents are induced in the metal which produce an equal and opposite field outside the box, so that the effective field outside is zero. The effect is, in fact, the same as if the electric field had been constrained to flow within the material of the box. This is the usual principle of screening with which the reader will be familiar.

Let us see what happens if we enclose this dipole within a metal tube having its axis along the length of the dipole. This has been illustrated on the right-hand side of Fig. 13.26, and it will be seen that the effect is to restrict the outward propagation of the electric field in every direction except along the axis of the tube. We further see that a short distance away from the dipole the configuration of the electric field has settled down to a regular series of practically rectangular closed loops, and that these loops will travel along the tube by the same mechanism as waves travel in free space.

The arrangement shown would produce radiation in both directions along the tube, but it is clear that one could terminate the left-hand side of the tube in some manner which either absorbed the energy or reflected it in the correct phase, so that it strengthened the fields progressing towards the right, and in this case all the energy would be concentrated in the right-hand direction. This is the more usual arrangement, and in Fig. 13.26 the portion to the left of the dipole has been ignored.

It will be clear that for this device to be practicable it is necessary for the dipole to be small compared with the dimensions of the tube,

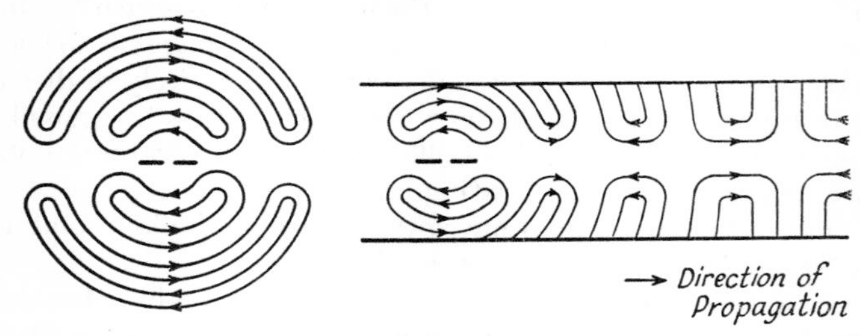

FIG. 13.26. ILLUSTRATING CONFINEMENT OF ELECTRIC FIELD WITHIN A GUIDE

so that it was not until it became practicable to generate wavelengths of centimetre order that this form of transmission came within the bounds of practice.

Types of Wave

It will be noted that the configuration of the electric field in a waveguide differs in an important manner from that in a coaxial or parallel-wire feeder. In the latter case the field in the dielectric is at right angles to the direction of propagation whereas in a waveguide the field is mainly in the direction of propagation. In the diagram of Fig. 13.26 this axial component is electric and this type of wave is therefore known as an E wave.

FIG. 13.27. MAGNETIC WAVE CORRESPONDING TO ELECTRIC WAVE OF FIG. 13.26

Associated with these electric fields are the usual magnetic fields, just as with waves in free space. The magnetic field in Fig. 13.26, for example, is a series of circular fields coaxial with the tube. These magnetic fields will not be in an axial direction but will be transverse to the direction of propagation.

Clearly, if we rearrange the manner in which the wave is started it is possible to transpose the fields, so that we have axial loops of magnetic field, as shown in Fig. 13.27, in which case the electric field will be transverse.* Such a wave is called an H wave.

* Note that these magnetic loops complete themselves within the guide, whereas the electric fields of Fig. 13.26 complete themselves through the material of the guide walls.

The waves in fact, are designated as E or H type according to whether it is the electric or the magnetic field which is axial. There is an alternative nomenclature based on the *transverse* field. As we have seen, the magnetic field in a E wave is transverse to the axis of the guide, so that this type of wave is sometimes called a TM (transverse magnetic) wave. Similarly an H wave, in which the electric field is transverse, is called a TE wave.

These designations are usually followed by certain suffixes because, as we shall see shortly, there are various possible modes of operation. Thus the waves of Figs. 13.26 and 13.27 are known as E_0 and H_0 respectively; we will discuss the significance of these designations later.

Rectangular Guides

Waveguides are not necessarily cylindrical and in practice it is customary to use a rectangular form which is not only more convenient mechanically, but also provides a better performance in certain particulars.

The most frequent form is illustrated in Fig. 13.28 in which the ratio of width to depth is 2:1. This form of guide can be excited in a very simple mode, in which the magnetic field is a series of lozenge-shaped loops along the axis, while the electric field is a simple, uniform transverse field across the guide.

We shall see that this form of guide has a number of advantages in practice, and it is, in fact, by far the most common.

Propagation in Guides

It will be realized that the field patterns shown in the figures are not stationary but travel along the guide. The wave will be propagated along the guide, and in the process the electric and magnetic fields will suffer attenuation and phase shift in the same way as occurs with a normal transmission line. We find, in fact, that the ratio of the field at a distance x to that at the input end is of the form

$$v_x/v_0 = \mathrm{e}^{-\gamma x}$$

where $\gamma = \alpha + j\beta$.

As with a normal transmission line, the first term α is the attenuation coefficient and is zero for a loss-free line. The second term is the phase-shift coefficient, representing a progressive change of phase along the guide. The phase changes by one complete cycle every time $\beta x = 2\pi$, so that $x = 2\pi/\beta = \lambda_g$, the wavelength of the field distribution inside the guide.

With a normal feeder the wavelength is virtually independent of the feeder parameters, but with a waveguide this is not so. In fact β is dependent upon the dimensions of the guide and the mode of operation. For the simple H_{10} mode of Fig. 13.28,

$$\beta = \sqrt{\left[\frac{\omega^2}{c^2} - \frac{\pi^2}{a^2}\right]}$$

where a is the width of the guide and c is the velocity of light.

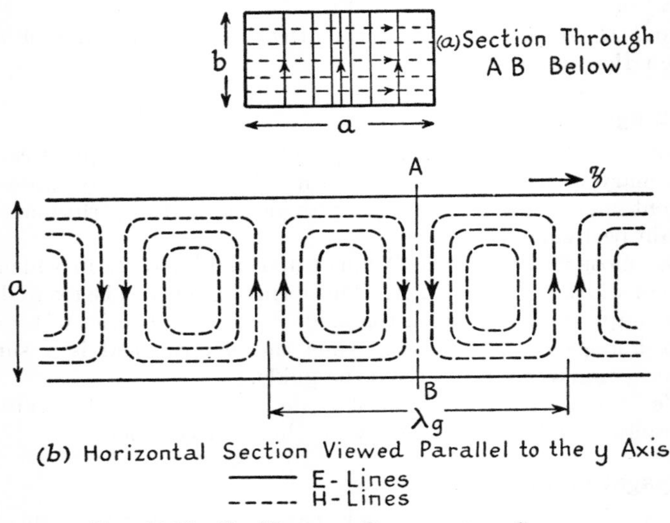

FIG. 13.28. H_{10} WAVE IN RECTANGULAR GUIDE

If this is to be real ω^2/c^2 must be greater than π^2/a^2, which means that there is a limiting frequency below which the wave will not be propagated along the guide. This is called the *cut-off frequency* and it will be seen to correspond to a wavelength *in free space* of $\lambda_c = 2a$.

At frequencies well above the cut-off value, β becomes simply ω/c and λ_g becomes c/f, which is the wavelength in free space. As the frequency is reduced λ_g gradually increases until at the cut-off point it becomes infinite. (It is necessary to distinguish clearly between the guide wavelength λ_g and the wavelength produced by the same frequency in free space.)

The effect may be understood physically by reference to Fig. 13.29, which illustrates the basic form of the magnetic fields in an H_{10} guide. If the frequency is such that the guide dimensions have an appreciable effect then reducing the width of the guide will have the effect of squeezing the loops of field longitudinally, so that the guide wavelength is correspondingly increased.

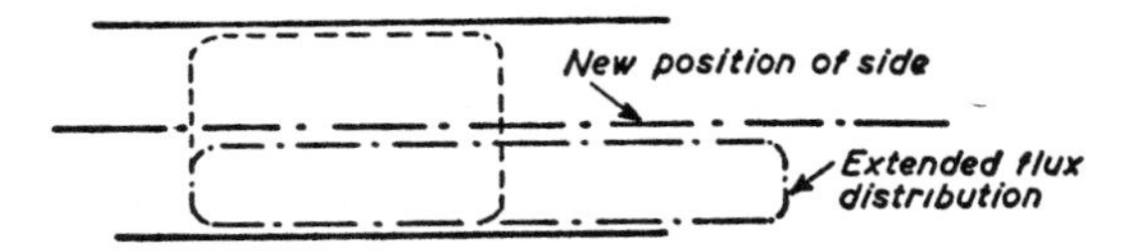

FIG. 13.29. ILLUSTRATING DEPENDENCE OF λ_g ON GUIDE DIMENSIONS

There is a simple relationship between the guide wavelength and the cut-off wavelength, namely

$$\lambda_g = \lambda/\sqrt{[1 - (\lambda/\lambda_c)^2]}$$

where λ_g = guide wavelength.
λ_c = cut-off wavelength.
λ = wavelength in free space = c/f.

This should be memorized as it is frequently used in practical calculations.

Orders of Wave

As stated earlier, it is possible for a waveguide to be excited harmonically. Normally the transverse component of the field varies sinusoidally across the guide, but it is clearly possible to have one or more complete cycles of variation in the space, and with a rectangular guide this may happen on both axes. The various types of wave are therefore distinguished by suffixes m and n, indicating the number of half-periods of field variation in the a and b directions respectively.

Thus the wave in Fig. 13.28 is an H_{10} wave, for the electric field has a sinusoidal distribution over the width of the guide while the magnetic field is uniformly distributed over the depth.

With H waves in rectangular guides it is possible to have a linear field distribution in either direction, which means that either m or n may be zero (but not both together). With E waves, however, only sinusoidal distributions are possible so that the lowest-order E wave (in a rectangular guide) is the E_{11}.

Similar harmonic distributions are possible with cylindrical guides, but because there are not two clearly-defined directions the nomenclature is not so easy to apply. We replace the dimension a by the semi-circumference and the dimension b by the radius. Hence the suffix m indicates the number of *full* periods of variation in the complete circumference, while the second suffix n indicates the number of maxima (or minima) between the centre of the guide and the circumference.

The wave of Fig. 13.27 is thus an E_{01} wave.

Figure 13.30 illustrates the field distributions in some representative guides. The left-hand diagram shows the field distribution

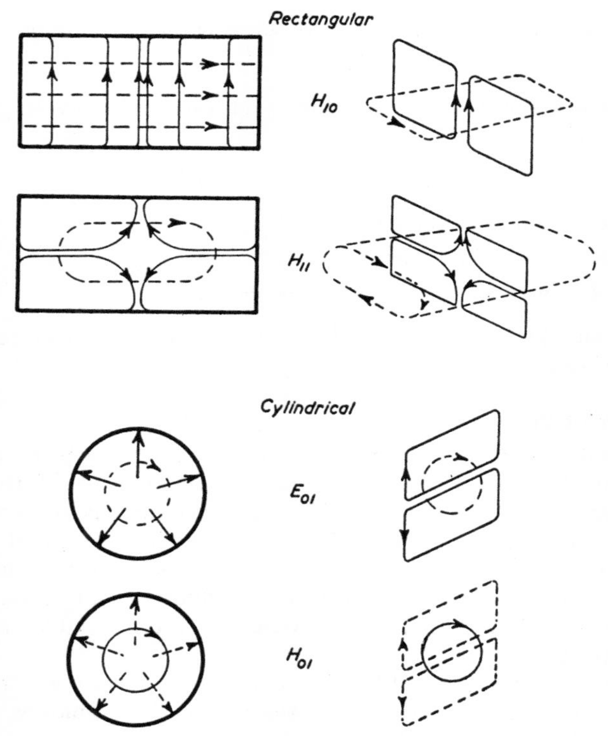

Fig. 13.30. Diagrams of Field Distribution in Typical Waveguides

——— electric field - - - - magnetic field

across a section of the guide at which this particular field is of maximum intensity, while the right-hand diagram gives a perspective view of the fields. Thus the top diagram represents a rectangular guide in its dominant (H_{10}) mode. There are longitudinal loops of magnetic field along the guide, uniformly distributed over its depth.

The transverse electric field is at right angles to the magnetic field and is of sinusoidal distribution.

The diagram on the right shows two completed loops of field. In the actual guide these fields are completed through the metal of the guide itself so that only the central vertical portion appears within the guide. Only two loops of field are shown for simplicity in the right-hand diagram, but, as just explained, the field is continuously spread over the width of the guide so that in the left-hand diagram a series of electric field lines is shown.

The second diagram shows the same guide excited in the H_{11} mode. Here the electric field strength is sinusoidal in both directions so that the cross-section is effectively divided into quadrants and the magnetic field takes the form of a closed tubular arrangement as indicated in the right-hand diagram.

(It should be noted, of course, that the right-hand diagrams only show one half-wavelength, and in fact the pattern repeats continuously, the adjacent half-wavelengths being of opposite sign.)

The third diagram shows a cylindrical guide operating in the E_{01} mode. Here the electric field is longitudinal, but at the end of each half-wave section it flows radially from the centre to the outside as shown by the arrows in the left-hand diagram. The right-hand diagram illustrates the arrangement in perspective, but only one pair of loops is shown, whereas in fact there is a continuous series of loops in a toroidal arrangement. The magnetic field is a circular field threading the electric field as shown.

The fourth diagram shows the same guide in the H_{01} mode, which is identical with the previous diagram except that the electric and magnetic fields are transposed.

These diagrams represent four of the simplest types of excitation. As already stated there are various other higher-mode excitations but these are normally avoided if possible.

Critical Wavelengths

The critical or cut-off wavelength depends upon the mode in which the guide is operating, the mode which gives the highest cut-off wavelength being known as the *dominant mode.* In general the higher the mode the lower the critical wavelength.

With a rectangular guide the critical wavelength is given by

$$\lambda_c = 2a/\sqrt{[m^2 + (na/b)^2]}$$

With cylindrical guides the expression is complex, but Table 13.1 gives the critical wavelengths for a variety of modes.

This dependence of critical wavelength upon the mode can be

turned to advantage in practice. Any mismatching or interference will give rise to harmonic modes which will absorb energy. By proportioning the guide so that it is large enough to transmit the dominant mode but too small to transmit any higher-order mode the spurious modes may be suppressed. From Table 13.1 it will be seen that a rectangular guide having $a = 2b$ provides a 2:1 range of frequency over which only the dominant mode is effective.

TABLE 13.1. CRITICAL WAVELENGTHS (AIR DIELECTRIC)

Mode	*Rectangular guide* $a = 2b$	*Cylindrical guide* *radius* $= r$
H_{10}	$2a$	—
H_{01}	a	$1{\cdot}64r$
H_{11}	$0{\cdot}89a$	$3{\cdot}42r$
E_{01}	—	$2{\cdot}61r$
E_{11}	$0{\cdot}89a$	$1{\cdot}64r$

Evanescent Modes; Piston Attenuators

If a waveguide is used at a frequency below the critical value there is no transmission of energy along the guide but merely a storage of energy near the source, the fields decaying exponentially. This can be considered as due to interference between imaginary or *evanescent* waves, somewhat like the standing waves in a mismatched feeder.

A pick-up device located within the guide will thus extract energy from the field to an extent exactly dependent on the distance from the source. If the wavelength is large compared with the critical wavelength, the attenuation is given by $\alpha = 54{\cdot}6/\lambda_c$, so that by moving the pick-up along the tube (or producing an equivalent effect by a reflecting barrier) a convenient attenuator can be produced. Such devices are known as *piston attenuators*.

Phase and Group Velocities

We have seen that the guide wavelength $\lambda_g = 2\pi/\beta$, which can be written in terms of the cut-off wavelength as

$$\lambda_g = \lambda/\sqrt{[1 - (\lambda/\lambda_c)^2]}$$

Hence, since the operating wavelength must always be less than λ_c, λ_g is always greater than λ. Now, the velocity of waves in free space $= c = f\lambda$. Hence it would appear that the velocity of the waves in the guide $v_p = f\lambda_g$, which is always greater than c, actually in the ratio $1/\sqrt{[1 - (\lambda/\lambda_c)^2]}$.

This is a somewhat surprising and apparently impossible state of affairs, but actually this quantity v_p, which is called the *phase velocity*, is not the velocity of propagation of energy along the guide. The configuration of the fields at any instant is such that, if we regard the field as having travelled from the input end, it has arrived sooner than it would have done if it had travelled in free space. But in fact the fields have not travelled in this way. Moreover, as we saw in Chapter 3, a steady wave cannot communicate any intelligence. To do this there must be variation, which we can regard as having been produced by two waves of slightly different frequency. These will produce beats which, because of the different frequency (and hence different wavelength) will travel along the guide with a finite velocity which is called the *group velocity*. This velocity v_g is slower than that of light, as one would expect, and in fact the two velocities are connected by the simple relationship $v_p v_g = c^2$.

Attenuation

In a loss-free guide the attenuation coefficient α is zero, but in a practical guide there is some loss. The attenuation, in fact, depends upon the wavelength relative to the critical wavelength for the particular conditions. At the critical wavelength, of course, the attenuation is infinite, i.e. the wave is not transmitted at all. Below the critical wavelength the attenuation drops rapidly and reaches a minimum at between one-third and two-thirds of the critical wavelength, depending on the type of wave. This minimum is fairly broad, as illustrated in Fig. 13.31, and thereafter the attenuation rises slowly as the wavelength is progressively decreased.

An exception to this rule is the H_{01} wave (or any H_0 wave), which has theoretically a continuously decreasing attenuation with decreasing wavelength. It has been found in practice, however, that the very slightest departure from true dimensions of the guide introduces rapid attenuation, and the attenuation for an H_{01} wave in a practical guide, while rather better than that for an E_{11} wave, is nevertheless of the same form and passes through a minimum.

Fig. 13.31 shows the order of attenuation obtained with a typical rectangular copper waveguide. The values for cylindrical guides of comparable area are similar. With brass walls the attenuation is approximately twice as great.

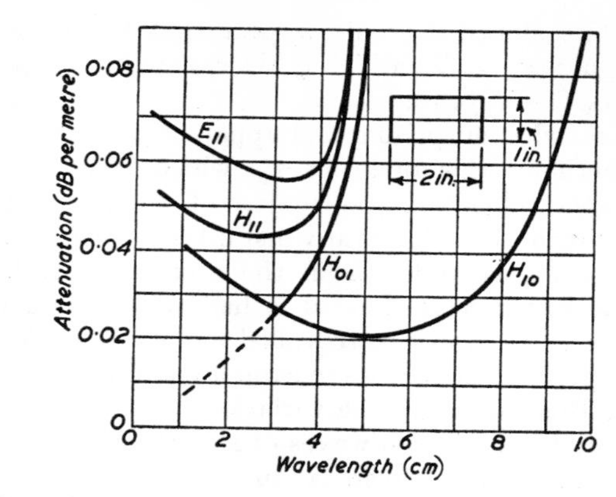

FIG. 13.31. ATTENUATION WITH RECTANGULAR COPPER WAVEGUIDE IN DIFFERENT MODES

Characteristic Impedance

The absence of any current flow of conventional form causes the usual concept of characteristic impedance to break down. However, Schelkunoff has suggested that it can be regarded as the ratio of the transverse electric field to the transverse magnetic field. On this basis it can be shown that for both rectangular and cylindrical guides, the impedance is

$$\text{For } E \text{ waves} \qquad Z_E = k\lambda/\lambda_g$$
$$\text{For } H \text{ waves} \qquad Z_H = k\lambda_g/\lambda$$

where $k = \sqrt{(\mu_0/\varepsilon_0)} = 120\pi$ for air dielectric. It is sometimes called the *intrinsic impedance of free space.*

We have seen that λ_g can be expressed in terms of λ and λ_c, the cut-off wavelength. Hence the characteristic impedance may also be expressed in these terms as

$$Z_E = k\sqrt{[1 - (\lambda/\lambda_c)^2]}$$
$$Z_H = k/\sqrt{[1 - (\lambda/\lambda_c)^2]}$$

Reflection; Standing-wave Ratio

It will be clear that any change of section, or obstruction, in the guide will cause reflections of the electric and magnetic fields which will travel back along the guide. The effect is analogous to that in a normal transmission line if one substitutes electric and magnetic fields for voltage and current, so that we obtain standing waves of field within the guide.

The ratio of maximum to minimum field intensity is known as the *standing-wave ratio.* It can be measured by inserting a probe detector into the guide (as explained in the section on reception) and noting the e.m.f. picked up as the probe is moved along the guide. Since the wavelengths involved are of the order of a few centimetres the movement from a node to an antinode is quite small; on the other hand the probe itself must be kept as small as possible or it will produce reflections of its own.

Ideally the standing-wave ratio should be unity, indicating no reflection, but this will only happen if the waveguide is correctly terminated by a load equal to its characteristic impedance. Alternatively we can deliberately introduce a reflected wave in opposite phase to that reflected by the load. In practice both methods are used. The load is designed to provide a reasonable match and an adjustable reflecting system is then introduced to obtain the final matching.

This reflecting system may take various forms. We can terminate the guide with a plunger located just beyond the reception point and vary its position; alternatively a small stub can be arranged at right angles to the guide, the reactance of the stub being variable by a plunger; while finally obstructions of various forms may be introduced within the guide which produce effects analogous to shunt capacitance or inductance in a transmission line.

Launching the Waves

The waves are first set up or launched in the guide by short rods or aerials arranged along the direction of the electric component of the wave required. Thus, in Fig. 13.26 where we launched an E_0 wave, the aerial was arranged along the axis of the tube. This particular aerial could be fed by a feeder coming in through the side of the guide at right angles to the plane of the paper.

A more usual arrangement is to feed the energy into the guide from a coaxial cable. In this case it is simply necessary to extend the central conductor of the coaxial cable along the guide for a short distance, as shown in Fig. 13.32 (*a*). Similarly if we wish to launch an E_1 wave, we can do this by having two rods protruding into the inside of the guide as shown in Fig. 13.32 (*b*). It is necessary that the phases of the voltages on the two rods be in opposition, which is conveniently arranged by introducing an additional half-wavelength of cable between one aerial and the other.

With an H_1 wave there is, as we have seen, a single electric field in the centre of the guide, and this may therefore be introduced by a single rod placed transversely to the guide as shown in Fig. 13.32 (*c*). With an H_0 wave we have two such fields, and these must be generated by two rods each extending across approximately half the diameter of the tube as in Fig. 13.32 (*d*).

There are various modifications of the essential technique which cannot be discussed here. Useful information will be found in a

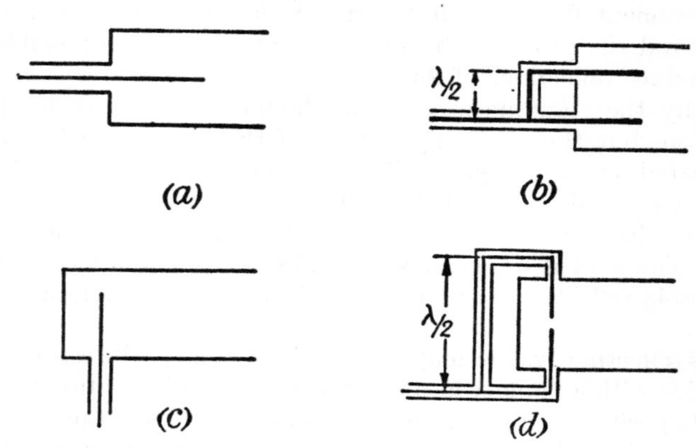

FIG. 13.32. METHODS OF LAUNCHING WAVES IN GUIDES

paper by Kemp,* which also discusses the effect of bends and obstructions which may introduce reflections and/or spurious modes. It may be noted that one type of wave may be transformed into another by interposing barriers or gratings within the guide. This is sometimes useful, for it may be that more than one type of wave is present at the same time, and the undesired type of wave may be trapped by the interposition of suitable barriers designed to coincide with the configuration of its electric field, but of such a nature that they do not seriously interfere with the substantially different field distribution for the wanted wave.

Radiation from Guides

We saw in Fig. 13.26 that a wave could be launched in a guide by locating the aerial within the guide, so that the electric fields instead of being able to spread out were confined within the guide. It is clear that if we cut a slot in the guide of length λ_g these fields

* Kemp, J., "Waveguides in electrical communications," *J. Instn. Elect. Engrs.*, **90**, Pt. III, p. 90.

will emerge from their constraint, and will again begin to radiate. A simple example of this is illustrated in Fig. 13.33, where a slot has been cut in a guide carrying an E_{10} wave and the manner in which the electric field will emerge is quite clear. If we have two slots a suitable distance apart as shown in Fig. 13.33 (*b*), we have two emergent loops of field which will join up and form one field of greater amplitude.

If the width of the slot is restricted the energy is radiated in the form of a relatively narrow beam, and this radiation may be assisted by the provision of external flanges. Alternatively radiation may be produced from the open end of a rectangular waveguide.

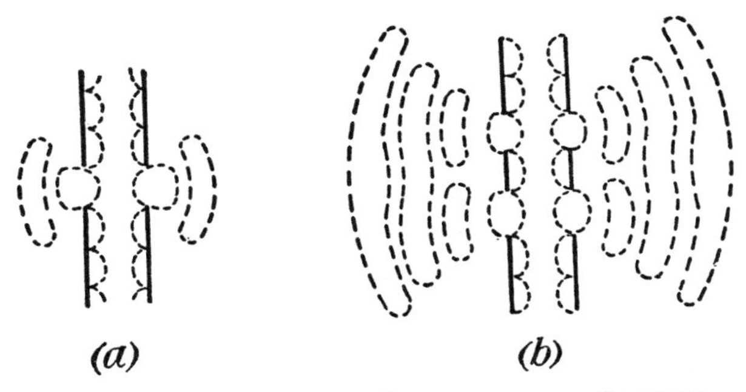

FIG. 13.33. ILLUSTRATING RADIATION FROM SLOTS IN RECTANGULAR WAVEGUIDE

The distribution of field in an H_{10} wave in a rectangular tube was shown in Fig. 13.28, and its suitability as a radiator will at once be apparent in view of the fact that the electric field lies straight across the guide from top to bottom, this travelling field being accompanied by a single configuration of magnetic field. If such a guide is left open at the far end the electric field will simply continue to travel out into space.

There will also be an appreciable directive effect, the waves being radiated within a fairly narrow angle depending upon the dimensions of the guide relative to the wavelength. This ratio is called the *aperture* and with apertures between 5 and 10 beam angles between 15 and 20 degrees are obtained.

Use of Horns

Still further improvement in the radiation may be obtained by flaring the mouth of the tube to form a horn. This may be done in a horizontal dimension only or in both directions according to the requirements. As might be expected, there is an optimum angle of flare which usually lies between 30° and 50°, while there is also a minimum length of horn. The horn acts as a resonating cavity, and thus amplifies the strength of the fields, but if the horn is less than the critical length the greater part of the energy is confined within the horn and does not radiate. Beyond the critical length, however, an increasing amount of energy is radiated, and because of the resonance effect just mentioned it will be seen that the radiated field is greater than it would be from a simple open-ended tube.

The classic references on waveguide theory are

Barrow, "Transmission of electromagnetic waves in hollow tubes of metal," *Proc. Inst. Rad. Engrs.*, **24**, p. 1298.

Southworth, "Some fundamental experiments with waveguides," *Proc. Inst. Rad. Engrs.*, **25**, p. 807.

Barrow and Chu, "Theory of the electromagnetic horn," *Proc. Inst. Rad. Engrs.*, **27**, p. 51.

13.5. Short-wave Receivers

The signals collected by the receiving aerial have to be amplified and subsequently demodulated. These operations are performed by the same basic processes as were discussed in Chapter 9, but the technique has to be modified to allow for the higher frequencies involved. There are two main factors to be considered, namely circuit configuration and transit time.

Circuit Configuration

As the frequency increases, the LC product decreases rapidly, being equal to $1/4\pi^2f^2$. The minimum practical value of C is of the order of 10 pF, which with a frequency of 30 MHz would require an inductance of 0·28 μH. This is comparable with the inductance of the circuit wiring and as the frequency is further increased the use of "lumped" inductances becomes increasingly impracticable.

At really high frequencies, therefore, the required inductance has to be provided by the circuit configuration, which is achieved by the use of *trough lines,* consisting of short lengths of rigid parallel wires (one of which may be the chassis or screen). The order of inductance and capacitance of such lines can be calculated from the expressions given in Chapter 1 (pages 23 and 38). As an example, two wires 1 mm diameter 5 cm long with 1 cm separation would have an inductance of 0·05 μH, which with a capacitance of 16 pF would tune to 175 MHz. The capacitance between the wires would be approximately 0·5 pF.)

If the length of the trough line is comparable with the wavelength the current and voltage distribution is no longer uniform and the arrangement becomes a form of resonant line, as discussed in Chapter 8, and at really high frequencies the calculations should be made on this basis. It is usual to make the lines a little over a quarter of a wavelength long. They then present an inductive reactance which can be tuned with appropriate series capacitors (as in Fig. 13.36).

Transit-time Effects

The second limiting factor arises from the fact that at the frequencies involved the period of the oscillation ($1/f$) becomes comparable with the transit time of the electrons in the amplifying devices used. Because of this the performance deterioriates rapidly as the frequency increases.

If the amplifier is a valve the effect of transit time is twofold. The transconductance g_m is no longer in phase with the input, but is subject to an increasing phase lag, while the input impedance begins to fall rapidly. The action may be explained broadly as follows.

Consider a valve working normally with a suitable negative bias on the grid. If this bias is suddenly increased it will prevent any further electrons from being emitted from the cathode, but we are still left with a quantity of electrons which have already left the cathode, and are prevented from reaching the grid due to the fact that it is now at a negative potential. Similarly, the electrons in between the grid and the anode are suddenly repelled, leaving a sort of electron vacuum which is filled by a flow of electrons from the external circuit.

A similar but opposite state of affairs results if the grid is made suddenly more positive, so that we obtain in actuality a momentary grid current, despite the fact that the grid is so biased that under static conditions no grid current flows. If the grid is being subjected to extremely rapid alterations of potential, a permanent alternating grid current will flow exactly as if there were a resistance across the grid and cathode. In other words, the input impedance of the valve falls from a very high value at normal frequencies down to something quite small—of the order of 10,000 ohms or even less—at frequencies comparable with the transit time of the electrons.

It can be shown that the input resistance is given by the expression

$$R = 1/kg_m f^2 \tau^2$$

where g_m is the transconductance of the valve,

f is the frequency,

τ is the electron transit time, and

k is a constant depending on the geometry and operating conditions of the valve.

An evaluation of this expression shows that with an ordinary valve the input resistance may be as low as 20,000 ohms even at 30 MHz (10 metres), and the results are confirmed by actual experiment.

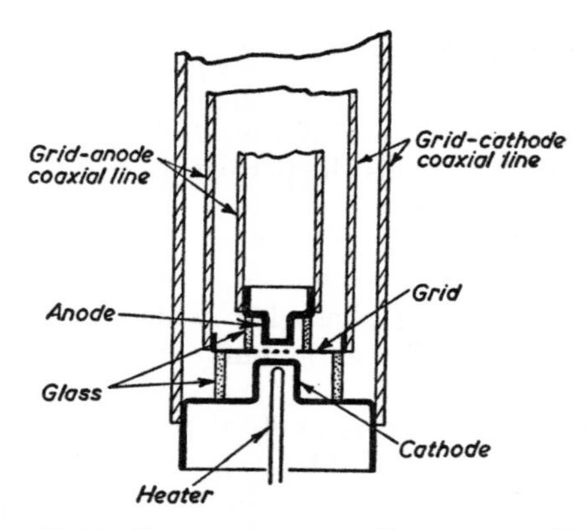

FIG. 13.34. CONSTRUCTION OF LIGHTHOUSE TUBE

It will be noted that, in a given valve, τ is constant, so that the grid resistance decreases as the square of the frequency, and thus very rapidly passes from a factor of negligible importance to one of major consideration.

The remedy is to reduce the transit time of electrons which is done by reducing the internal dimensions and increasing the density of the emission from the cathode. To minimize the inductance and self-capacitance of the leads, these are brought out through the sides of the valve by the shortest possible route, and with constructions of this type frequencies of several hundred MHz can be handled.

For still higher frequencies it becomes desirable to design the valve so that it can be built into a coaxial line as an integral part. Such valves have their electrodes in the form of flat discs sealed directly to the glass envelope. This also permits very close spacings, which reduces the transit time. Such valves are known as *disc-seal* or *lighthouse* tubes. (See Fig. 13.34.)

Grounded-grid Circuits

Valves using these close spacings are usually triodes and hence, when used as amplifiers, are subject to the usual limitations of feedback via the anode–grid capacitance. Since this is smaller than with a normal triode, and since also the slope is kept low to reduce transit-time effects, amplifications of 2 or 3 times can usually be attained without self-oscillation.

Somewhat better performance can be obtained by using the valve

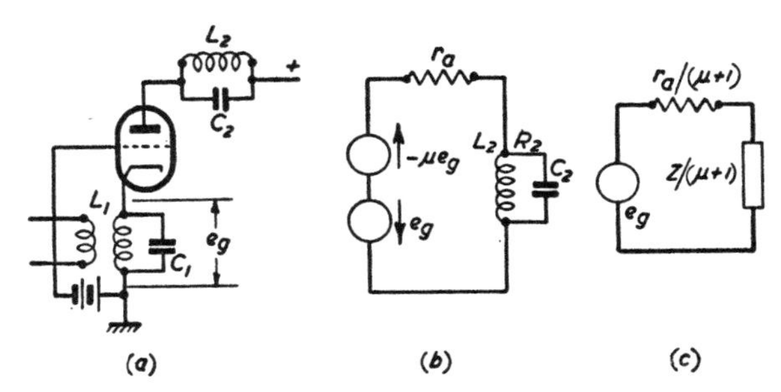

FIG. 13.35. GROUNDED-GRID CIRCUIT

in the *grounded-grid* connection illustrated in Fig. 13.35. Here the input is connected between grid and cathode, as usual, but the grid is earthed and thus acts as a screen between the output and input circuits.

The equivalent circuit is shown in Fig. 13.35 (*b*). The anode circuit contains an e.m.f. e_g and an amplified e.m.f. $-\mu e_g$ in opposition, so that the total e.m.f. is $e_g - (-\mu e_g) = (\mu + 1)e_g$ operating across r_a and the load $Z = L_2/C_2R_2$. The circuit may thus be simplified as at (*c*) where the input voltage e_g operates across $r_a/(\mu + 1)$ in series with $Z/(\mu + 1)$.

The stage gain is thus slightly greater than with the normal connection, being $(\mu + 1)Z/(r_a + Z)$ but the input impedance is $(r_a + Z)/(\mu + 1)$ which is relatively low.

Transistor Circuits

With transistor operation similar transit-time effects occur, but this is partially mitigated by the considerably smaller physical dimensions of the junctions. The behaviour was discussed in Chapter 6 (page 288), where it was shown that the principal effect of increasing frequency is a progressive reduction in α, with consequent modification of the gain and the circuit impedances.

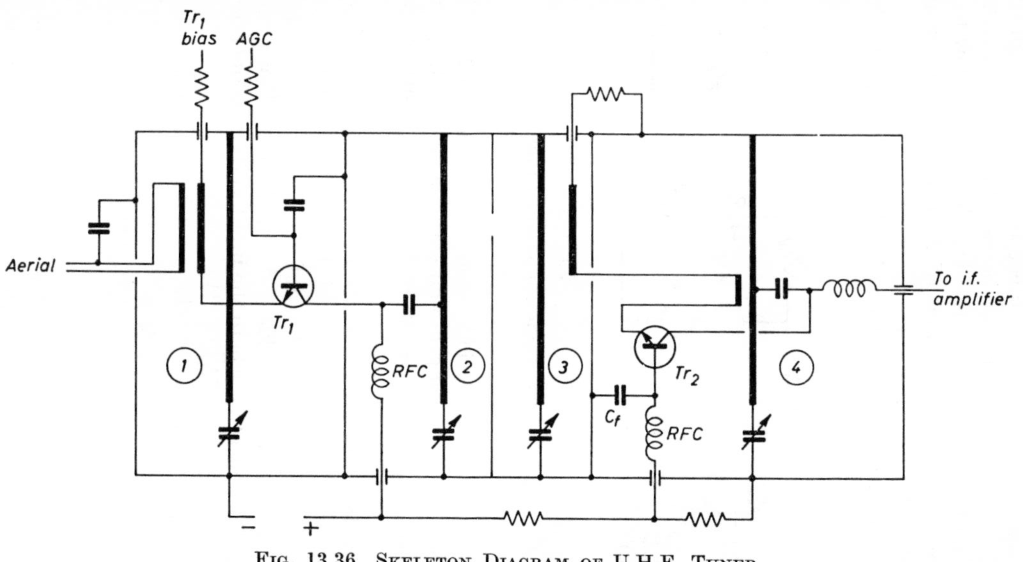

Fig. 13.36. Skeleton Diagram of U.H.F. Tuner
The Heavy lines represent resonant lines

Nevertheless, by suitable manufacturing techniques, transistors can be made which perform successfully at frequencies approaching the gigahertz region, often being made in a form which enables them to be built in as an integral part of the circuit, which must in any case conform with the configuration requirements mentioned earlier.

Fig. 13.36 illustrates the essential features of an r.f. amplifier and mixer suitable for use at frequencies in the u.h.f. bands adopted for television channels (470 to 854 MHz). It uses the trough-line technique mentioned earlier. A box chassis is employed, the partitions of which, with appropriate central conductors, form resonant lines which are tuned to the required frequency with small series capacitors.

In the first section two parallel wires constitute a transformer which matches the aerial to the emitter of Tr_1, which is operating in the common-base mode. The output from the collector is fed to a band-pass filter provided by the trough lines 2 and 3, the necessary coupling between them being provided by a small aperture in the intervening partition.

Tr_2 is a further common-base stage which feeds trough line 4, but is provided with feedback via the capacitor C_f in the base, so that it generates the required local oscillation. Since it also receives signal-frequency input via the coupling from trough line 3, it acts as an additive mixer, the output from the collector being fed to the i.f. amplifier.

The tuning capacitors are ganged to provide a single operational control.

Reception of Microwaves

Radio-frequency amplification is impracticable at the frequencies of several thousand MHz involved in microwave transmissions, so that it is necessary to utilize the signal as received. This is done by inserting into the waveguide a wire (or grating of wires in parallel) along the axis of the electric field and including in this wire a non-linear device. A point-contact silicon diode is found to be the best for this purpose, often designed to obey a square law so that appreciable gain may be obtained by heterodyning, as explained on page 438. It is customary, therefore, to use the crystal as a mixer, by injecting a local oscillation, when the output is the heterodyne product which may be fed to an i.f. amplifier in the usual way. Intermediate frequencies of the order of 60 MHz are employed, at which amplification can be provided by standard television technique.

Fig. 13.37 illustrates the essentials of a crystal mixing circuit.

The signal from the aerial is fed along one waveguide while a signal from a local oscillator is fed along a second guide. The crystal mixer is in the field of both these signals and thus delivers an output at the difference frequency to the i.f. amplifier. Both the waveguides and

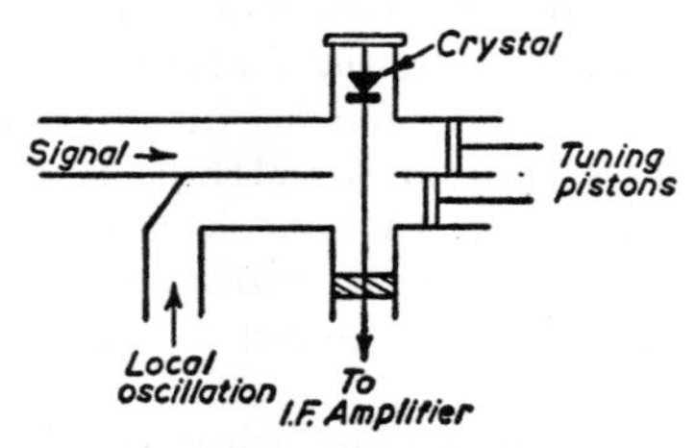

FIG. 13.37. DIAGRAM OF CRYSTAL MIXER

the crystal cavity are tuned to obtain optimum matching, though this need not necessarily be variable.

Cavity Resonators

The crystal housing in Fig. 13.37 is, in fact, a form of cavity resonator. It will be clear that if we have a short length of waveguide, closed at both ends, and of length $n\lambda_g/2$, we shall obtain the maximum reflection and standing waves will be set up within the

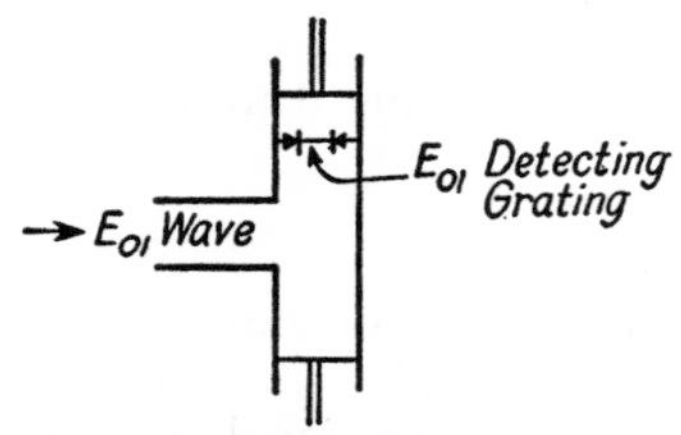

FIG. 13.38. ILLUSTRATING PRINCIPLE OF CAVITY RESONATOR

cavity. The arrangement is analogous to air vibrations in a closed organ pipe and so forms a type of resonant circuit.

The arrangement, in fact, produces substantial resonant amplification corresponding to a Q-value in a conventional tuned circuit of several thousand. Exact tuning is obtainable by making one of the ends in the form of a movable piston. It will be appreciated that, because of the high Q, the tuning will be very sharp, requiring considerable mechanical precision in the construction, and this is true of waveguide construction in general.

Fig. 13.38 illustrates the principle diagrammatically. The detector is located at a point of maximum intensity of the standing wave.

13.6. Short-wave Transmitting Valves

The production of the relatively large currents required in a transmitter poses increasing problems as the frequency is raised, partly because of the transit-time effects already discussed, but also because the connections themselves constitute a major part of the circuit. Hence the valves have to handle the full r.f. current, instead of the loss component only.

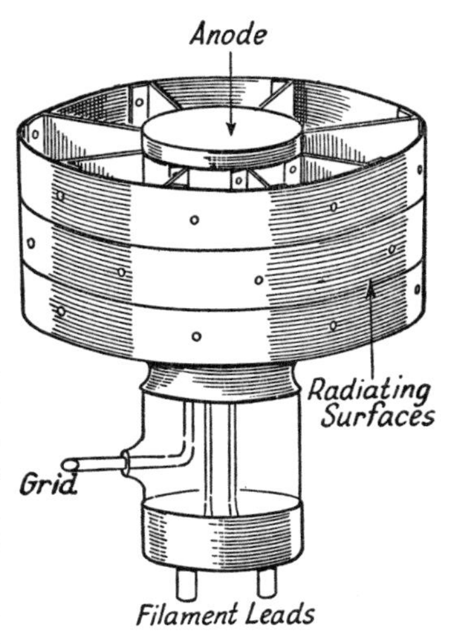

FIG. 13.39. COOLED-ANODE TRIODE

Up to frequencies of a few hundred megahertz the requirements can be met by appropriate modifications to the construction. Both grid and anode leads therefore are relatively massive, and are brought direct through the glass envelope at the side or top. The limit of output with a glass valve, however, is about 50 watts at 300 MHz.

Better results are obtainable by adopting the long-wave technique of making the anode the outer container, the grid and filament being housed on a glass support fixed to the anode with a glass-to-metal seal as shown in Fig. 13.39. Cooling may be assisted by fitting radiating fins as shown in the figure, with forced air draught or water cooling. By the last-mentioned technique up to 150 kW may be dissipated at 25 MHz and up to 2·5 kW at 300 MHz.

Conventional valves, however, are quite inadequate for the generation of the ultra-high frequencies required for microwave operation, and new types of valve have had to be devised in which, instead of attempting to limit the transit time, the operation is made to depend on this factor, so providing a radically different approach.

Barkhausen-Kurz Oscillations

In 1920 two German engineers, Barkhausen and Kurz, produced oscillations, using a circuit of the type shown in Fig. 13.40, at frequencies much higher than anything previously achieved. The grid was run at a positive potential with little or no voltage on the anode, and the explanation put forward was that electrons leaving the cathode were first attracted to the grid, shot through the spaces therein, and were brought to rest by the retarding field between grid and anode due to the lower potential of the latter electrode. They then commenced to return (due to the positive grid potential), shot through the grid again, and were once more brought to rest by the retarding field in the grid–cathode space. They then reversed again, and so an oscillation was built up independent of the external

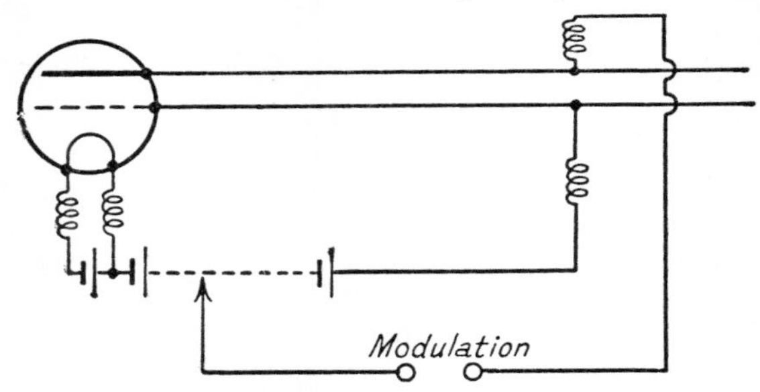

FIG. 13.40. BARKHAUSEN-KURZ CIRCUIT

circuit and having a frequency dependent on the dimensions and operating voltages of the valve.

The oscillation is extracted from the valve by connecting a tuned feeder or *Lecher wire* to grid and anode as in Fig. 13.40. The necessary d.c. voltages are fed in at nodal points, with h.f. chokes in the leads. In the original circuit, filament chokes were also used.

The technique is now obsolete, but it serves to illustrate the possibility of generating oscillations determined by the structure of the valve itself.

The Magnetron

A more practical arrangement employs what is called a magnetron which is a device comprising a cathode and a cylindrical anode, with a magnetic field applied axially down the tube by means of an external coil, as shown in Fig. 13.41 (*a*).

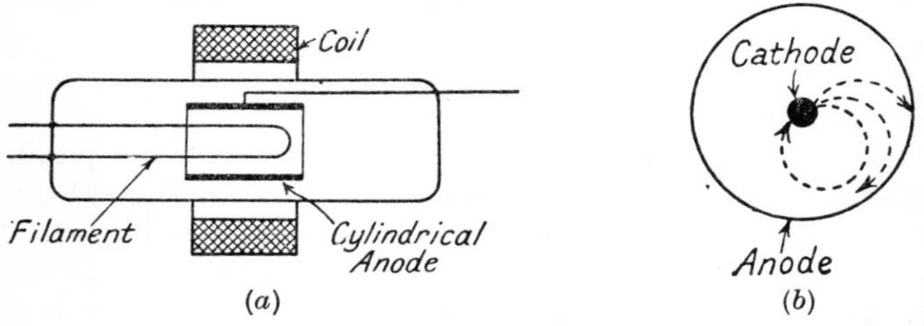

FIG. 13.41. SIMPLE MAGNETRON VALVE

Electrons leave the cathode in all directions in radial straight lines. The magnetic field, however, causes them to deviate to one side, so that they proceed in a curved path as indicated in Fig. 13.41 (*b*).

As the strength of the field increases, the curvature of the path becomes more and more until a critical value is reached, at which

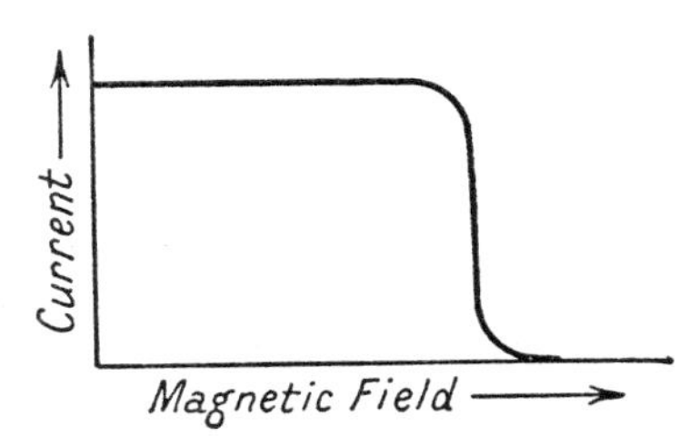

FIG. 13.42. MAGNETRON CHARACTERISTIC

the electrons fail to reach the anode altogether and return to the cathode region. Thus there will be quite a sharp dividing line between the condition of normal anode current and the condition of zero current. Up to this critical value the magnetic field has little or no influence on the actual current, the characteristic being of the form shown in Fig. 13.42.

Electron Oscillations with the Magnetron

It is clear that the electrons which leave the cathode and do not reach the anode must be describing roughly circular orbits in a definite time so that within the valve itself there are electronic oscillations. If we consider a valve operating around the cut-off region the actual anode current can be varied by altering the potential of the anode over a small amount. We are therefore able to extract energy from the valve by a process rather similar to that already described for the positive-grid electron oscillator.

If the anode potential decreases, the electrons in the vicinity will be retarded and therefore energy will be extracted from the system, while if the anode potential is increased the electrons will be accelerated and caused to reach the anode, where they will be drained out of the circuit, leaving us with a net acquisition of energy.

It is necessary that the variation of anode potential shall be suitably timed relative to the natural period of circulation of the electrons within the valve, so that the frequency of oscillations of this type is determined by the valve itself, but with the important advantage that it is possible to alter this natural frequency by altering the value of magnetic field.

It is found, in practice, to be most convenient to use an arrangement in which the anode is divided into two halves which are connected in push-pull fashion as indicated in Fig. 13.43. The external circuit is again made in the form of a resonant line, the

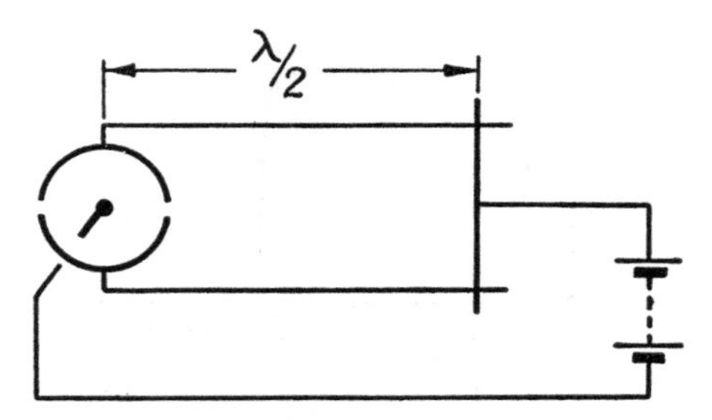

FIG. 13.43. SPLIT-ANODE MAGNETRON CIRCUIT

length being adjusted to be half a wavelength, and under such conditions the wavelength of the oscillation produced is given approximately by

$$\lambda = 1{\cdot}1/H \text{ cm}$$

where H is the field strength in amperes per metre.

This is an empirical expression based on calculation of the transit time of the electron in its orbit from cathode back to cathode again. It will be noted that it only depends upon the strength of the magnetic field, but, remembering that it is necessary to operate the valve around the cut-off condition, it will be clear that this, at the same time, necessitates a certain value for the anode potential, this value being determined by the physical structure of the valve. Clearly also, up to a point, the higher the anode potential the more the energy which can be extracted from the valve.

Dynatron Oscillations

The amount of power extracted from the electronic form of magnetron oscillation is small. Efficiencies of a few per cent only are quite normal. At longer wavelengths, however, it is possible to use the valve in a considerably more efficient manner by making use of a negative resistance effect which can be produced with the split-anode type of valve.

From Fig. 13.41 (*b*) it will be clear that by suitable choice of the magnetic field it is possible to arrange that the electrons make a half circuit of the circumference before reaching the anode. However, if the anode is split and connected to a source of r.f. potential which is varying at a suitably chosen rate the electrons will reach their respective targets at a time when the potential at that point is decreasing. There will thus be a negative resistance effect, the

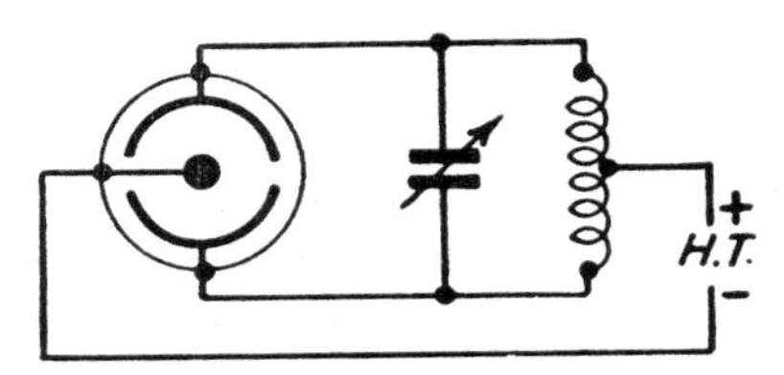

FIG. 13.44. DYNATRON MAGNETRON CIRCUIT

current increasing when the voltage is decreasing, so that energy will be delivered to the external circuit.

The frequency of such oscillations is controlled by the external circuit since it depends on the dynatron action just discussed and not on internal electronic movements, but the valve will only maintain this class of oscillation at relatively long wavelengths, such that the period of oscillation is long compared with the transit time.

As the wavelength is reduced the efficiency falls from about 50 per cent (at frequencies around 100 MHz), becoming progressively lower until the oscillations merge into the electronic type and cease to be controlled by the external circuit.

The Cavity Magnetron

Considerably higher frequencies may be generated, however, by what amounts to a combination of the two methods. The anode is made in the form of a ring having a number of circular cavities, as shown in Fig. 13.45. These cavities form resonators in which electrons will circulate at a frequency determined by the physical dimensions, in consequence of which there will be a stationary pattern of electric fields forming a series of loops from one cavity to the next.

This stationary field pattern can be regarded as being produced by two component fields rotating in opposite directions (at the frequency of the resonance in the cavities). But we have seen that the electrons emitted from the cathode follow a curved path due to the influence of the axial magnetic field. If the transit time is correctly adjusted, the electrons will arrive at the anode in phase with the rotating electric field pattern, and so will give up their energy to the field.

The effect is thus to cause the electrons emitted from the cathode to become concentrated in a series of "spokes" which rotate like a wheel round the "axle" formed by the cathode. In doing so they

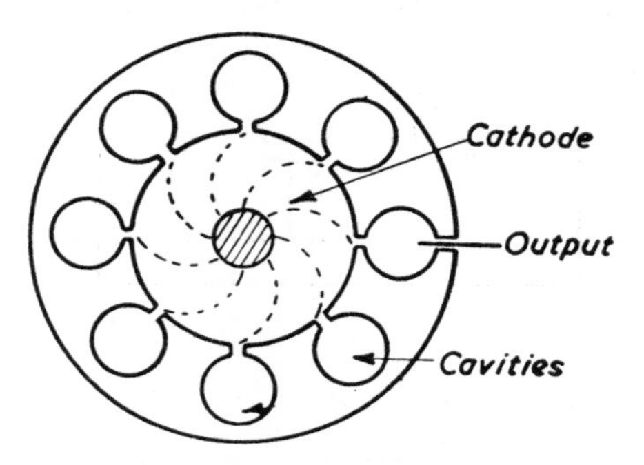

FIG. 13.45. CAVITY MAGNETRON

give up energy to the cavities in their passage so that the oscillations are sustained.

This can be shown to occur when

$$\frac{\lambda V}{H} = \frac{110\pi^3 D^2}{N\phi^2}$$

where V = anode voltage,
H = magnetic field strength,
D = diameter of anode hole,
N = number of resonators,
λ = wavelength,
and ϕ = phase difference between adjacent resonators.

The cavity magnetron is usually operated in what is called the π mode, for which $\phi = \pi$, in which case

$$\lambda = \frac{110\pi D^2}{N} \cdot \frac{H}{V}$$

Hence, although D and N are fixed for any given magnetron, it is possible to vary the wavelength over a small range by suitable adjustment of H/V.

By this means frequencies can be generated up to 30 GHz (1 cm) with considerable power output. When used for the production of very short pulses for radar and similar requirements, peak powers of the order of 100 kW have been achieved on these very short wavelengths, while at somewhat lower frequencies of the order of 3 GHz, peak powers of 3 MW (3,000 kW) can be obtained.

For further details reference may be made to a paper by Boot and Randall, "The cavity magnetron," *J. Instn. Elect. Engrs.*, **93**, Pt. IIIA, p. 928.

Velocity-modulated Oscillators

In a normal valve the velocity of the electron stream is constant. It is clear, however, that at very high frequencies such that the period of oscillation is comparable with the electron transit time, this is no longer true. In the B–K oscillator of Fig. 13.40, for instance, the electrons are continually changing their velocity as they oscillate to and fro past the grid.

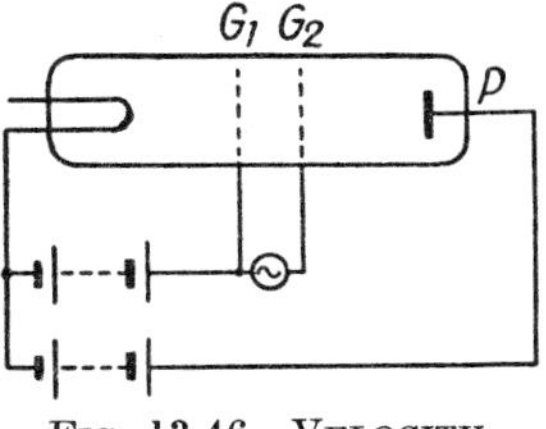

Fig. 13.46. Velocity-modulated Tube

This led various investigators to consider the possibility of controlling the electron stream by varying its velocity instead of its amplitude.

Such an arrangement is illustrated in Fig. 13.46. The grid G_1 is at a high potential and produces an electron stream from the cathode. G_2 is varied above and below the potential of G_1 by some small fraction, and this causes the electrons to be alternately accelerated and decelerated. In the space between G_2 and the anode P, therefore, we have a velocity-modulated stream.

Retarding Field Conversion

If we reduce the voltage on the anode P to a low value we can arrange that electrons travelling with normal or reduced velocity are turned back, so that only those electrons which have greater velocity than normal manage to reach the plate. Hence we obtain an anode current depending upon the modulation, thus converting the velocity modulation into an amplitude change.

Such an arrangement has been used with success. It is necessary to provide an additional electrode to collect the electrons which are turned back from the anode, while it is further essential for reasonable efficiency that the external impedance in the anode circuit shall be high at the oscillation frequency, and this is not easy to achieve.

Drift Tube Converters

A more successful method is to allow the electron stream to convert itself into an amplitude-modulated stream by simply allowing it to continue on its way. The faster moving electrons then overtake the slower moving ones, and since the velocity has been modulated periodically, there will be points along the *drift tube* (the space in which the electrons are allowed to proceed uninterrupted) where there will be accumulations or bunches of electrons.

Fig. 13.47 illustrates this graphically. The various lines represent electrons entering the drift tube at regular intervals. The slope of the

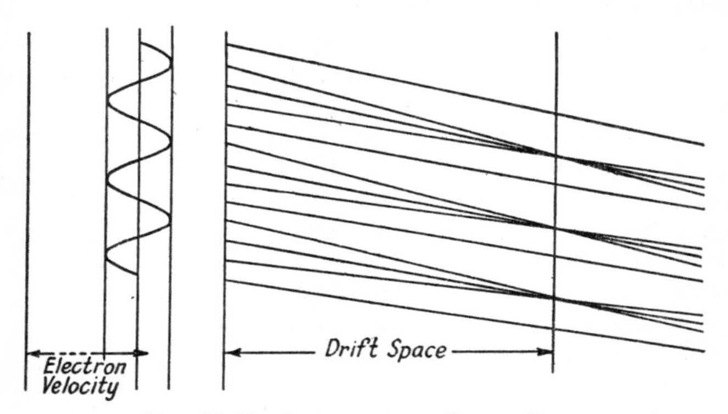

FIG. 13.47. PRINCIPLE OF DRIFT TUBE

lines represents the velocity which will be seen to increase and decrease rhythmically. It will be seen that the majority of the electrons arrive at the end of the tube together.

The process may be illustrated by a simple analogy. Suppose we dispatch a series of trains at five-minute intervals travelling at 30 m.p.h. If some time afterwards we dispatch another train on a parallel line running at 60 m.p.h. it will obviously catch up with each of the slower-moving trains in turn, and it is not difficult to see that if we had a series of the faster trains dispatched at five-minute intervals there would be certain points along the track at which the overtaking took place, and the inhabitants of any one of these places would always see two trains going through together, followed by an interval with no trains at all.

Rhumbatrons

The electrons will take a finite time in passing between the grids G_1 and G_2 of Fig. 13.46. Clearly this time must not be more than the period of one half-cycle of the modulating frequency. Otherwise an electron which had been accelerated immediately after passing G_1 would be decelerated again before it had got clear of G_2.

This requirement leads to the use of a combined oscillator and modulating grid structure of the form shown in Fig. 13.48, to which the name *rhumbatron* was given by the brothers R. H. and S. F. Varian, who first evolved a practical drift-tube type of oscillator. Fig. 13.49 illustrates the development of this form of oscillator from the more conventional form of circuit.

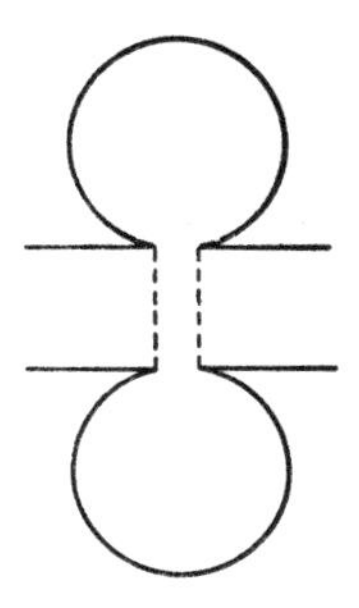

FIG. 13.48. RHUMBATRON

On the left we have a capacitor with two circular plates joined by a single-turn inductance. Such a circuit is limited in frequency by the inductance of the loop, while it is also of poor Q because the greater part of the energy is radiated (whereas a high Q implies that the greater part of the energy is *stored*). The second figure shows two loops in parallel, which reduces the effective inductance and also reduces the radiation since the fields from the two loops tend to cancel out. Hence both f and Q have been increased.

The more loops we add the greater the improvement, and the

FIG. 13.49. DEVELOPMENT OF RHUMBATRON FROM L-C CIRCUIT

ultimate configuration is a complete toroidal sheet of copper with the capacitor in the centre, which is the shape of Fig. 13.48. Such a device is an oscillatory circuit of very high Q since there is no radiation—the Q without external loading being of the order of 10,000. Even when appreciable power is taken from the circuit we can still maintain a Q of the order of 1,000, which is far better than is attainable by normal methods.

The Klystron

The brothers Varian, already mentioned, used such a device in the development of a practical oscillator. The centre section was made with a grid structure, and replaced the two grids G_1 and G_2 of Fig. 13.46. Oscillation is set up in this rhumbatron, which is called the *buncher*, by introducing a short rod or wire along a line of electric force. This is followed by a drift tube and a second rhumbatron, known as the *catcher*, where the arrival of the bunched electrons excites oscillations. To render the whole system self-maintaining, an electrode in the catcher extracts some of the energy, and feeds it back to the buncher, thus maintaining the oscillations therein.

When the device is working there is a strong field across the grids of the catcher, and this is in such a phase as to retard the electrons and thus extract most of the energy from them. They pass through the catcher at quite low velocity, and are collected by a final electrode

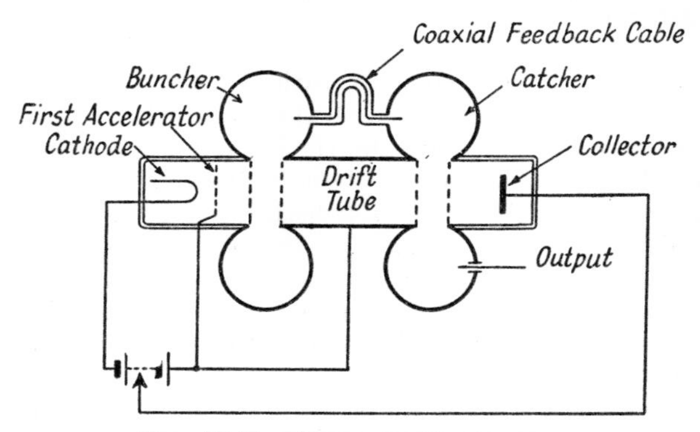

FIG. 13.50. KLYSTRON OSCILLATOR

at low potential. Thus the energy in the external circuit is quite small, most of the energy having been absorbed by the catcher, so that the arrangement is highly efficient. Powers of 300 watts at 10 cm are readily generated by this type of oscillator.

Multiple-cavity Klystrons

The velocity of the electrons must be such that the transit time across the grids of the buncher and also along the drift tube is such as to provide the correct phase relations. The velocity is controlled by the potential on the first accelerator, and hence there is a series of correct operating potentials at which oscillation is possible. At all other potentials the device does not operate. Moreover, since catcher and buncher are tightly coupled the system has two adjacent frequencies (as with any coupled circuit system), each with its own range of operating potential.

Hence for all but relatively low powers it is more convenient to use the device as an amplifier, supplying input to the buncher from an external source (which may be a solid-state circuit as described later) and taking the output from the catcher. This is equivalent to the drive oscillator technique used at low frequencies and avoids the double-tuning effect, while also providing appreciable power gain.

The performance may be still further improved, however, by providing an intermediate resonator, as shown in Fig. 13.51. Oscillations are excited in this intermediate cavity by the partially-bunched electron stream passing across the gap. These in turn develop voltages across the gap which react with the main beam. If the cavity is tuned to a frequency slightly higher than the signal

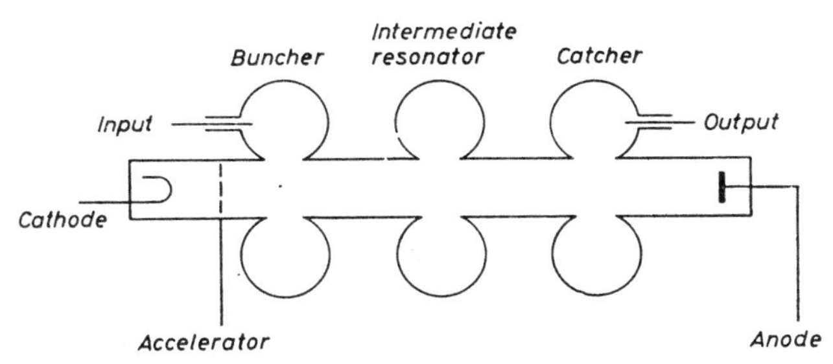

FIG. 13.51. THREE-CAVITY KLYSTRON AMPLIFIER

frequency, so that it presents an inductive impedance, the phase of the voltage across the gap is such as to increase the velocity modulation of the main stream. This considerably increases both the gain and the efficiency of the system. A three-cavity klystron will provide a power gain of 30 dB with an efficiency of the order of 40 per cent.

A further advantage is that, whereas the bandwidth of the simple klystron is relatively small, because of the high Q of the system, the introduction of an intermediate cavity mistuned by an appropriate amount produces a substantial broadening of the bandwidth. This is of value in handling wide-band transmissions such as are required for television programmes or frequency-modulated transmissions. Still further improvement is possible by adding further cavities, though the operation becomes more critical, but four- or five-cavity klystrons are commonly used in u.h.f. broadcast transmitters.

Continuous powers of 15 kW and more are available with such devices at frequencies of the order of 900 MHz, while pulsed powers of 30 MW have been obtained at frequencies as high as 3 GHz.

The fixed frequency of the klystron is perhaps its only serious disadvantage. This is partially offset by making the device mainly of metal with a glass portion containing the cathode and first accelerator sealed on to the input end. The rhumbatrons may then be made with corrugated sides enabling their shape, and hence their resonant frequency, to be changed over a small range.

Inductive Output Amplifier

A slightly different and more flexible arrangement is the inductive output amplifier illustrated in Fig. 13.52.

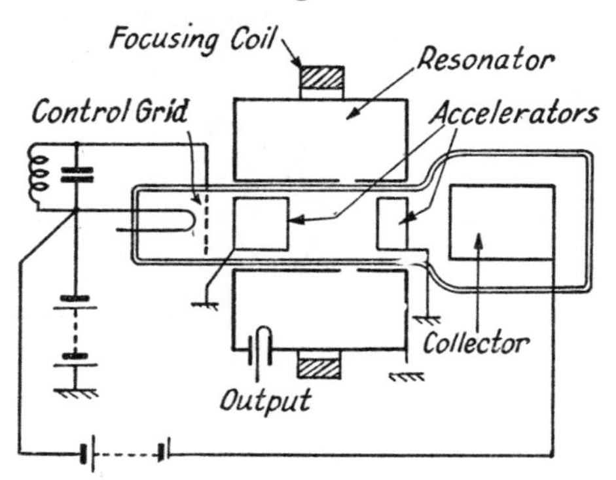

Fig. 13.52. Inductive Output Oscillator

A simple grid is used both as first accelerator and modulator. The arrangement is a modulator more of amplitude rather than of velocity, the effect being to generate bunches of electrons which are then accelerated down the tube by additional accelerators.

Outside the tube is a cylindrical resonator with an annular gap, across which strong electric fields appear when the resonator is oscillating. If the phase is correctly adjusted these fields will be of maximum intensity when a bunch of electrons is passing, and if they are in such a direction as to retard the electrons they will extract energy from them. In practice the energy can be almost wholly extracted, leaving only a small residual to be drained off by the final anode.

A further improvement results from the application of an axial magnetic field which prevents the electrons from scattering. It imparts a spiral motion tending to concentrate the electrons in a beam as in a magnetically-focused cathode-ray tube, and the arrangement as a whole forms a highly efficient mechanism for amplifying or generating oscillations between 100 and 1,000 MHz.

The fact that the resonator is external to the tube has obvious advantages, since the tube is separated from the associated circuits as with the more conventional forms of valve. There is, of course, still the need to adjust the operating potentials within critical limits to maintain the correct phasing.

Travelling-wave Tubes

The *travelling-wave tube*, is a device which makes use of the interaction between an electron beam and a travelling wave and is only applicable at very high frequencies in excess of several thousand MHz, at which the transit time is comparable with the velocity of a travelling wave.

The arrangement is illustrated diagrammatically in Fig. 13.53 and will be seen to consist of an electron gun which projects a beam of

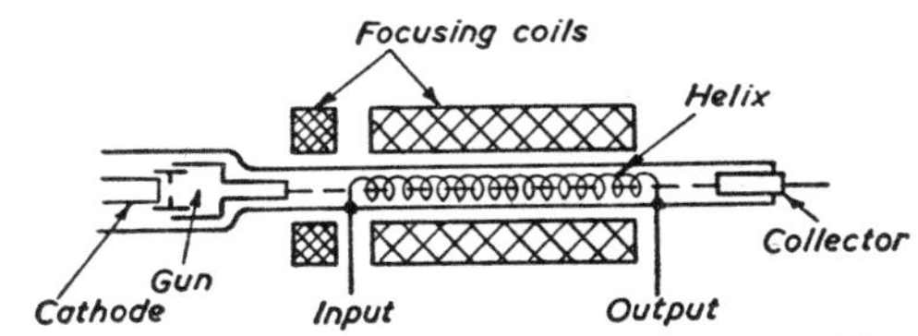

FIG. 13.53. DIAGRAM OF TRAVELLING-WAVE TUBE

electrons down a long tube with a collector at the far end. Within the tube is a helix and the signal to be amplified is applied to the end of this helix nearest the gun. This signal is then propagated along the helix at a finite rate, and if this can be arranged to coincide with the velocity of the electrons in the beam there is a cumulative extraction of energy from the beam so that the signal at the output end of the helix is considerably amplified.

It is not practicable here to do more than to refer to the general principle, but further information can be obtained from an article by J. R. Pierce and L. M. Field, "Traveling-wave tubes," *Proc. Inst. Rad. Engrs.*, **35**, p. 108.

13.7. Semiconductor Techniques

As said earlier, transistors are now available for use in the v.h.f. and u.h.f. ranges. To minimize transit-time effects the physical size has to be small, which limits the power output obtainable, but the efficiency of space-wave propagation is such that communications links, in general, do not require large powers. This does not apply to broadcast transmissions where a relatively large signal strength is required to permit the use of comparatively simple receiving circuitry, and for such purposes valves of the klystron or travelling-wave type have to be used. Radar transmissions also require high-power pulses in order to ensure adequate reflection from the target, for which purpose magnetrons are commonly employed.

It is nevertheless possible, and convenient, to use solid-state equipment in the early stages of transmitters, performing the generation and modulation at low power. Typical arrangements have already been discussed in Chapter 7, and these may be extended to

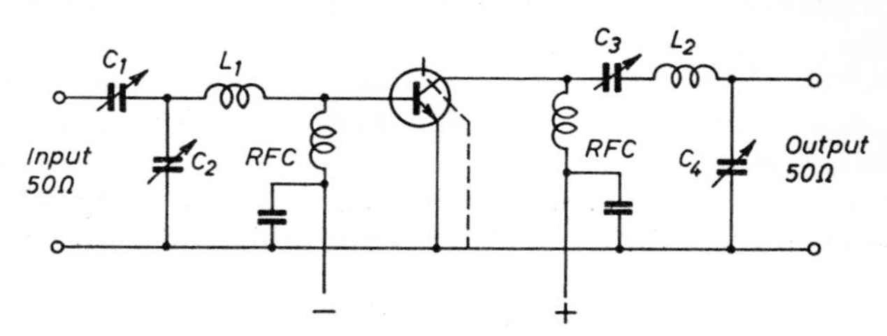

Fig. 13.54. 175 MHz Amplifier

the higher frequencies by suitable choice of transistors and circuit configuration.

A typical v.h.f. amplifier is illustrated in Fig. 13.54 which uses a silicon n–p–n transistor in a common-emitter mode. Transistors for r.f. applications are usually provided with an internal screen as shown dotted, to minimize the internal feedback, but apart from this the circuit is conventional. It will deliver an output power of

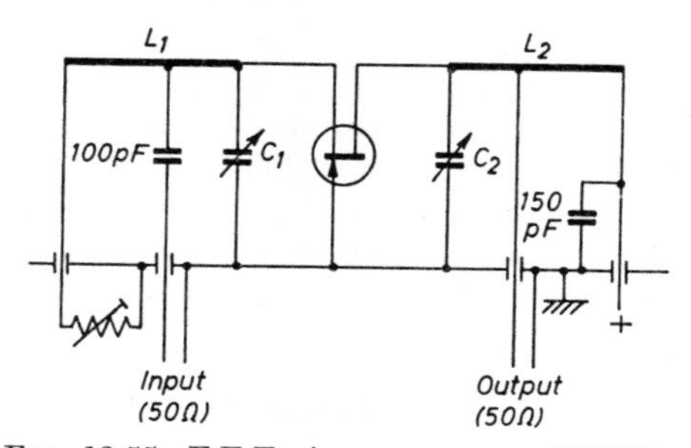

Fig. 13.55. F.E.T. Amplifier for 470 MHz

13·5 W at 175 MHz, both input and output being designed for coaxial or waveguide feed, for which the standard impedance is 50 ohms.

As said in Chapter 6, f.e.t.s are often employed at high frequencies, where they have the advantage of providing a better signal/noise ratio. With the common-source mode it is usually necessary to include suitable neutralizing circuitry to offset the internal feedback. This may be avoided by using the common-gate connection, which is thus often preferred despite its low input impedance. Fig. 13.55 shows a common-gate amplifier circuit suitable for operation at 470 MHz. L_1 and L_2 are resonant lines (approximately 6 cm long) tapped at appropriate points to provide the correct matching to the input and output lines. Tuning is provided by the capacitors C_1 and C_2, which would be trimmers variable over a range of 5–50 pF.

Frequency Multiplication

The use of a capacitance-diode to provide frequency modulation was discussed in Chapter 7, where it was also said that it is often convenient to generate (and modulate) the primary oscillation at a low frequency and then pass this through appropriate multiplying

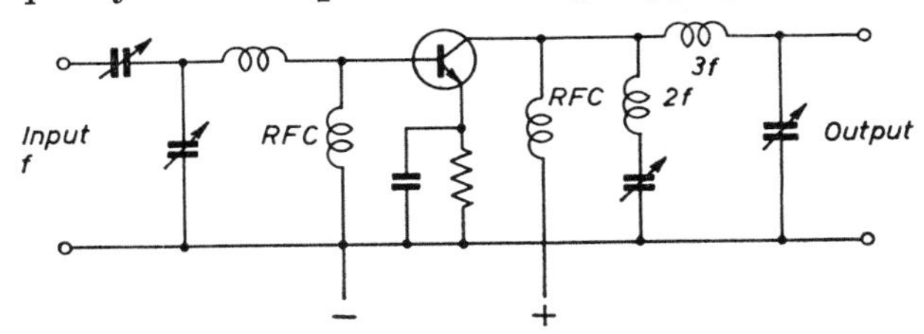

FIG. 13.56. FREQUENCY TRIPLER

stages. This multiplication can be performed simply and efficiently with transistor circuitry, by feeding in signals at the fundamental frequency, and tuning the output to a convenient multiple. Thus Fig. 13.56 shows a frequency tripler taking input at, say, 167 MHz and providing output at 500 MHz. There will be some second-harmonic component, and while this will not be accepted by the output circuit, it will absorb a certain amount of power so that an acceptor circuit tuned to 333 MHz is connected from collector to ground.

Diode Multipliers

Multiplication can be provided even more simply, and efficiently, by the use of r.f. diodes. These are usually of the varactor type characterized by a law of the form $I = kV^n$. A sinusoidal voltage across such a diode will produce a current which is rich in harmonics, so that by tuning the output to the appropriate harmonic as shown in Fig. 13.57 the required multiplication can be obtained.

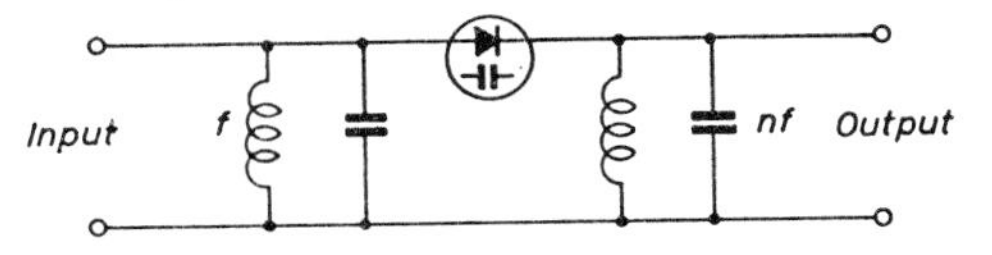

FIG. 13.57. DIODE MULTIPLYING CIRCUIT

The power output is determined by the input voltage swing, which in the forward direction is limited by the current-handling rating of the diode, and in the reverse direction must not exceed the reverse breakdown voltage. Subjects to these limitations appreciable powers can be handled at frequencies up to several GHz.

If the diode were ideal there would be no loss of power in the process. An actual diode, comprising a resistance with a small shunt capacitance, can be represented by an equivalent circuit containing a (modified) capacitance in series with a loss resistor r_s,

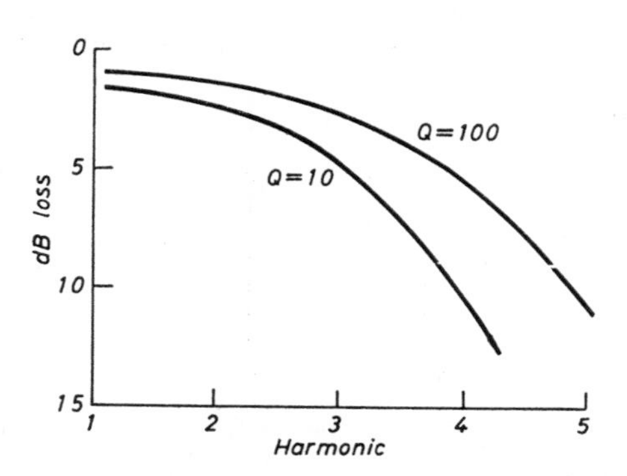

FIG. 13.58. DEPENDENCE OF LOSS ON Q AND HARMONIC

which will have a $Q = 1/\omega C r_s$. The capacitance C is dependent on the bias voltage, so that the Q value has to be specified for given operating conditions, but the feature of varactor diodes is that the losses are kept small so that Q values of the order of 100 are possible.

The losses incurred in the process are then dependent on the Q of the diode, but it can be shown that they are also influenced by the degree of multiplication, as indicated in Fig 13.58. It is evident that large multiplications are not efficient, and it is preferable to use a series of multipliers in cascade. Thus two doublers in cascade will be twice as efficient as a single stage quadrupler.

The efficiency falls off as the frequency increases, but even so satisfactory output is obtainable. A typical multipler operating from a transistor input stage delivering 10 W at 125 MHz followed by four stages of multiplication of 3, 2, 2 and 2 gives an output of 0·6 W at 3 GHz.

At these frequencies use would be made of cavity waveguide resonators of the form discussed earlier, and the diodes would be constructed to fit into such systems. The same applies to the tunnel and Gunn-effect diodes mentioned in Chapter 6, with which gigahertz oscillations can be generated direct, though the outputs obtainable are appreciably smaller.

P-I-N Diodes

A device which has applications at microwave frequencies is the Positive–Intrinsic–Negative or p–i–n diode. This is provided with a high-resistivity layer adjacent to the normal junction, in consequence of which its dynamic resistance is of the order of 10 kΩ. The value of the resistance is controllable by the bias across the device, and if an r.f. signal is superposed on the bias the effective impedance changes in a rhythmic and *linear* manner. Hence no rectification is

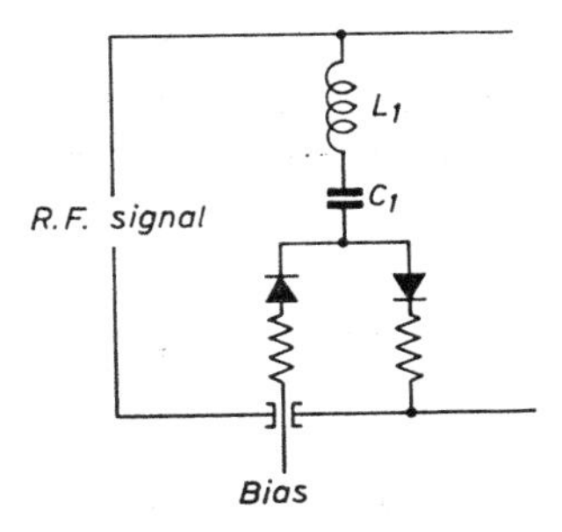

FIG. 13.59. FREQUENCY-SHIFT FILTER USING P–I–N DIODES

produced, nor are any harmonics generated. For this to happen the period of the r.f. signal must be small compared with the minority-carrier lifetime.

The device can be used as a variable-resistance amplitude modulator at microwave frequencies, while since the capacitance also changes with applied signal it can be used as a frequency-shift filter, as shown in Fig. 13.59. Bias is applied across the diodes in series, but to the r.f. signal they are in parallel and contribute a small capacitance in series with C_1. The network thus constitutes a rejector circuit of which the frequency can be switched very rapidly by suitable alteration of the bias voltage.

14

Filters and Attenuators

14.1. Passive Filters

THERE are many occasions in communications practice when it is desired to attenuate certain frequencies at the expense of others. A simple method of achieving this is to use a reactance and resistance

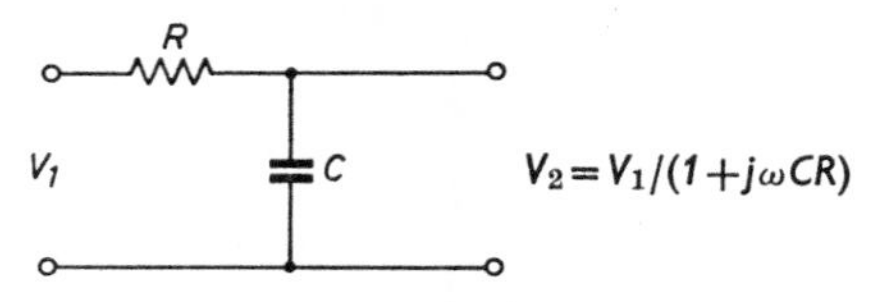

FIG. 14.1. R-C FILTER

in series, as shown in Fig. 14.1, and since the value of the reactance will change with the frequency the output voltage will rise or fall with frequency, depending on the type of the reactance. If the reactance is large compared with the resistance a simple calculation shows that the output voltage varies by 2:1 (6 dB) per octave.

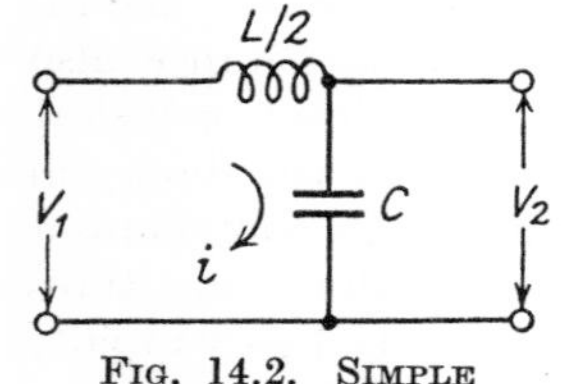

FIG. 14.2. SIMPLE L-SECTION FILTER

A simple linear variation of this nature, however, is of limited application; a much more interesting type of network is that shown in Fig. 14.2, in which two *inverse* reactances are employed, one increasing with frequency while the other decreases. Such an arrangement is known as a *filter* and has the property of transmitting a certain range of frequencies with relatively little attenuation, followed by a rapid fall to complete cut-off.

Consider the performance of the circuit in Fig. 14.2, which is known as a *low-pass* filter. (The value of the inductance is taken as $L/2$ because the circuit shown is actually the input end of a complete filter in which the *total* series arms have a value L, as explained later.) The current through the network is

$$i = \frac{V_1}{j\omega L/2 + 1/j\omega C}$$

The output voltage is the voltage across the capacitor, which is

$$V_2 = \frac{i}{j\omega C} = \frac{V_1}{j\omega C(j\omega L/2 + 1/j\omega C)} = \frac{V_1}{1 - \omega^2 LC/2}$$

Hence the ratio of V_2 to V_1 is $2/(2 - \omega^2 LC)$.

If $\omega^2 LC = 2$ this expression becomes infinite, indicating that the output voltage is more than the input voltage due to normal resonant

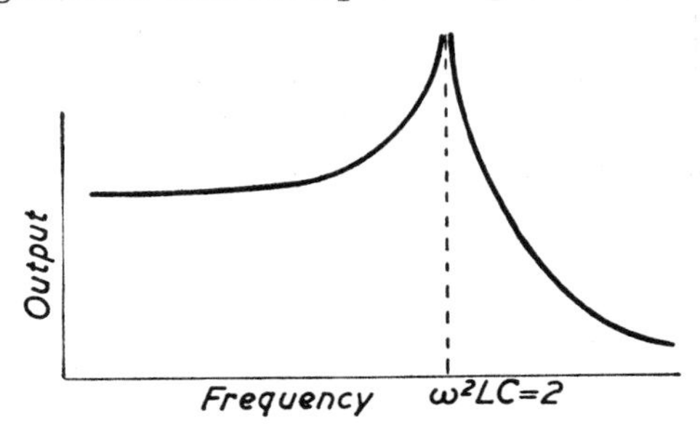

Fig. 14.3. Response of Fig. 14.2 Circuit

action. If $\omega^2 LC = 4$ then $V_2/V_1 = -1$ indicating that the output is the same as the input but with a 180° change of phase. Beyond this point the output falls off continuously as shown in Fig. 14.3.

In practice, of course, the output voltage does not rise to infinity because there is always resistance in the circuit, partly in the components and partly in the load into which the network operates, so that the conditions would be more as shown in Fig. 14.4. There would be a rise in output at the frequency for which $\omega^2 LC = 2$, at which point there would be a phase change of 90°, followed by a sharp fall, the phase changing rapidly to 180° when $\omega^2 LC = 4$, and thereafter remaining constant. This point at which complete reversal of phase occurs is called the *cut-off point.* The output voltage is (numerically) equal to the input so that up to this point there is no attenuation and the filter is said to be operating in its *pass band.* Beyond the cut-off point there is a progressive attenuation as is clear from Fig. 14.4.

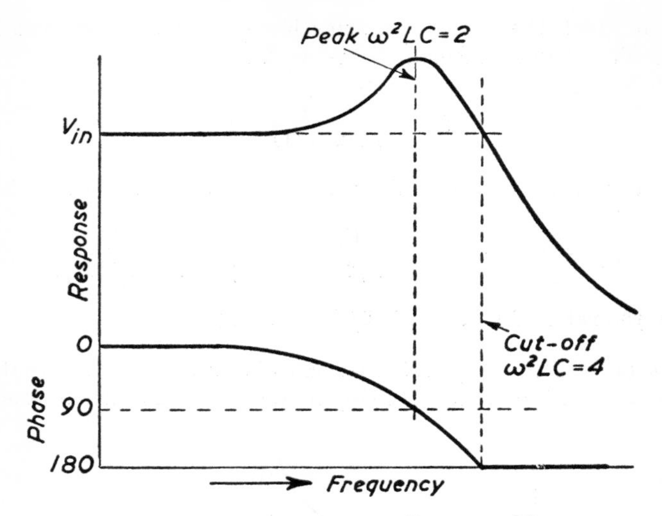

Fig. 14.4. Characteristics of Low-pass Filter

T- and π-Sections

Now a simple L-section filter such as this is of limited use. For one thing the cut-off is only gradual and a more effective filter can be obtained by connecting several sections in cascade. If this is done, we obtain a network rather like that of the transmission line discussed in Chapter 8, and for the reasons there explained it is

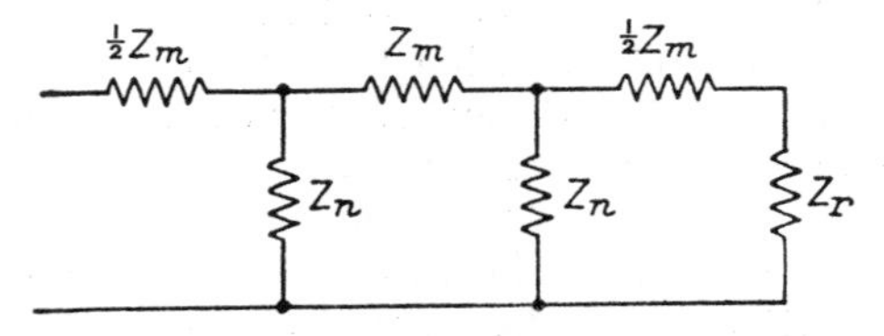

Fig. 14.5. General Filter Network

necessary to see that the various sections are suitably matched to one another and to the terminal load. Otherwise, reflection will occur, with consequent waste of energy, resulting in imperfect transmission.

Let us examine the impedance of a few typical filter sections. Fig. 14.5 shows a general case of a composite filter. This can be split up into two types of section. One is the T-section as shown in

Fig. 14.6 (*a*). Here the series impedance Z_m is split into two, one half on each side. A number of these sections placed in cascade will obviously be equivalent to the original network.

We can, if necessary, split the section into two half sections as at

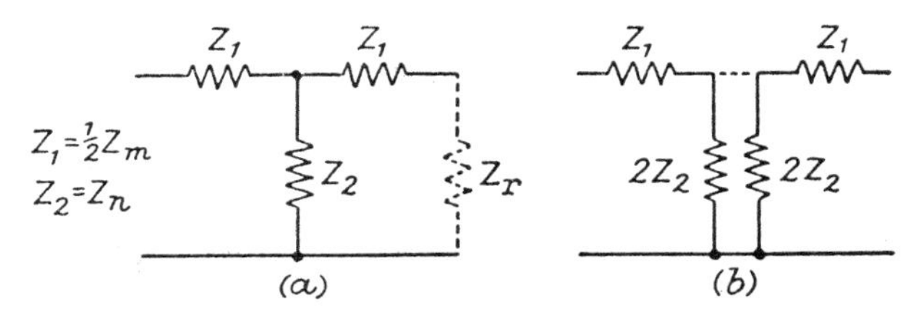

FIG. 14.6. T-SECTION FILTERS

Fig. 14.6 (*b*). To do this, the shunt impedance is replaced by two impedances each twice as great, so that the two in parallel give the original value.

The second type is the π-section shown in Fig. 14.7 Here the

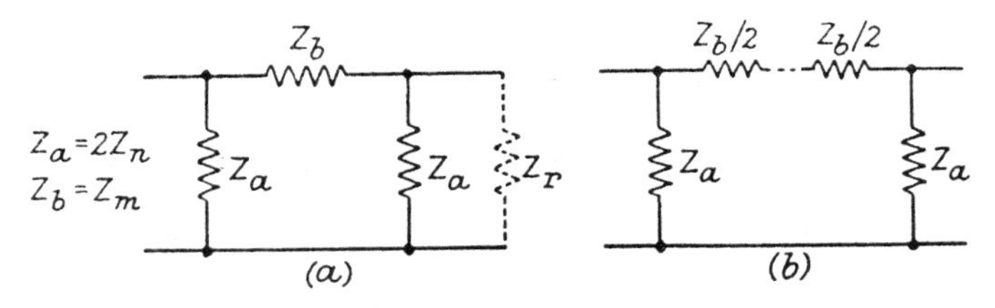

FIG. 14.7. π-SECTION FILTERS

series impedance is the full value, while the shunt impedances are each twice the original value. This network is, in fact, two half T-sections placed end to end. The π-section can also be split into two half sections which are identical with the T half-sections.

The T-section is also known as the *mid-series* termination and the π-section as the *mid-shunt* termination.

General Filter Equations

Let us now examine the impedance of these typical sections. For simplicity we will assume that the sections are symmetrical, as shown, but otherwise the impedances are unrestricted. Consider first the mid-series or T-section (Fig. 14.6). We can write down the following expressions for the impedance of the filter viewed from the input end:

$$Z_o = \text{impedance with far end open } (Z_r = \infty)$$
$$= Z_1 + Z_2 \quad . \quad . \quad . \quad . \quad . \quad . \quad . \quad . \quad (1)$$

$$Z_s = \text{impedance with far end short-circuited } (Z_r = 0)$$
$$= Z_1 + \frac{Z_1 Z_2}{Z_1 + Z_2} \quad . \quad . \quad . \quad . \quad . \quad . \quad (2)$$

$$Z = \text{impedance with impedance } Z_r \text{ across output}$$
$$= Z_1 + \frac{Z_2(Z_1 + Z_r)}{Z_1 + Z_2 + Z_r} \quad . \quad . \quad . \quad . \quad . \quad (3)$$

Iterative Impedance

A case of particular importance arises if $Z = Z_r$ so that the input impedance is equal to the load or terminal impedance. Let Z_k be the particular value of Z_r which satisfies this condition. Then, from (3) above, we have

$$Z_k = Z_1 + \frac{Z_2(Z_1 + Z_k)}{Z_1 + Z_2 + Z_k}$$

$$\therefore \quad Z_1 Z_k + Z_2 Z_k + Z_k^2 = Z_1^2 + Z_1 Z_2 + Z_1 Z_k + Z_1 Z_2 + Z_2 Z_k$$

Cancelling out $Z_1 Z_k$ and $Z_2 Z_k$ on both sides we have

$$Z_k^2 = Z_1^2 + 2Z_1 Z_2 \quad \text{or} \quad Z_k = \sqrt{[Z_1(Z_1 + 2Z_2)]}$$

This is known as the *iterative impedance* of the filter and is important for the following reasons.

1. If the network "looks like" Z_k, we can terminate it either with a load Z_k, or with *another filter section*, provided this is also terminated by the iterative impedance Z_k, and so on indefinitely.

2. Provided the network is terminated in its iterative impedance no reflection will occur at the receiving end. This aspect is discussed later (page 621).

It is worth noting, in passing, that $Z_k = \sqrt{Z_o Z_s}$, as the reader may easily verify.

Cut-off Frequency

The iterative impedance may be written in an alternative form which involves the ratio of the series and shunt impedances in a significant way. If Z_m is the *total* series impedance ($= 2Z_1$) and Z_n is the total shunt impedance ($= Z_2$), we can write

$$Z_k = \sqrt{[Z_m Z_n(1 + \rho)]} \tag{4}$$

where $\rho = Z_m/4Z_n$ (5)

Now, we saw earlier that with the low-pass filter section of Fig. 14.2 the cut-off frequency is given by $\omega^2 LC = 4$.

This may be written in the form

$$\frac{-\omega^2 LC}{4} = -1 = \frac{j\omega L}{4/j\omega C} = \frac{Z_m}{4Z_n} = \rho$$

Hence the cut-off point of a filter is obtained when $\rho = -1$.

This is not a rigid proof, but the statement is true and is in a convenient form. It applies, whatever the form of Z_m and Z_n.

Mid-shunt Termination

Let us now consider the mid-shunt or π-section. We treat this in the same way as before, and once again we can deduce the iterative impedance. This is found to be

$$Z_k' = \sqrt{\left[\frac{Z_a Z_b}{2 + Z_b/Z_a}\right]}$$

If again we write Z_m for the total series impedance ($= Z_b$), and Z_n for the *total* shunt impedance ($= Z_a/2$), we have

$$Z_k' = \sqrt{[Z_m Z_n/(1 + \rho)]} \tag{6}$$

Thus, the only difference between the mid-series and mid-shunt termination is in the position of the factor $(1 + \rho)$.

The cut-off frequency is unaltered, being given as before by the relationship $\rho = -1$.

These expressions in terms of Z_m and Z_n can easily be memorized, and are in the most convenient form for numerical calculations. It is easy to become confused, when working in terms of Z_1 and Z_2, by the factors of $\frac{1}{2}$ or 2 in the end sections. If one works in terms of Z_m and Z_n, which are the *total* series and shunt elements respectively, the cut-off frequency is always given by $\rho = -1$ (even for the high-pass filters considered shortly), while Z_k is given by the expressions (4) or (6) according to whether the filter is of T or π form.

Variation of Iterative Impedance

It will be clear from these expressions that the iterative impedance is not constant but varies with frequency. For example, with a mid-series low-pass filter, as in Fig. 14.6, having a *total* series element L and a shunt element C, Z_k varies from $\sqrt{(L/C)}$ when $\omega = 0$ to zero at the cut-off frequency, for which $\rho = -1$. With mid-shunt connection the variation is from $\sqrt{(L/C)}$ to infinity.

Over the majority of the pass band the variation is small, as shown in Fig. 14.8, and moreover within this band Z_k is purely resistive. Beyond the cut-off point, however, it becomes imaginary.

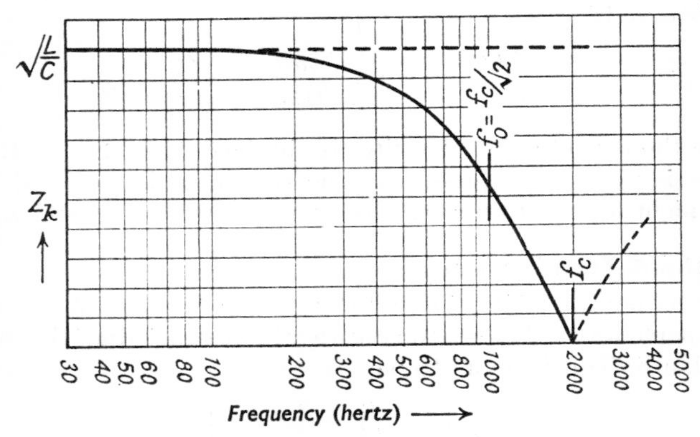

FIG. 14.8. VARIATION OF ITERATIVE IMPEDANCE WITH FREQUENCY (LOW-PASS FILTER)

In practice it is not usually practicable to terminate the filter in a load which varies with frequency, and it is customary to use a resistive load $= \sqrt{(L/C)}$, which means that the performance will depart slightly from the ideal. It may be noted, however, that if the filter were terminated in its iterative impedance at all frequencies within the pass band it would always "look like" its load impedance so that the input and output voltages would be the same and there would be no attenuation. The idealized curves often shown for filter networks are based on this not truly valid assumption. The effect of mismatching is discussed on page 623.

Constant-K Filters

If the product of the series and shunt impedances is constant at all frequencies so that $Z_m Z_n = K^2$, the filter is said to be a *constant-K* type. The simple low-pass filter just described is of this type, and it should be noted that for this law to hold good the series and shunt impedances must be inverse, i.e. if one increases with frequency the other must decrease at the same rate.

With the simple low-pass filter, $Z_m = j\omega L$ and $Z_n = 1/j\omega C$, so that $Z_m Z_n = (j\omega L)(1/j\omega C) = L/C = R^2$, where R is the iterative impedance. Hence, for a simple T-section filter terminated in a load R, we can define L and C exactly, as follows

$$L = R/\pi f_c \quad \text{and} \quad C = 1/\pi f_c R$$

where f_c is the cut-off frequency.

High-pass Filters

If we wish to attenuate the low frequencies we use a *high-pass* filter as shown in Fig. 14.9. This works in just the opposite manner from that just described. It will accept frequencies above a certain value, but will cut off below the critical frequency. Here, again, we can work out the action of the filter from simple resonant theory, or by making $\rho = -1$. Either method gives $4\omega^2 LC = 1$ (as distinct from $\omega^2 LC = 4$ for the low-pass filter). This corresponds to a frequency 0·7 times the resonant frequency f_o (and not $1{\cdot}4 f_o$ as before, which is obvious when one considers that we are running through resonance from a higher frequency this time).

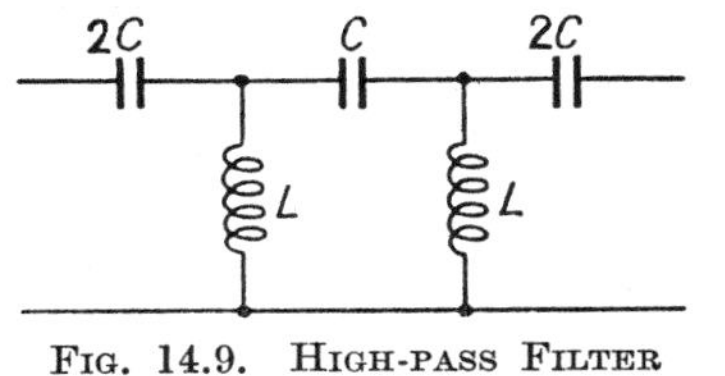

Fig. 14.9. High-pass Filter

The iterative impedance Z_k is calculated as before and is found to be

$$Z_k = \sqrt{Z_m Z_n (1 + \rho)} = \sqrt{\left[\frac{L}{C}\left(1 - \frac{1}{4\omega^2 LC}\right)\right]}$$

At cut-off, when $4\omega^2 LC = 1$, this is zero, while at frequencies *above* this (we are not interested in lower frequencies which are not in the pass range) the value rapidly rises and tends to a limiting value $\sqrt{(L/C)}$. This is similar to the low-pass case if the frequency is assumed to vary from infinity *down* to cut-off instead of from zero *up* to cut-off.

The filter is a constant-K one, and if we put $Z_m Z_n = R^2$ we have

$$L = R/4\pi f_c \quad \text{and} \quad C = 1/4\pi f_c R$$

Derived Filters

A series of sections of a simple filter network will provide a cut-off of increasing sharpness as the number of sections is increased. It is, however, possible to improve the cut-off with a smaller number of sections. For example, if we include a tuned circuit in the shunt arm of the filter, such as is shown in Fig. 14.10 (*a*), we shall obtain a point where the attenuation is complete, for when L' and C resonate, the voltage across the points AB will be zero (assuming that the resistance in the network is negligible). On the other hand, beyond this point the voltage at the output will commence to rise and we

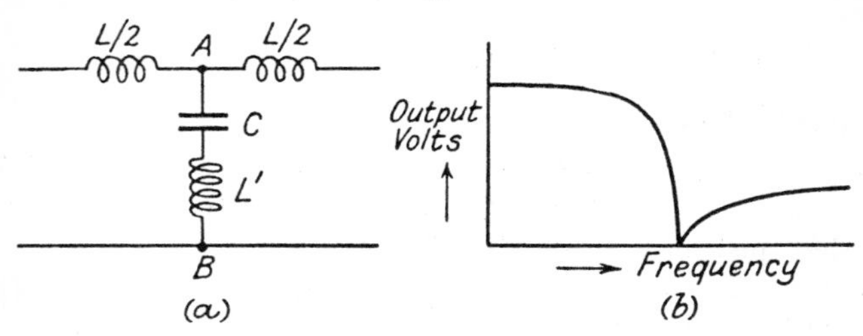

FIG. 14.10. DERIVED FILTER

shall obtain a characteristic somewhat similar to that shown in Fig. 14.10 (*b*).

It is obviously desirable that this *derived* filter, as it is called, should have the same cut-off frequency and the same iterative impedance as the *prototype* from which it was obtained. Assuming a mid-series termination again, if we multiply Z_1 by a constant, m,

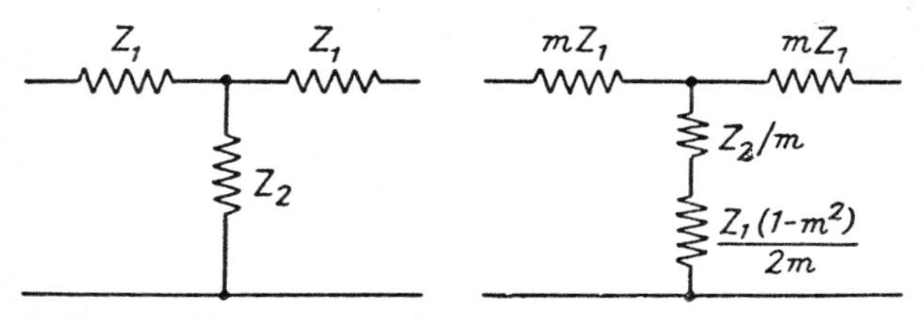

FIG. 14.11. MID-SERIES M-DERIVED FILTER SECTION

and divide Z_2 by the same constant, we have a filter having the same cut-off but different impedance. To comply with the matching condition, therefore, we make the additional impedance in the shunt arm equal to $Z_1(1 - m^2)/2m$, so that the total shunt impedance becomes

$$Z_2/m + Z_1(1 - m^2)/2m$$

as shown in Fig. 14.11.

This, in conjunction with the modified series impedance mZ_1, will give the same value of Z_k as for the simple filter, as the reader can verify for himself.

The additional impedance must, of course, be of the same type as Z_1 and the inverse of Z_2 (e.g. if Z_2 is a capacitance the additional impedance is an inductance), and the frequency of maximum attenuation is given by the resonance condition between the two shunt impedances. This can be shown to be α times the cut-off frequency f_c, where $\alpha = 1/\sqrt{(1 - m^2)}$ for a low-pass filter and $\sqrt{(1 - m^2)}$ for a high-pass filter.

Now, m may be any arbitrary number, but a case of special

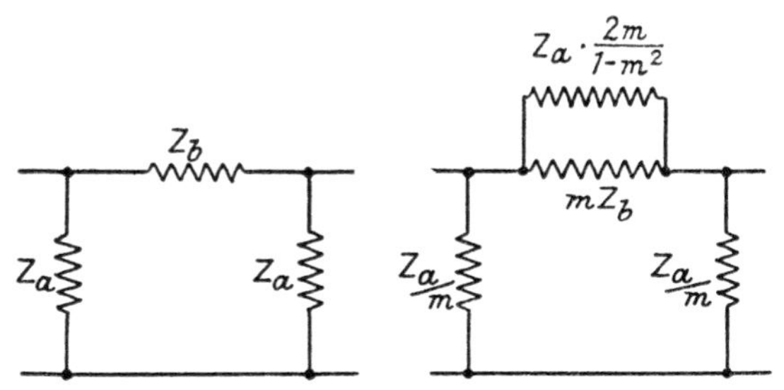

FIG. 14.12. MID-SHUNT M-DERIVED FILTER SECTION

importance arises if $m = 0{\cdot}6$, for then the *mid-shunt* iterative impedance is constant and equal to $\sqrt{(L/C)}$ for the greater part of the pass band. When $m = 0{\cdot}6$, $\alpha = 1{\cdot}25$ or $0{\cdot}8$ according to the type of filter, which results in a very sharp cut-off as shown in Fig. 14.14.

We can, if desired, obtain a derived filter from a mid-shunt or π-section as shown in Fig. 14.12. Here a similar procedure is adopted. We multiply the series impedance by m and divide the shunt impedance by the same amount. Then we connect *across* the series impedance an inverse impedance $= [2m/(1 - m^2)]Z_a$. With this filter, if $m = 0{\cdot}6$, the *mid-series* iterative impedance is constant over the greater part of the pass band.

Multi-section and Composite Filters

Filter sections of simple or derived types may be joined together provided they have corresponding terminations. Thus, a mid-series termination can be joined to another mid-series termination, but not to a mid-shunt termination.

Suppose now we wish to make up a filter consisting of one derived section and one simple section. Fig. 14.13 shows an example of how this may be done. First of all we derive a mid-shunt section. Then we split this into two half sections. Next, to the end of the derived section we connect a simple mid-shunt or π-section, and to the end

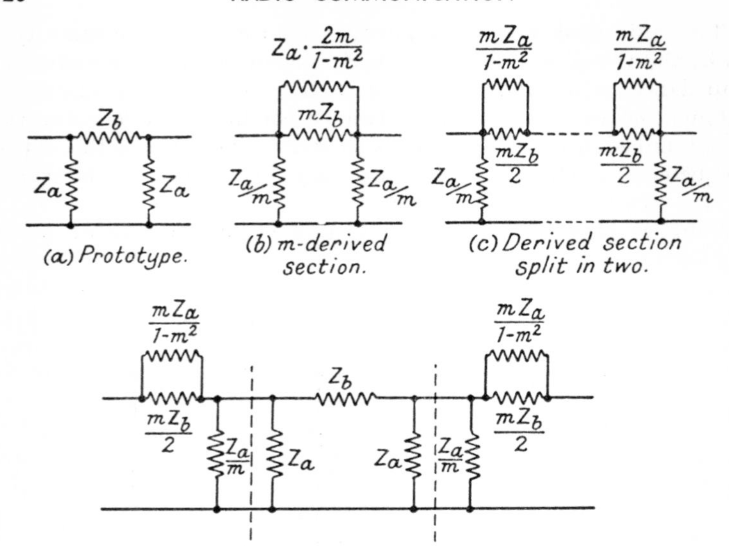

FIG. 14.13. ILLUSTRATING CONSTRUCTION OF COMPOSITE FILTER

of this we connect the front half of the derived filter. Thus internally we have mid-shunt terminations adjacent to one another, so avoiding reflection, while viewed as a whole the filter has mid-series termination, which as we have seen gives a practically constant iterative impedance.

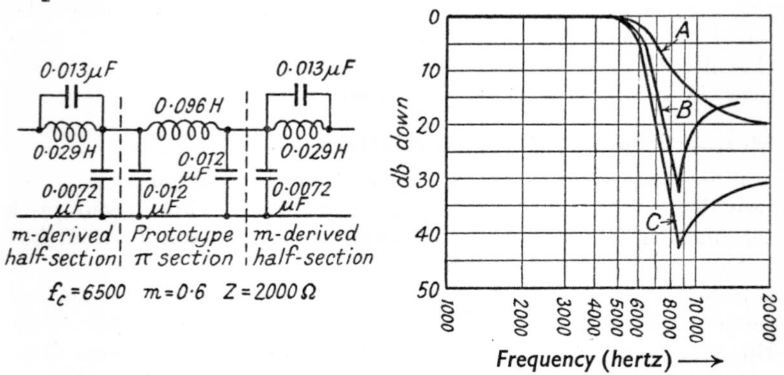

FIG. 14.14. TYPICAL COMPOSITE FILTER WITH TRANSMISSION CHARACTERISTICS

Fig. 14.14 shows the response curve of a filter of this type. Curve B is that of the derived filter giving a sharp cut-off but starting to pass frequencies above the maximum attenuation point. Curve A shows the gradual cut-off of the simple section with heavy attenuation later on. Curve C is the combination of the two, giving a sharp cut-off and over 30 dB attenuation thereafter.

Band-pass Filters

Filters are sometimes required which will accept a band of frequencies only. These are made up by using two filters in series, one a high-pass filter which is designed to cut off all frequencies below the lowest frequency in the band, and the second a low-pass filter which cuts off all frequencies above the highest desired frequency. Provided that the filters are suitably matched to one another so that their iterative impedances are equal, and equal to that of the input and the output circuits, frequencies within the band are transmitted practically unattenuated.

Propagation; Phase Angle

This treatment of filters has been in physical rather than mathematical terms, for a detailed discussion of the subject is beyond the present scope. It may be noted, however, that a series of filter sections is really only a special case of the transmission lines discussed in Chapter 8, with lumped circuit values instead of distributed impedances; and hence the same formulae will apply with $x = 1$.

Thus, for a filter correctly terminated with an impedance equal to the iterative impedance, $v = V_0 e^{\gamma}$. The exponent γ is the propagation coefficient and obviously $= \log_e v/V_0$. As we saw in Chapter 8, γ is a complex quantity containing a real and an imaginary component, and hence varies with frequency.

If we write $\gamma = \alpha + j\beta$ it is found that over the pass range of the filter ($\rho = 0$ to $\rho = -1$), the attenuation coefficient, α, is zero and β, the phase-change coefficient, is $2 \sin^{-1}\sqrt{-\rho}$. Thus when $\rho = 0$ there is no phase shift while at the cut-off frequency ($\rho = -1$) the phase shift is $2 \times \pi/2 = 180°$. This was clearly illustrated in considering the simple filter section in Fig. 14.3.

Effect of Resistance

In a practical filter, of course, the impedances are not pure reactances. The effect of resistance in the network is to cause some attenuation in the pass region, while the change of phase is delayed. At cut-off it does not reach 180° and may never do so. Fig. 14.15

illustrates the effect for a single section in terms of ϕ, the phase angle of the ratio $\rho = Z_m/4Z_n$.

With pure (inverse) reactances $\phi = 180°$, and under such ideal conditions the attenuation coefficient α is zero in the pass band (up to $\rho = -1$) and then rises sharply. Similarly the phase-change coefficient β shows a gradual change in the phase of the output from zero to π (complete reversal) after which there is no further phase shift but only attenuation as already mentioned.

The presence of even a small amount of resistance spoils this ideal action. The filter does not transmit signals without loss even in the pass band, and as will be seen with $\phi = 160°$ the loss at the cut-off

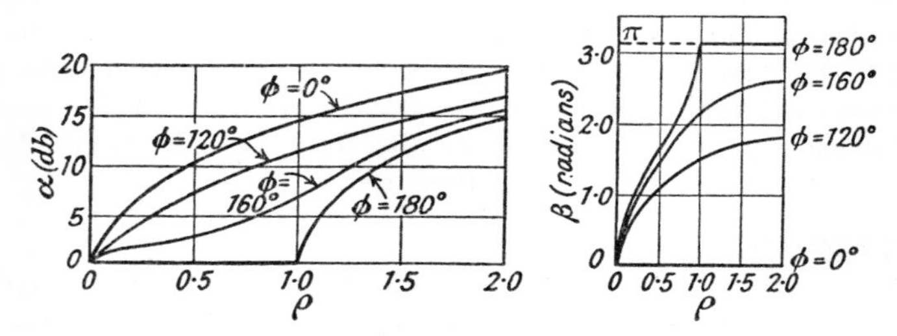

Fig. 14.15. The Effect of Resistance on the Attenuation Coefficient (α) and the Phase-change Coefficient (β)

point is not zero but 7 dB—more than 2:1 down so that we shall only obtain half the correct output. Moreover, the cut-off is no longer sharp but gradual, and as we introduce still more resistance, conditions become steadily worse until when $\phi = 120°$ there is no filter action but merely a progressive attenuation.

The phase no longer exhibits a discrete reversal when $\rho = 1$; there is a gradual change of phase which continues beyond the cut-off point. When $\phi = 0$, corresponding to a simple resistance network, there is no phase shift at all and we merely obtain a fixed attenuation determined by the value of ρ.

It is clearly essential to reduce the resistances to a minimum. Even when ϕ is 160° we have largely destroyed the filter action. A phase angle of 160°, however, is quite poor. If we assume the resistance to be all in the inductance it is easy to show that

$$\phi = \pi + \tan^{-1} Q$$

where $Q = L\omega/R$. On this basis $\phi = 160°$ is obtained with a Q of 2·7 only. It is easy to obtain a Q at least ten times as high, which makes $\phi = 178°$, under which conditions the performance will not be seriously different from the ideal. The figures in the pass band for values of ϕ near to 180° are shown in Fig. 14.16.

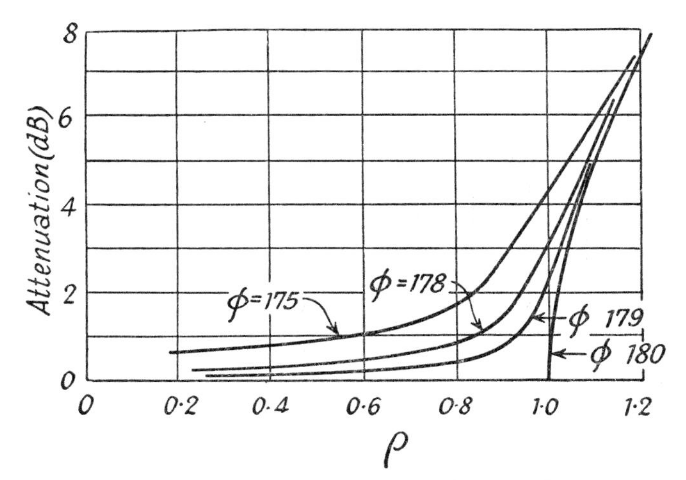

FIG. 14.16. ATTENUATION CURVES FOR FILTERS CONTAINING RESISTANCE

Effect of Mismatching

The performance both in and beyond the pass band is seriously affected if the matching is incorrect. A detailed discussion of this point is apt to become involved, though the calculations are simplified if we only consider symmetrical sections, as has been done here.

The curves of Figs. 14.15 and 14.16 are for a filter terminated in its correct (iterative) impedance, but we have seen (Fig. 14.8) that this is not constant. Z_k is actually zero at the cut-off point so that no fixed value of resistance can give correct matching over the whole of the pass band.

Fig. 14.17 shows the effect of terminating the filter with various values of resistance. There are four curves as follows—

1. Terminating resistance P equal to the surge resistance R. As explained below, this is a resistance equal to $\sqrt{(L/C)}$, which is constant, whereas the characteristic impedance Z_k varies with frequency. This gives negligible loss until $\rho = -0{\cdot}6$, after which a gradually increasing loss appears. There is no sharp transition at the cut-off point ($\rho = -1$), and the loss in the stop band is appreciably less than the ideal.

2. $P = R/10$. Here there is perfect transmission at the cut-off

point and a very steep cut-off thereafter, but there is a serious loss in the pass band.

3. $P = R/2$. This is a compromise giving a small loss in the pass band, and a rather steeper cut-off than with $P = R$.

4. $P = 10R$. Here all trace of filter action has been lost.

The ideal matching curve is shown dotted for comparison. With $P = Z_k$ there is zero loss up to cut-off and a rapidly rising loss thereafter.

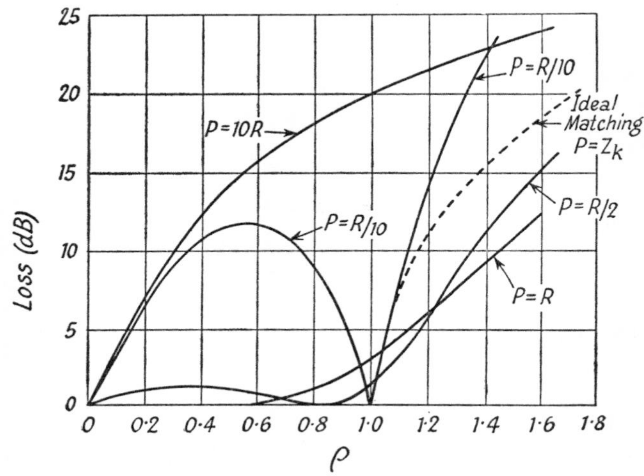

FIG. 14.17. EFFECT OF MATCHING ON PERFORMANCE

All these curves apply to a mid-series section, either low or high pass, and show that the terminating impedance should not exceed R and may often be made appreciably less with advantage. The effect of mismatching is reduced if several (correctly matched) sections are used in series because the intervening sections are then ideally terminated.

The tendency in practice is to design the filter for a cut-off somewhat short of the theoretical point—e.g. if we require to accept frequencies up to 1,000 Hz we should design the filter with a cut-off at about 1,300 Hz and use $P = R$ or possibly $R/2$.

For a mid-shunt section Z_k varies between $\sqrt{(L/C)}$ and infinity over the pass band. Hence with such a filter inverse reasoning will apply so that we should tend to the use of higher loads than R. The $R/2$ curve will be obtained with $P = 2R$ (very nearly) and so on, so that with this type of section the filter action is destroyed if R is too low.

Network Impedances

We have so far considered simple symmetrical networks in a virtually non-mathematical manner. In a full treatment these would be regarded as particular examples of a generalized four-terminal network having two input and two output terminals, and not necessarily symmetrical. When dealing with such networks certain expressions are used which, while not required in the present treatment, may be briefly defined for completeness, as follows.

Image Impedance

Consider the four-terminal network of Fig. 4.18. The input image impedance is the apparent impedance of the network when the

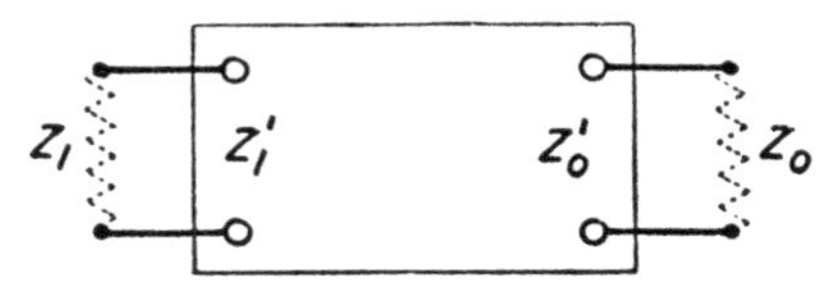

Fig. 14.18. Four-terminal Network

output is terminated with the output image impedance, and vice versa. When the network is not symmetrical these two image impedances are *not* equal.

Iterative Impedance

This, as we have seen, is the impedance such that if the network is terminated with this impedance it "looks like" this impedance. If the network is not symmetrical the iterative impedances are not the same looking at either end.

Characteristic Impedance

With a symmetrical network the image and iterative impedances are the same at either end. This is called the *characteristic impedance* of the network.

Wave Impedance

A long line is necessarily symmetrical and its characteristic impedance is often called the *wave* or *surge impedance*, and is equal to $\sqrt{(L/C)}$.

Insertion Loss

If we have a source of e.m.f. E having an internal impedance Z_s connected to a load Z_r, then the current which will flow is

$I = E/(Z_s + Z_r)$. Suppose now that we interpose a network between the source and the load. The current flowing will depend upon the matching of the various impedances.

Let us assume that the network is correctly matched at both ends. The network will then "look like" Z_s so that the input current $I_s = E/2Z_s$.

Now, neglecting losses, the volt-amperes at the input and output will be equal, so that $I_s^2 Z_s = I_r^2 Z_r$, whence

$$\begin{aligned} I_r &= I_s \sqrt{(Z_s/Z_r)} \\ &= \frac{E}{2Z_s}\sqrt{\frac{Z_s}{Z_r}} = \frac{E}{\sqrt{(4Z_s Z_r)}} \\ &= \frac{E}{Z_s + Z_r} \cdot \frac{Z_s + Z_r}{\sqrt{(4Z_s Z_r)}} \end{aligned}$$

This means that the presence of the connecting network modifies the received current in the ratio $(Z_s + Z_r)/\sqrt{(4Z_s Z_r)}$, which may be more or less than unity according to whether Z_r is greater or less than Z_s. The network in fact behaves like a transformer.

However, if the network is not correctly matched there is reflection at either or both ends, which further modifies the received current, while if the network is not loss-free there is attenuation due to the resistance. Taking all these factors into account it can be shown that the received current is

$$I_r = \frac{E}{Z_s + Z_r}\left[\frac{Z_s + Z_r}{\sqrt{(4Z_s Z_r)}} \cdot \frac{\sqrt{(4Z_0 Z_s)}}{Z_0 + Z_s} \cdot \frac{\sqrt{(4Z_0 Z_r)}}{Z_0 + Z_r} \cdot e^{-\gamma}\right]$$

where Z_0 is the characteristic impedance and γ is the propagation coefficient. The expression within the square brackets is called the *insertion loss.*

The input and output volt-amperes are the same only if the network is loss-free and is correctly terminated. Under any other conditions there will be some difference, and the (logarithmic) ratio of the two is called the *image transfer coefficient*, which is $\theta = \frac{1}{2}\log_e (V_1 I_1)/(V_2 I_2)$ when the network is terminated in its image impedances. In a loss-free network $V_1 I_1 = V_2 I_2$, so that $\theta = 0$.

Crystal Filters

It was shown in Chapter 7 that a quartz crystal, being a mechanical resonator, posseses properties analogous to inductance and capacitance, the equivalent circuit being shown in Fig. 7.8. It is evident that these can replace the elements of a conventional filter.

A single section is of limited application, but by providing a succession of small electrodes along the length of a crystal slice a compact multiple filter can be obtained. Each element operates virtually independently, a small degree of coupling being provided by fringe vibrations between one element and the next.

This is a highly specialized application of crystal technique which cannot be discussed here. It permits the production of very-sharply tuned band-pass (and other) filters in miniaturized form. Ceramic piezo-electric materials are also used for certain applications.

14.2. Active Filters

The filters described in the previous section all utilize *passive* elements, i.e. components which do not contain any inherent source of e.m.f. Moreover, it was shown that a simple RC network such as that of Fig. 14.1 only produces a progressive attenuation with frequency, and that to obtain true filter characteristics it is necessary to employ complementary reactances, so that resonant conditions can be produced to obtain a sharp cut-off. This involves the use of inductances, which are not only inherently less stable, but become unwieldy, particularly at lower frequencies.

However, if *active* elements can be introduced into the network, i.e. devices which provide a measure of amplification and/or phase reversal, it is possible to devise RC networks which behave like inductances, and in fact all the properties of LCR filters can be realized using only RC elements with appropriate associated active devices. The basic theory of such active filters has long been established, but it is only in recent years that the evolution of the integrated-circuit operational amplifiers described in Chapter 18 has permitted the practical realization of the possibilities by providing low-cost active elements of predictable characteristics.

Transfer Functions

The behaviour of active filters can most conveniently be assessed in terms of the transfer functions of the system. These are functions involving an operator p to denote differentiation with respect to time, so that $p = d/dt$.

Now $v = L\,di/dt$ and $i = C\,dv/dt$. These expressions may be written $v = Lip$ and $i = Cvp$, whence the transfer function

$$v/i = pL = 1/pC$$

(The analogy with the expressions for steady-state impedances $j\omega L$ and $1/j\omega C$ is readily apparent.)

For the simple RC network of Fig. 14.19 (a) the frequency response function is $(1/j\omega C)/(R + /j\omega C) = 1/(1 + j\omega CR)$. Expressing this in terms of the p operator we have

$$\frac{V_{out}}{V_{in}} = \frac{1}{1 + pCR} \tag{1}$$

Now, CR has the dimensions of time (as also has the corresponding term involving inductance, L/R), so that eqn. (1) may be written

$$V_{out}/V_{in} = 1/(1 + pT) \tag{2}$$

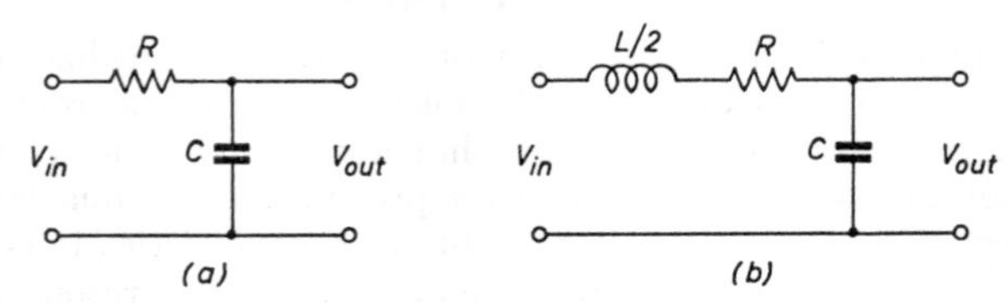

Fig. 14.19. Passive Low-pass Filter Sections

This is the fundamental expression for a *first-order* low-pass filter (i.e. a filter involving the first power of T). By similar reasoning the corresponding expression for a high-pass filter can be shown to be

$$V_{out}/V_{in} = 1/(1 + 1/pT) \tag{3}$$

Let us now consider a filter with complementary reactances, as in Fig. 14.19 (b). Using the same technique we can write

$$\frac{V_{out}}{V_{in}} = \frac{1/pC}{1/pC + R + \frac{1}{2}Lp} = \frac{1}{1 + pCR + p^2(\frac{1}{2}LC)} \tag{4}$$

Now, we saw earlier (page 611) that with this filter the cut-off frequency occurs when $\omega_c{}^2LC = 2$. Hence if we put $T = 1/\omega_c = \sqrt{(\frac{1}{2}LC)}$ the product RC is given by

$$RC = \frac{R}{\frac{1}{2}L\omega_c{}^2} = \frac{R}{\frac{1}{2}L\omega_c} \cdot \frac{1}{\omega_c} = \frac{T}{Q}$$

where Q is the conventional circuit magnification factor, which in this case is equal to $\frac{1}{2}L\omega_c/R$.

Eqn. (4) can thus be written

$$\frac{V_{out}}{V_{in}} = \frac{1}{1 + pT/Q + p^2T^2} \tag{5}$$

This is an expression involving the second power of T, so that this kind of circuit is called a *second-order* (low-pass) filter.

Similar reasoning can be used to derive the equation for the high-pass filter of Fig. 14.20 (*a*), which is

$$\frac{V_{out}}{V_{in}} = \frac{p^2T^2}{1 + pT/Q + p^2T^2} \qquad (6)$$

and for the band-pass filter of Fig. 14.20 (*b*), which is

$$\frac{V_{out}}{V_{in}} = \frac{pT/Q}{1 + pT/Q + p^2T^2} \qquad (7)$$

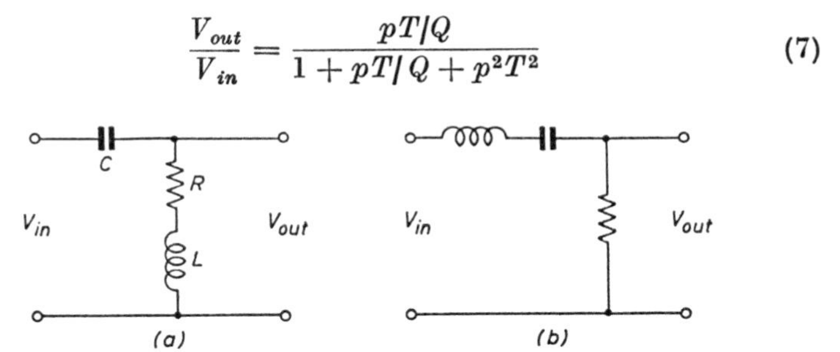

FIG. 14.20. PASSIVE HIGH-PASS AND BAND-PASS SECTIONS

Synthesis of Second-order Active Filter

Defining the transfer function in this form specifies the performance required from the active elements of the filter. It will be remembered that the equation for an amplifier with negative feedback was shown (page 473) to be

$$\frac{V_{out}}{V_{in}} = \frac{A}{1 + \beta A}$$

where A is the amplifier gain without feedback. If the feedback factor is unity, this expression reduces to

$$\frac{V_{out}}{V_{in}} = \frac{1}{1 + 1/A} \qquad (8)$$

This is of the same form as eqn. (5). Hence to provide a second-order low-pass filter we require an active element having a gain given by

$$1/A = pT/Q + p^2T^2$$

so that

$$A = \frac{Q}{pT} \times \frac{1}{1 + pTQ}$$

This can be achieved by using a two-stage amplifier as shown in Fig. 14.21 consisting of an integrator having a voltage transfer ratio Q/pT followed by a simple low-pass element having a transfer ratio $1/(1 + pTQ)$, as per eqn. (2). The time constant C_1R_1 would be made equal to T/Q, while $C_2R_2 = TQ$. This is known as an *integrator-and-lag circuit*. To approach perfect equivalence the gain of A_1 should tend to infinity, while A_2 should have unity gain with a high input impedance and low output impedance.

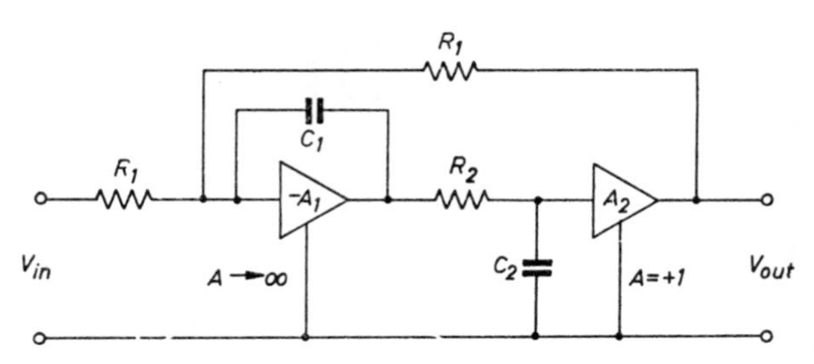

FIG. 14.21. INTEGRATOR-AND-LAG ACTIVE FILTER
The minus sign in A_1 denotes a phase reversal.

These conditions can be fulfilled with modern integrated-circuit operational amplifiers,* providing a practical circuit occupying small space and requiring no inductances. Moreover, since it has a low output impedance, any number of second-order sections can be connected in cascade to produce a filter of a higher order. Similar conditions can be derived for the high-pass or band-pass filters of eqns. (6) and (7).

Integrator-Loop Filter

A disadvantage of the simple integrator-and-lag circuit is that large ratios of component values are required to obtain optimum performance. This can be overcome, at the expense of additional circuit complexity, by replacing the passive *RC* network with its active equivalent. The resulting circuit, which is known as an *integrator-loop filter*, is illustrated in Fig. 14.22.

This configuration has the further advantage that it will tolerate reasonable variations in the component values of both passive and active elements.

* An *operational amplifier* is one which is designed to perform a specific function, such as differentiation or integration, as distinct from straightforward amplification (see Chapter 17).

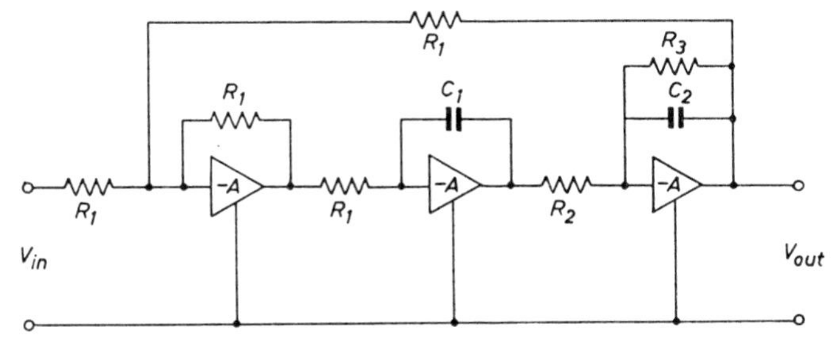

FIG. 14.22. INTEGRATOR-LOOP FILTER

Partitioned Feedback

An alternative approach is to adopt what is called *partitioned feedback* as shown in Fig. 14.23. Here the voltage transfer ratio is

$$\frac{V_{out}}{V_{in}} = \frac{AB}{1 - \beta A}$$

where $B = V_1/V_{in}$, which is the transfer ratio of a passive *RC* network at the input, and A is the gain of an active element having a feedback ratio β.

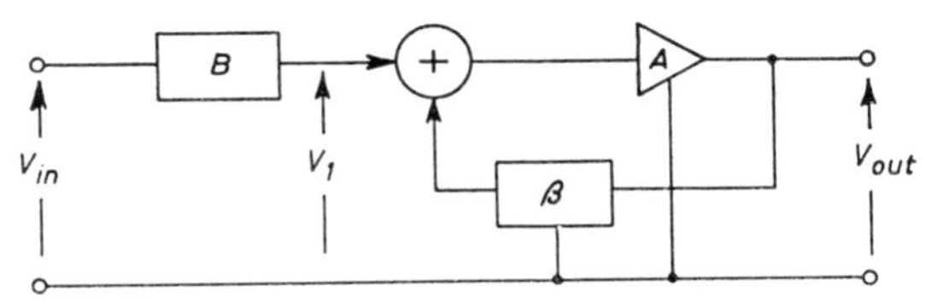

FIG. 14.23. PARTITIONED FEEDBACK NETWORK

The symbol between the passive and active stages in the diagram denotes a *buffer*, which will pass signals from input to output, but not in the reverse direction.

This form of circuit is frequently used, and is often called the Sallen and Key circuit, after its originators. Fig. 14.24 shows a low-pass filter using this configuration; the equivalent circuit may be drawn as at (*b*), since C_1 is earthed with respect to the input through the output impedance of the amplifier, and the resistor R_1 is effectively earthed through the (low) impedance of the input voltage source.

Hence the circuit is equivalent to that shown in Fig. 14.23 with a passive *RC* input network introducing a frequency-selective loss, followed by an amplifier with a frequency-selective feedback. Analysis of the transfer functions shows that this circuit gives the required overall filter characteristics, but the value of βA must be close to unity for high values of Q. This means that the circuit is susceptible to small changes in component values.

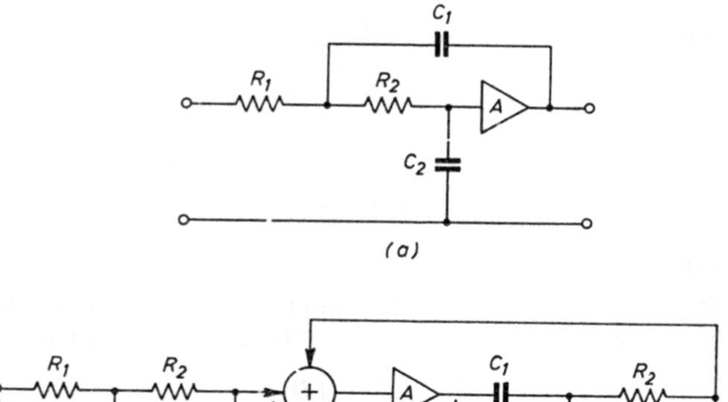

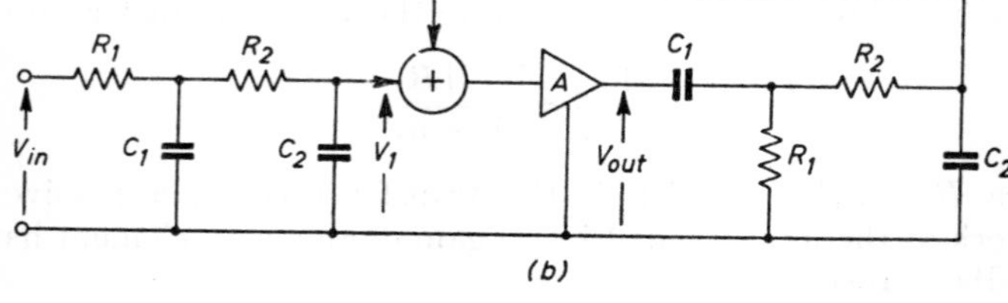

FIG. 14.24. LOW-PASS SALLEN AND KEY FILTER, WITH EQUIVALENT CIRCUIT AT (*b*)

Negative Impedance Convertors

Still another method of realizing the desired filter characteristics involves the use of circuits which transform impedances from one form to another. This may be done in various ways, using the sophisticated techniques now possible with integrated-circuit operational amplifiers.

One such device is the *negative-impedance convertor* (NIC), which is an active network in which any impedance placed across the input terminals will appear as the negative of that impedance when observed at the output. Fig. 14.25 illustrates a simple low-pass filter comprising two passive *RC* elements with a NIC in between. By

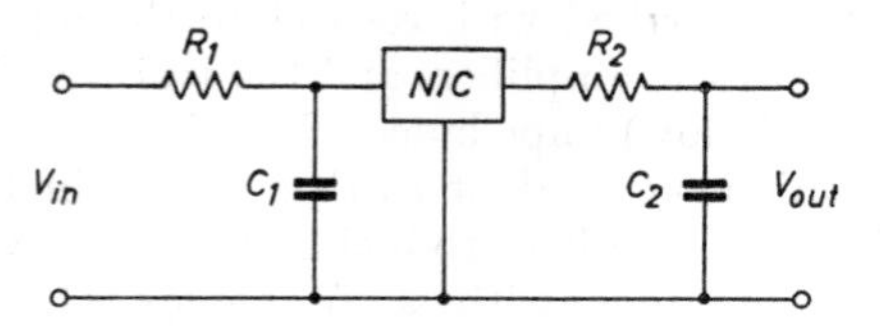

FIG. 14.25. NIC LOW-PASS FILTER

suitable choice of component values this can be arranged to provide the required filter characteristics. However, it has a high-impedance output so that a buffer stage is required to couple it to succeeding stages.

Gyrators

An alternative device is the *gyrator*, which is a network having an admittance matrix* of the form

$$\begin{vmatrix} 0 & -g \\ g & 0 \end{vmatrix}$$

Thus if a gyrator is terminated with a capacitor, the overall circuit is equivalent to an inductance of value $L = C/g^2$. (This is an

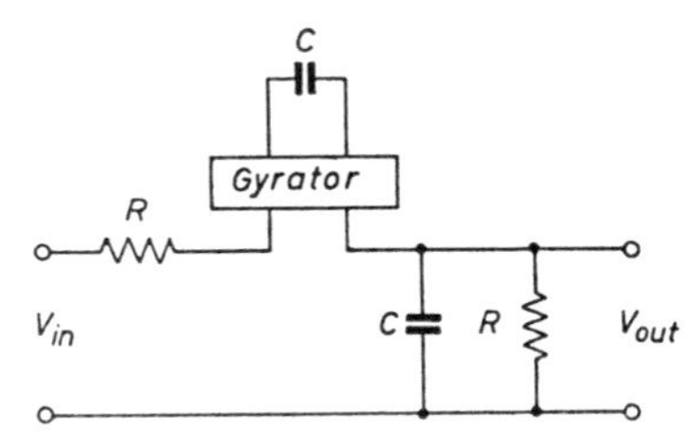

FIG. 14.26. LOW-PASS FILTER USING A GYRATOR

arrangement analogous to the reactance valve described on page 352.) With such a device the inductance in any of the standard forms of *LC* filter may be replaced with a gyrator and capacitor, as for example in the circuit of Fig. 14.26, which is equivalent to the low-pass half-section of Fig. 14.19 (*b*).

Filters using gyrators have a very good theoretical performance, the added complexity being more than offset by the considerable reduction in size. Variations in component values and/or ambient conditions can produce variations in the apparent inductance, but these can be minimized by appropriate circuit modifications.

The technique of active-filter design has, in fact, become highly sophisticated, and the foregoing discussion has necessarily been limited to consideration of the basic principles involved in the production of simple second-order filters.

It is apparent that higher-order filters can be realized by suitable extensions of these techniques. Practical higher-order filters are obtained by cascading appropriate integrator-loop sections, or by

* See page 87.

using gyrator-capacitor combinations as replacements for the inductors in higher-order passive *LC* networks. Much care has to be taken in the detailed design to minimize the effects of changes in component values which may occur in either passive or active elements.

14.3. Attenuators and Equalizers

Finally, we come to the consideration of *attenuator* networks. These are arrangements of resistances designed to cut down the voltage by a known and constant amount irrespective of frequency. This is usually done by a T-section network as shown in Fig. 14.27.

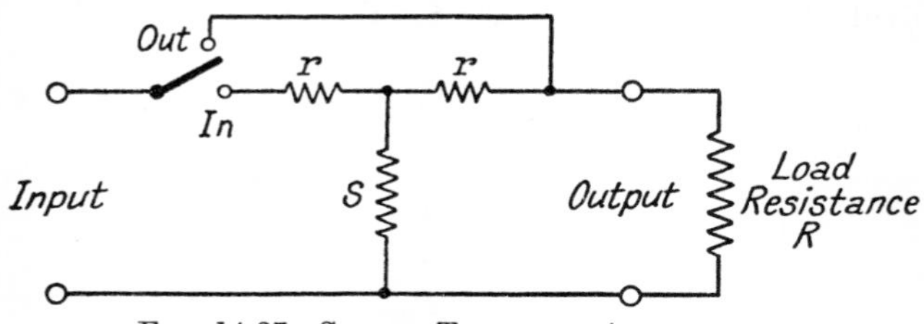

Fig. 14.27. Simple T-section Attenuator

The requirement is that the impedance of the attenuator shall always be the same and equal to the load resistance.

The attenuator is thus a special case of a T-section filter terminated in the iterative impedance. The load resistance R must be

$$\sqrt{[r(r + 2s)]}$$

The attenuation of a given section is easily deduced. Let P be the resistance of s in parallel with $R + r$. We have across the input a voltage divider made up of r and P in series. Across P is a further voltage divider consisting of r and R in series. Hence, the output voltage

$$V_2 = \frac{P}{P + r} \cdot \frac{R}{R + r} V_1$$

so that

$$\frac{V_2}{V_1} = k = \frac{PR}{(P + r)(R + r)}$$

If the network is terminated in its iterative impedance, the input resistance $= P + r = R$, so that $P = R - r$.

Hence

$$k = (R - r)/(P + r)$$

This expression and that previously given for the iterative impedance completely determine r and s. By rewriting the equations we have

$$r = R\left(\frac{1-k}{1+k}\right)$$

and

$$s = 2R\left(\frac{k}{1-k^2}\right)$$

Provided the resistances conform with these requirements, the effective resistance is always the same, and several sections, each having different attenuation factors, may be connected in series.

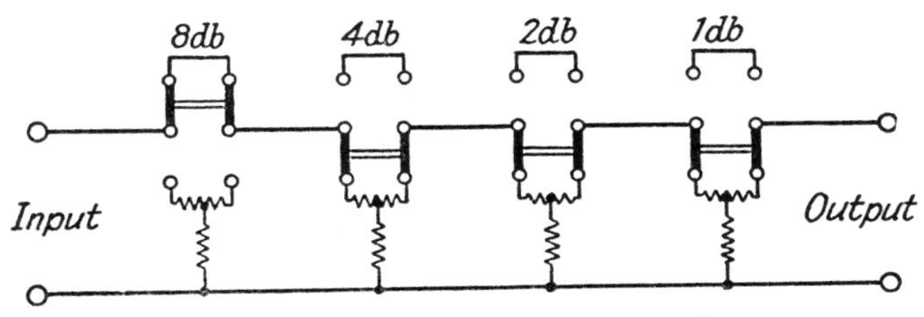

FIG. 14.28. T-PAD ATTENUATOR COVERING A VARIATION OF 1 TO 15dB SET TO GIVE 7dB ATTENUATION

Fig. 14.28 shows a network capable of giving 15 dB attenuation in steps of 1 dB.

Effect of Frequency

In all the foregoing expressions it is assumed that the resistors are non-reactive, i.e. that the reactance of any self-inductance and/or stray capacitance is negligible at the frequencies over which the attenuator is to be used. With high resistance values stray capacitances may well not be negligible, but with fixed attenuators this may be allowed for by connecting compensating capacitances across suitable portions of the network so that the capacitive reactances are in the same ratio as the resistances.

Variable Attenuators

Where correct matching is required an attenuator of the form of Fig. 14.28 is essential. Fine adjustment is obtained, if required, by adding further sections—e.g. 0·4, 0·4, 0·2, and 0·1 dB to provide fractions of 1 dB.

Occasions arise, however, when approximate matching is sufficient, as with an oscillator or amplifier which is designed to feed into a specified load. Here the performance will not suffer seriously if the impedance varies by 10 or even 20 per cent. In such cases it is possible to make up attenuators having a continuously variable section as shown in Fig. 14.29. Here a 20 dB section is provided

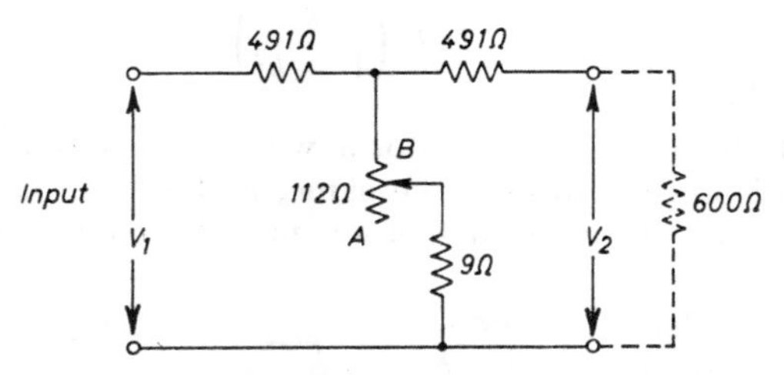

FIG. 14.29. VARIABLE ATTENUATOR SECTION

with a variable shunt element. In position A the shunt resistance is 121 Ω which with the 491 Ω series arms and 600 Ω load will look like 600 Ω, but $V_2 = 0{\cdot}1\,V_1$. In position B, $V_2 = 0{\cdot}01\,V_1$ but the input impedance is only 500 Ω, a variation which may be tolerable.

Ladder Attenuators

In r.f. practice use is made of attenuators which are required to work into a high impedance. Such a case occurs in the output of a signal generator which works into a low resistance of 10 or 20 ohms, but which feeds the tuned aerial circuit of a receiver or the grid circuit of a valve, either of which, when tuned, has an impedance of some thousands of ohms which is negligibly high compared with the attenuator impedance.

For such work a ladder attenuator of the form of Fig. 14.30 is used. The input, fed in at the point A, is progressively attenuated in fixed steps, the output being taken from the points A, B, C, or D. If the impedance of the output is high the input resistance of the network is clearly not affected by the position of output tap. Moreover, it is simple to arrange that the resistance of the network viewed from the output end is always the same (and equal to the input resistance) at any of the tapping points, so that the device behaves like a generator of constant source impedance.

Consider first the end section alone. The resistance to the right of C is $P + T$. If this step is to attenuate n times, $P + T = nT$.

To the right of B the resistance is $P + Z$, where Z is the resistance of S in parallel with $P + T$. But if we are to maintain the same attenuation per stage Z must be equal to T.

Hence $$Z = T = nTS/(nT + S)$$

whence $$S = nT/(n - 1)$$

Similar reasoning applies as we move progressively to the left, for the network to the right of B can be replaced by a simple end section $P + T$ which gives the same value for the shunt resistance

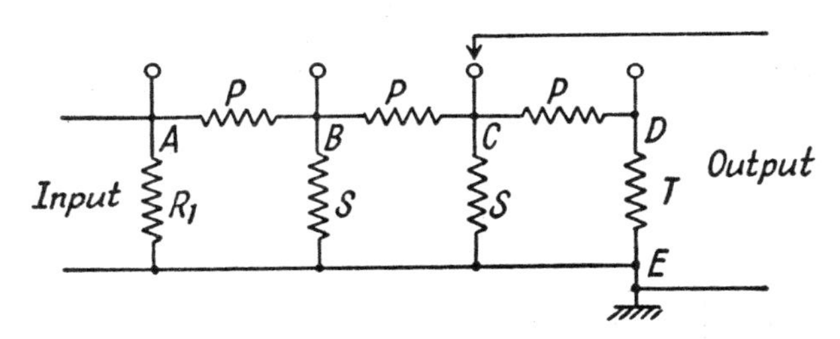

FIG. 14.30. LADDER ATTENUATOR NETWORK

at B, and again to the right of A we can replace the whole network by $P + T$.

The input resistance of the whole network is thus R_1 in parallel with $P + T$. If we make $R_1 = T$ the whole network becomes symmetrical. The resistance at any of the points A, B, C, or D then becomes

$$(P + T)T/(P + T + T) = nT^2/(nT + T)$$
$$= nT/(n + 1) = R$$

Hence $$T = (n + 1)R/n$$

If $n = 10$ and $R = 10$, $T = 11$ ohms, $S = 12{\cdot}2$, and $P = 99$ ohms, which are values commonly used in r.f. attenuator networks.

If R_1 is not made equal to T the input resistance will be

$$R_1(P + T)/(R_1 + P + T)$$

This will still remain constant whatever the position of the output tap, but the resistances at the points B, C, and D will not be the same. If this point is not of importance it may be found convenient to adopt a value of R_1 different from T.

Balanced Networks

The use of an unbalanced line (i.e. with one side earthy) may give rise to crosstalk, so that symmetrical arrangements are often employed. An attenuator or filter for use with such a system must be similarly "balanced."

This is achieved by splitting the series impedances in two and inserting one half in each line as shown in Fig. 14.31 (*a*). The calculations are as before, the two separate elements being considered as one. If the network includes a terminating impedance or transformer this will be centre-tapped to provide the earth point. Such terminating devices are themselves symmetrically constructed to

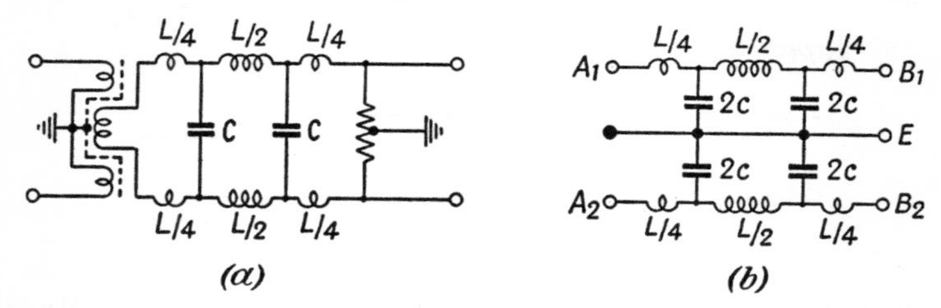

FIG. 14.31. EXAMPLES OF BALANCED NETWORKS

equalize the stray capacitances. An earth screen is often provided in addition, the device being known as a screened and balanced transformer.

If no such termination is provided the shunt impedances must be similarly split into two elements in series, as shown in Fig. 14.31 (*b*).

Equalizers

With many signal sources (e.g. a record-player pick-up) the e.m.f. is not uniform over the frequency range. The variation may be compensated by introducing a network having a complementary frequency characteristic. In many cases a simple RC network will suffice, but with such a circuit the input impedance and the phase of the output both vary with frequency.

Both these parameters can be held constant by the use of networks employing two complementary reactances as shown in Fig. 14.32. These are called *equalizers,* and it can be shown that if a resistance R is connected across Z_1 equal to the load resistance, then if $Z_1 Z_2 = R^2$

the whole network has a resistive impedance $= R$ at all frequencies. Z_1 and Z_2 must be complementary, which requires that their reactances must be inverse and also that any resistance in series with Z_2 must be compensated by an appropriate resistance *in parallel* with Z_1, and vice versa. The attenuation is then given by

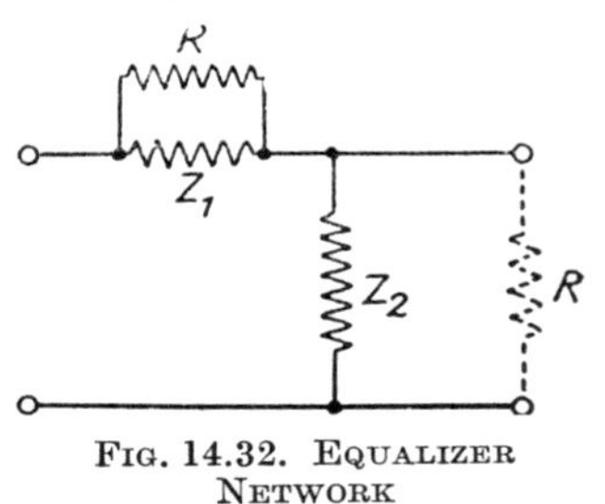

FIG. 14.32. EQUALIZER NETWORK

$$\alpha = (Z_2 + R)/Z_2 = (R + Z_1)/R$$

A typical example is illustrated in Fig. 14.33, which shows an equalizer designed to provide uniform output above 500 Hz but a rising output below this point, being 4 times up at 50 Hz. The shunt arm comprises a capacitor C in series with a resistor r_2. Some initial attenuation α_1 must be assumed at 50 Hz, say 1·25. Then at 500 Hz, α_2 must be 4 times as great $= 5$.

At 500 Hz, $1/j\omega C$ will be small compared with r_2 so that α_2 will be simply $(R + r_2)/r_2$, which with $R = 50\,\text{k}\Omega$ makes $r_2 = 12{\cdot}5\,\text{k}\Omega$. Using this figure the value of C can then be calculated to make $\alpha_1 = 1{\cdot}25$ at 50 Hz, which requires $C = 0{\cdot}04\,\mu\text{F}$.

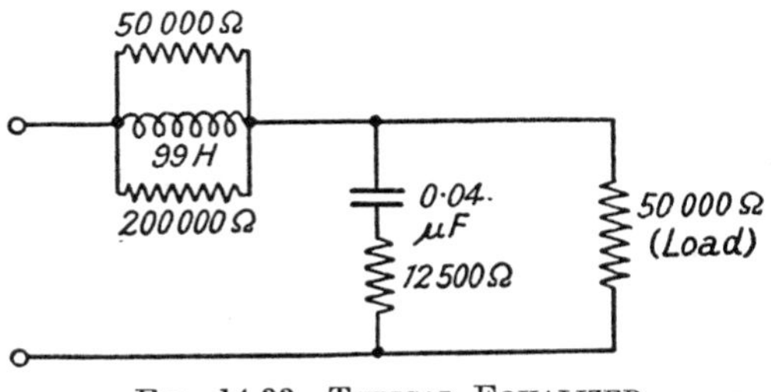

FIG. 14.33. TYPICAL EQUALIZER

To equalize the network, Z_1 is then calculated from the relationship $r_1 r_2 = L/C = R^2$, whence $L = 99\,\text{H}$ and $r_1 = 200\,\text{k}\Omega$. This will be in parallel with the $50\,\text{k}\Omega$ reflecting the load impedance, so that the two resistors would be replaced by a single resistor of $40\,\text{k}\Omega$. In practice it would probably be convenient to replace the inductance with a gyrator-capacitor combination.

It is desirable to verify at the outset that the required ratio of α_2/α_1 is practicable. If in whichever impedance has elements in series the ratios of reactance to resistance are Q_1 and Q_2 at the frequencies in question, the maximum ratio of α_2/α_1 is $Q_2/(1 + Q_1^2)$.

15

Measurements

The field of measurement in radio communication ranges from the determination of simple constants such as resistance, inductance and capacitance, and fundamental quantities such as current, voltage and power, up to tests on complete receivers for sensitivity, fidelity and selectivity.

15.1. Voltage and Current

For the measurement of direct currents and voltages a moving-coil meter is usually employed. This consists of a coil pivoted in a strong magnetic field, as illustrated in Fig. 15.1. As explained in Chapter 4 (page 215) the coil will take up a position depending upon the strength of the current through it.

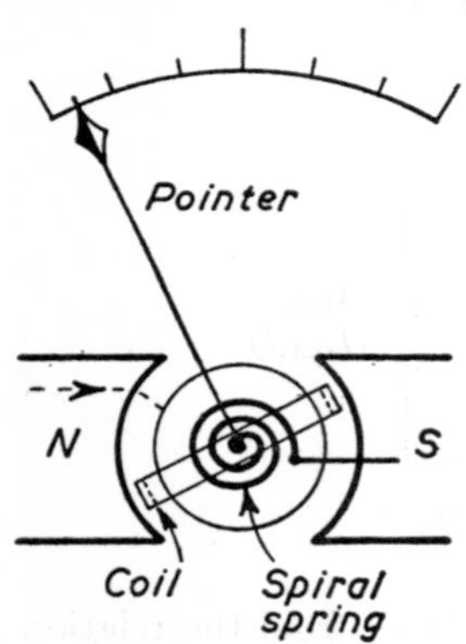

Fig. 15.1. Diagram of Moving-coil Meter

A similar indication can be obtained with a moving-iron mechanism, but as such movements require more power for their operation, their use is usually confined to measurements of alternating currents, as explained later.

Voltages may be measured by connecting a high resistance in series with the instrument. The current flowing through the instrument will then be directly proportional to the voltage (since $I = E/R$). The resistance is made high so that only a small current will be absorbed by the device.

The Potentiometer

Such instruments, of course, do not provide an absolute measurement and have therefore either to be individually calibrated or adjusted during manufacture by comparison with some suitable standard. This is a matter with which the communications engineer is not normally concerned, but reference should be made to the basic device used in all such fundamental calibrations, namely the *potentiometer*.

In its simplest form the potentiometer consists of a long wire AB (Fig. 15.2) which has a resistance proportional to its length and a tapping which can be moved along the wire. If a galvanometer is connected in series with the sliding contact, a balance will be obtained when $E_2 = (AC/AB)E_1$. This balance is indicated by zero

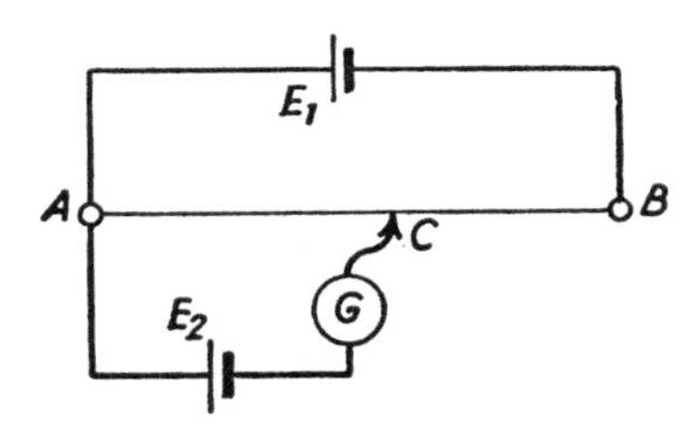

FIG. 15.2. THE SIMPLE POTENTIOMETER

current in the galvanometer, under which conditions the internal resistance of E_2 and G produce no voltage drop and the voltage $(AC/AB)E_1$ is the true e.m.f. of the battery E_2.

In practice, E_1 is a source of reasonably steady potential, while E_2 is a standard cell which, as described in Chapter 4.6, delivers an exactly-known e.m.f. under conditions of zero current, and the setting of the sliding contact is noted. The standard cell is then replaced by the source of e.m.f. to be measured and the new balance point is noted. The ratio of the two settings then accurately determines the unknown e.m.f. in terms of that of the standard cell.

For measurement of current, the unknown e.m.f. would be replaced with a standard resistance, R, through which the unknown current is passed from a suitable source. Balance would then be obtained when $IR = (AC/AB)E_1$ and this is compared with E_2 as before.

Effect of Meter on Circuit

The essential feature of potentiometer technique is that no current is taken by the test circuit at balance. In practical measurements this condition does not apply, but it is nevertheless a fundamental requirement in any measuring technique that the measuring device shall not appreciably alter the quantity being measured. Thus a voltmeter connected across a circuit should not affect the circuit impedance, but this is an ideal condition. In practice some variation in the conditions is inevitable and it is necessary to assess whether this can be tolerated or not. For instance, consider the circuit of Fig. 15.3, The voltage at the point A will be $V_{cc} - RI_z$, where I_z is the current through the load Z. If we try to measure this with a voltmeter across Z, the conditions will be modified by the current I_m through the voltmeter itself, so that V_A is now $V_{cc} - R(I_z + I_m)$. Hence if the conditions are not to be appreciably disturbed, I_m

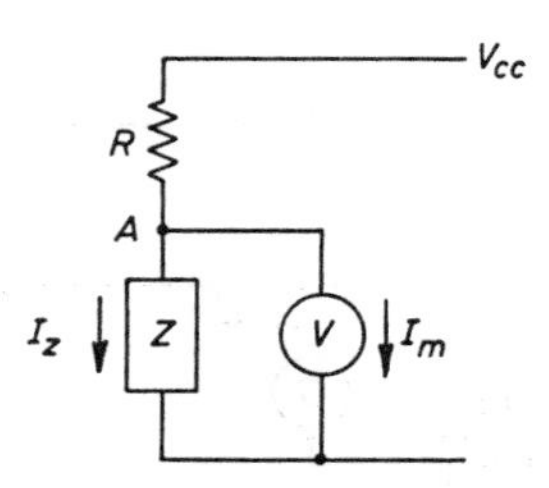

FIG. 15.3. ILLUSTRATING INFLUENCE OF METER ON CIRCUIT
The current taken by the voltmeter V may vitiate the voltage reading.

must be very small compared with I_z. In practice this is often not true. Many small test meters take 5 mA for full-scale deflection (f.s.d.) so that even a partial reading will draw several milliamperes, which may be comparable with I_z. (For example, Z may be a transistor taking only 1 mA.) Even with a meter taking 1 mA f.s.d. the reading will be inaccurate, and to avoid this difficulty meters are often used which take only 50 μA f.s.d.

Meters are often rated in terms of ohms per volt. Thus a meter having a 1 mA f.s.d. would require a series resistor of 1,000 ohms to read 1 volt, and *pro-rata*, so that it would be rated as 1,000 ohms/volt. A voltmeter using a 50 μA movement would be rated at 20,000 ohms/volt.

If a high-resistance voltmeter is not available it may be possible to determine V_A by connecting the meter across R. Provided that the meter resistance is large compared with R, this will not seriously disturb the conditions, and V_A is simply $V_{cc} - V_R$.

A.C. Measurements

The deflection of the pointer in a moving-coil instrument depends on the direction of the current through the coil. When we come to consider alternating currents or voltages, therefore, it is obvious that a device must be employed which is not affected by the direction in which the current is flowing, but only by the value. Since the square of a quantity is always positive, an instrument which gives an indication depending on the square of the current or voltage will be suitable for measuring a.c.

This requirement can be met by using a moving-iron instrument, such as was discussed in Chapter 4 (page 216), and for power-frequency measurements such instruments are often used. They are less sensitive than the moving-coil type, normally requiring some 5 mA to produce full-scale deflection. This means that such meters absorb appreciably more power, which cannot be tolerated in many instances, while they become increasingly inaccurate as the frequency increases.

Rectifier Meters

For audio-frequency voltages, therefore, a rectifier meter is normally used. This consists of a moving-coil meter, the current to which is supplied through a small bridge rectifier as shown in Fig. 15.4.

Such instruments will operate at frequencies up to 100 kHz; beyond this point the stray circuit capacitances produce an increasing deterioration in performance. They are not suitable for

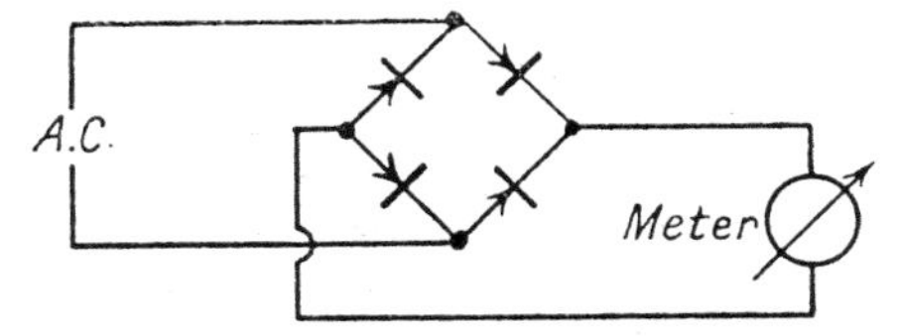

FIG. 15.4. CIRCUIT OF A.C. RECTIFIER METER

measuring small currents because the diode characteristics are non-linear (cf. Fig. 6.2) and are also dependent on temperature. Rectifier meters are therefore mainly used as voltmeters, with a series resistance large enough to swamp any variations in the resistance of the diodes, under which conditions a substantially linear scale can be obtained.

A limitation of the device is that it measures the mean rectified current. What we require to know is usually the r.m.s. current. The instrument can be calibrated on a sine wave to read the r.m.s. value, but this calibration will obviously not hold if the waveform is seriously distorted. In fact, quite a small amount of distortion is sufficient to introduce an error of 5 or 10 per cent, and as distortion is quite common in audio-frequency measurements, particularly during investigation, this inherent disability of the rectifier meter should always be borne in mind.

Current Measurements

The current in a circuit is usually assessed by measuring the voltage drop on a small resistance, known as a *shunt*. This avoids the necessity of passing heavy currents through the meter coils. The shunt must, of course, be capable of carrying the full current, and is designed to provide a suitable (small) voltage drop, which can be measured with an appropriate millivoltmeter, either d.c. or a.c., calibrated in terms of the current.

For large a.c. currents a current transformer is used. These consist of a primary winding of one or two turns only, again designed to carry the full current, with a large number of secondary turns. The primary voltage drop is thus very small and exercises a negligible effect on the circuit while the secondary voltage is large enough to be conveniently measurable.

Hot-wire Instruments

For radio frequencies, hot-wire instruments are sometimes used. These are based on the fact that a fine wire carrying a current will exhibit an appreciable extension of its length due to the heating effect of the current. While this is small, it can be arranged by suitable geometry to provide an indication of the current.

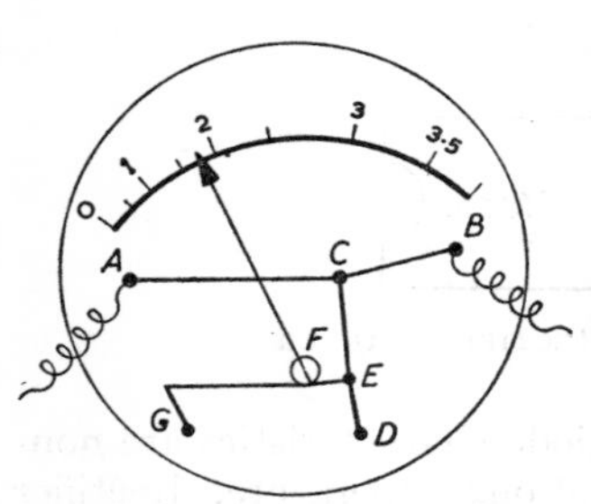

Fig. 15.5. Diagram of Hot-wire Meter

Fig. 15.5 gives a diagram of a hot-wire ammeter. If a current is passed through the wire AB it expands and the point C moves downwards, which causes a sag in CD; this is taken up by the spring G, operating through a thin silk thread GE passing round the drum F, so that the pointer rotates. By passing known currents through the wire AB the movement of the pointer can be calibrated to read the current direct in amperes.

A hot-wire instrument of this type can be made to measure currents from 0·5 ampere up to 10 or more amperes, but such instruments are sluggish in operation and have only a limited accuracy.

Thermocouples

Better results can be obtained by using thermocouples. As said in Section 1.8 (page 69), if the junction of two dissimilar metals (such as iron and constantan, or platinum and platinum-iridium) is heated, an e.m.f. is generated proportional to the temperature. Hence if a thermojunction of this type is constructed in close (thermal) contact with a fine-wire filament or heater, the heat generated by the passage of the current through the heater will produce an e.m.f. in the junction which can be measured by a suitable millivoltmeter.

For relatively large currents (100 mA or more) the wires can be fairly robust and need only be mounted in a suitable housing to protect them from mechanical damage. For smaller currents fine wires are required, and these are mounted in a partially-evacuated

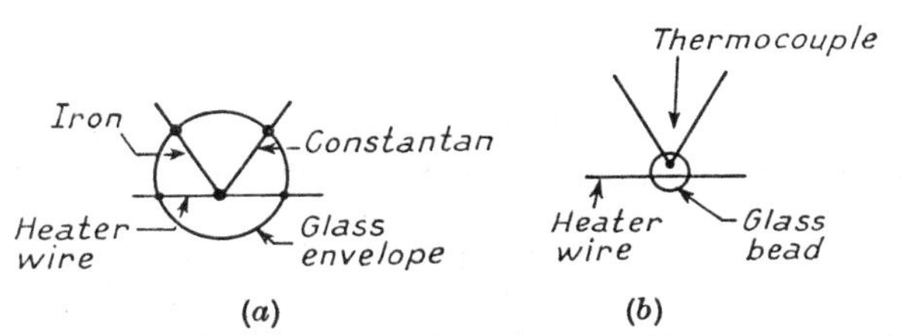

FIG. 15.6. ILLUSTRATING PRINCIPLE OF THERMOCOUPLE

glass bulb. This not only affords mechanical protection but increases the sensitivity. The heat developed in the heater is dissipated partly by radiation and partly by convection. If the bulb is evacuated, the convection loss is eliminated, so that for a given current the wire will attain a higher temperature and hence will produce greater e.m.f. in the thermocouple. Fig. 15.6 illustrates the principle.

Currents as low as 50 microamperes may be measured by this means, the e.m.f. produced at the thermojunction by such a current being of the order of 0·5 microvolt.

The thermojunction is usually welded to the heater wire as in Fig. 15.6 (*a*). However, if it is desirable to isolate the meter circuit, this can be achieved, at the expense of a certain loss of sensitivity, by insulating the junction from the heater, as shown in Fig. 15.6 (*b*).

Instruments are made containing built-in thermocouples. Their disadvantage is that they will carry little overload, and if the couple burns out the whole instrument requires recalibrating.

Electrostatic Voltmeters

For measuring high alternating voltages an electrostatic instrument is usually employed. If a voltage is applied to a capacitor a physical force exists between the plates tending to draw them together. At high voltages this force becomes appreciable, and by suitable arrangement it can be made to produce measurable deflections of a pointer. Two types of instrument are in use. One for moderately high voltages has a series of fixed and moving plates, similar to a variable receiving capacitor but with the plates farther apart. When the voltage is applied the moving plates are attracted into the space between the fixed ones.

For very high voltages two discs are erected on standards and placed some 10 cm apart. The actual pull on the discs is then measured by a mechanism similar to a spring balance.

Electrostatic instruments can be designed to read from about 200 volts up to 20,000 volts or more.

Dynamometer Instruments

For certain classes of work, *dynamometer* instruments are used. These employ an arrangement of two coils, one fixed and one moving. If current is passed through the whole arrangement, the magnetic fields produced interact and set up a force between the coils, causing the moving one to rotate. This movement is indicated by a pointer, and is proportional to the square of the current in the circuit.

Dynamometer instruments are considerably less sensitive than conventional types because of the weaker magnetic fields involved. Hence they absorb appreciable power and are only used for certain special (low-frequency) applications.

Power Measurement

As shown on page 48, the power in a circuit is $EI \cos \phi$. The measurement of power thus requires a device which provides an indication proportional to the product of the instantaneous values of voltage and current. This can be done with a dynamometer instrument in which the current is passed through one set of coils, usually the fixed one, since this can the more easily carry a larger current, while the other set is connected in series with a high resistance across the source of supply so that it measures the voltage. The resultant force is therefore proportional to the product of the current and the voltage, with due regard to their phase relationship.

This is only practicable for relatively large powers in which the consumption of the meter itself can be neglected. The more usual practice for audio frequencies is to measure the current in the circuit. The useful power is then I^2R, where R is the resistive component of the load impedance. (With the average loudspeaker this is practically the same as the ohmic impedance because only a small proportion of the total energy is converted into sound energy.) If the load is resistive the power is also given by E^2/R, where E is the voltage across the load, but if there is an appreciable reactive component, $P = (E^2/R) \cos \phi$, where ϕ is the phase angle.

If a true measurement of power is required, one must use some form of multiplicative circuit which responds to the instantaneous product of two inputs. This can be achieved in a variety of ways, some of which are referred to in Chapter 17 (page 719).

At high frequencies the power is sometimes estimated by measuring the heat generated in the load by means of a calorimeter, but this is cumbersome and is only used in laboratory investigations.

Valve Voltmeters

The range of usefulness of the simple meter can be considerably extended by the addition of suitable amplifying circuitry. One such arrangement, called a *valve voltmeter*, is illustrated in Fig. 15.7.

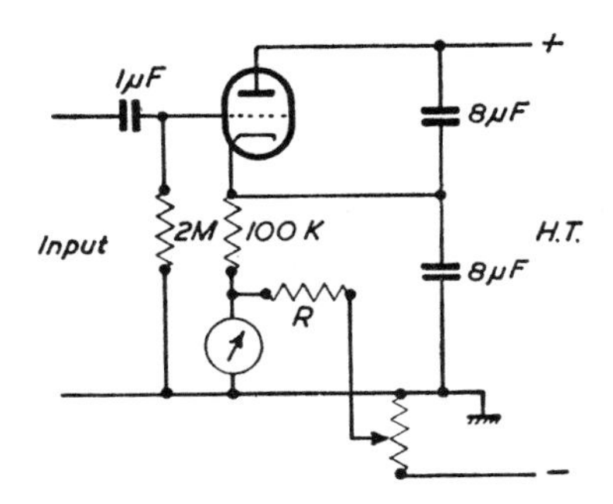

FIG. 15.7. REFLEX VALVE VOLTMETER CIRCUIT

The valve is biased nearly to cut-off and so acts as an anode-bend rectifier so that any signal applied to the input causes the anode current to increase. By deriving the bias from a resistor in the cathode circuit, as shown, substantial negative feedback is provided and the anode current is made proportional to the input signal.

Such an arrangement will handle signals at frequencies well into the megahertz region. To reduce Miller effect a large capacitance is connected from anode to cathode, so preserving a high input impedance. The small zero current through the meter is backed off by a current through a high resistance R from a negative supply.

This simple form of circuit is only occasionally used, having been superseded by the more sophisticated circuits described later, but it serves to illustrate the basic principle. It measures the mean value of the applied signal and is thus subject to "turnover error", since with an asymmetrical waveform the mean values of the two halves of the wave are different.

Diode Voltmeter

Fig. 15.8 illustrates an alternative form of valve voltmeter which measures the peak voltage of a signal. Here a conventional diode circuit is used to rectify the signal and the d.c. voltage across the

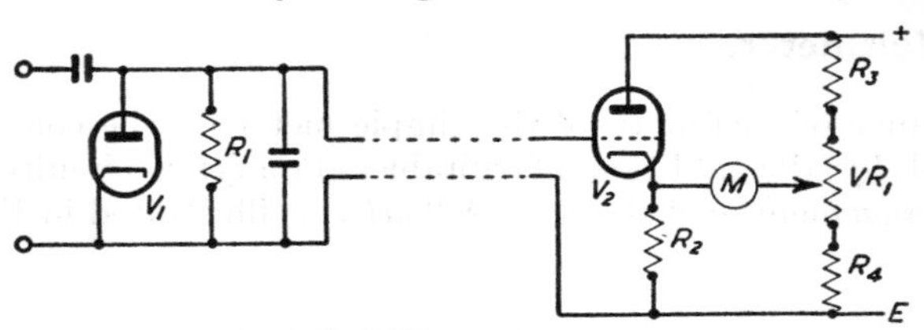

FIG. 15.8. DIODE VOLTMETER CIRCUIT

high load impedance R_1 is converted to a low-impedance signal in the cathode follower V_2.

The valve V_2, its cathode resistor R_2 and the h.t. voltage divider R_3, VR_1, and R_4 constitute a bridge circuit in which the h.t. supply is the bridge voltage source and the microammeter M is the null indicator. The bridge is balanced by means of the zero-setting control VR_1 so that no current flows through the meter in the absence of a signal. The zero setting is then inherently stable against changes in h.t. voltage and the current through the meter is proportional to the a.c. signal applied to the diode. Different ranges of sensitivity can be achieved by varying the microammeter series resistance.

The diode itself, being small, can be accommodated with its circuit elements in a small probe at the end of a flexible cable. This enables the actual terminals to be located quite close to the voltage to be measured, while the instrument and its power supplies, which are necessarily bulky, can be located some feet away.

Solid-state Voltmeters

It is clear that the basic principles used in valve-voltmeter techniques can be utilized with semiconductor circuitry. The diode voltmeter of Fig. 15.8 can be considerably improved by using a semiconductor diode in place of the valve diode; whereas with a valve diode the lowest practicable reading is about 0·1 V, signals as low as 1 mV can be successfully rectified with a semiconductor diode. The rectified current is, of course, correspondingly smaller, so that an amplifier is required, and since d.c. amplifiers cannot easily be constructed with the high stability needed in a measuring instrument, the d.c. signal is usually fed to a "chopper" which delivers an a.c. signal which can be amplified by conventional circuitry.

Such voltmeters can be designed with a sensitivity of 1 mV at frequencies up to 1,000 MHz. They read the peak value of the

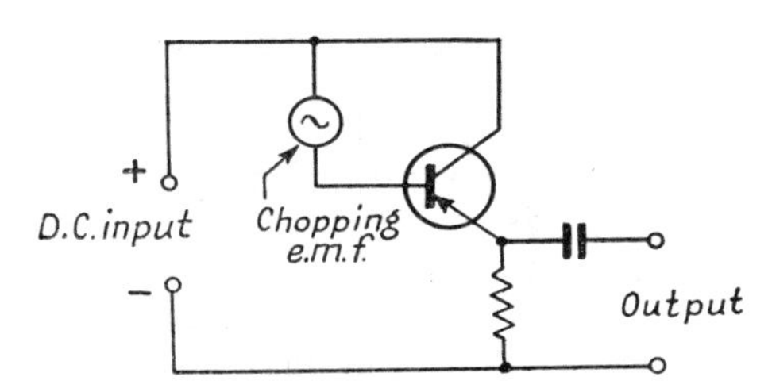

FIG. 15.9. TRANSISTOR CHOPPER

signal but are usually calibrated in r.m.s. terms. There is thus a liability to error if the waveform is not pure, and this can be quite large. The chopper used may be a mechanical device, but is more usually transistor operated. Fig. 15.9 shows a typical circuit which is essentially a common-collector circuit (having a high input impedance), but which is swung between bottoming and cut-off by the chopping e.m.f. between base and collector, thus providing a square-wave output proportional to the d.c. input. The chopping e.m.f. may be obtained from the mains, or from a suitable local oscillator.

Amplifier-rectifier Voltmeters

In the foregoing discussion, instruments have been described which incorporate no amplification before rectification. However, for the measurement of small a.c. voltages, input amplification may usefully be employed, either with an untuned circuit or a tuned circuit where it is required to measure only voltages of a given frequency. The untuned version is of more general application and, by using a number of stages and incorporating negative feedback, instruments are made in which it is possible to measure voltages in the range 100 μV to 1,000 V over the wide frequency range of 10 Hz to 30 MHz.

The amplified signal is usually applied to the meter through a bridge rectifier using semiconductor diodes of the type discussed in

Chapter 6. This gives a linear scale shape, and since it reads the mean value of both halves of the waveform, it eliminates the turn-over error mentioned above.

Transistorized versions of the conventional amplifier-rectifier voltmeter are also used, a block diagram of a typical instrument being shown in Fig. 15.10. A high-impedance input circuit using a field-effect transistor is housed in a probe to minimize the input capacitance and the output is passed at low impedance to an attenuator followed by an amplifier with heavy negative feedback

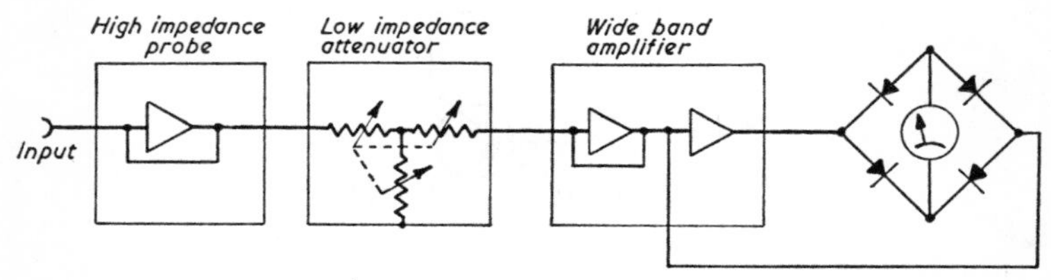

FIG. 15.10. BLOCK DIAGRAM OF ELECTRONIC VOLTMETER

using diffused-base transistors. The rectifier and meter circuitry is included in the feedback loop. Such instruments can read signals as low as 10 μV at 30 MHz. The meter reads the mean current, but is calibrated in r.m.s. input. Alternatively, the bridge rectifier may be replaced by a thermocouple bridge, in which case the meter gives a true r.m.s. reading.

Digital Voltmeters

The reading accuracy of the normal measuring instrument is limited to about 1 per cent of the full-scale reading. However, if the voltage to be measured can be presented in a digital form, considerably greater accuracies can be obtained (better than 0·05 per cent), the limiting factor being the accuracy of analogue/digital conversion. This can be achieved in various ways, one typical method being to generate a linear "ramp" voltage—i.e. a voltage which increases uniformly with time—and compare this with the voltage to be measured. A timing oscillator is connected to a "gate" which is opened when the ramp voltage commences and is closed when the voltage equals the unknown d.c.; during the interval a specific number of cycles of the timing oscillator have been passed to a counter and by suitably co-ordinating the frequency of the timing pulses with the rate of rise of the ramp voltage, the number of pulses shown on the display can be a direct measure of the unknown voltage.

The ramp voltage is generated by a form of integrator circuit similar to that of Fig. 15.22. When the ramp voltage equals the unknown voltage, the capacitor is discharged and the process recommences, but the display on the counter is left set up until the next coincidence is attained. The sampling rate is usually made adjustable, being normally of the order of two or three per second. A decimal point in the display is positioned by the setting of the input attenuator, which can be adjusted manually or, more usually, automatically by suitable sensing circuitry in the instrument.

It is clear that the rate of rise of the ramp voltage must be extremely precise, which requires highly sophisticated circuitry. In many instruments the ramp voltage does not rise continuously but increases in a series of discrete steps, since this makes it easier to achieve the strict proportionality required. The time interval between steps can be made equal to one cycle of the clock oscillator so that no ambiguity results.

Since in addition to the integrating system there are the counting and sensing circuits the equipment is necessarily much more expensive than a normal pointer-type instrument, but this is often justified where continuous monitoring is required with an accuracy of a few parts in 10,000.

15.2. Resistance, Inductance and Capacitance

The second main division of measurement technique is concerned with the basic circuit constants. The (d.c.) resistance of a circuit can be assessed by noting the current produced by a given e.m.f., so that a normal milliammeter can be adapted to measure resistance. For general-purpose testing multi-range meters are used which provide for a range of voltage and current measurements, and also one or more resistance ranges. For these a small internal battery is provided, with a pre-set series resistor which is adjusted to give full-scale reading with the meter terminals short-circuited. Any finite resistance across the terminals will then result in some lesser deflection, and an auxiliary scale is provided calibrated directly in resistance. The scale, of course, is non-linear, having approximately a reciprocal law, but the method is convenient for approximate assessment.

The method is not suitable for measuring high (e.g. insulation) resistance, partly because of the very small currents involved but mainly because insulating materials are normally required to withstand high voltages and should therefore be measured under similar conditions. Special types of instrument are used for the purpose, containing a hand-driven generator producing an e.m.f. of the order of 500 volts in association with a double moving-coil movement, one operated by the voltage across the test sample and the other by the current through it. The coils are so arranged that their combined torque is determined by the ratio V/I, so providing a direct reading of the external resistance.*

The Wheatstone Bridge

Any direct-reading method, however, is of limited accuracy owing to the inherently non-linear scale and for more precise measurement use is made of a bridge technique in which the unknown resistance is compared with a standard resistance of known value. Such an arrangement is the *Wheatstone bridge* illustrated in Fig. 15.11. The

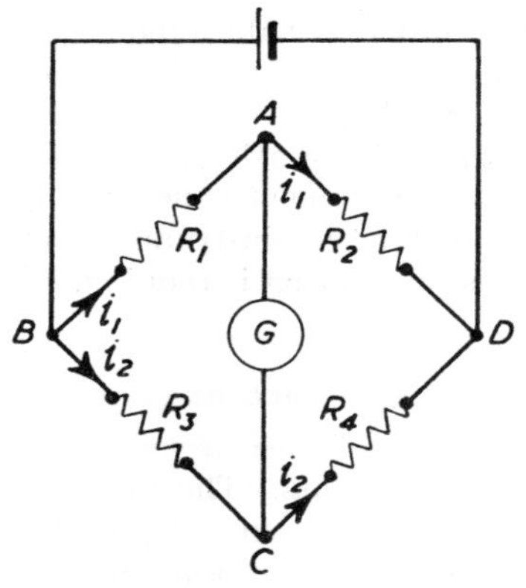

FIG. 15.11. THE WHEATSTONE BRIDGE

current from the battery divides through the arms BAD and BCD. Normally some current will flow across the arm AC but a condition can be found when this current is zero. The bridge is then said to be balanced and a meter connected across these points will thus give a zero reading.

For d.c. measurements this meter is a sensitive moving-coil instrument known as a galvanometer, and it will be clear that for a condition of balance, indicated by zero deflection on the galvanometer,

$$i_1 R_1 = i_2 R_3 \quad \text{and} \quad i_2 R_4 = i_1 R_2$$

Hence

$$\frac{R_1}{R_2} = \frac{R_3}{R_4}$$

* Further details of ohmmeters of various types may be found in *Telecommunication Principles* by R. N. Renton.

Thus an unknown resistance R_1 (say) can be measured if the value of R_2 and the ratio R_3/R_4 are known.

In practice, therefore, R_2 would be a resistance box containing a series of decades such as 10×1, 10×10, 10×100 and 10×1000 ohms, so that any value between 1 and 11,111 ohms can be obtained by suitable setting of the dials. R_3 and R_4, known as the *ratio arms*, could each have values of, say, 1, 10 or 100 ohms permitting the ratio R_3/R_4 to be set in steps of 10:1 between 100:1 and 1:100. The bridge would then measure values of R_1 between 0·01 Ω and 1·1111 MΩ.

Since the current in AC may be in either direction the galvanometer is arranged to have a central zero so that the direction of the deflection will indicate the direction in which R_2 has to be varied to approach balance, while to avoid overloading the meter when the bridge is seriously off balance a variable shunt is provided which is progressively removed as the balance point is approached. In its most sensitive condition the galvanometer will indicate small fractions of a microampere, and for the highest precision special forms of instrument known as *mirror galvanometers* are used in which the pointer is replaced by a small mirror which reflects a spot of light onto a scale.

It will be noted that the condition for balance may be expressed in the form $R_1R_4 = R_2R_3$—i.e. the products of the opposite arms are equal. This is a fundamental relationship which applies to any bridge network.

A.C. Bridges

The Wheatstone bridge relationships obviously apply equally well to networks containing impedances instead of pure resistances, and hence if in Fig. 15.11 we replace R_1 and R_2 with capacitances C_1 and C_2 then, remembering that the capacitive reactance is inversely proportional to the capacitance, the bridge will balance when $C_2/C_1 = R_3/R_4$. Similarly if R_1 is replaced by an unknown inductance and R_2 is a standard inductor, a balance could be expected when $L_1/L_2 = R_3/R_4$. This would only apply, however, provided that the resistances of the inductors were also in the same ratio, which would not normally be the case, and in general for reactance measurement it is necessary to adopt a modified form of bridge which permits independent balances to be obtained on both reactive and resistive components.

It is not practicable here to discuss the many forms of bridge network used for impedance measurements and the reader should refer to *A.C. Bridge Methods* by B. Hague (London, Pitman, 1971).

A typical example is the *Hay bridge* shown in Fig. 15.12, where an unknown inductance is balanced against a standard capacitor (which is often more conveniently obtainable than a standard inductor). Balance is obtained, as in the case of any other bridge, when the products of the opposite arms are equal. Hence

$$(r + j\omega L)(R_3 + 1/j\omega C) = R_1R_2$$

Expanding this and separating the resistive and reactive terms we have

$$R_3 = r/\omega^2 LC$$

and

$$L = CR_1R_2 - rR_3C$$

Now $R_3 = r/\omega^2 LC = 1/QC\omega$ where $Q = \omega L/r$. The expression for L may then be rewritten as

$$L = CR_1R_2/(1 + 1/Q^2)$$

which, if Q is greater than about 10, reduces simply to $L = CR_1R_2$.

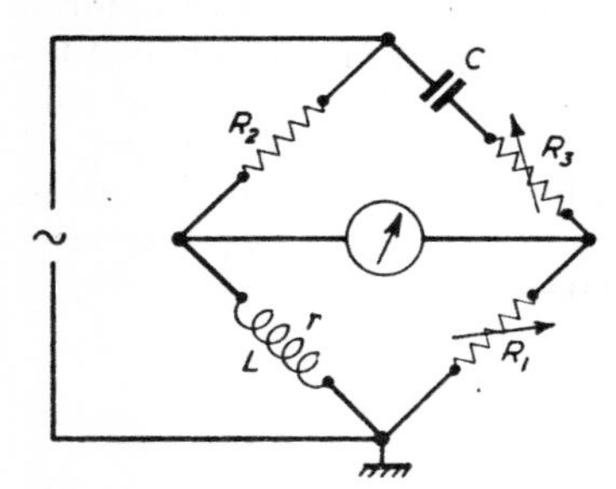

FIG. 15.12. HAY BRIDGE

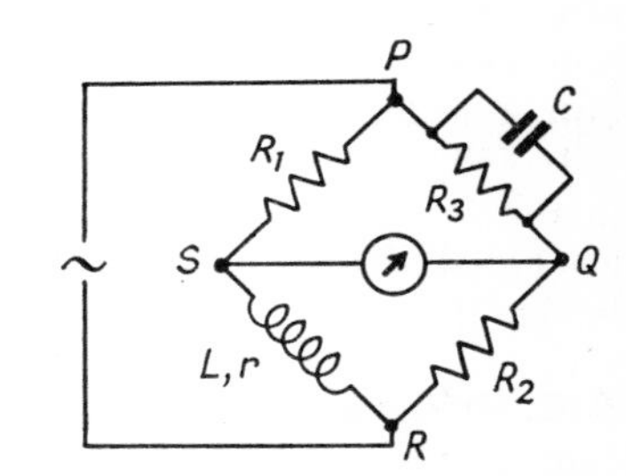

FIG. 15.13. MAXWELL BRIDGE

Both the balance conditions must be satisfied together. It will be noted that they are dependent on frequency. This is normally no problem, since a fixed frequency is being used. However, if instead of a series CR combination in the standard arm, a parallel circuit is used, as in Fig. 15.13, the balance conditions become independent of frequency. For this arrangement, which is called a *Maxwell bridge*,

$$R_1R_2 = (r + j\omega L) \,.\, R_3/(1 + j\omega CR_3)$$

whence, by expanding and separating the real and quadrature terms as before,

$$L = CR_1R_2$$

$$r = R_1R_2/R_3$$

These conditions are independent of frequency because the impedance of the arm PQ is the inverse of the arm RS, the reactance of L being balanced by the inverse reactance of C, the series resistance r being compensated by a parallel resistance R_3 across C.

Similar compensation may be obtained by omitting R_3 in Fig. 15.13 and including a capacitor C_1 in series with R_1. This is known as an *Owen bridge*, the balance conditions being

$$L = CR_1R_2 \qquad \text{and} \qquad r = R_2C/C_1$$

If the inductor is iron-cored it must be remembered that the *incremental inductance* depends on the operating condition of the

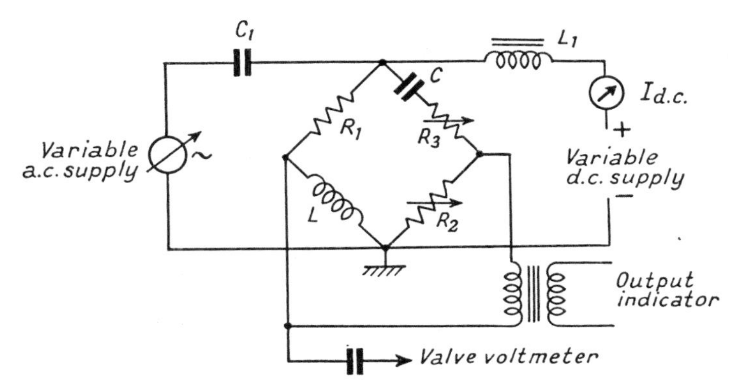

FIG. 15.14. HAY BRIDGE FOR MEASURING IRON-CORED INDUCTORS

core as was explained in Chapter 4 (page 185). It is therefore necessary to adjust the a.c. input to the bridge so that the voltage across the inductor corresponds with that in the circuit in which the inductor is to be used, and also to arrange for a suitable d.c. current to pass since this also affects the incremental permeability. This may be done in various ways, a suitable arrangement being shown in Fig. 15.14. The d.c. polarizing current is supplied through the choke L_1, while the a.c. is fed through the isolating capacitor C_1. The reactance of L_1 must be large compared with the bridge impedance. Otherwise an appreciable proportion of the a.c. feed current passes through the d.c. circuit, causing an unnecessary waste of power.

For the measurement of capacitance, particularly if the phase angle and/or loss is to be determined, the *Schering bridge* illustrated in Fig. 15.15 is used. Here by equating the products of the impedances of the opposite arms, as before,

$$C = C_1R_2/R_1$$

$$r = C_2R_1/C_1$$

These balance conditions are independent of frequency.

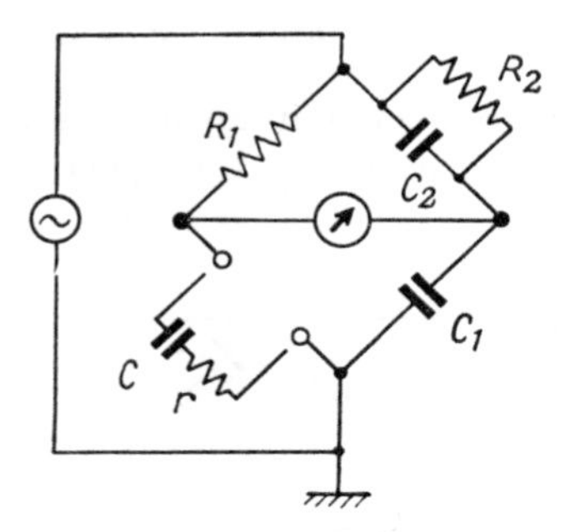

Fig. 15.15. Schering Bridge

15.3. Frequency and Wavelength

One of the most common measurements in radio communication is that of the frequency of an oscillation. This may vary from a few hertz in power and audio-frequency applications up to many megahertz in radio-frequency transmitters. Such measurements may be made with the *Wien bridge* illustrated in Fig. 12.12, but are more usually carried out by a comparison method, the unknown frequency being compared by some suitable means with an oscillation of known frequency. One method of comparison is the heterodyne principle already referred to (page 434). Beats will be obtained between the two frequencies which will be audible if the difference is within the audio-frequency range, and as the two frequencies are brought nearer and nearer to coincidence the frequency of the beat note will fall continuously until it disappears when the two frequencies are identical.

The actual method depends on circumstances. In some cases it is the unknown frequency which is adjusted to coincidence with the standard frequency, while in other cases the standard source is made capable of variation over a suitable range so that it can be adjusted to be equal to the unknown frequency. It should be noted also that beats can be obtained between harmonics of one or other of the frequencies, a method which is frequently convenient, while other methods of frequency comparison may be used, as for example by using the Lissajous figures obtainable with cathode-ray tubes (see page 677).

A more fundamental method is to count the number of cycles over a given period. Various forms of electronic counter circuits exist, some of which are discussed in more detail in Chapters 17 and 18. If the input to such a counter is "gated" so that it remains operative for exactly one second, the total count indicates the number of cycles per second—i.e. the frequency. The result can be displayed on suitable indicator tubes as described in Chapter 6, so providing a visible record of the frequency in question.

Alternatively, though less accurately, the pulses can be integrated, and the output used to charge a capacitor which is being discharged at a predetermined rate by suitable control of the time constants. The net voltage on the capacitor will be a function of the input frequency and may be calibrated accordingly.

The provision of an accurate standard for comparison methods (and/or a known time-interval for integration methods) is usually accomplished with a quartz crystal oscillator of the type discussed in Chapter 13. By setting up a crystal oscillator and arranging by some convenient method to count the number of oscillations over a long period of time, it is possible to determine the frequency of the oscillations very exactly, the standard in this case being astronomical time deduced from the movement of heavenly bodies. Of recent years, however, it has been found that the period of the earth's rotation on which our conventional notions of time are based is not sufficiently constant for modern requirements, and standards have been evolved depending upon atomic or molecular structure. Under certain conditions absorption of energy can occur at certain calculable and extremely precise frequencies and this can be used as a primary standard of frequency.

Wavelength

For measurement of the wavelength of a given circuit, an instrument known as a *wavemeter* is employed. This simply consists of a circuit which is tuned to resonance with the circuit under test. The measurement is thus really a check on the frequency to which the circuit is tuned and the wavelength is deduced from this.

The wavemeter may be arranged to generate oscillations which are received and detected in the circuit under test. Alternatively the test circuit may be arranged to oscillate, the oscillations being detected on the wavemeter. In either case the wavemeter tune is altered until a maximum response is obtained, which indicates that the two circuits are in tune.

In a transmitting plant there is plenty of spare energy, and the currents set up in the wavemeter may be made to light a lamp. For receiving purposes, however, it is necessary to detect the absorption of energy, not on the wavemeter, but on the receiving circuit.

For example, with an oscillator the d.c. feed current depends upon the amplitude of the oscillation. Hence, if a milliameter is inserted in the anode (or collector) circuit, it will be found that the absorption of energy by the wavemeter as it is tuned to resonate with the oscillator will produce a change in the milliammeter reading. The wavemeter should gradually be moved farther and farther away, tuning it slowly over the point where the absorption takes place, until the absorption only occurs over one degree or so on the wavemeter dial, when it can be said with reasonable accuracy that the wavelengths of the oscillating circuit and the wavemeter are both the same.

None of these methods is really a measurement of wavelength. They are all frequency measurements, but an actual measurement of wavelength is very rarely carried out except at short wavelengths, where a Lecher wire may be used as described in Chapter 13.

15.4. R.F. Measurements

Measurements are sometimes required to be made at radio frequencies, as for example the determination of the power-factor of a capacitor. This can be done with the Schering bridge of Fig. 15.15 provided that due allowance is made for the effect of the inevitable stray capacitances. The same proviso is necessary for any form of r.f. bridge and special techniques have to be adopted which are beyond the present scope.

There are, however, certain commonly used types of r.f. measurement which should be reviewed briefly.

High-frequency Resistance

It is sometimes desired to measure the high-frequency resistance of a given circuit. There are various methods of doing this, but one of the most reliable is the resistance variation method. Briefly the method is this.

The circuit under test is supplied with r.f. current from a convenient source and contains a sensitive thermocouple so that the actual current in the circuit can be measured. The circuit is tuned to maximum current. A resistance is now inserted and the current is again noted. Now, provided that the voltage induced in the circuit under the two conditions is the same,

$$E = IR = I_1(R + R_1)$$

where I = original current in the circuit,

I_1 = current in circuit after the introduction of resistance R_1,

R = resistance of circuit.

Hence
$$R = R_1 \frac{I_1}{I - I_1}$$

Fig. 15.16 gives the circuit arrangements. It is essential that the configuration of the circuit shall not be altered *in any way* during

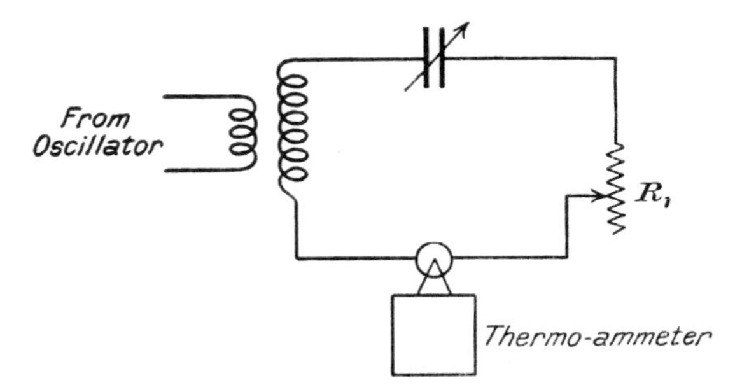

FIG. 15.16. CIRCUIT FOR H.F. RESISTANCE MEASUREMENT

the test. Stray radiation from the oscillator induces voltages in the circuit, and if the layout is appreciably altered the voltage will not be the same throughout the test.

In any case the test is repeated several times with different values of added resistance, and it is convenient to plot the reciprocal of the current against the added resistance. A graph is then obtained similar to that shown in Fig. 15.18, $1/I$ being plotted horizontally and added resistance vertically. The points should lie on a straight line and the negative intercept (corresponding to the self-capacitance in Fig. 15.18) gives the h.f. resistance of the circuit itself. If the points do not lie on a straight line, errors are being introduced and the circuit must be laid out more carefully and perhaps removed farther from the oscillator. The coupling from oscillator to test circuit may be too strong.

The resistance obtained is that of the whole circuit, including the leads and the capacitance. If one is interested mainly in the coil the leads must be kept short and a high-grade capacitor employed of a capacitance such that it is used towards the maximum position.*

Reactance Variation Method

An alternative method, useful if a calibrated capacitor is available, is to tune the circuit to resonance and then mistune it until the current is $1/\sqrt{2}$ times the peak value. This is done above and below resonance, and the capacitances C_1 and C_2 are noted. Now, by the usual circuit laws, if $I_1 = I_o/\sqrt{2}$ the resistance and reactance are equal (numerically).

When C is increased to C_1,

$$X = L\omega - 1/C_1\omega$$

When C is reduced to C_2,

$$X = 1/C_2\omega - L\omega$$

$\therefore$
$$2X = 1/C_2\omega - 1/C_1\omega$$

whence
$$X = R = (C_1 - C_2)/2\omega C_1 C_2$$

Measurement of Q

Direct measurement of the h.f. resistance, however, is both cumbersome and difficult and the more usual practice is to measure the Q of the circuit. This is the circuit magnification $= \omega L/R$. The obvious method is to inject a known voltage into the circuit and measure the voltage developed across the coil when the circuit is tuned to resonance. The ratio of output to input voltage is the figure required.

There are practical difficulties in this technique. The only convenient way of injecting the voltage is by inserting a small resistance in series as shown in Fig. 15.17. To avoid disturbing the circuit under test this resistance must be a small fraction of an ohm, and to develop a reasonable voltage across this a large current is required. But unless the oscillator and the leads to the injection resistance are carefully shielded, voltage will be induced into the circuit by direct induction which will invalidate the results.

Instruments known as Q-meters are made in which these matters are attended to. The valve voltmeter is built-in and calibrated

* So that the dielectric shall be mainly air. As the capacitance of the usual parallel-plate type of variable capacitor is reduced, an increasing proportion of the effective dielectric is occupied by the solid insulating supports, so that the relative dielectric loss increases.

directly in Q, and reliable measurements can be made up to frequencies of the order of 50 MHz.

If such an instrument is not available it is preferable to use an indirect method. The circuit under test is energized by any convenient means and the voltage developed at resonance is measured with a valve voltmeter. The circuit is then mistuned until the reading on the voltmeter falls to 0·71 times its resonant value.

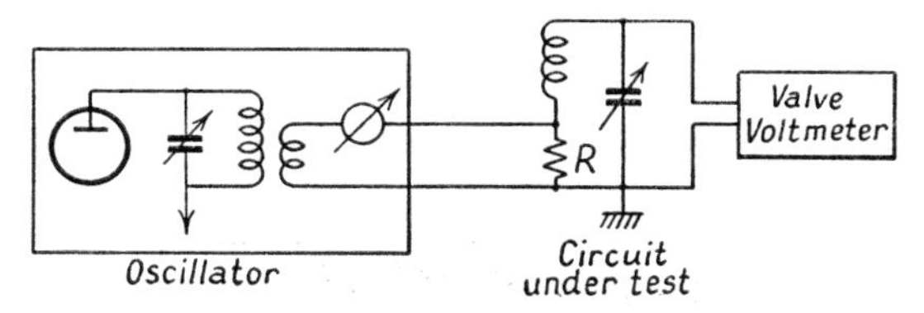

FIG. 15.17. CIRCUIT FOR MEASURING Q

The mistuning may be carried out by varying the frequency of the injected signal. Then, if f_1 and f_2 are the frequencies, on either side of the resonant frequency f_0, at which the output is 0·71 times the maximum,

$$Q = f_0/(f_1 - f_2)$$

(This is another way of expressing the relations quoted on page 96.)

Alternatively the circuit may be mistuned by altering the capacitance. In this case, if C_1 and C_2 are the capacitances at which the voltage is 0·71 times the maximum, and C_0 is the resonant value,

$$Q = 2C_0/(C_1 - C_2)$$

Self-capacitance

By a development of the foregoing method it is possible to determine the self-capacitance of a coil at radio frequency. To do this the coil in question is connected in parallel with a calibrated capacitor, and the wavelength measured for various values of parallel capacitance. The wavelength is proportional to the square root of the capacitance, so that if the capacitance is plotted against the wavelength squared, the various points will all be found to lie on a straight line.

If the coil had no self-capacitance, this straight line would pass through the zero point, but due to the self-capacitance the line cuts the zero wavelength axis at a point below the zero capacitance mark. The length of the negative intercept determines the value of the self-capacitance of the coil to the same capacitance scale as was used in plotting the curve. The method is illustrated in Fig. 15.18.

If it is more convenient to measure the frequency the same technique can be used by plotting the *reciprocal* of f^2 against the capacitance.

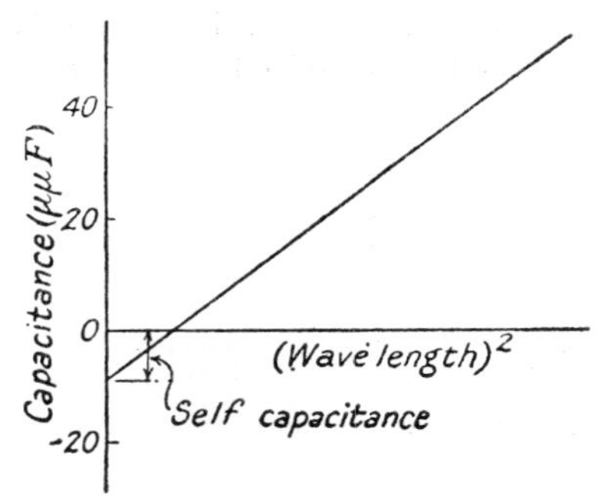

FIG. 15.18. ILLUSTRATING ESTIMATION OF SELF-CAPACITANCE

15.5. Performance Tests

In addition to measurements of individual components it is necessary to assess the performance of a circuit, to determine how closely it meets the design requirements in respect of gain, selectivity, fidelity, etc. For this purpose special assemblies are built up using equipment of the type already discussed. The subject is one about which a great deal can be (and has been) written and it is only possible here to indicate the basic methods in common use.

Gain Measurement

A type of measurement often required is that of the step-up or gain of a stage. The principle adopted here is usually the same whether at high or low frequencies. The circuit is supplied with e.m.f. at the required frequency and the voltages developed at appropriate points in the circuit (e.g. the input and output) are measured with an electronic voltmeter. Since stage gains are often more than 100, the voltmeter will need to be a multi-range instrument.

Alternatively the valve voltmeter or other measuring device may be left connected to the output of the amplifier and the input transferred from one point to another. Thus in Fig. 15.19 the input is first connected to the point A and the output reading noted. The input is then transferred to B and increased until the same output is obtained. The increase in input required is the gain of the valve V_1.

As in any measurement, care must be taken during the test that the introduction of the voltage at any point does not alter the operation of the circuit seriously, and that the connection of the voltage measuring device (e.g. an electronic voltmeter) across the output or input does not in itself affect the circuit at the particular frequency at which the measurement is being carried out.

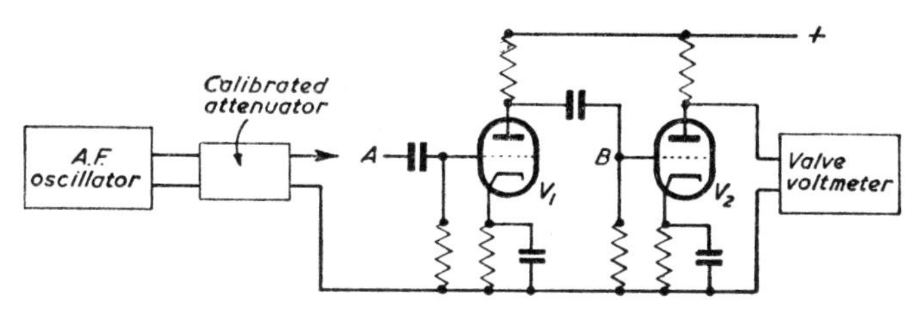

FIG. 15.19. TYPICAL SET-UP FOR MEASURING STAGE GAIN

Signal Sources

An essential requirement in such testing is a generator capable of delivering an input of known and constant amplitude over the range of frequencies to be investigated. For audio-frequency tests this involves frequencies within the range 10 to 20,000 Hz, for which conventional *LC* oscillators are unsuitable owing to the large values of inductance which would be required. It is usual to employ *RC* oscillators of the phase-shift or Wien-bridge type, as described in Chapter 12, which with suitable precautions can be made to deliver a constant and stable output over the range required.

The range of frequency variation, however, is limited to about 10:1 and is not linear. The frequency is proportional to $1/CR$, so that if C, or more usually R, is varied the frequency scale follows a reciprocal law. The linearity may be improved by using a graded resistor in which R is proportional to $1/\theta$, but this usually limits the overall frequency variation obtainable. Whichever method is used, the full a.f. range can only be covered in a series of steps, the circuit values being appropriately changed from range to range.*

Alternatively the whole range may be covered in a single sweep by using a *beat-frequency oscillator*. This employs two r.f. oscillators (usually of the conventional *LC* type) one of which is fixed and the other variable over a small range, the outputs being mixed in a conventional frequency changer. Thus if f_1 is 100 kHz and f_2 is variable

* This only applies to the simple circuits described in Chapter 12. It is possible to use more elaborate RC networks which permit a considerably extended range of frequency to be covered in a single sweep, but these are not in common use.

from 100 to 120 kHz, the output from the mixer will vary over a range of 20 kHz in a single sweep. The system has two limitations. Firstly, if the oscillators are too close in frequency they will pull into step. This causes the beat-frequency output to become subject to increasing distortion below about 25 Hz and to cease altogether at about 10 Hz. The other limitation is that of drift. A quite small variation in the frequency of either of the r.f. oscillators will produce a large change in the beat-frequency output, particularly at the low frequencies. This is minimized by making the oscillators as stable as possible, and arranging that they shall drift in the same direction.

Transient Response

Faithful reproduction requires an adequate response not only to the fundamental components of the waveform, but also to the important harmonic components. This may often be more conveniently assessed by feeding the input with a step-function or square wave, and observing the output on an oscilloscope, as described later. Any shortcomings in the transient response will then be shown up as a departure of the output waveform from the true square wave. For this purpose many a.f. oscillators incorporate a shaping circuit, such as the Schmitt trigger (cf. Fig. 17.14), which converts the initial sine wave into a square wave, again of known amplitude and frequency, but now in this more convenient form for transient testing.

Receiver Measurement

A development of this simple gain measurement is the complete measurement of the performance of a receiver. Here a small dummy transmitter is employed known as a *signal generator*. It usually consists of an oscillator followed by a modulating stage (the modulation usually being employed on an amplifying stage instead of on the oscillator itself to avoid frequency modulation as explained in Chapter 7) and a radio-frequency attenuator. This provides a source of modulated voltage which can be varied in output from about a tenth of a volt down to a few microvolts. It is obvious that the most careful shielding must be adopted to ensure that energy can only be obtained from the generator via the output terminals, where it is under proper control, and not by any direct radiation which would, of course, invalidate the measurement. In any case, it is always desirable to keep the receiver under test several feet away from the generator, to be on the safe side.

Receiver measurements are carried out by injecting into the aerial terminal a signal having a standard modulation (usually 30 per cent at 400 Hz), and adjusting the input until a given a.f. output is obtained (usually 50 mW). The power output is measured by a suitable (thermal) meter in series with a resistance equal to the optimum load for the particular output stage in use.

This measurement is carried out at various frequencies within the range of the receiver and the sensitivity curve obtained. Fig. 15.20 shows a typical layout.

Selectivity is measured by mistuning the oscillator (leaving the receiver unaltered) and increasing the input until the receiver again gives 50 mW output. This can be done for a series of frequencies gradually getting farther away from the tune of the receiver, and a selectivity curve plotted. It will, in fact, be an inverted resonance curve.

Fidelity measurements are taken by tuning the receiver to some

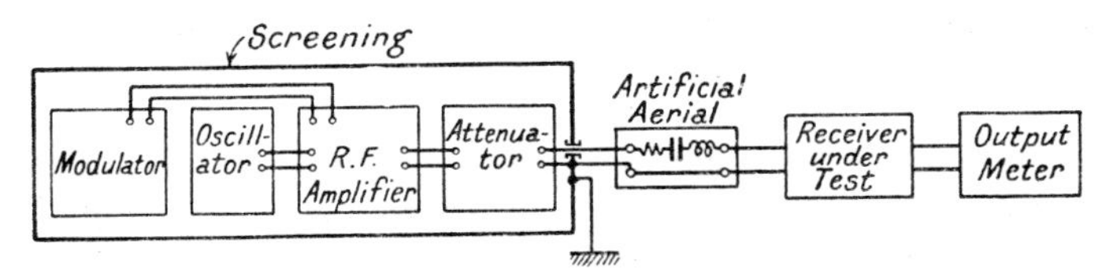

Fig. 15.20. Set-up for Measuring Receiver Performance

suitable radio frequency and then varying the frequency of the modulation, keeping the percentage constant. The output power is then noted. If the receiver were perfectly faithful, of course, the output power would be the same irrespective of the modulation frequency. In practice, this is very far from being the case.

These are the standard measurements. Numerous other measurements can be made. 50 mW, for example, does not represent full power output by any means, and one can examine the behaviour of the receiver as the input is gradually increased, or the relative output with constant input but differing depths of modulation. Special tests are also made using two signal generators to determine the degree of cross-modulation present (see page 421). One generator is adjusted to give normal output and its modulation then switched off. The other generator is then mistuned by a certain amount and modulated in the normal manner. By cross-modulation, this modulates the wanted carrier and the input of the interfering signal is adjusted until normal output from the wanted carrier is obtained. This is then repeated for varying amounts off tune, the interfering signal input becoming progressively larger as it becomes more and more mistuned.

Modulation Measurement

An important factor in these measurements is the estimation of the modulation, and various methods are employed here. Where it is not possible to obtain access to the low-frequency modulating voltage, the modulation depth may be estimated by noting the change in the operating current when the circuit is modulated. Although the mean value of a modulated current is the same as the normal carrier wave, the r.m.s. value increases, the rise being approximately 23 per cent with full (sine-wave) modulation. Actually, the depth of modulation is given by the expression

$$M = \sqrt{\left[2\left(\frac{I_2^2 - I_1^2}{I_1^2}\right)\right]} \times 100 \text{ per cent}$$

where I_1 is the unmodulated current, and I_2 is the modulated current.

It may be noted that this method can be applied to measure the effective modulation depth of a transmitter.

In a signal generator, however, it is usually practicable to measure the actual audio-frequency voltage across the modulating choke, and the peak a.f. voltage divided by the H.T. voltage gives the percentage modulation rather more accurately than can be deduced from the current rise.

Still another method is to apply the modulated signal to a detector valve and measure the change in d.c. anode current and the alternating anode current. The latter may be estimated by measuring the modulation-frequency voltage across a resistance in the anode circuit. The a.c. component divided by the change in d.c. current gives the modulation depth.

The diode circuit of Fig. 9.27 can also be used by inserting a microammeter in series with R_1, and also measuring the a.f. voltage across R_1 with a valve voltmeter (an r.f. choke being inserted in the voltmeter lead to keep out the carrier).

Then $$M = \frac{(\sqrt{2})\beta V_{af}}{I_R R_1} \times 100 \text{ per cent}$$

where β is the rectification efficiency of the diode.

A visual check on modulation depth can be made with an oscilloscope such as is described in the following section. This enables the modulated waveform to be displayed, and the heights of the peaks and the troughs can be measured. If these heights are a and b respectively, the modulation depth is $(a - b)/(a + b)$. It is also possible, by a modified form of connection, to examine not only the extent but also the linearity of the modulation, as is described on page 678.

Acoustic Measurements

The loudness of any tone is a function not only of its intensity but also of its frequency, and a frequency of 1,000 Hz is taken as the standard. To estimate the loudness of any tone it is therefore first compared with a 1,000 Hz tone which is adjusted to give the same apparent loudness. The intensity is then stated in *phons* which is the intensity level (of the 1,000 Hz tone) expressed in decibels above a standard reference level of 10^{-16} watt per cm^2.

A logarithmic unit is used because relative sound levels are dependent on the ratio of the sound pressures, and in fact the least variation in intensity which can be detected by the ear is produced by a change of 1 dB in level.

For purposes of noise measurement a loudness unit has been standardized, this being an empirical unit derived from subjective observations by Fletcher and Munson. The scale is based on the principle that doubling the number of loudness units is equivalent to a sensation to the average ear of twice the loudness.

Sound in radio communication includes wanted sound—speech and music—and unwanted sound or noise. Sound level measurements may be carried out by a subjective method, by estimating the equivalent loudness of sounds by ear, but an objective method is more usually employed.

A typical *sound level meter* consists of a microphone, amplifier, calibrated attenuator and indicating meter. The microphone is generally designed to be non-directional and to have a flat frequency response over the audio range of 20 Hz to 20 kHz, while the frequency response of the amplifier is controlled by negative feedback to have a flat characteristic over the same range. However, the readings of an instrument such as described above would bear little relationship to true subjective aural effects, since the average ear has neither a flat frequency response nor a linear amplitude response. Consequently sound level meters are provided with weighting networks so that various response characteristics can be obtained. Thus the instrument can be calibrated in phons or loudness units or in accordance with various "equal loudness" contours which have been subjectively determined.

Noise and interference on telephone and broadcast circuits are measured with a *psophometer*. The psophometer is essentially similar to the sound level meter but since it is used to measure electrical noise, no microphone is used; it consists only of a calibrated valve voltmeter which contains networks which weight every frequency in accordance with its interference relative to that at 800 c/s. Usually the instrument can be used either with a flat response or with one of two internationally agreed weighting networks. One of these networks is used for measurements on commercial telephone circuits, and the other with circuits used for the transmission of broadcast music.

15.6. The Cathode-ray Oscilloscope

In addition to the generators and measuring devices so far discussed, an essential instrument in the modern laboratory is the *cathode-ray oscilloscope*. This is a device which permits the visual examination of the performance of a circuit under working conditions. It is based on an electronic structure known as a cathode-ray tube

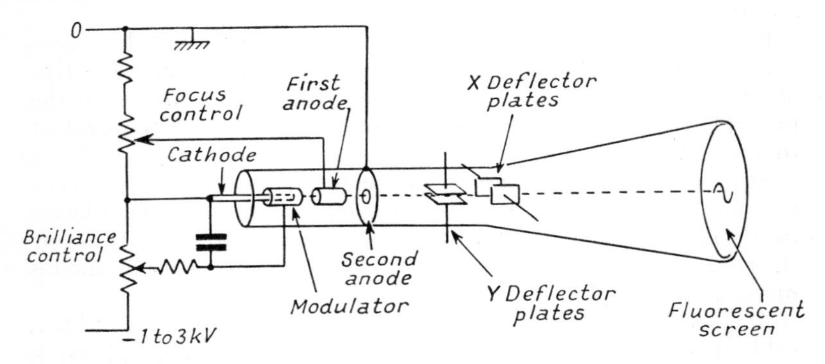

FIG. 15.21. BASIC CIRCUIT OF CATHODE-RAY OSCILLOSCOPE

which consists, in essence, of a cathode emitting electrons, and an anode in the form of a plate with a hole in the centre which is maintained at a steady positive potential. Some of the electrons emitted from the cathode shoot through the hole in the anode in the form of a pencil or stream of electrons which is allowed to impinge on a fluorescent screen at the end of the bulb. The point where the electron beam strikes the screen is shown by the appearance of a small spot of light, usually of a greenish or bluish colour depending upon the material of the screen.

In order to concentrate the electrons in the beam a control cylinder is placed round the cathode, this being connected to a negative potential rather like the grid of an ordinary triode. This also enables the number of electrons in the beam to be varied, so that the intensity of the spot can be varied, if required. For this reason this cylinder is often called the *modulator*. An auxiliary anode is also provided prior to the main anode which, in conjunction with the modulator and main anode, produces an electric field distribution which helps to produce a sharp focus of the beam at the screen, this focusing action being largely independent of the intensity control produced by the modulator. The whole assembly thus acts as an electron lens, and is termed an *electron gun*.

The same result can be achieved magnetically by an axial magnetic field produced by a focusing coil round the neck of the tube. This has no effect on electrons which are pursuing a strictly axial path but deflects any divergent electrons into a convergent spiral so that again they all meet at a sharp focus. This type of focusing is usually adopted with the television tubes described later.

The electron beam may be deflected by electrostatic or magnetic means. By placing a pair of plates on either side of the beam and applying to these plates suitable voltages, the spot of light may be deflected from side to side. A similar pair of plates set at right angles will cause a deflection in a vertical direction, and between the two a composite motion can be obtained which is very useful for analysing waveforms and similar phenomena. Fig. 15.21 shows the essential parts of a cathode-ray oscilloscope.

Waveform Analysis

For example, the application of an alternating voltage to the vertical deflecting plates will cause the spot of light to lengthen out into a thin vertical line. If, at the same time, a voltage of lower frequency is applied to the horizontal plates, this vertical line will be spread out so that a waveform will be traced on the screen.

The horizontal deflecting voltage must obviously be of a special form. If the voltage slowly and regularly increases until the spot has been deflected over the full width of the screen, and then rapidly returns to zero and commences the process again, it is possible to cause the successive traces of the wave to lie one on top of the other so that a stationary picture of the actual waveform is obtained. The only conditions required are that the time taken for the spot to travel across the screen shall be an exact sub-multiple of the waveform period, and this is easily arranged.

Time Bases

The circuit which produces the horizontal deflection is called a *time base*. As said, it is required to provide a linearly-rising voltage (or current) followed by a sharp return to zero after a given level has been reached, as shown in Fig. 15.22 (*a*). This is known as a sawtooth waveform, and can be produced in a variety of ways, the simplest being as in Fig. 15.22 (*b*). Here a capacitor C is charged through a high resistance R, producing a rising voltage which is applied, either directly or through an amplifier, to the X plates of the oscilloscope.

Across the capacitor is a discharge device D which is normally non-conducting but which rapidly passes to a conducting state when the voltage across it reaches a predetermined value. When this occurs a rapid *flyback* is produced and the cycle recommences.

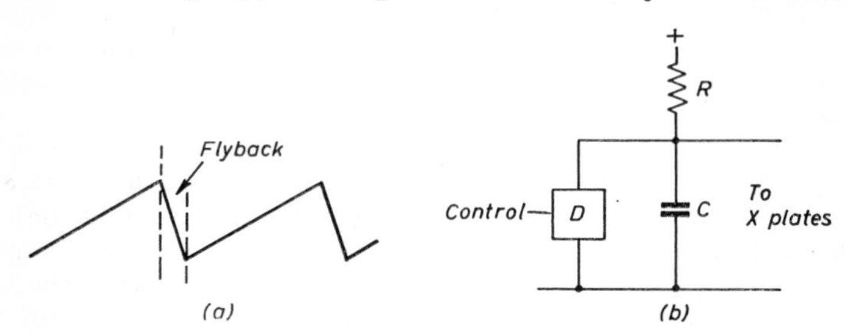

FIG. 15.22. BASIC TIME-BASE CIRCUIT

The time of the cycle is controlled by the product CR, which determines the time required for the voltage to rise to the chosen value. It is customary to make R continuously variable in association with a series of switched values of C to cover a wide range of frequency.

It only remains to ensure that the time-base frequency is exactly related to that of the signal under examination. This synchronization is achieved by applying a small fraction of the signal voltage to the control electrode of the discharge device, which triggers the discharge at the correct instant.

The discharge device may be a gas-filled triode of the type described in Chapter 5 (page 251) or its semiconductor equivalent, the discharge voltage being determined by the bias on the control electrode.

Practical forms of time base incorporate various modifications of this basic circuit. A disadvantage of the arrangement of Fig. 15.22 is that the rise of voltage on the capacitor is not linear, but is exponential in form, as was explained on page 19. As long as the sweep is not more than about $0{\cdot}1E$ the departure from linearity is only about 5 per cent, but this clearly imposes limitations in the design. One solution is to replace the resistor with a constant-current device such as a pentode or transistor, but it is better to use a modified form of circuit known as a *Miller integrator*, as discussed in Chapter 17.

A further limitation is that the simple discharge device described above will only function satisfactorily at relatively low frequencies, so that for high-frequency investigations it is necessary to use more sophisticated forms of circuit, usually based on multi-vibrator techniques which can provide a very rapid transition from one state to another; and these circuits can also provide a more precise form of synchronization which can be arranged to operate at variable points in the waveform, and not just at one end.

A third consideration is that for many applications it is convenient to use time bases which are not "free-running" but are driven by specific and controllable input pulses.

These refinements do not affect the basic principles of waveform analysis and will therefore not be pursued at this point. They fall within the category of wave-shaping techniques and are more conveniently discussed in Chapter 17.

Astigmatism

The deflection of the spot at the screen depends upon the axial length of the tube and also upon the gun voltage (as will be shown shortly). Hence if the deflector voltage is an appreciable fraction of the gun voltage, the sensitivity of the tube will vary as the deflector voltage changes, which will cause distortion of the image.

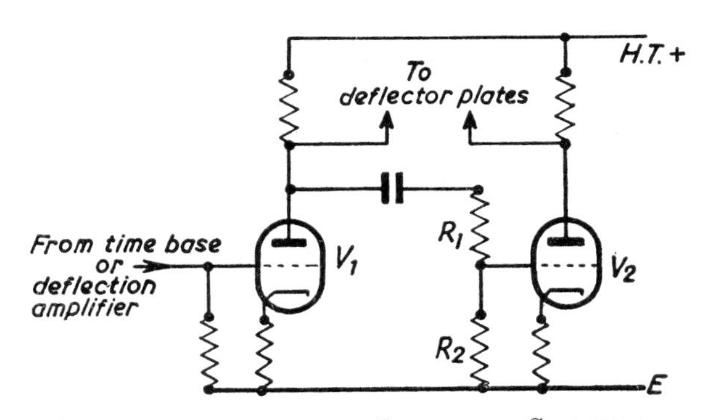

FIG. 15.23. PUSH-PULL DEFLECTION CIRCUIT

It will also cause a certain defocusing of the spot. These defects are called *astigmatism* and may be minimized by using a push-pull deflection system so that when one plate is being driven negative the other is driven positive. A typical push-pull output stage is shown schematically in Fig. 15.23.

The output of the time base or deflection amplifier is applied to the grid of V_1 and a small proportion of the amplified waveform at the anode of V_1 is fed to the grid of V_2. R_1 and R_2 are so arranged that the waveform at V_2 anode has the same amplitude but is 180° out of phase with the waveform at V_1 anode.

In many instances an individual fine adjustment of the potentials is provided to correct any residual astigmatism.

Magnetic Deflection

It is possible to deflect the beam by applying a magnetic field at right angles to the beam. The direction of deflection of the beam is then at right angles to the direction of the magnetic field, and by arranging two pairs of coils at right angles to one another we can produce the desired deflections exactly as in the electrostatic case. The disadvantage of this method of working is that power is consumed by the coils.

There are, however, compensating advantages in using magnetic deflection, the most important of these being that the spot can be deflected through a wide angle without the defocusing encountered when using electrostatic deflection. Consequently, the screen area can be made large without the tube being excessively long. In general all tubes with a screen of more than 6 in. diameter are designed for magnetic deflection, and deflection angles of more than 45° from the straight line have been achieved. This form of deflection is used in all television tubes, and tubes having a total deflection angle of 110° are now being used.

The possibility of deflecting the beam accidentally must always be borne in mind, because stray magnetic fields from nearby apparatus (including the power supply to the oscilloscope itself) may distort the traces or the screen. To avoid this interference it is customary to enclose the tube in a cylinder of Mumetal, which provides a magnetic shield and prevents stray fields from influencing the electron beam.

Deflection Sensitivity

In an electrostatically-deflected tube the deflection at the screen is determined primarily by the geometry of the tube and is not affected by the speed of the electrons in the beam *as such*. However, the deflector plates have a definite length, which means that the electrons in the beam remain in the deflecting field for a finite time that is inversely dependent on their speed. Since the speed is directly proportional to the gun voltage, V, the deflection sensitivity is inversely proportional to V. As already mentioned, the actual deflection is dependent on the tube construction, and the manufacturers quote figures for the particular types, in the form $S = k/V$ mm/volt.

With magnetic deflection the displacement of the spot *is* dependent on the speed of the electrons. The electrons are accelerated during the period when the beam is in the magnetic field and the deflection is $\frac{1}{2}\alpha t^2$, where α is the acceleration. The time t is inversely proportional to V, as before, but since the deflection is proportional to t^2, the deflection is inversely proportional to $\sqrt{V}$. Hence with magnetic deflection the sensitivity is of the form $S = k/\sqrt{V}$.

Measuring Oscilloscopes

From the foregoing principles it is apparent that, by using deflection amplifiers of known gain and linear time bases of known sweep speeds, the oscilloscope can be used to perform a wide variety of measuring functions. The measuring function is generally carried out in one of two ways. If a graticule is used in front of the tube face, the oscilloscope can be calibrated directly in volts applied to the input to produce 1 cm deflection and the time base can be similarly calibrated in sweep-time/cm. Alternatively known d.c. voltages can be applied (indirectly) to the tube and calibration effected by comparing the deflection so produced with the signal being measured. Measuring oscilloscopes are made to perform a wide range of specialized functions, and voltages as low as 100 μV can be measured while the frequencies covered range from d.c. to 50 MHz and higher.

Special Types of Oscilloscope

The design of oscilloscopes has been brought to a fine art, and it is probably the most universally used tool in communication engineering. Not only have there been developments in the construction of the tubes themselves, but highly sophisticated time-base and amplifier circuitry has been evolved to permit the examination of phenomena over a wide range of frequency, from d.c. to 50 MHz and above. Techniques have also been devised for the examination of the many forms of pulsed waveform now in common use, many of which have rise times measured in nanoseconds (10^{-9} sec). Such circuitry is really ancillary to the tube and is discussed in more detail in Chapter 17, but there are certain developments in tube construction which may be noted briefly.

P.D.A. Tubes

As just said, the deflection sensitivity of an oscillography tube is inversely proportional to the gun voltage. On the other hand, the brilliance of the trace is directly influenced by the gun potential, and if the screen is to be adequately actuated with very rapidly moving traces, gun potentials as high as 10kV may be required.

The requirements of high sensitivity and adequate brilliance are thus mutually conflicting, to mitigate which many tubes use *post-deflection acceleration* (p.d.a.). Here the normal electron gun is operated at a relatively low potential so that a reasonable sensitivity is obtained. Beyond the deflector plates, however, a third anode is provided, often in the form of a graphite coating on the inside of the tube, which is maintained at a substantially higher potential. This accelerates the already-deflected electrons, which thus acquire sufficient energy to produce an adequate trace on the screen.

Double-beam Tubes

For many purposes it is convenient to display two traces on the screen simultaneously—e.g. voltage and current, or the voltages at two different parts of a network. This can be achieved by providing two electron beams, either by the use of two separate guns or by means of a splitter plate following a conventional single gun. The deflecting system is arranged so that the beams have a common X deflection but are provided with individual (and independent) Y deflections so that two traces can be displayed simultaneously.

This is of considerable convenience in circuit analysis and is in common use.

Storage Oscilloscopes

Information is often required concerning phenomena of a transient nature which produce traces of too short a duration to be comfortably observed visually. In some instances a solution can be provided by using a tube having a special long-persistence phosphor which continues to glow for an appreciable time after it has been actuated. The fluorescence decays fairly rapidly, but with an adequate gun voltage may persist for several seconds. The persistence is confusing, however, if the tube is used on repetitive phenomena.

The trace may, of course, be recorded photographically, for which purpose a tube is used having a phosphor which provides a trace of high actinic value, but the process is necessarily cumbersome.

An alternative is the use of a storage tube. In this a special storage mesh is provided adjacent to the phosphorescent screen. The normal electron beam activates the appropriate elements of this storage mesh which is then irradiated with electrons from an auxiliary "flood gun". Over those parts of the mesh which have not been actuated these flood electrons are repelled, but in the activated portions they are permitted to pass, and are then accelerated to the screen, where they produce a visible trace.

The activated elements of the storage mesh hold their charge indefinitely so that the trace remains displayed on the screen until the mesh is cleared.

When the storage facility is not required the flood-gun electrons are suppressed. The higher-velocity electrons in the writing beam then penetrate the mesh and reach the screen in the normal manner so that the tube behaves conventionally.

Storage oscilloscopes are necessarily more expensive than conventional types, but for transient investigations the extra cost is well justified.

Other Uses of Oscilloscopes

The oscilloscope is used for many purposes apart from waveform examination, some of which are reviewed briefly as under.

Response Curves

If the time base of an oscilloscope is coupled to, or operated by, a circuit which generates a variable frequency, and the vertical deflection is made to indicate the output voltage from a circuit which is fed from this variable-frequency source, the oscilloscope will display the frequency-response curve, and this method is often used for the alignment of i.f. amplifiers.

The variable frequency may be produced either mechanically or electrically. In the former case a conventional oscillator is designed to cover the frequency range required. The operating spindle of this oscillator is coupled to a voltage divider across which is connected a suitable source of d.c. voltage. The spindle is continuously rotated by a motor so that, as the frequency varies over the range, the voltage divider delivers a steadily increasing d.c. voltage which is applied to the X-plates of the oscilloscope, producing a deflection which is proportional to the frequency.

Alternatively, the normal time-base voltage in the oscilloscope may be applied to a reactance valve of the form of Fig. 7.22, where it will produce a variable-frequency output proportional to the time-base voltage. The maximum variation of frequency with this type of circuit is limited, but by heterodyning the output with a suitable fixed frequency (as in a normal beat-frequency oscillator) the range of the frequency variation may be extended as required.

Curve Tracing

Alternatively the oscilloscope can be used for a variety of static presentations such as valve or transistor characteristics or B-H

curves of magnetic materials as shown in Fig. 15.24. The X-plates are fed from a small resistor in series with the coil, so giving a deflection proportional to H; the voltage across the coil (which is

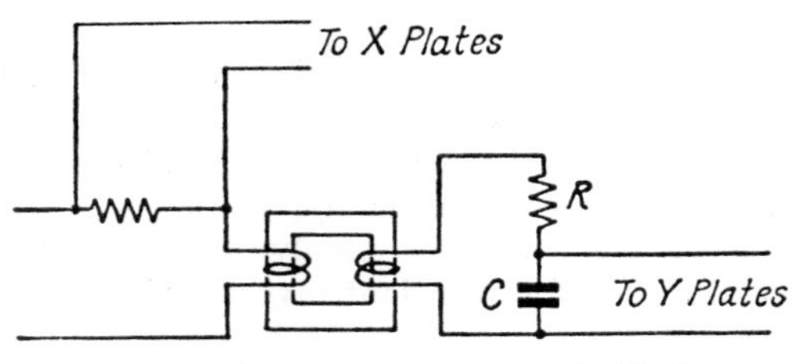

FIG. 15.24. CIRCUIT FOR TAKING B–H CURVE

proportional to dB/dt) is fed to the Y-plates through an integrating circuit (cf. Fig. 17.17) so giving a deflection proportional to B.

Phase Angle and Distortion

An oscilloscope can be used to determine the phase difference between two signals. If the plates are fed with two equal voltages of the same frequency the display will be a diagonal line if the e.m.f.s

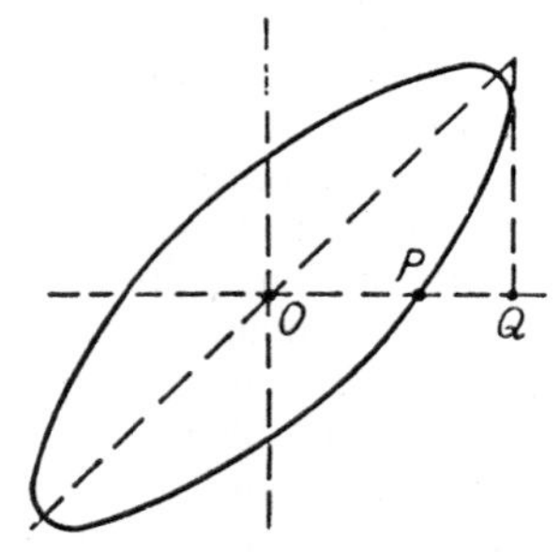

FIG. 15.25. METHOD OF DETERMINING PHASE ANGLE FROM PHASE ELLIPSE

are in phase and a circle if they are 90° out of phase. At any intermediate phase angle the display will be an ellipse, Fig. 15.25, and from the eccentricity of this the phase angle can be calculated from the relationship $OP/PQ = \sin\phi$. This may be proved as follows. Let the vertical deflecting voltage be $V_1 \sin \omega t$ and the horizontal voltage be $V_2 \sin(\omega t + \phi)$. At the point P the vertical voltage is zero so that $\sin \omega t = 0$. Hence the horizontal voltage at that instant $= V_2 \sin \phi$, which is represented by OP. The maximum value of the horizontal voltage $= V_2$, represented by the maximum horizontal travel OQ. Hence $OP/PQ = \sin \phi$, from which ϕ can at once be determined.

If the X- and Y-plates are fed from the input and output of an amplifier (with suitable adjustment of the sensitivity) the trace will disclose any distortion present. If the amplifier is linear the trace will be a diagonal line, but any amplitude distortion will result in a curvature of some part of the trace, while if there is any phase shift present the trace will be a distorted ellipse.

Frequency Comparison

If the voltages are of different frequencies a confused rapidly changing pattern will be obtained, but when the frequencies come into an integral relationship the pattern becomes stationary. There

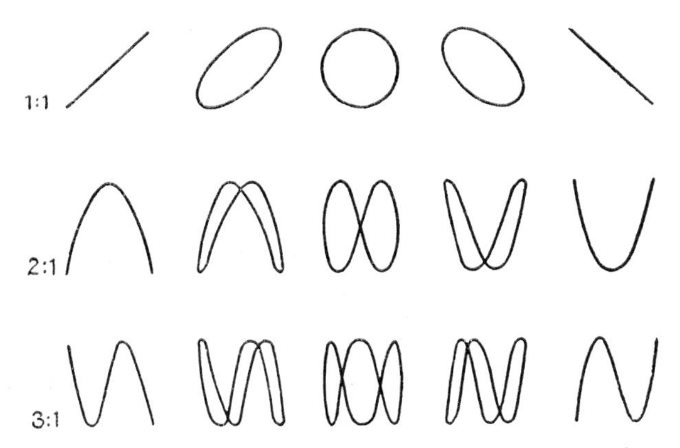

FIG. 15.26. LISSAJOUS FIGURES

is a range of distinctive patterns, corresponding to simple relationships between the frequencies, which are known as *Lissajous figures*, and by the use of these patterns frequencies may be compared very accurately.

Some typical patterns are shown in Fig. 15.26. The actual pattern depends on the relative phase of the signals, the five patterns illustrated corresponding to phase angles of 0, 45, 90, 135 and 180 degrees. In practice, unless the phase angle (or frequency coincidence) is exact, the pattern will slowly change from one form to the next and back again.

Circular Time Base

If the X- and Y-plates are supplied with voltages of equal amplitude but with a 90° phase displacement between them, the trace on the screen will be a circle, the radius of which will be controlled by the amplitude of the applied voltage. The basic circuit is shown in Fig. 15.27, the phase bridge being so arranged that $R = 1/2\pi fC$, where f is the frequency of the scan voltage. This produces a 45° lag on the X-plates and 45° lead on the Y-plates, giving the required 90° displacement.

This arrangement is used in various ways. If a signal (of appreciably higher frequency than the scan frequency) is superposed on

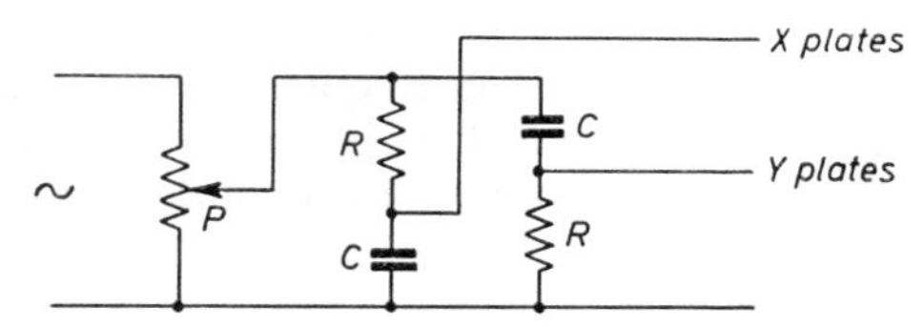

Fig. 15.27. Circuit for Circular Time Base

the input, a waveform will appear round the circular trace, which is sometimes useful in examining a repetitive waveform in which there is a periodic change.

Alternatively, the trace may be normally blacked out but momentarily brightened by signal pulses applied to the modulator grid. This will produce blips at appropriate points round the circle, which may be made stationary by synchronizing the time base to a suitable sub-multiple of the pulse frequency.

If the input pulse is also arranged to control the amplitude of the time base, the radius at which the blips appear will also vary, and this technique is used in the P.P.I. radar display referred to in Chapter 19.

Modulation Measurement

If a modulated r.f. signal is applied to the Y-plates and the X-plates are fed with a suitable sub-multiple of the modulation frequency a pattern will be obtained similar to that of Fig. 15.28; from which the depth of modulation can be seen, and calculated from the ratio $(a - b)/(a + b)$, where a and b are the maximum and minimum amplitudes.

Alternatively, by applying the modulating frequency to the X-plates, patterns of the form shown in Fig. 15.28 will be obtained from which any amplitude or phase distortion can be discovered. The modulation depth, as before, is $(a - b)/(a + b)$.

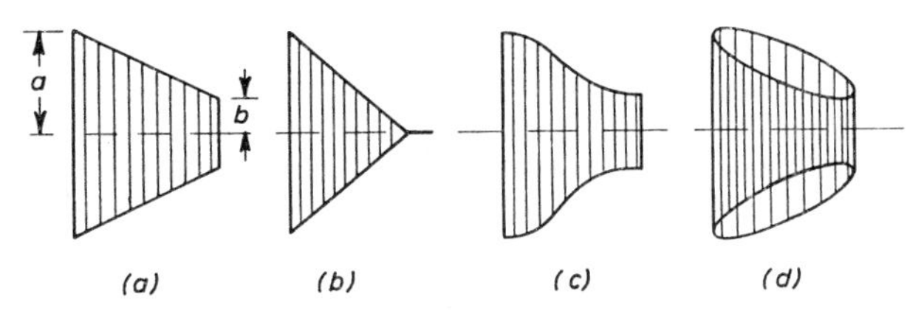

Fig. 15.28. Modulation Patterns

(a) Linear modulation, 50 per cent.
(b) Over-modulation.
(c) Amplitude-distorted modulation.
(d) Phase-distorted modulation.

15.7. Ballistic Technique

Reference may be made, in conclusion, to certain types of measurement of a quasi-instantaneous nature. The normal moving-coil movement has a restoring torque provided by a light coil spring which restrains the movement of the pointer, which comes to rest when the torque produced by the current in the coil is balanced by the restraining torque of the spring. Such a system has a natural frequency of (mechanical) oscillation and it is therefore provided with some form of damping so that the movement shall be "dead beat."

The arrangement may be modified to produce what is called a *ballistic* movement by providing a very weak restoring spring and omitting the damping. If such a movement is supplied with a very short pulse of current it will take up a position proportional to the *quantity* of electricity and can thus be used to measure the charge in a capacitor, and hence the capacitance. For example, an unknown capacitor may be charged from a battery and then discharged through a ballistic galvanometer. The reading is noted and the test repeated with a known capacitor, when the ratio of the two readings will give a direct indication of the relative capacitances.

A similar technique may be used to measure magnetic flux. If a search coil in a magnetic field is suddenly removed the induced e.m.f. will produce a pulse of current which can provide an indication on a ballistic meter, and by using a standard search coil the meter can be calibrated directly in flux.

A further, though not strictly ballistic, application is in the measurement of time intervals. If the restoring spring is entirely omitted a current through the coil will produce a continuous movement which will stop when the current ceases. The device may thus be calibrated directly in time interval and is known as a millisecondmeter.

16

Picture Transmission and Television

16.1. Principle of Scanning

The transmission of pictures by wire or radio link is in everday use, while television has long emerged from the experimental stage and now takes its place with standard engineering developments in radio communication.

The basic method of transmitting either a still picture, for press or similar purposes, or a succession of pictures at very short intervals, to constitute a television image, is carried out by a process known as *scanning*.

Scanning

The picture is divided up into a series of thin strips or lines, and the point under examination moves along each line in turn. Each individual line is thus composed of a number of successive elements which may be light, dark, or half-tone, according to the nature of the image, and if this light and shade can be translated through a photo-electric cell into electric currents it is possible to obtain a varying voltage having a modulation which follows the variations of light and shade in the original strip of the picture.

Having analysed one line, the transmitter then proceeds to an adjacent line and so on until the whole picture has been covered. In this way the picture or image is divided up into a series of successive picture elements, each one of which will produce its corresponding modulation voltage, and clearly such currents can be transmitted either by wire or by a radio link to a distant point.

At the receiving end it is necessary to re-convert the current into light, the intensity of which is directly proportional to the current at each particular instant. The light must be focused on to or otherwise caused to illuminate the receiving screen in a similar series of lines and in synchronism with the transmitter. Thus the spot of light at the receiving end will travel along one line and the intensity of the illumination at each successive point will be controlled by the current received from the distant end, so that a series of light and dark picture elements is built up. The spot then proceeds with the second line and so on until the various picture elements have all been reassembled and a complete picture is obtained which should be a duplicate of the original scene.

With picture telegraphy the whole process can take several minutes since only one copy is required. With television, on the other hand, the whole picture must be built up in a small fraction of a second and must then be repeated over and over again, the picture frequency being sufficiently rapid to enable the visual persistence of the eye to blend the whole impression into a continuous image as in the case of the cinema.

Facsimile Transmission

For the transmission of pictures by wire or radio, the original is scanned by a source of light operating through some convenient mechanical arrangement, which may be a mirror drum system similar to that described for television later, or the simple arrangement of wrapping the picture round a drum, causing the drum to rotate and at the same time tracking the light axially down the picture. This will obviously scan the picture in the manner described, and a somewhat similar arrangement can be used at the receiving end.

For the reception photo-sensitive paper is usually employed so that the variation in the intensity of the light spot affects the paper in the same way as the varying light transmitted through a photographic negative, and when the whole picture is developed the image comes up in the usual way and is then fixed.

There are numerous matters of technique which can only be referred to briefly here. In particular, since the whole transmission takes some minutes, the frequencies of the signals involved are very low. This would necessitate d.c. amplifiers at the transmitting end to augment the very small voltages produced by the photocell, and it would also demand characteristics from the line over which the transmission was sent far in excess of those which are satisfactory for normal telegraphy or telephony. It is customary, therefore, to chop up the impulses by an interrupter, usually interposed in the path of the light beam itself, so obtaining a rapid succession of impulses, the amplitude of which is varying more slowly in accordance with the light and shade of the picture points of the image being scanned.

It is, of course, necessary to ensure that the mechanism at the receiving end revolves at exactly the same speed as that at the transmitting end. This is ensured by the transmission of synchronizing signals in somewhat similar manner to that adopted in television practice.

Television Scanning

For television purposes the image has to be scanned in a small fraction of a second, which requires electronic techniques. The earliest systems used a Nipkow disc (named after its Polish inventor) in which a series of small holes were arranged in a spiral towards the periphery of the disc. As the disc rotated these holes passed across the image in turn, so providing the necessary scanning. With any reasonable size of disc the holes have to be very small so that the illumination is very weak and the method is quite impracticable for the high-definition systems now in use.

Variants of the method which, while still quite inadequate for high-definition requirements, are still employed for certain low-definition applications are the *mirror drum* and *mirror spiral*. In the former a series of mirrors is arranged round the periphery of a drum, while in a mirror spiral they are mounted on an axial shaft. Either arrangement, in association with a suitable lens system, will scan the image in the same way as a disc. Alternatively the mirrors may be illuminated by a powerful light source, so that as they rotate the successive portions of the image are illuminated. This is known as the *flying-spot* system.

Cathode-ray Scanning

As said, mechanical methods are quite inadequate for modern high-definition systems using 405, 625 or even more lines, which can only be handled by an electronic technique. The cathode-ray tube provides an obvious solution since it is possible to deflect an electron beam at speeds far higher than are demanded for picture scanning.

The essentials of a cathode-ray tube were described in the previous chapter. If the spot is deflected horizontally by a suitable time base, and is simultaneously deflected vertically by a similar time base operating at a much lower speed, it will produce a *frame* of the form shown (with a very small number of lines) in Fig. 16.1. This is exactly what is required, and by suitable relationship between the time-base speeds any desired number of lines may be produced.

Interlacing

Although visual persistence lasts for approximately 1/10th second it is necessary to use a more rapid repetition rate if objectionable

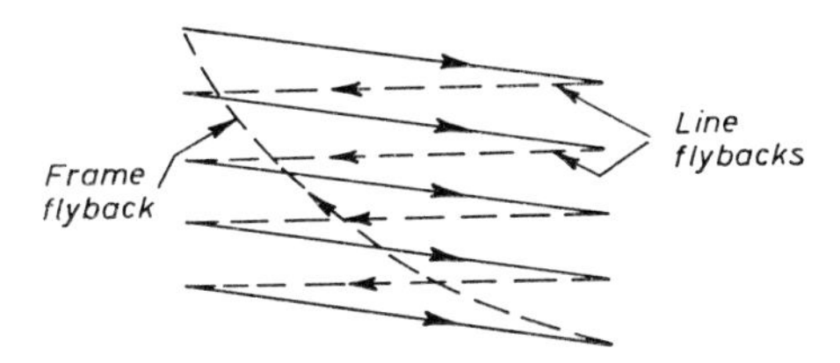

FIG. 16.1. PICTURE "FRAME" PRODUCED BY TWO TIME BASES

flicker is to be avoided; but the more rapid the rate, the shorter is the time occupied by each picture element, which involves a higher bandwidth. A convenient compromise can be achieved by the process of *interlacing*.

An odd number of lines is required, of which half are transmitted during each frame, and the time of the vertical or frame scan is adjusted so that the lines are spaced apart by twice the normal width. Since there is an odd number of lines the scan will finish in the middle of a line, and the succeeding scan will then occupy the

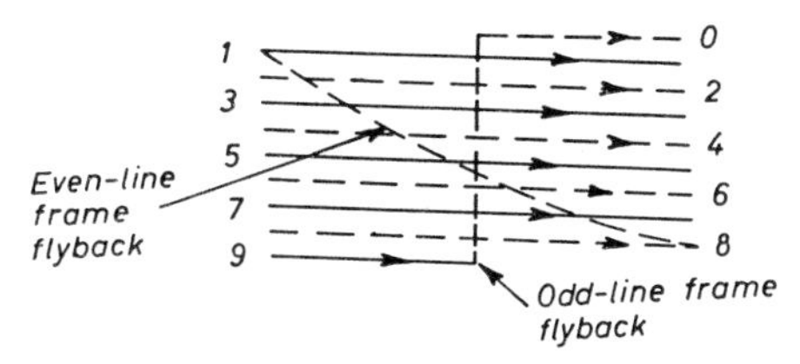

FIG. 16.2. ILLUSTRATING INTERLACED SCANNING
(The line flybacks are omitted for clarity)

space between the lines of the first scan. The process is illustrated in Fig. 16.2, from which the interlacing can be seen clearly.

Thus with a 625-line picture, $312\frac{1}{2}$ lines are transmitted in 1/50th sec, followed by the remaining $312\frac{1}{2}$ lines in the next 1/50th sec. Visual persistence causes both frames (and several more) to be seen together, but the high repetition rate virtually eliminates any sensation of flicker; and since only half the picture elements are transmitted during each scan the overall bandwidth is the same as if the complete picture had been scanned in 1/25th second.

Synchronism

It is clearly essential that the movement of the spot at the receiver shall be exactly in synchronism with that at the transmitter. This is achieved by the provision of synchronizing pulses between each line, and at the end of each frame. These are applied to the scanning system at the transmitter and are also radiated as part of the video waveform and then separated at the receiver from the picture modulation and used to synchronize the line and frame scans.

To facilitate this separation and prevent interference with the picture the synchronizing pulses are made "blacker than black." Thus in the British system the picture modulation is from 30 per cent (black) to 100 per cent (full white) while the synchronizing pulses reduce the modulation to zero, as shown in Fig. 16.3. (In

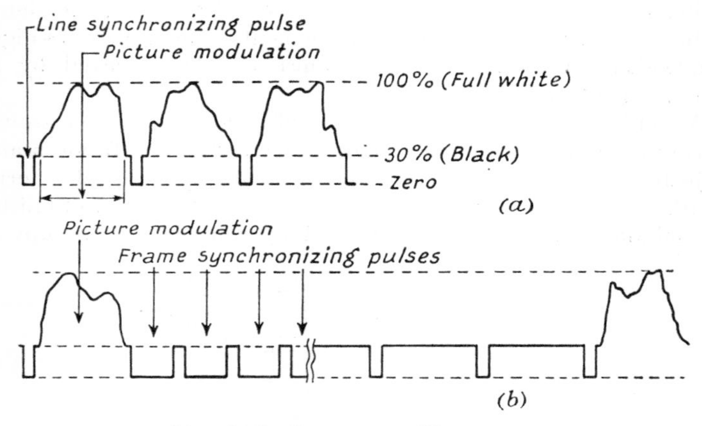

Fig. 16.3. Television Waveform

America the reverse procedure is used, with full white corresponding to zero modulation, with black at 75 per cent and blacker than black from 75 to 100 per cent, but the principle is the same.)

To distinguish line from frame synchronization the frame pulses are made much longer, so that they may be separated from the line pulses by suitable choice of time constants. It is clear, however, that some time interval must be allowed for these pulses, and also to allow for the flyback of the spot between successive lines and frames. It is customary therefore to allow a period of about 15 per cent of the total line time for the flyback and sync. pulse, and a similar blank period between successive frames, usually between 5 and 10 per cent of the frame-scan time. This blank frame period reduces the effective number of lines available for the picture.

A short period of black is allowed at the beginning and end of the blank periods to allow the circuits to settle before the sync. pulse is transmitted. The form of the frame sync. pulse is complex. Not only has the line synchronization to be maintained during the idle period, but the frame scan has to be triggered alternately at the beginning and half way along a line. These two requirements can be met by breaking up the frame pulse into a series of shorter pulses of (total) duration equal to half the line time, as shown in Fig. 16.3.

16.2. Television Cameras

As said, the requirements of a high-definition television system can only be met by electronic scanning methods. The camera tube is essentially a modification of the cathode-ray tube described in the previous chapter, in which an electron beam is deflected in the horizontal and vertical axes by appropriate saw-tooth generators. The beam, however, is not directed on to a fluorescent screen but is arranged to scan an image of the scene to be televised in such a manner as to generate an e.m.f. corresponding to the intensity of the illumination at each successive picture point.

This involves the use of some form of photo-electric mosaic which can be provided in a variety of ways.

The Iconoscope

The most widely used of the early tubes was the Iconoscope of Zworykin. In this arrangement the usual fluorescent screen at the end of the oscillograph tube is replaced by a plate containing a mosaic of very small photocells. The ordinary photocell contains an emitting surface of cæsium on a thin film of silver which is deposited on the glass or emitting surface first. By suitable manufacture it is possible to deposit this emitting surface in a series of very small

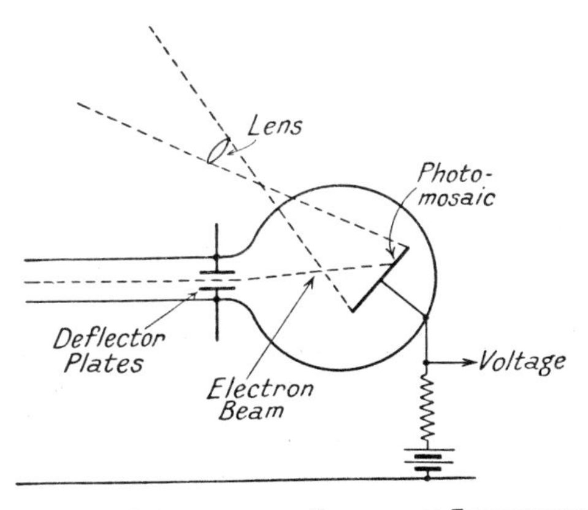

FIG. 16.4. PRINCIPLE OF ZWORYKIN ICONOSCOPE

globules, each one of which therefore becomes an individual and independent photocell. This mosaic is deposited on (but insulated from) a suitable metal plate and is situated at such an angle that it can be viewed conveniently through the side of the tube, as shown in Fig. 16.4.

The scene to be televised is focused down on to this plate, and each of the tiny photocells in the mosaic will thus respond individually to the amount of light in that particular part of the scene. Since there is, as yet, no circuit, the cells will acquire a charge proportional to the intensity of the light. A cathode-ray beam generated in the usual way is now caused to scan over the plates by deflecting voltages of the usual type, and as it reaches each individual photocell in the mosaic the latter is discharged, producing a current in the external circuit proportional to the charge.

This method, therefore, combines the flexibility of the cathode-ray tube with the necessary photo-electric properties required to convert light to electric current and this was, for a number of years, the best camera tube available.

Electron Multipliers

The great difficulty in television is to obtain sufficient light on the photocell, for the amplification which can be used is limited by the background noise which is produced due to electronic disturbances

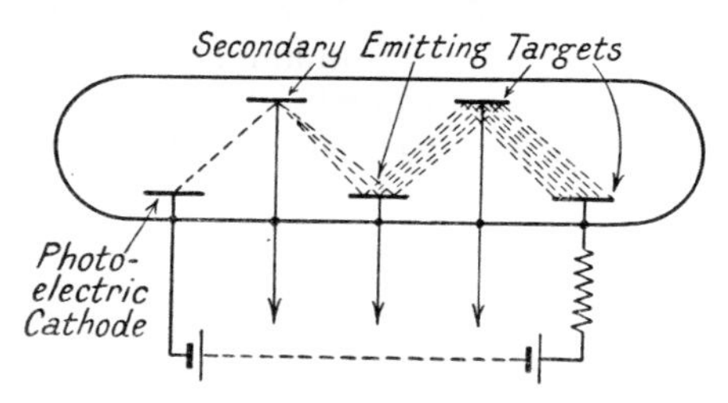

FIG. 16.5. SIMPLE ELECTRON MULTIPLIER

as mentioned on page 418. Considerable improvement has been possible as the result of the development of *electron multipliers*, which are devices using secondary emission.

A simple form of multiplier is shown in Fig. 16.5. The electrons emitted from the photo-electric cathode are focused on to a plate which is coated with material which readily emits secondary electrons. This means that, if it is struck by electrons moving with sufficient rapidity, it will in turn emit electrons of its own in the proportion of perhaps five or ten electrons for each one which strikes it. Hence, there will be emitted from this secondary plate a current five or ten times as great as the original current, and this process can be repeated by successive emission from subsequent plates all arranged at a higher potential than the preceding one until an amplification of hundreds of thousands of times is obtained in the one tube.

Since the circuit contains no coupling impedances or subsidiary source of emission, the question of background noise does not arise and very weak illuminations on the photocell cathode are capable of giving several milliamperes emission at the far end.

The Image Orthicon

The most widely used camera tube at the present time is the *Image Orthicon*, which can produce a usable picture at reasonably low levels of illumination. The structure of this device is shown schematically in Fig. 16.6.

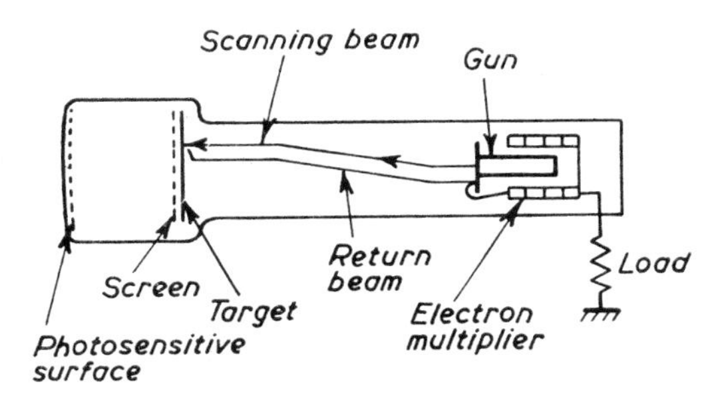

FIG. 16.6. SCHEMATIC IMAGE ORTHICON

The image section consists essentially of a photosensitive surface operated at a high negative potential, a target plate of thin, low-resistivity glass and a very fine screen very close to the target. When an image is focused on the thin photosensitive film, electrons are emitted from the rear of the film in proportion to the light intensity. These electrons are attracted towards the screen and the majority pass through it and strike the target. As these electrons strike the target, they produce secondary electrons which are attracted to the screen leaving positive charges on the target. The focusing arrangements in the image section are so arranged that all the electrons emitted from a given point on the photosensitive surface are focused at a corresponding point on the target, so that the charge distribution on the target reproduces the light distribution of the optical image focused on the photosensitive film.

The back of the target plate is now scanned in the normal way and the beam deposits just enough electrons on the rear of the target to neutralize the charge that has been built up on the image side, the remaining electrons returning to the gun. The current in the returning beam thus varies in accordance with the variation in light intensity of the optical image being (indirectly) scanned. Since the target is thin and of low resistivity the electrons deposited by the scanning beam will neutralize the optically-derived positive charges before the target is scanned again. Most of the electrons in the return beam strike the surface of the electron gun, which is treated to be a good emitter of secondary electrons; the secondary electrons thus produced are deflected into an electron multiplier system and the output of the electron multiplier is a fairly large current which varies linearly with light intensity between the limits of blackout (when all the electrons striking the target are returned) and peak white (when the positive potential of the target equals the potential on the screen).

Other camera tubes in use at the present time include the *Vidicon* and the *C.P.S. Emitron.* The Emitron uses a development of the mosaic principle described in the section on the Iconoscope; a very low-velocity scanning beam is used and a good sensitivity obtained, though in this respect the C.P.S. Emitron suffers in comparison with the Image Orthicon.

The Vidicon

The Vidicon camera tube is based on the photoconductive properties of semiconductors. The construction of the Vidicon is outlined schematically in Fig. 16.7. The signal plate is a metallic

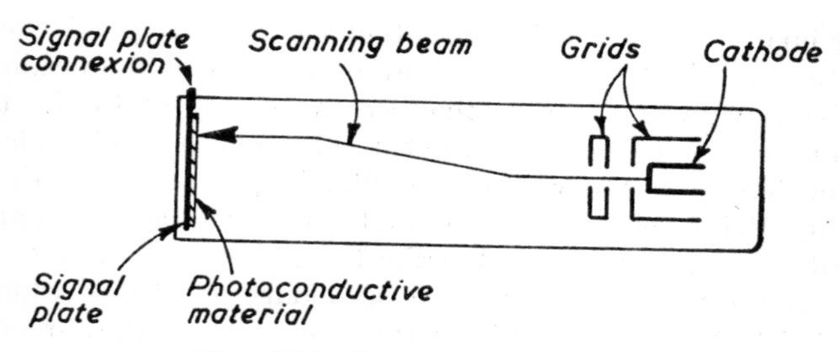

FIG. 16.7. SCHEMATIC VIDICON

plate, so thin as to be transparent, coated on one side with an even thinner layer of a photoconductive material. The optical image is focused on the other side of the signal plate and the photoconductive side is scanned in the normal way with an electron beam from a cathode which is some 30 volts negative to the signal plate.

As the scanning beam passes over the photoconductive material it deposits electrons so as to reduce the potential of that side of the film to that of the cathode. However, the other side of the film is still at its original potential so that the charge leaks through the film at a rate determined by the conductivity of the material and hence on the incident light. Consequently the charge deposited by the electron beam the next time it scans the plate will vary over the surface of the plate according to the variation of light intensity in the optical image. Therefore the current through the load, and hence the output voltage, will reproduce the variations in light intensity as successive points in the optical image are scanned.

The Vidicon has the advantages of low cost and simplicity of operation but at present suffers from minor defects which make it unsuitable for entertainment use. It has the same order of sensitivity as the Image Orthicon, but the response is non-linear and it suffers from a slight time lag in adjusting itself to different levels of light intensity. These problems, however, stem from properties of the photoconductive material and will no doubt be overcome in the near future. Meanwhile the Vidicon is extensively used in small industrial closed-circuit television system where extremely good picture quality is not of paramount importance.

The Flying-spot Scanner

Reference has already been made to the early mechanical flying-spot system. An extension of this principle is used for reproducing slides and film. The spot is produced by a cathode-ray tube with very short persistence, and focused on to the transparency. The light transmitted through the transparency is focused on to a photocell of the electron multiplier type. Scanning is achieved by deflecting the cathode-ray beam in accordance with normal scanning practice.

The same technique may be used to scan the successive frames of a motion-picture film, but it is found more convenient to move the film through the scanner at a constant speed and to reflect the flying spot onto the film from a series of rotating mirrors which cause the scanning lines to "pursue" the film to the exact extent required to scan the whole frame in the allotted time.

Resolution and Bandwidth

The vertical resolution of a television picture is determined by the line width and hence by the number of lines in the picture. The horizontal resolution is determined by the ability of the system to respond to abrupt changes of light intensity and hence to the bandwidth of the system.

The finest pattern which can be reproduced can be considered as being made up of alternate black and white squares each having the dimensions of the scanning spot. The bandwidth required can then be estimated as follows:—Let N be the number of lines and N_a the effective number of lines (deducting those required for the frame

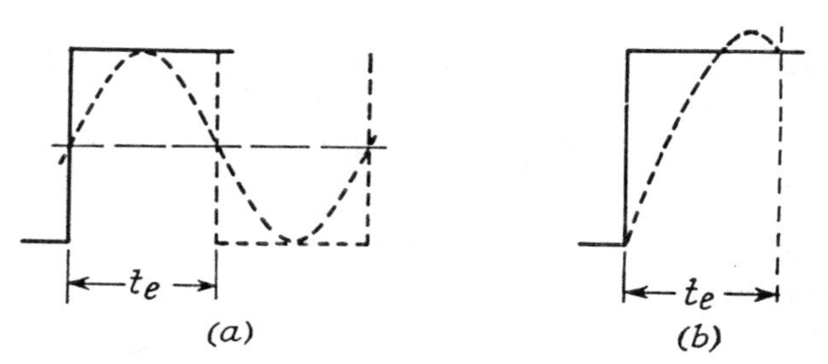

FIG. 16.8. WAVEFORM OF TRANSITION FROM BLACK TO WHITE

blanking period). Let F be the picture repetition frequency, and p the picture ratio (width/height, usually 4/3).

The effective line time, assuming a 15 per cent line blanking period, is then $0{\cdot}85/NF$. The number of picture elements per line is pN_a. Hence the time per element $t_e = 0{\cdot}85/pNN_aF$. This time may then be treated in two ways. It may be regarded as the time of one half-cycle of a square wave as shown in Fig. 16.8 (*a*), in which case the highest (fundamental) modulation frequency is

$$f_m = 1/2t_e = 0{\cdot}59pNN_aF$$

If the response of the video amplifier falls by 3 dB at the upper limit the bandwidth required is $f_m/\sqrt{2} = 0{\cdot}42pNN_aF$.

Alternatively the waveform may be regarded as a pulse having a rise-time not greater than the time t_e, as shown in Fig. 16.8 (*b*), in which case the bandwidth required (see page 717) is approximately $0{\cdot}4/t_e = 0{\cdot}47pNN_aF$.

The bandwidth requirements thus rise approximately as the square of the number of lines. With a 405-line transmission $B = 2{\cdot}4$ MHz while for a 625-line transmission $B = 5{\cdot}85$ MHz. The i.f. and video-frequency amplifiers must be able to accommodate bandwidths of this order, both at the transmitter and the receiver.

Actually some correction has to be introduced to allow for the finite size of the scanning spot. This will, in general, be comparable in size with the picture elements so that it cannot produce a sharp transition as in Fig. 16.8. Circuitry is therefore introduced which magnifies the rate of rise and so preserves the definition. There are also other corrective networks which are necessary to preserve uniformity of response over the scan, but these are matters of detail which do not affect the basic operation.

Transmitters

Obviously it is necessary to use a very high frequency carrier wave to accommodate this modulation bandwidth, and the B.B.C. transmitters operating in Band I cover the frequency range 41·5 to 66·75 MHz while the I.T.A. transmitters in Band III range from

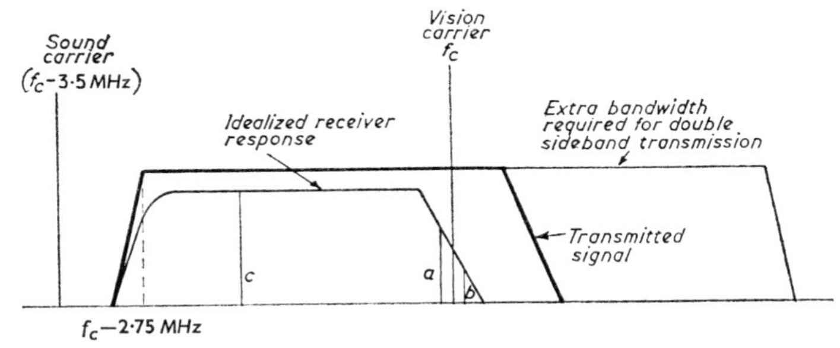

FIG. 16.9. VESTIGIAL TRANSMISSION AND RECEPTION

Two sidebands at a and b close to f_c add to give the same amplitude as one sideband at c.

186·25 to 199·75 MHz. Still higher frequencies are used for the 625 line transmissions, which operate in the U.H.F. band 470 to 890 MHz.

The transmitters use the techniques already described in Chapter 13, but the large bandwidth involved introduces certain problems in both transmitting and receiving equipment. Using normal amplitude modulation, a 625-line transmission would require a total bandwidth of over 11 MHz, which not only complicates the design but increases the background noise (which is directly proportional to bandwidth).

Television transmitters therefore use the vestigial-sideband system described in Chapter 7 (page 345), in which the upper frequencies in one sideband are suppressed, thereby reducing the overall bandwidth by some 35 per cent. The sound channel uses normal double-sideband modulation on a carrier displaced slightly beyond the maximum video side frequency, as shown in Fig. 16.9.

16.3. Television Receivers

For the reception of the pictures, electronic methods are again employed. A cathode-ray tube is employed, the electron beam being arranged to scan the screen in synchronism with the transmitter, as already described, the brilliance of the spot at any point being controlled by the video modulation.

However, in order to cover a large screen area a wide deflection angle is required, which is best achieved with magnetic deflection and focusing, so that the gun is a simple structure of cathode, anode and a modulator by which the intensity of the spot is caused

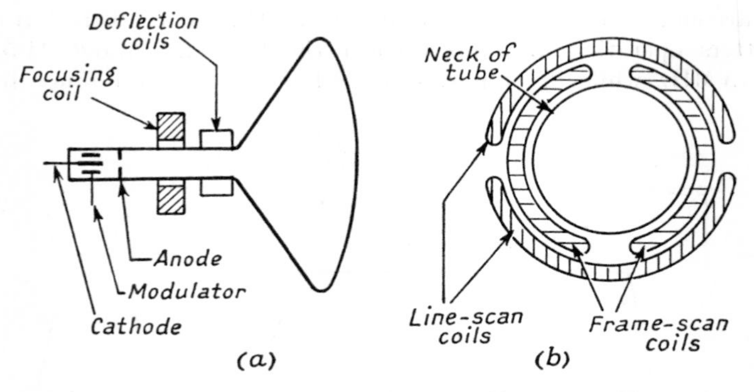

FIG. 16.10. DIAGRAM OF TELEVISION RECEIVING TUBE
(*b*) Shows a cross-section through the deflection coil system

to vary in accordance with the received video signal. A diagram of a television tube is shown in Fig. 16.10.

The time bases are basically relaxation oscillators of the type already described, but additional circuitry has to be provided to convert the sawtooth voltages into corresponding current swings for feeding the deflection coils. The coils are formed round the neck of the tube as shown in Fig. 6.10 (*b*).

The design of the receiver is based on the basic principles which have already been discussed in Chapters 9 and 13, but these have clearly to be adapted to meet the special requirements. It is not practicable here to discuss these in detail, but the principal features can be summarized by reference to the block diagram of Fig. 16.11.

The receiving aerial is an important part of the system. The short wavelengths used are subject to many reflections from obstacles in the path. These create standing-wave patterns so that the location of the aerial can be quite critical. These reflections can also

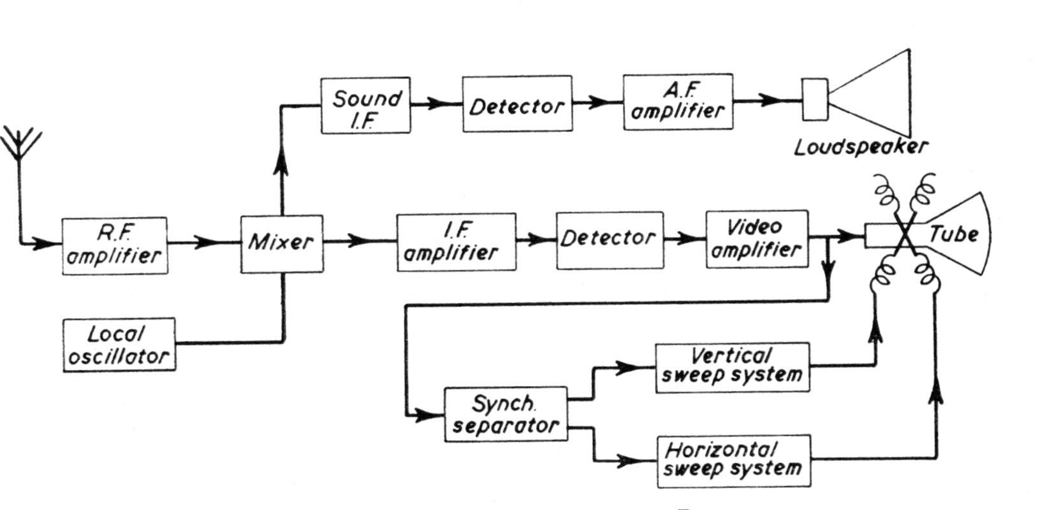

Fig. 16.11. Block Diagram of Television Receiver

produce "ghost images", which are the result of signals weaker than the true signal, but slightly delayed in time so that they do not correspond exactly with the true image.

In areas of high signal strength the aerial may be a simple rod a few feet long, the position (and angle) of which must be adjusted for optimum results, but beyond this range directive arrays of the Yagi type are required, as discussed in Chapter 13.

It is customary to use an r.f. amplifier prior to the mixer. The tuned circuits have a bandwidth of 5 MHz disposed asymmetrically with respect to the carrier as shown in Fig. 16.9. This is followed by a video i.f. amplifier which again has a bandwidth of 5 MHz, but here symmetrically disposed around the centre i.f., which is usually of the order of 35 MHz. To achieve this bandwidth at this relatively lower frequency limits the permissible Q (and hence the gain) so that it is usual to adopt the staggered tuning technique discussed in Chapter 9.

The actual frequencies are so arranged that the same local oscillator beats with the sound carrier to provide a second i.f. of the order of 500 kHz which is handled by a conventional i.f. amplifier.

An alternative arrangement employs a frequency-modulated sound channel. Only one (video-frequency) i.f. amplifier is then required, of slightly extended bandwidth. The sound modulation, being of constant amplitude, does not affect the vision modulation, and can be extracted from the output with a ratio detector.

The time constants of the video detector have to be such as to provide full output at the highest video frequency (5 MHz) at which 100 per cent modulation is required for sharp transition from black and white. (In a sound channel the highest modulation frequencies are harmonics which are of relatively small amplitude.)

The output of the second detector is amplified and applied to the picture tube in such a way as to control the beam current of the tube. The output of the video amplifier is also fed to the synchronization separator circuits which extract the synchronizing signals from the picture information and then separate the line and frame synchronizing pulses from each other. These pulses are then used to synchronize the horizontal and vertical scanning circuits.

As said earlier, magnetic deflection is used in television picture tubes since the beam can be deflected through a wide angle without undue deflection distortion. The sawtooth current waves for the scanning coils are derived from sawtooth generators, synchronized by the appropriate pulses. These low-power sawtooth waveforms are then used to drive an output stage, consisting of a pentode or beam tetrode whose anode is transformer-coupled to the scanning coils.

The power developed by the line output valves during the flyback period is often used to develop the e.h.t. voltage for the tube anode, as discussed in Chapter 11.

16.4 Colour Television

Any given colour can be produced by a suitable blend of three primary colours, namely red, blue and *cyan* (which is a yellowish green). Since phosphorescent materials are available which will glow with these colours when activated by an electron beam, it is

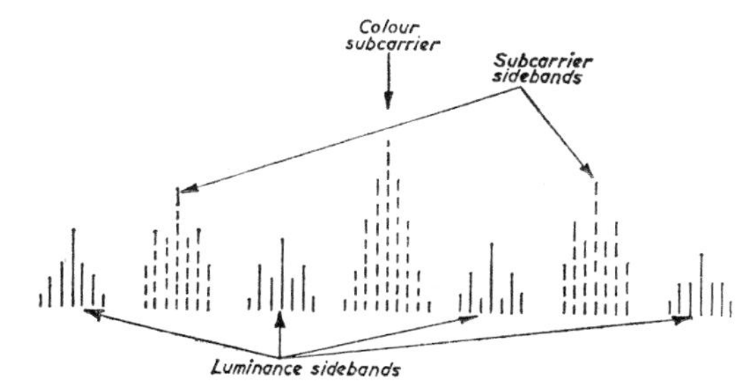

FIG. 16.12. ILLUSTRATING INTERLACING OF CHROMINANCE SIDEBANDS

clearly possible to reproduce television pictures in colour if suitable arrangements can be devised to superpose a succession of three-colour images.

This has involved much research, one of the main problems being that, since colour processes are necessarily considerably more expensive, any system adopted for general use must be *compatible*, i.e. it must still provide adequate black-and-white pictures on a normal television receiver.

Various practical solutions have been devised, which cannot be discussed in detail here. It will suffice to summarize the basic technique of one of the systems in common use. The first requirement is to provide at the transmitter three separate signals corresponding to the relative illumination at the three primary colours. This involves, basically the use of three co-ordinated cameras each viewing the scene through an appropriate colour filter.

These signals are then combined to provide a *luminance* signal which is proportional to the overall brightness of the scene and acts in the same way as the normal intensity modulation with a black-and-white picture. Now, it is found that with this normal modulation the side frequencies within the total band occur in a series of groups, as shown in Fig. 16.12 (because the successive picture elements produce voltages which repeat at intervals along the line). It is therefore possible to introduce auxiliary modulation in the spaces between these groups which can be used to transmit appropriate *chrominance* signals corresponding to the individual colour components.

The chrominance components are therefore arranged to modulate a sub-carrier located towards the upper end of the luminance spectrum, as shown in Fig. 16.13, producing interlaced sidebands

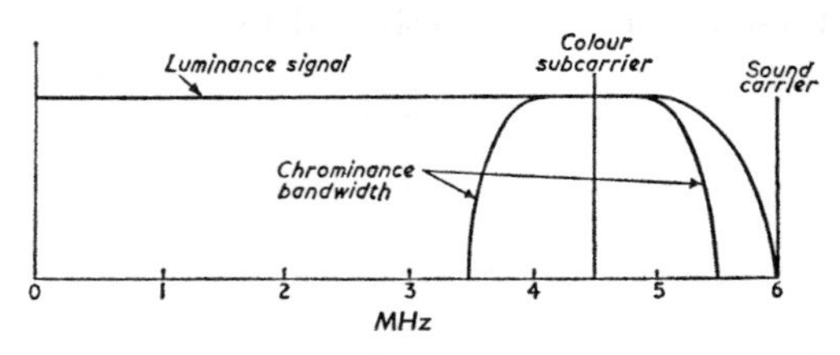

Fig. 16.13. Illustrating Use of Sub-carrier to Convey Chrominance Signals

which do not interfere with a black-and-white picture but provide the required colour information.

At the receiver this subcarrier is filtered out and the composite modulation which it carries is resolved by a complex decoding matrix into the three basic colour components which are then applied to the colour tube.

Colour Receiving Tubes

The various signals must then be combined at the receiving end within a single tube, since the use of three separate tubes would involve prohibitive optical problems. This requirement has also been met in a variety of ways, one of the most successful being the *shadow-mask* tube, illustrated in Fig. 16.14.

This uses an assembly of three separate guns, one for each of the primary colours. The tube face is coated (internally) with a mosaic of three different phosphors which when activated glow red, cyan and blue respectively. Behind this screen is located a fine mesh mask, the beams from the three guns reaching the screen through the holes in this mask.

By accurate alignment of this mask, together with the correct focusing of the electron beams, it is arranged that the beam from, say, the red gun impinges exactly on the red phosphor dots and does not activate the adjacent spots. Similarly the cyan gun activates the cyan spots, and the blue gun activates the blue spots. It is evident that considerable precision in manufacture is involved, particularly since the overall area of the spot clusters must be of the same order as the normal spot size on a black-and-white screen, but it has proved practicable in use.

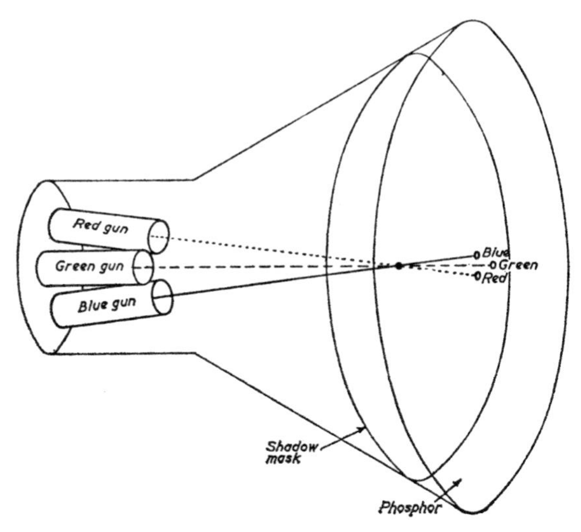

FIG. 16.14. SHADOW-MASK TUBE

The intensity of each of the beams is then controlled by the decoded chrominance signals, resulting in an effective colour picture. In other respects the tube behaves like a normal tube, being scanned by deflecting voltages derived from sawtooth generators, synchronized in the normal way. If the chrominance signals are omitted all three guns act together to produce white, the intensity being controlled by the luminance signal.

This description is, of course, greatly simplified. There are other methods of transmission which differ in detail, and alternative types of receiving tube. One such variant is the *trinitron* developed by the Sony Corporation of Japan. This uses a single gun assembly having three cathodes, which is claimed to provide more accurate focus and a 50 per cent increase in beam current and hence increased brilliance. The shadow mask is replaced by an *aperture grille* containing a rigid assembly of closely-spaced vertical wires associated with vertical stripes of phosphor which provides improved colour registration.

17

Wave Shaping and Pulse Techniques

17.1. Wave Shaping

A large number of the waveforms used in communications practice are non-sinusoidal In addition to square and sawtooth waveforms, pulses are extensively employed, constituting the basis of all digital techniques.

The desired waveform may be obtained by employing a circuit such as the multivibrator or blocking oscillator which generates the waveform directly, or alternatively we may start with some waveform such as a sinusoid and operate on it with appropriate circuits to obtain the required waveshape.

Clipping

One of the simplest waveshaping circuits is the clipping or limiting circuit shown in Fig. 17.1 (*a*). This circuit operates so as to prevent the positive and negative peaks of the wave from exceeding the value

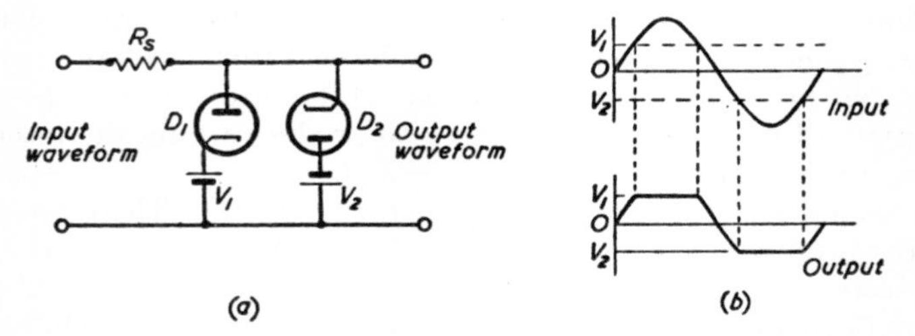

FIG. 17.1. DIODE CLIPPING CIRCUIT AND ASSOCIATED WAVEFORMS FOR SINE-WAVE INPUT

set by the clipping circuits. If the peak value of the input waveform is less than V_1, the input waveform is transmitted unchanged, but if the peak positive value exceeds V_1, the diode D_1 will conduct, holding the output voltage very nearly constant at this value until the level of the input waveform falls below the limiting voltage. For the clipping action to be effective, the resistance R_s must be much greater than the dynamic resistance of the diode. The same action takes place on the negative peak due to the diode D_2 and the output waveform is clipped as shown in Fig. 17.1 (*b*). Clearly either positive or negative peaks alone may be clipped by omitting the appropriate diode.

In this circuit thermionic diodes are shown, but semiconductor

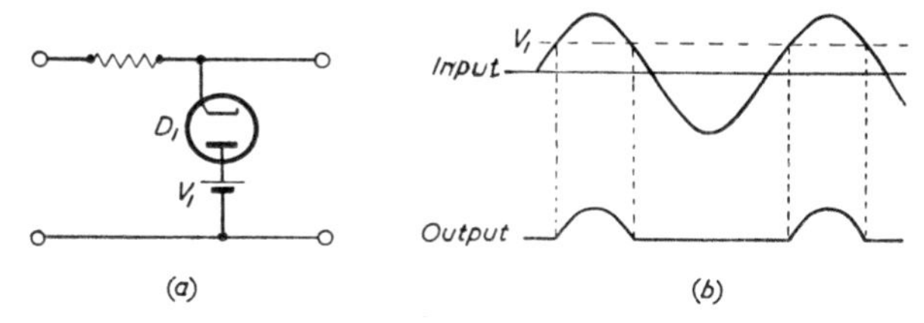

FIG. 17.2. DIODE BASE-CLIPPING CIRCUIT

diodes may be used equally well. The same applies to the circuits of Figs. 17.2 to 17.6.

Alternatively we may require to suppress all voltages below a certain level. This can be achieved by using the circuit shown in Fig. 17.2 (*a*). Here the diode is biased so that the output voltage remains at the level of the clipping voltage V_1 until the peak value of the input waveform exceeds V_1; and the resultant waveform is as shown in Fig. 17.2 (*b*).

The diode circuits described above are the simplest clipping

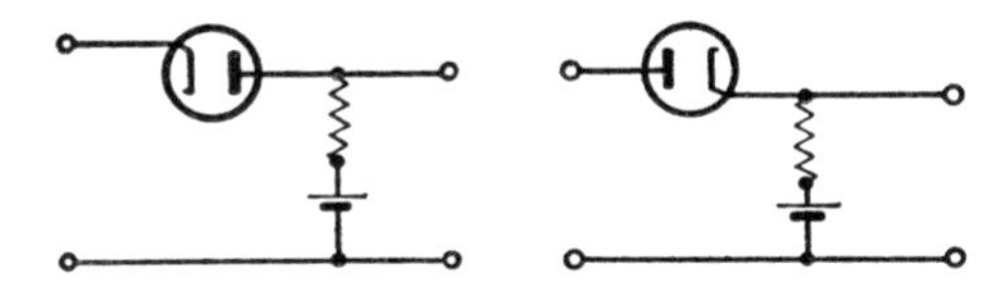

FIG. 17.3. ALTERNATIVE FORMS OF CLIPPING CIRCUIT

circuits used, but possible variations can be effected by taking the output voltage across the resistor rather than from the diode, as shown in Fig. 17.3, or by other placing of the reference voltage.

Valves or transistors are often used in clipping circuits where the limitation is obtained due to bottoming on the one hand and cut-off on the other. The amplification obtained in such a circuit is sometimes useful.

Clipping circuits may be combined in various ways; in particular by combining two circuits such as shown in Fig. 17.3 together with appropriate choice of the voltages V_2 and V_3, we can arrange that the output waveform consists of a slice of the input waveform, as shown in Fig. 17.4.

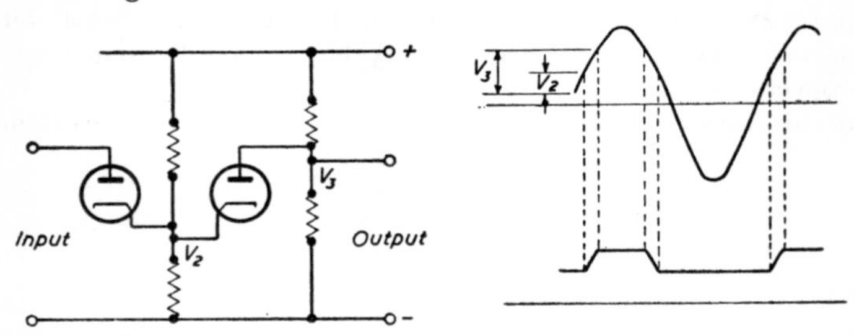

Fig. 17.4. Slicer Circuit and Waveforms

D.C. Restoration

Many unsymmetrical waveforms possess a d.c. component which will be lost if the waveform is passed through a capacitor or a transformer. A typical example is the pulse train shown in Fig. 17.5. The waveform originally consists of a series of positive-going pulses (*a*) based on 0 V; after passing through an a.c. coupling, the level is as shown in Fig. 17.5 (*b*). It is often important to restore the original d.c. level of the waveform.

Such a function, which is known as *d.c. restoration*, is a particular case of a general process known as *clamping* in which any desired d.c.

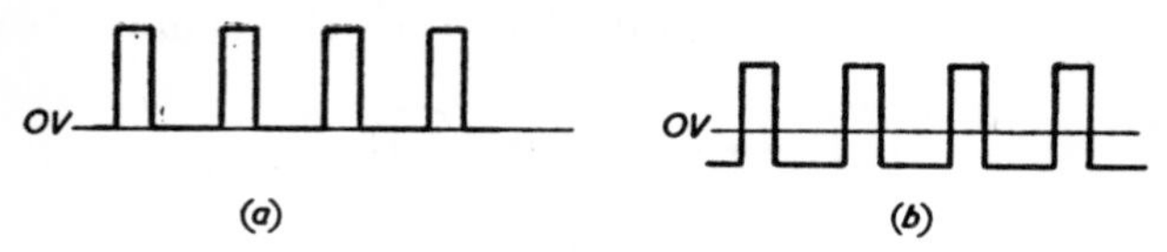

Fig. 17.5. Unsymmetrical Waveform
(*a*) With d.c. component; (*b*) With d.c. component removed

reference level is introduced into a waveform. The simplest form of clamping circuit (Fig. 17.6) consists of a diode, a resistance and a capacitance. If the time constant *RC* is so large compared with the period of the incoming wave that it can be considered infinite, then the diode *D* in Fig. 17.6 (*a*) acts as a rectifier which charges the capacitor *C* to a voltage V_1 which just prevents the voltage at the output from going positive. The d.c. voltage V_1 has thus been re-inserted in the output waveform or, in alternative phraseology, the waveform has been clamped onto the positive peak.

Figs. 17.6 (*b*) and (*c*) show alternative arrangements in which the negative peaks of the waveform are clamped to zero and the positive peaks clamped to V_2. In practice the time constant CR can seldom be made so large as to be considered infinite, and if this ideal condition cannot be realized some distortion of the output waveform will occur. However CR can usually be made large enough for the distortion to be negligible.

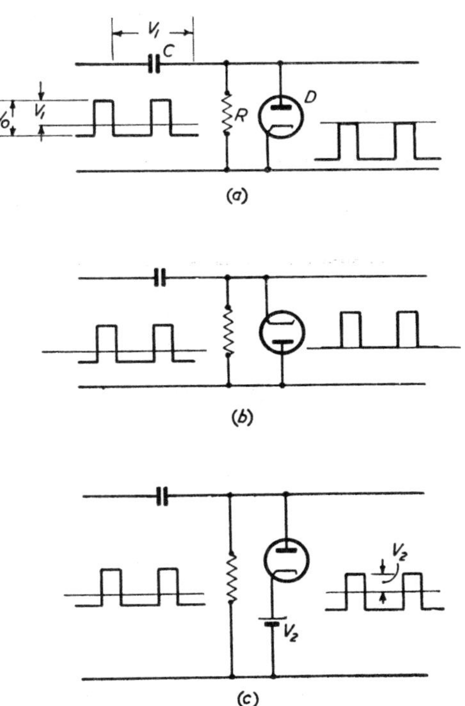

FIG. 17.6. DIODE CLAMPING CIRCUITS

(*a*) Positive peaks clamped to zero; (*b*) Negative peaks clamped to zero. (*c*) Positive peaks clamped to V_2

17.2. Pulse Generation; the Multivibrator

A form of pulse generator which has many applications in communications engineering is the *multivibrator*. This is a form of relaxation oscillator originally devised by Abraham and Bloch in 1918 as a square-wave generator; and since a square wave is rich in harmonics it was called a multivibrator.

In its original form (which is still used) it produces a continuous square-wave oscillation at a frequency determined by the internal time constants. This form is known as the free-running or *astable* multivibrator. The circuit may be modified, however, to a form in which it does not oscillate continuously but can be caused to emit a single pulse when it is triggered by an external impulse. Two possible conditions exist, known as *monostable* and *bistable*

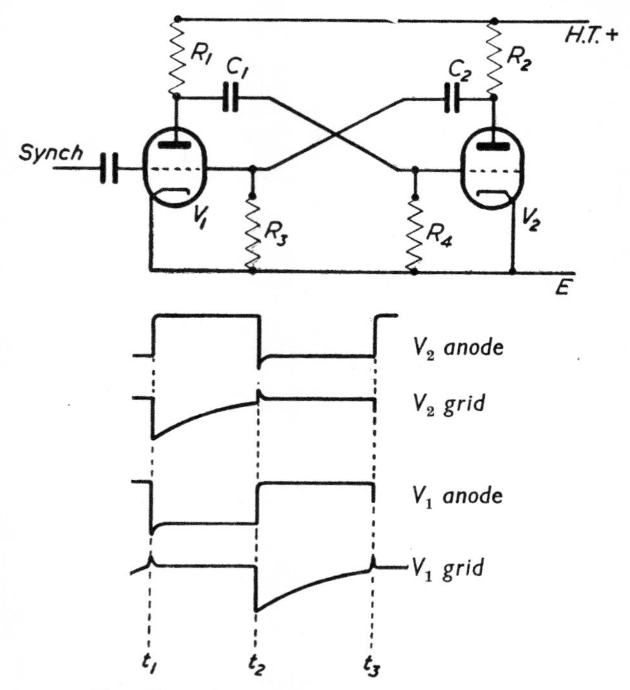

FIG. 17.7. ANODE-COUPLED MULTIVIBRATOR AND WAVEFORMS

respectively, according to the nature of the quiescent state, as will be explained shortly.

Fig. 17.7 shows the circuit of a valve-operated astable multivibrator. It may be regarded as a two stage *RC*-coupled amplifier in which the output of the second valve is fed back to the input of the first valve. Oscillations will take place because each valve produces a phase shift of 180°, so that the fed back voltage is in phase with the input. The operation of the circuit may be illustrated by reference to the waveforms shown in Fig. 17.7. Oscillations are started by random noise which drives V_1 grid positive (say). This

voltage is amplified and fed back positively so that the grid potential of V_1 rises, causing the valve to take heavy current and the anode voltage to fall. This action continues until V_1 anode (and hence V_2 grid) falls to such a value that V_2 is cut off; amplification then ceases. The action described above takes place practically instantaneously and the waveforms at this time are as shown at $t = t_1$.

The voltage on V_2 grid then rises at a rate determined by C_1R_4 until V_2 becomes conducting. The switching action of the previous paragraph is then reversed and V_2 is turned hard on, and V_1 is turned off. This point is represented by t_2 in the figure. The voltage on V_1 grid then rises at a rate determined by C_2R_3 until V_1 is switched on again at time $t = t_3$, and this cycle of events then repeats indefinitely.

The waveforms in Fig. 17.7 are for the case when $C_1 = C_2$, $R_3 = R_4$, and $R_1 = R_2$ so that the total period $t_3 - t_1$ is divided into two equal periods, $t_2 - t_1$ when V_2 is conducting and $t_3 - t_2$ when V_1 is conducting. This is known as 1:1 mark-space ratio. Obviously both the period and the mark-space ratio can be altered by changing the circuit constants.

The frequency of the oscillation depends on the extent to which C_1 (or C_2) has to discharge before the change-over takes place. It is of the order $f = 1/CR$. (See page 706.)

Synchronization

The operating condition of the multivibrator described above is known as a "free running" condition, i.e. the period of operation is determined solely by the circuit constants. If, however, positive pulses are applied to V_1 grid, the switchover point can be made to occur at a time determined by the input pulse. This action is illustrated in Fig. 17.8.

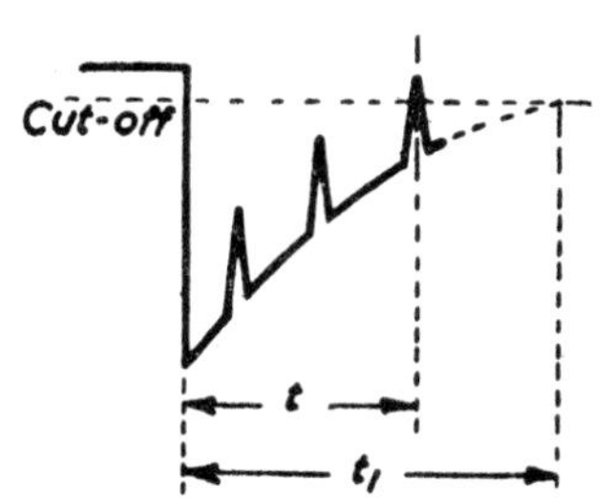

FIG. 17.8. GRID WAVEFORM OF TRIGGERED MULTIVIBRATOR

If the multivibrator were free running V_1 grid would reach the cut-off voltage after a time t_1, but the positive pulse triggers the multivibrator action after a shorter time t. This process is known as *synchronization* and has many uses in electronics.

Frequency Division

One of the most important uses of synchronization of an oscillator is in frequency division. If pulses with a repetition interval of $t/3$

are applied to V_1 grid in Fig. 17.8 we see that the multivibrator triggers on every third pulse, i.e. after a time interval t. A similar action can be obtained by feeding the pulses to V_2 grid so that the total period of the multivibrator can be made exactly $2t$, i.e. the output frequency of the multivibrator is exactly one-sixth of the frequency applied to the grids. This principle can be carried on in a series of stages to produce a low frequency exactly related to a much higher frequency and is clearly of great importance.

Monostable Multivibrator

In certain applications it is required that the multivibrator should remain quiescent until a trigger pulse is applied. The multivibrator then goes through one cycle of operation and remains quiescent until a further trigger pulse is applied. Such a circuit is called a *monostable multivibrator*, a typical example being shown in Fig. 17.9. This circuit differs from the multivibrator in Fig. 17.7 in that cathode coupling replaces the anode-to-grid coupling of one of the amplifier

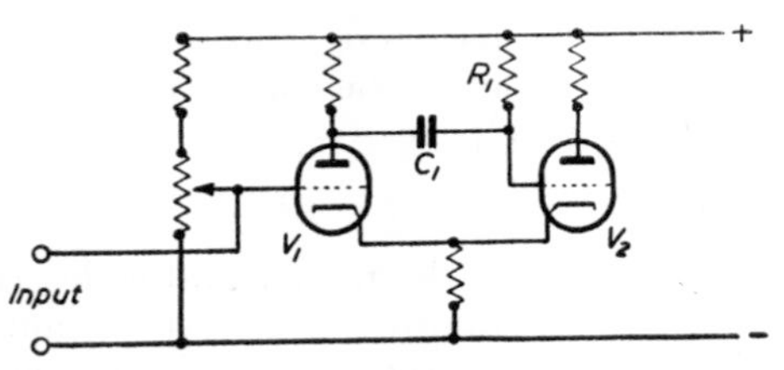

FIG. 17.9. MONOSTABLE MULTIVIBRATOR

stages. The monostable action is obtained by returning the grid of the second valve to the positive potential whilst the grid of the first valve is returned to a variable positive bias. In the quiescent condition V_2 carries a large current which raises the common cathode potential sufficiently to cut V_1 off. However, if a pulse is applied to V_1 grid large enough to turn this valve on, multivibrator action is initiated as in the circuit of Fig. 17.7 and V_2 is cut off. The grid voltage on V_2 then slowly increases at a rate determined by C_1R_1, until the grid–cathode voltage on V_2 is such that the valve can conduct. The circuit then switches over to the original quiescent condition. The period of operation is clearly dependent on C_1R_1 but it is also dependent on the (variable) bias on V_1, since this bias affects the common cathode voltage when V_1 is on and hence the voltage to which V_2 grid must rise before switchover occurs.

Bistable Multivibrator

The transition from one state to the other in an astable or monostable multivibrator results from the operation of the internal time-constants. If these are omitted, however, so that the circuit becomes direct-coupled, it becomes a stable two-state device, as illustrated in Fig. 17.10.

If V_1 commences to conduct, its anode (and hence V_2 grid) will go negative. V_2 anode will go positive which will be fed back to V_1

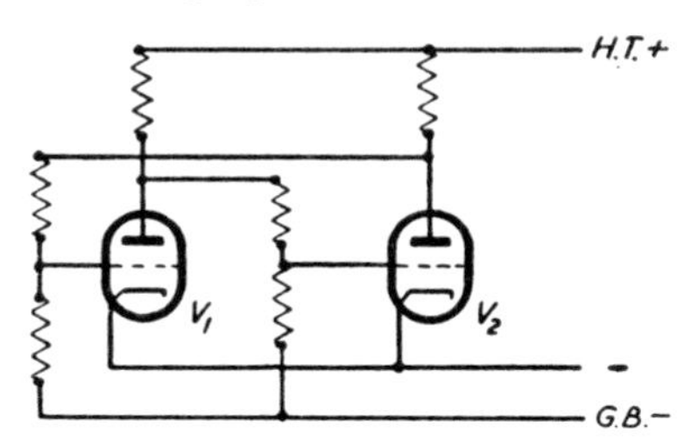

FIG. 17.10. BISTABLE MULTIVIBRATOR

grid and produce a rapid increase in current. At the same time V_2 will be cut off, and *the circuit will remain in this condition*.

A positive pulse applied to V_2 grid, however, will cause V_2 to conduct, and the same action will occur with V_2 as the operative valve. The circuit thus exhibits a *bistable* characteristic and will change rapidly from one state to the other by the application of a positive pulse to the non-conducting valve.

A second pulse applied to the *conducting* valve has no effect. Hence it is usual to apply the pulses to both grids together. Each successive pulse then causes the circuit to change its state.

This circuit is often called the Eccles–Jordan circuit, after its originators, or alternatively, a "flip-flop" circuit.

Transistor Multivibrators

Any of the circuits just described may be constructed with transistors instead of valves, the three types being illustrated in Fig. 17.11. The action is similar to that for a valve circuit, but the performance is slightly modified by the finite resistance of the transistors. Thus in Fig. 17.11 (*a*) the capacitor C_1 has two discharge paths, through R_2 and through Tr_2, and since the latter current includes the collector leakage current, which is dependent on temperature, the behaviour of a transistor multivibrator is somewhat less stable than its valve equivalent and this must be allowed for in a practical design.

The performance of the monostable and bistable types may be improved by connecting small capacitors across R_2 and R_3 as shown dotted, since this assists in providing a rapid transition from one state to the other.

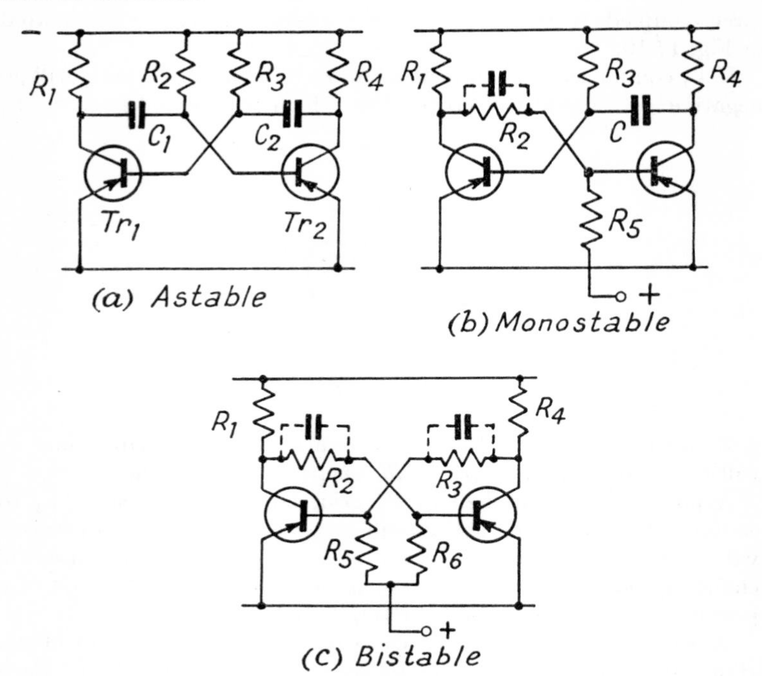

FIG. 17.11. TRANSISTOR MULTIVIBRATOR CIRCUITS

Switching Period

As said in Chapter 12, the frequency of the oscillation in a free-running (astable) multivibrator is given by $f = 1/2T$, where T is the switching period.

Consider the circuit of Fig. 17.11 (*a*). When Tr_1 is off and Tr_2 is on, the capacitor C_1 is charged to a voltage (very nearly) equal to the battery voltage V_B. When the multivibrator changes state, the collector voltage of Tr_1 falls very nearly to zero so that the base of Tr_2 is taken to a potential V_B positive to zero due to the charge on C_1. The capacitor now commences to discharge through R_2 towards the supply voltage $-V_B$, so that the voltage on the base of Tr_2 is given by

$$v_b = 2\,V_B(1 - e^{-T/C_1R_2}) - V_B$$
$$= V_B(1 - 2e^{-T/C_1R_2})$$

The transistor Tr_2 will switch on again when $v_b = 0$ (very nearly), which occurs when $2e^{-T/C_1R_2} = 1$. Hence

$$T/C_1R_2 = \log_e 2 = 0{\cdot}69.$$

The switching period T is thus $0{\cdot}69\ C_1R_2$.

A similar action occurs on the reverse cycle, the time being determined by C_2R_3. If the time constants are the same, so that $C_1R_2 = C_2R_3$ the waveform is symmetrical (unity mark-space ratio). If the time constants are not equal the two periods are different and the mark-space ratio is modified accordingly.

In a valve circuit the period is somewhat less because the valves do not fully "bottom," and in practice $T \simeq 0{\cdot}5\,CR$.

The Emitter-coupled Multivibrator

So far we have dealt with collector–base coupled multivibrator circuits. However, in some applications there are advantages in replacing one collector–base coupling with feedback via the emitters. The circuit of an emitter-coupled monostable device is shown in Fig. 17.12. Let us assume that the potential at the base of Tr_1

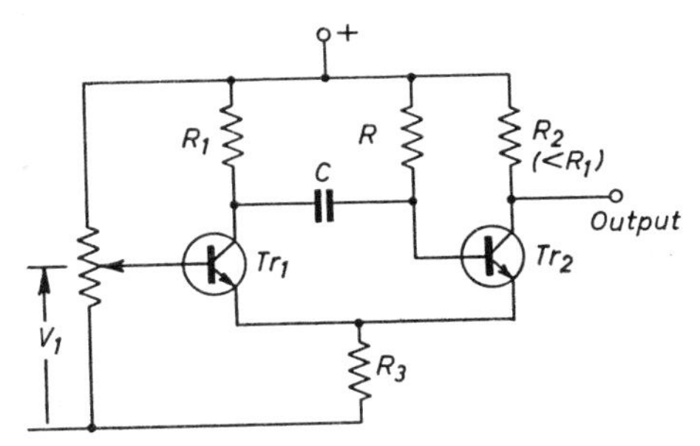

FIG. 17.12. EMITTER-COUPLED MONOSTABLE CIRCUIT

is adjusted so that in the normal state Tr_1 is cut off and Tr_2 is saturated. Application of a positive trigger to this base turns Tr_1 on and Tr_2 is turned off via C. Tr_1 now acts as an emitter-follower and stabilizes the common emitter potential at $V_1 - V_{be}$. C now charges through R until the base voltage of Tr_2 reaches V_1, when Tr_2 turns back on. Since R_2 is less than R_1, the emitter voltage now rises, cutting off Tr_1, and the consequent rise in collector voltage drives Tr_2 into saturation via C.

One advantage of this circuit is that R_2 is relatively isolated from the coupling paths and the output may therefore be conveniently taken from the collector of Tr_2. Another advantage is that the common emitter coupling allows a greater flexibility in choosing the operating points of the transistors than is possible in collector–base coupled circuits. This is particularly useful in high-speed circuits where the recovery time of a saturated transistor is significant. It is possible, by adjusting the circuit conditions, particularly making $R_1 \simeq R_2$, to make the emitter-coupled multivibrator "free run" i.e. operate in an astable mode. The analysis of the operation is somewhat complex, but it is possible to obtain a very unequal mark-space ratio, and this circuit is therefore often used when such a waveform is required.

Delay Circuit

Multivibrator circuits are used in many ways, apart from the generation of pulses. For example, the monostable circuit may be used to provide a variable delay, as shown in Fig. 17.13. Here an

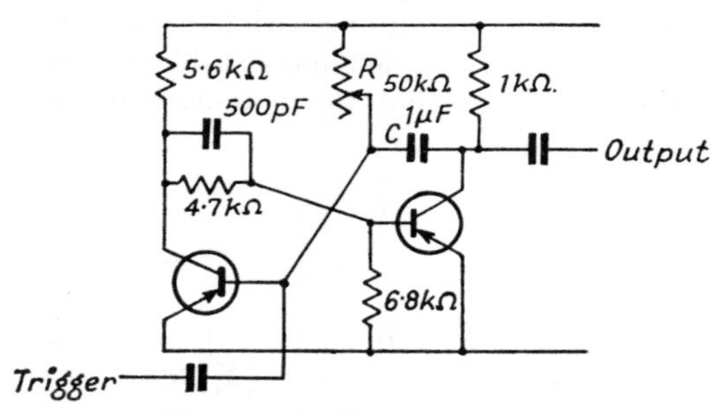

FIG. 17.13. DELAY CIRCUIT

input pulse is employed to trigger a circuit of the type of Fig. 17.11 (*b*), which then generates a single square-wave pulse of which the duration can be controlled by CR. The collapse of this pulse, when the circuit reverts to its normal condition, may be differentiated, so providing a second pulse delayed beyond the original pulse by a known amount.

Schmitt Trigger

It is sometimes required to have a circuit which will change its condition when the input exceeds a certain value and will revert to normal when the input falls below this critical value again. Such a circuit is called a *Schmitt trigger*, and Fig. 17.14 illustrates one way of producing the action.

The circuit is an asymmetrical monostable amplifier of which the operating state depends on the d.c. input level. Tr_1 is normally cut off by reason of the negative emitter voltage produced by the

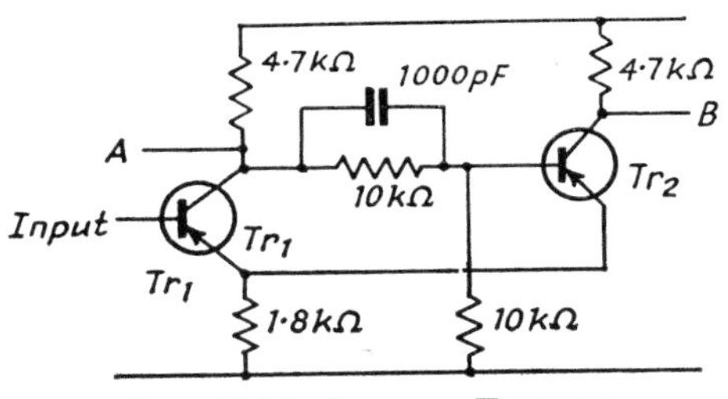

FIG. 17.14. SCHMITT TRIGGER

current in Tr_2. As the input rises Tr_1 begins to conduct and a rapid transition occurs to the alternative state with Tr_2 cut off. This persists until the input falls below the critical level again. The transition is rapid but does not require a pulsed input; it takes place when the input exceeds the critical value, even if slowly. The output may be taken from A (positive going) or B (negative going). The circuit may be modified so that the ON and OFF input levels are not identical, providing a (controlled) backlash which is sometimes useful.

Blocking Oscillators

An alternative form of pulse circuit is the *blocking oscillator* illustrated in Fig 17.15. Here the anode is coupled back to the grid through a transformer, so connected as to produce positive feedback. This causes a large pulse of anode current, but at the same time the grid is driven positive, resulting in a large pulse of grid current which charges the capacitance C negatively and so cuts off the anode current. The valve remains paralysed until this charge has leaked away, when a further pulse occurs, the successive pulses following one another at a rate determined mainly by the time constant CR.

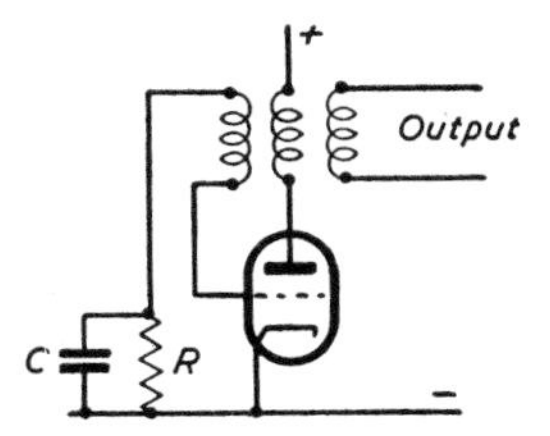

FIG. 17.15. BLOCKING OSCILLATOR

The repetition rate is not precise because it depends on the grid current and also on the transformer characteristic, but with proper design the circuit will produce very sharp pulses of fairly constant duration at repetition rates which can be controlled by varying R.

It is also clearly possible to synchronize the repetition rate from an external source, as with a multivibrator, by applying suitable pulses to the grid of the valve, and blocking oscillators are frequently used in frequency-dividing networks. The circuit can also be made to operate as a *single-shot* device by applying a large steady bias which cuts off the anode current. A trigger pulse large enough to overcome this bias will then produce a single pulse of operation, after which the

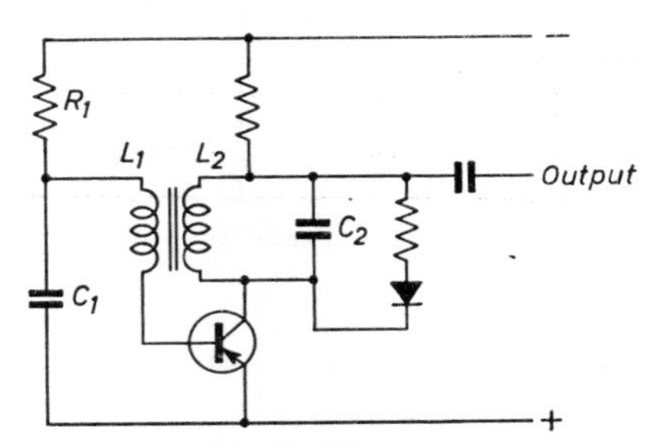

FIG. 17.16. TRANSISTOR BLOCKING OSCILLATOR

circuit resumes its quiescent state. Used in this way the circuit will respond accurately to incoming pulses whether they arrive regularly or at random.

A transistor version of the circuit is shown in Fig. 17.16. At switch-on the feedback via L_2L_1 produces a rise in base current which charges the capacitor C_1 causing V_b to rise and cutting the transistor off. The capacitor then discharges through R_1 until the transistor again conducts. The circuit thus generates bursts of collector current at a repetition frequency determined by the time constant R_1C_1, the duration of the pulses being controlled by the resonant frequency of the circuit L_2C_2. The windings must be tightly coupled, and a diode is often connected across L_2 to suppress the reverse voltage (overshoot) generated when the collector current is cut off. With suitable choice of constants the circuit will provide substantially rectangular pulses of (mean) collector current.

17.3. Operational Circuitry

For many purposes it is required to develop e.m.f.s which are derivatives of the original signal. With suitable circuitry a wide range of mathematical functions can be performed, some of the simpler examples being described below.

Differentiation and Integration

Differentiation or *integration* of a waveform consists in producing another waveform which is approximately proportional to the derivative de/dt or the integral $\int e dt$ of the original wave. The required accuracy varies considerably with the application; the closer the waveform has to be to the true derivative or integral, the more elaborate is the circuit required.

Fig. 17.17 (*a*) shows a differentiating circuit in which the voltage across R is proportional to the rate of change of the voltage applied.

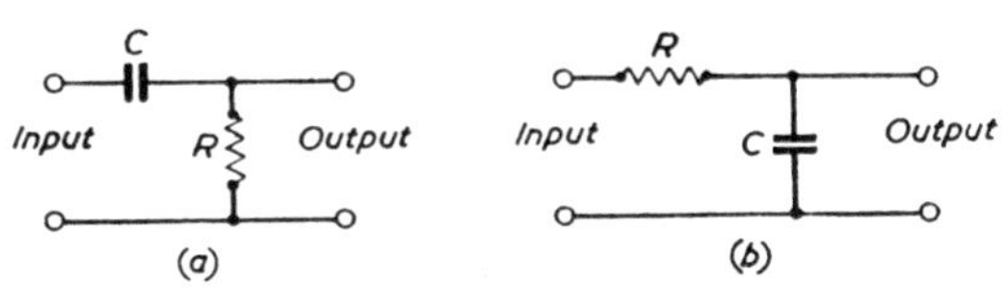

FIG. 17.17. SHAPING CIRCUITS
(*a*) Differentiating; (*b*) Integrating

For this to be strictly true the resistance R must be negligible in comparison with the reactance of C at the frequency of the highest (important) harmonic of the input wave.

This may be proved as follows. The charge on the capacitor is $q = Cv_c$. The current through the circuit is the rate of change of charge and is $C dv_c/dt$, so that the voltage across R is $CR dv_c/dt$. But if R is small, the voltage on the capacitor is practically equal to that across the whole circuit, so that the output voltage $= CR dv/dt$ very nearly.

Clearly this stipulation that R is small compared with the reactance of C means that the circuit will attenuate considerably, but this is often offset by the fact that the rate of change is large. Nevertheless an amplifier is usually required following the resistance to provide a satisfactory signal.

With a pure sine wave, R must be small (less than about 10%) compared with $1/C\omega$. The rate-of-change (differential) wave is then a second sine wave lagging 90° behind the input wave. With a wave having a steep front (such as a synchronizing impulse or a transient), R must be still smaller, for a steep-fronted wave can be analysed into a fundamental and a series of high-order harmonics, and the reactance of C to these high harmonics is much less than it is to the fundamental. R will generally need to be about one-tenth of the value which would suffice for a sine wave.

Fig. 17.17 (*b*) shows an integrating circuit. Here the voltage on the capacitor $= \frac{q}{C} = \frac{1}{C}\int i\, dt$. But if the reactance of C is small compared with R (the reverse of the previous condition), $i = v/R$ nearly, so that the voltage across the capacitor $= \frac{1}{CR}\int v\, dt$.

With a sine-wave input, making $1/C\omega = 0{\cdot}1R$, or less, the output voltage is a sine wave leading by nearly 90°. It will be noted that this form of circuit is the same as is used for filtering in power supply circuits (see Chapter 11).

Operational Amplifiers

As said, the basic circuits of Fig. 17.17 only give approximate differentiation and integration. However, if an amplifier with very high negative voltage gain is included in the circuit, substantially perfect operation can be achieved.

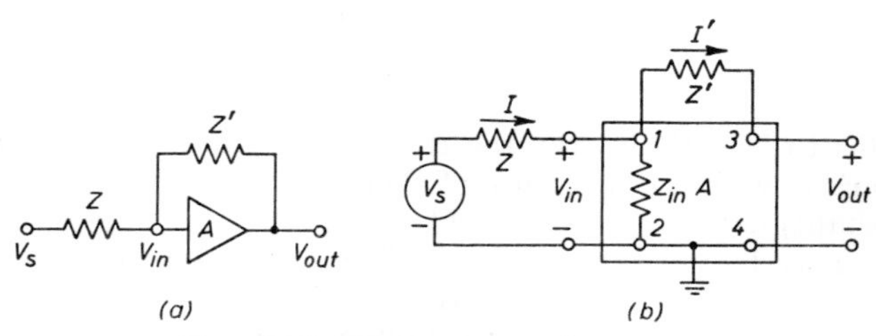

FIG. 17.18. OPERATIONAL AMPLIFIER
(a) Basic circuit (b) Expanded circuit

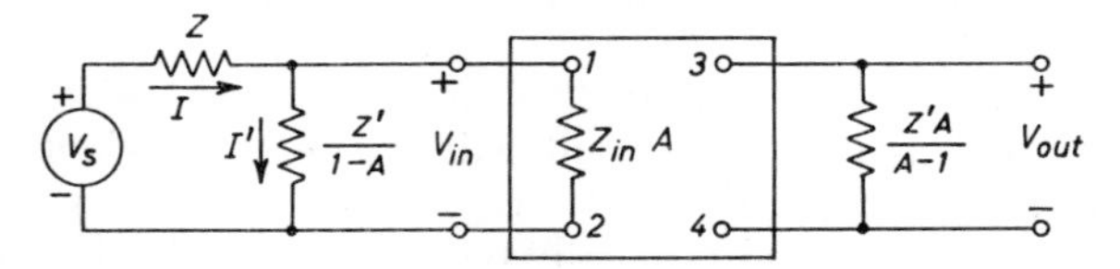

FIG. 17.19. EQUIVALENT CIRCUIT OF FIG. 17.18.

If we consider such an amplifier connected as shown in Fig. 17.18 (a,) the circuit may be expanded as shown at (b) where Z_{in} is the input impedance of the amplifier and A the voltage gain with Z' connected. We may now draw an equivalent circuit as shown in Fig. 17.19, where the same input current I is drawn from the source and the same input and output voltages are obtained. Now, if $Z'/(1 - A)$ is much larger than Z_{in}, $I' \simeq I$, so that the output voltage is

$$V_{out} = AV_{in} = AIZ'(1 - A)$$

As A tends to infinity, the impedance across terminals 1 and 2 tends to zero, and it is said that a "virtual earth" exists at terminal 1. Under these conditions $I \simeq V_s/Z$ and $V_{out} \simeq -IZ'$, so that the overall voltage gain is

$$V_{out}/V_{in} = -Z'/Z$$

Now, if Z is a capacitor C and Z' is a resistor R, as shown in Fig. 17.20 (*a*), and V_{in} is an alternating input $v(t)$, the input current

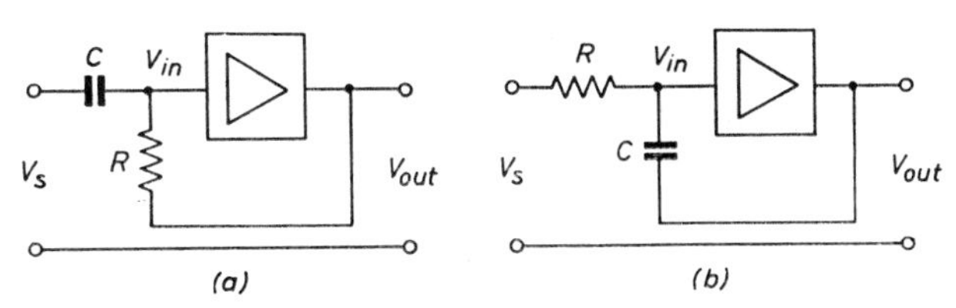

FIG. 17.20. FORMS OF OPERATIONAL AMPLIFIER
(*a*) Differentiation (*b*) Integration

$i = C\,dv/dt$ and $V_{out} = -Ri = -RC\,dv/dt$. Hence the output voltage is the differential of the input.

Integration is performed with the arrangement shown in Fig. 17.20 (*b*), where Z is a resistor and Z' is a capacitor. A particular case of the integrator occurs when the input voltage, V, is constant;

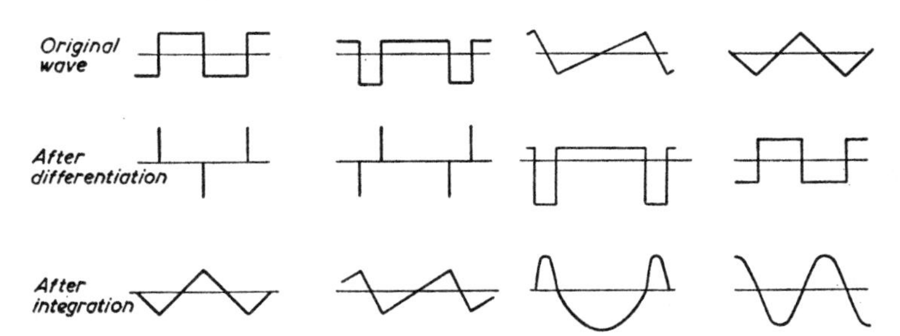

FIG. 17.21. DIFFERENTIATION AND INTEGRATION OF SOME IDEAL WAVEFORMS

the output is then given by $V_{out} = -Vt/RC$. This is a voltage rising linearly with time, or a *linear ramp*. This circuit is known as the *Miller integrator* and is discussed in more detail later.

Circuits of this type can be used to perform a variety of mathematical operations. They are therefore known as *operational amplifiers*, and are often made up, either singly or in combination, in the integrated-circuit form discussed in Chapter 18. Fig. 17.21

illustrates the effect of differentiation and integration on some simple waveforms.

Differential Amplifiers

In the previous section we have considered the operational amplifier to have a single-ended input, that is to have one side of its input referred to earth. It is often convenient, however, to use an amplifier which amplifies the difference voltage between its two inputs, neither of which is necessarily at earth. This allows greater flexibility in both input and feedback conditions and most commercially

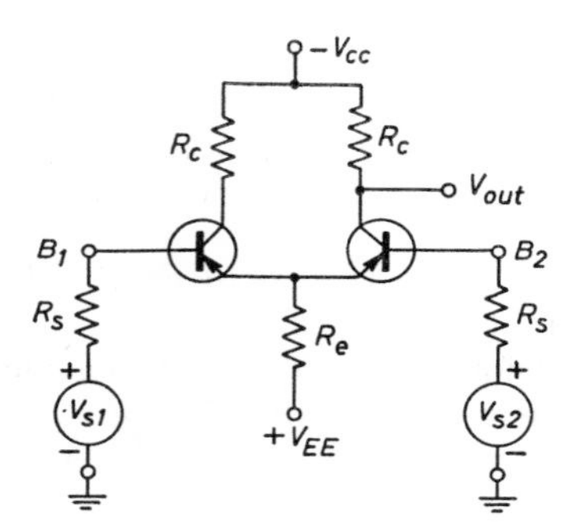

FIG. 17.22. LONG-TAILED PAIR

available operational amplifiers use this principle, though they can, of course, be used in the single-ended mode by connecting one input to earth.

The majority of differential amplifiers are based on the principle of the *long-tailed pair* illustrated in Fig. 17.22. Analysis of this circuit shows that, if the emitter resistance R_e is very large and the circuit is symmetrical, the rejection of common-mode voltages is very high and the output voltage V_{out} is a true amplified representation of the difference between V_{s1} and V_{s2}. (Common-mode voltages are those which affect both inputs in the same direction.)

There are, however, practical limitations on the value of R_e, since the supply voltage V_{cc} is usually limited, and if the operating currents of the transistors are reduced too much, the characteristics will be degraded. In practice, therefore, R_e is frequently replaced by a transistor circuit connected as shown in Fig. 17.23. This transistor, Tr_3, acts as a close approximation to a constant-current source so that a very high effective emitter resistance is presented to Tr_1 and Tr_2.

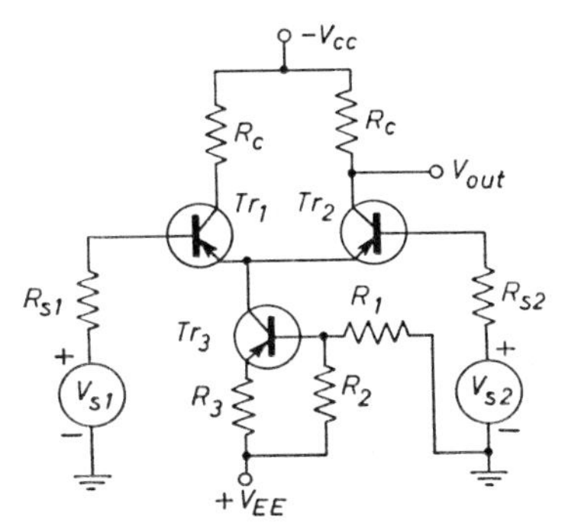

Fig. 17.23. Constant-current Emitter Circuit

Average Values

It is sometimes required to find the r.m.s. and/or mean values of a square wave or a pulse. This can be calculated from the relevant Fourier series, but this is apt to be cumbersome and it is usually easier to obtain the values by simple proportion.

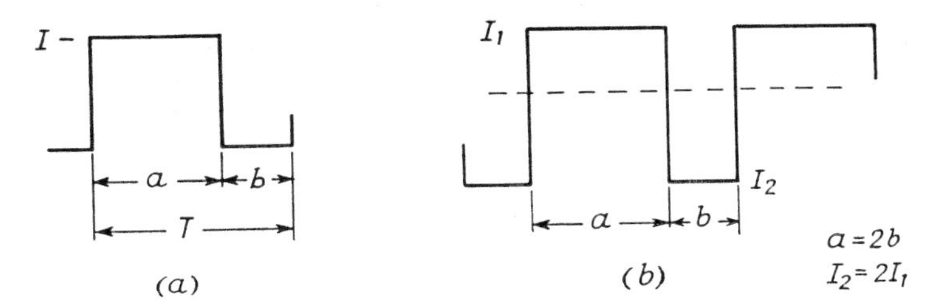

Fig. 17.24. Illustrating Calculation of Average Values

Thus for the waveform in Fig. 17.24 (a) the current flows for $a/(a + b)$ of the total period. Hence

$$\text{Mean value} = Ia/(a + b)$$
$$\text{R.m.s. value} = \sqrt{[I^2a/(a + b)]} = I\sqrt{[a/(a + b)]}$$

If the current is negative for part of the cycle, as in Fig. 17.24 (b) the two portions must be considered separately. Thus

$$\text{True mean} = I_1a/(a + b) - I_2b/(a + b)$$
$$\text{This is usually} = 0\text{, which makes } I_2 = I_1a/b$$

The rectified mean is $I_1a/(a + b) + I_2b/(a + b)$, which for the case shown is $4I_1/3$.

$$\text{The r.m.s. value} = \sqrt{[I_1^2a/(a + b) + I_2^2b/(a + b)]}$$

which for the case shown is $I_1\sqrt{2}$.

Rise Time

The pulses produced by the various circuits discussed (and others) are usually required to have a steep wavefront with a similarly sharp tail, though this is often less important. The shape of the pulse can be controlled by suitable attention to the time constants. Thus the connection of suitable capacitances across R_2 and R_3 in Fig. 17.11 will assist a rapid transition from one state to the other.

The maintenance of the full amplitude throughout the duration of the pulse usually presents no problem, but if the pulse is long, and the circuit contains a.c. couplings, attention to the low-frequency time constants may be necessary.

If the pulses are to be amplified, however, it is important to ensure that the amplifier shall reproduce the shape of the pulse as nearly as possible, and in particular that it shall not slow up the rise time. This clearly requires a frequency response much greater than the pulse repetition rate, and the circuits have to be designed specifically to accommodate the sharp transients involved. A waveform which suddenly changes its level is known as a *step function*, and circuits to handle such functions must be related not to the pulse rate, but the rise time.

It was shown in Chapter 1 that a pulse contains a converging series of odd harmonics which, in a square wave, diminish in amplitude in inverse proportion to the order of the harmonic. With a short-duration pulse the amplitudes of the harmonics are different, but a reasonable reproduction of the wavefront can be achieved if the amplifier response is capable of handling harmonics of the fifth order relative to the *equivalent* fundamental (i.e. an imaginary fundamental having a period equal to twice the pulse duration).

For a more precise calculation the behaviour of the circuit must be analysed with specific reference to the step function involved. The various capacitances (both external and internal) can only acquire their altered potential after a finite time determined by the (small) series resistance present. Moreover with fast rise times of the order of a few nanoseconds* the inductance of the leads becomes important and can produce an overshoot as in Fig. 17.25.

The start may also be slightly delayed, so that the rise time is

* A nanosecond is 10^{-9} second, i.e. one thousandth of a microsecond.

usually taken as the period from 10 to 90 per cent of full amplitude. A convenient empirical criterion is given by the expression

$$\text{Rise time in seconds} = k/B$$

where B is the (3 dB) bandwidth and k is a factor which lies between 0·35 and 0·45, depending on the characteristics of the amplifier. With a critically-damped amplifier (in which the response falls off naturally due to the stray capacitances) $k = 0{\cdot}35$. With a maximally-flat amplifier (where the response is artificially prolonged, followed by a rapid cut-off) $k = 0{\cdot}45$, in which case there is a small overshoot as shown in Fig. 17.25.

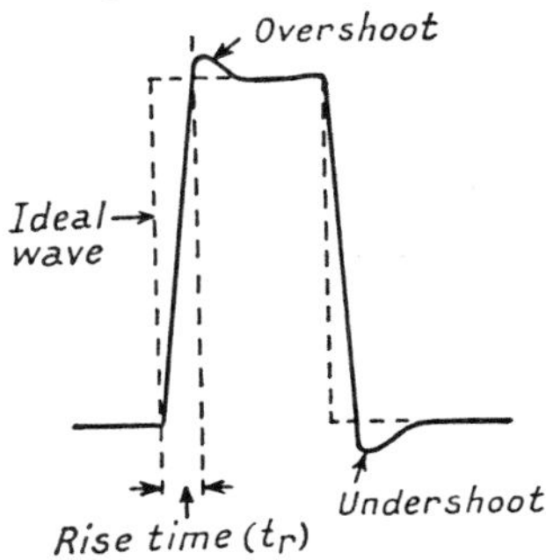

FIG. 17.25. ILLUSTRATING DEVIATION OF PRACTICAL PULSE FROM THE IDEAL

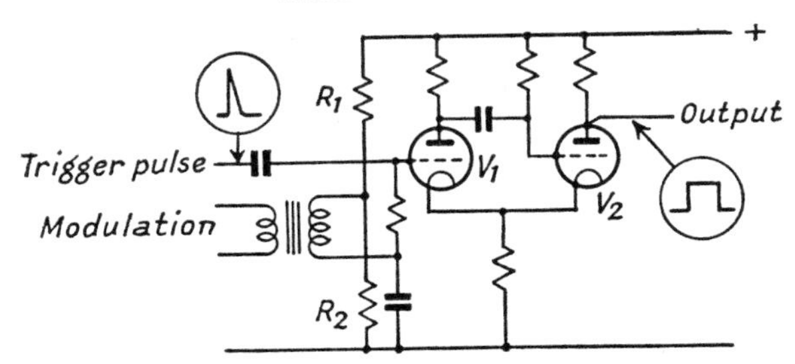

FIG. 17.26. CIRCUIT FOR PRODUCING CONTROLLABLE PULSE WIDTH

Pulse-width Control

For many purposes pulses are required having a flat top and an accurately-controlled duration. A convenient technique here is to generate a trigger pulse from a relaxation oscillator arranged to deliver sharp spikes—i.e. pulses having a steep wavefront with a rapid tail. This is then used to trigger a monostable multivibrator of the type shown in Fig. 17.9. The duration of the pulse can then be accurately controlled either by the time constant C_1R_1, or by varying the grid bias on V_1, as stated on page 704.

By using grid-bias control it is possible to make the pulse width proportional to an input signal (as is required for pulse modulation). Fig. 17.26 shows a circuit for doing this. The standing bias on V_1 is set by suitable choice of R_1 and R_2. The modulation then varies

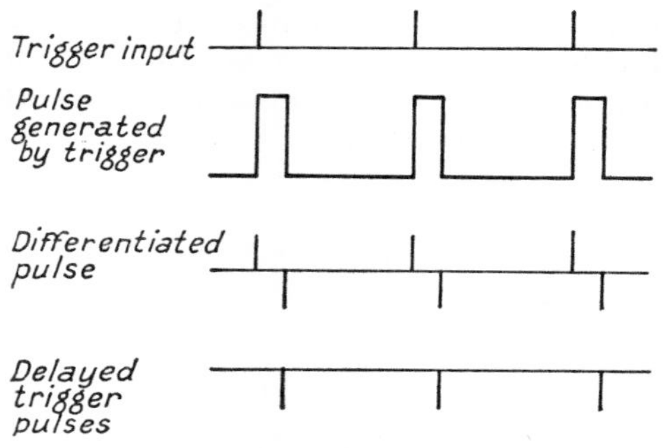

FIG. 17.27. ILLUSTRATING PRODUCTION OF DELAYED TRIGGER PULSES

this bias and so alters the pulse width. The modulation input may be d.c. coupled if necessary.

A similar circuit can be developed from the transistor multivibrator of Fig. 17.11 (*b*).

Subsidiary trigger pulses may be derived from such a circuit for pulse-position modulation or other requirements (such as the delay circuit of Fig. 17.13). If the pulse is differentiated, as in Fig. 17.27,

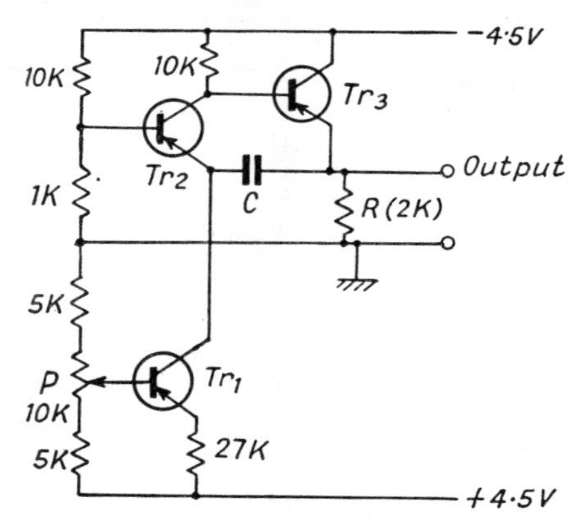

FIG. 17.28. CIRCUIT FOR PRODUCING TRIGGER PULSES

spikes will be produced (in opposite directions) by the leading and trailing edges. The positive spikes are suppressed (e.g. by arranging that they produce bottoming or cut-off) leaving only the delayed negative spikes.

Various circuits may be used to produce the initial trigger spikes. The arrangement in Fig. 17.28 shows a suitable form of relaxation oscillator. Tr_2 is normally cut off. As the capacitor C charges (at constant current) through Tr_1 the emitter voltage of Tr_2 falls until it becomes positive to the base when Tr_2 conducts which turns Tr_3 on, causing rapid discharge of C which produces a spike across R. Tr_2 is thus immediately cut off again for a relatively long period while C recharges. The repetition frequency is determined by the current through Tr_1, which is controlled by the voltage divider P.

Multiplying Circuits

Occasions arise which require an output proportional to the product of two (or more) inputs, a typical instance being the measurement of power which is the product of the instantaneous voltage and current. A frequency-changer and a square-law detector both

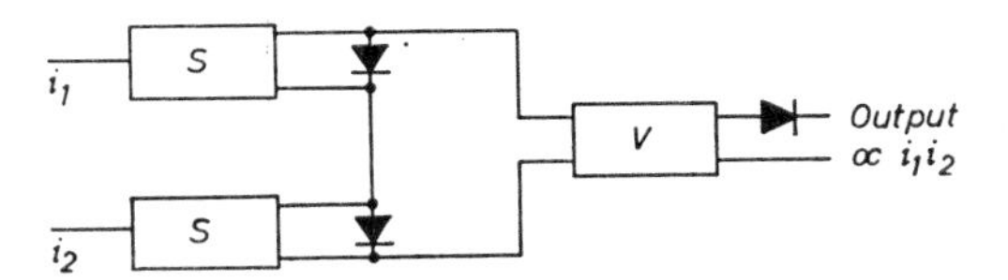

FIG. 17.29. DIODE MULTIPLIER CIRCUIT
S. Constant-current source *V*. Buffer amplifier

contain a product term in the output, but where the inputs are of the same frequency this product term cannot be suitably isolated.

A simple multiplier is shown in Fig. 17.29, which uses a combination of diodes. These will have a characteristic of the form $i = i_o e^{kv}$, so that $v = (1/k) \log_e (i/i_o)$. If i_o can be considered (or held) constant, this simplifies to $v = k_1 \log_e i$. If the diodes D_1 and D_2 are then fed (through suitable constant-current sources) with currents i_1 and i_2, and the voltages across them are added together, the overall voltage is proportional to

$$\log_e i_1 + \log_e i_2 = \log_e i_1 i_2$$

This voltage can then be applied across a third diode which will deliver a current proportional to $i_1 i_2$.

Alternatively, the inputs may be applied to the long-tailed pair circuit of Fig. 17.30. This circuit is usually used with a fixed "tail" resistance, in which case it produces a differential amplifier as previously described, but if the tail current is variable, as shown, the

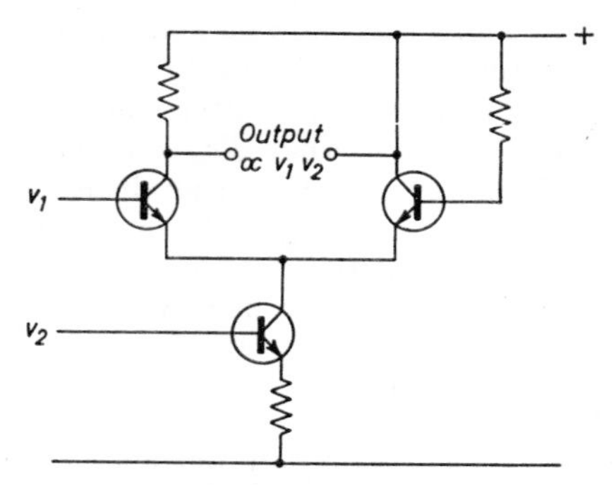

FIG. 17.30. LONG-TAILED PAIR MULTIPLIER

output is proportional to $v_1 v_2$. Circuits of this type will operate from d.c. to 100 kHz with accuracies of the order of 2 per cent.

Pulse-width Multipliers

The pulse circuitry previously discussed can be adapted to provide multiplication by arranging that the height of the pulses is controlled by one of the inputs, while the second input controls the width of the pulses by one of the standard width-control circuits such as that of Fig. 17.26. If the output is then rectified, the mean value will be directly proportional to the product of the two inputs.

The arrangement is more elaborate, but with adequate precautions it can provide a high degree of reliability and accuracy.

Hall-effect Multiplier

If a slab of semiconductor material carrying a normal (ohmic) current is subjected to a magnetic field at right angles to the direction of the current flow, an e.m.f. is developed across the material in the Z axis, as shown in Fig. 17.31. This effect is used in certain specialized applications, and can obviously be applied to provide a multiplier.

The main disadvantage is in the additional circuitry involved, particularly since the e.m.f. developed is only a few microvolts so that substantial subsequent amplification is required, but it operates in all four quadrants and so provides output of the correct polarity with input signals of any phase relationship.

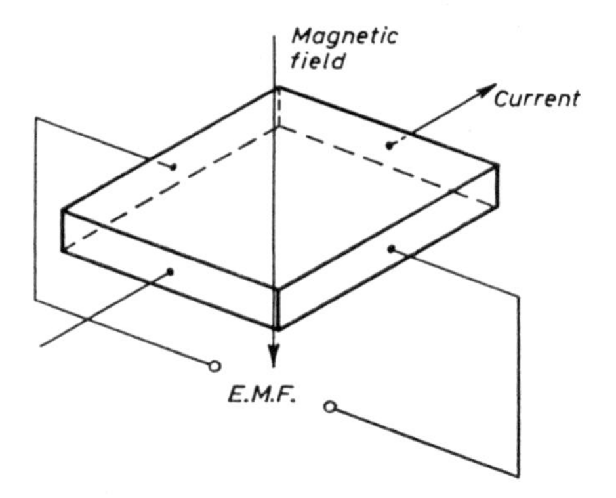

FIG. 17.31. ILLUSTRATING THE HALL EFFECT

17.4. Sawtooth Generation

A particular waveform which finds widespread application is the linear voltage sweep or sawtooth. This waveform is the basis of the display systems used in television, in oscilloscopes and in some radar systems, and is also used in various control systems to time-programmed functions.

Some types of circuit used for the generation of sawtooth waveforms have been mentioned in Chapter 15, and unijunction transistors can also be used as described in Chapter 18. However, when highly linear voltage sweeps are required, the two most commonly used circuits are the *bootstrap* and the *Miller integrator*.

The Bootstrap

Bootstrapping is the term used to describe a positive feedback process which linearize the sweep by applying a correction to overcome the effect which the increasing voltage on the capacitor being

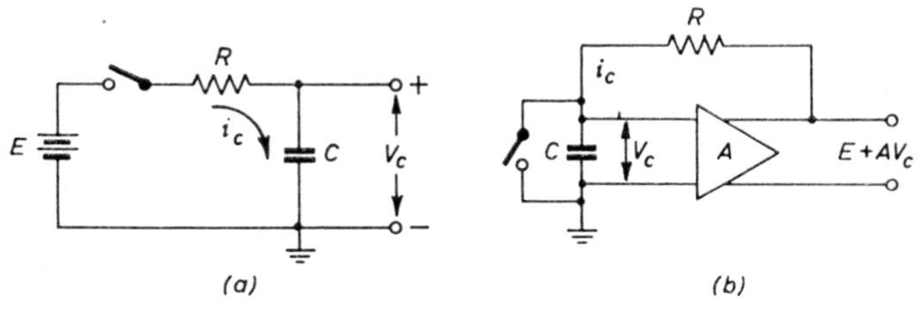

FIG. 17.32. SAWTOOTH GENERATION

charged would otherwise have on the sweep linearity. If we consider the circuit of Fig. 17.32 (*a*), we see that the charging current i_c is dependent on the capacitor voltage V_c, since $i_c = (E - V_c)/R$. However, if we now add an amplifier as shown in Fig. 17.32 (*b*), we see that $i_c = (E + AV_c - V_c)/R$, where A is the amplifier gain. If A is exactly unity, perfect constant current charging of C is obtained. This circuit represents the general form of bootstrap linear voltage sweep in which positive feedback is employed to increase the effective charging-source voltage at the same rate as the increase in the sweep voltage V_c.

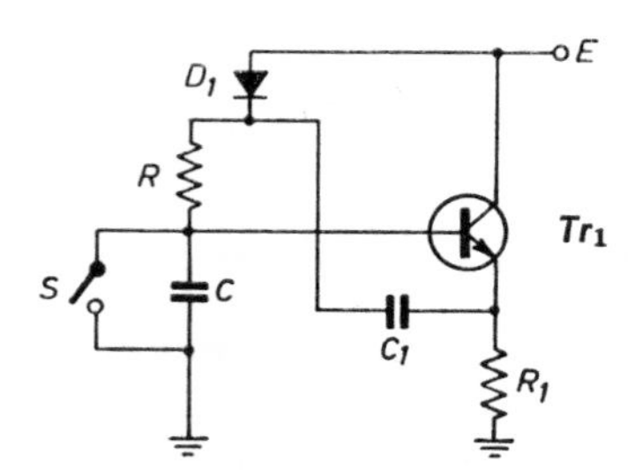

FIG. 17.33. PRACTICAL BOOTSTRAP CIRCUIT

A practical form of bootstrap circuit is shown in Fig. 17.33. Here the energy for the voltage sweep is stored in the capacitor C, which is charged through D_1 and R_1 during the period when the switch S is closed. When the switch is opened, the capacitor C begins to charge through R and the base voltage on Tr_1 rises. Since Tr_1 is connected as an emitter follower, the emitter voltage also rises, and since C_1 is fully charged, the diode D_1 is reverse biased and all the current charging C is drawn from C_1. If the current gain of Tr_1 is high, the base current drawn can be neglected and we can consider that charge is simply transferred from C_1 to C through R, so that if C_1 is very much larger than C, substantially constant current charging will be obtained.

At the end of the sweep switch S is closed, C discharges and allows D_1 to conduct so that the charge drawn from C_1 is restored. It should be noted that, since C_1 is very much larger than C, a significant time will be required to recharge C_1 through R_1. The switch S may be formed by a transistor which is normally saturated and is cut off during the sweep period. The total recovery time of the circuit would then include the time constant and switching time of the discharge path established through this transistor.

The Miller Integrator

The Miller integrator, like the bootstrap, incorporates an amplifier in the charging circuit in order to linearize the sweep. However, in this configuration negative voltage feedback is employed and the linearity is substantially independent of amplifier gain, provided the voltage gain is high. (The principle of the use of high-gain amplifiers to obtain ideal integration was discussed earlier, in the section on "Operational Amplifiers".)

A practical form of Miller integrator is shown in Fig. 17.34. The switch S is normally closed, the transistor is cut-off, and the capacitor C is charged to $+E$. When the switch is opened, the transistor conducts and collector current flows through R_2 causing the collector voltage to fall; this voltage drop is fed back to the base, and the

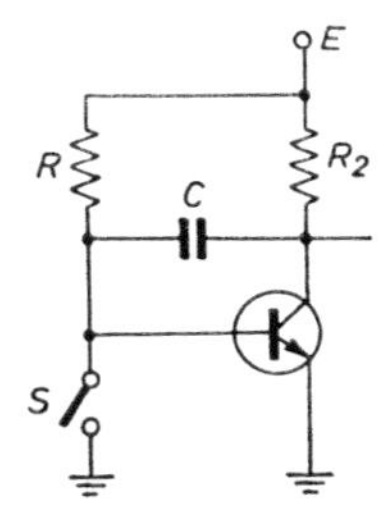

FIG. 17.34. MILLER INTEGRATOR

resulting change in base voltage is due to a combination of this voltage feedback with the effect of the current flow through R. If the stage had infinite voltage gain, the base voltage would remain constant whilst capacitor C charged through R, and the collector current would also remain constant, so that a linear voltage run-down would be obtained at the collector. In practice, the finite gain of the stage leads to some non-linearity, but the voltage gain of the stage can be made so high that this non-linearity is insignificant, particularly if Tr_1 is replaced by a Darlington pair as described in Chapter 6.

As in the bootstrap, the switch S may be formed with a transistor, since the collector saturation voltage will be sufficiently below the base cut-off voltage. It should be noted that, when the switch is opened, allowing the sweep to commence, a small voltage step will be seen at the collector before the linear run-down commences. This is due to the fact that there will be a small change in voltage at the base when the switch is opened, and this step will be reflected at the collector. The effect is not serious unless low-voltage supplies

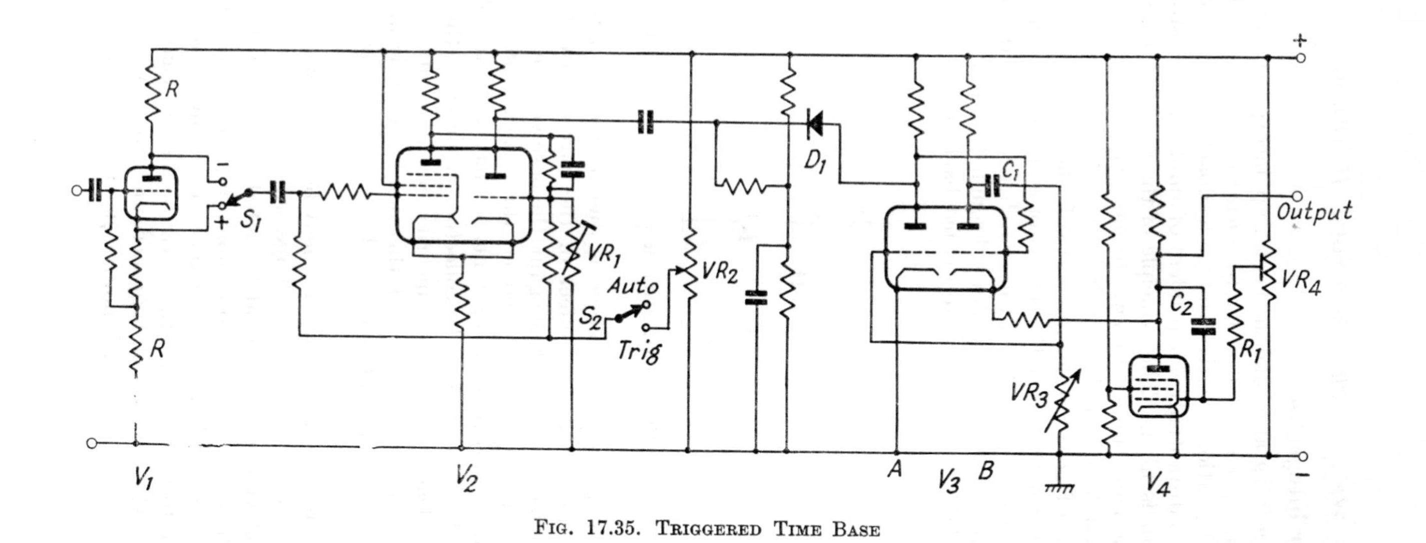

FIG. 17.35. TRIGGERED TIME BASE

are to be used, but the step can be eliminated if necessary by adding a small resistor $r = 1/g_m$ in series with the capacitor.

Triggered Time Bases

Mention was made in Chapter 15 of time bases which are normally quiescent, but which produce a single sweep in response to a suitable trigger impulse. Various forms of circuit are employed for this purpose, one particular arrangement which uses a Miller integrator as the sawtooth generator being shown in Fig. 17.35.

The trigger signal is fed to the grid of V_1 which produces antiphase signals at anode and cathode; the switch S_1 selects either the anode or cathode waveform and enables a positive pulse to be fed to V_2 irrespective of the polarity of the input pulse. This valve is a monostable multivibrator with the triode section normally on; a positive pulse turns the pentode section on and with S_2 switched to AUTO the turn-on level is preset by VR_1. When S_2 is closed, the triggering level may be adjusted by VR_2.

The output waveform is differentiated and the resultant negative pulse causes the diode D_1 to conduct. V_{3A} is normally cut-off and V_{3B} is conducting so that the anode of V_4, the Miller integrator, is held at a high potential. When a negative pulse is fed through D_1, the anode voltage on V_{3A} falls, causing the valve to conduct, the threshold being determined by the stability control VR_3. V_{3B} is then cut off, allowing anode current to flow in V_4. The anode voltage of the Miller integrator now decreases linearly at a rate determined by C_2R_1 and the positive return voltage set by VR_4. The linear run-down of V_4 anode voltage continues until the anode bottoms. The anode voltage then remains at near earth potential until V_{3B} starts to conduct again after a period determined by the capacitor C_1; the gate valve V_3 is then reset ready to operate the next time a pulse triggers V_2.

A range of sweep speeds can be selected by switching C_1 and C_2, and intermediate steps can be obtained by variation of R_1.

The principle may be applied equally well to transistor circuitry.

18

Switching and Control Circuitry

SINCE semiconductor devices in general exhibit a sharp transition from the non-conducting (reverse-biased) condition to the conducting (forward-biased) state they can be used to perform a wide variety of switching operations. Fig. 18.1 shows an idealized diode characteristic. If the voltage across the diode is less than OA the current will be negligibly small, apart from any leakage current. When the

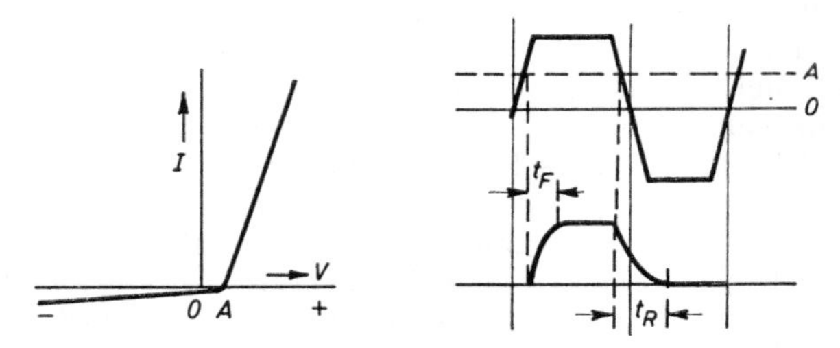

FIG. 18.1. ILLUSTRATING FORWARD AND REVERSE RECOVERY TIMES

voltage exceeds OA the diode passes to the conducting state and appreciable current flows.

The transition, however, is not instantaneous because it is necessary to dissipate the small charge stored in the internal capacitance of the diode. In the "off" state this charge is small because of the small current so that there is only a minimal delay in changing to the "on" state; but in this state the charge is appreciably greater so that the reverse transition occupies a longer time interval. The intervals, which are known as the *forward* and *reverse recovery times*, are very small, being measured in nanoseconds, so that they are only significant in high-speed switching operations.

A similar effect occurs when the diode constitutes the emitter–base junction of a transistor. The collector current of a transistor in common-emitter mode will exhibit an abrupt transition from "off" to "on" when the base voltage exceeds the hop-off value, but there is again a small, but finite, delay in the reverse transition. Both forward and reverse recovery times are somewhat greater with a transistor.

There is also the effect of leakage current, which increases the effective "off" current. With germanium transistors this may be appreciable and increases rapidly with temperature, so that switching transistors are usually silicon types.

The relative importance of these various factors depends upon the circuit requirements. There are two main classes of use. The first is for relatively low-speed control circuitry in which the switching time is of minor importance. The second application is concerned with the use of diodes and transistors as two-state devices in what is called *logic* circuitry, by means of which highly sophisticated functions can be performed. The requirements are appreciably different, so that they are considered separately in the sections which follow.

Where high-speed operation is of importance special manufacturing techniques are adopted, to which some reference will be found in Appendix 4. With such constructions the reverse recovery time can be reduced to a few nanoseconds with simple diodes; with transistors the figure is somewhat greater.

18.1. Switching Devices

A simple control circuit is illustrated in Fig. 18.2 in which a transistor in common-emitter mode is used to operate a relay. Assuming a silicon transistor, which has negligible leakage current, the collector current will be virtually zero with zero input on the base. A positive input greater than the hop-off (approximately 0·6 V) will turn the

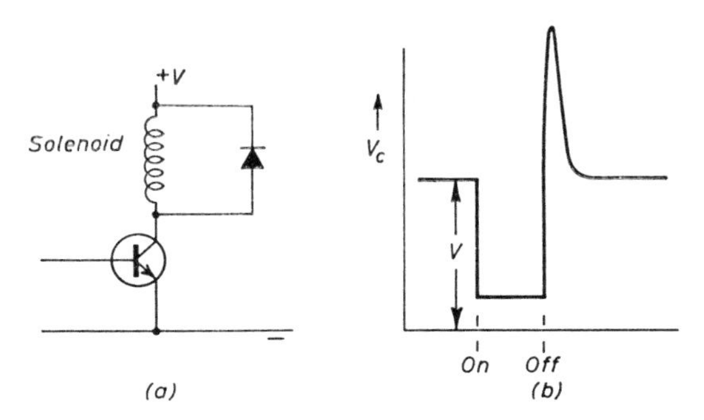

FIG. 18.2. TRANSISTOR SWITCHING CIRCUIT

transistor on, and since quite small transistors will handle currents of up to 1 A, the arrangement constitutes a sensitive relay requiring negligible input power. For example, if $\beta = 100$, an output current of 0·5 A will require a base current of 5 mA, which with an input voltage of one volt only involves an input power of 5 mW.

The use of the protective diode across the relay should be noted. In the "on" condition the collector voltage will fall nearly to zero because of the voltage drop on the relay, but when the transistor is turned off, the voltage will rise momentarily above the supply voltage by reason of the transient e.m.f. generated by the collapse of the magnetic field in the relay coil, as shown in Fig. 18.2 (*b*). This over-voltage may destroy the transistor so that a diode is connected across the coil to absorb the surge and limit the voltage rise to a safe figure. The diode must be capable of dissipating the stored energy ($\frac{1}{2}LI^2$).

This kind of protection is essential with any transistor circuit having an inductive load, since even a momentary voltage in excess of the rated maximum will destroy the transistor. It is also important to ensure that during the "on" condition there is no risk of thermal runaway (cf. page 297).

Thyristors

Where appreciable power is to be handled, a more convenient form of control can be obtained by using a modified form of device known as a *thyristor*, or silicon controlled rectifier (s.c.r.). This

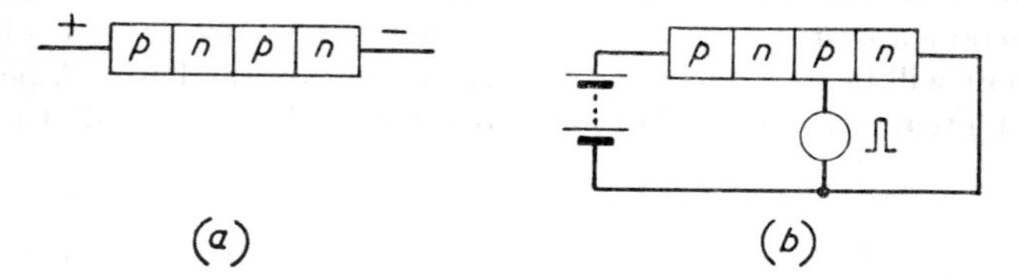

FIG. 18.3. BASIS OF SILICON THYRISTOR

consists essentially of a four-section p–n–p–n assembly as shown in Fig. 18.3. If a small e.m.f. is applied across this circuit in the direction shown, only leakage current will flow because the central section is reverse biased. As the voltage is increased the current increases slowly until a point is reached when the junction between the central n–p regions breaks down permitting a rapid and multiplicative formation of electron-hole pairs (avalanche effect), which results in a large increase in current, limited only by the saturation of the material and/or the resistance of the external circuit. In this condition the voltage drop across the device becomes very small (of the order of a volt). The behaviour is illustrated in Fig. 18.4; once the breakdown has occurred it can only be stopped by reducing the current below the critical value at which the n–p barrier breaks down. This is called the *holding current*.

Now, if instead of increasing the voltage, electrons are injected into the central region as in Fig. 18.3 (*b*), this will trigger the avalanche effect, which can thus be controlled by an external pulse. The *gate current* required is only small, and in practice a current of 50 A can be initiated with a gate pulse of the order of 30 mA only.

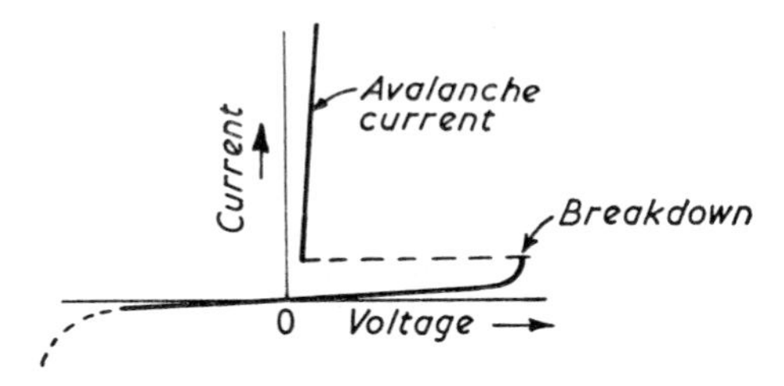

FIG. 18.4. THYRISTOR CHARACTERISTIC

Once the current has been initiated it will continue to flow as long as the supply voltage is maintained (above a small holding value, usually of the order of 2 V), irrespective of the gate potential. Hence a thyristor can be triggered by a single pulse lasting only a fraction of a millisecond.

Regenerative Action

The current builds up by a regenerative action which may be analysed more specifically by considering the device as two interconnected transistors, as shown in Fig. 18.5. The collector current of Tr_1, $I_{C1} = \beta I_{B1} + I_{co}/(1 - \alpha)$, as was shown in Chapter 6 (page 272).

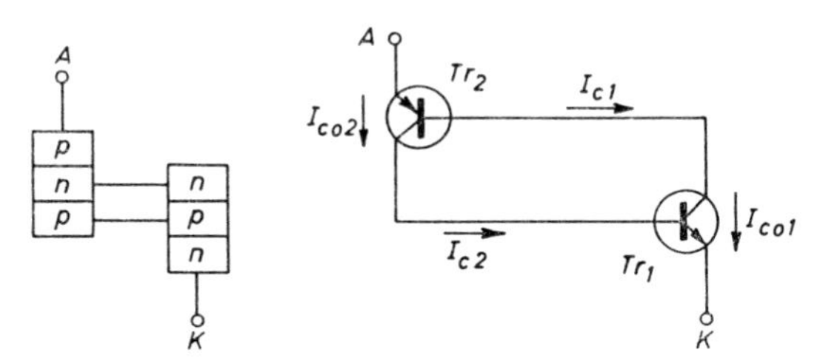

FIG. 18.5. EQUIVALENT CIRCUIT OF THYRISTOR

But $I_{B1} = I_{C2}$, while for $1/(1 - \alpha)$ we may write $1 + \beta$. Hence

$$I_{C1} = \beta_1 I_{C2} + (1 + \beta_1)I_{co1}$$

and

$$I_{C2} = \beta_2 I_{C1} + (1 + \beta_2)I_{co2}$$

The total current is $I_{C1} + I_{C2}$ which from these two equations simplifies to

$$I_A = \frac{(1 + \beta_1)(1 + \beta_2)(I_{co1} + I_{co2})}{1 - \beta_1\beta_2}$$

In the "off" condition β_1 and β_2 are very small and I_A is virtually simply the total leakage current $I_{co1} + I_{co2}$. A forward bias on Tr_1, however, produces an appreciable current which will also flow through Tr_2, so that both β_1 and β_2 increase cumulatively and the term $1 - \beta_1\beta_2$ approaches zero, producing a theoretically infinite current.

In practice, the resistance of the external circuit must be chosen to limit the current to a safe value. This may be simply the saturation current of the transistors, in which case the device is called a non-avalanche type, but for larger powers the avalanche effect may be utilized, permitting currents of several hundred amperes to flow under suitably controlled conditions.

Practical Thyristor Circuits

Fig. 18.6 illustrates a basic d.c.-controlled thyristor circuit. When the gate voltage exceeds a specified value (usually 3 V) the thyristor "fires" and a current will flow limited by R_L (which must be of a

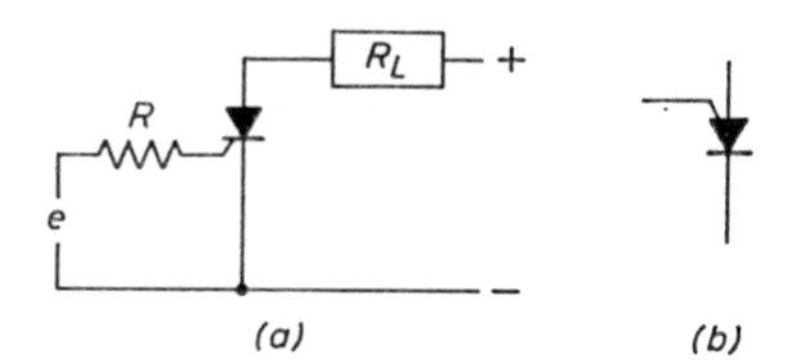

Fig. 18.6. Basic Thyristor Circuit

value sufficient to keep the current within the rated limit). The gate resistance is high in the "off" condition but falls rapidly when the circuit triggers. Hence it is desirable to include a resistance R in the gate circuit to limit the gate current.

For reliable operation the gate current must exceed a specified value, which is determined by the junction temperature. Fig. 18.7 shows the gate characteristics for a typical thyristor which requires a minimum gate voltage of 3 V and a gate current of 30 mA (at $T_j = 25°C$). If the input voltage is, say, 5 V the voltage at the gate when the circuit fires will be less than this because of the voltage drop on the resistor R, which must therefore be chosen such that the circuit will always supply the minimum gate voltage *and* current. This may conveniently be determined by drawing a load line on the

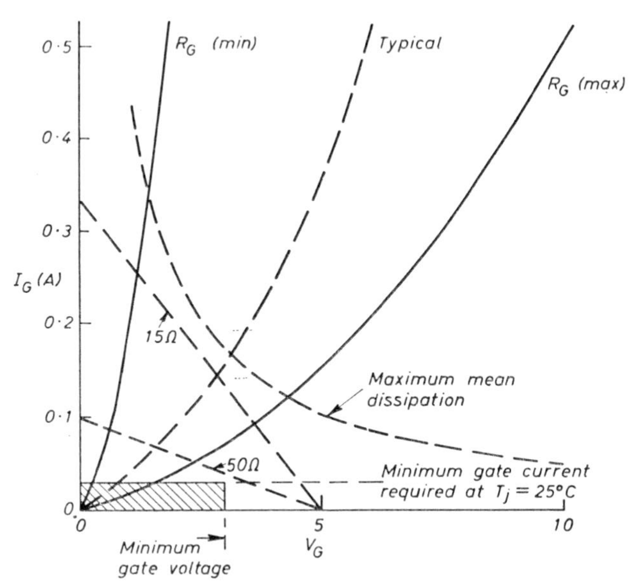

FIG. 18.7. TYPICAL GATE CHARACTERISTICS

characteristic as shown, and choosing the value such that this line is clear of the shaded area bounded by the limits of 3 V and 30 mA. A value of 50 Ω meets these requirements. The actual final gate current is then given by the intersection of this load line with the I_g–V_g curve, which with an average sample would be 70 mA. However, the production spread may result in a curve anywhere between the limits shown, so that in practice the current might be between 50 and 85 mA, but would always be above the 30 mA limit.

A lower value of R (e.g. 15 Ω) would comply with the requirements equally well, but the gate current would then be more than necessary—perhaps 250 mA with a sample having a low-limit characteristic—which would be wasteful.

When the thyristor is of the form of Fig. 18.3 it is known as a p-gate or cathode-controlled type, the symbol for which is shown in Fig. 18.6 (*a*). A similar action can be obtained with an n–p–n–p sandwich, in which case a negative trigger pulse is required. This form is called an n-gate or anode-controlled device, the symbol being as shown in Fig. 18.6 (*b*).

A.C. Operation

The circuit of Fig. 18.6 may be used with an a.c. supply, in which case the anode current is a series of half-wave pulses (since the thyristor will not conduct when the anode is negative). If the gate voltage is derived from the same supply the amplitude and/or phase may be adjusted to trigger the anode current over a part of the cycle only. Thus in Fig. 18.8 (*a*) the gate voltage reaches the

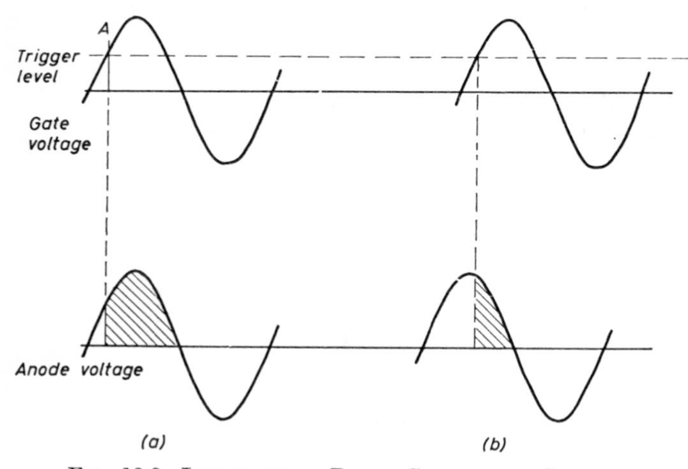

FIG. 18.8. ILLUSTRATING PHASE CONTROL OF CURRENT

trigger level at *A*, so that the thyristor conducts over most of the positive half-cycle. If the phase of the gate voltage is delayed, however, by 90° as shown at (*b*) the trigger level is not reached until later in the cycle so that the current pulse is restricted. The angle over which the device is conductive is called the *conduction angle* and may be varied by adjusting the phase of the gate voltage from virtually the full 180° (allowing for the small time required for the gate voltage to reach the trigger level) down to nearly zero.

Fig. 18.9 illustrates the application of this principle to the control of motor speed. The motor develops a back-e.m.f. at *K* proportional to the speed. The gate voltage at G is so arranged that for a given motor speed it becomes sufficiently positive to trigger the thyristor at the appropriate point in the cycle. This could be done with a simple voltage divider, but by using the network R_1R_2C the trigger point can be adjusted in both amplitude and phase, which provides

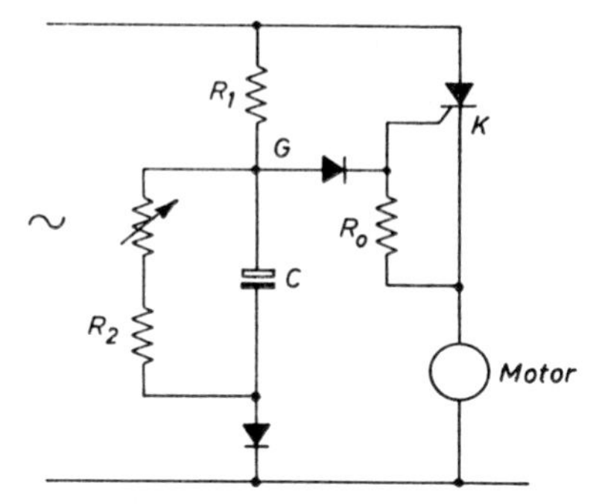

FIG. 18.9. MOTOR SPEED CONTROL CIRCUIT

an extended range of control, the conduction angle being variable over a range of 20° to 140°. The resistor R_o serves to set the minimum speed.

Triacs

By using two thyristors back to back current can flow throughout the full cycle. These may be combined in a single unit, as shown in Fig. 18.10, constituting what is called a *triac*. It has the advantage of requiring only a single gate terminal, which may be supplied with suitably-phased a.c. signals, but the device will function (on both halves of the wave) with a d.c. gate pulse, provided that this meets the requirements previously discussed. This permits relatively large a.c. powers to be switched by a d.c. gate signal of a few volts only.

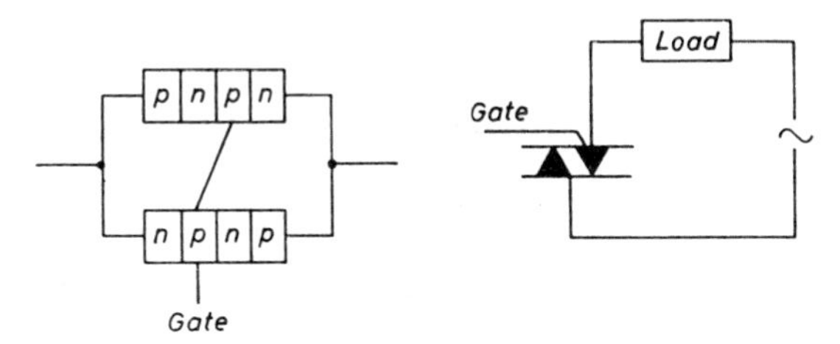

FIG. 18.10. TRIAC CIRCUIT

The Unijunction Transistor

A switching device which has a variety of applications is the double-base diode, or *unijunction transistor*. This consists essentially of a (short) rod of n-type silicon with a p-type junction near the centre, as shown diagrammatically in Fig. 18.11 (*a*). The customary symbol for the device is shown at (*b*), with the equivalent circuit at (*c*). With no voltage on the emitter, the device is simply a rod of resistive material, with ohmic connections at the two ends, called base 1 and base 2, the interbase resistance $R_{B1} + R_{B2}$ being of the

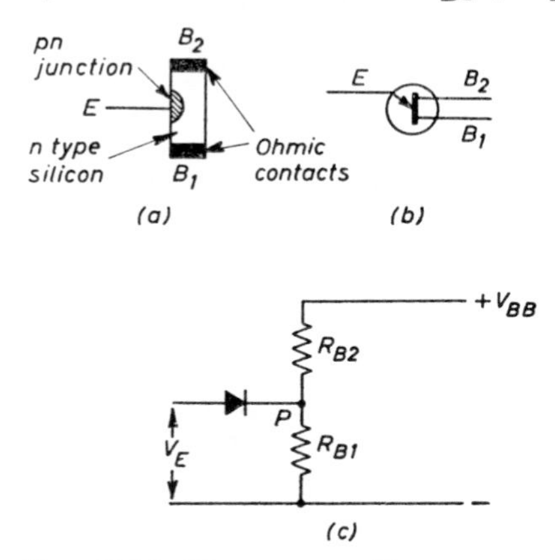

FIG. 18.11. UNIJUNCTION TRANSISTOR

order of 10kΩ. Hence a voltage V_{BB} applied across the circuit will produce at the point P a voltage $V_P = \eta V_{BB}$, where $\eta = R_{B1}/(R_{B1} + R_{B2})$. This is known as the *stand-off ratio* and is typically about 0·6.

If a voltage is now applied to the emitter, the p–n junction will initially be reverse biased so that only a small leakage current flows. However when V_E exceeds V_P the diode becomes forward biased and current flows into the region R_{B1}, causing an increased conductivity so that R_{B1} falls. This initiates a cumulative effect because the reduction in R_{B1} causes the interbase current to increase, producing a greater voltage drop on R_{B2}, so that the voltage at P falls causing a further increase in the emitter current until R_{B1} becomes negligibly small and the device becomes simply a normal diode.

The static $V_E - I_E$ characteristic is shown in Fig. 18.12. As V_E is increased there is at first only a small leakage current proportional to V_E, but at the critical voltage V_P there is a rapid increase of current as just described and the voltage necessary to sustain it *falls*, producing a negative-resistance action until a "valley" point is reached beyond which the characteristic is effectively simply that of the p–n junction alone.

If V_E remains equal to V_P the final current may be very large so that in practice it is usual to include a suitable resistance in the

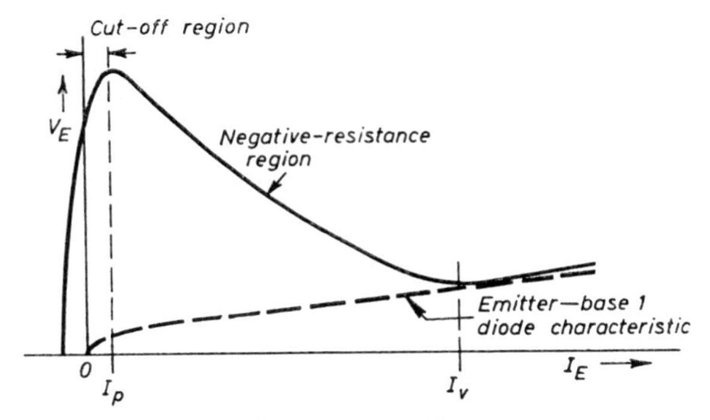

FIG. 18.12. UNIJUNCTION CHARACTERISTIC

external circuit, and/or to arrange that when emitter current flows V_E is allowed to fall appropriately. With these precautions the unijunction provides a convenient trigger device having a very rapid transition from the "off" to the "on" state which has many applications.

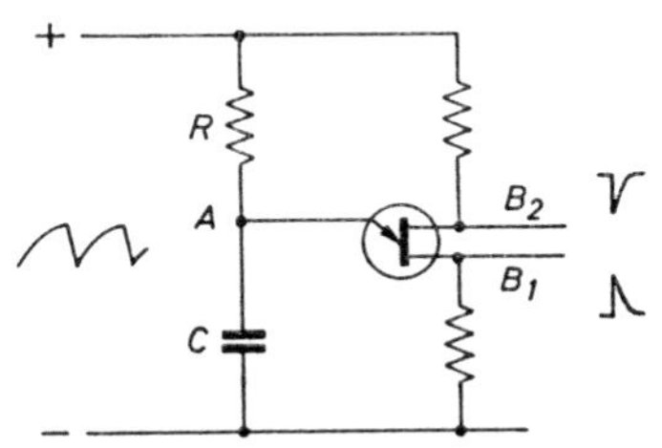

FIG. 18.13. RELAXATION OSCILLATOR USING UNIJUNCTION TRANSISTOR

For example, it can be used as a relaxation oscillator as shown in Fig. 18.13. Here the capacitor C charges exponentially through the resistor R. When the voltage reaches the critical value the unijunction conducts, producing a rapid discharge of the capacitor, and the process recommences. This generates a sawtooth voltage at the point A, but also produces steep-fronted pulses of positive or negative polarity at B_1 or B_2 which can be used to trigger further devices, such as a driven time-base as discussed in the previous chapter, or a thyristor.

The frequency of the oscillation is controlled by the time constant CR and the stand-off ratio η, being given by

$$f = 1/CR \log_e 1/(1 - \eta)$$

Programmable Unijunctions

The performance of a normal unijunction is determined mainly by the stand-off ratio, which is inherent in the device. It is often convenient to be able to control the parameters externally, which may be done by using a modified form of device known as a *programmable unijunction transistor* (PUT). This employs a four-element p–n–p–n assembly as shown in Fig. 18.14. This will conduct when

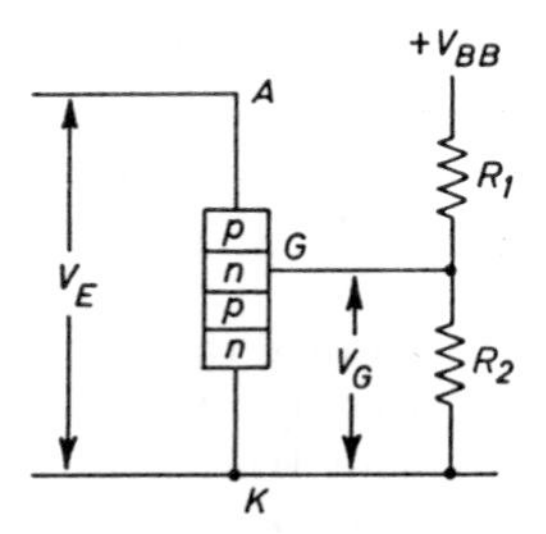

FIG. 18.14. BASIC PUT CIRCUIT

V_E exceeds V_G by the regenerative action described earlier, but V_G is now determined by the external voltage divider R_1R_2, so that the stand-off ratio is controllable.

18.2. Logic Circuitry

A significant development of switching technique is the evolution of a wide range of logic circuitry. This is a system based on the use of two-state devices which are either on or off. The action is thus basically different from that of a normal amplifier in which the output is proportional to the input, and a different approach is required. Solid-state devices are particularly suitable for this type of operation, because of their low power consumption and well defined on and off states.

Gates

The circuit possibilities are considerably increased if the response can be controlled by more than one input. This can be achieved by suitable combinations of diodes and/or transistors which are known as *gates*. While these can be made up with individual components, it is usual to use integrated units in which the semiconductor elements and their respective loads are assembled in one package, often in monolithic form as described in the next section. Gates are commonly available having as many as eight inputs, but to understand the behaviour we need, at the outset, only consider two-input types.

AND Gate

The simplest form of gate uses ordinary (switching) diodes as shown in Fig. 18.15. If both inputs are open-circuited (or preferably connected to $+V$) the diodes are non-conductive and the output

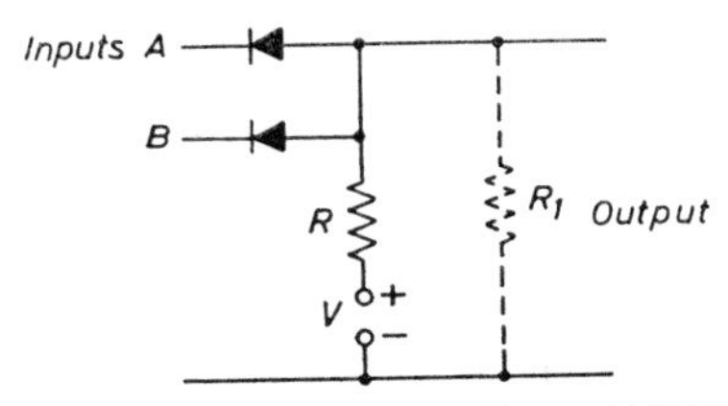

FIG. 18.15. SIMPLE DIODE GATE (AND/OR)

(neglecting the voltage drop caused by any current in R_1 due to leakage or a subsequent circuit) will be $+V$. If either input is negative, however, the relevant diode conducts and the output falls virtually to zero. Hence a high output is only obtained if *both* inputs are high together, and the device is therefore called an AND gate.

Truth Tables

The actual output depends on the supply voltage and the load but the relevant factor is the *state* of the output, which will be either *high* (approaching the supply voltage) or *low* (approaching zero). The relationship between the input and output conditions is called the *truth table* for the particular circuit, which for the AND gate of Fig. 18.15 can be written as below:

Input A	*Input B*	*Output*
High	High	High
High	Low	Low
Low	High	Low
Low	Low	Low

Truth tables like this can be written down for any type of gate, or combination of gates, and if expressed in these terms they are fundamental.

Logic Conventions

Alternatively, the conditions may be expressed in terms of *Logic* 1 and *Logic* 0. This is often convenient, but has a major disadvantage in having two possible interpretations according to which is regarded as the operational state.

With positive logic, 1 represents the high state and 0 the low state.

With negative logic, 1 represents the low state and 0 the high state.

Hence it is essential in designing gate circuits (and interpreting data) to specify which kind of logic is applicable.

With positive logic, the truth table for Fig. 18.15 is Table 18.1.

TABLE 18.1. AND FUNCTION ((POSITIVE LOGIC)

A	*B*	*Output*
1	1	1
1	0	0
0	1	0
0	0	0

OR Gate

It will be seen from Table 18.1 that the circuit of Fig. 18.15 will give a low output if either input is low (or if both are low together). This constitutes an OR gate, i.e. a gate which responds if either *A* or *B* is actuated. But in this condition the actuating signal, and the output, must be *low*, so that the circuit is operating under conditions of negative logic.

The truth table can thus be written in these terms, using 1 for low and 0 for high, as in Table 18.2.

TABLE 18.2. OR FUNCTION (NEGATIVE LOGIC)

A	*B*	*Output*
0	0	0
0	1	1
1	0	1
1	1	1

It should be noted that this is *not* the inverse of the AND function. To provide an OR function with positive logic, the output must be

1 if either or both inputs are 1, the required truth table being as shown in Table 18.3.

TABLE 18.3. OR FUNCTION WITH POSITIVE LOGIC

A	*B*	*Output*	*Inverted AND*
1	1	1	0
1	0	1	1
0	1	1	1
0	0	0	1

It will be seen that this is not the same as an AND gate with the output fed through an inverting stage, as shown in the fourth column, and in general a function with negative logic cannot be converted to positive logic by simply adding an inverter. It is necessary to invert the inputs as explained later (Fig. 18.19).

Transistor Gates

If the inputs are fed to a transistor as in Fig. 18.16, the magnitude of the output is independent of the input. If both inputs are high,

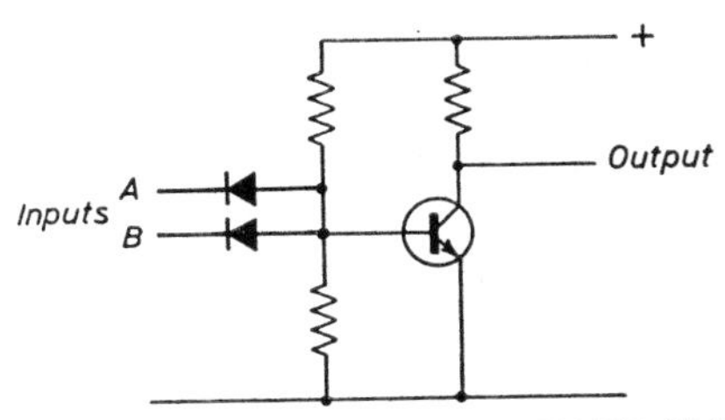

FIG. 18.16. TRANSISTOR GATE (NAND/NOR)

the transistor is turned on and the output goes low. Any low input turns the transistor off so that the output goes high. The truth table is thus as in Table 18.4.

TABLE 18.4. TRUTH TABLE FOR FIG. 18.16 (NAND GATE)

A	*B*	*Output*
1	1	0
1	0	1
0	1	1
0	0	1

If this is compared with an AND gate (Table 18.1) it will be seen that the outputs are inverted. Hence this type of gate is called a NAND (not AND) gate. It can be converted to an AND gate by feeding the output through an inverter, which may be a simple common-emitter transistor. Conversely an AND gate can be converted to a NAND gate by the addition of an inverter.

The circuit will also perform an OR function *if negative logic is used,* but because of the 180° phase shift the output will be inverted, as shown in Table 18.5. Hence this is said to be a NOR function.

TABLE 18.5. NOR FUNCTION (NEGATIVE LOGIC)

A	*B*	*Output*
0	0	1
0	1	0
1	0	0
1	1	0

NOT Gate

If the inputs of a gate are commoned the arrangement becomes simply a common-emitter transistor. It thus behaves as a simple inverter, and does not act as a gate, so that it is often called a NOT gate. This may be seem wasteful, but with the integrated-circuit units described later, which contain several elements in the same unit, this is often convenient and economical.

Gate Symbols

Gates are represented in circuit diagrams by an enclosure containing a symbol denoting the function. There is some lack of uniformity as to the shape of the enclosure, three typical forms being

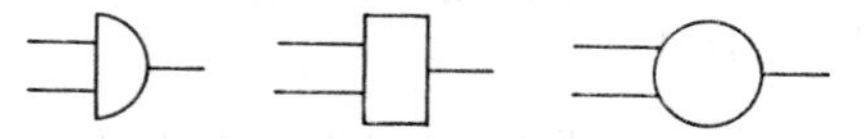

FIG. 18.17. ALTERNATIVE FORMS OF GATE SYMBOL

shown in Fig. 18.17. The number of inputs is denoted by appropriate input lines.

The complete symbol with the function indication is then as shown in Fig. 18.18. For an AND gate the function symbol is &, inversion being signified by a bar over it, so that for a NAND gate the symbol is

$\bar{\&}$. With an OR gate the symbol denotes the number of inputs which must be altered to change the state of the output. This is usually 1, as in the circuits so far described, but special forms of gate are made in which 2 or more inputs have to be changed together.

An inverter (NOT gate) is designated by a horizontal line, but if this is included as part of the gate it is shown as a small circle around the appropriate lead, as shown in Fig. 18.19.

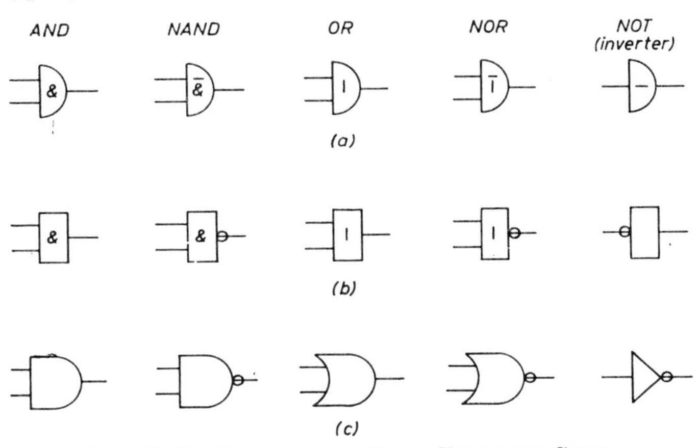

FIG. 18.18. SYMBOLS FOR BASIC FORMS OF GATE

(*a*) Used in this book.
(*b*) Adopted by British Standards Institution, July 1971.
(*c*) American and Continental forms.

An alternative symbol for a NAND gate is the AND symbol followed by a small circle. Similarly a NOR gate may be represented by an OR gate followed by an inverting circle.

Many American and Continental circuits employ a slightly different convention as shown in Fig. 18.18 (*c*). The symbols do not include any internal character, but are differentiated by the fact that AND and NAND gates have a straight input side, while for OR and NOR gates the input side is curved.

An alternative symbol is also used for an inverter, consisting of a triangle followed by a small circle, denoting an amplifier (in this case of unity gain) with an inverted output.

Transformations

The transistor gate has certain advantages over the simple diode gate of Fig. 18.15, which is only occasionally used. The standard form of gate is thus the NAND gate of Fig. 18.16 (with certain refinements described later, which do not modify its basic function).

From what has been said so far it will be clear that certain simple transformations are possible, namely

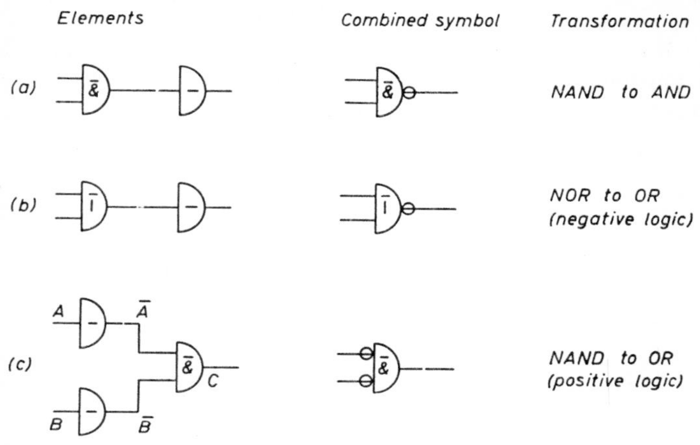

FIG. 18.19. TRANSFORMATIONS OF FUNCTIONS

1. A NAND gate can be converted to an AND gate (with positive logic) by the addition of an inverter as in Fig. 18.19 (*a*).
2. A NOR gate can be converted to an OR gate by a similar addition as at Fig. 18.19 (*b*).
3. A NAND gate can be transformed to an OR gate with positive logic, if inverters are included in the inputs as shown at (*c*). The relevant truth table is then as in Table 18.6.

TABLE 18.6. NAND/OR TRANSFORMATION WITH POSITIVE LOGIC (FIG. 18.19 (*c*))

A	B	$\bar{A}$	$\bar{B}$	C
1	1	0	0	1
1	0	0	1	1
0	1	1	0	1
0	0	1	1	0

Wired-OR Connection

It is often convenient to connect the outputs of two or more gates in parallel. Then, if the normal condition of the outputs is high, the combined output is high, but if any output goes low, the combined output is low. Thus in Fig. 18.20, if both inputs to *either* gate are high, the output is low irrespective of the condition of the other gate. The circuit thus performs an OR function, and is called a *wired*-OR connection but see p. 751.

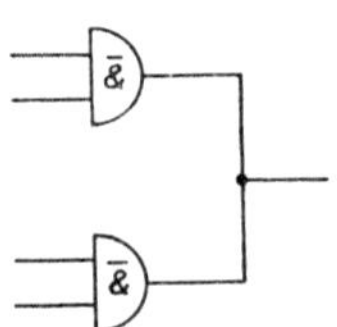

FIG. 18.20. WIRED-OR CONNECTION

Exclusive-OR Connection

Circuits are often required to respond to a change of one or other of the inputs, but not both together. This is known as an *exclusive*-OR connection, and may be provided by using two NAND gates as in Fig. 18.21. The second gate is fed with signals which are the inverse of those on the first, and the outputs are commoned so that the combined output is 0 if either gate is giving output 0. The truth table is as in Table 18.7.

TABLE 18.7. EXCLUSIVE OR (POSITIVE LOGIC)

A	B	C	$\bar{A}$	$\bar{B}$	D	*Output* ($C \times D$)
1	1	0	0	0	1	0
1	0	1	0	1	1	1
0	1	1	1	0	1	1
0	0	1	1	1	0	0

In many instances the input circuitry contains points at which the necessary inverse signals already exist, in which case the inverters may be omitted as in Fig. 18.21 (*b*). A pair of gates connected in this manner is often represented by a single symbol as shown at (*c*).

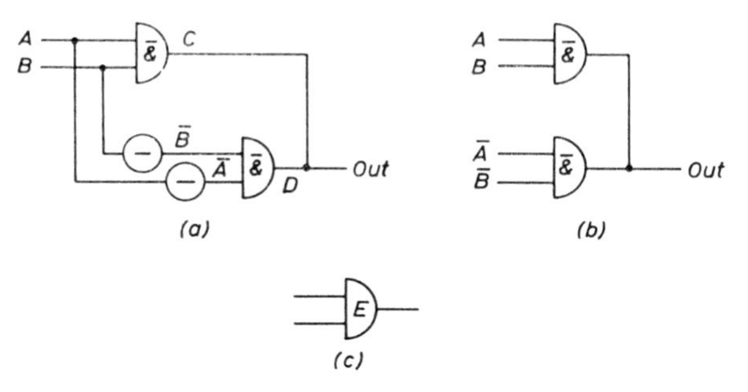

FIG. 18.21. EXCLUSIVE-OR CIRCUIT

Comparator

The exclusive-OR circuit provides an output if the inputs are different, but not if they are both the same. A modification of this is the *comparator*, or *equivalence circuit*, of Fig. 18.22, which provides

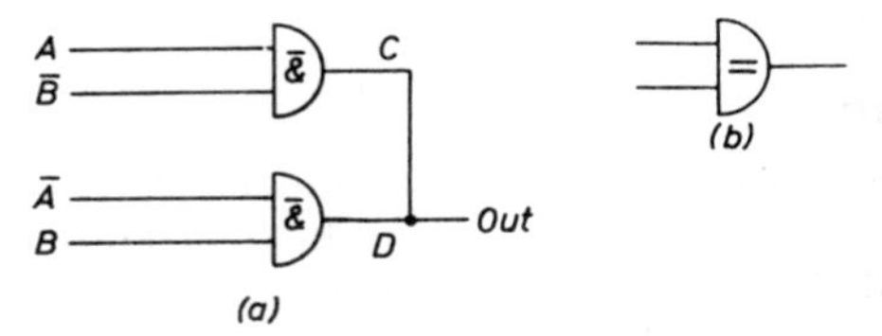

FIG. 18.22. COMPARATOR (EQUIVALENCE CIRCUIT)

an output when the inputs are both 1 or both 0. Two NAND gates as used as before with slightly different cross-connections, the truth table being shown in Table 18.8.

TABLE 18.8. TRUTH TABLE FOR EQUIVALENCE CIRCUIT

A	B	$\bar{A}$	$\bar{B}$	C	D	*Output*
1	1	0	0	1	1	1
1	0	0	1	0	1	0
0	1	1	0	1	0	0
0	0	1	1	1	1	1

As with the previous circuit, complementary inputs are required. If these are not available from the preceding circuitry they must be provided through inverters. The symbol for an equivalence circuit is shown in Fig. 18.22 (*b*).

Latch

It is often required to hold a condition until it is subsequently cancelled. This can be achieved by using two cross-connected NAND gates as in Fig. 18.23, which is called a *latch*. A momentary earth pulse on input A_1 will send output A high. This makes input B_2 high, and since B_1 is also high, output B goes low. Input A_2 thus goes low, so that gate A is held cut off and the output is "latched" high, with output B held low.

An earth pulse on input B_1 will reverse the conditions so that gate B is latched with its output high, with output A held low. This circuit responds to input pulses of very brief duration, and may thus be vulnerable to random interference. Hence it is often convenient to slow down the operation by connecting a small capacitor (e.g. 0·047 μF) across one of the outputs as shown dotted in Fig. 18.23.

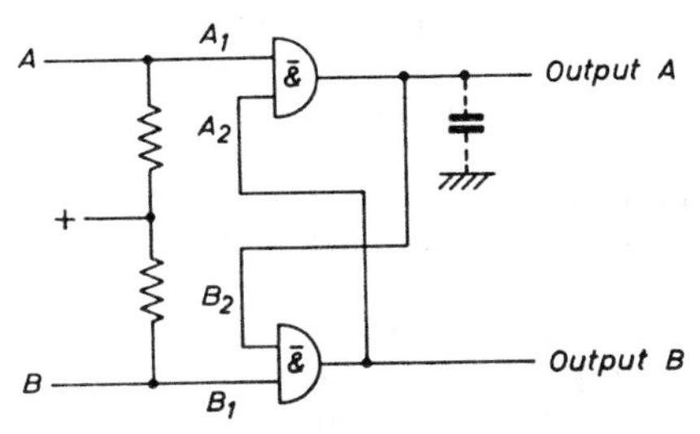

FIG. 18.23. LATCHING CIRCUIT

Calculating Circuits

Logic circuitry is extensively applied to arithmetical calculations but since the circuit elements have only two possible states, 0 and 1, it is necessary to use a binary system of counting. As explained in Appendix 3, the ordinary system of numbering is based on powers of ten, the successive digits, from right to left, indicating the number of units for successive powers of ten. Thus the number 57 $= (7 \times 10^0) + (5 \times 10^1)$.

The number can be expressed equally well in terms of powers of two, in which case it would be built up as

$$\begin{aligned}(1 \times 2^5) + (1 \times 2^4) + (1 \times 2^3) + (0 \times 2^2) + (0 \times 2^1) + (1 \times 2^0) \\ = 32 + 16 + 8 + 0 + 0 + 1 \\ = 57\end{aligned}$$

so that in binary code it would be written as 111001.

Now, although this looks a little cumbersome it is clearly in a form which can be interpreted by two-state devices which can indicate whether any given power of two is present or not, and it is then possible to perform any of the customary mathematical processes by successive operations of addition or subtraction.

Suppose we want to add $1 + 1 = 2$ in conventional terms. But $2 = 1 \times 2^1$, so that in binary terms the expression is written $01 + 01 = 10$. In other words, two unit characters combine to give a 0, plus a carry 1 to the next register.

To extend the process, let us add 9 to 3. The sum is then as below, and we simply add the successive columns, from right to left, remembering that $1 + 1 = 0 +$ a carry 1 to the left. If this carry

1 adds to an existing 1, it will again = 0, plus a carry 1 stage further to the left. Hence we have

1001	(9)
0011	(3)
1100	(12)

The operation can be extended to provide multiplication by successive additions. Thus to multiply 1 by 5, the process is

$$0001 + 0001 = 0010$$
$$0010 + 0001 = 0011$$
$$0011 + 0001 = 0100$$
$$0100 + 0001 = 0101 \; (= 2^2 + 2^0 = 5)$$

This may appear clumsy, but can be performed with modern high-speed circuitry in far less than one microsecond.

Adders

To perform these operations circuits are required which will provide the required binary addition. The simple addition of two

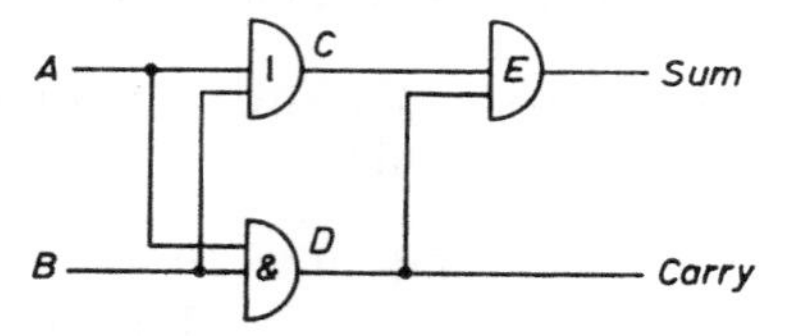

FIG. 18.24. BINARY ADDER (HALF ADDER)

digits can be achieved by the circuit of Fig. 18.24. This is a combination of an AND and an OR gate with an exclusive-OR unit, the truth table being shown in Table 18.9. It will be seen that the sum output is the correct binary sum of the inputs, with a carry 1 when required.

TABLE 18.9. TRUTH TABLE FOR HALF ADDER

A	*B*	*C*	*D*	*Sum output*	*Carry output* (*D*)
0	0	0	0	0	0
1	0	1	0	1	0
0	1	1	0	1	0
1	1	1	1	0	1

The circuit of Fig. 18.24 will only handle one digit, but it can be shown that any number of digits can be handled by the use of two such circuits in association with suitable delay networks which provide the correct routing of the carry signals. Such an arrangement is called a *full adder*, the Fig. 18.24 circuit being shown as a *half adder*.

Subtraction

To complete the basic calculating processes it is necessary to be able to subtract, and hence, by successive subtractions, to divide. This can be arranged by similar combinations of gates which provide an output corresponding to the difference between the inputs, but in this case provision must be made to *borrow* from the adjacent register when necessary.

Thus 2 — 1 in binary code will be

$$\begin{array}{cc} 1 & 0 \\ -0 & 1 \\ \hline 1 & (-1) \end{array}$$

But we have no means of representing —1 in a two-state device, so we borrow a digit from the next stage to the left, i.e. $2^1(=2)$. This converts the calculation to the form

$$\begin{array}{ccl} 1 & 0 & (=2) \\ (-1) & (2) & \text{Borrow} \\ 0 & 1 & (=1) \\ \hline 0 & 1 & (=1) \end{array}$$

In other words $0 - 1 = 1$ with a *borrow* from the adjacent register.

Thus 9 — 3 in binary code would be written

$$\begin{array}{r} 1001 \\ -0011 \\ \hline 0110 \end{array}$$

There are numerous refinements of these techniques which involve the whole art of computer technology and are beyond the present scope, which is necessarily limited to the basic principles.

Stores

To perform extended calculations the individual items of information, which are called *bits*, are fed into a *store*, which holds the data in readiness for further operations. Any detailed discussion of these assemblies again comes within the ambit of computer technology, and is beyond the present scope. In broad terms, a store consists of an assembly of two-state devices together with a series of *shift registers* by means of which the information received (or extracted) is routed to the appropriate part of the store, the function of the carry or borrow signals being to actuate the appropriate shift register. Some brief examples of the circuitry are given in Section 18.5.

Stores can be arranged to hold the information in many different ways, depending on the particular requirements. Moreover they do not necessarily use semiconductor elements, and for large stores it is often more economical to employ magnetic elements, such as are discussed in Section 18.5.

The initial information is prepared in suitable binary form on punched tape or magnetic tape. The required answers are then obtained by interrogating the store and so deriving a sequence of similar (binary) signals which are then translated into suitable form to operate a print-out device or actuate appropriate control equipment.

18.3. Monolithic Circuits

The gates described in the previous section can be assembled with individual component elements, but they are more usually made up in monolithic form. With this construction the diodes and transistors, together with the appropriate resistors and internal connections, are formed by suitable processing of a single silicon chip, so forming what is called an *integrated circuit* (i.c.). With this technique a single unit can include a number of interconnected gates, providing a considerably extended range of possibilities.

Details of the construction of these, and other forms of composite circuit, are given in Appendix 4. The technique is not limited to switching circuits, and a wide range of amplifiers and control devices is available in this form, as described in Chapter 17. The present section is concerned with some of the basic forms of integrated-circuit logic devices.

Types of Circuit

There are three main types of circuit, namely

RTL CIRCUITS (Resistor-Transistor Logic)

These are circuits using simple transistors with their associated resistors. They are rarely used today.

DTL CIRCUITS (Diode-Transistor Logic)

Improved performance can be obtained by the addition of diodes in suitable positions. Fig. 18.25 shows a two-input DTL gate, which is similar to Fig. 18.16 but includes four diodes. Two of these are included in the inputs, so that they can operate independently to produce an off condition. The diodes D_1, D_2 (which in series provide a forward voltage drop of approximately $0{\cdot}7 + 0{\cdot}7 = 1{\cdot}4$ V) are

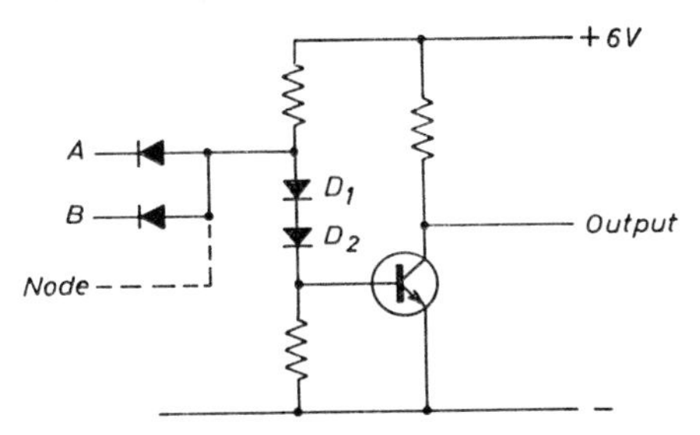

FIG. 18.25. TWO-INPUT DTL GATE

included to improve the signal/noise immunity as described later.

Sometimes a "node" input is also provided which feeds the base of the transistor directly, as shown dotted. This gate is a NAND gate, like Fig. 18.16, and has a similar truth table.

TTL CIRCUITS (Transistor-Transistor Logic)

Here the input transistor is of a special form having several emitters on a common base. Up to eight emitters are normally available, a simple circuit having three inputs being shown in

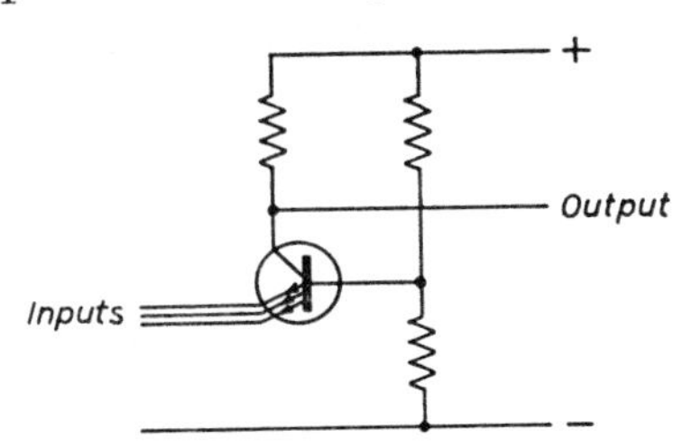

FIG. 18.26. MULTIPLE-EMITTER GATE

Fig. 18.26. In this case the transistor is operating in the common-base mode, so that earthing any of the inputs will cause the transistor to conduct. The output will then go low, so that the circuit behaves as an AND gate.

This simple arrangement has certain disadvantages, and moreover is not compatible with DTL circuitry, which provides NAND characteristics, as previously described. Hence it is usual to follow the input transistor with a cascode amplifier as shown in Fig. 18.27, which provides a greater output and a more rapid transition from one state to the other. Very broadly, the action is that with all inputs high the collector voltage of Tr_1 is also high, which turns Tr_2 on. This cuts off Tr_3 and turns Tr_4 on so that (since this is in common-emitter mode) its output goes low (less than 0·4 V), and can sink appreciable current from succeeding circuitry (typically up to 16 mA).

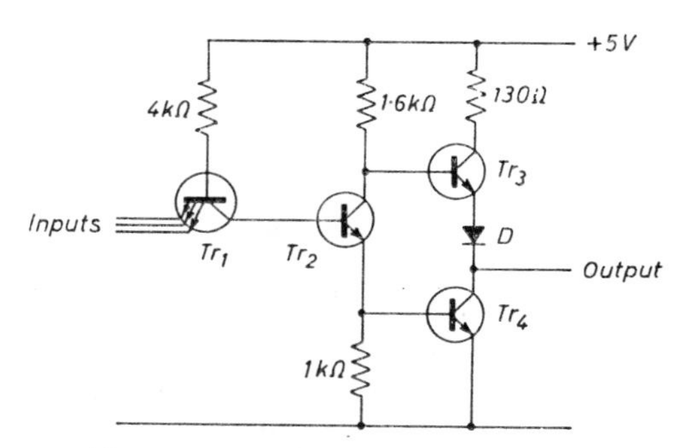

FIG. 18.27. THREE-INPUT TTL GATE

With any of the inputs low the conditions are reversed and the output goes high. The circuit thus behaves as a NAND gate.

Multiple-gate Assemblies

The monolithic technique permits DTL and TTL circuits to be produced in very small space so that it is often possible to include several gates, either separate or interconnected, in a single unit, so providing a very wide range of possible circuit configurations.

Fan-out and Fan-in

A gate with its input low will take a small but specific current, often called the *sinking* current. Hence if a gate is required to feed a number of subsequent gates in parallel it must be able to provide sufficient current to sink all the gates which it feeds. The maximum number of gates which can be so fed is called the *fan-out* and is typically of the order of 6 to 8. Similarly the permitted number of separate sources which may be connected to the input of a given gate is called the *fan-in*.

Note that the fan-out does *not* indicate the number of *outputs* which may be commoned in a wired-OR connection (page 743). This is determined by the collector circuitry, and some types of gate (notably those with a "pull-up" circuit as in Fig. 18.27) will not tolerate wired-OR. It is necessary to use "open collector" types with a suitable (common) resistor connected externally.

Propagation Delay

The transition from one state to the other occupies a small but finite time. The manufacturing processes are designed to keep this as small as possible, and in many instances the transition time is determined by the external circuit rather than the gate itself. Where very rapid switching is required, however, the *propagation delay* may be significant.

The delay is slightly different for the rise and fall transitions, as was mentioned earlier, and it is customary to specify the mean value. With DTL gates it is of the order of 40 nanosec (0·04 microsec). TTL gates are two or three times faster in operation.

Noise Margin

A logic 0 condition in a gate does not imply zero voltage. There is, in fact, a threshold input level (usually of the order of 0·4 V) below which the gate is in a low state, but above which it will switch

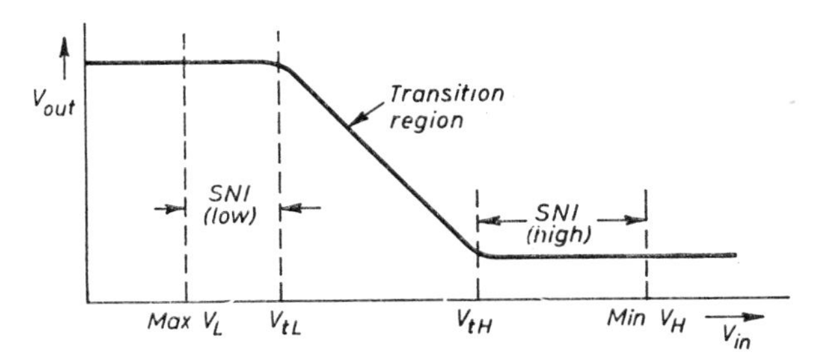

FIG. 18.28. ILLUSTRATING SIGNAL/NOISE IMMUNITY

to the high state. Similarly there is a corresponding high threshold which must be exceeded if stable logic 1 conditions are to be obtained. This is, in general, appreciably less than the supply voltage, as it must be if gates are connected in cascade, since the output voltage of any gate will be reduced by any load connected across it.

The characteristic of a gate is of the form shown in Fig. 18.28. If the input voltage is below V_{tL}, the output will be high. If it is above V_{tH}, the output will be low. In between these two thresholds the output will be indeterminate.

If the input is noise-free it is sufficient to ensure that it is beyond these limits, but if there is noise present (either external or internal), the operating conditions must be so chosen that the effective input is always clear of the threshold conditions as shown at $V_{L(max)}$ and $V_{H(min)}$ in Fig. 18.28.

It is usual, therefore, to specify the input conditions in conjunction with a *signal/noise immunity factor* (S.N.I.). There are two such factors relating to the low and high conditions respectively. These are defined as

S.N.I. (low) The difference between the low input threshold V_{tL} and the *maximum* instantaneous input (which with gates in cascade is the output of the preceding gate in the low condition).

S.N.I. (high) The difference between the high input threshold V_{tH} and the *minimum* instantaneous input (which with gates in cascade is the output from the preceding gate in the high condition).

As an example, the permissible noise levels for a typical DTL gate are 0·4 and 1·6 V respectively. The input conditions are then specified as

		V
Maximum low input voltage	(S.N.I. = 0)	0·8
	(S.N.I. = 0·4V)	0·4
Minimum high input voltage	(S.N.I. = 0)	2·3
	(S.N.I. = 1·6V)	3·9

Emitter-coupled Logic

The gates so far described are all saturated types, the transistors being driven between the states of cut-off and saturation. For many purposes this is no disadvantage, but since the time taken for a transistor to change from its on to its off condition depends on the current it is possible to obtain an increase in operating speed by using circuits which operate over limited current swings.

This is known as *emitter-coupled logic*, usually called ECL (or sometimes E²CL), a simple circuit being shown in Fig. 18.29. This will be seen to be a variant of the long-tailed pair discussed in Chapter 17 (Fig. 17.22). Tr_2 is held at a steady reference potential. If the potential on either of the transistors $Tr_{1(a)}$, $Tr_{1(b)}$ is increased above this reference level, the potential at A rises and that at B falls, while if the input level falls the conditions are reversed. Hence a gating action can be provided by arranging to swing the input wihint prescribed limits, producing corresponding changes of output state which do not run into saturation or cut-off.

This results in appreciably faster operation, with operating times of the order of a few nanoseconds, which is comparable with that of a diode, and is useful for very-high-speed switching circuits. The operating conditions can be controlled by varying the reference potential, while the emitter-followers Tr_3 and Tr_4 serve as isolators,

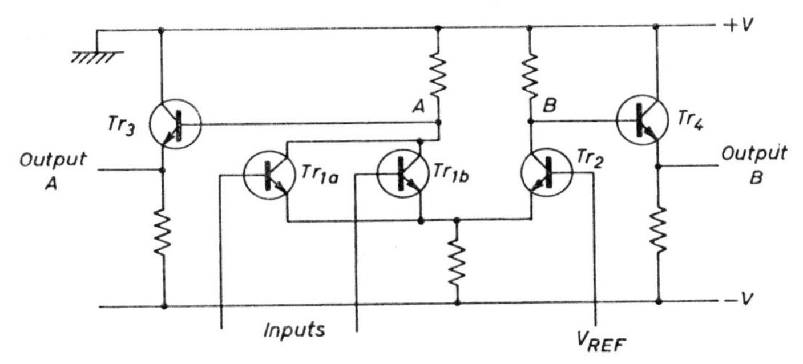

FIG. 18.29. ECL GATE

and provide outputs having one side grounded. Because of the smaller range of input signals the signal/noise immunity is reduced and precautions have to be taken accordingly.

M.O.S. Integrated Circuits

As said, monolithic construction, using the planar technique described in Appendix 4, permits a variety of circuits to be included on a single chip, including the specialized devices discussed in the next section. With the usual bipolar transistors, however, the possibilities are necessarily restricted by the limitations imposed by the interconnections and the isolation between the circuits.

The difficulty can be overcome in some cases by using suitable arrays of m.o.s. transistors, which permits many more devices to be incorporated in a given chip size. This is due in part to the smaller size of m.o.s. devices in comparison with their bipolar equivalents, but mainly because no isolation diffusion is required (cf. Appendix 4), and since fewer steps are entailed, closer tolerances can be maintained.

The disadvantages of m.o.s. devices are mainly related to speed, since the problems of high-impedance operation limit the system speed to below 10 MHz, and to the difficulties involving interfacing with external heavy-current loads. However, where there is a large amount of logic to be done between input and output at relatively low speed, m.o.s. arrays have a considerable advantage and are being increasingly used in such applications as multiple counter/dividers or shift registers, read-only stores or memories and read-write stores.

The limiting factor determining the complexity of a circuit on a single chip is usually power dissipation, but m.o.s. arrays have been developed in which information is effectively stored as charge on the gate capacitances within the array, and if this information is refreshed at intervals with a clock pulse, static power consumption is greatly reduced. Indeed, dynamic logic circuits using trains of either two or four clock pulses (two- and four-phase logic) have been developed in which the total power for the arrays is drawn from the clock pulses and no static d.c. supply is needed.

M.O.S. arrays are likely to become increasingly important for large-scale logic applications, but detailed discussion is outside the scope of this book. For further information the reader is referred to *MOS Integrated Circuits and their Applications* (Mullard, 1970).

18.4. Bistable Circuits

The circuit element most widely used for storing a bit of information is the bistable circuit which has been described in terms of discrete

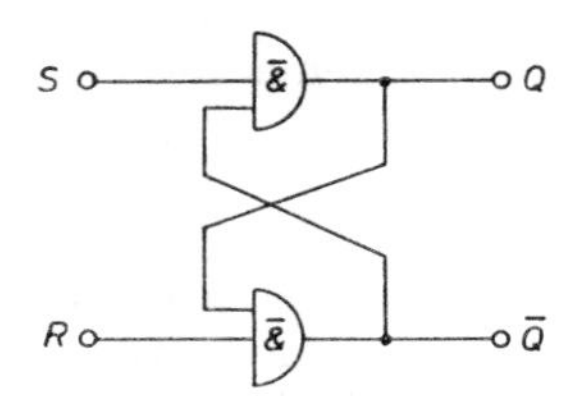

FIG. 18.30. R–S BISTABLE CIRCUIT

components in Chapter 17. This circuit is often referred to, particularly in American parlance, as a *flip-flop*, although the British Standard definition of a flip-flop refers to a monostable circuit.

This simple bistable circuit can also be realized by cross-connection of a pair of gate circuits. Any of the gate configurations described in Section 18.2 can be used, but in this section reference will only be made to NAND configurations since these are most widely used. Fig. 18.30 shows a simple bistable obtained by cross-coupling two NAND gates. This circuit is known as an *R–S bistable* since the output Q is *set* to 1 by applying a 0 signal to S and is *reset* to 0 by applying a 0 to R. The circuit is stable in either state in the absence of triggering signals and can therefore be used to store either a 0 or a 1, i.e. one bit of information. The conditions of the circuit after the receipt of an input are shown in Table 18.9, where $Qn + 1$ means the state of the output Q after the input conditions shown have been applied (possibly only momentarily).

TABLE 18.9. TRUTH TABLE FOR R–S BISTABLE

S	R	$Qn + 1$
1	1	Qn
0	1	1
1	0	0
0	0	Not permitted

Both the R and S inputs are normally at the 1 level (or open circuit), and if they remain at this level the output Q will not change state. If, however, the S input is changed to 0 with R remaining at 1, output Q is set to 1 and is reset when R changes to 0 with S at 1. The condition with R and S both at 0 makes both Q and $\bar{Q}$ outputs high, and if both inputs are removed simultaneously the state of the circuit is unpredictable. This input condition is therefore usually not permitted.

The R–S bistable is often more useful as a circuit element if the change of state takes place only when a clock pulse occurs. The circuit of a clocked R–S bistable is shown in Fig. 18.31, with the truth table in Table 18.10.

TABLE 18.10. TRUTH TABLE FOR CLOCKED R–S BISTABLE

CP	R	S	Q_{n+1}
1	0	0	Q_n
1	0	1	1
1	1	0	0
1	1	1	Not permitted
0	0	0	Q_n
0	0	1	Q_n
0	1	0	Q_n
0	1	1	Q_n

Another useful bistable element is the binary divider or *T bistable*, which is required to change state in sympathy with every input pulse. This circuit can be realized by cross-connecting two gates as shown in Fig. 18.32. Change of state occurs on the leading edge of a negative triggering signal at point T.

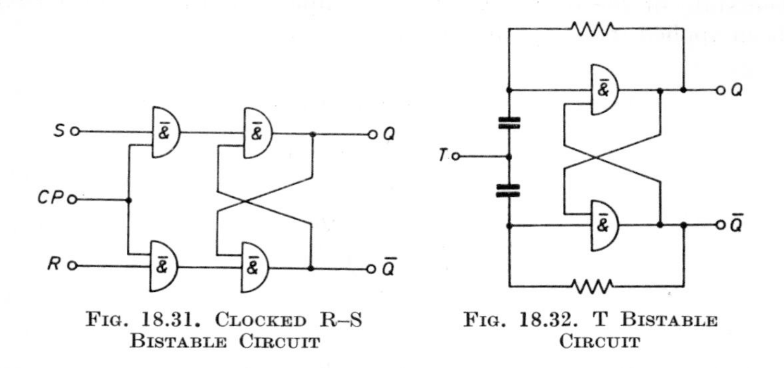

FIG. 18.31. CLOCKED R–S BISTABLE CIRCUIT

FIG. 18.32. T BISTABLE CIRCUIT

The J–K Bistable

More powerful bistable elements can be obtained by using more gates as control elements. The most widely used complex bistable element is the *J–K bistable,* which has the following truth table.

TABLE 18.11. TRUTH TABLE FOR *J–K* BISTABLE ELEMENT

J	K	Q_{n+1}
0	0	Q_n
1	0	1
0	1	0
1	1	$\bar{Q}_n$

This result may be realized in a number of ways, and large numbers of gates are often used in integrated-circuit J–K bistables in order to obtain elements which are easy to use. There are three basic kinds of J–K elements in use, namely

(*a*) Master-slave type.
(*b*) Capacitive-memory type.
(*c*) Edge-triggered type.

In the first two of these types, the input information is stored when the leading edge of the clock pulse occurs, and the output changes state at the trailing edge of the clock pulse. In the edge-triggered type, the output changes at the leading edge of the clock pulse. As said, the actual realization of these elements is often complex and will not be considered here. However, a possible configuration of a master-slave J–K bistable using only NAND gates is shown in Fig. 18.33. When the clock pulse is high, the outputs A and B in the master will latch in a condition determined by the J and K inputs, but the slave will not operate because the input at C is low. When the clock pulse goes low, C goes high and the slave will then respond to the inputs A and B.

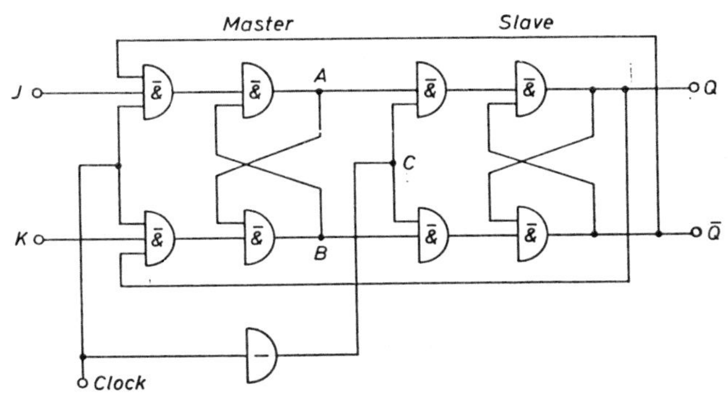

FIG. 18.33. MASTER-SLAVE J–K BISTABLE CIRCUIT

In practice, a number of J and K inputs are usually provided, and since the truth table of most other bistable elements can be realized by suitable connection of a J–K element, these are most widely used. One other element which is often used is the *D bistable*, which is a single-input clocked-memory element in which the state of the output is that of the input one clock pulse earlier. The truth table is thus

D	Q_{n+1}
0	0
1	1

Both J–K and D bistables are usually provided with *preset* and *clear* inputs to enable the output state to be set independent of the clock pulse when necessary.

Shift Registers

J–K bistables are widely used either as interconnected circuit elements or combined in single medium or large-scale integrated circuits to produce more complex digital circuits such as shift registers and counters. Shift registers are used to store numbers in binary form and are much used in arithmetic operations for temporary storage. The binary number may be entered serially or in parallel, and the size of the register can be built up to cater for the largest number to be stored. Fig. 18.34 illustrates the principle,

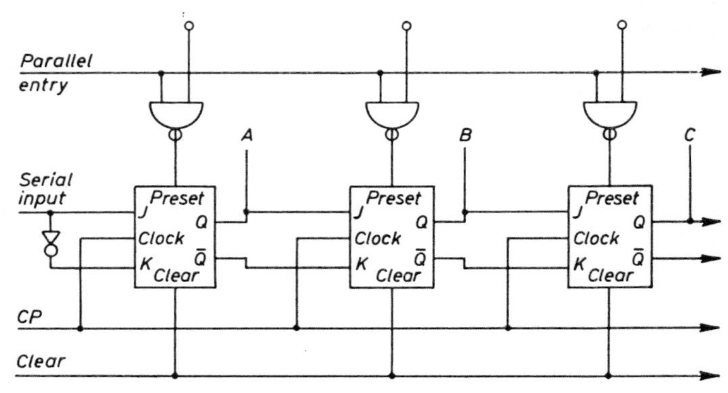

FIG. 18.34. THREE-BIT SHIFT REGISTER

showing three J–K elements and associated gates connected to allow either serial or parallel entry to a 3-bit register.

Counters

Counting may be defined as the process of successive addition and memory. The principle of counting may be illustrated by reference to the simple circuit of Fig. 18.35 consisting of three J–K

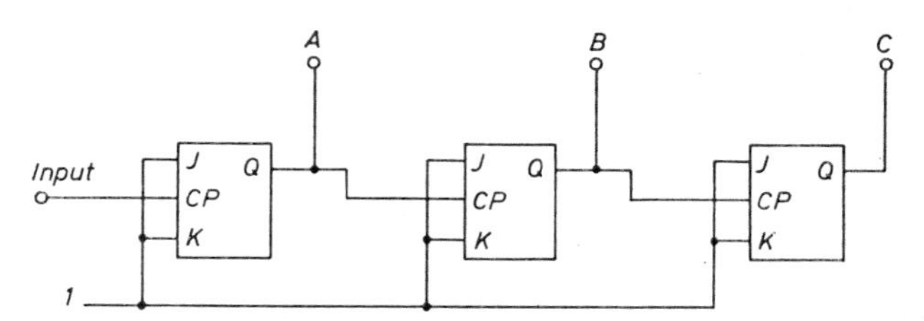

FIG. 18.35. SIMPLE BINARY COUNTER

bistables in cascade. Here the first clock pulse received at the input is stored as a 1 at A, the next clock pulse causes A to go to 0 and B to go to 1; the next causes A to go to 1 but B remains at 1 since the second bistable only responds to change from 1 to 0 at A, and the process is continued so that the circuit is a true 3-bit binary counter operating in accordance with Table 18.11. This counter is called a *ripple-through* or *asynchronous* counter, since, if a number of successive stages are all in the 1 state when an input pulse is received, the last stage will not change to a 0 until each preceding

TABLE 18.11. TRUTH TABLE FOR BINARY COUNTER OF FIG. 18.35

Digit	*A*	*B*	*C*
1	1	0	0
2	0	1	0
3	1	1	0
4	0	0	1
5	1	0	1
6	0	1	1
7	1	1	1
8	0	0	0

stage has changed to a 0. Since each bistable has a finite operating time, these changes will give rise to a ripple running through the counter.

This cumulative delay gets worse as the number of stages increases and is often a disadvantage, particularly when certain counts have to be decoded for driving other functions at certain intervals.

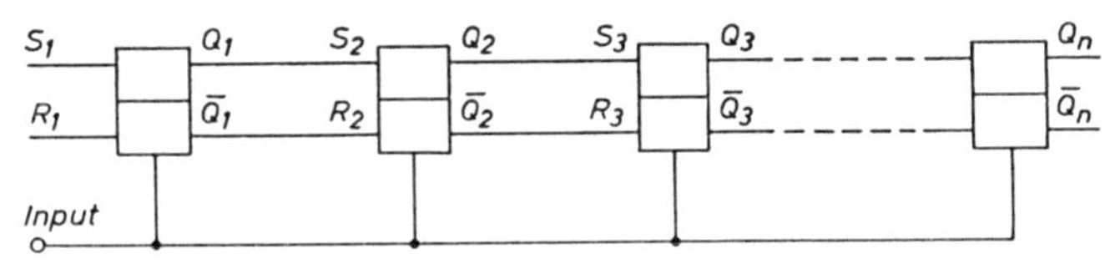

FIG. 18.36. RING COUNTER

Synchronous counters, in which all the outputs change state at the same time, are therefore often used. The simplest synchronous counter is a ring counter in which a 1 is made to recirculate through a ring of bistables connected as shown in Fig. 18.36. The instantaneous position of the 1 will indicate the state of count. Many more complex synchronous counters can be devised, using suitably gated J–K bistables to count in binary or other codes.

Because the bistable element has, by definition, only two stable states, all counters using bistables count basically in a binary code. It is often more convenient, however, for the counter to operate on the decimal system; that is to return to the starting state after 10 counts. This can be accomplished by the addition of suitable gates to a pure binary counter so that the state corresponding to the change from 9 to 10 is recognized and used to reset the counter to 0000. A typical ripple-through decade counter using four J–K bistables

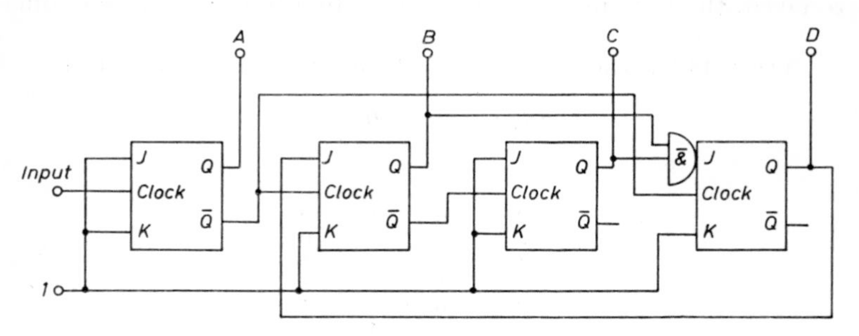

FIG. 18.37. BCD COUNTER

is shown in Fig. 18.37. This counter operates as a ÷2 followed by three bistables which would normally divide by 8 but are modified by the gating to divide by 5. The truth table for this counter is

Digit	*A*	*B*	*C*	*D*
0	0	0	0	0
1	1	0	0	0
2	0	1	0	0
3	1	1	0	0
4	0	0	1	0
5	1	0	1	0
6	0	1	1	0
7	1	1	1	0
8	0	0	0	1
9	1	0	0	1
10	0	0	0	0

The counter therefore operates in a pure binary code except that it returns to 0000 after 10 counts. It is therefore known as a 1–2–4–8 *binary-coded decimal counter*. Decimal counters can be made to operate using many other codes, but BCD counters are most common, and are often made as a single integrated circuit.

If a decimal readout is required, the output of the decade counter must be decoded by some means that recognizes the state of all four stages. This can be done with an array of diodes or multi-input NAND gates or similar gate configurations. Here again single integrated circuits are available which carry out the decoding function and provide outputs suitable for driving cold-cathode or other display devices.

Clearly more and more complex counting functions can be built onto a single integrated-circuit chip as the demand for these devices grows, and multiple counter/memory/decoders and multibit shift registers are already available.

18.5. Core Stores

As mentioned earlier, the stores required in computer work do not necessarily utilize semiconductor elements, and with units of medium or large size there are advantages in size and cost in using magnetic storage techniques.

As was mentioned in Chapter 4 (page 203), ferrites with rectangular hysteresis loops can be used to store information in a binary code. If an array of small toroidal cores of this material is constructed in the form of a matrix plane and the cores are threaded by suitable windings, random access can be obtained to the entire field of stored binary digits.

The simplest form of matrix plane consists of rows and columns parallel to the X and Y axes. Each core is threaded by two drive windings X and Y common to all cores in *one* row or column and by an output winding common to *all* the cores. If the current required to produce a sufficient field H_m to change the direction of magnetization is called I_m, positive information, known as a "one", can be written into one particular core, say X_3, Y_4 by feeding coincident current pulses of $+I_m/2$ to the X_3 and Y_4 drive wires. All the cores in the X_3 and Y_4 column are then subject to a field $+H_m/2$ which is not sufficient to cause a permanent change in direction of magnetization; core X_3, Y_4, however, receives two $I_m/2$ pulses and, if these are arranged to add, the resultant field $+H_m$ will drive the core onto the square loop, leaving remanent flux density $+B_r$ after the current pulses have been removed.

Information is thus stored in one unique location and can be read out by applying pulses $-I_m/2$ to the drive wires. If a "one" has been stored at the coincident location, the flux density will be changed from $+B_r$ to $-B_m$ and a large voltage will be induced in the output winding. If, on the other hand, a "zero" ($-B_r$) has been stored, the voltage resulting from the change in flux density from $-B_r$ to $-B_m$ will be small. It should be noted that this read-out destroys the information in the store and that a "one" digit must be rewritten if it is required for subsequent operations.

The store may be extended to a three-dimensional stack in which information can be stored as binary *words*. The most usual way of incorporating the additional address information is to add an inhibit winding common to all cores in a plane, which is driven at $-I_m/2$ when the Z address of the plane is a "zero".

Thus, if the word X_3, Y_4 is to be written 1001, write pulses of $+I_m/2$ are applied to the X_3 and Y_4 drive wires of all planes, and inhibit pulses of $-I_m/2$ are applied to the Z_2 and Z_3 planes. Only the required "one" address then receives a cumulative current of $+I_m$, all other addresses in the stack receiving $+I_m/2$, 0 or $-I_m/2$, none of which is sufficient to change the direction of magnetization.

The size of core store is limited by the problem of core disturbance, which is the noise generated on the output line by the small voltages induced when cores are fed with $I_m/2$. While these voltages are small in themselves, the cumulative effect of many such voltages is very significant. This effect is reduced by arranging that the output wire links half the cores in the matrix in one sense and half in the other, thus giving approximate cancellation of unwanted signals while true "one" outputs appear as either large positive or negative pulses. Other methods of discrimination include using two cores per bit and arranging that both a true "one" and a true "zero" give a definite output but with opposite signs.

Core stores are basically simple and reliable and remain a preferred choice for medium-size stores. They have the further advantage that they can be driven by high-speed diodes, which are appreciably faster than transistors.

19

Specialized Communication Techniques

MODERN communication practice involves a wide range of specialized applications beyond the scope of the present work, which is necessarily confined to discussion of basic principles. Brief reference may be made, however, to some of the more important techniques not hitherto mentioned.

19.1. Radar

A widely-used development of pulse circuitry is the system known as *radar*. The word is a contraction of "Radio Detection and Ranging", and relates to the utilization of the reflection of short pulses of radio waves from distant objects. It was originally used by Appleton to conduct experiments on the ionosphere, and was later developed for military needs, in which field it is extensively employed today; but many of its techniques have since been profitably adapted to civil requirements.

Brief pulses of radio waves, of suitable (short) wavelength, will be reflected by distant objects in their path, and since they travel with a constant and known velocity, the total time elapsing in the go-and-return travel provides a measure of the range, while by using highly-directive aerial systems the direction can also be determined with considerable precision. For an object at a distance d metres the echo time t is $2d/c = d/(1{\cdot}5 \times 10^8)$ seconds. For close-range work d may be as little as 150 m, for which $t = 1\,\mu$sec, while for ranges of several kilometres it will be upwards of $10\,\mu$sec.

Clearly the interval between the pulses must be sufficient to allow the echo to be received, and is thus conditioned by the requirements. Subject to this limitation, the higher the *pulse-repetition frequency* (p.r.f.) the better the signal/noise ratio, as explained later, typical values ranging from 200 to 10,000 pulses/sec. The pulse duration must be short compared with the interval time, and is again a compromise. The shorter the pulse the greater the discrimination between neighbouring objects; on the other hand, the bandwidth of the transmission is inversely proportional to pulse duration, as explained in Chapter 17, so that a compromise is necessary. Typical pulse durations range from 0·1 to 10 μsec.

Such brief pulses require the use of very high frequencies, usually in the S or X bands, and are customarily generated by magnetrons, as described in Chapter 13. Because of the small duty cycle (e.g. 1 μsec on, 100 μsec off), peak powers of several megawatts are practicable. They are radiated by parabolic aerials, often of the *cheese*

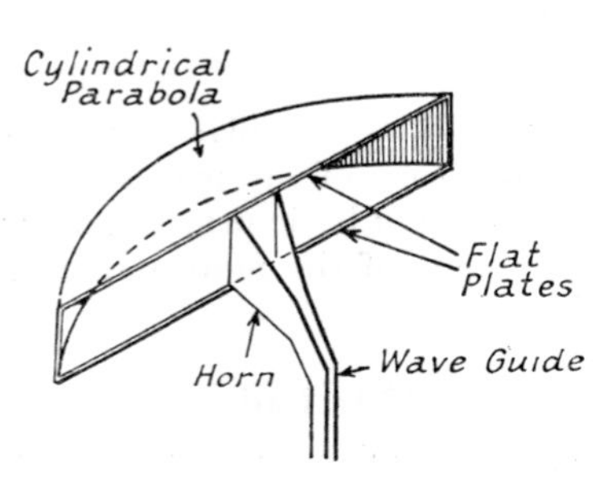

FIG. 19.1. CHEESE AERIAL

type, which are sections of a paraboloid cylinder as shown in Fig. 19.1. These are then rotated, or oscillated, over a prescribed arc in order to scan the required area.

Types of display

The same aerial is used to receive the echo, which is then amplified and displayed on a cathode-ray tube. In the type A display the horizontal scan is provided by an accurately-known linear sweep, calibrated in distance (e.g. 1 μsec = 150m). This is triggered by the transmitter pulse when the echo appears slightly later as shown in Fig. 19.2. The display is usually fed through a delay network, so that the (suitably attenuated) transmitter pulse and the echoes are seen together. Near-by objects may produce a random collection of echoes, known as *ground clutter*, and there may be false echoes from large fixed obstructions, but it is not difficult to select the true echo. The successive echoes from the rapidly-recurring pulses are superposed, producing a virtually stationary *blip*, while the inevitable background noise is not so integrated, so that a reasonable signal/noise ratio is obtained, which improves as the p.r.f. is increased, as said earlier.

If the aerial is oscillated over a suitable arc, the echo may also be displayed on a second tube of which the time base is synchronized with the movement of the aerial, thus providing an indication of the direction, in both horizontal and vertical planes.

It may be noted that this type of display is used in such applications as echo sounding—plotting the contour of the sea bed, or locating shoals of fish—and it has been adapted to the detection of flaws in materials. In these applications, however, radio waves are

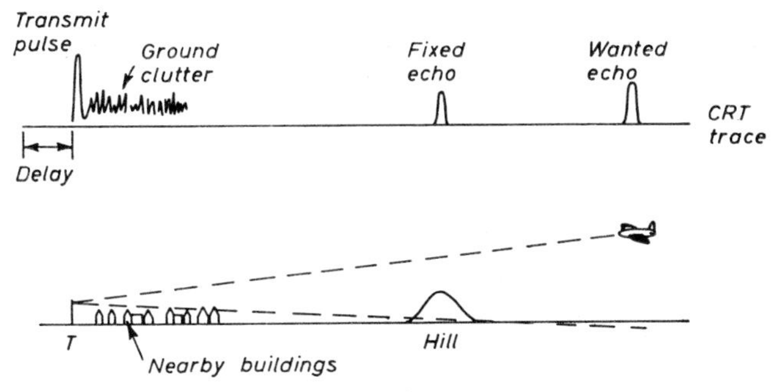

FIG. 19.2. RADAR A DISPLAY

neither necessary nor practicable, and pulses of supersonic waves are used.

An alternative display, known as the *plan position indicator* (p.p.i.) provides a radar map of the area being explored, and is used where visual observation is impossible (e.g. at night or through fog or cloud). Here a special form of time base is used in which the electron beam is rotated in synchronism with a rotating aerial, but the distance from the centre is determined by the echo time. Thus as the scan rotates a series of traces is produced which records the distances of the various obstacles encountered and so provides a map of the territory. A long-delay screen is used which holds the pattern and is reactivated at each successive revolution of the beam. Thus, for example, a map can be produced of a harbour showing the movement of any vessels in the area. It can also be used to provide weather charts, since areas of rain or heavy cloud can be arranged to provide suitable echo patterns.

T–R Switch

The use of the same aerial for both transmission and reception involves some high-speed switching arrangement to isolate the receiver during the radiation of the (high-power) pulses. This is known as a *T–R switch*, illustrated in Fig. 19.3. It consists of a resonant cavity housing a cold-cathode tube containing gas at a low pressure. The high-energy transmitter pulse ionizes the gas, which imposes a heavy load on the cavity and destroys its resonant properties so that there is no path between the aerial and the receiver.

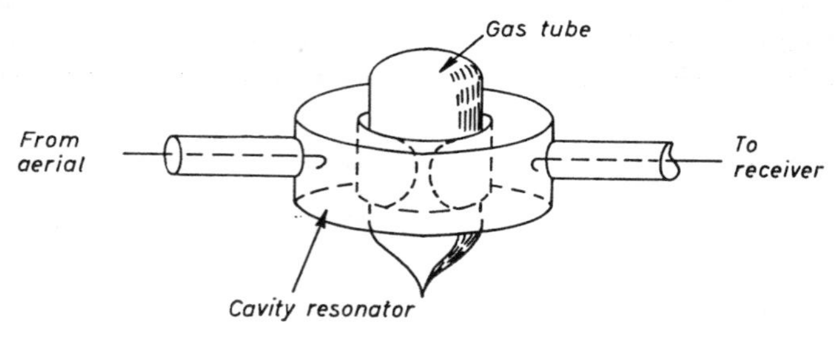

FIG. 19.3. T–R SWITCH

It will be evident that there are many detailed refinements in practical radar systems, but it is not possible here to do more than review the basic principles.

19.2. Weak-signal Amplifiers

One of the most spectacular of recent achievements has been the facility with which communication has been possible, notably in the space programmes, over distances measured in millions of miles. This is the more remarkable since in spacecraft the power available is necessarily limited so that the signals to be received and amplified are of the order of 10^{-9} volt or less. This is well below the level of background noise in any conventional form of amplifier, and entirely new forms of resonant system have had to be devised.

Parametric Amplifiers

One of the methods used for the reception of extremely small signals utilizes what is known as a *parametric amplifier*. The basic principles of this system may be explained by considering a simple parallel-resonant circuit fed from a source of sinusoidal current as shown in Fig. 19.4. This current produces a sinusoidal voltage

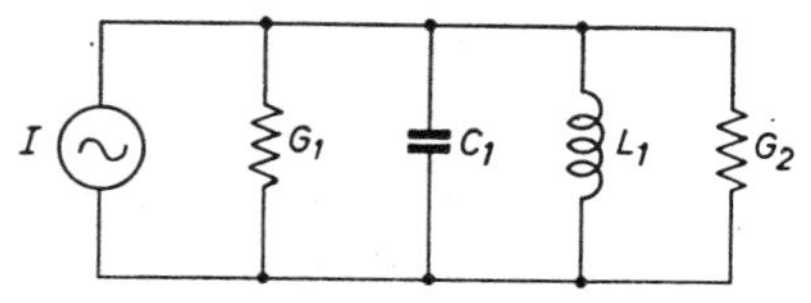

FIG. 19.4. SIMPLE RESONANT CIRCUIT FED FROM CURRENT SOURCE

across the capacitor; if the capacitance can be decreased at the instant when the voltage is a maximum, the voltage will be increased since for a given charge the product of voltage and capacitance is constant. If the capacitance is returned to its original value one-quarter period later, no voltage change is introduced, since both voltage and charge are now zero. The process can now be repeated, and progressive amplification of the voltage across the circuit will

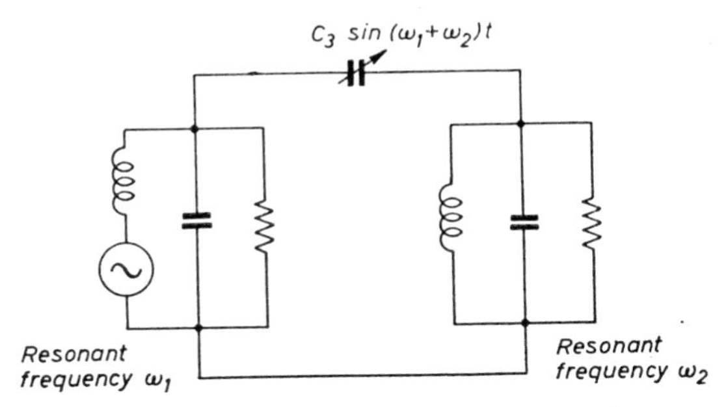

FIG. 19.5. GENERAL CIRCUIT OF PARAMETRIC AMPLIFIER

be obtained if we can provide a means of varying the capacitance at a frequency twice that of the input signal.

This simple model shows that amplification is possible in a circuit having a variable reactance element, usually a varactor diode (page 266). An increase in energy in the signal circuit is obtained by transferring energy from the circuit which modulates the varactor, known as the *pump circuit*.

In the circuit described above, the varactor is pumped at exactly twice the signal frequency and with an exact phase relationship. This is, however, a particular case and the basic general circuit of a parametric amplifier is shown in Fig. 19.5. This consists of a circuit tuned to the signal frequency and provided with input and output connections with another circuit tuned to a second frequency and known as the *idler*. The varactor couples the two circuits and is pumped at the sum of the signal and idler frequencies. Power enters the system at the signal frequency and mixes with the pump power to produce power at the idler frequency, which in turn mixes with the pump signal to produce signal power. The effect of this double mixing in a variable reactance can be shown* to be that of introducing a negative resistance across the signal and idler tuned circuits, energy from the pump source being converted to energy at the

* J. M. Manley and H. E. Rowe, *Proc. I.R.E.*, **44**, p. 904.

signal and idler frequencies. As the pump power is increased, the effective negative resistance rises until the resultant circuit resistance is negative and the amplifier oscillates at signal frequency.

In a practical amplifier, the signal, idler and pump frequencies must be separated, and the input must be separated from the output. A typical arrangement is shown in Fig. 19.6, where a ferrite *circulator* is used to buffer the input from the output. The circulator is a device in which energy can only pass from the input terminal to the second terminal and from the second to the third. Thus, the parametric amplifier is protected from any noise generated in subsequent

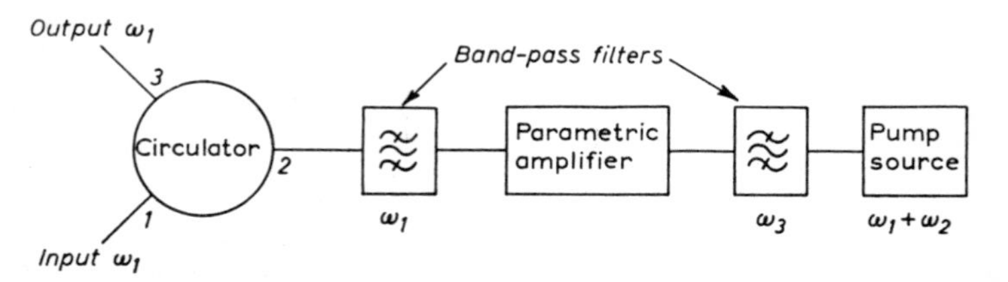

FIG. 19.6. TYPICAL ARRANGEMENT OF PARAMETRIC AMPLIFIER

stages, while the pump and idler frequencies are excluded from subsequent stages by a band-pass filter.

The advantage of the parametric amplifier is that a very low noise figure can be obtained, and these amplifiers are used as the first stage in microwave receivers for such applications as satellite communication and radio astronomy. The upper frequency of operation is limited by the pump, since the pump power required increases as the square of the pumping frequency. However, parametric amplifiers have been made for frequencies above 10 GHz.

The Maser

Where very low noise microwave amplification is necessary, the *maser* is superior to the parametric amplifier for the first stage preceding a conventional microwave amplifier. The name "maser" is derived from the initial letters of "Microwave Amplification by Stimulated Emission of Radiation".

The operation of the maser is based upon the quantum theory that atoms or molecules exist in certain well defined energy levels. If an atom can be made to drop to a lower energy by the application of an electromagnetic wave of appropriate frequency, the energy of the emitted radiation can be made to add to the stimulating radiation, thus causing amplification. Maser systems have been devised using gases such as ammonia or carbon monoxide as the maser material, but most systems use solid paramagnetic materials such as artificial ruby.

One form of maser amplifier uses a cavity resonator. The crystal is contained in a cavity mounted in a magnetic field and cooled with liquid helium to reduce noise to a minimum. Energy is fed into the cavity from a pump oscillator, and some molecules emit radiation. This radiation is reflected from the cavity walls and induces further emission so that amplification is obtained at a frequency determined by the resonant frequency of the cavity. Input and output are coupled to the cavity through the same waveguide, using a ferrite circulator to effect separation as described above.

A disadvantage of the cavity resonator for radio communication is the limited tuning range and narrow bandwidth. Greater flexibility can be obtained if the maser material is enclosed in a travelling-wave tube, such as was described in Chapter 13 (page 605). Here pump and and signal frequencies are fed in at one end of the tube, and after energy transfer using the maser material, the amplified output is obtained at the other end of the tube.

Since the maser must be supercooled for best noise performance, maser amplifiers are necessarily expensive and bulky. However, in sophisticated applications such as radio astronomy or high-precision radar where the received signal is very small, the noise performance may be the overriding factor.

The Laser

The name *laser* is derived from the initial letters of "Light Amplification by Stimulated Emission of Radiation". The laser functions on the same basic principle as the maser, but as the name implies, it is concerned with frequencies in the visible spectrum. Its significant property is that coherent light is generated which is plane polarized, monochromatic and emitted as nearly parallel rays.

As with the maser, the material used can be either a gas or a solid, the most usual materials being carbon dioxide, gallium arsenide or artificial ruby. In one form of ruby laser, the ends of a cylindrical rod of ruby are coated with reflective material except for a small hole at one end. The rod is then placed in a magnetic field to stimulate energy transfer and excited with green light from a xenon source. The ruby absorbs this green light and emits red light, most of which is reflected from the coating at each end. This reflection from end to end of the rod produces highly selective regeneration so that the light emitted through the small exit hole is monochromatic and can be focused into an extremely narrow beam with negligible divergence.

It is this concentration of the laser beam which is the property that leads to the majority of applications. Lasers are used in many range-finding and surveying applications, including measurement of distance in space, and in communication, where the beam can be modulated with a large number of information channels like a radio wave, the advantage being that the narrow beamwidth conserves energy and enhances secrecy. Other applications are those in which the heating properties of the concentrated beam are employed, including such diverse examples as metal forming and eye surgery.

19.3. Telemetry

Telemetry literally means "measuring at a distance", but the term is now generally applied to the transmission of the results of physical measurements from a remote location, and sometimes to the transmission of data which is to perform a control function at a remote location. The telemetry information can be transmitted over directly wired systems or over radio systems. Wired systems are limited in distance, flexibility and capacity, and the following considerations relate mainly to radio systems.

Radio telemetry systems are usually classified by the methods of multiplexing and modulation involved. Data signals from the separate measuring instruments at the remote location must be kept separate from one another by a *multiplexing* process which may take the form of a frequency division or a time division. In *frequency-division* multiplexing, the various data signals are separated by allocation of independent frequency channels to each variable, while in *time-division* multiplexing the data is sampled on a time basis and referenced to a synchronizing signal. The composite multiplex signal is then modulated on to a carrier using one of the techniques of amplitude, frequency or phase modulation described in Chapter 7.

Thus frequency-division multiplexing is effected by modulating sinusoidal sub-carriers of different frequencies with the various data messages and then, in a radio telemetry system, using these sub-carriers in turn to modulate the main carrier at a higher frequency. Since each stage of the modulation can use any of the three basic techniques, the complete modulation can be effected in nine ways. In fact, phase-modulation techniques cannot be used for signals having a d.c. component, which must be preserved in transmission, so that six possible systems are used. These are designated by first describing the multiplexing process and then the final carrier modulation. Thus an AM/FM system uses amplitude modulation of the data on to sub-carriers and frequency modulation of the sub-carriers onto the main carrier.

Time-division multiplexing is accomplished by sampling the various data channels at slightly different times and using these samples for one of the pulse-modulation techniques described in Chapter 3; these modulated pulses are then used to modulate the main sinusoidal carrier. The combination of PAM, PDM, PPM and PCM with AM, FM or PM would give twelve possible systems, but the use of PPM with a continuous carrier defeats the main power-saving advantage of PPM so that only ten systems are usually used.

Detailed discussion of the techniques employed in preparing physical data in a form suitable for telemetry is beyond the scope of this book, but a large variety of active and passive transducers have been developed for translating physical data into electrical form. Active transducers generate a voltage or current, either directly as in piezo-electric transducers or using an external power source as in photoelectric transducers, whilst passive transducers exhibit a change in some passive electrical parameter such as resistance, capacitance or inductance due to a change in the physical parameter to be measured.

Development in the telemetry field is mainly concerned with microminiaturization and the achievement of a high degree of reliability of both transducers and circuits for such purposes as space research. This activity is reflected in the industrial field, where pulse-modulation techniques are compatible with the use of computers for direct digital control of industrial process.

19.4. Opto-electronics

Developments in solid-state technology have led to an increase in the use of optical paths for the transmission of information. Many applications involve the use of infra-red sources such as gallium arsenide light-emitting diodes combined with photosensitive detectors as described in Chapter 6. These optical links may be used over very short distances to produce switches with very high isolation, or over considerable distances using lens systems to concentrate the beam. Transmission is possible over very long distances using laser techniques.

Amplitude modulation may be employed to impress information onto the carrier, but other modulation techniques have also been developed. An interesting system involves the use of the YIG modulator. Yttrium-iron-garnet is a crystalline substance which exhibits interesting properties when subjected to a magnetic field and has been used in microwave as well as optical applications. In an optical modulation system, the property of interest is that the plane of polarization of light is rotated after transmission through the YIG by an amount which depends on the applied magnetic field. Thus information can be impressed onto a plane-polarized carrier

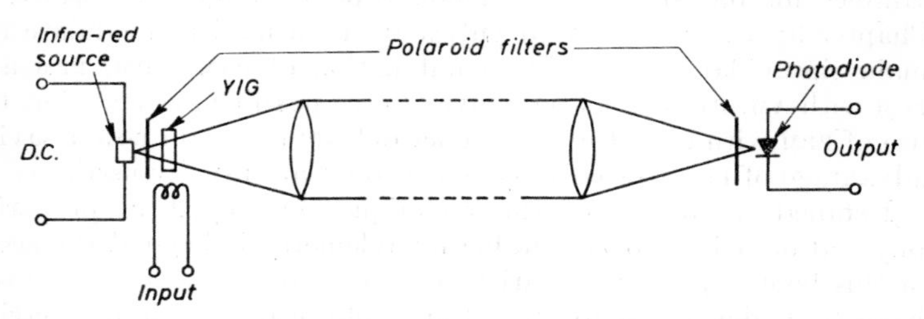

FIG. 19.7. SIMPLE OPTICAL LINK USING YIG MODULATOR

as a varying angle of polarization. Demodulation at the receiver may be effected over a limited bandwidth by passing the signal through a polaroid filter, the resultant output varying in amplitude with the angle of polarization. Fig. 19.7 shows a simple optical link using an infra-red source and a YIG modulator.

Appendix I

Atomic Structure

THE first practical atomic theory was propounded by Dalton in 1803. According to this formulation all physical substances were composed of suitable combinations of certain essential *elements* such as carbon, copper, silver, oxygen, etc. Experimental evidence indicated that these elements could combine with certain other elements but could not themselves be analysed into constituent parts, so that they appeared to be the smallest possible particles of matter.

The various elements did not all behave alike, particularly in respect of the relative proportions in which they combined with other elements. Moreover, whereas some elements combined very readily, others would not combine at all; and there were other peculiarities which were observed and recorded but not understood.

Towards the end of the nineteenth century the researches of various physicists into electrical phenomena led to the idea that matter itself was electrical in character. It began to appear as if the very atoms of matter were composed of incredibly small electrical particles, assembled in some sort of miniature solar system. These supposed particles were given the name *electrons*. Yet while this theory accounted for many of the observed effects there was much that still remained unexplained.

In 1909, however, Nils Bohr, then a young student at Manchester University, put forward the inspired hypothesis which forms the basis of modern atomic theories. Some years earlier Max Planck had discovered that, when dealing with very small quantities, changes could only occur in distinct jumps, corresponding to what he called *quanta* of energy. Bohr now suggested that if atoms were in fact miniature solar systems, the orbits in which the planetary electrons rotated would be limited in number and disposition to correspond with certain definite quantum energy levels, such that the angular momentum of the electron was always an integral multiple of $h/2\pi$, where h is Planck's constant.

To comply with this requirement the permissible orbits must lie within a series of shells at increasing distances from the nucleus, known as the *K*, *L*, *M*,. . . shells respectively. Moreover as the diameter of the shell increases it is found that the requirements can be met in several ways (e.g. by various forms of elliptical orbit), so that increasing numbers of orbits become possible as the diameter of the shell increases. In fact the maximum number of electrons in the successive shells is $2n^2$ where n is 1, 2, 3, . . ., etc.

Now, as far back as 1869 Mendeleev had suggested that if the various elements were arranged in order of their atomic weights, they fell naturally into a series of "octaves" in which elements having the same relative position had similar characteristics. The quantum requirements produce just such a grouping, indicating that the behaviour of an atom is primarily dependent upon the number and disposition of its planetary electrons.

Thus the *K* shell can contain two electrons, which provide the two simplest and lightest elements, hydrogen with one electron and helium with two. Then follows a series of 8 further elements from lithium to neon, having from 1 to 8 electrons in the next (*L*) shell (in addition to the two in the *K* shell).

A further octave then follows with the elements sodium to argon, having from 1 to 8 electrons in the *M* shell. This shell, however, can have 18 electrons, but it is found that the 10 additional electrons are accommodated in a slightly different manner, which does not come into operation immediately. The next two elements, potassium and calcium, have one and two electrons respectively in the *N* shell. Then follows a series of ten *transition elements*, which utilize the ten vacant possibilities in the *M* shell, after which the octave in the *N* shell is completed by the elements gallium to krypton.

The process continues with increasing complexity until the structure becomes, so unwieldy that the atoms break down spontaneously by the process known as *radioactivity*, but we cannot here pursue the matter further.* The structure of the first 36 elements is shown in Table A.1, from which we can see the reasons for many of the physical effects which have been discussed in the preceding chapters.

First it will be noted that the "alkali" metals, lithium, sodium and potassium, each have only one electron in their outer shell, while the acid gases fluorine, chlorine and bromine have seven. These two types of element have a great affinity for each other, combining to

* An excellent discussion of the subject will be found in *Intermediate Electrical Theory* by H. W. Heckstall-Smith (London, Dent).

form salts like sodium chloride (common salt) in which the combined total of outer electrons is eight. This, in fact, is the basis of chemical combination, the elements grouping themselves, often more than two at a time, in such a manner as to achieve a combined total of eight outer electrons.

TABLE A.1. ELECTRON STRUCTURE OF THE FIRST 36 ELEMENTS

Element	*Period*	*Atomic Number*	*Electrons in shells*				
			K	*L*	*M*		*N*
					M_1	M_2	
H	1	1	1				
He		2	2				
Li	2	3	2	1			
Be		4	2	2			
B		5	2	3			
C		6	2	4			
N		7	2	5			
O		8	2	6			
F		9	2	7			
Ne		10	2	8			
Na	3	11	2	8	1		
Mg		12	2	8	2		
Al		13	2	8	3		
Si		14	2	8	4		
P		15	2	8	5		
S		16	2	8	6		
Cl		17	2	8	7		
A		18	2	8	8		
K	4	19	2	8	8		1
Ca		20	2	8	8		2
Sc		21	2	8	8	1	2
Ti		22	2	8	8	2	2
V		23	2	8	8	3	2
Cr		24	2	8	8	4	2
Mn		25	2	8	8	5	2
Fe		26	2	8	8	6	2
Co		27	2	8	8	7	2
Ni		28	2	8	8	8	2
Cu		29	2	8	8	9	2
Zn		30	2	8	8	10	2
Ga		31	2	8	8	10	3
Ge		32	2	8	8	10	4
As		33	2	8	8	10	5
Se		34	2	8	8	10	6
Br		35	2	8	8	10	7
Kr		36	2	8	8	10	8

Secondly, it will be noted that elements which are good conductors have a structure containing relatively few outer electrons so that these atoms are more susceptible to the process of electron loss and recapture which constitutes an electric current. It will be noted also that the familiar metals, including copper and iron, are all transition elements, having only two outer electrons.

Thirdly, the energy level in an atom depends upon which of the possible orbits are occupied. The greater the radius of the orbit the greater the energy, so that if an electron jumps from one orbit to a closer one, energy will be released (in the form of vibrations of heat, light or X-rays), while conversely the atom can absorb energy causing an electron to jump to an orbit of greater radius.

The electrons in the inner shells are confined to their respective orbits, but the outer electrons can occupy not only their normal orbit, but any orbit beyond, so that various possible "excited" states exist, including the possibility, in conductors, of complete escape under the influence of an external e.m.f. The energy required to cause a transfer from one orbit to another is measured in *electron-volts*, one eV being the work done in changing the potential of any electron by one volt.

It is not practicable to discuss the mechanism in greater detail here, but enough has been said to indicate the existence of different energy levels in the atom, and the possible transitions from one level to another.

The Nucleus

So far we have said nothing about the positively-charged nucleus around which the electrons travel in their orbits. This nucleus is now believed to consist (mainly) of *protons* and *neutrons*. Protons are particles carrying a positive charge equal to that on an electron, but having a mass some 1,800 times as great. The mass of an atom is, in fact, mainly in the nucleus.

Neutrons are particles which behave like a proton and an electron bound together in some way. Thus a neutron has practically the same mass as a proton but no electric charge, and all atoms (except hydrogen) contain several neutrons in their nucleus.

These neutrons, however, have no effect on the electrical properties of the atom, and very little effect on its chemical behaviour, which is determined almost entirely by the number of protons, normally balanced by an equal number of planetary electrons. The atom can, as we have seen, be deprived of one of its outer electrons, and so become ionized, without losing its chemical identity. If it loses or gains a proton in the nucleus, however, it becomes a different element, but this can only happen under the very special circumstances which are the province of nuclear physics, and in ordinary electrical engineering it is only the electrons which are involved.

Atomic Weight

The presence of neutrons in the nucleus means that the atomic weights of the various elements do not increase uniformly, like the atomic number. Thus the atomic weight of helium is not twice that of hydrogen, but (nearly) four times, because its nucleus contains two protons *plus* two neutrons. Even so, the ratio is not exactly four because there appears to be a slight loss of mass (or *mass defect*) when a number of protons and neutrons are in close association, as in a nucleus.

Some elements which have the same atomic number (same number of protons) have different mass numbers (different numbers of neutrons). They are similar in chemical and electrical behaviour, and differ only in atomic weight. They are called *isotopes*.

It has now been agreed to take the unit of atomic weight as one-twelfth of that of the isotope carbon-12. On this basis the atomic weight of hydrogen is very nearly 1·008.

Appendix 2

Units and Dimensions

THE units by which numerical values are assigned to physical phenomena are based on the three fundamental units of length, mass and time. From these, derived units can be devised to measure various related quantities. Thus velocity is the distance travelled in a given time, or in symbols, LT^{-1}. The principal physical quantities are

Length	$= L$
Mass	$= M$
Time	$= T$
Velocity	$= LT^{-1}$
Acceleration	= rate of change of velocity
	$= LT^{-2}$
Force	= mass × acceleration
	$= MLT^{-2}$
Work (energy)	= force × distance
	$= ML^2T^{-2}$
Power	= work/time
	$= ML^2T^{-3}$

In the now accepted International System (SI), the fundamental units of length, mass and time are the metre, kilogram and second. The unit of force, which is mass × acceleration, is the *newton*, which is the force developed by a mass of 1 kilogram falling under the influence of gravity (9·81 m/sec²).

The unit of energy is then the newton-metre which is called the *joule*, while the unit of power, which is the expenditure of one joule per second, is called the *watt*.

Electrical Units

The units adopted for electrical quantities must clearly be compatible with the above mechanical units, which is achieved by defining them in terms of the forces exerted by electrical charges or currents. For many years two systems were in use, one employing electrostatic units based on the forces developed between charges, and the other employing electromagnetic units based on the forces between current-carrying conductors. Both systems used the centimetre, gram and second as fundamental units, and they were therefore known as CGS systems.

Unfortunately the units in the two systems were of widely different magnitudes, the electromagnetic unit of current being 3×10^{10} times the electrostatic unit (this factor being the velocity of light in cm/sec), while neither was of a convenient size; and there were other inconveniences.

In 1950 it was decided to rationalize the whole system. The fundamental electrical unit adopted was the ampere, and at the same time the fundamental units of length and mass were changed to the metre and the kilogram, constituting what was for a time known as the MKS system.

It can be shown that the force between two conductors of length l, separated by a distance d in a vacuum, and carrying currents I_1 and I_2 respectively, is $\mu_o I_1 I_2 l/2\pi d$. With unit quantities, this should equal 2 newtons (since I_1 and I_2 each contributes to the force). This can be achieved by assigning an arbitrary value to $\mu_0 = 4\pi \times 10^{-7}$, and defining the current accordingly, as below.

One must then assign a value to the permittivity of free space, ε_0. Now, Clerk Maxwell showed that in any medium the velocity of wave propagation is constant, and is equal to $1/\sqrt{(\varepsilon\mu)}$. In a vacuum, $1/\sqrt{(\varepsilon_0\mu_0)} = c = 3 \times 10^8$ metres/sec, whence, since $\mu_0 = 4\pi \times 10^{-7}$, $\varepsilon_0 = 1/(36\pi \times 10^9)$. We thus have the two essential constants—

$$\text{Magnetic space constant } \mu_0 = 4\pi \times 10^{-7}$$

$$\text{Electric space constant } \varepsilon_0 = 1/(36\pi \times 10^9)$$

The *ampere* is the current which, when maintained in two straight parallel conductors of infinite length, of negligible circular cross-section, separated by one metre in vacuo, produces a force between the conductors of 2×10^{-7} newton per metre length.

From this basic unit a series of derived electrical units have been developed, which were defined in their appropriate context in Chapter 1. They are summarized in Table A2 below.

TABLE A.2. PRINCIPAL S.I. ELECTRICAL UNITS

Quantity	*Symbol*	*Unit*	*Abbreviation*
E.M.F. or p.d.	E or V	Volt	V
Current	I	Ampere	A
Quantity	Q	Coulomb	C
Energy	W	Joule	J
Power	P	Watt	W
Resistance	R	Ohm	Ω
Reactance	X	Ohm	Ω
Impedance	Z	Ohm	Ω
Resistivity	ρ	Ohm-metre	Ω-m
Conductance	G	Siemens	S
Magnetic flux	Φ	Weber	Wb
Magnetic flux density	B	Tesla	T
M.M.F.	F	Ampere turn	At
M.M.F. gradient	H	Ampere turn per metre	At/m
Reluctance	S	Ampere turn per weber	At/Wb
Inductance	L or M	Henry	H
Electric flux	Ψ	Coulomb	C
Electric intensity	E	Volt per metre	V/m
Displacement (electric flux density)	D	Coulomb per square metre	C/m²
Capacitance	C	Farad	F

Logarithmic Units

Many physical effects are dependent on ratios. For example, in order to produce a discernible change in the loudness of a sound it is necessary to vary the intensity not by any absolute amount but by a specific ratio—actually a little over 10 per cent. Hence it is frequently convenient to express the ratio between two quantities in logarithmic units, because if this is done the combined effect of two ratios can be assessed by simply adding (or subtracting) the logarithmic values.

The neper

One such unit is the *neper*, which is the natural logarithm of the ratio between two currents, *irrespective of the impedances* in which they are flowing. Hence the ratio between two currents I_1 and I_0 can be expressed as $\log_e (I_1/I_0)$ nepers.

The unit can also be used to express relative voltage levels, but cannot be applied to power ratios unless the impedances are the same; in which case a ratio P_1/P_0 can be expressed as $2 \log_e (I_1/I_0)$.

The decibel

Power ratios, however, are more conveniently expressed in terms of common logarithms (i.e. logarithms to base 10 rather than to base e). The corresponding unit would then be $\log_{10} (P_1/P_0)$ which was called the *bel*, after Alexander Graham Bell, the inventor of the telephone. This unit, however, is inconveniently large and today the unit in use is the *decibel*, which is defined by saying that the power P_1 is at a level of x decibels with respect to P_0, where

$$x = 10 \log_{10} (P_1/P_0)$$

Thus if $P_1 = 2P_0$, $x = 10 \log_{10} 2 = 10 \times 0{\cdot}3010 = 3{\cdot}01$ dB. Similarly if $P_2 = 3P_1$, P_2 is 4·77 dB above P_1, and is $4{\cdot}77 + 3{\cdot}01 = 7{\cdot}78$ dB above P_0 (which is the same answer as would be obtained by saying that $P_2 = 6P_0$, so that $x = 10 \log_{10} 6 = 7{\cdot}78$).

The decibel can only be used for expressing voltage or current ratios if the impedances are the same; if they are, then the powers are proportional to the square of the voltage (or current) and the ratio in decibels is $20 \log_{10} (V_1/V_2)$ or $20 \log_{10} (I_1/I_2)$.

If the impedances are the same, one neper is equal to 8·68 dB.

dBm

In communications circuits it is customary to refer power levels to an arbitrary datum level of 1 mW. Hence if we use this level for P_0 we can say that a given power level P_1 has a level of $10 \log_{10} (P_1/10^{-3})$ dB with respect to 1 mW. This is often expressed by saying that P_1 is so many *dBm*. Thus 1 watt = 30 dBm.

Photometric Units

The performance of photosensitive devices is customarily rated in terms of the incident illumination, the relevant units for which are defined below.

Luminous Intensity

The original unit of light intensity was the candle, from which illuminations were expressed in candle-power. Such a unit was far too indefinite, and it has been replaced by an exact unit called a *candela* which is defined as the luminous flux intensity of $1/60 \text{ cm}^2$ of black body at a temperature of 2042 K.

In practice, one is only indirectly concerned with the intensity as such, since it varies with the nature and temperature of the source, but as an indication of the magnitude of this unit, an ordinary tungsten-filament lamp develops (very roughly) one candela per watt.

Luminous Flux

A source of light is considered to radiate a luminous flux in all directions, analogous to the electrostatic flux radiated by an electric charge. It thus obeys similar laws, notably the inverse square law.

A sphere of unit radius will have a surface of 4π. The solid angle which encloses unit area on such a sphere is thus $1/4\pi$ of the total, and is called a *steradian*.

The unit of luminous flux is then taken as the flux emitted in one steradian by a point source having a uniform intensity of one candela, and is called the *lumen*.

Illumination

The illumination is defined as the flux per unit area, the unit being the *lux*, which is a flux density of one lumen per square metre.

A former unit, still used but not now preferred, is the *foot-candle*, which is one lumen per square foot. It is equal to 10·75 lux.

Irradiance

Alternatively the illumination may be expressed in terms of the total incident radiated power, usually expressed in mW/cm^2. This necessarily depends upon the temperature of the source and is thus not directly compatible with the lux.

For purposes of comparison, however, for a tungsten incandescent source at a temperature of 2,870 K, an illumination of 1,000 lux corresponds to a radiant power of $4{\cdot}8\,mW/cm^2$.

Obsolescent Units

References will inevitably be found in the literature to units which are obsolescent. For example, torque may be expressed in *oz-in* or *g-cm*, whereas in SI units it is expressed in newton-metres. In such cases the conversion is simple, but there are other units where the conversion is less straightforward, and a brief reference to these will be useful.

Energy

Mechanical work was formerly expressed in *ft-lb*. This is not a true unit because energy involves not only mass and distance but also acceleration. The work done in lifting a mass M through a distance L is thus MLg, where g is the acceleration due to gravity $=$ 9·81 m/sec² (32·2 ft/sec²).

Energy expressed in *ft-lb* may be converted to *joules* by multiplying by 1·356.

Power

The former unit of mechanical power was the *horsepower*, which was defined as the expenditure of 550 *ft-lb/sec.*

The SI unit of power is the *watt*, the requisite conversion being 1 *hp* = 746 *watts.*

Magnetic Flux Density

This is occasionally expressed in the old CGS unit, the *gauss.* The SI unit, the *tesla* = 10^4 gauss.

Conductance

This is merely a change of name. The unit of conductance corresponding to a resistance of 1 ohm, formerly called the *mho,* is now called the *siemens.*

Appendix 3

The Binary Scale

MANY electronic devices (particularly in the computer field) use elements possessing only two possible states. If such devices are to be used to handle ordinary numbers it is necessary to convert the number into a scale-of-two or binary nomenclature.

The ordinary notation is based on a scale of ten, the respective digits from right to left representing increasing powers of 10. Thus 123 is 3×10^0, plus 2×10^1 plus 1×10^2. In a binary scale the digits represent powers of 2, but each digit can only have two values, usually chosen as 0 and 1.
Thus

$$57 = \underset{32}{1 \times 2^5} + \underset{16}{1 \times 2^4} + \underset{8}{1 \times 2^3} + 0 \times 2^2 + 0 \times 2^1 + \underset{1}{1 \times 2^0}$$

so that on a binary scale 57 = 111001

The maximum number with six (binary) digits is 63 (or 64 counting 0) $= 2^6$, so that in general the maximum scale-of-ten number which can be written with n binary digits is 2^n.

Thus if a wave is quantized into 8 regions (see page 148) the amplitude can be expressed in a binary code with 3 digits only, viz:

0	000
1	001
2	010
3	011
4	100
5	101
6	110
7	111

Certain modifications of this basic scale are sometimes used, but the principle remains the same.

Appendix 4

Transistor Constructions

As explained in Chapter 6, semiconductor materials are basically tetravalent elements in Group 4 of the Periodic Table, the most commonly used being germanium and silicon, which are intrinsic semiconductors. These materials, however, have a relatively high resistivity, so that to obtain an improved, and *controllable* conductivity, accurately determined amounts of suitable impurity atoms are introduced, the process being known as *doping*. If pentavalent impurities are used, the resulting combination contains free electrons, so that it is called an n-type (impurity) semiconductor. With trivalent impurities there are regions having a deficiency of electrons (holes) which are effectively positive current carriers, so providing a p-type material.

It is clearly essential that the original material shall be virtually free from any intrinsic impurities so that the ultimate performance is accurately controllable by the deliberate impurity content introduced. This is achieved by first preparing bars of nominally pure material by conventional chemical processes. This, however, still contains impurities, which are removed by a *zone-refining* process. This utilizes the fact that the impurities present are more soluble in the liquid form of the element than they are in its solid condition. The bar is therefore heated at one end by r.f. induction, in an inert atmosphere, and is then moved slowly through the heating zone, collecting the impurities as it goes. The process is repeated until virtually all the impurities have been concentrated at the exit end of the bar, which is then cut off. The refined material is then melted and controlled impurities added to provide the required n or p characteristics.

Grown Transistors

The earliest forms of transistor were made by a growing or pulling process. By dipping a small seed crystal into a bath of molten material and slowly withdrawing it, a long single crystal is produced. For an n–p–n transistor the material would be suitably doped to provide a resistivity of the order of 1 to 2 ohm-cm, which would be suitable for the collector. When a sufficient length has been achieved a p-type impurity is added to the bath, sufficient to neutralize the existing n-type impurity and provide an overall p-type characteristic having a similar order of resistivity. This forms the base region and the growing continues for a very short space—about 0·025 mm—after which n-type material is reintroduced, this time in sufficient quantity to produce a much lower resistivity of around 0·01 ohm-cm for the emitter region.

The composite crystal is then cooled and cut into thin wafers along the growth axis, so producing several hundred n–p–n elements to which fine wire connections are then made to the three regions. (The thin base region cannot be seen and an electrical probe has to be used to locate it.)

By reversing the order of the materials, p–n–p transistors can be produced.

Alloy-junction Transistors

A more convenient process is the alloy-diffusion technique illustrated in Fig. A.1 for a p–n–p transistor. To a thin wafer of n-type base material (typically 0·1 in square and 0·003 in thick) pellets

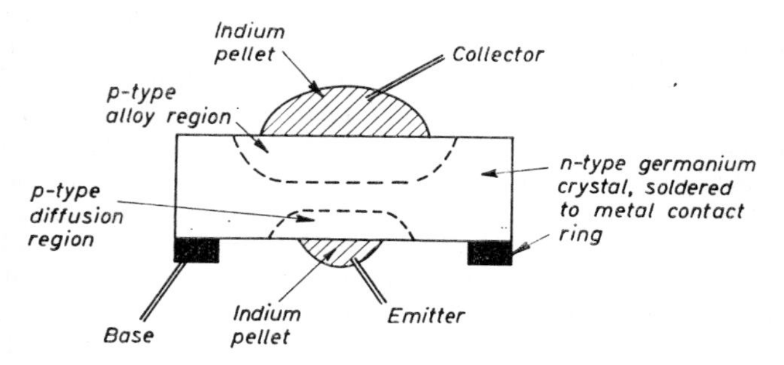

Fig. A.1. Germanium p–n–p Alloy Junction Transistor

of appropriate p-type impurity are fixed on each side. The wafer is then heated to a temperature above the melting point of the impurity but well below that of the base. The impurity then diffuses into the base, forming emitter and collector regions (of appropriate resistivity) separated by an n-type base region which can be very thin and can be controlled by the length of time for which the diffusion is allowed to proceed. Alpha cut-off frequencies in excess of 5 MHz can be attained with this method.

The method is applicable to germanium and silicon types, of either polarity, though with silicon p–n–p types, where the impurity is usually aluminium, some difficulty arises because of the different coefficients of expansion of silicon and aluminium.

Diffused Transistors

The high-frequency performance can be improved by reducing the thickness of the base. One method of achieving this is to etch the faces of the base wafer by a sophisticated electrolytic process

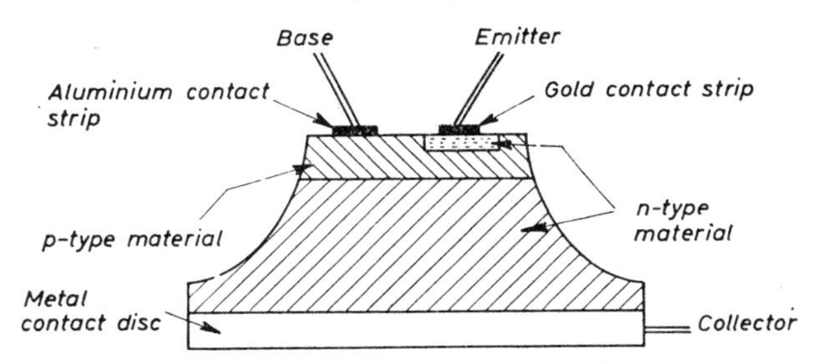

FIG. A.2. DIFFUSED SILICON N–P–N (MESA) TRANSISTOR

and then by reversing the direction of the current to deposit the required impurity material.

An alternative method is to vary the concentration of the impurity in the base region. If the concentration is a maximum near the emitter and a minimum near the collector, an internal electric field is produced which aids the passage of charge carriers across the region. This technique is applied in a variety of ways, one of which is illustrated in Fig. A.2. Here the starting point is the *collector* material, which for the example shown is n-type silicon. This is heated to around 1250°C in an atmosphere containing p-type impurity, which diffuses into the surface to provide a controllable base region. The surface is then masked except for a small area, and is then heated in an atmosphere of n-type impurity, which diffuses into the base to provide the emitter. Ohmic contacts are provided by depositing thin strips of aluminium or gold, which are alloyed to the base and emitter regions by suitable heating.

Since it is only the material between the emitter and collector junctions which is essential to the transistor action, the material outside these regions is etched away to reduce the collector capacitance, leaving the active portion standing proud of the collector in a pedestal or *mesa*. In practice, a slab of material is used sufficient for several hundred transistors. After the base region has been formed it is covered with a mask containing a lattice of small apertures through which the emitter regions can then be formed and the whole slab is then cut up into individual transistors.

There are variants of this diffusion process which need not be discussed here. It provides an effective method of controlling the behaviour and permits alpha cut-off frequencies to be attained as high as 500 MHz.

Epitaxial Construction

To provide a satisfactory reverse breakdown potential the collector region must have a reasonable thickness, which conflicts with the requirements just discussed, and also increases the effective collector resistance. This difficulty can be resolved by making the collector of low-resistance material (e.g. $\rho = 0{\cdot}002$ ohm-cm) but providing on the surface a thin *epitaxial* layer of higher resistivity (e.g. 1 ohm-cm) by suitable choice of the impurity element. Such an epitaxial layer can be made very thin (of the order of 10 microns only), but provides the necessary characteristics with a greatly reduced collector resistance.

Planar Transistors

A considerable improvement in performance and robustness is obtainable with silicon transistors by providing a thin oxide layer on the wafer. This effectively seals the transistor against the ingress of moisture and substantially reduces the leakage current. This is known as *planar* construction and is illustrated in Fig. A.3.

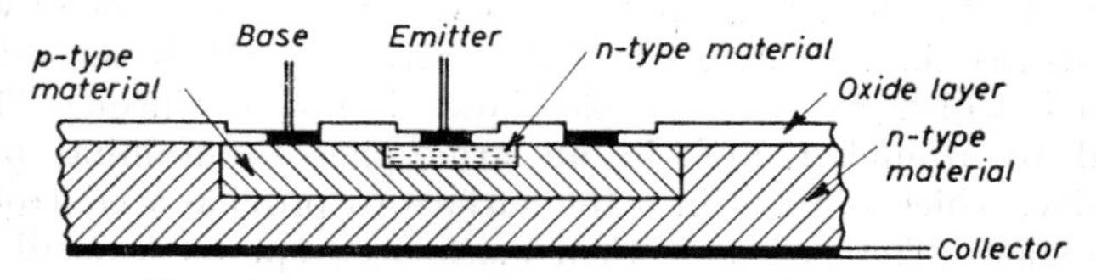

FIG. A.3. SILICON N–P–N PLANAR TRANSISTOR

A slab of n-type silicon (having a superposed epitaxial layer if desired) is heated to about 1200°C in an atmosphere of oxygen or water-vapour, which forms a protective layer of silicon dioxide on the surface. The slab, which can be of appreciable size, say 25 mm square, is then coated in a dark room with photo-sensitive material, and is then exposed to light through a mask containing perhaps

1,000 apertures of the size of the individual transistors required. The photo-sensitive coating is then developed, by normal photographic technique, and the slice is then etched to remove the oxide coating under the exposed areas. It is then heated in a boron-rich atmosphere, which provides a series of p-type base regions. The slice is then recoated with photo-sensitive material and exposed to light via a matrix of smaller apertures corresponding to the emitter regions. It is then heated in a phosphorus-rich atmosphere, which produces the required n-type emitter regions, and finally the whole slice is reheated in an oxidizing atmosphere. Holes are then made in this (new) coating by a similar masking process which permits an aluminium layer to be deposited on the base and emitter regions, providing ohmic contacts, and the slice is then cut up into individual transistors.

The process is well suited to mass production and provides transistors of considerable consistency and robustness, having alpha cut-offs within the gigahertz region. It is obviously equally applicable to the production of field-effect transistors, of both junction and insulated types, resulting in constructions of the form illustrated in Figs. 6.40 and 6.41.

Integrated Circuits

By suitable development of this planar technique it is possible to produce assemblies containing a variety of circuit elements on a single silicon chip. Such devices are called monolithic or *integrated circuits* and are widely used in communication practice today.

It is evident that by appropriate control of the forming processes, diodes and transistors of various types can be produced, while resistors can be provided by suitable areas of n- or p-type material having impurity concentrations which produce the required resistance value. Ohmic connections can be provided between the component elements, so that on a single chip of the order of 5 mm square it is possible to provide operational amplifiers, multivibrators, multiple-input gates and a variety of other applications as discussed in Chapters 17 and 18.

By way of illustration, Fig. A.4 shows a cross-section through part of such a unit, incorporating a double-emitter transistor, a diode and a resistor. Because the individual circuits are mounted in close proximity on a common substrate it is necessary to surround them with isolating walls which are provided by diffusing into the substrate barriers of suitable p- or n-type material which prevent the current carriers from drifting between the circuits.

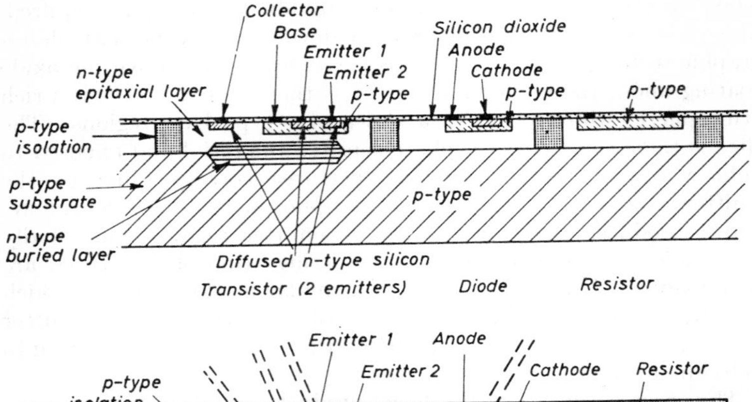

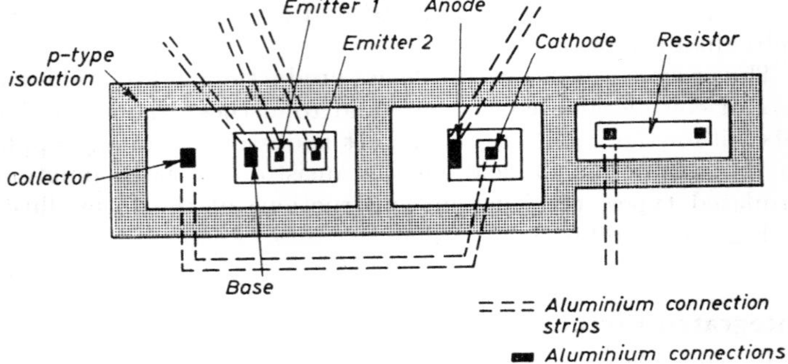

FIG. A.4. CROSS-SECTION AND PLAN OF PART OF A MONOLITHIC CIRCUIT

M.O.S. Integrated Circuits

Integrated circuits using bipolar transistors are necessarily limited in complexity by the size of the chip. Appreciable improvement can be obtained in many cases by using arrays of m.o.s devices. This arises partly because such devices can be smaller than their bipolar equivalents, but mainly because with this class of circuitry isolation barriers are frequently not required. Moreover, since fewer stages are entailed in the manufacture, closer tolerances are possible.

A further advantage is that, as explained in Chapter 18, the circuitry can be arranged to utilize internal capacitance storage effects which provide the necessary operating potentials for brief periods, and are periodically replenished from an external source.

Examples

THE following examples are reproduced by kind permission of the City and Guilds of London Institute and the Institution of Electrical Engineers, from papers in the subjects covered by the present scope. The examples have been selected to provide practical experience in the application of the techniques discussed in the various chapters and where it seems desirable a brief indication of the method of approach has been given. (The student should refrain from referring to this until he has attempted to solve the problem for himself.)

Chapter 1

1. Distinguish between the terms *power* and *energy*.

A 50-V secondary battery used to supply power to two bays of equipment gives on discharge only 60 per cent of the energy that is required to charge it to its initial condition. The first bay requires 8 A at 50 V. The second bay requires 10 A at 40 V and an 8-ohm 40-V lighting load is also connected in parallel with this bay. A resistor is connected in series with the total 40-V load so that this load can be supplied from the 50-V battery.

Calculate the electrical energy required to charge the battery to replace the charge drawn from it in supplying the whole load for one hour. Assume that the battery voltage is constant at 50 V throughout the discharge period.

(*C. & G.*, *T.P.A*, 1964)

[Lighting load $= 40/8 = 5$ A; hence total load $= 400 + 750$ W. Recharging energy $= 1150/0{\cdot}6 = 1920$ Wh]

2. (*a*) Explain *power* and *energy*, naming units in which each may be measured.
   (*b*) An electrically operated lifting device raises a load of 10 tons through a vertical distance of 1 ft in 4 s. If the efficiency of the lift mechanism is 60 per cent and that of the driving motor 85 per cent, calculate
      (i) the brake (output) horse-power of the electric motor,

(ii) the current taken by the motor if the supply voltage is 240 V d.c.,

(iii) the cost of electrical energy used in 900 such lifting operations if the cost of electrical energy is 2d/kWh.

(*c*) If the load had been 10,000 kg and the vertical distance travelled in 1 s had been 0·5 m, would the above answer have been greater or smaller? Give, with approximate calculations only, the reason for the answer.

(*C. & G., T.P.A.*, 1970)

[$(22{,}400 \times 1 \times \frac{1}{4})/(0{\cdot}6 \times 0{\cdot}85 \times 550) = 20$ b.h.p. $= 14{\cdot}92$ kW; 62·17A; 29·48 d; more]

3. In the circuit of Fig. E.1, determine the current that would be taken from a d.c. supply that maintains 25 V across AB.

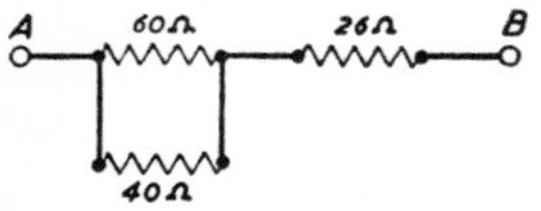

Fig. E.1

Calculate the current that would flow in the 40-Ω resistor.

If the source of supply is a battery which has an internal resistance of 8 Ω, what is the e.m.f. of the battery?

What would be the voltage across the terminals of this cell when the connection to A is broken? (*C. & G., Tel. A*, 1959)

[0·5 A; 0·3 A; 29 V; 29 V]

4. The speed of an electric motor is to be regulated by an adjustable resistor connected in series. The current taken from a 200-V d.c. supply is 4 A when the resistor is set at 15 Ω.

Calculate (*a*) the voltage drop in the resistor,

(*b*) the power dissipated by the resistor,

(*c*) the potential difference across the terminals of the motor.

In what form is the power dissipated in the resistor?

What fraction of the power drawn from the source is dissipated in the resistor? (*C. & G., Tel. I*, 1959)

[60 V; 240 W; 140 V; heat; 30 per cent]

5. State Kirchhoff's laws. Are these laws applicable for varying as well as for steady currents?

The circuit of Fig. E.2 is arranged for the measurement of very small changes in the current I. G is a sensitive galvanometer having a resistance R_g and E is a battery of constant e.m.f. 2·02 V.

Calculate the value of the resistance R necessary to reduce the galvanometer current to zero when the current $I = 0{\cdot}100$ A. If I increases by a small amount δI, deduce an expression for this increase in terms of the galvanometer current and the resistances R and R_g. (*E.I.E., Elec.*, Oct. 1957)

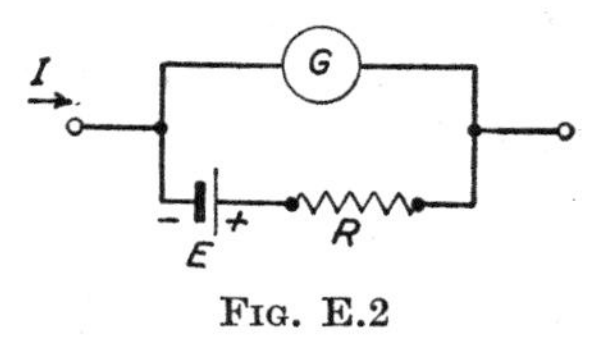

FIG. E.2

[At balance $V_g = 0 = E - IR$; whence $R = 20{\cdot}2\ \Omega$. If I increases by δI, $V_g = R_g I_g = R(\delta I - I_g)$.]

6. What do you understand by the term *charge on a capacitor*? How is this charge related to the voltage that it produces across the plates?

A capacitor of 10 μF is connected in a circuit that supplies it with a steady charging current of 75 μA. Calculate the rise in voltage that occurs across its plates during 3 sec.

Plot the rise in this voltage with time, using the vertical axis to represent voltage and the horizontal axis time.

(*C. & G., Tel. I*, 1959)

[22·5 V]

7. What do you understand by the term *permittivity*? A parallel-plate air capacitor consists of eleven equispaced plates. The plates are interleaved, alternate plates being connected together. The effective area of one side of each plate is 20 cm^2 and the distance between adjacent plates is 2 mm. Calculate—

(*a*) the charge held by the capacitor when it is connected across a 100-V battery,
(*b*) the energy stored in the capacitor in this condition.

The supply is disconnected and the capacitor is immersed in insulating oil which has a relative permittivity of 5. No leakage of electricity occurs in the process. Determine—

(*c*) the new value of capacitance,
(*d*) the charge held by the plates,
(*e*) the potential difference across the terminals,
(*f*) the energy stored.

How do you account for the difference between (*b*) and (*f*)?

(*C. & G., Tel. II*, 1959)

[(*a*) $8 \cdot 85 \times 10^{-3}$ μC; (*b*) $4 \cdot 425 \times 10^{-7}$ J; (*c*) 442·5 pF; (*d*) same as (*a*); (*e*) 20 V; (*f*) $8 \cdot 85 \times 10^{-8}$ J. The lower energy in condition (*f*) is because the dielectric is less severely stressed, the difference being released as mechanical energy (and/or radiation). Conversely, when the plates are withdrawn from the oil, part of the work done in moving the plates is transformed into increased energy stored in the capacitor.]

8. In the last question what would be the p.d. across the plates if the capacitor were only half-immersed in the oil?

[33·3 V]

9. A capacitance of C μF is charged to a potential difference of V volts. Write down expressions for

(*a*) the charge Q on the capacitor,
(*b*) the energy stored in the capacitor.

State the units employed in each case.

A 10-μF capacitor is charged from a 100-V battery. The battery is then removed. Calculate the energy stored in the capacitor.

Two other capacitors, of 5 and 3 μF joined in series, are now connected across the terminals AB of the charged capacitor. Determine—

(*c*) the total capacitance across AB,
(*d*) the p.d. across each of the three capacitors.

Assuming that no leakage of charge occurs find—

(*e*) the energy stored in each capacitor.

(*C. & G., Tel. II*, 1958)

[(*a*) CV; (*b*) $\frac{1}{2}CV^2 = 0 \cdot 05$ J; (*c*) 11·875 μF; (*d*) 31·6 V across 5 μF, 52·7 V across 3 μF, 84·3 V across total combination; (*e*) 0·0025 J in 5 μF, 0·0042 J in 3 μF, 0·035 J in 10 μF. Note that these do not total the original figure of 0·05 J because the overall voltage is now less, due to the redistribution of the charge over a greater volume of dielectric.]

10. Derive expressions for the inductance and capacitance per metre of a transmission line consisting of two parallel conductors in free space. Each conductor is a hollow thin-walled cylinder of radius r metres and the centre-to-centre spacing is d metres. State

clearly any assumption used. Find the numerical values when $r = 0{\cdot}6$ cm and $d = 30$ cm.

(*I.E.E., Elec. Eng. I*, April 1965)

[9·25 pF/m; 1·57 μH/m]

11. A concentric cable has an inner conductor of 2·0 cm diameter; the insulating material, of thickness 1·0 cm, has a relative permittivity of 3·5 and a resistivity, of 8×10^6 MΩ-m. Calculate the capacitance and the insulation resistance of the cable per kilometre length.

(*I.E.E., Elec. Eng. I*, April 1959)

[$C = 2\pi\varepsilon_0\varepsilon_r/\log_e(r_2/r_1)$ μF/m. In the example, $r_2/r_1 = 2$, whence $C = 0{\cdot}28$ μF/km. The tube of insulating material may be regarded as a block 1 cm thick and 1 km long, of width equal to the mean circumference $= 3\pi$ cm. The insulation resistance $R = \rho l/A$, where l is the length of insulation between the conductors (which here is the thickness of insulation), while A is the area, here equal to the mean circumference multiplied by the length of cable. Hence

$$R = (8 \times 10^6)/(10^{-2})(3\pi \times 10^{-2})(10^3) = 850 \text{ M}\Omega\text{/km}]$$

12. A concentric cable has an inner conductor of 16 mm diameter, the insulating material being 8 mm thick. The resistivity of the insulating material is 10^7 MΩ–m.

The dielectric stress is not to exceed 50 kV (r.m.s.)/cm. Determine the highest r.m.s. voltage to which the cable may be subjected and calculate the current taken per kilometre of cable at this voltage due to the conductance of the insulation.

(*I.E.E., Elec. Eng. I*, April 1960)

[The maximum potential gradient in a concentric cable is

$$V/r_1\log_e(r_2/r_1) = 5 \times 10^6 \text{ (volts per } metre\text{); whence}$$

$$V = (5 \times 10^6)\ (8 \times 10^{-3} \times 0{\cdot}693) = 27{\cdot}2 \text{ kV (r.m.s.)}.$$

The insulation resistance (calculated as in the previous example) is 1,060 MΩ/km; whence the conductance current $= 25{\cdot}7\mu$A.]

13. Two parallel metal plates, each of area A square metres, are separated a distance of D metres by a dielectric which has a relative permittivity ε_r and restivity ρ Ω-m.

Obtain expressions for—

(*i*) the capacitance C between the plates,
(*ii*) the leakance G between the plates.

What determines the maximum voltage which can be applied across the plates?

If a sinusoidal voltage having an amplitude of 1 V at an angular frequency of ω radians/sec is applied across the plates, what would be the magnitude and phase of the current supplied?

(*C. & G., T.P.C*, 1963)

[$8{\cdot}85A\varepsilon_r/D$ pF; Leakage = 1/Resistance, so $G = A/\rho D$ siemens; G is in parallel with C, so circuit admittance is $G + C\omega$, whence $i = (G + C\omega) \sin \omega t$ leading by arc tan $C\omega/G$.]

14. A parallel-plate capacitance is charged from a 200-V battery which is then removed and the capacitor is lowered into a bath of oil. Assuming no loss of charge, what change occurs in—

(*a*) the p.d. across the plates,
(*b*) the electric flux density between the plates,
(*c*) the energy stored in the capacitor.

Take the effective area of each of the two plates as 200 cm² and their distance apart as 0·1 cm. The relative permeability of the oil is 5·0.

The experiment is repeated but this time the 200-V battery remains connected. What happens in the battery circuit as the capacitor—

(*d*) enters the oil,
(*e*) is withdrawn from it?

(*C. & G., Tel. B*, 1960)

[(*a*) The p.d. decreases 5 times; (*b*) flux density Q/A is unchanged; (*c*) C increases fivefold and hence the energy ($= \frac{1}{2}CV^2$) decreases 5 times.

In the second experiment the p.d. is constant; hence the charge will increase as the capacitor enters the oil, requiring a momentary current from the battery; on withdrawal, the capacitor voltage will rise above that of the battery and the surplus charge will be dissipated by a momentary reverse current.

Note that the dimensions of the capacitor do not influence the answer so that the capacitance need not be evaluated.]

15. Explain the terms *induction density*, *permeability*, *reluctance* and *magnetomotive force*.

Estimate the value of each of the above quantities in the case of an iron ring which has a mean circumference of 1 m and a section of 10 cm². It is uniformly wound with a coil of 1,000 turns, and a current of 1 A through the coil produces a flux of 0·001 Wb in the ring. (*E.I.E., Elec. I*, Oct. 1957)

[1 T; $\mu_r = 796$; 10_p At/Wb; 1,000 At]

16. Write down a law by which the direction of the force on a conductor carrying a current in a magnetic field can be determined. Illustrate your answer by a diagram.

Calculate the force acting on a conductor 10 cm long carrying a direct current of 10 A at right angles to a uniform magnetic field of 10 mT. Hence find the turning moment on a coil 10 cm square of 500 turns carrying a current of 10 A when the plane of the coil is parallel to the above magnetic field. The coil is pivoted on an axis perpendicular to the direction of the field and in the plane of the coil, on a line joining the midpoints of opposite sides.

What is the turning moment when the coil is rotated by 90° from this position? Explain your answer.

(*C. & G., T.P.A*, 1964)

[Force $= BIl = 10^{-3}$ newton; with coil parallel to field force is zero; when rotated through 90°, turning moment
$= BIlNr = 10^{-3} \times 500 \times 5 \times 10^{-2} = 0{\cdot}025$ newton-metre.]

17. A coil having an inductance of 250 H and a resistance of 200 Ω is connected to a 50-V supply through a milliammeter and switch. The milliammeter coil resistance is 50 Ω.

Explain the reason for the slow rise of current when the switch is closed. State, in full, any law of electromagnetism that you quote in your answer.

Evaluate—

(*a*) the time constant of the circuit and explain its meaning,
(*b*) the value of the current at the instant of switching-on,
(*c*) the rate of increase of current at this instant,
(*d*) the energy stored in the coil when the current reaches a steady value.

Write down an expression for the current–time relation and sketch its form. (*C. & G., Tel. II*, 1958)

[(*a*) The current $i = (E/R)\,(1 - e^{-Rt/L})$; this is an exponential function, the rate of rise being determined by the time constant $L/R = 250/(200 + 50) = 1$ sec.

(*b*) When $t = 0$, $e^{-Rt/L} = 1$ so that $i_0 = 0$.

(*c*) The rate of increase of current is

$$di/dt = (-E/R)\ (-R/L)\ e^{-Rt/L}$$
$$= E/L \text{ when } t = 0.$$

Hence the rate of increase of current at switch-on = 50/250 = 0·2 A/sec.

(*d*) The energy stored in the coil when the current is steady is $\frac{1}{2}LI^2$. $I = E/R = 50/(200 + 50) = 0{\cdot}2$ A; hence the energy is $\frac{1}{2} \times 250 \times 0{\cdot}04 = 5$ J.]

18. A constant voltage V is maintained across an inductance of L henrys in series with a resistance of $R\ \Omega$. Write down an expression for the current that flows in the circuit t seconds after switching on.

On what factors does the rate of rise of current depend?

What is the initial rate of rise of current?

A relay coil of resistance 200 Ω and inductance 8 H is connected in series with a 100-Ω resistor and 60-V battery. The relay operates when the current in its coil is 31·6 mA. How long does it take to operate?

The operation of the relay armature increases the inductance of the coil to 20 H. Sketch the current/time curve from the moment of switch-on, showing the effect of this increase in inductance.

How much energy is stored in the magnetic field when the current has reached a constant value?

[$i = 0{\cdot}2(1 - e^{-37{\cdot}5t}) = 31{\cdot}6 \times 10^{-3}$, whence $t = 4{\cdot}6$ millisec; up to this time, i increases exponentially at the above rate but then drops abruptly on to a lower curve (also starting from zero) given by $i = 0{\cdot}2(1 - e^{-15t})$; energy stored when current is steady is $\frac{1}{2} \times 20 \times (0{\cdot}2)^2 = 0{\cdot}4$ J]

19. A square coil of side 20 cm has 1,000 turns and rotates on an axis through the centre points of two opposite sides, at a speed of 50 rev/s. The axis is perpendicular to a uniform magnetic field of intensity 0·02 T. Plot a curve showing how the e.m.f. generated in the coil varies with time and explain how the curve is derived.

For which condition would the torque required to rotate the coil be greater—

(*a*) when its terminals are open-circuited, or
(*b*) when they are connected to a resistor?

Give the reason for your answer. (*C. & G., Tel. A*, 1960)

[$E_{max} = Blnv$, l being the *total* length of conductor which cuts the flux; hence $e = 80\pi \sin 100\pi t$; condition (*b*) requires more torque because power ($= I^2R$) is being dissipated in the resistor.]

20. Give an account of the phenomenon of self-induction and discuss the factors which determine the self-inductance of a coil.

A uniformly-wound toroidal coil has a resistance of 20 Ω and takes a current of r.m.s value 5 A when connected to a source of voltage given by $v = 200 \sin 314t$.

(*a*) Derive an equation for the instantaneous value of the current.
(*b*) Calculate the maximum value of the flux linking the coil if it has 750 turns. (*E.I.E.*, *Elec. I*, April 1960)

[$Z = 200/5\sqrt{2} = \sqrt{(R^2 + L^2\omega^2)}$; whence $L\omega = 20$ and $L = 0{\cdot}0635$ H. $L\omega = R$, so that $\phi = \pi/4$.
Hence $i = 7{\cdot}07 \sin (314t - \pi/4)$.
Maximum flux $= L\hat{I}/N = 0{\cdot}0635 \times 7{\cdot}07/750 = 6 \times 10^{-4}$ Wb.]

21. What is the meaning of the term *root mean square value* of an alternating wave?

Write down the equations representing sinusoidal voltages of—

(*a*) 50 V r.m.s. and frequency 100 Hz,
(*b*) the same frequency and amplitude as (*a*) but leading it in phase by 90°.

Plot these waveforms on the same pair of axes and hence draw the waveform of the current that would flow in a 100-Ω resistor, if the two voltages were applied in series across it. (*C. & G.*, *Tel. B*, 1960)

[$e = 70{\cdot}7 \sin 628t$; $70{\cdot}7 \sin (628t + \pi/2)$]

22. A resistance of 100 Ω, a capacitance of 2 μF and an inductance of 0·02 H are connected in series across an alternator of negligible internal impedance. If the e.m.f. of the alternator is $1{\cdot}0 \sin 5{,}000t + 0{\cdot}5 \sin 10{,}000t$ calculate the r.m.s. values of the applied voltage and circuit current.

What power is being dissipated in the circuit and what is the power factor? (*C. & G.*, *Tel. III*, 1959)

[R.M.S. voltage $= 0{\cdot}707\sqrt{(1^2 + 0{\cdot}5^2)} = 0{\cdot}79$.

At $\omega = 5{,}000$, $X_L = -X_C = 100$, so that circuit resonates and $Z = R = 100$. Hence $I_{r.m.s.} = 1/(100\sqrt{2}) = 7{\cdot}07$ mA.

At $\omega = 10{,}000$, $X_L = 200$ and $X_C = -50$, so that $Z = \sqrt{(100^2 + 150^2)} = 180$ and $I = 0{\cdot}5/(180\sqrt{2}) = 1{\cdot}96$ mA.

Total r.m.s. current is thus $\sqrt{(7{\cdot}07^2 + 1{\cdot}96^2)} = 7{\cdot}32$ mA.

Power dissipated $= I^2R = 5{\cdot}385$ mW.

Power factor $= P/VI = 5{\cdot}385/(0{\cdot}79 \times 7{\cdot}32) = 0{\cdot}93$.]

23. Explain the meaning of the term *electrochemical equivalent* of a metal.

An electrolytic tank filled with a solution of silver nitrate is to be used to silver-plate a copper article. If a steady current of 1 A deposits 6·71 g of silver in 1 h 40 min, calculate the electrochemical equivalent of silver.

On which electrode is the silver deposited?

(*C. & G., Tel. I*, 1958)

[$Q = 1 \times 100 \times 60$, hence e-c equivalent $= 6{\cdot}71/6{,}000 = 1{\cdot}118$ mg/coulomb; metal is deposited at the cathode.]

24. Explain what is meant by the terms *reactance* and *impedance* of an a.c. circuit.

A 50 Hz alternating voltage applied to a 40-μF capacitor has a triangular waveform which rises to a maximum of 100 V and falls to zero at a uniform rate in each half-cycle. Plot one complete cycle of this voltage waveform, and calculate and plot, on the same graph, the waveform of the current.

Calculate the r.m.s. values of the voltage and current and the reactance of the capacitor under these conditions.

(*E.I.E., Elec. I*, April 1960)

[The current is a square wave of peak value 0·8 A; $V_{rms} = 57{\cdot}7$ V; $I_{rms} = 0{\cdot}8$ A; $X_C = 72{\cdot}2\ \Omega$.]

25. A rectifying device produces a square-topped wave of current of equal conducting and non-conducting periods of 10 millisec. The value during the conducting period is 10 A. Determine the r.m.s. values of the two harmonic components having frequencies nearest to 500 Hz. (*I.E.E., Pt. III, Elec. Meas.*, 1957)

[9th harmonic $= 20/9\pi\sqrt{2} = 0{\cdot}5$ A (r.m.s.); 11th harmonic $= 20/11\pi\sqrt{2} = 0{\cdot}41$ A; no 10th harmonic.]

26. A d.c. circuit contains a 40 Ω resistor connected in parallel with a second resistor having a current/voltage characteristic, over the operating range, of

$$I = 8 \times 10^{-4} V + 6 \times 10^{-5} V^3$$

where I is in amperes and V is in volts.

If the total current taken by the two resistors is 2 A, determine, graphically or otherwise, the current in each resistor and the voltage across them. If the total current is increased to 10 A, determine the percentage increase in the power dissipation in the 40 Ω resistor.

(*I.E.E. III, Adv. Elec. Eng.*, June 1964)

[Total current

$$= V/40 + 8 \times 10^{-4} V + 6 \times 10^{-5} V^3 = 258 \times 10^{-4} V + 6 \times 10^{-5} V^3$$

whence by plotting I against V, a value for V can be determined which makes $I_t = 2$ A, $= 27{\cdot}8$ V. Whence power in 40 Ω resistor = $(27{\cdot}8)^2/40 = 19{\cdot}3$ W.

If $I_t = 10$ A, $V = 52{\cdot}5$, whence $W_{40} = 69$ W. Percentage increase is $(69 - 19{\cdot}3)/19{\cdot}3 = 257$ per cent]

27. A direct voltage E is applied to an uncharged capacitor C in series with a resistor R. Derive from first principles expressions for the time variation of the power supplied to R and C. If the values of R and C are respectively 1,000 Ω and 2 μF, determine the time at which the power supplied to C is a maximum. Sketch the time variation of the power dissipation in R, the power supplied to C and the total power supplied.

(*I.E.E. III, Adv. Elec. Eng.*, June 1965)

[$v_C = E(1 - e^{-t/CR})$; $v_R = Ee^{-t/CR}$; $i = (E/R)e^{-t/CR}$. Instantaneous power in $C = v_C i = (E^2/R)(e^{-t/CR} - e^{-2t/CR})$, which is a maximum when $dP_C/dt = 0$ which occurs when $e^{-t/CR} = 2e^{-2t/CR}$, for which $t/CR = 0{\cdot}693$, so that with the values given

$$t = 0{\cdot}693 \times 2 \times 10^{-6} \times 10^3 = 1{\cdot}386 \text{ millisec}]$$

Chapter 2

1. State the Thévenin network theorem.

Use the theorem to find the current in a 50-Ω resistor connected across the output terminals PQ of a network comprising impedances and a.c. generators (all of the same frequency). Measurements on the network itself gave the following results—

(1) Voltage across PQ on open circuit, 100 V.
(2) Current in a conductor short-circuiting PQ, 2·0 A.
(3) Current in a 10-Ω resistor across PQ, 1·77 A.

(*I.E.E., Elec. Eng. II*, 1958)

[Conditions (2) and (3) cannot both apply unless the internal impedance of the network is reactive.

If $Z = R + jX$ then, from (2), $R + jX = 50\ \Omega$, while from (3), $(R + 10) + jX = 56{\cdot}5\ \Omega$. From this, $R = 30\ \Omega$ and $X = 40\ \Omega$.

With 50 Ω across PQ, $I = 100/(80 + j40) = 1{\cdot}12$ A.]

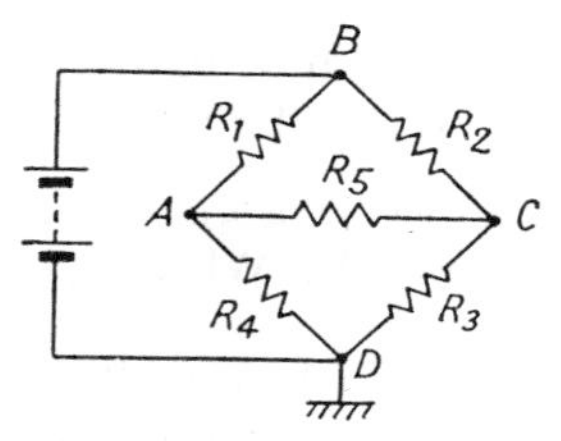

FIG. E.3

2. Fig. E.3 shows a Wheatstone bridge used for temperature measurement. It is balanced at 15·5°C when all the resistances are equal, i.e. $R_1 = R_2 = R_3 = R_4 = R_5 = R$, and the battery p.d. is V volts.

The resistances R_1, R_3 and R_5 (which includes the resistance of the galvanometer) are independent of temperature. The resistances R_2 and R_4 are both made of material having a resistance temperature coefficient of α (ohms per ohm per deg C) referred to 15·5°C.

Obtain an expression in terms of V, R and α relating the galvanometer current I_5 and the temperature rise T deg C above 15·5°C.

Find the temperature at which the galvanometer current would be 1 mA if the resistances were each 500 Ω at 15·5°C, and R_2, R_4 were of copper having $\alpha = 1/250$. The battery p.d. is 10 V.

(*I.E.E. Elec. Eng. II*, April 1964)

[Let the currents BAD, BCD and BACD be i_1, i_2 and i_3. By symmetry $i_1 = i_2$. $R_1 = R_3 = R_5$ and $R_2 = R_4 = R_1(1 + \alpha T)$. $V_{AC} = V_{AD} - V_{CD} = R_5 i_3$, whence $R_1(\alpha T i_1 - i_3) = R_1 i_3$, and $i_3 = (\alpha T/2) i_1 = \alpha T V/2R_1(2 + \alpha T)$. With the values given

$$i_3 = 1\ \text{mA when } T = 55{\cdot}5°]$$

3. State Kirchhoff's Laws relating currents and voltages in a network of resistances.

In the bridge network of Fig. E.4 the current in the meter M is 1mA in the direction shown when a 2-V supply is connected to the points AC. Calculate the resistance of R. (*C. & G., Tel. B*, 1959)

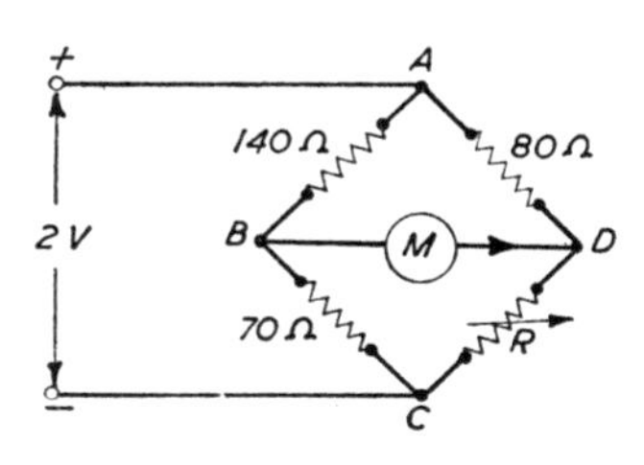

Fig. E.4

[This appears at first to be a simple exercise in Kirchhoff's Laws, which can be solved by writing down the usual mesh equations; but in fact the resistance R cannot be uniquely determined without knowing the value of the meter resistance M, and the solution must therefore be in the form of a relationship between R and M.

One method is to assume currents i_1, i_2 and i_3 in the branches ABC, ABDC and ADC, respectively. For convenience, express all currents in mA and V_{AC} as 2,000 mV. Then, since $i_2 = 1$, $140\ (i_1 + 1) + 70i_1 = 2{,}000$, whence $i_1 = 62/7$.

From the mesh ABDA we have

$$140(i_1 + 1) + M - 80i_3 = 0$$

whence $i_3 = (1{,}380 + M)/80$.

For the branch ADC we have

$$80\ i_3 + R(i_3 + 1) = 2{,}000$$

Substituting the value of i_3 we find that

$$R = 80(620 - M)/(1{,}460 + M)$$

which is satisfied by values of R between zero ($M = 620$) and 34 Ω ($M = 0$).]

4. What is the meaning of the Q-factor of (a) a reactor, and (b) a resonant circuit?

A capacitor of 1 μF and an inductor of 0·01 H are connected in parallel. At the resonant frequency of the combination the capacitor has a Q-factor of 100 and the inductor has a Q-factor of 50. Calculate the resonant frequency of the combination, its Q-factor and impedance at this frequency. (*C. & G., Tel. III*, 1959)

[$f_r = 1{,}590$ Hz; $R_L = 2\,\Omega$; $R_C = 1\,\Omega$; $Q = 33{\cdot}3$; $Z = 3{\cdot}33 \times 10^3\Omega$]

5. A series tuned-circuit consists of a 4 mH coil having a Q-factor of 250 and a loss-free capacitor. It is connected to a 5 mV source, having negligible impedance, whose frequency is set to 80 kHz.

Calculate the value of the capacitor for maximum current flow and the value of that current.

If the capacitor is changed for one of equal capacitance but with a loss angle of 0·001 rad, calculate the new value of the maximum current and the frequencies at which the current will fall to 0·707 of its maximum value. (*C. & G., T.P.C.*, 1970)

[1,000 pF (approx); 0·625 mA; 0·5 mA; 79·8 and 80·2 kHz]

6. A circuit consisting of an inductor, of resistance R and inductance L, and an uncharged capacitor of capacitance C in series is connected through a switch to a battery of e.m.f. V volts and of negligible impedance.

Derive the current–time equation for the circuit from the time of closing the isolating switch, assuming that the current is oscillatory.

What relationship between R, L, and C satisfies this condition? Determine the periodic time of oscillation when these quantities are respectively 50 Ω, 1·25 mH and 1·0 μF.

(*I.E.E., Elec. Eng. II*, April 1959)

[$i = V\{1/R - (1/2\beta L)e^{-\alpha t}(e^{\beta t} - e^{-\beta t})\}$ where $\alpha = R/2L$ and $\beta = \sqrt{(\alpha^2 - 1/LC)}$. Current is oscillatory if $\alpha^2 < 1/LC$, when $\omega = \sqrt{\{(1/LC) - \alpha^2\}}$; with values given $f = 3{,}185$ Hz.]

7. A capacitor of capacitance C is charged to a voltage V. It is allowed to discharge into an inductive resistor of resistance R and inductance L.

Assuming a non-oscillatory discharge, derive an expression for the voltage of the capacitor at time t after the beginning of discharge and determine the voltage after 0·1 msec in the case when C is 1·0 μF, R is 250 Ω, L is 10 mH, and V is 1,000 V.

(*I.E.E., Elec. Eng. II*, April 1960)

[$v = Ve^{-\alpha t}(e^{\beta t} - e^{-\beta t})$ where $\alpha = R/2L$ and $\beta = \sqrt{(\alpha^2 - 1/LC)}$; for example given, $v = 472$ V after 0·1 msec.]

8. A source of internal resistance of 200 Ω is loaded by an 800 Ω resistor in parallel with a 15·92 μF capacitor. The source e.m.f. is $(159 + 250 \sin \omega t - 106 \cos 2\omega t)$ volts and the fundamental frequency is 50 Hz. Determine the r.m.s. and average voltages across the load and the power dissipated in the load resistor.

(*I.E.E. Elec. Eng. II*, April 1965)

[$V_{out}/V_{in} = R_2/(R_1 + R_2 + j\omega CR_1R_2) = 800/(1{,}000 + j\,16f)$.

When $f = 0$, $V_{out} = 159 \times 0{\cdot}8 = 127$

$f = 50$, $V_{out} = 250 \times 800/(1{,}000 + j \times 16 \times 510) = 156$

$f = 100$, $V_{out} = -106 \times 800/(1{,}000 + j\,1{,}600) = -45$

Hence $V_{out} = 127 + 156 \sin 314t - 45 \cos 628t$

Average value $= 127$

$$\text{Rectified average} = 127 + \frac{2}{\pi}(156 - 45) = 198$$

$$\text{R.M.S. value} = \sqrt{[127^2 + \tfrac{1}{2}(156^2 + 45^2)]} = 171$$

Power in load $= V_{out}{}^2/800 = 36{\cdot}6$ W]

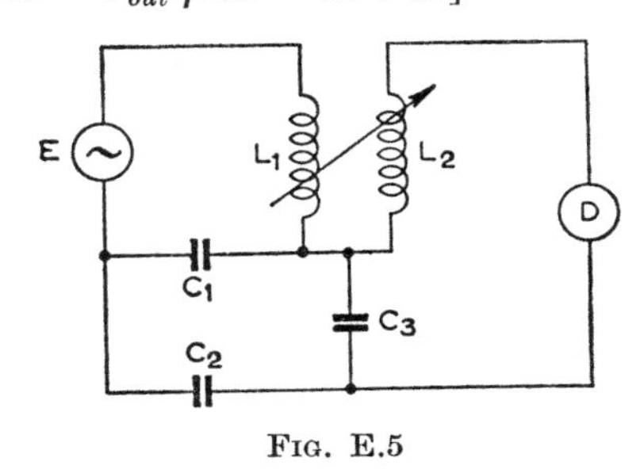

FIG. E.5

9. The circuit shown in Fig. E.5 is used to measure the frequency of the source E.

Determine the necessary condition for zero detector current. If $L_1 = L_2 = 100$ mH, and $C_1 = C_2 = C_3 = 1{\cdot}0$ μF, find the range of frequency measurable when the coefficient of coupling can be varied between 0·10 and 0·95.

(*I.E.E. III Adv. Elec. Eng.*, June 1965)

[Assume cyclic currents i_1, i_2 and i_3 in the networks L_1C_1, $C_1C_3C_2$, and C_3L_2D. At balance $i_3 = 0$, so that $M\omega i_1 = i_2/C_3\omega$, whence by writing the expression for the network $C_1C_3C_2$, and substituting $i_3 = 0$ and $MC_3\omega^2 i_1$ for i_2, an equation is obtained involving ω, namely $\omega^2 = C_3/M(C_1C_2 + C_2C_3 + C_3C_1)$. Since $L_1 = L_2 = 0{\cdot}1$ H, $M = 0{\cdot}1k$; whence with $k = 0{\cdot}1$, $f = 920$ Hz; and with $k = 0{\cdot}95$, $f = 297$ Hz]

10. Explain what is meant by *self-inductance* and *mutual inductance*. Define the unit in which self-inductance and mutual inductance are measured.

Two identical coils are mounted coaxially a short distance apart. When one coil, having a resistance of 35 Ω, is connected to a 230-V 50-Hz supply the current taken is 4 A and the e.m.f. induced in the second coil on open circuit is 115 V. Calculate the self-inductance of each coil and the mutual inductance between the coils.

(*E.I.E.*, *Elec. Eng. I*, April 1959)

[0·145 H; 0·0915 H]

11. Two coils are connected in parallel to an a.c. supply of angular frequency $\omega = 2\pi f = 200$ rad/sec. One coil has a self-inductance of 45 mH and a resistance of 3 Ω, the values for the other coil being 15 mH and 1 Ω. The coils have a mutual inductance of 25 mH. What is the input impedance of the combination, and why are there two possible answers? (*I.E.E.*, *Elec. Eng. II*, 1957)

[$$V = (3 + j9)\, i_1 \pm j5i_2 = (1 + j3)\, i_2 \pm j5i_1$$

from which $i_1(3 + j4) = i_2(1 - j2)$ with M positive

or $i_1(3 + j14) = i_2(1 + j8)$ with M negative

Using these relationships the initial equations may be written in terms of i_1 *or* i_2; but $i_1 = V/Z_1$ and $i_2 = V/Z_2$ while

$$1/Z = 1/Z_1 + 1/Z_2$$

Whence $Z = 2 + j\,3{\cdot}5$ ohms (M positive); or $0{\cdot}8 + j0{\cdot}1$ ohms (with M negative).

Note: It is simpler not to rationalize until the final stage.]

12. Each of two circuits A and B consists of a coil and a variable capacitor. Each coil has an inductance of 1,500 μH and a resistance of 30 Ω, and the two circuits have a mutual inductance of 100 μH. Circuit A is tuned to $750/2\pi$ kHz with B on open circuit, and circuit B is tuned to $810/2\pi$ kHz with A on open circuit. Calculate the current in B when an e.m.f. of 2 V, having a frequency of $780/2\pi$ kHz, is induced in circuit A.

[10·35 mA]

13. A voltage of 300 V r.m.s. at 159 kHz is applied across a 1,000 pF capacitor whose power factor is 0·05.

Find (*a*) the current, (*b*) the power and (*c*) the equivalent series resistance of the capacitor.

If a 1·3 mH inductor of Q-factor equal to 10 is placed in series with the capacitor calculate the conductance and susceptance of the combination. (*C. & G., Tel. IV*, 1959)

[(*a*) 0·3 A; (*b*) 4·5 W; (*c*) 50 Ω; conductance $= 1/R = 5{\cdot}55 \times 10^{-3}$ S; susceptance $= 1/X = 3{\cdot}33 \times 10^{-3}$ S.]

14. The primary winding of an air-cored mutual inductor has a resistance of 100 Ω and an inductance of 0·1 H. The secondary has a resistance of 1,000 Ω and an inductance of 0·4 H. When an alternating current of isosceles-triangular waveform, having a peak value of 1 mA and frequency 800 Hz, flows through the primary, the peak value of e.m.f. induced in the secondary winding is 0·5 V. What are the values of the mutual inductance and coupling coefficient between the windings? Sketch the waveform of the secondary e.m.f.

Would the magnitude and waveshape of the voltage across the secondary winding have been affected if a non-reactive resistance of 1,000 Ω had been connected across its terminals? Give reasons for your answer. (*C. & G., Tel. III*, 1959)

[$M = 0{\cdot}157$ H; $k = 78$ per cent; e_2 is a square wave varying between $\pm$ 0·5 V; with the 1,000-Ω resistor across the secondary, the output voltage is halved and the leading edge is delayed by the self-inductance of the secondary.]

15. A primary circuit of self-impedance Z_1, coupled by mutual inductance M to a closed secondary circuit of self-impedance Z_2, is connected to a supply of sine waveform and angular frequency ω. Show that the effective input impedance of the primary circuit is given by $Z_1 + (\omega M)^2/Z_2$.

A generator of e.m.f. 20 V r.m.s. and internal impedance 500 $+ j500$ Ω is connected to a coil of impedance $100 + j1{,}000$ Ω. The coil is magnetically coupled to a second coil, of impedance 40 $+ j600$ Ω, loaded with an impedance $800 - j200$ Ω to form a secondary circuit. The mutual inductive reactance is $j500$ Ω. Find the primary and secondary currents, and draw (not necessarily to scale) a vector diagram of currents and voltages in the combination. (*I.E.E., Elec. Eng. I*, 1958)

[$i_1 = 12{\cdot}4$ mA$\underline{/\overline{59°}}$; $i_2 = 6{\cdot}67$ mA$\underline{/\overline{15°}}$]

16. Explain what is meant by resonance in a tuned circuit.

A tuned circuit comprises an inductor and capacitor in series. The inductor has a value of 50 μH and the frequency of resonance is 1,500 kHz.

If the capacitance is increased by 100 pF, at what frequency does the circuit resonate? *(C. & G., R.L.T. A, 1959)*

[1,250 kHz]

17. Sketch the admittance–frequency relation for an *LCR* series circuit for the frequency range in the neighbourhood of resonance. Explain the significance of the frequencies for which the admittance is $1/\sqrt{2}$ of that at resonance.

The impedance of an *LCR* series circuit is 225 Ω both at 8·0 kHz and at 12·5 kHz, while at the resonant frequency it is 10 Ω. Find the resonant frequency, the *Q*-factor of the circuit, and the value of the inductance. *(I.E.E., Elec. Eng. I, 1958)*

[Neglecting R, $L\omega_1 - 1/C\omega_1 = 1/C\omega_2 - L\omega_2$. Whence $LC\omega_1\omega_2 = 1 = LC\omega_0^2$, so that $\omega_0 = \sqrt{(\omega_1\omega_2)}$. Hence $f_0 = 10$ kHz; by substitution $Q = 50$ and $L = 7{\cdot}96$ mH.]

18. A series-tuned circuit, resonant at 1 MHz, has a coil of inductance 100 μH and resistance 15·9 Ω. Another series-tuned circuit has a 3 dB bandwidth of 10 kHz and a resonant frequency of 2 MHz, when tuned by a capacitance of 50 pF. The two tuned circuits are then connected in series. Calculate the resonant frequency, 3 dB bandwidth, resonant impedance and effective *Q*-factor of the combination assuming that the coil resistances do not change with frequency.

(C. & G., R.L.T. B, 1964)

[For the second circuit $Q = f_r/B = 200$, whence $L = 127$ μH and $R = 8$ Ω. For the first circuit $C = 252$ pF. Hence for the combined circuit $L = 227$ μH, $R = 23{\cdot}9$ Ω and $C = 41{\cdot}7$ pF; whence $f_r = 1{\cdot}64$ MHz; $B = 16{\cdot}8$ kHz; $Z_r = 228$ kΩ and $Q = 97{\cdot}5$]

19. The primary winding of a mutual inductor has a resistance of 300 Ω and a self-inductance of 350 μH, and is supplied from a sinusoidal source at 796 kHz. When the secondary winding, which has a resistance of 20 Ω and a self inductance of 16 μH, is connected to a 0·005 μF capacitor, a current of 2·5 mA flows in it. The mutual inductance between the windings is 12 μH.

Find from first principles the supply voltage and current, and verify that the power supplied is equal to the copper losses in the two windings.

(*I.E.E. Elec. Eng. II*, October 1964)

[3·16 V; 1·86 mA]

20. Show how the insertion of a transformer can increase the transfer of power from a source to a load. What is the meaning of *insertion gain*?

Calculate the insertion gain when a step-up transformer with a turns-ratio of 1:3 is connected between a source of resistance 200 Ω and a load of resistance 1,800 Ω. Express the result in decibels. Does this value of turns-ratio produce the maximum power transfer?

(*C. & G., T.P. C*, 1964)

[Without transformer $V_L = (1{,}800/2{,}000)V_o$. Power $= (0{\cdot}9V_o)^2/1{,}800 = 4{\cdot}5 \times 10^{-4}V^2_o$. With transformer, effective $R_p = 1{,}800/9 = 200\ \Omega$, which equals source impedance, producing maximum power transfer. $V_1 = 0{\cdot}5V_o$ and $V_L = 1{\cdot}5V_o$; Power in load

$$= (1{\cdot}5V_o)^2/1{,}800 = 12{\cdot}5 \times 10^{-4}V^2_o.$$

Insertion gain $= 12{\cdot}5/4{\cdot}5 = 2{\cdot}78$ (4·44 dB)]

21. When is a transformer said to be ideal? Explain why the magnetizing current of a transformer increases as the frequency of a constant applied voltage is reduced.

An ideal transformer has a turns-ratio of 1:2. What is the impedance measured at the primary when a resistance of 2,512 Ω is connected to the secondary? How is this value modified at a frequency of 200 Hz if the inductance of the primary winding is 0·5 H?

[Ideal transformer is assumed to have windings with infinite reactance and no losses (either copper or iron); magnetizing current increases because of fall in primary reactance; 628 Ω; $628\sqrt{2}$ Ω.]

Chapter 3

1. Sketch the waveforms of a radio-frequency carrier wave, amplitude-modulated by a sinewave tone, when the depth of modulation is (*a*) 100 per cent, and (*b*) 25 per cent.

If a radio-frequency carrier wave is amplitude-modulated by a band of speech frequencies, 50 Hz to 4,500 Hz, what will be the bandwidth of the transmission and what frequencies will be present in the transmitted wave, if the carrier frequency is 506 kHz?

(*C. & G., R.L.T. A*, 1959)

[9 kHz; 506·05 to 510·5; 506; 505·95 to 501·5 kHz]

2. The field strength F (volts/metre) at a distance d (metres) from a transmitter operating on a wavelength λ (metres) and producing a current I (amperes) in an aerial of effective height h (metres) is given approximately by—

$$F = \frac{120\pi h I}{\lambda d}$$

State the conditions under which this formula is applicable. Deduce the effective height of the aerial, given that a field strength of 1·5 mV/m is produced at a distance of 50 km by a transmitter operating on 150 kHz with an aerial current of 25 A.

(*C. & G., Radio III*, 1951)

[15·9 m]

3. The equivalent series circuit of a long-wave transmitting aerial consists of the following: inductance, 100 μH; capacitance, 850 pF; loss resistance, 1·5 Ω; radiation resistance, 0·3 Ω. The aerial current is 50 A (r.m.s.) and the operating frequency is 100 kHz. Calculate (*a*) the radiated power, (*b*) the total power input, (*c*) the efficiency and (*d*) the voltage at the aerial terminals.

(*C. & G., Radio III*, 1951)

[750 W; 4·5 kW; 16·7 per cent; 90·35 kV]

4. With reference to amplitude-modulation explain the terms *modulation envelope* and *depth of modulation* and distinguish between *side frequencies* and *sidebands*.

The amplitude of a 310-kHz wave is modulated sinusoidally at a frequency of 5 kHz between 0·9 V and 1·5 V.

Determine the amplitude of the unmodulated carrier, the depth of modulation and the frequency components present in the modulated wave.

(*C. & G., R.L.T. A*, 1965)

[1·2 V; 25 per cent; 305 and 315 kHz]

5. What are the advantages and disadvantages of single-sideband suppressed-carrier operation, as compared with double-sideband plus carrier operation, of long-distance point-to-point radio telephone links in the h.f. band?

The power output of a radio-telephone transmitter when modulated to a depth of 100 per cent is 100 kW. If the transmitter is now amplitude-modulated to a depth of 50 per cent, what will be the power in each of the sidebands? (*C. & G., Radio III*, 1959)

[$P_{mod} = P_{carrier}\ (1 + m^2/2)$, whence $P_c = 100/1{\cdot}5$ kW; power in sidebands $= P_c m^2/2$; hence with 50% modulation, power in *each* sideband $= (100/1{\cdot}5)(0{\cdot}5^2/4) = 4{\cdot}167$ kW]

6. State the meaning of the following terms when applied to a frequency modulated transmission—

(*a*) *modulation index*,
(*b*) *frequency deviation*,
(*c*) *deviation ratio*.

When the modulation index of a certain f.m. transmitter is 6, and the bandwidth in use is 140 kHz, what is its frequency deviation?

Explain the effect of altering the deviation in an f.m. system on the output from a receiver due to an interfering signal whose frequency is close to that of the carrier of the wanted signal.

(*C. & G., Radio C*, 1970)

[60 kHz]

7. Explain clearly the meanings of the terms—

(*a*) frequency swing,
(*b*) frequency deviation,
(*c*) deviation ratio,
and (*d*) modulation index

as applied to a frequency-modulated wave.

A carrier wave of amplitude 1 mA and frequency 100 MHz is received together with an interfering c.w. signal of amplitude 25 μV and frequency 100·1 MHz. Calculate the magnitude of the peak phase and the frequency deviations.

(*C. & G., Radio C*, 1965)

[The second part of this question appears to involve frequency modulation but is, in fact, an example of the intermodulation discussed on page 141. Using the expressions developed there the peak phase deviation is $\pm\ 0{\cdot}039$ radian ($2{\cdot}25°$); the frequency deviation is the difference between the two signal frequencies, i.e. ± 50 kHz]

Chapter 4

1. Describe briefly how the resistance of conducting materials commonly used in electrical circuits varies with change of temperature.

Define temperature coefficient of resistance.

An electric heater made from 10 ft of resistance wire which has a resistance of 0·6 Ω/ft at room temperature shows a temperature rise of 500°C when connected to a 27-V supply. The temperature rise can be assumed to be proportional to the power dissipated and the temperature coefficient of the resistance wire can be taken as constant at 0·001 per °C. Calculate the supply voltage needed to run the heater with a temperature rise of 400°C. (*C. & G. Tel. A* 1960)

[23·3 V]

2. Distinguish between the *loss angle* and the *power factor* of a dielectric.

Why are the requirements in telecommunications such that these two quantities should be nearly equal?

Mention some modern materials which satisfy this condition.

(*C. & G., Tel. C*, 1960)

3. An alternating voltage of 10 V r.m.s. at a frequency of 159 kHz is applied across a capacitor of 0·01 μF. Calculate the current in the capacitor.

If the power dissipated within the dielectric is 100 μW, calculate

(*a*) the loss angle,
(*b*) the equivalent series resistance,
(*c*) the equivalent parallel resistance.

(*C. & G., T.P. C*, 1965)

[$I = 0{\cdot}1$ A; $\phi = 10^{-4}$ radian; $R_s = 0{\cdot}01\ \Omega$; $R_p = 1\ \text{M}\Omega$]

4. An iron-cored choke consists of a magnetic circuit built up of steel laminations and a magnetizing winding of 400 turns. The magnetic circuit has a constant cross-section of 20 cm² and a mean length of 80 cm in which there is an air-gap of 2 mm.

The magnetization curve of the steel is given in the following table:

H (At/m)	100	200	300	400	500
B (T)	0·37	0·67	0·84	0·94	1·02

Determine (*a*) the current required to produce a magnetic flux of $1{\cdot}7 \times 10^{-3}$ Wb in the magnetic circuit, stating any assumptions made, and (*b*) the energy stored in the air-gap when this current is flowing. (*E.I.E., Elec. I*, April, 1960)

[From the curve, H to produce the required flux in the steel $= 310$, so that At $= 310 \times 0{\cdot}8 = 248$. To produce the same flux in the air-gap the m.m.f. required = $\Phi l/A\mu_0$ = 1,360. Hence total At = 1,608 and the current = 4·02 A. Energy in gap = $\frac{1}{2}BH$ × volume of gap = 1·05 × 10^{-2}J.]

5. Explain briefly under what conditions it is advantageous to use in a magnetic circuit—

(a) a laminated iron core,
(b) a granulated iron (iron dust) core.

A magnet core consists of a stack of 20 circular ring laminations each 0·5 mm thick with outer diameter 5·5 cm and inner diameter 3·5 cm. The relative permeability of the iron is 2,000. A radial air gap of 2 mm is cut in this core. Calculate the direct current that will be required in a coil of 1,000 turns uniformly distributed around the core to produce a magnetic flux of 0·3 mWb in the air gap.

Assume that magnetic leakage is negligible.

The permeability of free space is $4\pi \times 10^{-7}$ henry per metre.

(*C. & G.*, *T.P. B*, 1963)

[4·97 A]

6. The magnetic circuit of a simple cut-out is shown in Fig. E.6. Each coil has 100 turns and the spring exerts a force of 100 g in the unoperated position. What current must flow through the coils, when connected in series assisting, to operate the cut-out?

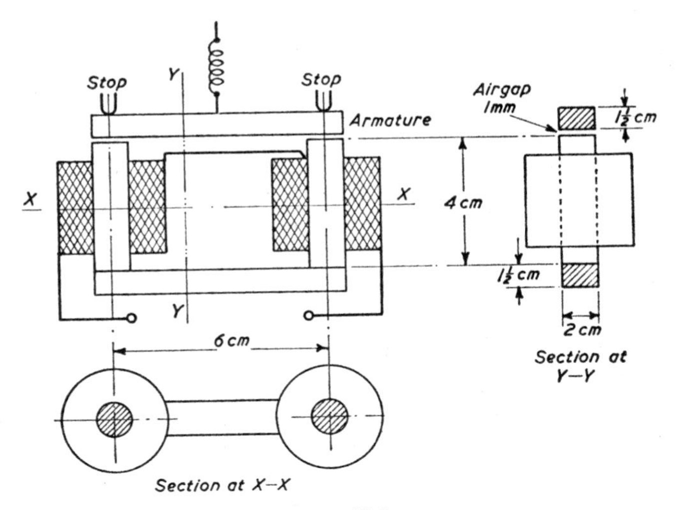

FIG. E.6

The relative permeability of the iron μ_r can be taken as 100.

How would you ensure that the armature returns to its original position when the current is switched off?

What would be the effect of an external magnetic field on the operation of the cut-out? (*C. & G., Tel. III*, 1959)

[$F = \frac{1}{2}B^2A/\mu_0$ newtons; 100 g = 0·98 newton; $B = NI\mu_0/[l_a + l_i/\mu_r]$; l_a = *total* airgap = 2 mm; whence operating current = 1·53 A]

7. Explain the meaning of *relative permeability* for a magnetic material.

A stalloy core has relative permeabilities of 6,000 at a flux density of 0·6 Wb/m² (tesla) and 2,000 at 1·2 Wb/m² (tesla); the $\mu_r - \beta$ characteristic is a straight line between these points. Calculate the current needed in a coil of 1,000 turns on a core of this material to produce a flux density of 0·9 Wb/m² (tesla). The mean flux path is 0·2 m long.

(*C. & G., T.P.B.*, 1970)

[18 mA (approx)]

8. A 100-kVA, 3,300/230-V single-phase transformer takes a no-load current and power of 1·5 A and 1·1 kW at rated voltage and frequency. On short circuit with full-load secondary current flowing, it takes 1·4 kW at a primary terminal voltage of 150 V and rated frequency.

Estimate the efficiency and primary power factor for 75 per cent of full-load secondary current at 0·9 lagging power factor. Ignore the effects of temperature rise. (*I.E.E., Elec. Eng. I*, April 1959)

[75 per cent f.l. at 0·9 p.f. requires a current equivalent to 83 per cent at unity p.f.; hence copper loss = $(0{\cdot}83)^2 \times 1{\cdot}4$ kW = 0·97 kW, and total loss = 2·07 kW; efficiency = 97 per cent; primary load current = 25·3 A, of which 11·1 A is reactive; adding magnetizing current of 1·5 A gives total primary current of 26 A at 0·875 p.f.]

9. Discuss the proportionate relation between iron loss and copper loss in a transformer, and show that this must be considered in the selection of a transformer to be used for a specific purpose.

The efficiency of a transformer at ambient temperature is a maximum at three-quarters full load and is then 97·0 per cent. What will be the efficiency at one and a quarter times full load at the same temperature and power factor? (*I.E.E., Elec. Eng. II*, April 1959)

[96·5 per cent]

10. Describe briefly, with the aid of a sketch, the principle of operation of a moving-coil milliammeter.

A meter of this type has a coil resistance of 50 Ω and gives a full-scale deflection when a current of 10 mA flows through it. Show how you would convert this into a direct-current voltmeter which gives a full-scale deflection for 10 V. Calculate the value of any additional component required. (*C. & G., Tel. A*, 1959)

[Series resistor of 950 Ω]

11. Briefly discuss the changes and improvements in the magnetic system of a moving-coil instrument which have resulted from the introduction of modern permanent-magnet materials.

A permanent magnet has the following demagnetization curve—

Flux density, T	1·25	1·21	1·1	0·93	0·63	0
Magnetizing force (or coercive force), At/m	0	−5,000	−10,000	−15,000	−20,000	−25,000

A flux density of 0·45 T is required in two air gaps in series each 0·3 cm long. The pole dimensions are 5 × 3 cm. If the magnet is of section 3 × 3 cm, find the magnet length required, neglecting leakage and allowing 10 per cent for the m.m.f. in the rest of the iron circuit.

(*I.E.E. Elec. Eng. II*, October 1964)

[Air gap is 15 cm² while magnet area is 9 cm²; hence magnet flux density must be 0·45(15/9) = 0·75 T, which from data given requires 18,500 At/m. Hence 18,500 $l_m = B_g l_g/\mu_o$, giving l_m = 6·5 cm. Adding 10 per cent, l_m = 7·2 cm]

12. Describe the constructional features of a lead-acid secondary cell and outline its principles of operation.

How does the voltage of such a cell depend on the load current drawn from it?

What factors limit the life of such a cell?

(*C. & G., Tel. A*, 1960)

13. Describe with the aid of a sketch, the construction and operation of a typical telephone receiver.

Explain the need for the permanent magnet in such a receiver.

(*C. & G., R.L.T. A*, 1959)

14. Explain with the aid of a sketch the principle of operation of a moving-coil loudspeaker. What is the function of a baffle-board?

The optimum load of a particular output valve is 4,900 Ω. What turns-ratio is required in the output transformer to match the valve to a 4·0-Ω moving-coil loudspeaker?

(*C. & G., R.L.T. A*, 1965)

[35:1]

15. Discuss briefly the factors which determine the choice of dielectric in the electrolytic, fixed and variable types of capacitors used in a radio receiver.

Quote typical capacitance values for (*a*) a preset trimming capacitor as used in a tuned radio circuit, and (*b*) an electrolytic capacitor. Mention two uses of an electrolytic capacitor.

What determines the shape of the plates required in a variable tuning capacitor?

(*C. & G., R.L.T. A*, 1965)

16. Draw an equivalent circuit to represent an electrolytic capacitor at low frequencies and sketch the reactance/frequency graph for this circuit. Indicate on this graph the frequency for which the capacitor would be most effective as a decoupling device and that for which it would be least effective, giving reasons.

Give two reasons why you would not expect to find an electrolytic capacitor used as the tuning capacitor in a fixed frequency r.f. oscillator.

(*C. & G., R.T.L B*, 1965)

17. Explain clearly how the loss components in the iron core of an inductor vary with the applied voltage and frequency. What practical steps are taken to reduce these losses to an economic low value?

An inductor has a winding of resistance 20 Ω; with an applied voltage of 120 V at 25 Hz the total loss is 60 W and the current is 1·1 A. When the voltage is changed to 240 V at 50 Hz the current becomes 1·2 A and the loss 120 W. Estimate each of the main iron loss components at both 25 and 50 Hz.

(*I.E.E. Elec. Eng. I*, April 1965)

[Deduct I^2R at each frequency; remainder is

$$\text{iron loss} = k_1 f + k_2 f^2$$

whence at 25 Hz $P_h = 26$ and $P_e = 9{\cdot}8$; at 50 Hz $P_h = 52$ and $P_e = 39{\cdot}2$ watts.]

18. Discuss the main differences in the constructional features of inductors used in the audio and radio frequency stages of a radio receiver.

A tuned circuit comprises an inductor and capacitor in series and is tuned to a resonant frequency of 1 MHz.

If a value of the inductance is increased by 21 per cent, what is then the frequency of resonance?

(*C. G., R.L.T. A*, 1964)

[0·91 MHz]

Chapter 5

1. What emissive materials are used for the cathodes of receiving valves and why?

Explain what is meant when the anode current of a valve is said to be (*a*) space-charge-limited, (*b*) temperature-limited.

(*C. & G., Radio II*, 1958)

2. An audio-frequency output stage uses a triode valve with an anode slope resistance of 2,000 Ω, an amplification factor of 7, and an anode load resistance of 3,000 Ω.

What is the value of the mutual conductance of the valve used?

What value of audio-frequency voltage must be applied to the grid of the valve if the power output is to be one watt?

(*C. & G., Radio I*, 1959)

[3·5 mA/V; 13 V]

3. Three points on the static characteristic of a triode valve are, in terms of the grid voltage and the anode current: − 1 V and 9 mA; − 2 V and 4 mA; − 3 V and 1 mA. Assuming that over this range the current i and the voltage v are related by the expression $i = (a + bv)^2$, find the components of the anode current when the grid bias is − 2 V and a signal 1·0 sin 5,000t volts is applied to the grid. Roughly sketch the components, and their sum, as functions of time.

(*I.E.E., Elec. Eng. I*, 1960)

[If $i = (a + bv)^2$ then when $v = 0$, $i = 4$ so that $a = 2$. Similarly when $v = \pm 1$, $i = 9$ or 1 respectively, so that $b = 1$. Hence $i = (2 + \sin 5{,}000t)^2 = 4 + 4 \sin 5{,}000t + \sin^2 5{,}000t$]

4. A diode valve has the following anode-current/anode-voltage characteristic—

Anode volts (V_a)	Anode current (mA) (I_a)
0	0
20	0·42
40	1·56
80	6·48

Plot the characteristic, using anode voltage as abscissae. If the equation for the curve is $I_a = aV_a^2$, find a by plotting the curve of I_a against V^2 or otherwise.

A resistance of 100,000 Ω and a battery are connected in series with the diode. Construct a graph to a suitable scale, showing the variation of circuit current with battery e.m.f.

Thence, or otherwise, compare the performance as a rectifier of the diode alone and of the diode with the 100,000 Ω resistance connected in series. (*C. & G., Tel. III*, 1958)

[$a = 10^{-6}$; for $i = 1, 2$ and 4 mA, the battery voltage must be 132, 245 and 463 V]

5. An audio-frequency amplifier uses a triode valve with an anode resistance of 10,000 Ω and an amplification factor of 50. What value of anode load resistor will be required to give a voltage amplification of 40 times?

If the direct current taken by the valve is 5 mA, what should be the power rating of the anode load resistor?

(*C. & G., Radio I*, 1958)

[40,000 Ω; 1 W]

6. Show, with reference to characteristic curves, that an ideal pentode valve when used as an amplifier can be treated as a source of constant current.

A pentode amplifier valve has the static characteristics given below—

	$V_g = -2$ V	$V_g = -4$ V	$V_g = -6$ V
$V_a = 200$ V	$I_a = 25{\cdot}0$ mA	$I_a = 18{\cdot}6$ mA	$I_a = 12{\cdot}8$ mA
$V_a = 250$ V	$I_a = 25{\cdot}2$ mA	$I_a = 18{\cdot}8$ mA	$I_a = 13{\cdot}0$ mA
$V_a = 300$ V	$I_a = 25{\cdot}4$ mA	$I_a = 19{\cdot}0$ mA	$I_a = 13{\cdot}2$ mA
$V_a = 350$ V	$I_a = 25{\cdot}6$ mA	$I_a = 19{\cdot}2$ mA	$I_a = 13{\cdot}4$ mA

For the static operating point $V_a = 300$ V; $V_g = -4$ V, find the mutual conductance and anode resistance of the valve.

If the anode is choke-coupled to a resistive load of 8,000 Ω, find the voltage amplification of the stage. (*C. & G., R.L.T.B*, 1959)

[$g = 3{\cdot}05$ mA/V; $r_a = 0{\cdot}25$ MΩ; voltage gain = 24·4]

7. A single-valve triode amplifier has a resistive anode load. Explain how the anode-current/anode-voltage (I_a/V_a) characteristics of the valve can be used to calculate the voltage amplification of the stage.

The characteristics of a particular valve are given in the following table—

Grid voltage V_g (V)	0	−4·4	−8·8
Anode current I_a (mA)	40, 54, 62, 70	30, 45, 49, 54	27, 34, 36
Anode voltage V_a (V)	40, 90, 150, 250	75, 150, 200, 300	100, 200, 300

This valve is to operate from a 350-V anode supply with a grid bias of − 4·4 V. From your curves determine the value of anode load resistance that will give as high a gain as possible consistent with low distortion, when the value of the grid swing is ± 4·4 V peak-to-peak. (*C. & G., Tel. A*, 1960)

[6,000 Ω approx.; $V_a = 110$ V]

8. Why is it necessary to use a valve with an exceptionally small grid–anode capacitance in a radio-frequency amplifier of a receiver?

Explain clearly how this small grid–anode capacitance is achieved in the pentode valve.

Why is this type of valve fitted with a suppressor grid?

(*C. & G., Radio I*, 1959)

9. Explain, with the aid of sketches, the principle of operation of a thermionic triode valve in an amplifier. Mention in your answer the effect of negative grid bias on the action of the valve.

A triode is working as an amplifier with a resistive anode load of 5,000 Ω, and an anode supply battery of 200 V. Draw the load line for this circuit with the anode currents as ordinates.

The anode characteristics for the valve are such that the curves for grid voltages of − 3, − 2 and − 1 V, respectively, cut this load line at anode currents of 12·0, 19·5 and 27 mA.

Deduce the voltage amplification of the circuit if the peak input to the grid is 1 V. (*C. & G., Tel. II*, 1958)

[37·5]

10. Explain the operation of a neon tube voltage stabilizer.

A neon tube and resistor are connected in series to provide a 180-V stabilized output from a 220-V d.c. supply. If the load resistance is 15,000 Ω and the current through the neon stabilizer is 12 mA, determine the value of the series resistor and the power dissipated in it.

If the load resistance is reduced to 10,000 Ω, determine the new current which flows through the neon tube.

Assume that the operation is ideal. (*C. & G., R.L.T.B*, 1960)

[Load current = 180/15 mA = 12 mA; hence total current is 24 mA. Series resistor must drop 40 V at this current, requiring $R = 1{,}667\ \Omega$; power dissipation = 0·96 W. If load = 10,000 Ω, stabilizer current = 6 mA.]

11. By reference to a sketch, explain the purpose of each of the three grids in a pentode valve.

Briefly discuss the reasons for using pentode valves rather than triode valves in the high-frequency stages of medium-wave radio receivers.

(*C. & G., R.L.T. A*, 1964)

12. What is a constant-current generator?

A pentode valve has a mutual conductance of 3·0 mA/V and an anode slope resistance of 600,000 Ω. It is transformer coupled to a non-reactive load of 10 Ω. If the transformer has a turns ratio of 30, draw the equivalent anode circuit showing the valve as a constant-current generator.

What will be the r.m.s. current in the load and the power supplied to it when a voltage $1{\cdot}0 \sin \omega t$ is applied to the grid of the pentode?

What is the source of output power?

(*C. & G., T.P. C*, 1963)

[62·7 mA; 39·5 mW]

13. Draw a sketch to show the constructional features of the triode-hexode valve.

Draw a circuit diagram to show how such a valve is used in the frequency-changer stage of a superheterodyne receiver, and briefly explain the operation of the stage.

(*G. & G., R.L.T. B*, 1964)

Chapter 6

1. A single-stage, common-emitter, transistor amplifier has a 4 kΩ collector load resistor. The effective hybrid parameters of the transistor are: $h_{ie} = 1$ kΩ, $h_{fe} = 60$, whilst h_{re} and h_{oe} are both negligible. A signal source having an open-circuit e.m.f. of 2 mV and a 250 Ω resistive internal impedance is connected across the amplifier input.

Calculate—

(*a*) the signal current at the amplifier input,
(*b*) the signal voltage across the input terminals,
(*c*) the signal voltage across the collector load.

(Assume the effects of biasing and coupling components on the signal to be negligible.)

Sketch a practical circuit for the amplifier giving typical values for the resistors used in the circuit and the power supply voltage.

(*C. & G., T.P.C.* 1970)

[1·6 μA; 1·6 mV; 384 mV]

2. Calculate, after deriving any necessary formulae, the input resistance of a common-collector transistor stage with emitter load resistance of 5 kΩ. The working point is such that the equivalent-T parameters are: $r_e = 40$ Ω; $r_b = 500$ Ω $r_c = 1$ MΩ; $\alpha = 0{\cdot}98$. Justify any approximations made.

Show an arrangement for obtaining the necessary base bias current that will not result in excessive lowering of the input resistance.

(*I.E.E., Elect. Eng.*, Nov. 1958)

[201·5 kΩ, using complete expression. Approximate expression gives $R_{in} = 250$ kΩ]

3. Draw circuit diagrams and briefly explain how a junction transistor may be used in a single-stage audio-frequency amplifier (*a*) with common-base connection, and (*b*) with common-emitter connection. Discuss the order of magnitude of the input and output impedances and the current gain characteristic for each type of amplifier.

(*C. & G., R.L.T. A*, 1965)

4. Explain briefly what is meant by each of the following when applied to a transistor:

(*a*) cut-off frequency,
(*b*) collector to base capacitance.
(*c*) thermal runaway,
(*d*) donor impurity.

(*C. & G., Radio C*, 1963)

5. Draw curves to show how the current gain of a transistor varies with the resistance of its load when it is used in the common-collector, the common-base, and the common-emitter configurations.

A transistor, having a short-circuit current gain α, and input impedance of R_i, is connected in the common-base configuration with a load of R_L Ω. Derive expressions for the voltage gain and the power gain of the circuit, and calculate their values for $\alpha = 0{\cdot}98$, $R_i = 40$ Ω, and $R_L = 6{,}000$ Ω.

(*C. & G., Radio C*, 1965)

[$A_v = \alpha R_L/R_i = 147$; $A_p = \alpha^2 R_L/R_i = 144$ (21·58 dB)]

6. What restrictions are placed on the use of equivalent circuits for valves and transistors?

Draw the *constant-voltage* equivalent circuit for a valve and show that—

$$i_a = g_m v_g - \frac{v_a}{r_a}$$

where v_g is the signal voltage applied between the grid and the cathode and v_a and i_a are the corresponding changes produced in the anode-cathode voltage and the anode current.

Draw an equivalent T-circuit for a transistor and show that if the base resistance r_b is negligible—

$$i_c = \alpha i_e - \frac{v_c}{r_c}$$

where i_e is the signal current in the emitter lead and v_c and i_c are the corresponding changes produced in the collector–base voltage and the collector current.

(*C. & G., T.P. C*, 1964)

7. Explain the nature of the amplifying process in a junction transistor.

A transistor can be connected as an amplifier with any one of its three electrodes connected as the common element in the circuits of the other two. Sketch the basic circuit of one of these three amplifiers and explain briefly how it operates.

Why must excessive temperature rise in transistors be avoided?

(*C. & G., T.P. B*, 1963)

8. Draw an equivalent T-circuit for a transistor and explain the effect on each component of this circuit when a second identical transistor is connected in parallel with the first.

Define a set of h-parameters for a transistor and carefully explain, by setting down the initial equations, how these could be used to determine the small-signal gain of the transistor when inserted between a specified source and load.

(*C. & G., T.P. C*, 1965)

9. Draw circuit diagrams showing how a junction transistor may be used in a single-stage audio-frequency amplifier (*a*) with common-base connection, (*b*) with common-emitter connection. Mark clearly the input, output and biasing arrangements and the currents flowing. State the input and output impedances and current gain characteristics for each type of amplifier.

Derive an expression showing how, for small input signals, the current gain of a transistor connected in a common-emitter circuit is related to the gain of the same transistor connected in a common-base circuit.

A transistor connected in a common-emitter circuit shows changes in emitter and collector currents of 1·0 mA and 0·98 mA respectively. What change in base current produces these changes and what is the current gain of the transistor?

(*C. & G., R.L.T. A*, 1963)

[0·02 mA; $\alpha' = 50$)]

10. Sketch typical characteristics and explain briefly the physical principles underlying the operation of a p–n–p junction transistor. Define the parameters α and I_{co}.

A transistor has $\alpha = 0{\cdot}98$. The base–emitter potential difference is adjusted to make the emitter current 5 mA. Calculate the base and collector currents if I_{co} is 10 μA. For what magnitude of emitter current will the base current be zero? Mention all assumptions.

(*I.E.E. III Applied Electronics*, June 1964)

[Assuming common-base mode, $I_c = \alpha I_e + I_{co} = 4{\cdot}91$ mA whence $I_b = 0{\cdot}09$ mA; I_b will be zero when $I_e(1 - \alpha) = I_{co}$; whence $I_e = 0{\cdot}5$ mA]

Chapter 7

1. Describe briefly, with the aid of a circuit diagram, a method of producing an amplitude-modulated wave.

An amplitude-modulated wave has a maximum value of 3 V and a minimum value of 1 V. What is the depth of modulation?

(*C. & G., Radio I*, 1958)

[50 per cent]

2. Sketch the waveform of a carrier which is amplitude modulated by a sinusoidal signal and show that this is equivalent to a carrier and two side frequencies. Show how these components can be represented by vectors.

Draw the circuit of a modulator and explain its action.

A carrier is modulated 100 per cent by a square wave and the carrier component is then suppressed. Sketch diagrams showing the waveform and spectrum of the resulting signal.

(*C. & G., Tel. IV*, 1959)

3. Give the circuit diagram of a radio-frequency Class C amplifier stage suitable for operation with an anode supply voltage of 10 kV, and state the peak voltages experienced by any capacitors used in the circuit.

A Class C r.f. amplifier valve, which has a peak emission of 10 A, is supplied from an anode supply voltage of 10 kV. If the angle of current flow is 100°, the minimum anode voltage is 10 per cent of the anode supply voltage, $\mu = 40$, and the ratio of peak current to peak r.f. fundamental current is 2·95, determine—

(*a*) the output power,
(*b*) the anode direct current,
(*c*) the anode dissipation, and
(*d*) the grid-bias voltage,

assuming that the valve characteristics are linear.

(*I.E.E., Radio Comm.*, 1958)

[15·25 kW; 1·77 A; 2·45 kW; 1,670 V]

4. The anode current in a Class-C amplifier may be regarded as triangular pulses having a peak value of 2·5 A and angle of flow 90°. The grid voltage varies sinusoidally and the I_a/V_g characteristic is linear. If the h.t. supply is 2,500 V and the r.m.s. current delivered to a 750-Ω load is 0·8 A, what is the efficiency of the amplifier?

If the peak-to-peak amplitude of the anode voltage is 4,000, what is the instantaneous anode voltage when anode current commences to flow?

(*C. & G., Radio C*, 1963)

[Mean of triangular wave = $\frac{1}{2}$ peak, but this only flows for $\frac{1}{4}$ cycle, so mean anode current is $2{\cdot}5/8 = 0{\cdot}3125$ A. Input power = $2{,}500 \times 0{\cdot}3125 = 780$ W; output power = $0{\cdot}8^2 \times 750 = 480$ W; efficiency = 61·5 per cent.

Standing current is 0·3125 A at 2,500 V; peak current is 2·5 A at 500 V; by proportion, a line through these points will cut the zero current axis at a point x volts above 2,500 such that

$$x/0{\cdot}3125 = (2{,}000 + x)/2{\cdot}5$$

whence $x = 285$ so that anode current will begin to flow at 2,785 V.]

5. Describe the action of a reactance-valve frequency modulator, illustrating your answer with a circuit diagram.

An oscillator operating at 100 MHz is tuned by a 50 pF capacitor. What capacitance change must be introduced by the reactance valve to cause a 75 kHz change in frequency? Derive any expressions you use.

(*C. & G., Radio C*, 1964)

[$\pm 0{\cdot}075$ pF]

Chapter 8

1. An aerial has an effective height of 100 m and the current at the base is 450 A r.m.s. at 40 kHz. What is the power radiated? If the total resistance of the aerial circuit is 1·12 Ω, what is the efficiency of the aerial?

[57 kW; 25·1 per cent]

2. The aerial at a transmitting station has an effective height of 150 m and the current at the base is 700 A at 16 kHz. What is the power radiated? If the capacitance of the aerial is 0·04 μF what is the voltage across the lead-out insulator? Give a sketch.

[50 kW; 174 kV]

3. A transmitting station has an aerial of 100 m effective height carrying 200 A at 100 kHz. A receiving station 80 km distant has

an aerial of 10 m effective height which is tuned with a series inductance of 10 mH. The resistance of the receiving aerial circuit is 100 Ω. What voltage would appear across the inductance coil?

[19·6 V]

4. If the effective series inductance and capacitance of a vertical aerial are 100 μH and 100 pF respectively, what is the resonant frequency? If a capacitor of 200 pF capacitance is connected in series with the aerial, what is the new value of the resonant frequency?

(*C. & G., Radio I*, 1948)

[1·592 MHz; 1·950 MHz]

5. Sketch and explain the construction of (*a*) an inverted-L aerial and (*b*) a T aerial for the reception of medium-wave signals.

Indicate the manner of coupling these aerials to the receiver. Briefly discuss the directional properties of each of these aerials.

What type of aerial (other than a frame aerial) is now often used in portable broadcast receivers? Briefly discuss the construction of such an aerial and its orientation for the best reception.

(*C. & G., R.L.T. A*, 1964)

6. Explain the basis of operation of the mast-radiator used at a medium-wave broadcasting station. What factors limit the useful service range (*a*) by day and (*b*) by night?

(*C. & G., Radio III*, 1959)

7. Explain how, in direction-finding, a radio-frequency signal is picked up by a loop aerial.

A loop aerial for use at 500 kHz is of height 0·5 m, breadth 0·5 m and has 25 turns. When directed to receive a maximum signal the e.m.f. induced in the loop is 150 μV. What is the field strength in μV/m of the signal picked up? (*C. & G., Radio II*, 1958)

[2,300]

8. What is meant by *night effect* in the operation of a direction-finding loop aerial and how may it be eliminated?

A loop of 20 turns and area 1 m^2 lies in a plane making an angle of 30° with the direction of propagation of an incoming signal of frequency 150 kHz. What is the field strength of the signal if the e.m.f. induced in the loop is 1·5 mV? Prove any formula you may use.

(*C. & G., Radio III*, 1958)

[27·6 mV/m]

9. Describe the construction and principles of operation of the spaced-loop direction-finding aerial system. What advantage is gained with this system of direction finding?

(*C. & G., Radio III*, 1959)

10. A long transmission line having a characteristic impedance of $2{,}400 \, \overline{|30^\circ} \, \Omega$ is connected to a termination by means of a step-down transformer which has a turns ratio of 2 : 1.

What must be the impedance of the termination

(*a*) if the line is to be correctly terminated, and
(*b*) if maximum power is to be transferred from the line to the termination? (*C. & G., Tel. C*, 1960)

[$600 \, \overline{|30^\circ}$; $600 \, \underline{|30^\circ}$]

11. A transmission line of negligible loss has the following primary constants at a frequency of 0·5 MHz—

$$L = 0{\cdot}4 \text{ mH/km}$$
$$C = 0{\cdot}08 \text{ } \mu\text{F/km}$$

Calculate the characteristic impedance, propagation coefficient, wavelength and velocity of propagation at this frequency.

(*C. & G., Tel. III*, 1948)

[70·7 Ω; j17·7; 0·353 km; 176,500 km/sec]

12. Explain the meaning of the terms (*a*) *reflected wave*, and (*b*) *standing-wave ratio*, with reference to a transmission line.

A low-loss transmission line of 600 Ω characteristic impedance is connected to a transmitting aerial of impedance $400 + j200$ Ω. Calculate the amplitude of the reflected wave relative to the incident wave and the standing-wave ratio. (*C. & G., Radio IV*, 1949)

[0·277; 1·77]

13. Define the *reflection coefficient* and give an expression for it in terms of the impedances of the line and termination. Show that with a line of non-reactive characteristic impedance any purely reactive termination gives complete reflection.

A low-loss line of 500 Ω non-reactive characteristic impedance is terminated in an impedance of $500 + j577$ Ω. The voltage across the termination is 1·32 V r.m.s. Calculate the amplitude of the voltage across the line (*a*) at a point one-twelfth of a wavelength from the termination, (*b*) at a point one-third of a wavelength from the termination. (*C. & G., Tel. IV*, 1954)

[1·5 V; 0·5 V. These are actually max. and min. points.]

14. A 600-Ω transmission line supplies a terminating impedance at a frequency of 100 MHz. Measurement discloses a standing wave, of ratio 3, with a voltage minimum at 60 cm from the termination. Calculate the effective values of the resistive and reactive components of the terminating impedance. Assume that the wavelength on the line is equal to the free-space value. (*I.E.E., Radio Comm.*, Dec. 1959)

[$1{,}000 - j800\ \Omega$]

15. A 4-metre length of air-spaced transmission line is terminated in its characteristic resistance of 100 Ω and fed at the input with an alternating voltage given by the expression $\sqrt{2} \sin (2\pi \times 10^8 t)$ volts. Assuming the line to be loss-free and taking the free-space velocity of light as 3×10^8 m/s, calculate

(i) the wavelength on the line,
(ii) the power dissipated in the termination,
(iii) an expression for the instantaneous current in the termination,
(iv) the phase difference between the input and output voltages,
(v) the total energy transmitted along the line in 1 minute.

(*C. & G., T.P. C*, 1965)

[$\lambda = c/f = 3$ m; Since feeder is correctly terminated there will be no standing wave, but a travelling wave which will arrive at the receiving end leading by a small amount. The line is 4 m long so the angle of lead will be $2\pi(4 - 3)/3 = 2\pi/3$, whence

$$i_r = 0{\cdot}01\sqrt{2} \sin (2\pi \times 10^8 t + 2\pi/3);$$

the r.m.s. power is not affected by the phase difference and is simply $V^2_{r.m.s.}/R$; $V_{r.m.s.} = \sqrt{2}/\sqrt{2} = 1$, so power is 10 mW, and energy in 1 minute = 0·6 J]

Chapter 9

1. The valve used in an r.f. amplifier has $g_m = 5$ mA/V, $r_a = 500$ kΩ. The anode circuit contains an r.f. transformer having a tuned secondary in which $L_1 = 80\ \mu$H, $L_2 = 150\ \mu$H, $Q_2 = 100$ at 600 kHz and $k = 0{\cdot}15$. Calculate the gain in dB at 600 and 1,600 kHz.

Discuss the factors which determine the manner in which the gain may vary over the frequency range, and indicate a possible method of reducing the gain variation. (*I.E.E., Radio Comm.*, 1958)

[Gain $= g\omega MQ = 30{\cdot}9$ or $29{\cdot}8$ dB at 600 kHz. Assuming Q unchanged, gain at 1,600 kHz is 82·5 (38·3 dB).]

2. An i.f. transformer has identical primary and secondary circuits inductively coupled; each winding has an inductance of 500 μH and is tuned by a capacitor. If the bandwidth (3 dB loss) of the i.f. stage is 16 kHz centred on 450 kHz, calculate for critical coupling conditions—

(i) the coefficient of coupling,
(ii) the mutual inductance between the windings,
(iii) the Q-factor of each winding, and
(iv) the value of the tuning capacitor.

(*C. & G., R.L.T. B*, 1960)

[0·025; 12·5 μH; 40; 250 pF]

3. Explain what is meant by the term *signal-to-noise ratio* in connection with a communication system, and discuss its importance.

A carrier which is amplitude-modulated by a sine-wave tone to a depth of 30 per cent is applied to a receiver and the output signal-to-noise ratio is 20 dB. Assuming that the whole of the noise arises in the receiver input circuit, and that a linear detector is used, how is the signal-to-noise ratio affected—

(i) when the transmitter path loss is decreased by 3 dB,
(ii) when the depth of modulation is increased to 60 per cent, and
(iii) when the receiver audio-frequency gain is increased by 6 dB?

(*C. & G., R.L.T. B*, 1960)

[(i) + 3 dB; (ii) + 6 dB; (iii) no change]

4. A pentode frequency changer has a pair of critically coupled circuits. The resonant impedance of the primary circuit when not coupled to the secondary is known to be 50,000 Ω, and primary and secondary circuits are identical. An i.f. peak voltage of 3·54 V is measured across the primary when an input signal of 100 mV r.m.s. is applied. What is the conversion conductance of the valve?

(*I.E.E., Radio Comm.*, 1958)

[1 mA/V]

5. Draw a block diagram of a communications type superheterodyne radio receiver and briefly explain the function of each stage.

Discuss any inherent disadvantage of the superheterodyne principle.

(*C. & G., R.L.T. A*, 1964)

6. Explain the meaning of the terms *automatic gain control* (*a.g.c.*) and *delayed a.g.c.* as applied to a communications receiver. Sketch a typical a.g.c. circuit and indicate possible values of those components that determine the a.g.c. time constant.

(*C. & G., Radio III*, 1959)

7. Sketch the circuit diagram of a two-stage i.f. amplifier employing transistors, suitable for use in a superheterodyne receiver. Give the reasons for your choice of transistor configuration and state features and components included which contribute towards gain stability.

In a receiver, each i.f. transformer stage reduces the power of a signal in an adjacent channel by a factor of 20 relative to the wanted signal. Calculate the net reduction of the adjacent channel signal if the i.f. amplifier has three such similar stages.

(*C. & G., R.L.T. B*, 1970)

[20^3]

8. The table below gives three points on a valve characteristic which may be assumed to follow a square law.

Grid voltage (V)	—3	—2	—1
Anode current (mA)	1	4	9

The valve has a grid bias voltage of —2 and a 1,000 Hz signal of 1 V peak amplitude is applied to the grid. Calculate the mean value of the anode current.

If a 1,500 Hz signal of the same amplitude is applied in addition to the 1,000 Hz signal, calculate the mean value of the anode current and the peak value of the component of 500 Hz in the anode current.

(*C. & G., Tel. IV*, 1959)

[With the 1,000-Hz signal, i_a will swing between 1 and 9 mA with a mean value of 5 mA (approximately).

With the additional 1,500-Hz signal the peak input is doubled so that i_a swings between 0 and 16 mA. The peak value of the 500-Hz component is thus 16 mA and the mean value (again approximately) is 8 mA.

For accurate assessment of the mean values it is necessary to write $i_a = (2 + v)^2$ and then to integrate this between the limits 0 and 2π, when $v = \sin 2\pi(1{,}000)t$ and $2 \sin 1{,}500\pi t \cos 500\pi t$ respectively, which need only be done if a rigid analysis is required.]

9. A Class A radio-frequency amplifier operates at a frequency of 1 MHz and employs a pentode valve with a mutual conductance of 8 mA/V. The anode load is a parallel resonant circuit the inductor of which has a value of 50 μH. If the bandwidth of the amplifier, for a reduction in output of 3 dB, is 9 kHz what is the Q-factor of the inductor?

If a sinusoidal signal of amplitude 0·5 V r.m.s. at a frequency of 1 MHz is applied to the grid, calculate—

(*a*) the power dissipated in the inductor,
(*b*) the peak voltage developed across the load.

(*C. & G., Radio II*, 1958)

[$Q = 111$; power in inductor = 0·55 W; peak voltage = 196 V]

10. A simple diode detector, having a square law characteristic, is used for demodulating a sinusoidal carrier which has been amplitude-modulated by a signal having a rectangular waveform.

By projecting from the diode characteristic show how the signal is distorted by demodulation.

How can the distortion be reduced?

(*C. & G., T.P. C*, 1963)

11. Draw a circuit diagram of a triode-hexode frequency-changer stage for use in a communication receiver, indicating typical component values.

A triode-hexode frequency-changer stage has as its anode load a 354-μH inductor of $Q = 120$ in parallel with a loss-free capacitor, resonating at an I.F. of 450 kHz. If the conversion conductance of the hexode is 0·5 mA/V under the working conditions, calculate the I.F. voltage developed across the load when a signal of 100 μV is applied to the grid.

Assume that the internal resistance of the hexode is much larger than the load impedance.

(*C. & G., R.L.T. B*, 1965)

[6 mV]

12. Draw the circuit diagram of *one* stage of a transistorized intermediate-frequency amplifier normally used for a superheterodyne broadcast radio receiver. Neutralizing and stabilizing circuits should be shown. Describe the function of each of the components in the design.

Why is this particular transistor circuit configuration normally chosen for this application?

(*C. & G., Radio C*, 1964)

13. Draw the circuit diagram and explain the action of an amplitude limiter and frequency discriminator for detecting a frequency-modulated wave.

Draw curves of the discriminator characteristic and the relationships between the input and output voltages of the limiter.

(*C. & G., Radio C*, 1965)

14. Define the *decibel* and give three reasons why its use is convenient in transmission problems.

The input signal to an amplifier varies between 23·5 mW and 1·25 W. Express each power in dB relative to 1 mW and state the fluctuation in the level of the signal in dB.

(*C. & G., R.L.T. A*, 1965)

[13·71; 30·97; 17·26]

15. Discuss briefly the significance of the various types of distortion which may occur in amplifiers.

A pentode tuned-anode amplifier stage has an anode load of magnification factor Q tuned to the fundamental frequency of a non-sinusoidal signal. Show that the amplification of the fundamental is $Q(n^2 - 1)/n$ times that of the nth harmonic if Q is large.

(*I.E.E. III Applied Electronics*, June 1964)

[Gain $= gZ$. At the fundamental frequency $Z_f = L/CR = QL\omega$, but this is not applicable at the harmonic frequency, for which the circuit *admittance* (with Q large) is $\dfrac{1}{L\omega_1}\left[\dfrac{R}{L\omega_1} + j(\omega_1{}^2LC - 1)\right]$. Writing $n\omega$ for ω_1 the ratio $Z_f/Z_h = Z_fA_h$ can be shown to be $Q(n^2 - 1)/n$.]

16. A data transmission system uses an amplitude-modulated 1 kHz carrier. The modulating signal is retrieved in a detector comprising a diode in series with a capacitance C with 1 kΩ in parallel.

Determine a suitable value for C to avoid envelope distortion. The signal is sinusoidal and has a frequency range of 10 to 50 Hz and modulates the carrier to a maximum depth of 60 per cent. Derive all formulae used.

(*I.E.E. III Applied Electronics*, June 1965)

[Time constant must be such that V falls to 25 per cent peak in 0·01 sec. This requires a t.c. of $1{\cdot}4CR = 0{\cdot}01$; whence $C = 7{\cdot}2\mu F$]

17. Derive an expression for the input admittance of an amplifier stage due to the anode–grid capacitance of the valve.

The tuned circuit which determines the frequency of a master oscillator consists of a 500 pF capacitor in parallel with an inductor. The voltage across this circuit is applied to the grid of a tuned-anode buffer amplifier whose gain at resonance is 100. The frequency of oscillation is 500 kHz.

Calculate the change in oscillator frequency produced by detuning the amplifier from resonance to a condition where its gain is 70·7. The anode–grid capacitance of the pentode used in the amplifier is 0·05 pF.

(*I.E.E. III Applied Electronics*, June 1965)

[Frequency increases from 500 to 505 kHz]

Chapter 10

1. A triode valve with an amplification factor of 20 and an anode slope resistance of 7,000 Ω is used as a voltage amplifier. The anode load is a pure 1-H inductor. Calculate the voltage gain in decibels and the phase-shift produced by the amplifier at a frequency of 1,000 Hz.

Draw a vector diagram showing the anode current and the various voltages involved. (*C. & G., Radio III*, 1958)

[25 dB; Output voltage lags input voltage by 132°]

2. Explain briefly why the response of an audio-frequency output transformer falls at very low and very high audio frequencies.

The valve of an a.f. output stage has a grid resistor of 0·5 MΩ. What value of capacitor must be used to couple the valve to the preceding stage if the output voltage at 50 Hz is to be not less than 75 per cent of the output voltage at high audio frequencies?

(*C. & G., Radio II*, 1958)

[0·0072 μF]

3. With the aid of both a circuit diagram and an equivalent circuit diagram, explain the operation of a push-pull audio-frequency output stage working under Class A conditions.

State the advantages of such a circuit over a circuit which uses the same pair of valves in parallel. Mention the essential circuit conditions needed in order to realize these advantages.

(*C. & G., R.L.T. B*, 1960)

4. What are the special characteristics of the cathode-follower circuit? Give an example of its use.

A triode valve, amplification factor μ and anode slope resistance r_a, is connected as a cathode-follower; the value of the cathode resistor is R. Prove that the voltage gain of the stage is

$$\frac{\mu R}{r_a + (\mu + 1) R}$$

(*C. & G., Radio III*, 1958)

5. Draw a sketch showing the waveforms of anode voltage, anode current, grid voltage and grid current in a push-pull Class B amplifier.

In such an amplifier each valve draws a peak current of 0·5 A. If the supply voltage is 5,000 V and the peak value of the alternating component of anode voltage is 4,500 V calculate—

(*a*) the power output of the amplifier,
(*b*) the anode dissipation in each valve, and
(*c*) the turns ratio of the output transformer if the secondary supplies a load of 600 Ω. (*C. & G., Tel. IV*, 1959)

[(*a*) 1·125 kW; (*b*) 232·5 W; (*c*) 7·75 overall]

6. Explain carefully the meaning of the following terms as applied to amplifiers—

(*a*) input impedance,
(*b*) source impedance,
(*c*) output impedance, and
(*d*) load impedance.

Explain two possible meanings of the term *optimum load impedance*.

Draw the circuit of a cathode follower and explain its useful properties. Deduce an expression for its output impedance.

(*C. & G., Tel. IV*, 1959)

7. A triode valve has an anode slope resistance of 5,000 Ω and a mutual conductance of 3 mA/V. It is used in an amplifier with an anode load of 20,000 Ω and a cathode resistor of 600 Ω by-passed by a 1-μF capacitor. There is a total capacitance of 40 pF across the output.

Draw the equivalent circuits of the amplifier at high and at low frequencies. Calculate the amplification at frequencies in the middle of the band and the fall in amplification, in decibels, at 1 MHz and at 20 Hz. (*C. & G., Tel. IV*, 1958)

[At 1 MHz the effective anode load is greatly reduced by the 40-pF shunt; at 20 MHz the 1 μF does not effectively by-pass the cathode resistor, so that feedback results. At mid-frequencies $A = 12$; at 1 MHz and 20 Hz the gain falls by 2·25 and 2·42 dB respectively.]

8. A valve used as an audio-frequency class-A amplifier draws 30 mA anode current at an anode voltage of 200 V with no input signal applied. When supplied with a sinusoidal input signal the anode voltage varies between 50 and 350 V and the corresponding anode current change is 50 to 10 mA.

Calculate—

(*a*) the power drawn from the h.t. supply,
(*b*) the output power from the stage,
(*c*) the efficiency of operation.

(*C. & G., R.L.T. B*, 1965)

[6 W; 1·5 W; 25 per cent]

9. Draw a circuit diagram of an audio-frequency transistor amplifier comprising a common-emitter stage in tandem with an emitter-follower stage. Show clearly the biasing arrangements and briefly explain their operation.

State the order of magnitude of the input and output impedances of the amplifier.

Say what factors will affect the low-frequency response of the circuit that you have drawn.

(*C. & G., R.L.T. B*, 1965)

10. An idealized triode valve of $r_a = 500\ \Omega$, $g_m = 10$ mA/V takes an anode current of 200 mA at 100 V with zero grid bias. Construct the anode-voltage/anode-current characteristics over the ranges v_a(0, 400 V), i_a(0, 300 mA), v_g(0, −70 V). Determine, for class-A conditions, the power delivered to a resistive load of 6·667 Ω

coupled to the anode circuit through an ideal transformer of ratio 14·14/1; the d.c. supply voltage is 235 V, bias −35 V and signal amplitude ±35 V. Find also the average anode dissipation.

Explain briefly how to modify the procedure if the transformer winding resistances are not negligible.

(*I.E.E. Elec. Eng. II*, April 1965)

[Characteristic with $v_g = 0$ is a line through zero and 200 mA, 100 V (2 mA/V); $de_a/de_g (= \mu) = 5$; so with $v_g = -35$, v_a is everywhere $5 \times 35 = 175$ V to the right, and so runs from 0 mA, 175 V through 200 mA, 275 V. Similarly for $v_g = -70$ the characteristic is from 0 mA, 350 V through 200 mA, 450 V.

The primary load is $6{\cdot}667 \times 14{\cdot}14^2 = 1{,}333\ \Omega$. A load line of this value through 235 V on the $V_g = -35$ characteristic ($i_a =$ 125 mA) cuts the zero and −70 V characteristics at 110 V, 22 mA and 365 V, 30 mA. Hence the r.m.s. power output is

$$(225 \times 190)/8 = 6\ \text{W}$$

The anode dissipation is $235 \times 125 \times 10^{-3} = 30$ W approximately and the efficiency is thus $6/30 = 20$ per cent. (Exact figures are not necessary because of the possibility of graphical errors)]

11. Show that the gain of an amplifier with negative feedback may be given by the expression—

$$\mu_F = \frac{\mu}{1 + \mu\beta}$$

where μ_F = gain with feedback,
μ = gain without feedback,
β = fraction of output voltage fed back to input.

An amplifier has a gain without feedback of 54·8 dB. What will be the gain when a feedback path, having a β of 0·0082, is connected? If the gain without feedback rises by 6 dB, what will be the new gain with feedback, expressed in decibels?

(*C. & G., R.L.T. B*, 1964)

[The gain in dB must be converted to voltage gain; $54{\cdot}8 = 20 \log \mu$ whence $A_v = 550$; μ_F is then 100 (40 dB). If μ rises by 6 dB, A_v will be twice as great (= 1,100); μ_F then becomes 110 (42·8 dB)]

12. An amplifier without feedback has an internal gain M and an output resistance R_A of 50 kΩ. It is coupled by a capacitor C_c of 0·08 μF to a load consisting of a resistor R_L of 50 kΩ in parallel with a capacitor C_L of 500 pF.

Find (*a*) the frequencies at which the overall gain will fall to 0·707 times the mid-band gain; (*b*) the mid-band frequency.

Describe briefly the probable effects on this amplifier of negative feedback from the load voltage.

(*I.E.E. Elec. Eng. II*, April 1965)

[At low frequencies, neglect shunt capacitance. Gain then falls to $1/\sqrt{2}$ times normal when $\omega = 1/RC_c$, for which $f = 40$ Hz.

At high frequencies, neglect series capacitance. Gain is then $1/\sqrt{2}$ times normal when $RC_L\omega = 1$, giving $f = 6{\cdot}4$ kHz.

Mid-band frequency is thus $(6{\cdot}4 + 0{\cdot}04)/2 = 3{\cdot}22$ kHz]

13. A multi-stage amplifier has a gain tolerance of ± 6 dB due to component and transistor tolerances. A fraction of the output voltage is fed back in series with the input so that the overall gain is reduced by 20 dB with respect to its mean value without feedback.

Calculate the new tolerance in overall gain in decibels.

(*I.E.E. III Applied Electronics*, June 1965)

[Let full gain be A. Then gain with feedback $= 0{\cdot}1A$, from which $\beta A = 9$. 6 dB tolerances will give gains of $0{\cdot}5A$ and $2A$ which are reduced by feedback to $0{\cdot}11A$ approximately. Hence tolerance with feedback $= 0{\cdot}11/0{\cdot}1 = 1{\cdot}1 = 1$ dB approximately.]

14. Explain briefly the term 'class-B push-pull' as applied to the output stage of an audio-frequency amplifier. Illustrate the answer by sketches of waveform and a simple circuit diagram. Derive an expression for the theoretical maximum efficiency of such a stage.

Two transistors, each having a rated dissipation of 500 mW, are used in a class-B push-pull output stage. Calculate the maximum power output if the input is sinusoidal and the actual efficiency is 75 per cent. What would be the maximum power output if the efficiency were 50 per cent?

(*I.E.E. III Applied Electronics*, June 1965)

[3 W; 1 W]

15. The silicon transistor used in the circuit shown in Fig. E.7 has negligible leakage current, a current gain $\beta = 50$, an output resistance of 10 kΩ and a maximum dissipation of 200 mW.

Draw the I_C/V_{CE} characteristics. Plot the maximum dissipation curve. Determine the operating point and draw the load line. Calculate the maximum "undistorted" power in the load for a sinusoidal input. Indicate, with the aid of circuit diagrams, how the stability of the amplifier could be improved.

(*I.E.E. III Applied Electronics*, June 1965)

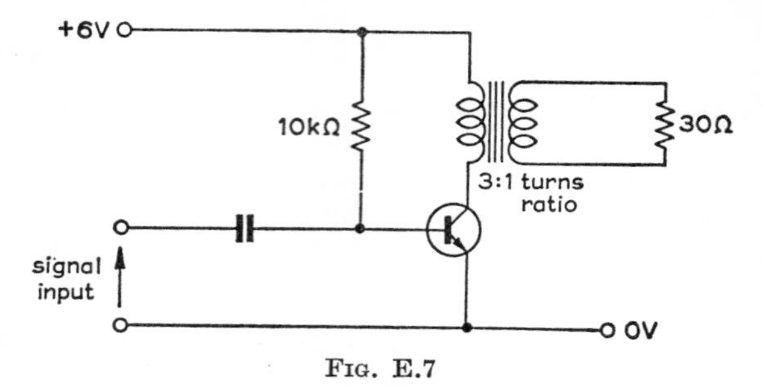

Fig. E.7

[The standing value of I_b is determined mainly by the 10 kΩ resistor, so that $I_b = 6/10^{-4} = 0{\cdot}6$ mA. At this point

$$I_c = \beta I_b = 30 \text{ mA},$$

and since r_c is at least $R_{out}/(1 - \alpha)$ the characteristic will be virtually a horizontal line through the point 6 V, 30 mA. Similar horizontal lines through other values of I_c will correspond to values of $I_b = I_c/\beta$.

The maximum dissipation curve will be such that at each point $V_c I_c = 0{\cdot}2$. The effective load is $30 \times 3^2 = 270$ Ω. A load line of this value through the working point will lie just below the dissipation line. The peak value of V_c cannot exceed 12, at which point $I_c = 7{\cdot}5$ mA. I_b would be 0·15 mA corresponding to a swing of 0·45 mA so that $I_{b(\max)} = 1{\cdot}05$ mA which cuts the load line at $I_c = 52{\cdot}5$ mA and $V_c = 0{\cdot}5$, which would be just about the knee.

Hence the power output would be

$$(52{\cdot}5 - 7{\cdot}5)(12 - 0{\cdot}5)/8 = 65 \text{ mW.}]$$

Chapter 11

1. With the aid of sketches explain the operation of single-phase power rectifiers supplying—

(*a*) half-wave rectification,
(*b*) full wave rectification using (i) only two rectifiers, and (ii) bridge-connected rectifiers.

In each case sketch the approximate waveforms of the voltages across a non-reactive load when no smoothing circuit is connected between the rectifier and the load.

Compare the current and peak inverse voltage ratings required for the individual rectifier arms in each of these circuits, for the same power delivered to a given load. (*C. & G., Tel. B*, 1960)

2. A double-wave valve rectifier supplies a load resistor R. If each rectifier has a resistance r when conducting, and the d.c. component of the load current is I amperes, calculate—

(*a*) the peak value of the current flowing in the transformer secondary,
(*b*) the r.m.s. value of the current flowing in the transformer secondary, and
(*c*) the total power dissipated in the circuit if the resistance of the transformer is neglected.

Hence show that the efficiency of rectification is

$$\frac{8}{\pi^2}\frac{R}{R+r} \times 100 \text{ per cent}$$

(*C. & G., Radio II*, 1958)

[(*a*) $\pi I/2$; (*b*) $\pi I/2\sqrt{2}$; (*c*) $(\pi^2/8)\, I^2(R+r)$]

3. An H.T. power supply unit comprises a full-wave valve rectifier followed by a conventional π-type inductor-capacitor filter. What factors must be taken into account when determining the rating of

(*a*) the valve,
(*b*) the reservoir capacitor, and
(*c*) the inductor?

A typical π filter has shunt capacitors of 8 μF and a series inductor of 20 H. Assuming a sinusoidal ripple of 150 V, 100 Hz applied to the input, determine the ripple voltage at the output terminals.

(*C. & G., Radio II*, 1959)

[2·42 V]

4. In a two-section inductor-input smoothing circuit for a high-tension rectifier, each inductor is 5 H and each capacitor is 8 μF. If the ripple voltage at the input consists of 8 V (r.m.s. at 50 Hz and 1·2 V (r.m.s.) at 150 Hz, calculate the values of these components of the ripple voltage at the output of the smoothing circuit. The load resistance may be neglected. (*C. & G., Radio IV*, 1950)

[1·69 V; 1·04 mV]

5. Draw the circuit diagram of a high-voltage rectifier suitable for operation from a three-phase a.c. supply via a delta-star transformer and incorporating a single-stage *L-C* smoothing circuit. Given that the three-phase 50-Hz supply is at 450 V (r.m.s.) between lines, and the transformer primary to secondary turns ratio is 1:5, determine the direct output voltage on no-load. The smoothing circuit consists of a 2-H series inductor and one 8-μF shunt capacitor; find the output ripple voltage at 150 Hz, given that the 150-Hz component of the ripple voltage at the input to the smoothing circuit has a peak value 0·25 times the no-load output voltage.

(*C. & G., Radio IV*, 1953)

[3,180 V; 60 V]

6. Give the circuit diagram of a voltage-doubling rectifier for the high-tension supply to a receiver. If the rectified output voltage is 250 V and the peak amplitude of the 100-Hz ripple is 15 per cent of the rectified voltage, what is the peak amplitude of the ripple after passing through a single smoothing stage consisting of a 10-H inductor and an 8-μF capacitor? (*C. & G., Radio II*, 1954)

[1·2 V]

7. Sketch and describe the operation of—

(*a*) power-units suitable for the h.t. and heater supplies in an a.c.-d.c. broadcast receiver.

(*b*) a transistorized d.c. to d.c. converter for use with mobile equipment.

In the case of (*a*), state suitable component values.

(*C. & G., R.L.T. A*, 1964)

8. A full-wave power-supply unit consists of a centre-tapped transformer and two semiconductor junction diodes whose forward resistances may be considered equal and constant at 300 Ω. If the unit supplies a mean current of 100 mA to a 2 kΩ load, calculate—

(*a*) r.m.s. transformer voltage
(*b*) diode peak current
(*c*) diode peak inverse voltage
(*d*) diode dissipation
(*e*) a.c. input power neglecting transformer losses.

(*I.E.E. III Applied Electronics*, June 1964)

[Each diode supplies current in turn; hence peak diode current = $0{\cdot}1 \times \pi/2 = 0{\cdot}157$ A; peak transformer voltage (per half) is then

$0{\cdot}157 \times 2{,}300 = 361$, and $V_{r.m.s.} = 255$; peak inverse voltage = 361 — voltage drop on conducting diode = $361 - (300 \times 0{\cdot}157) = 314$ V; total power = $255^2/2{,}300 = 28{\cdot}3$ W; load power (r.m.s.) = $0{\cdot}11 \times 0{\cdot}1 \times 2{,}000 = 24{\cdot}6$ W; hence total diode loss = 3·7 W = 1·85 W per diode]

Chapter 12

1. Draw a circuit diagram of an *L-C* oscillator suitable for use at a frequency of 1 MHz. Describe clearly how oscillations are maintained in such a circuit.

If the tuning inductor used in a 1 MHz oscillator has an inductance value of 200 μH, what value of tuning capacitance will be required?

(*C. & G.*, *Radio I*, 1959)

[127 pF]

2. Explain, with the aid of a circuit diagram, the operation of a transistor oscillator.

What is the source of output power?

(*C. & G.*, *T.P. C*, 1963)

3. Draw the circuit diagram and describe the action of an oscillator in which the frequency is determined by a resistance-capacitance network.

Deduce the conditions necessary for sustained oscillation, and the frequency of oscillation.

(*C. & G.*, *Radio C*, 1963)

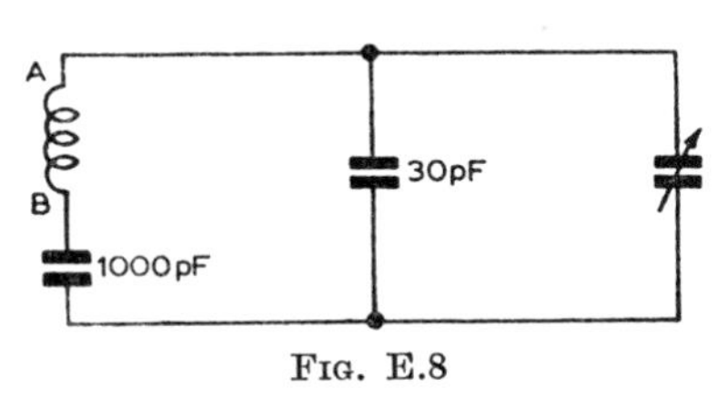

Fig. E.8

4. Fig. E.8 shows the arrangement of part of the oscillator circuit of a superheterodyne receiver. The variable capacitor has a minimum value of 30 pF and a maximum value of 500 pF. Calculate the inductance required so that the minimum resonant frequency is 1,000 kHz.

What will be the maximum resonant frequency?

If the Q-factor of the inductor is 100 and the capacitors are loss-free, what is the circuit impedance between points A and B at the resonant frequencies corresponding to minimum and maximum setting of the variable capacitor?

(*C. & G., R.L.T. B*, 1963)

[73 μH; 2·47 MHz; 113·5 kΩ; 46 kΩ]

5. Draw the circuit diagram of a tuned-grid oscillator incorporating an automatic biasing circuit. Explain the purpose of each component and describe with the aid of sketches how the automatic biasing circuit functions when the valve is operating under Class C conditions.

(*C. & G., Radio C*, 1965)

6. Sketch the circuit diagram of a simple mutual-inductance-coupled transistor oscillator. Briefly explain—

(*a*) the biasing and stabilization arrangements,
(*b*) how the oscillations are maintained.

If such an oscillator has a fixed inductance of 100 μH and it is required to tune over the band 2–4 MHz, calculate the range of the variable capacitor used.

(*C. & G., R.L.T. A*, 1970)

(Take $\pi^2 = 10$)

[62·5 to 250 pF]

7. Explain what is meant by the term "relaxation oscillator." Give two examples producing different waveforms and mention their uses.

A free-running multivibrator uses two p–n–p transistors in the common-emitter connection with 1 kΩ collector loads and 0·1 μF coupling capacitors. Calculate the approximate resistances to be connected between the base and the —10 V supply if the collector output is to be a square wave of frequency 1 kHz. What would be the effect on the frequency of changing the supply to —20 V? Mention all approximations made in the calculations.

(*I.E.E. III Applied Electronics*, June 1964)

[$T = 500\ \mu\text{sec} = 0{\cdot}69CR$, whence $R = (500 \times 10^{-6})/(0{\cdot}69 \times 10^{-5}) = 7{,}250\ \Omega$; assumptions are that v_b varies between 0 and V_B, and that I_{co} is negligible; if R is returned to battery rail, increasing V_B will have negligible effect on frequency]

Chapter 13

1. Explain what is meant by the *critical frequency* of an ionospheric layer. How is it related to the maximum usable frequency (M.U.F.) of a high-frequency radio link?

How would you expect the M.U.F. to vary throughout the day and night for a transatlantic radio link (*a*) in summer, (*b*) in winter?

(*C. & G.*, *Radio III*, 1958)

2. Explain briefly the following terms in connection with radio wave propagation—

(*a*) ionosphere
(*b*) ground wave
(*c*) sky wave
(*d*) skip distance

(*C. & G.*, *R.L.T. B*, 1963)

3. Sketch the polar diagrams, in the horizontal plane, of the following types of aerial—

(*a*) a horizontal dipole,
(*b*) a half-wave vertical dipole with reflector,
(*c*) an inverted-L,
(*d*) a ferrite rod as used in a portable broadcast receiver.

Sketch and explain the construction, including the lead-in to the receiver, of

either (i) an inverted-L aerial for the reception of medium-wave signals,
or (ii) a dipole and reflector type of aerial commonly used for the reception of television signals.

(*C. & G.*, *R.L.T. B*, 1965)

4. A centre-fed halfwave dipole is mounted vertically. Sketch the radiation pattern for this aerial

(*a*) in a horizontal plane
and (*b*) in a vertical plane,

and explain their significance.

Why is the wave said to be vertically polarized? Show the directions of the electric and magnetic fields relative to the direction of transmission.

If the electric field at a point 5 km distant from such an aerial is 10 mV/metre, what is the value at a point 10 km in the same direction? Assume free-space conditions.

(*C. & G.*, *T.P. C*, 1965)

[5 mV/m]

5. Sketch and describe a type of aerial commonly used for the reception of television signals. Give approximate dimensions.

Explain what is meant by the *radiation pattern* (polar diagram) of an aerial. Sketch the radiation pattern, in the horizontal plane, of the aerial described.

(*C. & G.*, *R.L.T. A*, 1963)

6. Draw an end-fire or broadside array, showing how the radiating elements are fed from the balanced feeder and indicating the approximate dimensions for an aerial working at 15 MHz.

What is the polarization of the wave radiated from this aerial?

How should the aerial be constructed so that it can work with wide bandwidths?

How can the direction of radiation from the array be reversed?

(*C. & G.*, *Radio C*, 1965)

7. Explain carefully what is meant by the *radiation pattern* (polar diagram) of an aerial array.

Give a dimensioned sketch of a dipole aerial with reflector suitable for the reception of a television transmission radiated at a frequency of 45 MHz.

Assuming the transmission to be horizontally polarized, sketch the horizontal polar diagram for the aerial and show, by a *dotted* line on the same diagram, the effect of adding a director element.

(*C. & G.*, *R.L.T. B*, 1959)

8. Two vertical, transmitting dipoles, separated by a quarter of a wavelength, are fed in anti-phase. Sketch the resulting polar diagram in the horizontal plane.

Suggest one method of connecting the aerials to the source in order to produce the 180° phase difference between them.

(*C. & G.*, *Tel. C*, 1960)

9. What is meant by the H_{01} mode of transmission in a rectangular waveguide? If the free-space wavelength of a wave is λ and the

width of the broad face of the guide is b, show that for the H_{01} mode the "wavelength in the guide" λ_g is given by

$$\frac{1}{\lambda^2} = \frac{1}{\lambda_g^2} + \frac{1}{(2b)^2}$$

Explain the significance of this equation and the limitation it places on the width of guide suitable for a wave of given frequency. In what other ways are the dimensions of the guide limited?

The end of a section of waveguide is completely closed by a perfectly conducting plate. Draw a diagram of the resulting electric and magnetic field distributions and the current paths in the guide walls and end plate. Show the significance of λ, λ_g and b.

(*I.E.E., Radio Comm.*, 1958)

10. Draw the electric and magnetic field distributions in a rectangular waveguide carrying the TE_{01} mode (H_{01} mode) and show how the energy can be launched into the waveguide.

A rectangular waveguide carries an electromagnetic wave having a frequency of 4 GHz. A standing wave indicator shows that the wavelength of this wave in the waveguide is 11·4 cm. What are the cut-off wavelength of the waveguide and the velocity at which energy is propagated along the guide? Take $c = 3 \times 10^{10}$ cm/sec.

(*C. & G., Radio C*, 1965)

[13·2 cm]

11. Give a diagram of a triode-hexode frequency-changer stage for use in a short-wave communication receiver, indicating typical component values.

Briefly explain the principle of operation of this type of frequency-changer and list its advantages relative to one other type.

(*C. & G., R.L.T. B*, 1960)

12. Why is automatic gain control necessary in a short-wave receiver?

Draw circuits, and explain their operation, for—

(*a*) the derivation of an a.g.c. voltage in the detector stage of a receiver,

(*b*) the use of this voltage for gain control.

Draw a circuit diagram to show how manual volume control is obtained in a radio receiver. Explain why the effect of this control is not offset by the action of the a.g.c.

(*C. & G., R.L.T. B*, 1963)

13. Use a diagram to explain the special characteristics of a varactor diode. Describe, with the aid of a simple circuit diagram, the operation of a varactor diode frequency modulator.

An oscillator operating at 100 MHz has a 75 pF capacitor in its tuning circuit. What total capacitance swing must the varactor supply to allow the modulator to have a 80 kHz peak deviation?

(*C. & G., Radio C*, 1970)

[0·24 pF]

Chapter 14

1. It is desired to insert a T-type low-pass filter having a cut-off of 3,000 Hz in a line of 600 Ω characteristic impedance. Compute the values of the series and shunt elements of a suitable filter section, assuming non-dissipative elements.

[Series elements 31·8 mH; shunt element 0·177 μF]

2. What is meant by a wave filter of (*a*) the constant-*k* type, (*b*) the derived type?

A high-pass filter for 600 Ω terminations is required to cut off below 20 kHz. Compute the values of the shunt and series elements.

[Series elements (T-section) 0·0133 μF; shunt elements 2·39 mH]

3. A low-pass filter, comprising one constant-*k* T-section, terminated at each end by a derived half-section ($m = 0{\cdot}6$) has a theoretical cut-off frequency of 4 kHz and a nominal image impedance in the pass range of 600 Ω. Determine the values of the components and give a circuit diagram. (*C. & G., Tel. V*, 1952)

[Shunt $C = 0{\cdot}132$ μF; series $L = 38{\cdot}1$ mH; terminating $L = 25{\cdot}4$ mH; terminating $C = 0{\cdot}0396$ μF]

4. Derive an expression for the characteristic impedance of a T-section low-pass filter in which each of the series arms contains an inductance L and the shunt arm consists of a capacitor C. Hence obtain an expression for the cut-off frequency. If each of the series arms of such a section contains an inductance of 6 mH and the shunt capacitance is 0·0333 μF, determine the cut-off frequency of the section. Determine also the value of the characteristic impedance at zero frequency, and at frequencies which are (*a*) 0·1 and (*b*) 0·9 of the cut-off frequency.

(*I.E.E. III, Line Communication*, June 1965)

[$\omega^2 LC = 2$; 15·9 kHz; 600 Ω; 596 Ω; 261 Ω]

5. State Thévenin's theorem.

A T-network consisting of a resistance of 60 Ω in each series arm and a shunt resistance of 100 Ω is connected to a source of 1 V e.m.f. and 40 Ω resistance. Find the value of the load resistance which will absorb maximum power and the value of that power.

Explain what is meant by insertion loss and calculate the insertion loss of the network under these conditions.

(*C. & G., Tel. III*, 1953)

[110 Ω; 0·57 mW; 9·35 dB]

6. Show that the maximum power which can be obtained from a source having an internal impedance $R + jX$ and open-circuit voltage E is $\frac{E^2}{4R}$.

An a.c. source having a non-reactive internal impedance of 300 Ω is connected across the input terminals of a T-network in which the series arms on the input and output sides are 158 Ω and 632 Ω respectively while the shunt arm is 158 Ω. What would be the value of the load resistance across the output terminals for maximum output power and the insertion loss of the network with this load?

(*C. & G., Tel. III*, 1952)

[749·4 Ω; 14·92 dB]

7. A network has two parallel arms consisting of

(*a*) a resistance R in series with a capacitance C,
(*b*) a resistance r in series with an inductance L.

Derive an expression for the impedance Z of the network.

Thence determine the frequencies at which Z is non-reactive and find its value at these frequencies.

What relationship must exist between R, r, L and C for Z to be independent of frequency? (*C. & G., Tel. III*, 1958)

[The combined impedance is

$$Z = \frac{(r + j\omega L)(R - j/\omega C)}{(R + r) + j(\omega L - 1/\omega C)}$$

Rationalize this and arrange the numerator in the form $A + jB$. Z then becomes non-reactive when $B = 0$ which occurs when $\omega^2 LC = (r^2 - L/C)/(R^2 - L/C)$.

In this condition the impedance is

$$Z = \frac{Rr(R + r) + R\omega^2L^2 + r/\omega^2C^2}{(R + r)^2 + (\omega L - 1/\omega C)^2}$$

For the impedance Z to be independent of frequency, the resistances must be equal and the reactances must be inverse, so that $R = r = \sqrt{(L/C)}$.]

8. What do you understand by the terms *insertion loss* and *insertion gain*?

Explain why an insertion gain can be obtained when a suitable transformer is connected between a source of resistance R_1 and a load of resistance R_2.

What features in the design of the transformer affect the magnitude of the gain?

Obtain an expression for the maximum gain.

What would be the effect of reversing the windings across which the source and load are connected? (*C. & G., Tel. III*, 1958)

9. A telecommunication circuit consists of four items of equipment connected in tandem by line and radio links.

The table below gives the input power to each item of equipment and the output power from items 1, 2 and 3. The power gain of item 4 is 23 dB.

	Item 1	*Item* 2	*Item* 3	*Item* 4
Power in (mW)	1,000	316	500	251
Power out (mW)	25,000	12,600	15,800	–

Determine—

(*a*) the input power to item 1 in decibels relative to 1 mW,
(*b*) the power gain in decibels for each of items 1, 2 and 3,
(*c*) the power loss in decibels for each link,
(*d*) the output power from item 4 in milliwatts,
(*e*) the overall power gain of the circuit in decibels.

(*C. & G., R.L.T. A*, 1970)

[(*a*) 30 dB; (*b*) 14, 16 and 15 dB; (*c*) 19, 14 and 18 dB; (*d*) 50,000; (*e*) 17 dB]

Chapter 15

1. Describe the principle of the thermocouple milliammeter and mention one application for which this type of meter is particularly suitable.

Is the following statement correct—

> "The reading accuracy of a thermocouple milliammeter is independent of the temperature of its surroundings"?

Give reasons for your answer. (*C. & G., Tel. II*, 1958)

2. Describe the principle of operation of a moving-coil wattmeter suitable for the measurement of power at frequencies of the order of 50 Hz. Give sketches of the circuit and of the general form of such an instrument. (*C. & G., Tel. II*, 1959)

3. Describe the principle of operation of an electrostatic voltmeter. What factors determine the sensitivity of this type of instrument? Give two applications of this class of instrument.

(*C. & G., Tel. B*, 1960)

4. A battery of e.m.f. 6 V and internal resistance 2·5 Ω, an adjustable resistor and an ammeter are connected in series. When the resistor is adjusted to 45 Ω the meter reads 100 mA. Find the resistance of the ammeter.

What value of shunt must be added across this ammeter in order that it still reads 100 mA when 0·5 A flow in the external circuit?

With this shunt in position, the adjustable resistor in the circuit is altered until the meter reads 50 mA. Calculate this new value of the adjustable resistor and the p.d. across the battery terminals.

With the shunt removed, how could the ammeter be converted into a voltmeter to measure 100 V for an indication of 100 mA on the scale?

(*C. & G., T.P. A*, 1963)

[12·5 Ω; 3·125 Ω; 19 Ω; 5·375 V; 1 kΩ]

5. Draw a diagram and explain the operation of a peak-reading diode valve voltmeter.

Such instruments are frequently calibrated to indicate the r.m.s. values of sine waves. What precautions must be taken when using instruments so calibrated? (*C. & G., R.L.T. B*, 1960)

6. Describe with the aid of a circuit diagram, a wavemeter of the absorption type. How would such a wavemeter be calibrated?

An absorption wavemeter is used to measure the frequency of an oscillator, and it is found that readings are obtained at a number of frequencies of which two consecutive values are 1 MHz and 1·5 MHz. What is the reason for this, and what is the frequency of the oscillator? (*C. & G., Radio I*, 1958)

[0·5 MHz]

7. Explain how a potentiometer and standard cell can be used to measure the e.m.f. of a battery. How could the arrangement be adapted and used to measure the direct current in a circuit?

A potentiometer consisting of a 1-m slide-wire gives a balance against the standard cell at a point 21·5 cm from the end to which the cell is connected. When the voltage to be measured has one terminal connected to this same end, the balance point is obtained 80 cm along the wire. What is the value of the unknown voltage?

The e.m.f. of the standard cell should be taken as 1·43 V.

(*C. & G., Tel. A*, 1960)

[5·32 V]

8. The a.c. bridge circuit shown in Fig. E.9 is used to measure the inductance L_1 and resistance R_1.

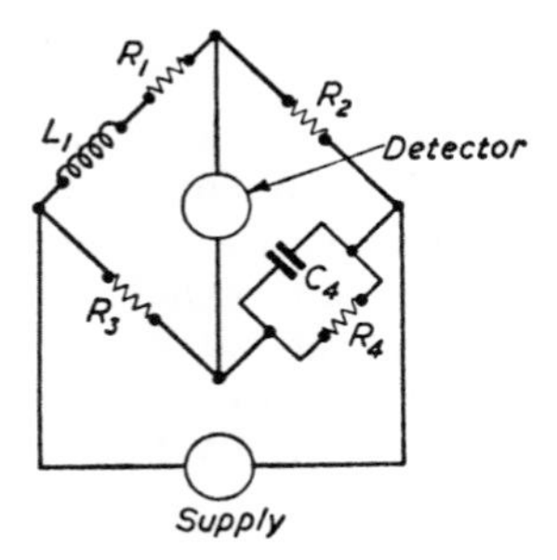

FIG. E.9

Determine the conditions of balance and sketch a vector diagram (not to scale) representing the conditions at balance.

(*I.E.E., Elec. Eng. II*, 1960)

[$R_2R_3 = Z_1Z_4$; whence by expanding and equating the resistive and reactive terms we get $L_1 = C_4R_2R_3$ and $R_1 = R_2R_3/R_4$; whence $L_1\omega/R_1 = Q_L = \omega C_4R_4$]

9. A test oscillator is very loosely coupled to a tuned circuit consisting of an inductance L tuned with a variable capacitor C_1. A resistor R is shunted across the circuit and is adjusted until the voltage, as indicated on a valve voltmeter across the circuit, is 1 V. An additional capacitor C_2 is now placed in parallel with L, and C_1 is readjusted to give maximum reading on the voltmeter. R is then adjusted so that the voltmeter again reads 1 V. If the new values of C_1 and R are 200 pF and 3·78 MΩ respectively, calculate the capacitance, reactance, equivalent r.f. shunt resistance and power factor of C_2. The inductance L is 80 μH and the oscillator frequency is 1 MHz. *(C. & G., R.L.T. B, 1959)*

[$C_1 = 1/\omega^2 L = 317$ pF; hence $C_2 = 317 - 200 = 117$ pF and $X_{C2} = 1{,}360\ \Omega$. Since in the second case the voltmeter reading is the same, the effective shunt resistance is still 1 MΩ; hence the shunt resistance of C_2 in parallel with 3·78 MΩ = 1 MΩ, which requires $R_2 = 1{\cdot}36$ MΩ. The power factor of C_2 is then $1/\omega C_2 R_2 = 0{\cdot}001$.]

10. Give a brief description of the principle of operation of a moving-coil milliameter. In particular explain the meaning of *control torque*.

A moving-coil meter has a coil of 1 cm square containing 100 turns. The coil is pivoted about an axis in the plane of the turns and bisecting two opposite sides. It moves in a radial uniform magnetic flux density of 2 mT. Calculate the torque on the coil when a current of 10 mA is flowing in its turns.

(C. & G., T.P. A, 1965)

[2×10^{-7} newton-metre]

11. A 50 Hz Schering bridge ABCD, used to determine the capacitance and power factor of a piece of insulating material, is balanced when the components are as follows—

AB air capacitor of capacitance 230 pF
BC 407 Ω non-inductive resistor in parallel with a 0·5 μF loss-free capacitor
CD 100 Ω non-inductive resistor
DA insulating material under test.

Derive the balance condition for the bridge and hence determine the capacitance and power factor of the specimen.

Explain why this type of bridge is suitable for measurements at high voltages and describe the techniques used to minimize errors.

(*I.E.E. III, Elec. Meas.*, June 1964)

[$C = 936$ pF; $r = 0{\cdot}217$ MΩ; power factor $= 0{\cdot}064$]

12. Describe the circuit and principle of operation of an a.c. bridge suitable for measuring the inductance and effective resistance of a coil at a frequency of about 1 kHz. Show clearly how these values can be determined using the bridge.

What type of detector and source of alternating current would you employ? What precautions must be taken in connecting these to the bridge itself?

(*C. & G., T.P. B*, 1963)

13. With the aid of a diagram, and starting with the electron emission from the cathode, describe the complete path taken by the current in a cathode-ray tube circuit.

Make a list of the important properties of the fluorescent material which is used to coat the screen of the tube.

Why, in practice, is a saw-tooth the most common of all time-base waveforms?

(*C. & G., T.P. C*, 1964)

14. Draw a block diagram of a cathode-ray oscilloscope and briefly explain the purpose of each portion. Detailed circuit diagrams are not required.

An oscilloscope is to be used to examine waveforms occurring in a high-quality audio amplifier. Briefly discuss the requirements for the bandwidth of the deflection amplifier and the velocity of the time base.

(*C. & G., R.L.T. B*, 1965)

15. With the aid of careful sketches describe briefly the construction and operation of a cathode-ray tube employing electrostatic focusing and deflection.

Obtain an expression for its sensitivity.

Give a circuit diagram showing the essential details of one form of time base which is approximately linear and explain its action.

(*C. & G., T.P. C*, 1963)

16. Explain how a cathode-ray oscilloscope may be used to measure

(i) the phase difference between two voltages at the same frequency,

(ii) the depth of modulation of an a.m. transmitter when it is sinusoidally modulated.

Show the expected trace in each case.

(*C. & G., Radio C*, 1970)

17. Draw a block schematic diagram of the equipment needed to produce a visual display of the amplitude/frequency responses of the i.f. and audio sections of a radio receiver. Explain the function of the items in the diagram.

Draw sketches of the displays that would be expected when making these tests on a communication receiver.

(*C. & G., Radio C*, 1965)

18. The output from a beat-frequency oscillator is to be automatically swept over the frequency range 0–20 kHz by the application of a low-frequency saw-tooth waveform to a simple reactance valve connected across the tank circuit of one of the radio-frequency oscillators. The required frequency range is 1 MHz to 1·02 MHz and the inductor in the tank circuit has a value 100 μH. The pentode used in the reactance valve circuit has its mean operating point at $V_g = -20$ V and its mutual conductance is given by the expression $g_m = (12 + 0{\cdot}1v_g)$ mA/V. A 5-pF capacitor is connected between anode and grid and a 500-Ω resistor between grid and cathode.

Derive an approximate expression for the effective reactance of the valve and calculate the amplitude of the saw-tooth voltage to be applied between grid and cathode to give the required frequency swing.

(*I.E.E. III, Applied Electronics*, June 1964)

[The effective reactance of the valve is $1/C'\omega$, where $C' = CRg_m$. To obtain change of f from 1 to 1·02 MHz a capacitance change of 10 pF is required; with $V_g = -20$, $g_m = (12 - 0{\cdot}1 \times 20) = 10$. Whence $C' = 25$ pF; to obtain $C' = 15$ pF, V_g must be -60; hence peak grid swing required is 40 V (negative)]

Chapter 16

1. With the aid of sketches explain briefly the reasons for providing the following in connection with the transmission and reception of 405-line television signals—

(*a*) synchronizing pulses,
(*b*) interlaced scanning.

(*C. & G., Radio III*, 1959)

2. Show how the bandwidth of the amplifiers required to handle the video signals of a television system may be determined.

Sketch the waveforms of the signals transmitted in a television system, indicating the form and duration of the various synchronizing pulses. Show that the video bandwidth is also adequate for the transmission of these pulses.

3. Develop an expression for the bandwidth required in a high-definition television system.

Explain how the bandwidth is affected by

(*a*) The number of lines in the picture,
(*b*) The aspect ratio,
(*c*) The frame frequency.

4. What is meant by vestigial-sideband transmission? What are the advantages of this system for television reception? How, in practice, is the receiver adjusted for optimum reception?

Chapter 17

1. Discuss the conditions in relation to input voltage waveform for the output of a simple *CR* circuit to be a differentiated or integrated version of the input voltage.

Draw the circuit diagram of a Miller integrator circuit which, when suitably driven, will give an output consisting of a succession of linear saw-tooth voltages.

Indicate the components of the circuit which control the slope of the saw-tooth and explain how to make the relevant calculations.

(*I.E.E. III*, *Radio Communication*, June 1964)

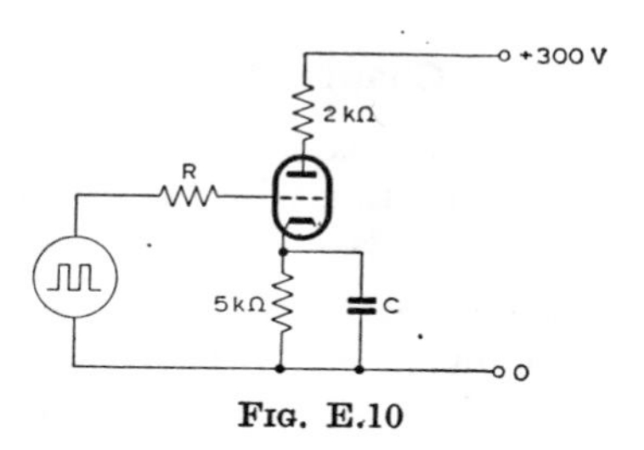

FIG. E.10

2. Describe briefly the operation of a circuit for producing positive rectangular pulses of voltage of duration 100 microseconds and repetition frequency 500 per second.

Such pulses are applied to the circuit shown in Fig. E.10. The capacitance C is sufficiently large to maintain an effectively constant p.d. across the cathode resistor. The input causes the anode current to be switched between 0 and 40 mA. The resistance R is very large and carries grid current during the "on" periods. Determine the ranges of the anode–cathode and the grid–cathode potential differences.

(*I.E.E. III, Applied Electronics*, June 1964)

[Voltage across C will remain steady at the peak signal value, i.e. $(40 \times 10^{-3} \times 5 \times 10^{3}) = 200$ V; hence when $i_a = 0$, $V_{ak} = 100$, $V_{gk} = -40$; when $i_a = 40$ mA the anode load drops 80 V so that $V_{ak} = 20$ V, $V_{gk} \rightarrow 0$]

3. A repetitive rectangular waveform has a mark-to-space ratio a/b. Show that when a/b is 81/19, the form factor of the wave is 1·11.

Compare this value with that obtained for a fully rectified sine wave, and criticize the use of form factor for indicating harmonic distortion in repetitive waveforms.

(*I.E.E., Elec. Eng. II*, April 1964)

Index